Third Edition

Modern Ceramic Engineering

Properties, Processing, and Use in Design

MATERIALS ENGINEERING

Third Edition

Modern Ceramic Engineering

Properties, Processing, and Use in Design

David W. Richerson

 Taylor & Francis
Taylor & Francis Group

Boca Raton London New York

A CRC title, part of the Taylor & Francis imprint, a member of the
Taylor & Francis Group, the academic division of T&F Informa plc.

Published in 2006 by
CRC Press
Taylor & Francis Group
6000 Broken Sound Parkway NW, Suite 300
Boca Raton, FL 33487-2742

1004547857

International Standard Book Number-10: 1-57444-693-2 (Hardcover)
International Standard Book Number-13: 978-1-57444-693-7 (Hardcover)
Library of Congress Card Number 2005048601

Library of Congress Cataloging-in-Publication Data

Richerson, David W., 1944-
   Modern ceramic engineering : properties, processing, and use in design / David W. Richerson.--3rd ed.
     p. cm. -- (Materials engineering ; 29)
   Includes bibliographical references and index.
   ISBN 1-57444-693-2
   1. Ceramic engineering. I. Title. II. Materials engineering (Marcel Dekker, Inc.) ; 29/

TP807.R53 2005
666--dc22
                                        2005048601

Taylor & Francis Group
is the Academic Division of Informa plc.

Visit the Taylor & Francis Web site at
http://www.taylorandfrancis.com

and the CRC Press Web site at
http://www.crcpress.com

Preface

Ceramic materials have become increasingly important in our modern world, yet most engineers and technologists receive little or no training in ceramics. This third edition (and the prior two editions) was prepared to provide engineers, students, teachers, and technicians with an introduction to the structure, properties, processing, design concepts, and applications of advanced ceramics. As in the earlier editions, emphasis is placed on developing an understanding of why ceramics are different from metals and organics and then applying this understanding to optimum material selection. Subsequent paragraphs describe how this edition is different from prior editions, review factors that affected progression from one edition to the next, and acknowledge the many individuals and organization that have supported or influenced all three editions.

The first edition was published in 1982 and was written primarily to provide a reference and text for mechanical engineers working on structural materials for such applications as gas turbine engines. The content was compiled based on a combination of personal engineering experience, preparation and teaching of a three-semester-hour ceramics course at Arizona State University starting in 1975, and a series of half-day (part of a three-day course on high-temperature structural materials) and three-day courses ("High-Strength Ceramics for Engineering Applications" and "Impact of Ceramics on Modern Engineering") taught for the American Society for Metals (ASM, now renamed the American Society for Materials International).

Even though the first edition was not written in a textbook format, a number of universities and organizations chose to use it as a textbook. By the mid-1980s, several professors suggested that a second edition be prepared in a textbook format and expanded to include a broader treatment of ceramics. In parallel, efforts were in progress to prepare short courses on structural ceramics and processing of ceramics for the American Ceramic Society, the American Institute for Chemical Engineers, and NASA plus an extensive home study course, "Introduction to Modern Ceramics" (Course 56), for ASM International. An arrangement was negotiated to share content being prepared for the second edition with ASM International for their home study course and to use content under preparation for the ASM course in the second edition. The second edition was published in 1992 and included substantially expanded content on ceramic structures and crystal chemistry, thermal behavior, electrical behavior, magnetic behavior, and all aspects of processing of ceramics. Completely new chapters were added on phase equilibrium and toughening of ceramics; problems also were added for each chapter.

Important advances have occurred in ceramic engineering since the second edition was prepared. In addition, I have now used the second edition for a number of years as a textbook for my "Introduction to Ceramics" course in the Materials Science and Engineering Department at the University of Utah and have a better idea of "what works" and "what doesn't work" in terms of student interest and comprehension. As a result, there are some significant changes in this third edition. Rather than starting with a discussion of the fundamentals, the book now begins a new Part I with three chapters oriented toward applications and engineering that encourage the student to think about how the needs of an application must be matched by the material and design. Many important applications of ceramics are described in this context to the extent that the key combination of properties and characteristics required is identified.

Part II, "Structure and Properties," then provides the fundamental discussion of why specific ceramics have the right combination of characteristics for specific uses. As in the second edition, this section contains chapters on atomic bonding and crystal structure; crystal chemistry and specific crystal structures; phase equilibrium and phase equilibrium diagrams; physical and thermal behavior; mechanical behavior; time, temperature and environmental behavior; electrical behavior; and dielectric, magnetic, and optical behavior.

What was Part II in the second edition, "Processing of Ceramics," is now Part III in the third edition. The major changes include updates in technology and materials examples, a study guide at the end of each chapter, and additional references or suggested reading. Similarly, Part III in the second edition is now Part IV in the third edition. The major changes here are an update of the chapter on toughening of ceramics (ceramic matrix composites), the addition of study guides for each chapter, and the deletion of the final chapter on applications (which is now covered in Part I).

I deeply appreciate the many individuals and organizations that have contributed directly and indirectly to the preparation of the first, second, and third editions: ASM International for collaboration on the second edition that resulted in many drawings and several chapters, especially Nick Jessen, Paul Urban, Dominic Youngross, and Brooke Willis; Greg Brigham for preparing line drawings for the first edition; Floyd Brown, Nelson Hope, Denise Birnbaum and others at Garrett Turbine Engine Company (now Honeywell Engines, Systems & Services) for review of draft manuscript of the first edition; Rachel Goeckeritz, Christy F. Johnson, Judith A. Martindale, Angie F. Peters, and Michael Anne Richerson for help with word processing of the first two editions; Norman Hecht, Dale Wittmer, and R. Nathan Katz for suggestions of content of the second and third editions; Dr Robert Shane who suggested writing the first edition and worked with Marcel Dekker to publish it; the Defense Advanced Research Projects Agency, the Air Force Materials Laboratory and Air Force Aero-Propulsion Laboratory, NASA, and Department of Energy for sponsoring many of the programs that produced many of the examples and photographs included within the three editions; and Vasantha R.W. Amarakoon, James F. Benzel, Alastair N. Cormack, Van Derck Frechette, Andrew Herczog, W. E. Lee, William G. Long, Anna E. McHale, Malcolm G. McLaren, James S. Reed, A. Safari, Daniel J. Shanefield, James R. Varner, and Thomas J. Whalen for reviewing the second edition manuscript and providing feedback prior to final revisions for publication.

David W. Richerson

The Author

David W. Richerson received degrees in ceramic science and engineering from the University of Utah (1967) and The Pennsylvania State University (1969). He conducted research on boron carbide armor, silicon nitride, and composites at Norton Company from 1969 to 1973; coordinated materials efforts from 1973 to 1985 at Garrett Turbine Engine Company to integrate ceramic materials into gas turbine engines; and conducted and managed a wide range of materials programs while Director of Research and Development and later Vice President at Ceramatec, Inc. from 1985 to 1991. From 1991 to the present Mr. Richerson has worked as a consultant, taught at the University of Utah, and planned and conducted volunteer science outreach projects in schools and in the community. Mr. Richerson has authored or co-authored 5 books, 13 book chapters, 21 government program final reports, 5 patents, and 59 technical publications. He has edited or co-edited four additional books, made numerous technical and educational presentations, and presented two-day to four-day short courses worldwide. Mr. Richerson is a Fellow and past Board member of the American Ceramic Society, a member of the National Institute of Ceramic Engineers and the Ceramic Education Council, and a past member of ASM International.

Introduction

The primary objective of this book is to provide an increased understanding of ceramic science and engineering in relationship to the overall field of materials science and engineering. The desire is for the reader to attain an appreciation of the wide range of properties and applications possible for ceramics, to understand the fundamental sources of behavior of ceramic materials compared to metals and organics, and to learn about the specific ceramics that have especially favorable combinations of properties to meet the needs of specific applications. To accomplish these learning objectives, the book is divided into four parts:

Part I: Ceramics as Engineering Materials
Part II: Structures and Properties
Part III: Processing of Ceramics
Part IV: Design with Ceramics

Part I is divided into three chapters, the first to help define in the reader's mind which materials are classified as ceramics, the second to review the history or evolution of ceramics, and the third to explore the wide range of engineering applications. The third chapter is especially important because it emphasizes the requirement of an engineer to match the material characteristics to the specific needs of an application. This then leads into subsequent sections of the book because the characteristics are controlled by factors such as the atoms present, the nature of atomic bonding, the crystal structure, the fabrication process, and even the component design.

Part II explores the physical, thermal, mechanical, electrical, magnetic, and optical properties of ceramics and their relationship to atomic bonding, crystal structure, and microstructure. Comparison with metals and organic materials is emphasized, and concepts of ceramic design begin to evolve.

Part III studies the ceramic fabrication processes, describing in detail each step from raw material selection to shape forming to quality control. Each process step is discussed in terms of its relationship to the properties and acceptability of the final ceramic component. Understanding how a ceramic component is fabricated can often help an engineer resolve an application problem.

Part IV applies the information covered in the first three parts to the design of ceramics. Emphasis is on the differences in design approach required for ceramics as compared to metals and plastics. The importance and techniques of fracture analysis are also described in Part IV, as well as techniques to increase toughness to resist fracture.

Ceramics are encountered in industry, aerospace, electronics, communications, medicine, transportation, construction, and virtually every other facet of our modern technology. It is hoped by the author that every reader of this book will gain an increased understanding of and ability to engineer with ceramic materials.

Contents

Chapter 7

Chapter 8

Chapter 9

Part I

Ceramics as Engineering Materials

The objective of Part I is to introduce the reader to ceramics as key engineering materials. Chapter 1 strives to define the field of ceramics and where ceramics fit into the overall field of materials. Chapter 2 reviews the history of ceramics and how they have expanded in scope from the invention of earthenware in ancient times to the slow evolution of traditional ceramic industries to an explosion during the last 75 years of high technology modern applications. Part I concludes with Chapter 3, which introduces examples of the wide range of applications of modern ceramics. The focus is on comparison of the engineering requirements of each application with the unique characteristics of the ceramic material best suited for that application.

1 What Is a Ceramic?

Individuals who have not previously studied ceramics typically ask: "What is a ceramic?" That is a good question, especially since this book is written to be used in an introductory course in Ceramic Engineering. Unfortunately, it is a difficult question. Do we categorize ceramics and other materials according to their behavior and properties or does another approach work better such as examining their chemical makeup or the nature of their atomic bonding? Or are materials categorized based on some historical perspective? This chapter provides a brief review of prior definitions of ceramics, discusses some problems with these definitions, identifies types of materials that clearly should be categorized as ceramics, and concludes with an explanation of the approach that will be used throughout the remainder of this book.

1.1 DEFINITIONS OF CERAMICS

The word *ceramic* was initially derived from the Greek word *keramos*, which means roughly "burnt stuff."[1] Webster defines *ceramics* as "of or having to do with pottery."[2] Both of these definitions go back to the earliest origins of ceramics when early peoples dug earthy clay, mixed in some water to achieve the consistency of potter's clay, crafted a shape, dried it in the sun, and placed it in a fire. The resulting hard, brittle material was our first ceramic that we now refer to as "earthenware." However, over the subsequent millennia many additional materials were invented that we also think of as ceramics. The definition broadened to "the art and science of making and using solid articles formed by the action of heat on earthy raw materials."[1]

By the latter half of the 20th century, scientists and engineers had learned to synthesize many new ceramics, sometimes by chemical methods that did not quite fit the older definitions. W. David Kingery, often regarded as the father of modern ceramic engineering, suggested in his classic text *Introduction to Ceramics*[1] a new definition: "the art and science of making and using solid articles which have as their essential component, and are composed in large part of, inorganic nonmetallic materials." This definition essentially says that a ceramic is anything that is not an organic material or a metal. It suggests three basic categories of materials: ceramics, organics, and metals. However, it still gives no information about how to distinguish between the three categories, so we are right back where we started. Also, it does not give us any hints about where salts, semiconductors, intermetallics, and metalloids fit into the scheme of materials.

Since we are not likely to come up with a precise or all-encompassing definition that really distinguishes ceramics from other materials, perhaps we can try a different approach. Let's look at some examples that are considered by most materials engineers to be in the category of ceramics.

1.2 MATERIAL TYPES GENERALLY CONSIDERED IN THE CERAMICS FAMILY

1.2.1 POLYCRYSTALLINE CERAMICS FABRICATED BY SINTERING

The first earthenware pottery and most ceramics that we use today are fabricated starting with a powder. The powder is mixed with water or other materials (such as a polymer and a solvent) and formed to the desired shape by processes such as pressing, slip casting, extrusion, and injection molding. This "greenware" is then dried, any organic materials removed, and the remaining particulate compact fired (sintered) at a high temperature such that the particles bond together to form a solid polycrystalline (made of many small crystals or grains) ceramic. This type of material certainly fits into the prior definitions and represents the majority of the discussion in this book.

1.2.2 GLASS

Glass typically starts out with powdered or crushed raw materials, but instead of sintering the powders to form a solid, the raw materials are heated to a high enough temperature that they melt to form a homogeneous liquid. The liquid, which is usually quite viscous, is formed into the desired shape and then cooled in such a way that no crystals or grains form. The resulting glass material can have the same chemical composition as a polycrystalline ceramic, but has much different properties. The difference between the polycrystalline ceramic and the glass is the arrangement of the atoms.

In the case of the polycrystalline ceramic (as well as a single crystal ceramic), the atoms are arranged in a repeatable three-dimensional pattern. For example, SiO_2 (silicon dioxide or "silica") is comprised of units of one silicon atom bonded to four oxygen atoms. Each unit is shaped like a tetrahedron and bonds to four adjacent SiO_4^{4-} tetrahedra to form a network structure that has both short-range and long-range repeatability (order) throughout the material. Silica glass involves the same SiO_4^{4-} units bonded to adjacent units, but not organized into as repeatable a structure. The glass is said to have short-range order but not long-range order. Should the SiO_2 glass and other glasses of typical ceramic chemical compositions be included in the broad materials category of "ceramics"? My vote is "yes."

Some metals also can be processed by extremely rapid cooling from a melt to result in a glassy structure. These are also sometimes described in the literature as "glass," but obviously would not fall into the category of ceramics.

1.2.3 GLASS CERAMICS

The atomic arrangement of glass is stable at high temperature, but not at room temperature. The atoms want to rearrange into a crystalline structure, but do not have enough mobility at room temperature. However, the glass can be induced to crystallize by adding a "nucleating" agent when the glass is initially formed and then by heat treating at a suitable temperature below the melting temperature. Enough energy is present at this temperature to allow the atoms to move (diffuse). The atoms in the glass rearrange around the atoms of the nucleating agent to form a polycrystalline ceramic called a glass ceramic. This also fits the prior definitions of a ceramic.

1.2.4 SINGLE CRYSTALS OF CERAMIC COMPOSITIONS

Single crystals of natural minerals such as quartz, ruby, and diamond are found occasionally in nature. They have special characteristics compared to polycrystalline forms of the same minerals, but are much more rare. At first thought, single crystals may seem to us to be too exotic to be considered for large-scale use. On the contrary, we have learned to synthesize single crystals of a wide variety of ceramic compositions and have established uses ranging from synthetic diamonds to ruby laser rods to cubic zirconia jewelry. For example, more than 500 tons of cubic zirconia crystals were produced in 1997.

Most single crystals are produced at high temperature from ceramic powders and clearly fit into the existing definitions as ceramics. However, some single crystals are produced at essentially room temperature by growth from a supersaturated solution. An example is sodium chloride. Is this a ceramic? NaCl certainly is not organic or metallic, but many people have trouble perceiving it as a ceramic, especially because it is soluble in water. But NaCl and other ionically bonded salts are best categorized as ceramics because of their mode of atomic bonding and crystal structures, as will be discussed in later chapters.

1.2.5 CHEMICAL SYNTHESIS OR BONDING

Some of the most widely used ceramics are formed by chemical reaction at room temperature. Two important examples are Portland cement and plaster. Both involve powdered ceramic compositions that react with water to form new ceramic compositions that either bond other materials into an aggregate (such as concrete) or form a stand-alone structure (such as plaster wallboard). Concrete is the most used material in the world. Approximately one ton of concrete is poured each year for every man, woman, and child on Earth. These chemically bonded materials fall within the new definition by Kingery, but not the older definitions.

1.2.6 NATURAL CERAMICS

The traditional definitions seem to exclude natural materials not synthesized by humans. Why should this be the case? If a natural nugget of copper or gold qualifies as a metal just the same as copper that has been smelted from an ore, then ceramic compositions produced by Mother Nature should be categorized as ceramics. In fact, nearly all of our modern synthesized ceramics are modeled after naturally occurring minerals. Quartz crystals, for example, were determined to have piezoelectric behavior in a particular crystal direction. Natural quartz crystals were (and still are) mined and sliced into devices such as oscillators. The demand has become so large, especially since the invention of the electronic quartz watch, that techniques were developed to grow quartz crystals in the laboratory. Should the synthetic crystals be categorized as ceramic and the identical natural crystals not?

Ceramic materials also are formed by living creatures. Sea creatures extract calcium and carbon dioxide from seawater and construct calcium carbonate shells and intricate ceramic–polymer composites. Our teeth and bones are mostly ceramic carefully constructed atom by atom at room temperature. An important new area of materials engineering is to try to duplicate these processes and structures.

1.3 SO WHAT IS A CERAMIC?

The above discussions and definitions lean in the direction of defining a ceramic in terms of chemical composition, but we still have the issue of which chemical compositions are ceramic. Is it the composition (the atoms present) or the way the atoms are bonded together? Diamond is a good example. Most people consider diamond as a ceramic. However, silicon has exactly the same crystal structure and nature of atomic bonding (covalent) as diamond and is rarely referred to as a ceramic. Instead, most people refer to silicon as a semiconductor or even sometimes as a metal. The term *semiconductor* semantically represents a property, but its use has been extended by people in the physics and electronics communities to represent a category of materials. This mixed use of semantics further confuses the picture.

The diamond and silicon example brings up the issue of properties. Can a ceramic be cleanly defined based on properties, characteristics, or behavior? Here again we have a problem. What properties are inherent to a ceramic? Most people have the concept or perception that a ceramic is brittle, has a high melting temperature, is a poor conductor of heat and electricity, and is nonmagnetic, and that a metal is ductile, is a good conductor of heat and electricity, and can be magnetic. These stereotypes may be true for many ceramics and metals, but not for all of them. For example, some ceramics are magnetic and are key constituents in a wide variety of products ranging from telephones to loudspeakers to linear accelerators. More than 600 million kilograms of ceramic magnets are produced each year. Other ceramics are electrically conductive and make possible products such as the oxygen sensors in our automobiles that increase our gas mileage and reduce pollution.

Coming up with an all-encompassing definition of a ceramic that everyone accepts and that really explains "what a ceramic is" appears even more difficult than we initially thought. There are just too many variables and conflicting personal perceptions. What is much more important than having a precise definition is to acquire a fundamental understanding of the relationship of the behavior of *any* material to its chemical composition, atomic bonding, crystal structure, fabrication process, and any other important variables. That will be the focus in subsequent chapters in this book: to understand the factors that control the behavior. As we proceed, hopefully you will come up with your own "working" perception of ceramics and other materials that will allow you to effectively select and design materials for the engineering challenges that you will face in your careers.

> **UNDERSTANDING MATERIALS**
>
> More important than having a precise definition of "ceramics" is to acquire a fundamental understanding of the relationship of the behavior of any material to its chemical composition, atomic bonding, crystal structure, fabrication process, and any other important variables. That is the focus and goal of this book.

REFERENCES

1. Kingery, W.D., *Introduction to Ceramics*, Chapter 1, John Wiley & Sons, New York, 1960.
2. *Webster's Dictionary*, Lexicon Publications, New York, 1988.

STUDY GUIDE

1. What are some of the factors one might consider when trying to categorize a material as ceramic, metal, organic, or some other designation?
2. Do all materials fit into neat, well-defined categories? Explain.
3. Explain the meaning of "polycrystalline."
4. Explain the difference in atom arrangement of a polycrystalline or single-crystal ceramic compared to the same chemical composition in the form of glass.
5. What is the term used to refer to firing of a compact of ceramic powder to achieve a polycrystalline ceramic?
6. Can a ceramic glass composition be converted into a polycrystalline ceramic? Explain.
7. Can a broad material category (such as ceramic, metal, or organic) be effectively defined based on properties? Explain.
8. What is a wise alternative to trying to "force" materials to fit into broad categories?

2 History of Ceramics

Chapter 1 attempted to define "ceramics" and to provide an initial glimpse into some of the materials that comprise the ceramics category. That was just a small first step toward reaching the ultimate goal of understanding the composition/property/processing relationships of ceramics and applying ceramics to engineer useful products. Chapter 2 takes a second step by reviewing how the use of ceramics progressed from prehistoric times to our modern times. Table 2.1 highlights some of the key milestones.

2.1 CERAMICS IN THE STONE AGE

2.1.1 USE OF NATURAL CERAMICS

Most of the earth's solid surface is made up of natural ceramic materials, either as solid rock or as fragments of rock eroded by the forces of nature. The earliest humans undoubtedly used these natural ceramics as they found them. Natural crevices between boulders or in hillsides probably provided shelter. Stones were likely thrown for defense and for hunting.

As time passed, people learned that they could chip rocks such as flint and obsidian to make tools that worked better than as-found rocks. Prehistoric sites along the shores of Lake Rudolf in northern Kenya have yielded deliberately shaped flint tools that have been dated to about 2.5 million years old.[1] These were crude, consisting of pebbles struck with another rock to form edges sharp enough to cut hide and meat.

During the next two million years, humans spread from Africa to Europe and Asia and continued to make simple stone tools. By 500,000 years ago *Homo erectus* became the dominant hominid species, with a slightly larger brain than his predecessors. Archaeological sites indicate that *Homo erectus* hunted and killed large game (bison, deer, wild boar, horse, and even elephant and rhinoceros), used fire for warmth and cooking, and prepared stone tools of increased sophistication. Of particular note were hand-held axes that were carefully chipped on all surfaces to produce a durable, versatile tool for skinning and cutting. These were in use until about 50,000 years ago and are usually referred to as Acheulian hand axes after the site where many of them were found.[2]

Homo sapiens (wise men) appeared around 250,000 years ago. Very little early archaeological information exists until about 70,000 years ago when the subspecies *Homo sapiens* Neanderthalis emerged. The Neanderthals developed their own style of stone tools, referred to as Mousterian after the cave site of Le Moustier in France where some of the best examples were discovered. The Neanderthals were masters at chipping flakes of flint of just the right size and shape for each desired use. Scrapers were especially important because many of the Neanderthals lived in cold northern climates and needed to prepare animal hides for clothing. However, they also made small spearheads, hand axes, long knives with one edge sharp and the other blunt, and notched blades for shredding food.

By 35,000 to 30,000 years ago Neanderthal man disappeared. Modern man, *Homo sapiens* sapiens, became the only remaining human subspecies and made significant advances in making tools of stone, bone, and ivory. One especially important technique they developed was pressure-flaking. First they would make a rough shape (such as a blade) by chipping and by the use of wood or bone punches struck with a stone hammer. Then they would refine the shape by carefully pressing pointed tools along the edges to remove small slivers of flint until the desired shape and edge

TABLE 2.1
Key Milestones in the Evolution of Ceramics

Approximate Time	Description of Milestone
24,000 BC	Clay mixed with water, molded into a shape, dried, and fired to produce the first man-made ceramic "earthenware"
10,000 BC	Early use of earthenware for useful articles such as containers
7000–6000 BC	Lime mortar used for filling spaces between stones in construction and also for making thick-walled containers
7000–5000 BC	Widespread use of earthenware for food storage and cooling
7000–5000 BC	Widespread use of plaster-like cementitious ceramics in floor construction and as decorated interior wall coatings
4500 BC	Ceramics used to line crude kilns to smelt copper from its ores and make possible the Chalcolithic, Bronze, and Iron Ages
4500–3500 BC	Fired brick shown to be more durable than dried mud bricks
4500–3500 BC	Development of early faience to produce white ceramics with bright blue glassy coatings; precursor to development of glass and glazes
3500–2800 BC	Evolution of first writing consisting of pictographs and "cuneiform" symbols inscribed in clay tablets
3300 BC	Potter's wheel invented, making earthenware pottery available and affordable for just about everyone
2500–1600 BC	From early isolated examples of glass to well-established craft
1750–1150 BC	Variety of glazes developed including lead glazes and colored glazes
1025–750 BC	Early developments in China of stoneware, which was lighter in color, stronger, and less porous than earthenware
500–100 BC	High-strength concrete developed by Greeks and Romans and used widely in construction especially by the Romans
100–50 BC	Glass blowing invented, making glass items available to other than royalty and aristocrats
600–800 AD	Evolution of high-quality porcelain in China
800–900 AD	Islamic refinements of the use of white tin-based glazes and lustrous overglazes
1740s	"Transfer printing" invented to greatly increase the rate of production of tiles
1840s	"Dust pressing" invented to further dramatically increase rate of production of tiles and eventually become the uniaxial pressing process and workhorse of modern high-volume production of polycrystalline ceramics
Late 1800s	Demonstration of successful laboratory synthesis of single crystals of ruby and sapphire
Early 1900s	Increase in temperature capability of production kilns to greater than 1400°C and later to above 1600°C
Mid-1900s	Establishment of alumina-based ceramics and introduction of a wide range of new engineered ceramics
Late 1900s	Development of ultra-high-temperature, high-strength, and high-toughness ceramics and ceramic matrix composites

sharpness were achieved. These techniques of flint smithing were employed to produce fine spearheads, axes, arrowheads with an extension for attachment to a wooden shaft, tiny pointed awls for punching holes in hide to sew clothing together with sinew, and even tools for cutting and engraving bone and making jewelry.

Homo sapiens sapiens also pioneered the use of other materials besides stone roughly between 30,000 and 15,000 years ago. They used their sophisticated tools to carve needles, fish hooks, harpoons, and spear throwers from bone or ivory.[3]

Some of the most important archaeological sites are graves.[4] At least as far back as 40,000 years, *Homo sapiens* sapiens buried their dead with ritual and symbolism. The bodies were buried wearing extensive ornamentation and were covered with red ochre (a natural ceramic powder rich in iron oxide). The ornamentation included ivory beads, seashell beads, bracelets, and necklaces. One skeleton was found with strands of over 3500 ivory beads. A few carvings of ivory and stone also have been found, a famous one being the "Venus of Willendorf." It is a 10 cm high limestone carving of a voluptuous female figure that was carved in Austria about 30,000 years ago. It probably had fertility significance.

At some distant time in the past, humans learned that they could mix water or perhaps animal oils with colored soil (such as red ochre or various clay minerals) to make pigments that could be used for body decoration. These pigments eventually were used to paint images of animals on cave walls. Chauvet Cave, discovered in France in 1994, is decorated with 420 animal images that date back as far as 35,000 years ago.[5,6] Some were painted with red ochre pigment, others with a yellowish clay pigment, and others with charcoal. Another important cave with wall paintings is Lascaux Cave, also in France. These paintings date to about 20,000 years ago.

Caves have preserved evidence of another important innovation of prehistoric artisans: three-dimensional sculptures molded in clay. The term "clay" represents a family of natural ceramic powders that result from the chemical alteration of minerals such as feldspar during weathering. The clay particles consist of tiny thin hexagonal crystals that have an affinity at their surface for water molecules. As water is added the clay–water mixture becomes pliable, similar in consistency to children's molding clay. A three-dimensional shape such as the image of an animal can be hand-formed and retain its shape due to the stiffness (high viscosity) of the clay–water mixture. The sculpture then becomes hard as it is dried. It would still soften and erode if exposed to rain or other source of water. However, if the sculpture was protected in a cave, it might be preserved. Such is the case for Tuc d'Audobert Cave in France. Two molded and dried clay bison are still as sharp and detailed as when they were sculpted 14,000 years ago.[7]

2.1.2 SYNTHETIC STONE: CLAY TRANSFORMED BY FIRE

No one knows when the first artisan crafted a three-dimensional shape from clay. No one knows when someone took the next crucial step of placing the dried clay article in a fire, but this was an enormous step for mankind. It represented the first time that a person invented a new material by using fire to transform a natural material.

The oldest evidence found so far for firing of clay articles comes from a 24,000- to 26,000-year-old site in what is now the Czech Republic.[3,8] This site consists of a fire pit that appears to be designed specifically for firing pottery. Fired clay figurines of animals and Venus-like female sculptures were found in and around the pit. Although no one knows for sure their purpose, archaeologists speculate that the fired clay figurines were used in rituals to assure a successful hunt or fertility.

This first man-made ceramic was a big improvement over dried clay because it did not fall apart in the rain. It had the characteristics of soft stone, but could be molded to complex shape rather than the tedious process of carving a shape from stone. We now refer to clay fired at low to moderate temperature as "earthenware." In spite of its ancient origins, earthenware is still made today in nearly every corner of the world.

> **EARTHENWARE**
>
> Major breakthrough; first man-made material, achieved by transforming a molded shape of soft clay into a hard stone-like article of ceramic.

2.1.3 FIRST PRACTICAL USE OF EARTHENWARE

Even though earthenware was invented at least 26,000 years ago, many centuries passed before it would reach widespread use for practical applications. Some pots were used in Japan about 12,000 years ago, but widespread use for storage and cooking did not occur until about 7000 to 8000 years ago. The rise in production in earthenware pottery coincided with the growth of permanent settlements and agriculture.[2,3]

Early agricultural settlements were established along the eastern coast of the Mediterranean Sea, north in Anatolia (present-day Turkey), and east in the regions presently occupied by Iraq and Iran. Most of these regions were semiarid. Dwellings were typically made of dried mud and clay and periodically required rebuilding. New construction was done on top of old construction. Thus many of the archaeological sites are mounds that are layered with the oldest strata at the bottom. Layers dated roughly between 8500 and 6300 BC contained unfired clay articles. This period is generally referred to as Pre-Pottery Neolithic. By 6300 BC some crude pottery that was only lightly fired began to be used. By 6000 BC fired clay articles were in use at most of the sites that have been excavated.[2]

One of the largest archaeological excavations has been at Çatal Hüyük in Anatolia.[9,10] Multiple layers were studied that were estimated to have been inhabited between about 6250 and 5400 BC. All of the layers contained fired pottery. The pottery in the oldest layer was cream colored and burnished (surface rubbed to provide smoothness and a little gloss), but had little or no decoration. Pottery in successively more recent layers was usually higher quality (typically fired at increased temperature so stronger and less porous) and contained surface decorations.

All of the evidence at the various archaeological sites indicates that the fabrication of earthenware pottery had become an important widespread craft by 5500 BC for preparation of vessels for food preparation and storage. All of the vessels were hand built using slabs or coils of damp clay, and some were greater than 1 m high.

2.1.4 OTHER NEOLITHIC CERAMIC INNOVATIONS

Other ceramic technologies were developed during the Pre-Pottery Neolithic period or perhaps even earlier. One was the use of plaster-like materials that would set like cement after water was added. Some of these materials were made with gypsum. Gypsum is $CaSO_4 \cdot 2 H_2O$.

> **GYPSUM PLASTER and LIME MORTAR**
>
> When a powder mixed with water, chemically reacted to produce at room temperature a rigid ceramic body.

If gypsum is carefully heated, a portion of the water can be removed to yield $CaSO_4 \cdot 0.5 H_2O$ (hemihydrate). When this is ground into a powder and water is added, it will rehydrate to grow tiny intertwining crystals of gypsum that form a solid (but porous) body. Today we call this material plaster of Paris and use it to make wallboard, molds, sculptures, and casts (such as for a broken arm).

Ancient people used plaster-like cementitious materials in construction. Excavations at Çayönü Tepesi in Anatolia, which was inhabited between about 7300 and 6500 BC by a population of 100 to 200 people, revealed several types of floor construction.[2] One floor was plaster-like. Another consisted of white limestone cobbles and crushed pieces bonded by a concrete-like ceramic. While the cement was still wet, salmon-colored pebbles and white pebbles were pressed into the surface to form a decorative pattern. After the concrete set, the whole floor was ground smooth and polished. A third type of floor consisted of limestone slabs interlaid like tiles.

Inhabitants of Çatal Hüyük between 6250 and 5400 BC had dwellings with plaster walls. The plaster was decorated using molded clay relief and incising to produce multicolored geometric patterns resembling woven rugs. Other decorations were paintings depicting scenes.

Plaster was also known further south in Jericho during the seventh millennium BC, but was used in a different way.[4] When a person died, the body was buried either beneath the floor of the house or in the space between houses, but the head was retained. When all of the flesh was gone and only the skull remained, the family would carefully mold plaster onto the front of the skull to depict the features of the departed person. The plastered skull would then be placed in a position of honor within the dwelling along with the plastered skulls of other departed family members.

Another early ceramics invention was lime mortar that was made using a source of lime (CaO) such as from pulverized heated (burned) shells or limestone plus the ashes of salty grasses. This was mixed with water, used to fill in the spaces between stones (such as in a wall), and hardened like cement without requiring firing. By around 6250 BC people along the eastern shore of the Mediterranean (a region know as the Levant) were making thick-walled vessels using mortar (*vaisselle blanche*; white ware). This predated the use of fired clay in that region.

Parallel to the developments of earthenware pottery and various forms of mortar, Neolithic people in the Near East learned that clay bricks became harder and more resistant to the weather by placing them in a fire. This became increasingly more important as civilizations were established, cities were built, and urban populations grew. Larger and larger buildings were needed. One of the first civilizations was established by the Sumerians in Mesopotamia in the fourth millennium BC. They constructed enormous temples called ziggurats. The core of the ziggurat was constructed of dried mud bricks. These were then covered by several layers of fired bricks cemented together with mortar made with waterproof bitumen (tarlike raw petroleum).[3]

The earliest pottery vessels and bricks were relatively porous because they apparently were fired at relatively low temperature. As time passed, though, potters learned how to build kilns (furnaces) that could achieve higher temperature and produce higher-quality ceramic vessels. They also developed techniques for applying decorations to their ceramic ware. As their ancestors had done for centuries for body decoration, cave paintings, and wall paintings, these potters prepared slips of colored clay and painted geometrical designs onto their pottery. Often they would burnish (polish by rubbing) the surface with a stone to achieve a pleasant sheen. Other techniques of decoration included incising (scratching) or pressing shapes into the damp clay before firing.

The concept of incising led to a completely different use of ceramic-based materials. By 8000 BC the first permanent settlements and towns had formed and Neolithic people were starting to domesticate crops and animals. Trade also was becoming important. To keep track of exchanges or shipments of grain and animals, people scratched markings onto damp clay tokens. As cities grew, this practice expanded and became increasingly more sophisticated and eventually evolved into the first written language by about 3500 BC in Sumeria.[11,12] The writing was scribed onto a damp clay tablet using a sharpened piece of reed. In the early stages of development (about 3500 to 3000 BC), the language was largely pictograph, but slowly transitioned (by about 2800 BC) into wedge-shaped script that we now refer to as cuneiform (from the Latin word *cuneus*, a wedge). Libraries of thousands of dried clay cuneiform tablets (such as the one shown in Figure 2.1) excavated from archaeological

FIGURE 2.1 Cuneiform tablet from the Utah Museum of Natural History collection. (Photo by D.W. Richerson.)

sites in the Near East have revealed to us in great detail information about commerce, law, daily life, and even early literature.

2.2 THE RISE OF TRADITIONAL CERAMIC INDUSTRIES

2.2.1 Ceramic Innovations during the Chalcolithic Period

Prior to about 4500 BC, nearly all tools and decorations were made from stone and ceramics. Metals were known from natural nuggets of copper and gold and from iron meteorites, but were very rare. Around 4500 BC people in the Near East learned how to extract copper probably from the green copper ore malachite.[13] This was the beginning of an era where people continued to use stone implements but progressively started to use more and more copper, thus the name Chalcolithic (copper-stone).

Although the exact details can only be speculated, archaeological evidence suggests that ceramics played a substantial role in the discovery of metals smelting and in the subsequent metal production processes. One possible scenario involves the decoration of ceramics. Ancient peoples prized colored stones such as lapis lazuli (blue), malachite (green), and azurite (blue). Lapis came all the way from Afghanistan, so was especially valuable. Malachite and azurite were more widespread in the Near East, but were not as hard and durable as lapis. Craftsmen apparently experimented with crushing malachite and azurite, preparing a slip or paste, and then painting this on the surface of pottery and also rocks such as soapstone. The painted pottery or rock was then placed in a pottery kiln with the hope that the ceramic or rock would take up the green or blue color. This approach did produce color and eventually did lead to the invention of glazes, but was not sufficient to convert the copper in the ore to metal.

A second condition was necessary to yield copper metal. The atmospheric conditions in the furnace needed to be reducing. Fortunately, by 4500 BC potters had already learned the importance of controlling the atmosphere inside the kiln.[8] They had learned by experience that the color of their pottery decorations was different if fired in a smoky atmosphere (black) vs. a nonsmoky atmosphere (red or brown). This was known as early as 5500 BC in Mesopotamia and was in widespread use by 4000 BC. Thus, it is likely that someone fired a ceramic article coated with malachite under smoky (oxygen-deprived reducing) conditions and observed little droplets of pure copper. This then led to designing clay-lined furnaces specifically for reducing copper ore to metallic copper. As the early metalworkers gained experience, they learned that they could increase the amount of copper they could produce by using charcoal for fuel, by forcing air onto the charcoal to increase the temperature in the furnace, and by designing better ceramic-lined furnaces. They also learned to make ceramic containers (crucibles) for holding and refining molten metal and to make ceramic molds for casting the metal into complex shapes. These early technologies were the ancient origin of the refractories industry.

Other ceramic innovations also occurred during the Chalcolithic period. One was faience.[8] Neolithic earthenware all had a dark color because the clays that were used contained iron as an impurity. Around 4500 BC someone began experimenting with white quartz pebbles. They crushed the quartz, ground it into a fine powder, and mixed it with malachite, natron (a sodium-containing salt found in dry lake deposits in the Near East), and other ingredients (possibly ground shells and ashes). The powders were mixed with water into a stiff paste and formed into a shape (probably in a clay mold). During drying, the water-soluble salt partially migrated to the surface. During firing to about 900°C, the natron melted at about 850°C to bond the quartz particles together and to act as a flux to react with the malachite to form a glassy blue surface coating. The temperature was not high enough to melt the quartz, so it remained as white grains. The result was that the beautiful blue color of the coating was not distorted by the white background as it would be if it were on dark-colored earthenware. This early faience is generally called Egyptian faience and represents perhaps the first example of a colored ceramic glaze.

Two other important ceramic developments occurred during the Chalcolithic period (~4500 to 3150 BC). One was the potter's wheel (probably around 3300 BC), and another was floor and wall tile. The potter's wheel allowed an individual to dramatically increase his output of ceramic vessels per day and thus make them available to every household.

> **FAIENCE**
>
> White ceramic coated with a glassy blue coating; early step towards development of glass and glazes.

2.2.2 CERAMICS AND THE METALS AGES

By ~3600 BC metal workers had learned the benefits of alloying copper with arsenic or tin. They didn't know anything about atoms, the chemical elements, solid solution strengthening, or work hardening, but they did observe that the mixture melted at a lower temperature, had greater hardness and strength, and could be crafted into superior tools and weapons. Ceramics continued to be a necessity in smelting and processing these metals. By about 1400 BC the Bronze Age came to an end and the Iron Age began. Iron required a substantially higher temperature to smelt and process, so more demands were placed on the ceramic furnace linings, crucibles, and molds. Steel required even higher temperatures. Evolution was very slow for the iron and steel industries. The iron industry peaked in the 1700s and ushered in the Industrial Revolution. The steel industry began to peak in the late 1800s. As these industries evolved and expanded, so also did the ceramic refractories industry.

> **REFRACTORIES INDUSTRY**
>
> Ceramics enabled the emergence and production of metals by comprising the furnace linings and containers (crucibles, molds) that could withstand the high temperatures and contact with the molten metals.

2.2.3 THE EMERGENCE OF GLASS

Glass is another of our important ceramic-based industries that can be traced back to ancient times.[14] Although we do not know exactly when glass was invented, most scholars believe it evolved out of faience. Faience had a glassy coating, but the temperature of the early kilns was not high enough to also melt the quartz (silicon dioxide) grains. Also the composition was not quite right to form a true glass. However, between 4500 BC and about 1600 BC, potters improved their kilns to achieve higher temperature and also explored a variety of raw materials. Someone found a composition containing natron, lime (CaO), and silica that completely melted into glass.

Early glass was highly valued and probably could only be possessed by royalty (such as the Pharaohs) and perhaps rich merchants. Two methods of fabrication were developed: casting and core forming. Casting was achieved by melting the glass in an earthenware vessel and pouring the viscous molten glass into a mold made of earthenware. Figure 2.2 shows a small cast glass article that was made in Egypt between 1436 and 1411 BC.

Although the Egyptians probably did not invent glass, they certainly embraced it and experimented with its fabrication. They pioneered core forming as a means of producing a hollow vessel such as shown in Figure 2.3. A core having the desired inner shape of the vessel was crafted by hand by compacting a mixture of dung and moist clay around a metal rod. After the core was dried, a glass worker would insert a rod of ceramic into viscous molten glass and draw out a strand (similar in consistency to taffy) and quickly wind it around the core. Generally strands of different colors would be alternated. Then the worker would reheat the vessel and roll the outer surface against a rounded stone to flatten and smear the strands into a smooth surface.

Casting and core forming were slow processes and not amenable to mass production of a low-cost affordable product. By around 50 BC a new technique of glass forming was invented, glass blowing. Now a glass worker could produce many vessels per day so that the product was affordable to a

FIGURE 2.2 Cast glass head of Amenhotep II. From the early 18th dynasty of Egypt, about 1436–1411 BC. Originally cast as blue glass, but weathered at the surface to a buff color About 4.1 cm high. (Photo reprinted with permission of The Corning Museum of Glass, Corning, NY.)

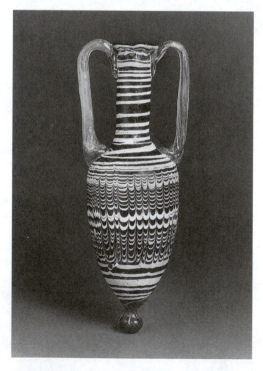

FIGURE 2.3 Core-formed glass amphoriskos. Probably from the Eastern Mediterranean Sea region in the first century BC and similar in style to the core-formed vessels from Egypt dating back to at least 1400 BC. About 23.9 cm high. (Photo reprinted with permission of The Corning Museum of Glass, Corning, NY.)

FIGURE 2.4 Blown glass from the Roman Empire. Examples from the first through fourth centuries AD that ranged in color from blue to greenish to yellowish to plumb. (Photo reprinted with permission of The Corning Museum of Glass, Corning, NY.)

much larger portion of the population. Figure 2.4 shows a variety of blown glass from the Roman Empire. Blown glass became so affordable that spectators attending events at the Colosseum could buy blown glass souvenirs commemorating the event.

2.2.4 CERAMICS IN BUILDING

All of the early civilizations used ceramics in building. By the time of the Greek civilization ceramic bricks, plaster and mortar, mosaic, and decorative tile all were in common use. The Greeks invented high-strength, water-resistant cement by mixing burnt lime (chalk and seashells heated to about 900°C to decompose to CaO) with volcanic rock. The Romans improved on this by adding volcanic ash called *pozzulana* from Mount Vesuvius and demonstrated that their cement could even set under water.[3]

The Romans were great engineers. They used their cement as a mortar and also as a base for adding various aggregates to make superior concrete. With these ceramic building materials they constructed aqueducts, bridges, and buildings with unprecedented spans. One bridge was 60 m high and stretched over 30 m across the Tagus River. Today it supports a four-lane highway. The aqueducts were equally impressive (see Figure 2.5 for one in Segovia, Spain). It was constructed of cut stone bonded with pozzulanic mortar and lined with hydraulic cement and is still in use today.

Perhaps the most spectacular example of Roman construction with concrete is the Pantheon, which was built between AD 115 and 125. It consists of a hemispherical roof supported only around the circumference by 20-foot-thick concrete walls. The dome spans 142 feet and contains 5000 tons of concrete.[3] To maintain maximum strength and minimize weight, the dome was poured in horizontal layers

FIGURE 2.5   Roman aqueduct in Segovia, Spain from the first century AD. (Copyright © 1998 by Stephen L. Sass. Reprinted from *The Substance of Civilization* by Stephen L. Sass, Published by Arcade Publishing, New York, NY.)

using three different grades of concrete. The bottom layer contained brick fragments as the aggregate, the middle layer volcanic lava rock pieces, and the top layer highly porous lightweight pumice.

Modern engineers have marveled at the quality of the Roman concrete. Besides the combination of ingredients, another "secret" to the quality was thorough mixing and ramming into place to decrease porosity. When the Western Roman Empire fell in the fifth century AD, the recipes and techniques for creation of high-quality cement and concrete were lost and were not rediscovered until ~1756. At that time John Smeaton in England systematically studied the chemistry of cement and concluded that clay or other suitable sources of alumina and silica must be added. This was the beginning of our traditional cement industry. Currently, ~1 ton of concrete is poured each year for each person in the world, which means that concrete is the most extensively used man-made material.

Parallel to the developments of mortar and cement, glazed tile and bricks became important building materials. Here the emphasis was more on beauty than structural benefits. Between about 1750 and 1170 BC new glazes were invented that contained lead. Lead acted as an effective flux that allowed the glaze to melt and freely flow over the surface at a much lower temperature than for prior glazes. This made possible brighter colors and a wider variety of colors.

By the end of the second millennium BC blue-glazed tile and brick were in use in Egypt, Mesopotamia, and Asia Minor. The scale of use of glazed tile and brick in the ancient world is illustrated by the Ishtar Gate built in Babylon under King Nebuchadnezzar II (604 to 562 BC). It was 10.5 m (34.5 feet) high and contained 575 glazed figures of real and mythical animals.

Lead glazes reached widespread use during the Roman Empire. After the fall of the Western Roman Empire, developments of ceramics and other technologies stagnated as Europe entered the Dark Ages, but continued in the Near East and after AD 632 in the Islamic territories. One technology that flourished was tiles coated with a glaze containing tin oxide. The tin oxide glaze was opaque white and could effectively hide the dark color of the underlying earthenware. When combined with a low-fired, lead-based overglaze, bright beautiful colors could be achieved. The interiors of many Islamic mosques are completely covered with colorful tiles (especially blue) made in this fashion.

Tile making continued as a labor-intensive handcraft until the 1740s. Up to that time each tile was formed from a slab of pottery clay and a design carefully crafted and glazed. Then an Englishman, Richard Prosser, invented transfer printing. This allowed a standardized design to be prepared separately with a wood-block tool and paper and then transferred to the surface of the tile. Two workers could now produce 1200 tiles in a day rather than requiring a workforce of 100 people. This was the first step to tile manufacturing becoming a major industry.

A second key innovation occurred 100 years later. It was "dust pressing." Rather than requiring a moldable clay mixture that took many hours to prepare and later dry, dust pressing involved adding just a little moisture. The nearly dry powder was placed between two metal plates and compacted by turning a large screw. The powder particles adhered together and the tile could be fired immediately without requiring a careful drying step. This enormously decreased the cost of making tile so that ceramic tile became afford-

> By Greek and Roman times, ceramics were widely used in building
>
> • Bricks
> • Plaster
> • Mortar
> • Concrete
> • Mosaic
> • Decorative tile

able to just about everyone. People began tiling virtually every room in their house. Besides beautification, a side effect was a big improvement in hygiene. As an illustration of the size of the traditional tile industry, enough tiles were manufactured in 1998 to pave a path 100 m wide completely around the world.[7]

2.2.5 Ceramic Whitewares

Another industry that can be traced to ancient times is whitewares, epitomized by fine porcelain.[8] The whitewares industry can be traced to three roots: faience, tin-glazed ware, and stoneware/porcelain. Faience and tin glaze have already been mentioned briefly and will be further discussed a little later. Stoneware and porcelain tech-

> Stoneware and porcelain were the beginning of the whitewares industry

nology began in China during the Shang Dynasty (1027 to 771 BC) and will be discussed first.

Chinese potters mixed finely ground "china stone" (petuntse, a mixture of quartz, feldspar, and mica) with white kaolin clays. This mixture required about 1200°C to fire, but yielded a dense "stoneware" pottery that was much lighter in color than earthenware pottery. Such a high temperature was beyond the capability of western potters, who could only reach about 1000°C, but was possible in China because of their special kiln designs. By the T'ang Dynasty (618 to 906 AD) Chinese kilns had been improved to reach ~1300°C, which made possible firing a true porcelain that was completely white and even translucent.

One type of kiln developed in southern China during the Song (Sung) Dynasty (960 to 1279 AD) was the "dragon kiln." A dragon kiln built in the 12th century as discovered and excavated near Longquan was ~30 m long and 2 m wide.[8] Constructed of brick arches covered with a thick mound of refractory earth insulation, this kiln was built climbing a 15 to 20° slope. The firebox was at the bottom and the flue at the top to encourage flow of heat through the furnace. Unfired porcelain items were each placed in a kiln furniture box (with lid) made of refractory clay such that ware could be stacked and essentially positioned along the whole length of the kiln. It has been estimated that a single kiln could hold as many as 100,000 pieces for a single firing, which represented a substantial level of production.

Firing the dragon kiln was done in stages. First, a fire was lighted in the main firebox. Ware near the firebox reached densification temperature first, while ware further up the kiln was preheated. Fire was then lighted in fire holes further up the kiln sequentially until all of the ware was successfully fired. Equivalent kiln technology would not be achieved in the West until the late 19th century.

The beautiful white porcelain of China generated much interest and led to efforts to duplicate it elsewhere. One approach was an extension of the faience concept. White quartz was crushed, mixed with ground up glass and white clay, and fired. This became popular in Persia by the 13th century AD.

Meanwhile, by the ninth century AD Islamic potters refined the art of the use of an opaque white tin glaze and developed techniques for applying an overglaze painting with metallic luster. This led by the 13th century to painting with cobalt blue pigments covered with a colorless glaze, which was modified by the Chinese for use with porcelain. The Islamic technology spread through Spain into Europe and became known as "maiolica." By the 16th and 17th centuries, maiolica (such as delftware in Holland) was the dominant form of luxury ceramics throughout Europe.

The faience-derived and maiolica ceramics were not comparable to Chinese porcelain. Efforts continued during the 16th and 17th centuries to duplicate Chinese porcelain, but, as in the past, were based on mixing a glass frit with other ceramic ingredients. Finally, in ~1708 in Meissen, Germany Count von Tschirnhaus and Johann Friedrich Böttger succeeded in duplicating Chinese porcelain using feldspar as the flux rather than lime or a glass frit. An important part of their success had involved reaching temperatures in the 1200 to 1300°C range.

Many technologies came together during the 1700s: increased understanding of chemistry, availability of the microscope to observe microstructure, introduction of machines and the assembly line approach, and improved kilns. White wares became an important industry in Europe and England and spread to colonial North America. Stoneware production was begun in the colonies in 1720 and porcelain in 1770.

2.3 FROM TRADITIONAL TO MODERN CERAMICS

The whitewares industry grew through the 1800s as population increased and as a larger portion of the population was able to afford fine ceramics. Then something new happened. People began to understand electricity and to build electrical devices. They discovered that porcelain was an excellent electrical insulator and that impurities such as iron were detrimental. Researchers began to improve existing ceramics by more carefully refining the raw materials and even synthesizing pure ceramic powders by chemical reactions. They also invented new ceramic compositions with special characteristics necessary to satisfy the demands of new high-technology applications.

A good example was the emerging automobile industry.[7] The reciprocating engine required a spark plug, which in turn required a reliable electrical insulator. The first spark plug insulators were made of porcelain and fabricated on a potter's wheel. They weren't very reliable; a driver would be lucky to travel 50 miles without having to replace a spark plug.

The early spark plugs were made from a traditional porcelain composition fabricated with naturally occurring clay, flint, and feldspar. Researchers determined that additions of aluminum oxide (alumina, Al_2O_3) resulted in improvements, but this caused two other problems. First, higher alumina compositions required higher firing temperature. For example, the baseline porcelain composition could be fired below 1300°C, while an 80% alumina composition required about 1425°C and a 95% alumina composition ultimately would require about 1620°C. New furnace technology had to be developed to achieve these higher temperatures.

The second problem was forming the complex shape needed for the spark plug insulator. The baseline porcelain composition included enough clay for the mixture of powders to stick together and be workable enough to shape simply by adding water. Water alone did not work when clay was replaced by alumina. A whole new technology of additions of organic binders and of high-rate shape-forming processes (such as injection molding and dry bag isostatic pressing) had to be developed. Figure 2.6 shows a modern high-alumina spark plug insulator (fabricated by dry bag isostatic pressing) next to a completed spark plug assembly.

One of the early spark plug developers was AC Spark Plug. By 1920 they were producing 60,000 spark plugs per day for World War I aircraft. By 1949 more than 250 million high-alumina automotive spark plugs were being produced per year, and today more than three million are produced worldwide each day.

> Evolution of alumina spark plug insulators played an important role in transition from traditional ceramics to highly engineered modern ceramics.

The dramatic rise of the automotive industry drove a rapid growth in ceramic science and technology. As the 20th century progressed, other favorable characteristics of alumina ceramics were recognized: high hardness and excellent corrosion resistance up to high temperature. This led to advanced applications as abrasives, laboratory ware (mortars and pestles, filters, etc., as shown in Figure 2.7), wear-resistant parts, and refractories. By 1995, ~5 million metric tons of high-alumina products were being manufactured each year.

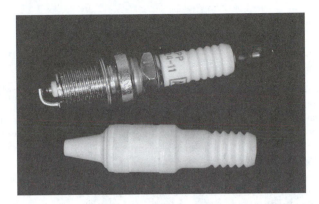

FIGURE 2.6   Alumina-based electrical insulator for an automotive spark plug plus an assembled spark plug. (Photo by D.W. Richerson.)

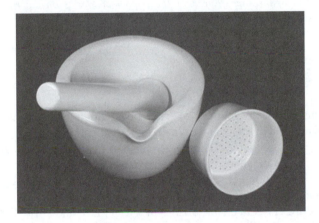

FIGURE 2.7   Alumina laboratory ware fabricated by Coorstek, Golden, CO. (Photo by D.W. Richerson.)

Alumina was the pioneer of synthesized, highly engineered modern ceramics. The science, techniques, and technologies learned for alumina during the first half of the 20th century were applied to a wide range of other ceramic compositions. Whole families of new ceramics were developed: magnetic ceramics, optical ceramics, high-temperature ceramics, high-strength ceramics, piezoelectric and other dielectric ceramics, semiconductor ceramics, electrochemical ceramics, bioceramics, single crystals, and specialty glasses.

2.4 SUMMARY

Ceramics is one of the oldest professions, dating back to the time that the earliest civilizations were forming. Traditional ceramics (earthenware, bricks, tile, refractories, glass, cement, and whitewares) have evolved over many centuries and continue to be indispensable today. Within the past century, though, there has been an enormous (virtually exponential) increase in the number and uses of ceramics. An engineer now needs to have a much broader awareness and understanding of ceramic materials. The next chapter identifies some of the key ceramic materials and families of materials and some of their applications and products. Subsequent chapters then focus on the relationships of composition, structure, processing, microstructure, and behavior that need to be known to effectively engineer with modern ceramic materials.

REFERENCES

1. *The Last Two Million Years*, The Readers Digest Association, London, 1973.
2. Redman, C.L., *The Rise of Civilization*, W.H. Freeman, San Francisco, 1978.
3. Sass, S.L., *The Substance of Civilization*, Arcade Publishing, New York, 1998.
4. *Tombs, Graves and Mummies*, Bahn, P.G., Ed., Barnes & Noble Books, New York, New York, 1996.
5. *La Grotte Chauvet: L'Art des Origines*, Clottes, J., Ed., Editions du Seuil, Paris, 2001.
6. Clottes, J., "Chauvet Cave," National Geographic Magazine, pp. 104–121, August 2001.
7. Richerson, D.W., *The Magic of Ceramics*, The American Ceramic Society, Westerville, Ohio, 2000.
8. Kingery, W.D. and Vandiver, P.B., *Ceramic Masterpieces*, The Free Press, a division of MacMillan, New York, 1986.
9. Mellaart, J., *Early Civilizations of the Near East*, Thames and Hudson, London, 1978.
10. Mellaart, J., *Catal Huyuk*, Thames and Hudson, London, 1967.
11. Claiborne, R., *The Birth of Writing*, Time-Life Books, Alexandria, Virginia, 1974.
12. Nemet-Najat, K.R., *Daily Life in Ancient Mesopotamia*, Greenwood Press, Westport, Connecticut, 1998.
13. Raymond, R., *Out of the Fiery Furnace: The Impact of Metals on the History of Mankind*, The Pennsylvania State University Press, University Park, 1986.
14. Zerwick, C., *A Short History of Glass*, H.N. Abrams, Inc., New York, 1980.

STUDY GUIDE

1. Explain the significance of learning to convert clay into a hard ceramic by firing.
2. What is gypsum plaster, and why was it an important early discovery?
3. What was the role of ceramics in the development of a written language?
4. What was the role of ceramics in establishing a metals industry?
5. What was the connection between lime mortar and the high quality cement later developed by the Greeks and Romans?
6. What are the differences in characteristics of a lead glaze and a tin glaze?
7. Why was "dust pressing" an important innovation?
8. What were some key factors that enabled a transition from traditional ceramic industries to modern ceramics technology, processing, and applications?

3 Applications: Engineering with Ceramics

Chapter 2 reviewed the history of ceramics and the evolution of some specific materials and applications, especially those that comprise our traditional ceramic industries: earthenware, whitewares, refractories, cement, bricks, tile, and glass. The traditional ceramic industries continue to be necessary to our modern society, but whole new categories of specialty applications have arisen as shown in Figure 3.1. For many of these applications, the ceramics have been carefully engineered to meet demanding and often narrow specifications of properties and configurations.[1] The objective of this chapter is to introduce you to the wide range of modern ceramic applications and the key characteristics that the ceramics must have to satisfy the application requirements.

> **TRADITIONAL CERAMIC INDUSTRIES**
>
> - Glass
> - Brick
> - Tiles
> - Earthenware
> - Whitewares
> - Refractories
> - Cement
> - Plaster

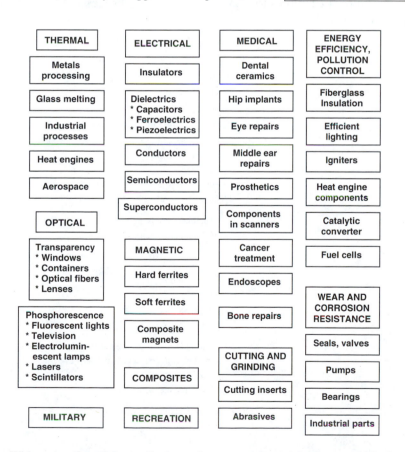

FIGURE 3.1   Wide range of specialty applications where ceramic materials play a critical role.

3.1 HIGH-TEMPERATURE APPLICATIONS

Throughout history ceramics have been the materials of choice for high temperature applications. They have a rich heritage as refractories for metal smelting and shaping, for glass production, for petroleum refining and chemical processing, and for energy conversion. They also have played an important role in the emergence of our aerospace industry and for key components in heat engines.

3.1.1 CERAMICS IN METALS PROCESSING

Our modern society is completely dependent on metals. By the beginning of the 21st century more than 725 million tons of steel[2] and 24 million metric tons of aluminum[3] were being produced per year worldwide. As shown in Table 3.1, ceramics are required in many ways in extracting and refining metals and converting these metals into

> Ceramic lining in BOF must survive severe conditions for thousands of cycles:
>
> - Thermal shock
> - High temperature
> - Slag corrosion
> - Molten metal attack
> - Erosion and impact

TABLE 3.1
Applications of Ceramics for Extracting, Refining, Processing, and Quality Assurance of Metals

Mining and ore processing	Diamond-studded drill bits for exploration and mining
	Wear parts for ore processing
	Vat linings for chemical treatment of some ores
Smelting, refining, melting, holding	Furnace linings (refractories)
	Heat sources (burners, heaters)
	High temperature lances and bubblers
	Cell liners and electrodes for electrolysis of aluminum
	Molten metal filters
	Crucibles
	Heat exchangers
	Oxygen sensors
Shape forming	Crucibles, molds, cores
	Molten metal filters
	Insulation for molds
	Nozzles
	Extrusion dies
	Molten metal pump parts
	Thermocouple protection tubes
Finishing	Cutting tool inserts
	Bearings for high-speed cutting tool machines
	Grinding wheels, sand paper, grit blasting abrasives
	Lasers, waterjet cutting
	Ultrasonic cleaners
Heat treatment	Furnace refractories, heat sources
	Setter plates, fixtures, conveyors
	Heat exchangers
Inspection	Transducers for ultrasonic nondestructive inspection
	X-ray radiography equipment components
	Dimensional measurement instruments

useful products. Note that often a combination of a material's characteristics is required: high temperature stability plus low thermal conductivity for some of the refractory liners; high hardness and high thermal conductivity for cutting tool inserts.

As we discussed in Chapter 2, the earliest metals that were extracted from ores were copper and lead. These are relatively low melting metals (copper, 1085°C; lead, 327°C) and did not require sophisticated ceramics. Simple clays and sand-based ceramics and even shaped molds carved from sandstone sufficed. Now metallurgists must deal with high temperature alloys such as the superalloys and titanium that require much higher temperature furnaces, crucibles, and molds. Thermodynamic stability of the ceramic becomes an important issue. For example, molten titanium has such a high affinity for oxygen that it will take oxygen away from most oxide ceramic refractories.

Another issue is complex shape. To support fabrication of the wide variety of metal shapes required by consumers and industry, a number of metal casting processes have been developed such as sand casting, lost wax casting (investment casting), and lost foam casting.[4] All of these require ceramics.

A good example of the role of ceramics is the casting of gas turbine engine rotor blades by the lost wax technique. Modern superalloy turbine blades are exposed to such high temperatures in the engine that they must be cooled with forced air flow through intricate internal passages. The only way to make these intricate air channels is during the casting process. Figure 3.2 shows a typical ceramic mold design to accomplish this difficult casting operation. Initially individual blades are fabricated from wax by injection molding. These wax blades are then attached to a cone-shaped solid piece of wax. This assembly is dipped into a slurry having the consistency of paint that is comprised of fine ceramic particles suspended in water. Upon removal from the slip, the wax model retains a thin coating of the ceramic particles. This is called the "dip" coat and is the ceramic layer that ultimately will be in contact with the molten metal during casting.

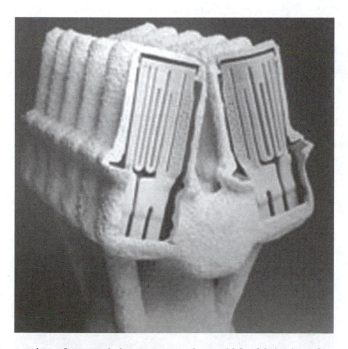

FIGURE 3.2 Cutaway view of a ceramic investment casting mold for fabrication of gas turbine engine rotor blades showing the ceramic cores for producing the complex internal cooling passages in the cast metal blades. Fabricated by Howmet Research Corporation, Whitehall, Michigan. (From Richerson, D.W., *The Magic of Ceramics*, The American Ceramic Society, Westerville, OH, 2000, p. 224. Reprinted with permission of the American Ceramic Society, www.ceramics.org)

The wax model is dipped a second time, but while still wet is immersed in a fluidized bed of dry coarser (sand size) ceramic particles that adhere to the damp dip coat. The sequence is repeated a number of times until the desired mold thickness is achieved. After thorough drying, the assembly is heated to a temperature at which the wax melts and flows out, leaving a hollow ceramic mold with internal pockets having the external shape of the desired rotor blades. The mold is now fired at a high enough temperature to give some strength, but not high enough to eliminate porosity or significantly change dimensions. Porosity is required to allow gases to escape during the casting process to avoid bubbles in the metal. Tight dimensional control is necessary to minimize expensive machining operations following casting.

Next a ceramic core is carefully positioned in each rotor blade cavity as shown in the cutaway mold in Figure 3.2. The core consists of a special mixture of ceramics that can withstand the temperature of casting and not react with the metal, yet can be removed completely by dissolving with a weak acid. The ceramic core is fabricated by injection molding, so it can be made to very complex shape and tight dimensional tolerances.

To complete the investment casting process, the ceramic mold, with core, is wrapped in fibrous ceramic insulation, a ceramic filter is positioned at the entrance to the mold, the assembly is preheated to a selected temperature, and molten metal is poured from a ceramic crucible through the ceramic filter into the mold. The ceramic filter technology was developed in the 1980s and was a big breakthrough, which substantially decreased the reject rate of blades, thus having a very favorable economic impact. This ceramic filter technology has now been extended to almost all metal casting operations. Figure 3.3 shows some ceramic molten metal filters.

After casting, the ceramic mold is removed by sand blasting, and the core is removed chemically. The blade is close to the final shape, but still requires some precision grinding to achieve the close tolerances required for the turbine application. This final machining is done using cermet (such as WC-Co) and ceramic (such as silicon nitride) cutting tool inserts (discussed later in this chapter).

Casting metal blades for gas turbines is just one example of the broad use of ceramics in metal processing. Ceramics are used on a much broader scale for smelting, refining, and forming iron and steel, and the conditions that the ceramic must survive are quite severe. Try to imagine the conditions inside a basic oxygen furnace (BOF). During each operation cycle, about 100 tons of scrap iron and 250 tons of molten crude iron are loaded into the BOF along with fluxes (to eliminate

FIGURE 3.3 Ceramic filters for screening debris from molten metals during casting and thus greatly reducing the percent of reject metal parts. (Ceramic filters from Hi-Tech Ceramics, Alfred, NY. Photo by D.W. Richerson. With permission.)

impurities) and other metals added to form the desired alloy. The crude iron contains 3 to 5% carbon. Oxygen forced through the molten metal reacts exothermally with the carbon producing vigorous mixing and an increase in temperature to between approximately 1590 and 1700°C (2900 and 3100°F). Each cycle lasts about 1 hour, and the furnace lining must survive for thousands of cycles. It is amazing that any material can withstand this abusive service.

3.1.2   GLASS PRODUCTION[2,5,6]

Production of glass exceeds 600 million tons per year worldwide, so there is nearly as much glass processed as steel. Molten glass is very chemically active and will dissolve most ceramics. The metals platinum, iridium, and

> World glass production is greater than 600 million tons per year

rhodium are resistant to most molten glass compositions, but are too scarce and expensive for lining the large furnaces necessary to meet the world glass production requirements. A typical furnace for making window glass has a melting tank larger than a tennis court and can produce more than 25,000 tons per day. The glass tank is lined with solid blocks of zirconium oxide (zirconia) or mixtures of zirconia, alumina, and silica. Even these blocks are attacked by the molten glass, but the rate of dissolution is slow enough that the ceramic lining provides an adequate lifespan.

3.1.3   INDUSTRIAL PROCESSES

3.1.3.1   Furnace and Reaction Vessel Linings

Many industrial processes involve high temperature and thus require materials that are stable under the temperature and chemical environment of the specific application. As in the case of metal processing and glass processing, high temperature ceramics are generally required for lining and thermally insulating high-temperature furnaces or reaction vessels. An obvious example is the ceramics industry. Ceramic linings are necessary for the furnaces involved in the fabrication of alumina spark plug insulators, cordierite catalyst supports for our automobile emission control systems, dinnerware, magnetic ceramics, electrical ceramics, and for nearly all other ceramics. Less obvious is that high-temperature furnace linings, thermal insulation, and other ceramics are required in petroleum and chemical processing, in producing cement, in a wide range of heat-treating processes, and even in papermaking.

Papermaking is an interesting example.[1,2] Moderate to high temperatures are involved in several stages of the papermaking process. The first stage is converting wood chips to pulp. Wood consists of cellulose fibers bonded together with organic compounds referred to as lignin. The wood chips are placed into a large pressure cooker up to about 75 meters tall where they are heated with steam at about 7 to 12 atmospheres pressure. Chemicals such as sodium sulfide and sodium hydroxide are added to help separate the lignin from the cellulose fibers. Although the temperature is not very high, the combined effects of temperature, pressure, and corrosive chemicals are best resisted by a ceramic lining.

Following pressure cooking, the chemicals and lignin are separated from the pulp as a fluid called "black liquor." This is thickened by evaporation to about 70% solids and is sprayed into a high-temperature boiler chamber where the combustible lignin is burned. Heat is extracted through metal heat exchanger tubes to run a steam turbine for generating electricity and to provide steam for other operations in the paper mill. Noncombustible inorganic salts and other solids in the black liquor fall to the bottom of the furnace chamber as a molten "smelt."

The sodium salts in the smelt are separated by dissolving the smelt in water and allowing insoluble materials to settle and be removed. Next, calcium oxide is added to the solution, resulting in a chemical reaction to form solid particles of calcium carbonate plus a solution of sodium hydroxide. The sodium hydroxide is pumped to the pulp pressure cooker and the calcium carbonate "mud" is

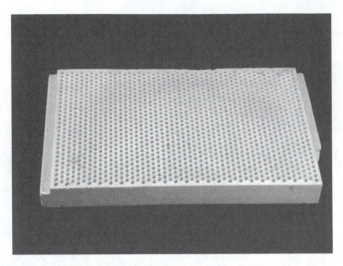

FIGURE 3.4   Perforated ceramic plate for a low-pollution, radiant surface burner. (Photo by D.W. Richerson.)

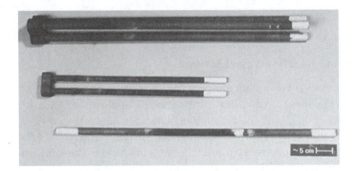

FIGURE 3.5   Variety of silicon carbide (SiC) heating elements.

fed into a lime kiln. The lime kiln, which is about 20 to 100 meters long and 2.5 to 4.5 meters in diameter and reaches temperatures of 980 to 1370°C, converts the calcium carbonate back to calcium oxide.

3.1.3.2 Heat Sources

Another category of high-temperature use of ceramics in industry is as a heat source. In one type of heat source, a fuel is mixed with air and combusted within a ceramic tube or liner to heat an enclosed chamber or to impinge directly onto the surface desired to be heated. In another type, the fuel and air are mixed and forced through an array of holes in a plate such that combustion occurs on the surface of the plate. This produces efficient and uniform heating that is especially low in nitrogen oxide (NO_x) pollutants. Figure 3.4 shows a ceramic surface burner. Ceramic radiant surface burners have been successfully used in such diverse applications as boilers, heat-treating furnaces, paint drying, and glass windshield shaping. They even work well for cooking pizza.

Another type of heating involving ceramics is resistance heating. Some ceramics have a limited degree of electrical conduction, but with substantial level of electrical resistance. As the electricity struggles to get through, much heat is produced. Important examples of ceramic resistance heating elements include SiC, $MoSi_2$, graphite, and doped zirconium oxide. Examples of SiC heating elements are shown in Figure 3.5.

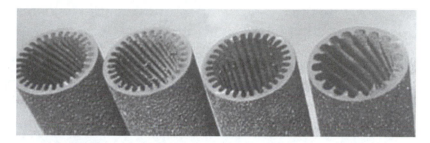

FIGURE 3.6   SiC tubular heat exchanger tubes with internal fins to increase heat transfer. Fabricated by Schunk-Inex Corporation, Holland, NY. (From Richerson, D.W., *The Magic of Ceramics*, The American Ceramic Society, Westerville, OH, 2000, p. 271. Reprinted with permission of the American Ceramic Society, www.ceramics.org)

3.1.3.3   Heat Exchangers

Much of the heat supplied to or produced in an industrial process is not used efficiently and goes up the stack as waste heat. For example, only 10% of the heat supplied to a combustion-heated glass-melting furnace is actually used to melt the glass; 67% escapes with the exhaust. The purpose of a heat exchanger is to reclaim a portion of this waste heat to preheat incoming air (such as combustion air) or process reactants.[7] Ceramic heat exchangers have been demonstrated that can salvage enough heat to reduce energy input or fuel consumption by nearly 50%.

Ceramic heat exchangers come in a variety of sizes and designs based upon the scale of the application and other factors. For the large scale glass furnace (which can have a melting tank the size of a tennis court), the heat exchanger consists of a massive refractory brick structure that extends along the length of the furnace.[5] Exhaust gases and inlet air are passed through this structure in such a way that heat is transferred from the exhaust gas to the inlet air. The more the inlet air can be preheated, the less fuel that needs to be burned to achieve the glass melting temperature.

Heat recovery in a glass melting furnace is essential to achieve a reasonable level of efficiency. Smaller scale industrial processes (such as metal heat treating) traditionally either did not use a heat exchanger or used a relatively low temperature metallic heat exchanger. As fuel and energy costs have increased and concerns about emissions of pollutants have increased, more and more industries have explored the use of heat exchangers. Two primary types of ceramic heat exchangers have been developed and tested, tubular designs and layered designs.

In a simple tubular design an array of tubes passes through the exhaust duct or chimney. Inlet air pumped through the tubes is heated by the hot exhaust gases. Figure 3.6 shows some silicon carbide heat exchanger tubes that have internal fins to increase heat transfer. A difficult engineering challenge with ceramic heat exchanger tubes is to deliver the cold incoming air to the tubes and the hot outlet air to be mixed with fuel in a combustion chamber. The tubes must be attached to an inlet manifold and an outlet manifold. Temperature gradients, heat transfer coefficients, and accommodation of thermally induced motion at the tube-manifold attachments all require careful design and material selection. The engineer must understand both the thermal and mechanical properties of each material involved. Figure 3.7 illustrates one way the challenge was approached in a government-funded program at the AiResearch Manufacturing Company of Arizona in the mid-1970s.[8] The SiC tubes were fabricated in a U-shape, which gave them more freedom to flex without breaking and allowed for a simple rigid attachment to the manifolds. While this approach minimized concerns about stress and fracture, it resulted in an increased challenge in fabrication. A straight ceramic tube is much easier and cheaper to fabricate than a U-shaped tube. This is another example of how an engineer must continually consider all of the tradeoffs including properties, performance, and cost.

FIGURE 3.7 U-shaped SiC heat exchanger tubes fabricated by the Norton Company in the 1970s and tested by the AiResearch Manufacturing Company under ERDA (Energy Research and Development Administration) and EPRI (Electric Power Research Institute) programs.

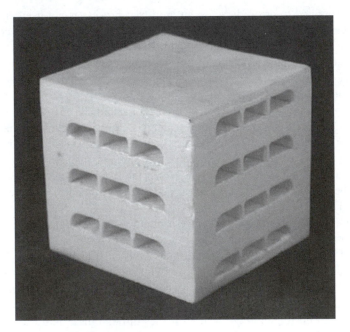

FIGURE 3.8 Small ceramic heat exchanger model illustrating a plate–fin layered cross-flow design. (Sample fabricated in the 1970s by GTE Sylvania. Photo by D.W. Richerson.)

The layered design involves a plate–fin concept such as illustrated in Figure 3.8. Note that alternating layers have flow channels at 90° to each other to provide a cross flow pattern between the hot gases and cold gases. The "plates" separate the hot-gas layers from the cold-gas layers. The "fins" provide high surface area to enhance heat transfer efficiency and also to provide structural strength. As an alternative to the rectangular channels, the fins can be in a sinusoidal pattern to yield a corrugated appearance similar to cardboard. A plate–fin heat exchanger can be relatively compact compared to tubular heat exchanger.

3.1.3.4 Heat Exchange for Chemical Processing

Heat exchange is conducted on a large scale in the chemical processing industries.[9] In many of these chemical processes, the precursor chemicals (usually gas or liquid) must be heated to a reaction or

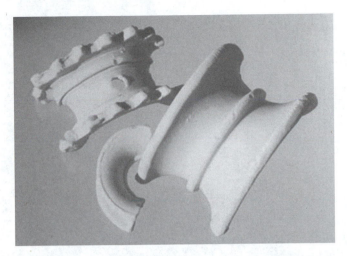

FIGURE 3.9 Ceramic "saddles" designed to provide heat transfer, temperature uniformity, and controlled flow of chemicals for chemical processing. (Samples manufactured by Saint-Gobain Norpro Corporation, Akron, OH. Photo by D.W. Richerson.)

distillation temperature. This is typically done in cylindrical chambers called towers or columns. These columns can be up to about 14 meters in diameter and accomplish chemical separations requiring as many as 128 theoretical stages.[10] The raw chemicals are typically fed into one end of the column and heated gases into the other end. The column is filled with packing materials that control the flow of the chemicals and heating and heat transfer, as well as act as surfaces to nucleate reactions.[10] Depending on the chemicals involved and the temperature, the packing material can be ceramic, metal, or polymer. Figure 3.9 shows ceramic "saddles" that absorb heat, provide open space for the chemicals to flow through, and provide chemical reaction sites.

Other chemical processes use ceramic packing such as shown in Figure 3.10 that are porous to provide an enormous amount of surface area to catalyze (and increase the efficiency of) chemical reactions. Sometimes these porous ceramic beads are coated with a catalyst such as platinum and can increase the rate of reaction exponentially. Ceramic catalyst supports are used extensively in petrochemical cracking processes, in ammonia synthesis, and in the processing of many polymer precursors (such as dehydrogenation of ethyl benzene to styrene).[2] Other applications of ceramic heat exchange media include removal of organic vapors, odor control, cooling and quench towers, and stills and fractionators.[10]

3.1.4 HEAT ENGINES

3.1.4.1 Gas Turbine Engines

A heat engine burns a fuel and converts the chemical energy to heat energy and the heat energy to mechanical energy (or with added equipment to electrical energy). Historically, the first heat engine was the steam engine invented in the early 1700s, but it was very inefficient, converting only 1% of the chemical energy in the fuel into useful work (the remaining 99% was wasted). The internal combustion engine invented in the mid-1800s and the gas turbine engine invented in the mid 1900s are much better, achieving efficiency approaching 30% in simple cycle designs. During the past 30 years there have been substantial efforts to increase the efficiency of heat engines by increasing the temperature and pressure inside the engine, especially for gas turbine engines. Temperature increase for gas turbine engines has been achieved through a combination of improved materials, elaborate cooling designs (as shown earlier in Figure 3.2), and thermal barrier coatings (TBCs).[11,12]

FIGURE 3.10 Porous ceramic beads with high surface area to provide catalytic sites to accelerate chemical reactions in chemical processing. (Samples manufactured by Saint-Gobain Norpro Corporation, Akron, OH. Photo by D.W. Richerson.)

Zirconium oxide (ZrO_2, zirconia) based ceramic TBCs are now widely used as a coating on hot section metal components (combustor liners, transition sections, nozzle guide vanes, and rotor blades) in gas turbine engines for aircraft, military, and power generation applications. Zirconia was selected for several reasons: (1) low thermal conductivity, (2) high-temperature stability in oxidizing and combustion atmospheres, (3) coefficient of thermal expansion similar to iron alloys and superalloys, and (4) ability to be applied cost-effectively as a coating onto metal surfaces. Because of the low thermal conductivity compared to metals, a layer only 0.6 to 0.7 mm thick can reduce the temperature to which the underlying metal is exposed by greater than 200°C. This allows the engine to be operated at a higher turbine inlet temperature, which results in increased efficiency (reduced fuel consumption and higher power output).

> **PRODUCTION CERAMIC COMPONENTS FOR ENGINES**
>
> - Seals
> - Spark plug insulators
> - Igniter insulators
> - Glow plugs
> - Swirl chambers
> - Cam follower rollers
> - Turbocharger rotors
> - Sensors
> - Fuel injection timing plungers
> - Fibrous thermal insulation
> - Electronic engine control modules
> - Catalytic converter honeycomb
> - Oxygen sensors
> - Thermal barrier coatings

For a TBC to act as a temperature barrier, cooling must be supplied to the back surface or interior of the metal component. If the metal and the TBC could be replaced with a high temperature ceramic, much of the cooling could be eliminated to provide a substantial increase in efficiency. Silicon nitride (Si_3N_4) and silicon carbide (SiC) ceramics have been under development for this purpose since the late 1960s and are just now reaching limited success in engine field tests. This development effort has been very difficult because the conditions inside a gas turbine engine (and most other heat engines) are severe. Consider a gas turbine engine being used to power oil pumps on the Alaska pipeline or a jet airplane parked at the gate in below zero weather during the winter. When the engine is started, the components in the hot section will be shocked by a change in temperature from ambient (in this case below zero) to well over 1000°C within a few seconds and will then have to operate often continuously at this high temperature for

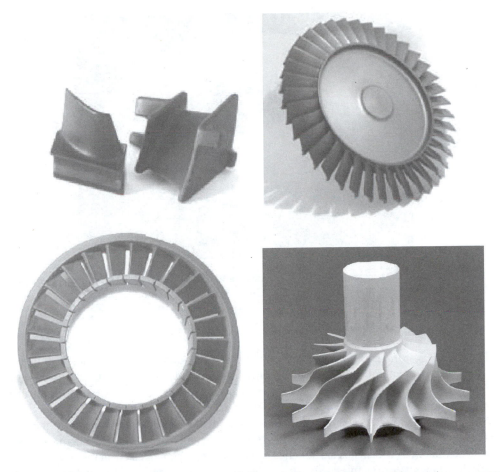

FIGURE 3.11 Examples of experimental silicon nitride and silicon carbide gas turbine engine components that have been fabricated since 1975 for various advanced gas turbine engine development programs. (Most photos by D.W. Richerson.)

thousands of hours. Material concerns include strength vs. temperature, stability in the combustion atmosphere, creep and stress rupture life, response to low and high frequency vibrations, coefficient of thermal expansion, and thermal conductivity. To achieve ceramics that can reliably survive these severe conditions has required extensive research and development including the improvement of composition, fabrication, microstructure and properties, extensive characterization and database generation, a whole new probabilistic design and life prediction methodology, and extensive rig and engine testing.[11,12] Figure 3.11 illustrates a variety of monolithic (noncomposite) ceramic turbine engine components that have been fabricated and tested.

In spite of all of the development efforts, though, accumulation of test hours in engines has been relatively slow. By the end of the 1970s only several hundred hours had been accumulated and only several thousand by the end of the 1980s. However, substantial field testing was begun in the late 1990s. Honeywell Engines and Systems accumulated more than 50,000 hours field testing of silicon nitride ceramic components in auxiliary power engines (including those in commercial aircraft service). The ceramic parts performed very well. For example, a set of silicon nitride nozzle guide vanes (a component exposed to the highest peak temperature in the engine) appeared nearly new after 7000 hours, while a set of metal nozzle guide vanes exhibited substantial erosion and corrosion after only 2000 hours.

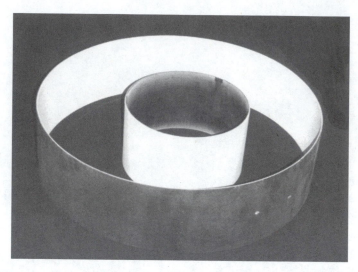

FIGURE 3.12 SiC–SiC composite gas turbine combustor liners with an oxide environmental barrier coating. (Photo courtesy of Solar Turbines, Inc., San Diego, CA, and the U.S. Department of Energy. With permission.)

The challenges of the gas turbine have provided a strong driver to the advancement in technology of advanced structural ceramics such as silicon nitride and silicon carbide. Dramatic improvements have been achieved in strength and complex shape fabrication technology. Examples are included in various chapters of this book. In addition, though, many other technology advances have evolved either directly or indirectly including improved understanding in the following areas: the relationship of chemical and microstructural properties (especially at high temperature), sintering of nonoxide ceramics, oxidation mechanisms and kinetics, the use of the scanning electron microscope and other analytic instruments for fracture analysis, and the nature and mathematical modeling of slow crack growth, creep, fracture toughness, thermal shock, and contact stress. The engineering significance of these also is discussed in subsequent chapters.

Monolithic ceramics fracture in a brittle mode. Major efforts began in the late 1980s and early 1990s to develop ceramic matrix composites that would fail in a more graceful nonbrittle fashion. By 1995 turbine engine testing of combustor liners of a SiC–SiC (silicon carbide fibers in a silicon carbide matrix) were in progress under a SolarTurbines Inc. and Department of Energy program. These tests identified material deficiencies that are still in the process of being incrementally overcome. By the end of 2002 more than 50,000 hours of engine operation were accumulated at an oil field location and a textile mill. Figure 3.12 shows a set of the SiC–SiC combustor liners. Note that they are coated with a white material. This white layer is an "environmental barrier coating" (EBC) consisting of oxide ceramics. Without the EBC, the SiC–SiC combustor liners exhibited oxidation degradation in the gas turbine combustor atmosphere that limited their life to less than 5000 hours. The EBC increased the life to greater than 15,000 hours, but further improvement is needed to get to the desired 20,000+ hours life. This is another example of how an engineering problem can sometimes be solved with a coating rather than a complete material change.

3.1.4.2 Internal Combustion Engines

Although internal combustion engines are presently constructed mostly of metals, there are some ceramic components and there are benefits that could be achieved with broader use of ceramics. The most obvious current use is the spark plug insulator. Here the ceramic must perform several engineering functions simultaneously. First, it must be an effective electrical insulator. Second, it must

FIGURE 3.13   Examples of ceramic turbocharger rotors, one illustrating attachment to a metal shaft. (Photo by D.W. Richerson.)

withstand high temperature pulses and severe thermal shock, as well as chemical corrosion. A high-alumina composition with a dense, fine-grained microstructure was developed to meet these combined challenges and to provide very high reliability.

Another current, but limited, use of ceramics is the turbocharger rotor. Here the optimum material selection is silicon nitride because of its low specific gravity, high strength, high toughness, low thermal expansion, and moderate thermal conductivity. The turbocharger extracts energy from the exhaust gases to provide an additional power boost such that a turbocharged four-cylinder engine performs more like an unturbocharged six-cylinder engine. For a turbocharger to provide a power boost when you want to accelerate from a stopped position, the rotor must accelerate from an idle speed up to about 140,000 rpm. Metal turbocharger rotors have a high specific gravity and must overcome a great deal of inertia. This results in "turbolag," a momentary hesitation from the time you step on the accelerator until you get the power boost. A silicon nitride rotor has less than half the specific gravity of metal rotors and accelerates much more quickly to eliminate turbolag.

Figure 3.13 shows some examples of ceramic turbocharger rotors that were developed in Japan where most ceramic turbocharger rotors have been produced and installed (starting in about 1988). Each ceramic rotor requires attachment to a metal shaft, as illustrated by one of the rotors in Figure 3.13. In this case the attachment was accomplished with the use of a braze alloy. Accommodating the large difference in thermal expansion between the metal and the silicon nitride was a difficult engineering challenge that had to be overcome for the brazed attachment or any other attachment design.

Ceramics are chosen for their temperature resistance for other components in conjunction with heat engines: silicon nitride glow plugs and swirl chambers for diesel engines, the doped zirconia oxygen sensor, the cordierite honeycomb substrate for the catalytic converter pollution control system, diesel particle traps, ceramic fiber insulation in the firewall between the engine compartment and the passenger compartment, and the alumina mounting substrates for the engine electronic control modules. In addition, ceramics serve many other functions in internal combustion heat engine systems where resistance to heat is not the primary issue. Examples include diesel cam follower rollers, diesel fuel injector timing plungers, water pump seals, air conditioner compressor seals, and a variety of sensors (intake air sensor, cooling water temperature sensor, oil temperature sensor, and knock sensor).[1]

One of the reasons for the low efficiency (~30%) of internal combustion engines is the requirement for cooling of the metals. Diesel engines lose about 30% of the energy through the cooling

FIGURE 3.14 Experimental ceramic parts evaluated for automotive engines plus a few production parts. (Photo courtesy of Kyocera Corporation. Richerson, D.W., *The Magic of Ceramics*, The American Ceramic Society, Westerville, OH, 2000, p. 203. Reprinted with permission of the American Ceramic Society, www.ceramics.org)

system and another 30% through the hot exhaust gases. Research has been conducted to replace the metal parts surrounding the cylinders and exhaust manifold with ceramic parts that can withstand higher temperature and also reduce through-the-wall heat loss. An early program demonstrated that cooling could be completely eliminated, but major long-term effort would be required to achieve high reliability and acceptable fabrication cost.[1] Reducing through-the-wall heat loss increases heat released through the exhaust system. Part of this heat can be converted to useful mechanical energy through turbocompounding, thus increasing the overall engine efficiency.

Figure 3.14 shows some experimental ceramic parts evaluated for automotive engines plus some production parts. The most successful experimental parts have been valves. Hundreds of thousands of miles have been successfully accumulated, especially by Daimler-Chrysler, using Si_3N_4 valves. The Si_3N_4 valves clearly out-perform metal valves, but are still considered too expensive for broad commercial use.

3.1.4.3 Aerospace

High temperature stability and strength are important for many aerospace devices and structures. Gas turbine engines were first developed for aircraft propulsion, and many gas turbine engine components already have been discussed. Other high-temperature ceramic applications for aerospace include rocket nozzle liners, thruster liners, afterburner components, igniter components, leading edge structures for hypersonic missiles, and thermal protection for the Space Shuttle.

SPACE SHUTTLE SURFACE PROTECTION MATERIAL DESIGN CHALLENGES

- Ultra-light-weight
- Attach to aluminum
- Withstand temperatures up to 1500°C
- Prevent heat from reaching underlying adhesive and aluminum
- Be 100% reliable for 100 missions
- Have high emissivity surface
- Not absorb water

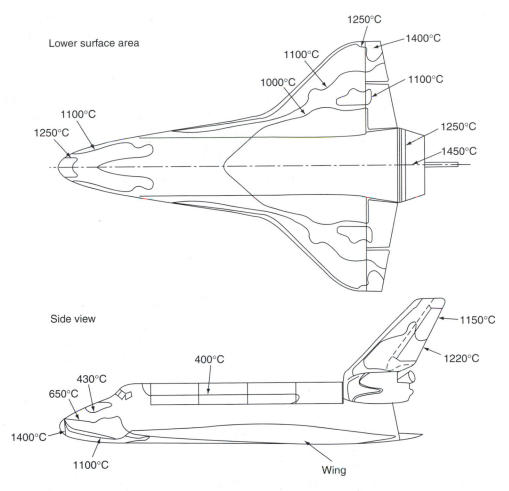

FIGURE 3.15 Approximate temperature distribution that the Space Shuttle Orbiter must withstand during ascent and re-entry. (From Svec, J.J., *Ceram. Ind.*, 107(4), 20–24, 1976. With permission.)

The Space Shuttle thermal protection is a good example of a difficult materials engineering challenge.[14,15] Figure 3.15 illustrates the temperatures that occur during ascent and re-entry. The temperatures are especially severe on the bottom, the nose, and on the leading edge of the wings. Materials engineers developed two material systems to satisfy the Space Shuttle design challenges, one based on highly porous silicon dioxide and the other based on a composite of carbon fibers in a carbon matrix.

The carbon–carbon composite was selected for the nose cap and the leading edge of the wings where the temperature can exceed 1400°C. The nose cap is about 1.4 m in diameter, and the leading edge of each wing totals about 36 m². While the carbon–carbon composite has excellent strength and thermal shock resistance, it is attacked by oxidation at high temperature. By itself it would be useless as protection for the Shuttle. To solve this problem, Vought Corporation developed a method of treating the carbon–carbon at high temperature with molten silicon to form a graded silicon carbide surface layer. They then applied a coating of silicon dioxide over the silicon carbide to perform as a diffusion barrier to oxygen. Another problem with the carbon–carbon is that it transports heat too well, so it required backing with insulation blankets to decrease the maximum temperature of exposure of the underlying metal to about 177°C (350°F).

Carbon–carbon is relatively light weight, but is still too heavy (and expensive) to cover the entire Shuttle. Another material was developed that contains about 80% pore space and only weighs about 144 kg/m^3 (9 lbs/ft^3). This material was developed by Lockheed Missile and Space Company and consists mostly of porous fused silica (silicon dioxide). The combination of the porosity (trapped dead air space) and the low thermal conductivity of the fused silica provides an effective barrier to the transfer of heat from the surface of the Shuttle through to the underlying elastomer adhesive and metal. In addition, the low thermal expansion and low elastic modulus (high strain tolerance) provide excellent resistance to thermal shock. However, this does not solve all of the problems. The porous silica is easily eroded and also susceptible to saturation with moisture. Furthermore it has low emissivity. That means that heat picked up during re-entry is not reradiated into space, but instead will slowly soak through to the inner side of the silica tile. Both of these problems were solved by applying a dense glassy coating onto the outer surface of the tiles. The composition of the coating was selected to provide high emissivity to rapidly reradiate heat.

As shown in Figure 3.15, different regions of the Shuttle are exposed to widely differing temperatures. Engineers have calculated the thickness of tile needed in each area. This has resulted in a requirement for about 33,000 different tiles ranging from about 0.5 to 11.4 cm thick and having various sizes and contours. So in addition to the technical challenges mentioned earlier, there are additional challenges of fabrication, quality control, and assembly.

3.2 WEAR AND CORROSION RESISTANCE APPLICATIONS

While many applications of ceramics are based upon temperature resistance, many others are based upon resistance to wear and chemical corrosion. Examples include seals, valves, pump parts, bearings, thread guides, papermaking equipment, and a variety of liners in equipment exposed to harsh chemical or erosive environments.

3.2.1 SEALS

A seal is an engineered interface designed to prevent leakage at a surface of contact between two materials. When someone asks for an example of a seal, most of us immediately think of one of the thousands of uses of gaskets. Gaskets are typically made of a flexible organic material (such as rubber, polyurethane, Teflon, nylon, or cork) or a soft ductile metal. Under a load the material deforms and fills spaces that otherwise would be paths for leakage. Good examples are the gaskets in our automobile engines and the gaskets at hose and plumbing junctions. Easily deformed materials are ideal for these seals. However, many applications require a seal between rapidly moving parts where a deformable or soft material would cause too much friction and would wear away quickly. Such a seal is usually referred to as a "face seal." Other seals must operate at too high a temperature for a polymer or in a severe chemical environment that would rapidly degrade a polymer or metal. These are the cases where a ceramic material is optimum.

Face seals essentially provide a seal at a rotating interface that prevents passage of liquids or gases from one side of the seal to the other. For instance, the compressor seal in an automobile air conditioner seals halogenated hydrocarbons and oil at pressures up to 250 psig and surface speeds of 1800 ft/min. A combination of graphite against Al$_2$O$_3$ has proven to be a low-cost, reliable seal for this application. Many millions of these seals are produced each year. Graphite against graphite has also been used.

Another important face seal fabricated by the millions for automobiles and trucks is the water pump seal. Examples of SiC and alumina water pump seals are shown in Figure 3.16. A more severe application is the main rotor bearing seal in jet engines. It must seal the oil lubricating system from 120-psig hot air at temperatures up to 1100°F and surface speeds up to 20,000 ft/min. Graphite impregnated with other materials to increase the strength and oxidation resistance has been used successfully for this application.[16] Recently, silicon nitride has been added as one of the seal elements for the main rotor seal and other critical gas turbine engine shaft seals in the Honeywell TFE731 turbofan propulsion engine. Since introduction in 1996, the silicon nitride–carbon-hybrid seals have

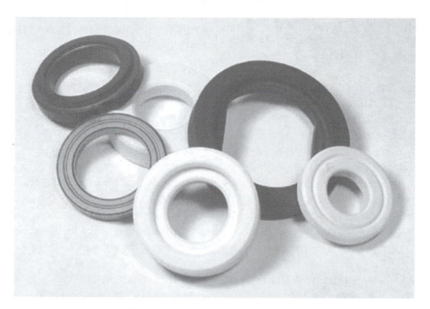

FIGURE 3.16 Examples of SiC and alumina seals, mostly for water pumps for cars and trucks. Many millions are produced each year. (Photo by D.W. Richerson.)

FIGURE 3.17 Silicon nitride "seal runner" that rotates at about 30,000 rpm in contact with a stationary carbon seal element to provide the main shaft seal in a turbofan business aircraft propulsion engine. The seal is 3.7 inches in diameter. (Seal fabricated by Honeywell Ceramic Components, Torrance, CA. Photo by D.W. Richerson.)

accumulated over 20 million service hours and have reduced unscheduled engine removal by 96%.[17] Figure 3.17 shows one of these silicon nitride "seal runners."

Another severe application that benefits from ceramic seals is the recovery of crude oil by the saltwater pressure system. Saltwater is pumped into the ground at about 2500 psi to force crude oil out of the rock formations so that it can be recovered in adjacent wells. The face seal in the pump must survive the 2500 psi pressure plus temperatures up to 600°F plus surface rub speeds of 5000 ft/min.

Face-type seals are used in many other applications including sand slurry pumps (which pump approximately 35% solids), chemical processing and handling, fuel pumps, torque converters, washing machines, dishwashers, garbage disposals, and swimming pool sand filter pumps. Figure 3.18 illustrates a grouping of SiC seals for industrial applications.

A ceramic seal works in a different mode than a flexible seal. Usually a ceramic material is selected that has high hardness and can be polished to tight tolerances of flatness and very smooth surface finish with low friction. For noncritical applications, flatness to within

---

APPLICATIONS OF CERAMIC SEALS AND VALVES

- Single lever faucet seal and valve
- Car and truck water pump face seal
- Air conditioner compressor seal
- Gas turbine engine main rotor bearing seal
- Slurry pump seal
- Experimental auto engine valves

0.002 inch (17 helium light bands) is usually acceptable. For higher pressure and more critical applications, flatness to within six or three light bands is usually required. Aluminum oxide, silicon carbide, transformation-toughened zirconia, and silicon nitride all have been successfully designed into a variety of face seal configurations. SiC is especially good because of its very high hardness, chemical inertness, and relatively high thermal conductivity (can conduct away heat produced at the rub face of the seal).

Sometimes a seal surface is desired that provides a degree of self-lubrication to sustain a low friction interface. Graphite is a good candidate. Many people assume this is due to its layer crystal structure. Graphite is comprised of carbon atoms that are strongly covalently bonded to each other in the crystal planes perpendicular to the c-axis, but are bonded to adjacent parallel planes only by weak van der Waal's forces, resulting in relatively easy shear between the weakly bonded planes. This may be a factor in some cases, but generally the lubrication behavior results from formation of a hydrodynamic or transfer film between the graphite and the mating surface. This film appears dependent on the presence of polar liquids, oxygen, or water vapor. Graphite operating in vacuum, dry gases, cryogenic fluids, or at high temperature does not form a suitable interfacial film. In these cases special grades of graphite containing impregnants are used.

Engineered graphite-based seals are effective as one surface running against a hard ceramic seal surface such as alumina, SiC, or silicon nitride.

Not all seals involve contact. A turbine tip seal for example strives to maintain minimum clearance between the tips of the rotating gas turbine engine rotor blades and the shroud that houses the rotor.

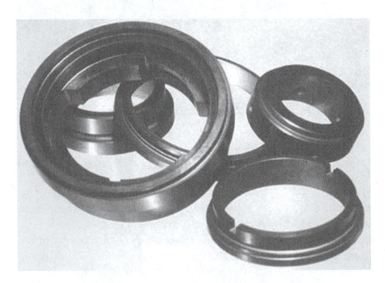

FIGURE 3.18   SiC seals for various industrial equipment applications. (From Richerson, D.W., *The Magic of Ceramics*, The American Ceramic Society, Westerville, OH, 2000, p. 245. Reprinted with permission of the American Ceramic Society, www.ceramics.org)

Any hot combustion gases that leak between the blade tips and the shroud decrease the efficiency of the engine. This is a difficult materials and engineering challenge. Depending on the size of the engine, the rotor can be turning at about 10,000 rpm for a large engine to over 100,000 rpm for a very small engine in a hot gas flow path that exceeds 1000°C. The engineer must carefully analyze temperature distribution and the thermal properties of the materials (thermal expansion and thermal conductivity) to achieve an optimum design. One option that has been explored has been to coat the surface of the metal shroud with a porous "abradable" ceramic to allow the blade tips to actually cut into the surface of the shroud and assure a minimum "form-fit" clearance.

In summary, seals are not as simple as they might first appear, but instead require substantial engineering. The engineer needs to be aware of material characteristics such as hardness, chemical activity, thermal properties, and even the tendency to form interfacial films. He or she also needs to thoroughly understand the conditions that the material will experience in the application environment.

3.2.2 VALVES

A valve performs the same function as a seal, but in an opening and closing mode rather than a continuous contact mode. There are many types of valves: rotary, gate, ball-and-seat, butterfly, and concentric cylinder, and each of these has many variations. One type of rotary valve consists of a ball with a hole through it and a rod or handle that protrudes from the side. The rod can be physically manipulated so that the hole either lines up with the pipe carrying the fluid (valve open) or so the hole faces the side wall (valve closed). This type of valve can control flow in both directions at any level between maximum and zero. A variant of this is the valve and seal for a sink faucet that has only a single handle. The valve and seal consists typically of alumina or silicon carbide pieces such as those shown in Figure 3.19. When you rotate the handle to the middle position, hot water comes in through one hole and cold through another and they mix to exit through another hole. If you rotate the handle all the way to one side, only one inlet hole is open so the water is either all hot or all cold. By control of the position of the handle you can achieve any mixture you desire of hot plus cold. A very smooth surface finish and excellent flatness between the ceramic contacting surfaces allows the valve to simultaneously perform as a seal, so no gasket is required. This is one of those hidden ceramic applications that you use all the time, but probably were never aware that ceramics were involved.

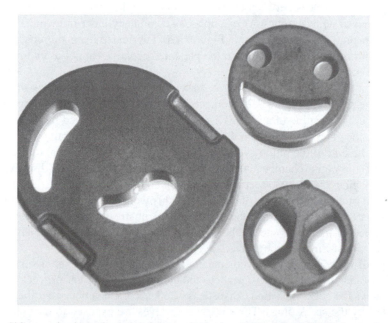

FIGURE 3.19   SiC ceramic pieces for a sink faucet valve and seal. (Photo by D.W. Richerson.)

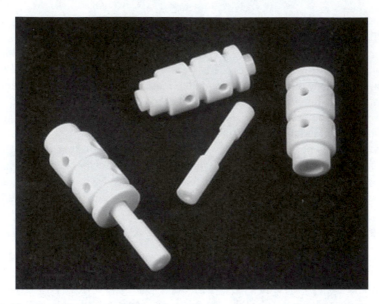

FIGURE 3.20   Alumina cylinder-in-cylinder valves for control of oxygen flow in a respirator. Fabricated by Coorstek, Golden, CO. (Photo by D.W. Richerson.)

A gate valve slides in and out of a passage to either block or allow flow. The gate in an irrigation ditch is a gate valve and is often made of metal or concrete. A gate valve that definitely requires ceramics is the slide-gate valve in a ladle holding molten steel. The slide-gate is opened to meter the flow of molten steel into a casting mold.

A butterfly valve, which is like a door that swings open and closed, is often used to control flow in large diameter passages such as air ducts. Like a rotary valve, it can allow flow in either direction at any level between maximum and zero. Variants also can be designed for use at a wide range of temperatures.

A cylinder-in-cylinder valve is just as the name describes. To be effective and not leak, it requires very smooth surfaces between the inner and outer cylinders and close tolerances, both of which can be achieved with ceramics. An example of a ceramic cylinder-in-cylinder valve is the oxygen valve for a hospital respirator shown in Figure 3.20. Fabricated from alumina, this valve cycles every time a breath of oxygen is metered to the patient. The valve cannot use a lubricant and must be 100% reliable for millions of cycles.

Another type of valve is the ball-and-seat, also often referred to as a check valve. A spherical ball, such as those shown in Figure 3.21, contacts against a tapered cylindrical opening, like placing a tennis ball on the opening of a glass. This type of valve is often used for pumps to allow flow only in one direction. The pressure from the pump cycle pushes the ball out of the seat and allows fluid to pass. At the end of the pump cycle, the ball drops back onto the seat to prevent fluid from leaking back into the pump chamber. An extreme application of this is the downhole pump for a deep oil well. Imagine a well 2000 feet deep and the weight of a column of fluid filling this well. At the conclusion of each pump cycle, the ball under this weight drops with great force against the seat. Very few materials can withstand the high mechanical stress involved, the erosion from rock particles suspended in the fluid, and the hot corrosive chemical environment. The exceptions are WC-Co cermets, transformation-toughened zirconia, and silicon nitride.

A final type of valve is familiar to you amateur auto mechanics. One set of valves allows a controlled amount of air–fuel mixture to enter into the cylinders of your car or truck engine. Another set of valves allows the hot combustion exhaust gases to leave the chamber after the air–fuel mixture has been ignited by the spark plugs. The ends of the valves are exposed to the full force and temperature of the explosion inside the cylinder. This type valve, which consists of a circular conical

FIGURE 3.21   Silicon nitride check-valve balls fabricated by Ceradyne, Inc., Costa Mesa, CA. (From Richerson, D.W., *The Magic of Ceramics*, The American Ceramic Society, Westerville, OH, 2000, p. 246. Reprinted with permission of the American Ceramic Society, www.ceramics.org)

FIGURE 3.22   Silicon nitride automobile valves. (Photo by D.W. Richerson.)

surface that seats against a matching conical surface, is typically fabricated from metal. Testing during recent years has demonstrated that experimental silicon nitride valves like the ones shown in Figure 3.22 perform with equivalent reliability to metal valves, but consume less energy because they are much lighter in weight than the metal. Also, the ceramic valves are more resistant to wear and to damage by heat. If these perform equivalently or better than metals, why are they not in common production use? The reason is cost. The best that has been achieved so far has been a cost about 15%

FIGURE 3.23   SiC pump parts fabricated by Saint-Gobain/Carborundum Structural Ceramics, Niagara Falls, NY. (From Richerson, D.W., *The Magic of Ceramics*, The American Ceramic Society, Westerville, OH, 2000, p. 247. Reprinted with permission of the American Ceramic Society, www.ceramics.org)

higher than current metal valves. The automotive companies have chosen to stick with metal valves, except in a few specialty vehicles. Here is another case where the engineer must address factors such as economics in addition to the technical material and design challenges. Economics are usually dominated by raw material costs and fabrication costs. Therefore, a significant portion of this book focuses on raw materials and fabrication processes.

3.2.3   PUMPS

Ceramic valves and seals are used extensively in pumps, but so also are ceramic bearings, plungers, rotors, and liners. Pumps in the mining industry must handle fluids containing large amounts of erosive solid particles, while pumps in the chemicals and petroleum industries are often subjected to corrosive chemicals sometimes up to about 800°F. Some pumps in the aluminum industry pump molten aluminum, and some pumps in the papermaking industry handle hot caustic soda (sodium hydroxide) solutions. A pump with ceramic parts will often last at least 10 times longer than pumps lined with other materials.

Figure 3.23 illustrates some SiC pump parts. Figure 3.24 shows a pump that is completely lined by ceramic parts (mostly alumina).

3.2.4   BEARINGS

Modern industry and transportation essentially run on bearings. Bearings provide rolling contact rather than sliding contact and thus greatly decrease wear and friction. To be viable for bearings, a material must be hard or tough to resist the high Herzian contact loads to which a bearing is exposed. It also must be capable of fabrication to high smoothness and close tolerances at reasonable cost and operate with a low coefficient of friction against a mating surface.[18] High performance steel, especially M-50, has dominated the market for industrial and transportation bearings.

FIGURE 3.24   Pump completely lined with ceramic materials. (From Richerson, D.W., *The Magic of Ceramics*, The American Ceramic Society, Westerville, OH, 2000, p. 247; photo courtesy of Kyocera Corporation. Reprinted with permission of the American Ceramic Society, www.ceramics.org)

Intuitively one would expect a ceramic material to be a good bearing. Ceramics can be hard, wear resistant, and surface finished to high smoothness. Furthermore, the major stress mode in a bearing is compressive. Ceramics have higher strength under compression than under tension and experience less elastic and plastic deformation than a metal, which should decrease the susceptibility to fatigue.

Some ceramics have been used for many years for bearings in low-load applications, in applications where lubrication is either limited or not possible, and in applications where the temperature is beyond the limits of metal bearings. A low-load example is the use of small spheres machined from single-crystal Al_2O_3 (ruby or sapphire) for older watches (before the electronic quartz watch was invented). You may have heard the term "17 jewel watch." The jewels were the single-crystal alumina bearings. Similar low-load bearings were important for many precision measurement instruments.

A variety of ceramics including alumina and SiC were tried over the years for high-load, high-speed bearings, but none were successful. They all failed catastrophically by brittle fracture. The first success was achieved in the early 1970s with a hot-pressed silicon nitride (Si_3N_4) ceramic.[19] Rolling contact fatigue tests showed that the Si_3N_4 material (NC-132 hot-pressed silicon nitride from the Norton Company) had an L_{10} life of 12,000,000 cycles at 700,000 psi Herzian stress loading. This was eight times the life of M-50 CVM steel. Subsequent engine testing verified the viability of the Si_3N_4 bearings.[20]

The Si_3N_4 bearings did not fail catastrophically like prior ceramics. Instead, they failed by slow development of surface spallation very similar to the failure mode of metals. What was different about silicon nitride compared to other ceramics that had been tried for bearings? The major difference was the fracture toughness. The hot-pressed silicon nitride had about 50% higher fracture toughness than alumina or SiC. So here is another example of the importance to the engineer of understanding the properties of materials and their relevance to performance in specific applications.

FIGURE 3.25   Silicon nitride ceramic bearing balls (fabricated by Saint-Gobain Advanced Ceramics, East Granby, CT) and an assembled hybrid bearing with silicon nitride balls in a metal race. The size is compared to a U.S. dime, which is about 1.3 cm across. (Photo by D.W. Richerson.)

Although hot-pressed silicon nitride was demonstrated as a superior bearing material in the early 1970s, the only way of fabricating a bearing ball was to use diamond abrasive tooling to cut a flat plate of hot pressed silicon nitride into cubes and to then tediously grind these cubes into spheres. This was prohibitively expensive. The fabrication problems would not be solved until the late 1980s when hot isostatic pressing was implemented for making hot-pressed silicon nitride balls with greatly reduced surface grinding. Even though the cost per ball was still about five times that of M-50 bearings, the performance improvement was so great for some applications that commercial production was started in about 1988. By 1996 more than 15 million Si_3N_4 bearing balls were being produced per year. Unit cost comes down as production volume increases. Figure 3.25 shows several individual balls and an assembly of balls mounted in a metal race. Table 3.2 lists some applications of silicon nitride bearings.

What other characteristics of silicon nitride besides the improved fracture toughness provide an engineering benefit? Because of its high hardness and fine-grained microstructure, hot-pressed silicon nitride can be surface finished to roundness and smoothness of about one-half millionth of an inch. This decreases friction, but also results in extremely low vibration operation at unprecedented speeds. Because of its low specific gravity (60% lighter than steel), silicon nitride bearings consume 15 to 20% less energy and can be run at up to 80% higher speed than an equivalent metal bearing.

3.2.5   THREAD GUIDES

Hardness and smoothness also are important in the textile industry, especially for the manufacture of thread and in automated weaving machines. Automated machines move thread at high speed under tension from spindle to spindle. The thread slides over and through many guiding surfaces referred to as *thread guides*. The thread guides must be very smooth to avoid fraying damage to the thread and must be very hard to resist wear. Alumina, porcelain, and glass-coated stoneware ceramics all have been used for thread guides. Figure 3.26 shows some examples of alumina thread guides.

3.2.6   CERAMICS IN PAPERMAKING

The logistics of papermaking are impressive. On a typical day, a paper mill produces around 1500 tons of paper, consumes enough electrical power to provide all the needs of a city of 40,000 people, and

---

TABLE 3.2
Equipment in Which Silicon Nitride Bearings Have Shown Benefits

High-speed machine tool spindles	Chemical processing equipment
Air-driven power tools	Food processing equipment
High-speed compressors	Textile equipment
Space shuttle main engine	Semiconductor processing equipment
Space shuttle liquid oxygen pump	Checkvalve balls
Gyroscopes	Pumps
Racing cars and racing bicycles	Gas turbine engines
Inline skates, skate boards, street luges	Helicopters
High-speed hand-held grinders	Aircraft anti-icing valves
Dental drills	Aircraft wing flap ball screws
Gas meters	Military missiles
Salt water fishing reels	CT scanners
High-speed train motors	X-ray tubes
Photocopier rolls	High-speed surgical saws
Turbochargers	MRI Equipment
Centrifuges	Disk drives

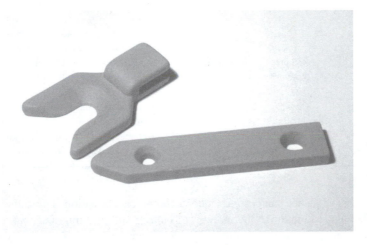

FIGURE 3.26 Aluminum oxide thread guides. The Y-shaped one is about 3.2 cm long. (Photo by D.W. Richerson.)

uses about 65 million gallons of water. In the final step, the paper machine, converts a slurry of 99.5% water and 0.5% pulp (wood fiber) into a 20-foot wide roll of dry paper in about 30 seconds through a sequence of water removal, pressing, and drying. The pulp and paper moves through this machine at nearly 80 miles per hour. With an operation of this size, shutdown time for maintenance is very expensive, so ceramics are used in many places because of their superior resistance to wear and corrosion. Table 3.3 lists some of the uses of ceramics in papermaking, and Figure 3.27 and Figure 3.28 show some examples.

3.3   CUTTING AND GRINDING

Most fabrication processes are not able to produce a part to the final shape, dimensions, and surface characteristics required by an application. A sequence of cutting and grinding is generally necessary,

TABLE 3.3
Ceramics in Papermaking

Stage of Process	Specific Application	Ceramic Material Use
Woodyard	Sawing of logs	6–8 foot diameter saws tipped with WC-Co cermet cutting edges
	Chipper to chop log into small chips	Composite consisting of TiC ceramic particles in a metal matrix
	Chip screening and conveying	Alumina tiles and blocks lining screens, pneumatic conveyor pipes, and sawdust-removing equipment
Pulp preparation	Chip "digestion"	Ceramic-lined pressure cooker 50–250 feet high, 100–180 psi pressure, sodium sulfide/sodium hydroxide
	Pulp cleaning and pulp bleaching	Ceramic-lined vats; ceramic valves, seals, cyclone separator liners, and pump parts
	Pulp storage and refining	Concrete tanks to contain up to 1500 tons of pulp/water slurry, lined inside and outside with ceramic tile
Paper machine	Water reduction from 99.5% to about 80%	Rows of alumina *suction box covers* about 2 feet wide and 20 feet long and rows of *foils* about 2 inches wide and 20 feet long made of segments of alumina, transformation-toughened alumina or zirconia, or silicon nitride
	Roll press to further remove water and compact wood fibers	Roll 20 feet long and 5 feet in diameter made of granite rock, followed by as many as 120 steel rolls
	Drying	Ceramic and other types of heaters to remove remaining water
	Slitting	Ceramic cutters made of transformation-toughened zirconia to cut paper to desired size

and sometimes this sequence can cost as much as the other fabrication steps combined. Industry is continuously striving to minimize the amount of cutting and grinding and to accomplish it more quickly and efficiently. Because they are the hardest materials available, ceramics have played an enormous role in increasing the rate of cutting and grinding.

Table 3.4 lists the hardness of a variety of ceramics. For comparison, most metals have hardness below 0.5 million psi (350 kg/mm^2), which is about one-two hundredth the hardness of diamond. The ceramics are used in a variety of ways. Sometimes they are bonded as small angular particles into or onto another material to form a cutting wheel or a grinding wheel. Other times they are used as fine particles bonded onto paper or cloth or as a free abrasive as loose particles mixed with a liquid for grinding, polishing, lapping, ultrasonic machining, or waterjet cutting. Perhaps the most amazing application is for cutting tool inserts where a solid polycrystalline piece of the ceramic is successfully used for high speed cutting of metals with a lathe or milling machine. The following paragraphs describe some of the cutting and grinding applications of ceramics that are very important to modern engineering.

3.3.1 CERAMIC CUTTING TOOL INSERTS

A study in the early 1970s estimated that there were 2,692,000 metal-cutting machines in the United States with an annual operating cost of about $64 billion. Those numbers are probably small compared to worldwide machines and costs today, especially considering the broad extent of use of numerically

FIGURE 3.27 Aluminum oxide suction box covers and foils installed on a papermaking machine. (From Richerson, D.W., *The Magic of Ceramics*, The American Ceramic Society, Westerville, OH, 2000, p. 253. Reprinted with permission of the American Ceramic Society, www.ceramics.org)

FIGURE 3.28 Close-up view of an alumina ceramic suction box cover. (From Richerson, D.W., *The Magic of Ceramics*, The American Ceramic Society, Westerville, OH, 2000, p. 254. Reprinted with permission of the American Ceramic Society, www.ceramics.org)

controlled (NC) machines. All of these machines have one thing in common. They force a sharp-edged tool (such as shown in Figure 3.29) against a metal surface in a turning, milling, or drilling operation. This results in high stress, high frictional temperature buildup, and severe erosion and corrosion.

Figure 3.30 illustrates the history of improvements in cutting-tool inserts.[21] Cutting-tool inserts in the early 1900s were made of high-grade tool steel. However, they could only be operated at a

TABLE 3.4
Hardness of Ceramics and Ceramic-Based Materials

Material	Hardness, in kg/mm^2	Hardness, in lbs/in^2
Single-crystal diamond	7000–9500	10.0–13.5 million
Polycrystalline diamond	7000–8600	10.0–12.2 million
Cubic boron nitride (CBN)	3500–4750	5.0–6.7 million
Boron carbide (B$_4$C)	3200	4.5 million
Titanium carbide (TiC)	2800	4.0 million
Silicon carbide (SiC)	2300–2900	3.3–4.1 million
Aluminum oxide (sapphire or polycrystalline)	2000	2.8 million
Tungsten carbide – cobalt "cermet" (94% WC-6% Co)	1500	2.1 million
Zirconium oxide (ZrO$_2$)	1100–1300	1.6–1.8 million
Silicon dioxide (SiO$_2$)	550–750	0.8–1.1 million
Borosilicate glass (such as ovenware)	530	0.75 million

Source: Richerson, D.W., *The Magic of Ceramics*, The American Ceramic Society, Westerville, OH, 2000, p. 240.

FIGURE 3.29   A variety of silicon nitride ceramic cutting tool inserts. Typical size is 1 to 2 cm. (Photo by D.W. Richerson.)

low speed of about 100 surface feet per minute (sfpm) or else they would overheat and deform or interact chemically with the workpiece. In the 1930s and 1940s, composites consisting of very hard tungsten carbide (WC) ceramic bonded with about 6% volume fraction of ductile cobalt metal were developed that could be operated up to about 300 sfpm. These composites, commonly referred to as *cermets* or *sintered carbides*, substantially increased the efficiency of metal machining and also were successful for many wear-resistance applications.

   The maximum working speed of the cermets was limited by softening and chemical degradation of the metal bond phase. Studies during the 1950s demonstrated that hot-pressed aluminum oxide could be used in some limited cutting operations up to 1000 sfpm, but tended to be unreliable

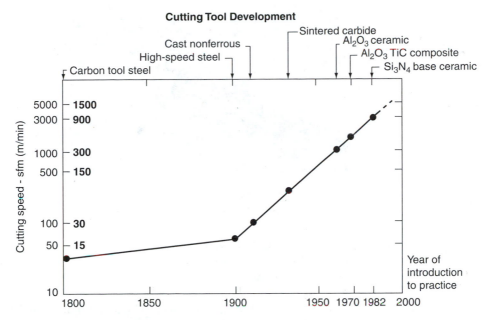

FIGURE 3.30 Historical improvements in the rate of metal cutting for cutting-tool inserts. (From Dow Whitney, E., *New Advances in Ceramic Tooling*, SME Tech Rept. MRR76-15, Society of Manufacturing Engineers, Dearborn, MI, 1976. With permission.)

due to brittle fracture and thermal shock. Addition during the 1960s of about 35% titanium carbide (TiC) particles to the alumina resulted in a ceramic–ceramic composite that was more reliable because it had increased fracture toughness and thermal conductivity and thus was more resistant to fracture and thermal shock.

Dramatic improvements in ceramic cutting tools were accomplished during the 1970s, which virtually revolutionized metal cutting by increasing cutting speed in some cases to over 3000 sfpm.[1] A wide range of ceramic cutting-tool inserts has emerged: hot-pressed silicon nitride, sintered *sialon*, hot-pressed silicon nitride with particle or whisker reinforcement, alumina with SiC whisker reinforcement, and transformation-toughened alumina. With these new ceramics, the time to machine a metal part could be reduced by 75 to 90%. In addition the ceramic tool lasted longer and could cut more metal parts per tool insert, which greatly enhanced the benefit to NC machines.

Although ceramic cutting-tool inserts have been dramatically improved, they still are not as tough as WC-Co cermets and cannot perform many machining operations. As a result cermets are still used extensively. Here again ceramics have proven useful. A thin coating – about two ten-thousands of an inch (5 micrometers) – of alumina, TiC, or titanium nitride (TiN) on the surface of a WC-Co cermet tool can increase tool life by up to five times. The cermet provides toughness and resistance to fracture and the ceramic coating provides resistance to erosive and corrosive chemical interaction between the tool and workpiece at the high temperature produced during machining.

3.3.2 SUPERHARD ABRASIVES

For many centuries naturally occurring ceramics such as quartz sand, garnet, and corundum have been used as abrasives. Natural diamond also has been an important abrasive. During roughly the past hundred years, researchers have learned to synthesize a variety of superhard ceramics including SiC, boron carbide (B_4C), TiC, cubic boron nitride, polycrystalline diamond, and single-crystal diamond.

Because diamond is the hardest material known, it can cut and scratch all other materials or conversely resist scratching by all other materials. Tiny single crystals were first synthesized by Howard

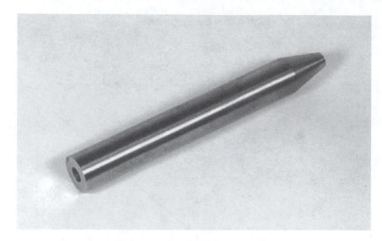

FIGURE 3.31 Waterjet cutting nozzle made from hard, wear resistant tungsten carbide. (Manufactured by Boride Products, Traverse City, MI. Photo by D.W. Richerson.)

Tracy Hall at General Electric Company in 1954 in an apparatus that could reach three million psi pressure and 5000°C (9032°F). About the same time, William Ebersole of Union Carbide Corporation demonstrated that diamond could be deposited from a carbon-containing gas, but the rate was way too slow to be of commercial interest. By the late 1980s scientists increased the rate of deposition by more than a hundred times so that a layer several thousands of an inch thick could be achieved in an hour. A diamond coating has been shown to increase the life of some cutting tools by a factor of 20 and to greatly increase the wear resistance of some industrial parts. This is a wide-open field of study and opportunity.

3.3.3 WATERJET CUTTING

Waterjet cutting is a relatively new method for rapidly and precisely cutting metals, plastics, composites, and ceramics. Fine ceramic particles (such as boron carbide) are mixed with water, and the resulting slurry blasted under about 60,000 psi pressure through the small bore of a ceramic nozzle. Figure 3.31 shows a nozzle fabricated from WC (not containing any cobalt). Waterjet cutting can slice through several inches of steel and has even been used to bore tunnels through solid rock.

3.4 ELECTRICAL APPLICATIONS OF CERAMICS

Most people do not associate ceramics with electrical applications, but the diverse electrical behavior of ceramic materials as indicated by Table 3.5 results in many important applications. In fact, surprisingly, these electrical applications entail the largest economic sector for ceramics. Perhaps this should not be surprising, considering the large number and variety of ceramic compounds possible. Most of these ceramic compounds are electrical insulators, but others are semiconductors, ionic conductors, and even electronic conductors and superconductors.

ELECTRICAL APPLICATIONS OF CERAMICS

- Electrical insulators
- Hybrid packages
- Ferroelectric ceramics
- Piezoelectric ceramics
- Semiconductors
- Ionic conductors
- Electronic conductors
- Superconductors
- Capacitors
- Varistors
- Thermistors
- Igniters
- Resistors
- Electro-optic ceramics

TABLE 3.5
Diverse Electrical Behavior of Ceramics

Insulators	Varistors
Ionic conductors	Resistors
Electronic conductors	Thermistors
Semiconductors	Ferroelectric materials
Superconductors	Piezoelectric materials
Capacitors	Electro-optic materials
Rectifiers	

When atoms bond together in a material, the outer electrons reside in discreet bands of specific energy level. If adjacent bands overlap, electrons can move between the bands and result in electrical conduction. This is the case for metals. In contrast, virtually all of the electrons in most ceramics are tightly held either within individual atoms or between only two adjacent atoms. The resulting bands do not overlap, but instead have an energy gap between them, i.e., additional energy would need to be added for an electron to jump across the gap and participate in conduction. If the energy gap is large, the material is classified as an electrical insulator. If the gap is small enough that electrons can be promoted across the gap by addition of energy in the form of heat, light, or an applied electric field, the material is classified as a semiconductor.

3.4.1 CERAMIC ELECTRICAL INSULATORS

An electrical insulator is a material that is very resistant to the flow of electricity, which seems pretty simple and straight-forward. However, every electrical insulation application has unique requirements, so the engineering team must have comprehensive knowledge and understanding of both electrical design principles and materials behavior and options. For example, the alumina insulator in an x-ray tube for a CT (computed tomography) scanner must withstand tens of thousands of volts, while the ceramic spacer or bracket in a light fixture or household on-off switch only has to perform in a low voltage and low current environment. The ceramic insulator for each spark plug in our car must survive many thousands of cycles of high voltage (about 8000 volts), rapid temperature change, and shock waves (from each air and fuel explosion). Equally important, the spark plug insulator must be low cost, extremely reliable, and able to be fabricated at a rate of about one million per day in a single factory (greater than three million per day worldwide) by relatively conventional powder compaction and densification processing. A ceramic spark plug insulator shaped by dry bag isostatic pressing plus green machining was shown earlier along with an assembled spark plug in Chapter 2 as Figure 2.5. Figure 3.32 shows a much larger alumina insulator for a high voltage application. Figure 3.33 illustrates a variety of much smaller ceramic electrical insulators for various electrical devices.

In contrast to the bulk ceramic x-ray tube and spark plug insulators, some applications require very thin films of ceramic insulation that are intricately integrated on a nanometer to micrometer scale with semiconductive, conductive, resistive, and dielectric materials to form arrays of miniature electrical devices. Perhaps the best example of this is the integrated circuit (the "silicon chip"). To achieve the necessary intricacy and precise control, the insulative ceramics such as SiO_2 and Si_3N_4 are deposited by processes involving microscopic scale chemical reaction, diffusion, physical vapor deposition, and chemical vapor deposition. A modern integrated circuit the size of a fingernail contains millions of transistors and diodes and is comprised of around 20 layers that require about 80 clean room fabrication steps to achieve.

The large number of devices in a silicon chip and the tiny size pose another engineering challenge. How does one transport and distribute electricity from a typical household electrical outlet to all of

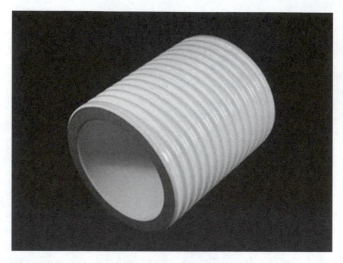

FIGURE 3.32   Example of an alumina electrical insulator (fabricated by Coorstek, Golden, CO) much larger than a spark plug. The insulator is about 12 cm long. (Photo by D.W. Richerson.)

FIGURE 3.33   Variety of small alumina electrical insulators. (Photo by D.W. Richerson.)

the tiny devices inside the integrated circuit? Engineers have learned to encapsulate hundreds of feet of metal electrical wiring in dense aluminum oxide (the electrical insulator) to form a small *hybrid package* into which the silicon chip fits. The technology and fabrication process are described in Chapters 10 and 13. Figure 3.34 shows an early hybrid package (vintage about 1975) with the silicon chip mounted inside. Note the microscopic wires that connect between the chip and the package. After the chip was mounted and connected, the assembly would be placed in a vacuum and a lid brazed on to hermetically seal the chip into the package.

The chip in Figure 3.34 contained less than 100,000 transistors. Modern chips contain many millions of transistors, so the package must be much more complex to support these chips. Figure 3.35 shows a variety of modern hybrid ceramic packages for individual silicon chips. Packages have also been designed for multiple chips such as required for a mainframe computer capable of handling all the transactions of a bank, an airline, or an Internet server. For example, IBM produces a hybrid ceramic package (illustrated in Figure 3.36) that can house 121 chips each containing millions

FIGURE 3.34 Close-up view of an early example (about 1975) of a silicon chip mounted in a hybrid package. This tiny chip is only about 3 mm long. (Photo by D.W. Richerson.)

FIGURE 3.35 Variety of modern hybrid packages, two containing a silicon chip mounted inside. The largest hybrid package in the photograph is 5 cm × 5 cm and contains 239 interconnection pins. (Photo by D.W. Richerson.)

FIGURE 3.36   View of the bottom and top of a modern IBM mainframe computer IC package that can hold 121 logic and array silicon chips. (Photo courtesy of IBM Microelectronics, East Fishkill, NY.)

of devices. The package is about 3.5 inches square and one-fifth inch thick and contains 33 layers, about 350,000 metal-filled holes that interconnect layers, 427 feet of internal wiring, and 1800 brazed-on pins for external connections. This IBM module contains over 100,000 separate electrical circuits.

3.4.2   DIELECTRIC CERAMICS

Even though electrons do not flow through a ceramic insulator, they still can respond to an electric field by slightly changing their position. This results in *polarization*, a slight shift in negative charge in the direction of the positive electrode and a slight shift in the positive charge in the direction of the negative electrode. Polarization can also result from movements of atoms or ions or a change in orientation of nonsymmetrical molecules.

The degree of polarization varies by several orders of magnitude from one ceramic to another. Alumina and silica have relatively low susceptibility to polarization, which is why they are such good insulators. At the other extreme, barium titanate ($BaTiO_3$) can polarize 500–1000 times more than alumina or silica, making them suitable for production of miniaturized capacitors. Capacitors are required in most electrical circuits and are produced at a rate of about one billion per day. Capacitors can be designed into the circuit to perform various functions: charge storage, blocking, coupling, decoupling, bypass, and filtering. These functions are described in Chapter 11.

The reason for the high polarization of the $BaTiO_3$ is linked to the atoms present and to the crystal structure. Each titanium ion (Ti^{4+}) is positioned between Ba and O atoms in the structure in a space that is a little too large for the Ti^{4+} ion. When an electric field is applied, the positively charged titanium ion can move very easily a tiny distance in the direction of the negative electrode. Because the Ti^{4+} has such a high charge (four times the charge of an electron), this tiny movement results in very large polarization. This is an example of the importance of understanding crystal structure and the factors that determine how individual atoms and ions fit into each structure. These factors are discussed in Chapters 4 and 5.

TABLE 3.6
Examples of Applications of Piezoelectric Ceramics

Quartz watches	Charcoal lighters	Smart skis
Ultrasonic cleaners	Hydrophone	Sonar
Medical ultrasound imaging	Buzzers, alarms	Fish finders
Underwater homing beacon	Ocean floor mapping	Emulsifiers
Dialysis air bubble detection	Wheel balancers	Motors
Nondestructive inspection	Accelerometers	Vibration sensors
Ultrasonic physical therapy	Homogenizers	Contact sensors
Autofocus camera	Underwater decoy	Impact sensors
Precise deflection measurement	Boat speedometers	Loud speakers
Zero-vibration tables	Musical greeting cards	Microphones
Transformers	Actuators	Printers
Telescope mirror distortion correction actuators	Transformers	Igniters

There are other variants to dielectric ceramics, each of which makes possible specialized applications.

- Ferroelectricity – The polarization goes to zero for most dielectric ceramics when the electric field is removed (reduced to zero). This is not the case for a ferroelectric ceramic. Instead, a portion of the polarization is retained and can only be removed by applying a reverse electric field of the appropriate intensity. Thus the degree and direction of residual polarization can be precisely controlled for each material. Furthermore, each material responds in a different way (such as ease of polarization and depolarization and level of retained polarization), so nearly infinite range of ferroelectric behavior is available to electrical device designers.
- Piezoelectricity – Some ceramics that have a crystal structure with no center of symmetry can polarize when a mechanical stress is applied. The stress elastically deforms the material in such a way that the centers of positive charge move in one direction and negative charge in the opposite direction. This results in a measurable voltage across the material. Conversely, application of a voltage causes the material to deform or to vibrate at its resonant frequency. A wide variety of applications and products have been developed based on piezoelectricity, some of which are identified in Table 3.6 and illustrated in Figures 3.37 and 3.38.
- Sensitivity to frequency – The different mechanisms of polarization respond at different rates to an electric field. Electron polarization happens very quickly because the electron only needs to move from one side of an atom to the other. Therefore electron polarization can occur over a wide range of frequency, including very high frequency. Conversely, molecular polarization requires time and also adequately high electric field for the molecule to re-align. Materials that depend on molecular polarization are very sensitive to frequency of the applied field, i.e., they exhibit a higher degree of polarization at low frequency and have a threshold above which the polarization significantly decreases.

3.4.3 SEMICONDUCTORS

Semiconductor technology is the basis for our modern electronics age. Most materials with *intrinsic* semiconductor behavior are covalently bonded elements and compounds from columns IIIB through VB of the periodic table (Si, Ge, GaAs, and GaP) as well as some from columns IIB and VIB (CdS, PbS, PbSe, Cu_2O, and ZnSe). These and many other compounds can become *extrinsic* semiconductors

FIGURE 3.37 Variety of ceramic piezoelectric parts and assemblies for many of the applications listed in Table 3.6. (From Richerson, D.W., *The Magic of Ceramics*, The American Ceramic Society, Westerville, OH, 2000, p. 150, photo courtesy of Edo Corporation Piezoelectric Products, Salt Lake City, UT. Reprinted with permission of the American Ceramic Society, www.ceramics.org)

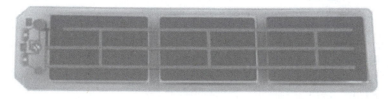

FIGURE 3.38 Piezoelectric module for a K2 "smart ski" that consists of three ceramic wafers imbedded in plastic and connected to high resistance electrical wiring. When the ski begins to vibrate as the skier goes faster and faster, some of the vibrations are converted by the piezoelectric ceramic into electricity that is then converted into heat by the high electric resistance wiring. The module manufactured by Active Control eXperts, Inc. (ACX), Cambridge, MA, and is 16.8 cm long. (Photo by D.W. Richerson.)

by doping (adding a tiny amount of selected elements). For further variation the extrinsic semiconductors can be designed to be either *n-type* (dopant results in extra electrons so the charge carriers are negative) or *p-type* (dopant results in missing electrons that behave as positive charge carriers). For example, doping silicon with aluminum atoms results in p-type while doping with phosphorus atoms results in n-type. This doping to p-type and n-type allows electrical engineers great flexibility in design of the miniature circuits and devices inside a silicon chip integrated circuit. It also makes possible semiconductor lasers and light emitting diodes (LEDs).

TABLE 3.7
Examples of Applications of Ceramic Semiconductors

Base Material with the Semiconductivity Controlled by Dopants or Heat Treatment	Device or Application	Function of Device/Ceramic Material
Si	ICs for computers, electronic watches, televisions, toys, breadmakers, automobiles, communications equipment	VLSI (very large scale integration) of transistors and other devices in integrated circuit (IC)
SiC, graphite, $MoSi_2$	Heating elements	Control of electrical resistance to produce heat
Si, CdS, InP, TiO_2, $SrTiO_3$	Solar cells	Directly convert energy from sunlight into electricity
Cu_2O	Rectifier	Allows current to only flow in one direction and can convert from alternating current to direct current
$BaTiO_3$, Fe_3O_4/ $MgAl_2O_4$ solid solution	Thermistor	Electric resistance varies as a function of temperature; used for thermal switches and other applications
ZnO, SiC	Varistor	Electrical resistance varies as a function of applied voltage; used for voltage surge protection
SnO_2, ZnO, Fe_2O_3	Sensors	Electrical output varies as a function of exposure to specific gases
GaAs, GaP, GN	LEDs and Lasers	Convert electrical input directly to bright light

Extrinsic semiconduction also can be achieved by formation of nonstoichiometric (excess cation or anion rather than ideal balanced formula) compounds. Examples include $Cu_{2-x}O$ and $Fe_{0.95}O$ (both p-type conduction) and $Zn_{1+x}O$ (n-type).

Semiconductors make possible the multi-trillion-dollar worldwide electronics industry. Table 3.7 identifies some important semiconductor applications and the key ceramic materials involved.

3.4.4 ELECTRICAL CONDUCTORS

Most people have the stereotypical view that metals are good electrical conductors and that ceramics are good electrical insulators. Actually ceramics can be excellent electrical conductors by two modes: electrons and ions. Electron conductivity occurs in some transition metal oxides due to formation of partially filled energy bands involving d or f electrons (conduction in metals involves s electrons). TiO, ReO_3, and CrO_2 all have conductivity greater than 10^2 $(ohm-cm)^{-1}$ with about 10^{22} to 10^{23} electrons available for conduction (comparable to metals). These have relatively flat conduction vs. temperature curves over a broad temperature range. VO_2, VO, and Fe_3O_4 have comparable conductivity at elevated temperature, but exhibit an abrupt drop to about 10^{-2} $(ohm-cm)^{-1}$ at a specific lower temperature for each material. NiO, CoO, and Fe_2O_3 have electron conductivity of 10^0 to 10^{-2} $(ohm-cm)^{-1}$ at elevated temperature with a rapid linear decrease over a several hundred degree lower temperature range to about 10^{-18} $(ohm-cm)^{-1}$. As was the case with semiconductor ceramics, this wide range of behavior with temperature provides the engineer with broad opportunity for design of electrical devices.

The second mode of conduction, ionic conduction, involves the movement of ions as charge carriers rather than electrons. Because ions are larger than electrons, ionic conduction generally requires extra energy (often in the form of heat) plus a path for the ions to move through the solid material. In the case of doped ZrO_2 (such as $Zr_{1-x}Ca_xO_{2-x}$), the charge carriers are O^{2-} ions and the path consists of oxygen ion vacancies in the crystal structure. In beta-alumina the charge carriers

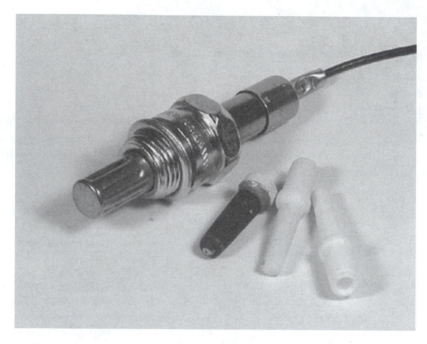

FIGURE 3.39 Oxygen sensor assembly, two as-fabricated doped zirconia solid electrolyte oxygen ion conductor parts, and a finished doped zirconia oxygen ion conductor coated with metal electrodes on the inside and outside and a porous wear-resistant spinel ceramic on the outside. This finished part is 2.8 cm long. (Photo by D.W. Richerson.)

are Na^{1+} and the path consists of directional channels in the crystal structure. Other ionic conductors include NaCl, AgBr, KCl, BaF_2, and various lithium compounds.

Some of the most important applications of ionic conductors include small solid state "lithium" batteries, oxygen sensors, and solid oxide fuel cells. An oxygen sensor assembly and the doped zirconia oxygen ion conductor, which is the key component inside the sensor, are shown in Figure 3.39. The zirconia part is essentially a tube with one end closed. The inner surface and the outer surface are coated with electrically conductive electrodes that are connected to electrical leads that go to a voltage measurement device. The inside of the tube is exposed to air, which has a relatively constant oxygen content. The outer surface is continuously bathed in the exhaust gases passing through the exhaust manifold of your car. Whenever there is a difference in the oxygen partial pressure between the inside and outside of the tube, oxygen ions begin to migrate across the ceramic. This results in a measurable voltage that varies precisely with the difference in partial pressure. Engineers have determined the ideal fuel to air ratio and the optimum residual oxygen content in the exhaust. Thus they know the optimum voltage to expect from the oxygen sensor if the automobile is operating at peak efficiency. This is programmed into the electronic control system. Any deviation from the optimum voltage automatically initiates a change in air to fuel ratio to get back to the ideal conditions. The ceramic oxygen sensor increases your gas mileage by about four miles per gallon. All automobiles with standard gasoline-burning engines in the United States, Western Europe, and Japan are required by law to have oxygen sensors.

The solid-oxide fuel cell works with the same principle as the oxygen sensor, except in this case air or oxygen is on one side and a fuel such as hydrogen or natural gas is on the other side.[22] The doped zirconia, referred to as a solid electrolyte, is a dense thin layer impermeable to gases, but doped to have oxygen vacancies that oxygen ions can use as a diffusion path to travel through. The zirconia electrolyte is coated with porous electrodes that allow access of the gaseous fuel and air but can also conduct electrons. The electrode on the air side takes electrons from an external circuit to convert air molecules to oxygen ions. These oxygen ions then migrate across the zirconia electrolyte until

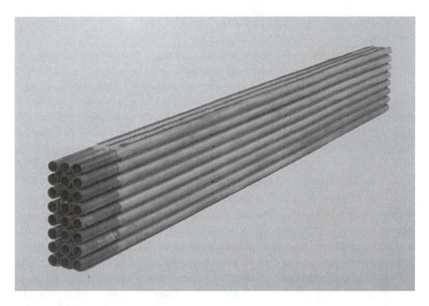

FIGURE 3.40   Tubular solid oxide fuel cell module manufactured by Siemens-Westinghouse, Orlando, FL. (From Richerson, D.W., *The Magic of Ceramics*, The American Ceramic Society, Westerville, OH, 2000, p. 279. Reprinted with permission of the American Ceramic Society, www.ceramics.org)

they reach the fuel side electrode. Here they react with the hydrogen in the fuel and give up electrons to the external circuit. This direct conversion of the chemical energy in the fuel to electricity is about 60% efficient, which is about double the efficiency of burning the fuel in a heat engine (such as your car engine) or production of electricity in a conventional coal power generation plant. If the waste heat produced by the reaction of the oxygen and hydrogen is harnessed, then the total efficiency can be around 80%. The solid oxide fuel cell (and other types of fuel cell) offers great potential for increasing energy efficiency, which will also reduce pollution emissions because of the decrease in fuel consumption. Furthermore, if the fuel is hydrogen the only emission is water vapor.

Fuel cells are being developed for home energy systems, automobiles, and larger scale systems including hybrids with gas turbines. There are many opportunities for materials engineers in this area. Figure 3.40 shows ceramic components for a tubular design solid-oxide fuel cell developed by Siemens-Westinghouse.

Engineers must deal with a broad spectrum of challenges to develop useful solid-oxide fuel cell products and other products based on ionic conduction. The following are some of the solid-oxide fuel cell issues: operation above approximately 700°C (ionic conductivity too low at lower temperature), requirement for a ceramic air-side electrode that is thermodynamically stable but also conducts electrons, thermal expansion match between the electrolyte and electrodes, noninteraction of the materials during the high temperature fabrication process and during more than 30,000 hours desired life, design to control flow of air and fuel and the heat produced in the reaction, sealing, and cost-effective fabrication process. Here is another example of the need for a broad understanding of the fundamentals controlling material's behavior and properties, design allowables for brittle materials and interfacing with other materials, and fabrication processes.

3.4.5   CERAMIC SUPERCONDUCTORS

Even more surprising than the electrical conduction of many ceramics is the fact that some ceramics are superconductors. A superconductor is a material that has zero electrical resistance. Even copper and aluminum, which we commonly use for electrical wiring, have significant electrical resistance at room temperature and convert part of the electricity flowing through them into heat. As much as

7% of the electricity produced in a central generation plant can be lost in transmission between the plant and your house.

In 1911 Heike Kamerlingh-Onnes discovered that mercury decreased in electrical resistance as the temperature was decreased and abruptly lost all electrical resistance at about 4.2 K (about −452°F). Between 1911 and 1986 many other materials were shown to be superconductive with transition temperatures as high as about 23 K. However, most applications were not practical because such low temperature could only be achieved by liquid helium, which was very expensive. Between 1986 and 1988 a series of ceramic compositions were identified that became superconductive at much higher temperature (90 to 120 K). The significance of this was that now a material could be superconductive with cooling by liquid nitrogen, which is much less expensive than liquid helium.

Much research and development has been conducted since 1988 to understand the behavior and fabrication issues of "high temperature" ceramic superconductors and to develop applications. The potential is enormous, but so are the challenges. An especially difficult challenge has been to fabricate ductile wire, since all of the ceramic superconductor materials are brittle at room temperature like most other ceramics. One solution has been to produce a thin, narrow ribbon consisting of a composite of a mixture of the superconductor powder with silver. As shown in Figure 3.41, this superconductive ribbon is being evaluated for electrical transmission cable. Many large cities (such as Detroit, MI) have underground conduit that carries the massive copper cables needed to supply the electrical demands of the city. Since personal computers and the Internet have became widely used, there has been a steady and rapid increase in the electrical requirement that is exceeding the capacity of the conduit, especially during the summer on hot days when everyone is using air conditioning. To replace the existing conduit would require digging up the city and would cost many billions of dollars. Some cities are considering replacing copper cable with ceramic superconductor cable that

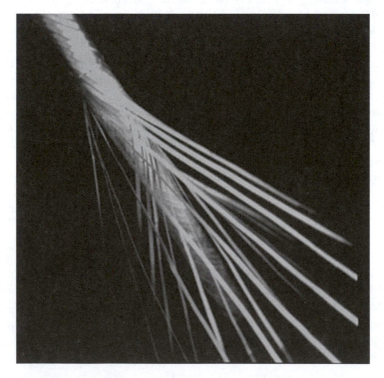

FIGURE 3.41 Ceramic superconductor ribbon being wound into a cable by American Superconductor Corporation, Westborough, Massachusetts for an experimental electrical transmission line. (From Richerson, D.W., *The Magic of Ceramics*, The American Ceramic Society, Westerville, OH, 2000, p. 144. Reprinted with permission of the American Ceramic Society, www.ceramics.org)

can carry the same current but with smaller cross section cable. Ceramic superconductors also are being evaluated for experimental motors, special antennas, microwave devices, magnetic shielding, and a variety of other applications.

3.5 MAGNETIC CERAMICS

Magnetism is another material property that can be traced to the electron structure within an atom and the arrangement of the atoms into a crystal structure. Most people do not think of ceramics being magnetic, but the first magnets were the naturally occurring ceramic magnetite (Fe_3O_4). Starting in the late 1940s, a variety of ceramic compositions with a wide range of magnetic characteristics began to be synthesized. In 1997 more than 1.3 billion pounds (nearly 300 million kilograms) of ceramic magnets were produced.

The general term for a ceramic magnet is a ferrite. Most ceramic magnets are categorized as *hard ferrites*; once they are magnetized they retain the magnetization and are excellent permanent magnets. Some of the applications for hard ferrites include speakers, household magnets, linear particle accelerators, and the motors for toothbrushes, electric knives, and all of the power accessories in automobiles. Other ceramic magnets are categorized as *soft ferrites*; they can be easily magnetized and demagnetized by changing the direction of an applied magnetic field. Soft ferrites are important in television, radio, touch-tone telephones, communications, electrical ignition systems, fluorescent lighting, high frequency welding, transformer cores, and high-speed tape and disk recording heads.

Ceramic magnets also can be combined with other materials (especially polymers) to produce a composite magnet. Examples are recording tape and the flexible magnetic strips that you use to post messages on your refrigerator.

Table 3.8 lists the applications of ceramic magnets in a typical automobile. Figure 3.42 illustrates the shape of the ceramic magnets for the windshield wiper motor and other accessory motors in your car. Figure 3.43 shows a ceramic magnet as an important component of a loudspeaker, while Figure 3.44 illustrates the force between two ceramic loudspeaker magnets.

3.6 OPTICAL APPLICATIONS OF CERAMICS

Optical behavior involves interaction of a material with light; not just the visible light spectrum, but also the much broader electromagnetic spectrum that includes infrared, ultraviolet, x-ray, microwave, and radio wavelengths and frequencies. Key properties that lead to numerous applications include transparency, refraction, color, phosphorescence and fluorescence, and electro-optics.

TABLE 3.8
Magnetic Ceramic Applications in a Typical Modern Automobile

Windshield wiper motor	Windshield washer pump
Window lift motors	Seat positioning motors
Tape drive	Cruise control
Automatic temperature control	Liquid level indicators
Economy and pollution control devices	Heating and air conditioning motor
Cooling fan motor	Air pump/suspension
Starter motor	Ignition
Antilock brake speed sensors	Speedometer, gauges and digital clock
Speakers	Antenna lift motor
Fuel pump motor	Door locks
Defogger fan motor	Tailgate or trunk latch
Sun roof motor	Recording tape

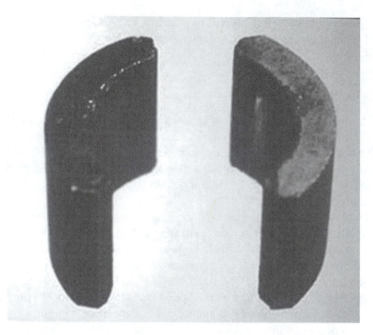

FIGURE 3.42   Ceramic magnets for a windshield wiper motor for an automobile. (From Richerson, D.W., *The Magic of Ceramics*, The American Ceramic Society, Westerville, OH, 2000, p. 206. Reprinted with permission of the American Ceramic Society, www.ceramics.org)

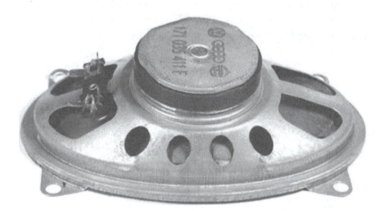

FIGURE 3.43   Ceramic magnet built into the backside of a loudspeaker. (Photo by D.W. Richerson.)

3.6.1 APPLICATIONS BASED ON TRANSPARENCY

Make a list of all of the different ways that we use transparent materials. Could we have achieved all of these application and products through the use of naturally occurring transparent solid materials? The answer is clearly "no." Transparency to visible light is actually quite rare for naturally occurring solid materials. The examples that are most obvious are single crystals such as quartz and mica, but these are not abundant enough or large enough in size to meet our needs. Until modern plastics were developed in the early 1900s, our primary source of a transparent material was soda-lime-silica glass, which is a noncrystalline form of ceramic. Glass remains so important to us that more than 600 million tons per year are produced worldwide, including a wide range of compositions beyond soda-lime-silica.[2,23] Examples of specialty glasses include high-lead optical glass, high-silica and

FIGURE 3.44 Two ceramic loudspeaker magnets mounted so they repel each other. The top magnet is suspended in mid air by the force between the magnets. (Photo by D.W. Richerson.)

TABLE 3.9
Applications of Ceramics Transparent to Visible Light

Glass		Synthesized Single Crystals
Windows	Eye glasses	Diamond
Drinking glasses	Storage containers	Ruby, sapphire gemstones
Mirrors	Cabinet and arcadia doors	Cubic zirconia gemstones
Computer screens	Fiber-optic cable	Sapphire bar code scanner plate
Light bulbs	Medical endoscopes	Specialty lenses
Car windshields	Microscope lenses	Cr-doped ruby laser host
Skylights	Nd-doped glass laser host	Nd-doped YAG laser host
Light guide fibers	Integrated optics	Many other laser hosts
Cooking ware	Ring laser gyroscope	Sapphire fibers
Decorations	Temperature measurement	Hi-temp. transparent parts
Art	Table tops	
Glazes	Protective coverings	
Marbles	Jewelry spheres	
Electrical insulators	Vacuum tube seals	
Telescope lenses	Photolithography lenses	

borosilicate low thermal expansion glasses, electrical grades, fibers for polymer matrix composites, optical fibers, doped laser glasses, and many colored glasses for art, decorative containers, and glazes. Table 3.9 lists some of the diverse applications of transparent glass along with transparent synthesized single-crystal ceramics.

3.6.1.1 Window Glass

Window glass was first produced by blowing a large bubble and rotating the blow tube until the bubble collapsed into a flat disk. This was then cooled, cut into somewhat irregular thickness panes, placed back in a "flattening" furnace for stretching or "ironing," and finally annealed. In 1902 an American, John Lubbers, designed and built a machine that could draw a long cylinder of glass directly from a pot of molten glass. This could produce a thin film of glass about 10 times larger than could be achieved by blowing. The cylinder could be cut along one side and flattened while heated. About the same time, Belgian Emile Fourcault invented a machine that drew a glass sheet five stories straight up. Unfortunately the surface was always marred, so the glass was not of high quality (for example not suitable for automobile windshields). Modifications were made to the Fourcault process by the U.S. firm Libby-Owens that doubled the rate and later by Pittsburgh Plate Glass Company that further increased production rate and achieved better quality glass.

By 1937 Ford Motor Company and British glassmaker Pilkington Brothers combined efforts to develop a large machine to produce high quality plate glass. Hot viscous glass was pressed between two rolls to form a thin sheet. This sheet was annealed and cooled, followed by surface grinding simultaneously on both surfaces. The glass was then washed, cut to the desired size, and polished on both sides. The machine was 1400 feet (427 meters) long. The Pilkingtons continued to experiment with methods to make high quality flat glass. During the 1950s they developed the "float glass process" that is now used worldwide. Molten glass flows onto a bath of molten tin, resulting in very flat glass that requires no grinding or polishing.

Corning Glass also developed a process for making high quality plate glass. This *fusion draw* process involved flow of glass over the edges of a trough. As the two flows met beneath the trough, they fused to form a pane of glass that subsequently made no contact with surfaces until it was fully cooled. This method was first used to make automobile windshields, later for photochromic glass lenses, and most recently for laptop computer screens for liquid crystal displays (LCDs).

Glass is normally susceptible to fracture by shattering when a local force is applied. This was not desirable for automobile windshields. In 1891 Otto Schott demonstrated that the strength of glass could be substantially improved by coating a high thermal expansion glass with a low thermal expansion glass. During cooling, the inner glass would shrink faster and pull the outer glass layer into a compressive stress state. Before the glass could be broken by a tensile force, the compressive stress would have to be overcome. By 1929 engineers had learned that compressive stresses could be induced into the surface of plate glass by quenching from a suitable temperature. However, only about one sheet out of ten survived the tempering procedure. In 1929 Achille Verlay of Saint-Gobain in France developed a successful technique. He hung glass sheets in an oven using tungsten-tipped tongs. At the appropriate temperature the glass was removed and exposed to an air blast simultaneously on both sides. This resulted in high yield and became the standard procedure. In 1962 Samuel Kistler of the University of Utah came up with an alternate approach. He exposed heated glass in a bath of a molten potassium salt. Potassium ions replaced smaller sodium ions near the surface of the glass. As the glass thermally contracted during cooling, the larger potassium ions resulted in a residual compressive stress state at the surface. This chemically tempered glass was 25 times stronger than untreated glass.

Many other innovations have occurred in glass technology for windows. Now most glass for buildings is laminated (using multiple layers of glass and polymers) to minimize thermal conduction and is coated with multiple thin layers of ceramics to reduce glare and transmission of infrared wavelengths of light. Engineers also have learned to reinforce glass with a metal wire mesh using a rolling process. Rolling is also used to emboss decorative designs onto the surface of glass.

3.6.1.2 Container Glass

Glass has many advantages as a container. It is very inexpensive, transparent, easy to clean, odorless, impermeable, hygienic, physically stable, and resistant to most acids and alkalis. It can readily be colored by small additions (usually less than 0.5%) such as chromium oxide for green glass and iron sulfide (pyrite) for brown glass. As long as colors are matched, glass can be easily recycled to reduce energy consumption and conserve raw materials.

Container glass has evolved along a similar path to window glass. The first major innovation was glass blowing around 50 BC. Another was pressed glass during the 1800s. The first automated glass blowing machine was developed by Michael Owens in 1903. Early machines could produce less than 20 bottles per hour per finishing mold. This was increased to about 90 and presently to about 900.

3.6.1.3 Optical Glass Fibers

Engineers learned to make thin glass fibers during the 1930s. These were initially used in wool-like bundles for thermal insulation (fiberglass insulation) for buildings and later adapted as aligned filaments as fiber reinforcement in polymer matrix composites (fiberglass composites). During the 1950s and 1960 communications scientists discovered that tremendous quantities of information could be transmitted by light through glass fibers. However, with the glass fibers and light sources available at the time, the information could only be transmitted about 10 feet due to scattering from impurities and bubbles in the glass. During the 1960s a variety of lasers were invented that could produce an intense monochromatic, coherent beam of light. By 1970 scientists at Corning, Inc. (Maurer, Keck, and Schultz) demonstrated an optical glass fiber that could transmit the light from a laser three fifths of a mile with retention of one percent of the initial intensity. This was adequate to launch the field of fiber optic communications.

Many improvements have been achieved in optical fibers since 1970. By the late 1990s the distance the fibers could transmit laser light or semiconductor LED light with one percent retained intensity had increased to more than 75 miles. This technology has had a dramatic impact on the amount of information that can be transmitted. According to a Corning brochure, two optical fibers, each thinner than a human hair, can carry 625,000 telephone calls at once. This means that about one-quarter pound of optical fiber can perform the same function as 2.5 tons of copper wire. Between 1976 and 1998 enough optical fiber has been installed worldwide to stretch back and forth between the earth and the moon 160 times. This fiber optic system has made the worldwide Internet possible and revolutionized the way we communicate.

> One-quarter pound of optical glass fiber can perform the same function as 2.5 tons of copper wire.

3.6.1.4 Lenses

Transparency combines with refraction and physical and chemical stability to make glass an effective material for lenses. Historically, glass lenses contributed immeasurably to our quality of life by making eye glasses, the microscope, and the telescope possible. The microscope in particular had a major impact by allowing scientists to see bacteria and thus understand the cause of disease.

The telescope also has been important to us and our understanding of science and our position in the universe. Hans Lippershey invented the first telescope in 1608 by placing two transparent glass lenses in a hollow tube. Galileo Galilei constructed a telescope the following year with larger lenses. However, as astronomers tried to make telescopes with still larger lenses, they were limited by image distortions resulting from the quality of the glass and other factors. In 1668 Isaac Newton invented the *reflecting telescope* as an alternative to the *refracting telescope*. He replaced the transparent lenses with lens-shaped mirrors of highly polished metal. This allowed much larger telescopes, but these also reached a limit. The metal had high coefficient of thermal expansion and the mirror distorted due to temperature fluctuations and gradients.

George Ellery Hale experimented during the early 1900s with fabrication of telescope mirrors from glass with a reflective coating on the surface. These mirrors had a lower thermal expansion than metals, but still too high to make the 200 inch diameter mirror that he desired. Here was another case of an improved design or revolutionary product not being possible because of a materials limitation.

It would not be until 1935 that a suitable near-zero expansion glass composition and other technologies would be developed to allow casting of a 200 inch diameter blank of glass for Hale's telescope mirror. And it would not be until 1949 before the mirror would be fully ground and polished and built into the telescope on Mt. Palomar. Once completed, the telescope allowed astronomers to peer much deeper into the universe than ever before.

Some lenses require transparency to other wavelengths besides visible light and in fact are not transparent to visible light. An example is the lens for night vision goggles or for an infrared (heat imaging) camera. Other examples of *electromagnetic windows* are discussed in a later section.

3.6.2 APPLICATIONS BASED ON PHOSPHORESCENCE AND FLUORESCENCE

Phosphorescence is the emission of light resulting from the excitation of a material by a suitable energy source. You have probably all seen a fluorescent mineral display at a natural history museum. Under daylight, incandescent light, or fluorescent light the minerals look normal and often drab. However, when the lights are turned off and the minerals are exposed to an ultraviolet light, they glow brightly each with a distinct color. Energy from the ultraviolet light is absorbed by specific electrons that then jump into a higher energy level within the atom. These electrons then drop back into their original position and give up the energy in the form of light of a specific wavelength. Many ceramics have phosphorescence and fluorescence and are important in fluorescent lights, oscilloscope screens, television screens, electroluminescent lamps, photocopy lamps, and lasers.

3.6.2.1 Fluorescent Light

The first electrical lights were *incandescent* and consisted of a delicate carbon filament sealed inside an evacuated transparent glass bulb. The electrical resistance of the carbon was high enough that considerable heat resulted when an electric current was passed through the thin filament, thus yielding bright incandescent light. Later the carbon filament was replaced by a tungsten filament. Although the incandescent light produces excellent white light (broad range of wavelengths), it also produces considerable heat and glare and is quite inefficient. The fluorescent light is much more efficient, cool, and has longer life.

The fluorescent light consists of a sealed glass tube coated on the inside with a ceramic phosphor and filled with a mixture of mercury vapor and argon. An electrical discharge from a capacitor stimulates the mercury vapor to emit radiation at a wavelength of 25.37 nm, which is in the ultraviolet range well below the visible range (400 to 700 nm). The ceramic phosphor then absorbs the ultraviolet and emits a broad band of wavelengths in the visible range that combine to form white light. The ceramic phosphor is a halogen phosphate such as $Ca_5(PO_4)_3(Cl,F)$ or $Sr_5(PO_4)_3(Cl,F)$ doped with Sb and Mn.

Fluorescent lights save an enormous amount of electricity compared to incandescent lights. Figure 3.45 illustrates some of the configurations now available for fluorescent lights.

3.6.2.2 Television and Oscilloscopes

While the phosphors in fluorescent lights are stimulated by ultraviolet light, the phosphors that coat the inside of a TV screen or oscilloscope screen are stimulated by a beam of electrons emitted by a cathode ray tube. For a black and white image only a single phosphor is required, but for a color TV three phosphors are necessary. One fluoresces blue, one green, and one red to combine to form all of the colors of the rainbow.

3.6.2.3 Electroluminescent Lamps

You have probably never heard of *electroluminescent lamps*, yet you probably use them every day. The light in your watch, your cellular phone, and the dash panel of your car all are electroluminescent.

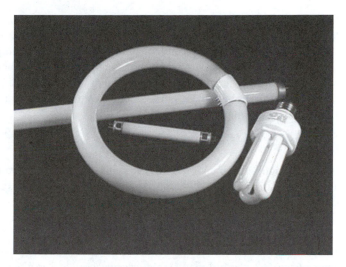

FIGURE 3.45   Variety of configurations of fluorescent lights. (Photo by D.W. Richerson.)

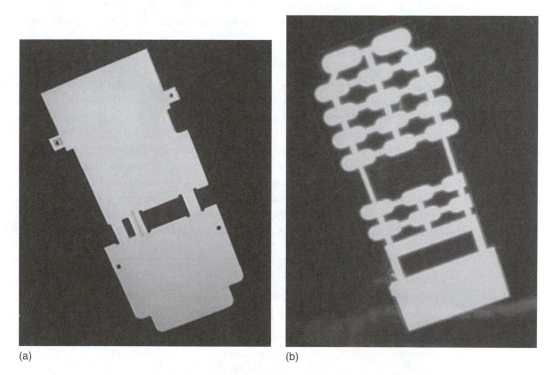

(a)   (b)

FIGURE 3.46   Electroluminescent lamp for a cellular phone. The image on the left shows the as-fabricated multi-layer sheet. The image on the right shows the sheet connected to the battery of the cellular phone to illuminate the keypad. (From Richerson, D.W., *The Magic of Ceramics*, The American Ceramic Society, Westerville, OH, 2000, p. 77. Reprinted with permission of the American Ceramic Society, www.ceramics.org)

The "lamp" is about as thick as the cover of a spiral notebook and consists of a laminate of the following thin film layers (from the surface inwards): transparent plastic electrical insulator, transparent electrically conductive front electrode (indium tin oxide ceramic), ceramic phosphor, ceramic dielectric layer, rear electrode (usually carbon), and rear plastic insulator. The layers can easily be deposited in any array such as the keyboard of a cellular phone, as shown in Figure 3.46. When the electricity is turned on, the phosphor is exposed to an alternating current electrical field and is stimulated to emit light of its own characteristic color.

Very little electricity is consumed, and very little heat is generated. For example, a wristwatch light operates off of a 1.5-volt battery at a current of only 3 mA and a cellular phone keypad off a 6-volt battery at 40 mA (and only consumes about one-quarter watt of electrical power). For comparison a single outdoor Christmas lightbulb consumes about 7 watts and an incandescent lightbulb for a bedside lamp uses about 40 to 60 watts.

3.6.2.4 Lasers

LASER is the acronym for "light amplification by the stimulated emission of radiation." The first laser, demonstrated by T.H. Maiman in 1960, was a ruby laser based on the phosphorescent behavior of Al_2O_3 doped with Cr^{3+} ions. Since 1960 thousands of applications have been found for lasers in communications, heat treating, medicine, guidance systems, cutting, weapons, powder synthesis, instrumentation, holography, entertainment, bar code readers, CD readers, and other fields and products. Each product requires specific laser characteristics, so many different types of lasers have been developed. Many of the lasers use doped single-crystal ceramics or glass as the laser host. Examples include tungstates, fluorides, molybdates, and garnet compositions doped with chromium, neodymium, europium, erbium, praseodymium, and other ions. Ceramics are also used for semiconductor lasers and as components in gas lasers, but these do not involve phosphorescence.

3.7 COMPOSITES

A composite is a mixture of two or more materials such that the combination has better characteristics than the individual materials. A good example is the addition of glass fibers to a polymer matrix to produce the composite most people refer to as *fiberglass*. The polymer alone is rather weak and easily deformed by bending, like a milk bottle. The glass fibers are very strong, but are like the thin filaments in thread. When woven into cloth and imbedded in the polymer, though, the fibers reinforce the polymer and provide great strength and stiffness. The polymer holds the fibers together and protects them from damage. The resulting composite is better than either material alone. Table 3.10 lists some of the many applications and products of fiberglass composites.

Glass fibers are very inexpensive, but have limited mechanical properties. Other ceramic fibers have been developed with higher elastic modulus (higher stiffness), higher strength, and retention

TABLE 3.10

Examples of Products Containing High Strength Ceramic Fibers as Reinforcement for Polymer Matrices

Pleasure boat hulls	Aircraft structures
Camper shells	Automobile parts
Skis	Windmill blades
Skateboards	Helicopter blades
Surfboards	Hot tubs
Kayak paddles	Cafeteria trays
Storage tanks	Fishing poles
I-beams	Tennis rackets
Pipe	Utility poles
Reinforced concrete	Bridge and pier decking
Roof shingles	Flooring
Golf club shafts	Prosthetics
Armor backing	Electrical mounting boards/substrates

of strength and stiffness to high temperature: carbon, SiC, alumina, and aluminosilicate. Carbon fibers are used extensively in fabrication of light-weight structures for aircraft and for stiff, light sporting equipment. A Boeing 757 aircraft contains about 3300 pounds of carbon–epoxy composite, which is a weight savings compared to aluminum metal alloys of nearly 1500 pounds. Weight savings is an important engineering criterion for aircraft and correlates with better fuel efficiency, longer range, and increased number of passengers. Composites also offer other engineering benefits. For example, a B-2 stealth bomber is estimated to contain about 50,000 pounds of carbon–epoxy composite as part of the stealth system.

A relatively new area of development is ceramic matrix–ceramic fiber composites. The most development has been conducted with carbon fibers in a carbon matrix (carbon–carbon composite), SiC-based fibers in a SiC matrix, alumina-based or mullite-based fibers in an alumina or aluminosilicate matrix. An example of a SiC–SiC ceramic matrix composite was shown earlier in Figure 3.12. Ceramic matrix composites are discussed in Chapter 20.

Ceramics are important in many other composites as a dispersion of particles, whiskers, or chopped fibers. As an example of the potential, addition of about 33% by volume of SiC whiskers to an aluminum alloy increased its strength from 10,000 psi to over 90,000 psi and allowed it to retain its strength to much higher temperature. This allowed the whisker-reinforced aluminum to perform successfully in an application that normally required a much heavier iron alloy.

Another interesting ceramic composite was developed in Japan and consisted of addition of 2% chopped carbon fibers to concrete. This small addition quadrupled the strength of the concrete and increased the amount of deflection the concrete could tolerate without fracturing by nearly 10 times. This proved to be very useful in building construction in a country that is prone to severe earthquakes.

3.8 MEDICAL APPLICATIONS OF CERAMICS[1]

A complete chapter could easily be written on the medical uses of ceramics. Biomaterials has been a rapidly growing field, and ceramics have been an important part. Ceramic involvement can be divided into three categories: (1) replacement and repair, (2) diagnosis, and (3) treatment and therapy.

> **BROAD MEDICAL APPLICATIONS OF CERAMIC MATERIALS**
>
> - Replacement and repair
> - Diagnosis
> - Treatment and therapy

3.8.1 REPLACEMENT AND REPAIR

Figure 3.47 illustrates many of the ways that ceramics are used as implants, repairs, and restorations in the human body. For a replacement to be successful, the material used must be able to perform the same function as the original body part and also must be compatible with the surrounding tissue. Many materials are toxic to tissue; others are attacked by our immune system and either destroyed or isolated by encapsulation in a special fibrous tissue. Some ceramic compositions are very similar to bone and are accepted by the body. Hydroxyapatite-based compositions actually stimulate bone and tissue growth. Other ceramics such as alumina and zirconia appear to be relatively inert and have a minimal effect on our body defenses.

3.8.1.1 Dental Ceramics

Perhaps the earliest significant use of ceramics for medical applications was for teeth replacement. Nicholas Dubois de Chémant in 1774 successfully used porcelain, which was a big improvement compared to the materials used prior to that time: ivory, bone, wood, animal teeth, and even teeth extracted from human donors. These earlier materials all quickly became stained and developed a bad odor.

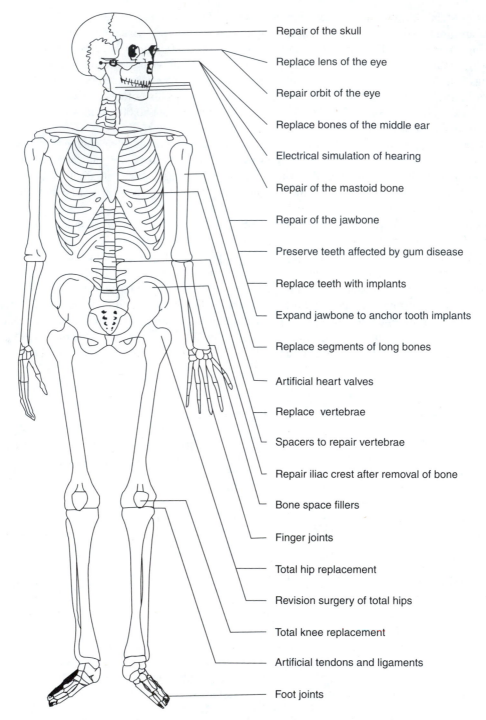

Repair of the skull

Replace lens of the eye

Repair orbit of the eye

Replace bones of the middle ear

Electrical simulation of hearing

Repair of the mastoid bone

Repair of the jawbone

Preserve teeth affected by gum disease

Replace teeth with implants

Expand jawbone to anchor tooth implants

Replace segments of long bones

Artificial heart valves

Replace vertebrae

Spacers to repair vertebrae

Repair iliac crest after removal of bone

Bone space fillers

Finger joints

Total hip replacement

Revision surgery of total hips

Total knee replacement

Artificial tendons and ligaments

Foot joints

FIGURE 3.47   Examples of the use of ceramics for implants, repairs, and restorations in the human body. (From Richerson, D.W., *The Magic of Ceramics*, The American Ceramic Society, Westerville, OH, 2000, p. 170 and courtesy of Professor Larry Hench, Imperial College, London. Reprinted with permission of the American Ceramic Society, www.ceramics.org)

Extensive advances have occurred in dental ceramics during roughly the past 40 years. Now ceramics are the standard for tooth replacement, dentures, and restoration techniques such as veneers, inlays, and crowns. Even cavities are now being filled with a composite of ceramic particles in a polymer that virtually duplicates the color of natural teeth and avoids the unsightly and potentially hazardous silver-tin-mercury fillings. Ceramics also are being employed for jawbone restoration in the form of porous hydroxyapatite implants into which bone and tissue grow. Similar porous ceramic implants have been successful in other parts of the body for bone restoration.

3.8.1.2 Hip Implants

Greater than 500,000 hip replacements are conducted each year. The technology was pioneered using a metal rod (stem) with a metal ball on one end. The metal stem was forced into the top of the femur and glued in place with a polymer adhesive. A polyethylene cap was then glued into the hip socket of the pelvis. These hip replacements were successful, but had a life of generally less than 10 years. The major problems were failure of the adhesive and wear particles causing irritation, inflammation, and fibrous tissue growth. Much progress has been made in extending the life and reliability of hip replacements, some involving ceramics. Placing grooves or wire mesh along the shaft of the stem and coating with bioactive ceramics results in a better bond and less chance of the stem breaking loose from the femur. Replacing the metal ball with a ceramic ball (of alumina, transformation-toughened alumina, or transformation-toughened zirconia), as illustrated in Figure 3.48, reduces wear particles and also increases compatibility with the body. About 250,000 hip replacements are now conducted each year using ceramic components.

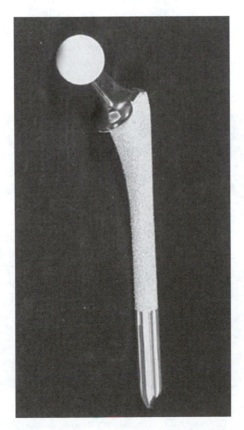

FIGURE 3.48   Ceramic ball on a metal stem for a hip replacement. (Photo by D.W. Richerson.)

Much progress also is being made in use of ceramics for other joint replacements including the thumb, knee, big toe, and shoulder. Many of the surgeries are still experimental as the doctors learn the most effective materials and procedures, but will probably become routine and common in the future.

3.8.1.3 Middle-Ear Implants

Hearing loss sometimes is caused by damage (often due to disease) of the tiny bones (ossicles) that transmit sound vibrations from the ear drum to the inner ear. Plastic and metal replacements were tried, but either failed to bond to the ear drum or were attacked by the body and coated with fibrous tissue that prevented sound waves from getting through. Ceramic *bioglass* implants, shown in Figure 3.49, solved these problems and resulted in successful restoration of hearing.

3.8.1.4 Eye Repairs

For centuries the only way of improving eyesight was corrective lenses. Since the laser was invented, new laser surgery options have been developed to correct vision. The laser also has made possible repair of problems inside the eye that can cause vision loss. One is the detached retina. The doctor can actually shine a special laser through the lens of the eye and "weld" the retina back in place. Another problem sometimes linked to hardening of the arteries and hypertension is the breakage of tiny blood vessels in the eye. The blood flows onto the retina and destroys the tiny vision receptors. A type of laser that produces a beam that is specifically absorbed by blood can stop the bleeding and remove excess blood, thus preventing further vision receptors from being destroyed.

Another cause of loss of sight is disease or damage to the lens of the eye or clouding over due to cataracts. Doctors can now remove the damaged lens and replace it with a transparent ceramic lens.

Ceramics also are used as artificial eyes, known medically as *orbital implants*. Natural eyes move together, but the original orbital implants simply filled the eye socket and did not move when the other eye moved. Now ceramic orbital implants are designed to bond to the tissue in the eye socket so they are connected to the muscle system and move synchronously with the other eye.

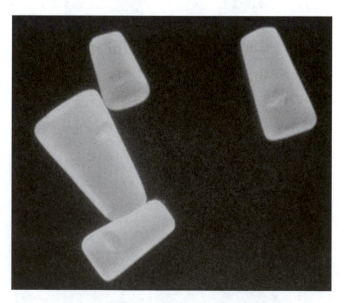

FIGURE 3.49   Bioglass middle ear implants manufactured by U.S. Biomaterials, Alachua, FL. (From Richerson, D.W., *The Magic of Ceramics*, The American Ceramic Society, Westerville, OH, 2000, p. 175. Reprinted with permission of the American Ceramic Society, www.ceramics.org)

3.8.1.5 Heart Valve Implants

A common heart ailment is an aortic valve that does not close completely with each beat of the heart, thus allowing backflow and decreasing the amount of blood pumped with each heart beat. This condition can now be repaired by implanting a mechanical heart valve made of pyrolytic carbon. This same carbon material has been successfully used as a bone replacement.

3.8.1.6 Prosthetic Devices

In spite of the progress that has been made in repair and implants, doctors still cannot implant complete limbs. However, much progress has been achieved in replacing limbs with prosthetic devices that provide at least some of the original function of the limb. Ceramics are important in two ways in these prosthetic devices: (1) high strength fiber reinforcement in polymer matrix composites and (2) sensors and integrated electronics systems that can provide an interaction between the person, the prosthesis, and the surroundings. This technology is at an early stage of development, but looks very promising. One exciting example consists of a probe based on an array of silicon electrodes (Figure 3.50) that can be inserted into the outer layer of the brain. These can either pick up neural signals from the brain or transmit signals from an external source to the brain. One possible application might be for a quadraplegic to be able to make mental commands to control a wheelchair or type a message into a computer. Another might be to provide a degree of artificial vision.

3.8.2 CERAMICS FOR MEDICAL DIAGNOSIS

Ceramics have supported medical diagnosis for many years through laboratory ware for sterile handling of medical samples for analyses (petri dishes, test tubes, etc.), through lenses for microscopes and for examining eyes and ears and for determining prescriptions for vision correction, and for electrical and optical components for x-ray machines, oscilloscopes, spectroscopes, computers, and a wide array of other specialized medical equipment. The role of ceramics has been especially dynamic for CT equipment, endoscopes, and ultrasonic imaging.

3.8.2.1 CT Scanner

A conventional x-ray radiography machine sends a broad beam of x-rays through your body and allows those that get through to expose film. Bone absorbs more of the x-rays than does soft tissue,

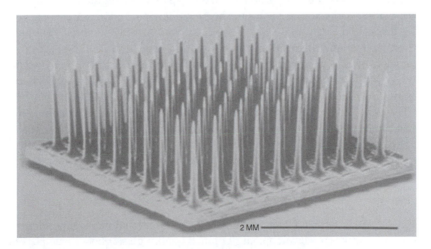

FIGURE 3.50 Brain probe developed by Bionic Technologies, Salt Lake City, UT. (Photo courtesy of Cyberkinetics, Inc.)

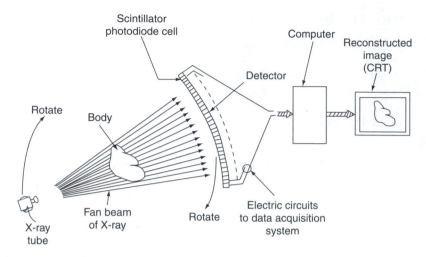

FIGURE 3.51 Schematic of a CT scan system. (From Richerson, D.W., *The Magic of Ceramics*, The American Ceramic Society, Westerville, OH, 2000, p. 179; courtesy of General Electric Medical Systems, Milwaukee, WI. Reprinted with permission of the American Ceramic Society, www.ceramics.org)

so the doctor ends up with a reasonable resolution picture of the bone structure but little information about soft-tissue organs. As illustrated in Figure 3.51, the CT unit passes x-rays through a collimator that separates the x-rays into many thin beams, each of which can be detected separately by an array of detectors. The x-ray source and detectors are rotated around the body to obtain many thousands of data points from every angle. The data is digitized and analyzed with computer programs that can reconstruct an image for any cross section in the body that the doctor wishes to examine. Examples of cross sections are shown in Figure 3.52.

The quality of the digital image is dependent upon the quality of the x-ray beam and the sensitivity of the detectors. Ceramics are critical elements of both. The detector consists of a miniature checkerboard of ceramic *scintillator* slices that are lined up to receive each narrow collimated beam of x-ray exiting from the body. The scintillator material is a transparent polycrystalline ceramic that phosphoresces when excited by x-rays. The greater the intensity of the x-rays, the more light given off by the specific grid of the scintillator. Each scintillator grid is backed by a silicon photodiode that produces an electrical output proportional to the intensity of the light, and each electrical pulse for each position of the x-ray source versus the body is digitized.

Early scintillators and electronics systems were not very sensitive and fast. The full body scan took at least half an hour and the resulting images did not have high resolution. New scintillators, photodiodes, and electronics systems are highly refined so that a scan can be done in a much shorter time (and expose you to less x-rays) and produce high-resolution digitized images such as those shown in Figure 3.52. Previously CT scanners were only in major hospitals and had to be run 24 hours per day just to keep up with prioritized patients. Now you can go to a clinic in a shopping mall in some cities and receive a complete virtual image of the inside of your body in about 15 minutes. This image can show tumors, deposits in blood vessels, and other abnormalities at an early enough stage so that they can be removed or treated before they become life threatening.

3.8.2.2 Endoscopy

Forty years ago the only way that a doctor could diagnose an internal medical problem was through major exploratory surgery. The development of optical quality glass fibers made possible an alternative approach, the *endoscope*. An endoscope is a flexible tube with an eyepiece on one end, a lens on the other end, and optical fibers inside to carry light in and an image out. The doctor can insert

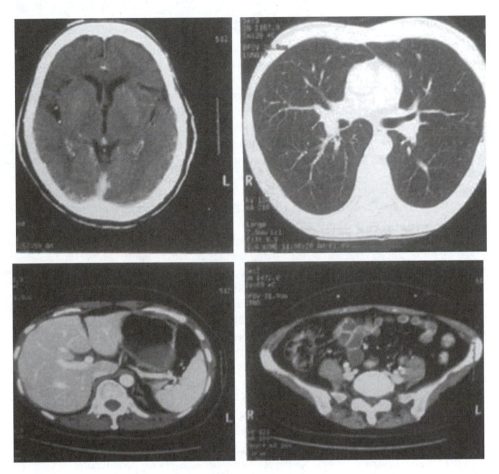

FIGURE 3.52 Examples of CT images of various cross sections of the body showing the high resolution that can be achieved with ceramic scintillators. (From Richerson, D.W., *The Magic of Ceramics*, The American Ceramic Society, Westerville, OH, 2000, p. 178; courtesy of General Electric Medical Systems, Milwaukee, WI. Reprinted with permission of the American Ceramic Society, www.ceramics.org)

the endoscope through natural openings in the body to search for signs of colon cancer or problems in the female reproductive system. Or the endoscope can be inserted through a small surgical incision to examine a damaged knee or to look for the source of abdominal pain such as might be caused by a ruptured appendix or a diseased gall bladder. Modern endoscopes are also designed to conduct surgery, as described in a later section.

3.8.2.3 Ultrasound Imaging

Ultrasound imaging is based on the use of piezoelectric ceramic transducers. An alternating current causes the transducer to vibrate at very high frequency. As the transducer is moved over the surface of the body, the waves penetrate into the body and bounce back whenever they intersect of new surface. The reflected waves stimulate the transducer to produce an electrical output that can be manipulated by a computer to show an image on a screen. This technology has become very popular for a mother to see an image of the baby developing in her womb. It also is used to scan the carotid artery (that passes though the neck) to look for narrowing due to buildups of deposits on the artery walls. This can provide early warning of the danger of a stroke or heart disease in time to take preventative measures.

3.8.3 CERAMICS IN MEDICAL TREATMENT AND THERAPY

Ceramics have enabled many important advances in medical treatment and therapy. The endoscope allows major surgery to be conducted with several small incisions rather than completely cutting the body open. As an example, endoscopic gall bladder surgery has been a major breakthrough by dramatically reducing risk, surgery time, and recovery time. Arthroscopic knee surgery is now widely practiced and has relieved countless people from pain and allowed them to return to normal activities and extraordinary activities such as top level athletics competition. Endoscopes also are used to prevent colon cancer. A precursor of colon cancer is the formation of polyps on the inner intestinal wall. These can be seen with the endoscope and immediately removed with a tool built into the endoscope. The tool consists of a tiny ceramic rod tipped with a coil of metal painted onto its circumference and connected to electrical wiring. The metal is compounded to have high electrical resistance. When a polyp is detected, the tool is touched to the polyp and the electricity turned on. The metal coil heats up rapidly and burns off the polyp.

Other applications of ceramics in surgery include ultrasharp scalpels of transformation-toughened zirconia, diamond-tipped cutting tools for sawing bone and drilling teeth, lasers (such as described earlier for eye surgery), and even high intensity light sources for operating rooms. A recent procedure for repair of sun-damaged skin is to vaporize the outer layer with a laser and allow undamaged skin and tissue to grow back. Another new procedure is to link a piezoelectric transducer to a scalpel to allow unprecedented control and precision in delicate operations.

One of the most amazing treatments involving ceramics is for liver cancer. Traditional treatments for liver cancer were chemotherapy or externally applied radiation therapy. Both of these caused severe side effects and were still relatively ineffective. Most patients died within 6 months. The ceramic treatment involves the use of tiny glass microspheres about one third the diameter of a human hair, as shown in Figure 3.53. By control of the chemical elements present (particularly yttrium and rare earth elements), the glass spheres can pick up different levels of radiation when placed in a suitable radiation source. Specifically, the half-life and the distance the radiation reaches from the spheres (about 1 to 10 millimeters) can be selected. The doctor conducts scans to determine the size and distribution of the tumors in the liver and places an order for the appropriate microspheres. When delivered, the microspheres are injected into a catheter in the femoral artery and carried by the blood flow directly into the liver. The size of the microspheres is such that they are trapped in the capillaries in the tumors. About eight times the radiation dose can be safely administered compared to use of external radiation. The treatment is done on an outpatient basis and the patient has no side effects. Some of the patients from

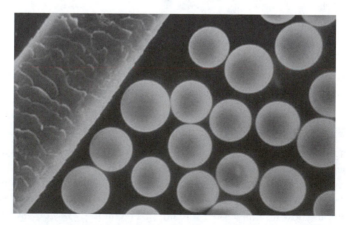

FIGURE 3.53 Scanning electron microscope image of radioactive glass beads compared in size to a human hair. (From Richerson, D.W., *The Magic of Ceramics*, The American Ceramic Society, Westerville, OH, 2000, p. 184; photo courtesy of Dr. Delbert Day, University of Missouri, Rolla. Reprinted with permission of the American Ceramic Society, www.ceramics.org)

the initial trials (who were all thought to be terminal within months and beyond treatment) survived for more than 6 years. The treatment has now been approved in many countries, and research is underway to allow larger doses and to explore treatment of other types of cancer.

Another interesting application of ceramics is for dialysis, which is a medical procedure that can keep you alive if your kidneys fail. Your blood flow is diverted through a polymer filter where toxins normally removed by the kidneys are removed. Before the blood passes back into your body, it must be monitored to make sure there are no bubbles. This is done with a set of piezoelectric transducers. One sends a continuous stream of vibrations through the tube containing the blood. Another transducer on the opposite side of the tube monitors these vibrations. As long as they are continuous, no bubbles are present. If a bubble comes through, the stream of vibrations is interrupted and protective action can be initiated.

Piezoelectric ceramics also are important for therapy. A high intensity of ultrasonic vibrations causes a slight increase in the temperature deep inside your muscles and in the ligaments that connect these muscles to your bones. This increases flexibility, decreases pain, and promotes healing. Figure 3.54 shows some piezoelectric ceramic parts for medical applications.

Ceramics are important in many other ways for medical treatment. Critical components of pacemakers (electrical insulators, capacitors, glass seals, and in some designs the outer case) and of respirators (the seal shown earlier in Figure 3.20) are ceramic. The valve for the respirator must move

FIGURE 3.54 Piezoelectric ceramic parts for medical applications. The powders from which the piezoelectric materials are synthesized are shown at the top. The samples at the lower left and right with the wires connected are for kidney dialysis bubble detectors. The dark samples are coated with a glaze and are built into hand-held physical therapy equipment. (From Richerson, D.W., *The Magic of Ceramics*, The American Ceramic Society, Westerville, OH, 2000, p. 185; photo courtesy of Edo Corporation Piezoelectric Products, Salt Lake City, UT. Reprinted with permission of the American Ceramic Society, www.ceramics.org)

TABLE 3.11
Historical Comparison of Energy Consumption and Pollution Emissions in the United States

	1950	1970	1990	2000	2020 (Projected)
Total energy consumption in United States, Quadrillion Btu	25	65	86	100	120
Carbon emissions (solid carbon equivalent from CO_2, soot, etc. from transportation, industrial, commercial, and residential sectors), Billion metric tons			1.346	1.585	1.975
Carbon emissions per person, metric tons				5.8	6.1
Electricity demand, billion kilowatt hours		1,392			4,345

Source: *Annual Energy Outlook 1999*, DOE/EIA-0383(99), Dec. 1998, published by the Department of Energy (DOE) Energy Information Administration, Washington, DC, and *National Energy Policy*, Report of the National Energy Policy Development Group, May 2001, U.S. Government Printing Office, Washington, DC.

smoothly and without the use of a lubricant, must have minimum clearance to avoid leakage, and must slide back and forth millions of times without sticking or wearing.

3.9 ENERGY EFFICIENCY AND POLLUTION CONTROL[1]

Energy and pollution are two of the most important issues in our world today. They affect our health, world economy, our standard of living, and perhaps the stability of our weather and environment. Statistics from the Energy Information Administration[26,27] are summarized in Table 3.11. They show that the United States consumed about 100 quadrillion (100 followed by 15 zeros) Btu of energy in the year 2000, compared to only 25 quadrillion Btu in 1950. The large increase in energy production has resulted in burning more and more fossil fuels and releasing more and more carbon dioxide, hydrocarbons, and other pollutants into the air. Of the six billion metric tons of carbon emissions in 1996, the United States accounted for 24%. By 2010 world carbon emissions are projected to be 8.3 billion metric tons. What does this mean to each of us personally? In 2000 each man, woman, and child in the United States was responsible for 5.8 metric tons of carbon emissions. Based on current trends, this number will increase to 6.1 metric tons by 2020. As engineers we have the responsibility to be aware of these issues and to lead efforts to address the issues.

Ceramics, as well as other advanced materials, have been key, in the past, in addressing energy and pollution challenges and will be even more important in the future. Table 3.12 summarizes some of the past successes and identifies opportunities for the future.

3.9.1 ENERGY SAVINGS IN THE HOME

3.9.1.1 Fiberglass Insulation

Fiberglass thermal insulation was developed between 1931 and 1938 through collaboration between Owens-Illinois and Corning Glass Works. Based on an assumption that 50% of homes have fiberglass insulation, it has been estimated that more than 25 quadrillion (25,000,000,000,000,000) Btu of heat energy has been conserved in single-family dwellings since 1938.[1]

Fiberglass insulation is fabricated by melting round marbles of refined glass in an electric furnace and flowing the molten glass across a platinum plate containing many tiny holes. As the tiny rivulets of molten glass flow from the bottom of the platinum plate, they are blasted with jets of high pressure steam. This stretches each flow of molten glass into a long filament of solidified glass

---

TABLE 3.12

Materials-Intensive Approaches to Decreasing Energy Consumption and Pollution Emissions

Improved thermal insulation
Emission cleanup technologies for coal-burning electricity generation systems
Use of heat exchangers
Automotive emission control (catalytic converter)
Replacement of less efficient lights with more efficient lights
Increased efficiency of heat engines
Fuel cells
Photovoltaic and solar thermal systems
Increasing efficiency of appliances
Hybrid and combined cycle energy systems
Replacement of pilot lights with igniters
Automobile electronic control systems
Lighter weight, stronger structural materials
Oxygen sensors and other control sensors
Improved recycling technology
Diesel particle trap

---

and results in a bundle of filaments with a large volume of air trapped between the filaments. This combination of glass filaments and "dead air space" is an excellent barrier to the flow of heat or cold. To give you a feeling for the size of the filaments, each five-eighths inch diameter marble yields about 97 miles of filament.

3.9.1.2 Efficient Light Sources

As we discussed earlier, the incandescent light is not very efficient. Fluorescent lights are much better, but for many years were only available in long slender form for overhead lighting. Recently a variety of circular and spiral shapes have been engineered that allow the fluorescent light to be screwed into a fixture designed for an incandescent bulb. Although these individual fluorescent lights are still expensive, replacing all of the incandescent bulbs in your home or office can conserve a lot of electricity and also substantially decrease the frequency of replacing burned out bulbs.

Have you flown on an airplane at night recently? Every street in virtually every city and town is lined with lights that stay on all night. These are high-pressure sodium vapor lamps, the lights that are mounted on long poles and give off golden or yellowish light. This type light was invented in the 1960s and is now manufactured at a rate of about 50 million units per year. While an incandescent light produces only 15 lumens per watt and usually lasts less than 1000 hours, a high-pressure sodium (HPS) vapor lamp emits 140 lumens per watt and lasts about 24,000 hours.

The key component in an HPS lamp is the *arc tube* or *discharge tube*, which is shown in Figure 3.55. This is the envelope that contains the high-pressure sodium vapor that emits light when an electric arc is passed through the vapor in the tube. The tube must be transparent and must withstand about 1200°C (2200°F) temperature and the corrosiveness of the hot vapors. Glass compositions were not able to withstand these conditions. A major development effort ensued to develop a transparent polycrystalline alumina material. In addition to being successful, the work provided enhanced scientific understanding of the sintering (densification) of pore-free polycrystalline ceramics that would help lead to other important products for transparent and translucent polycrystalline ceramics (x-ray scintillator, orthodontic brackets, and transparent ceramic armor).

The primary concern with HPS lighting is the golden color, which means that colors do not look the same in HPS lighting as in daylight. Halogen lamps have been invented that replace the sodium

FIGURE 3.55   High-pressure sodium vapor lamps showing the special translucent polycrystalline alumina arc tube inside. (From Richerson, D.W., *The Magic of Ceramics*, The American Ceramic Society, Westerville, OH, 2000, p. 267; photo courtesy of OSRAM SYLVANIA Products, Inc., Beverly, MA. Reprinted with permission of the American Ceramic Society, www.ceramics.org)

vapor with a metal halide vapor. These give off intense white light, but are more expensive than HPS lamps and don't last as long.

3.9.1.3   Gas Appliances

While electricity is the primary source of energy for many houses, natural gas is commonly used for space heating, water heating, cooking, and clothes drying. Until the early 1960s natural gas appliances had a pilot light that burned continuously, even when the appliance was not in use. A California study estimated that pilot lights were wasting 35 to 40% of the natural gas entering the home (about 6 million cubic feet per day) in the U.S. Ceramic igniters were developed during the 1960s as an alternative to the pilot light. A few became available for clothes dryers in 1968 and grew to a market of more than 750,000 per year by 1980. Igniters for gas ranges were introduced in 1974 and reached a market level of about 1.7 million per year by 1980.

The igniters were developed based on the earlier success of using semiconductor SiC as heating elements for furnaces. Figure 3.56 shows several configurations of SiC igniters. By control of the resistivity of the material, the cross section and length, and the applied voltage, a wide range of operating conditions can be achieved. For most appliance igniters, the electricity is turned on to the igniter about five to seven seconds before the gas flow valve opens, during which time the igniter goes from room temperature to over 1000°C, high enough to ignite the gas. The electricity then shuts off.

Electricity that currently enters our homes is not very efficient. A typical coal-burning power plant (that produces over 50% of our electrical energy in the United States) or a simple cycle gas turbine engine both have an efficiency of only 30 to 33%. That means that about 70% of the energy in the fuel is wasted. By the time the electricity is transmitted to our home, up to 7% more of the energy is lost due to resistance in the electrical transmission lines. As discussed earlier in this chapter, a solid oxide fuel cell (that is based on doped zirconia solid electrolyte) has potential for much higher efficiency. Engineers are trying to develop a small solid-oxide fuel cell (also working on a polymer electrolyte fuel cell) to produce electricity directly in your home initially with natural

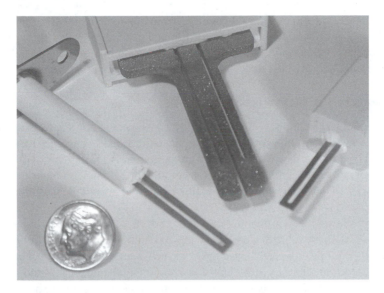

FIGURE 3.56 Silicon carbide igniters of various designs manufactured by Saint-Gobain/Norton Igniters, Milford, NH. (Photo by D.W. Richerson.)

gas fuel and later with hydrogen fuel. Initial systems likely will only produce electricity at 40 to 45% efficiency, but later ones with improved technology will provide our total energy requirement including electricity, heating, and cooling and operate at a total efficiency greater than 80%. This will save an enormous amount of fuel and at the same time greatly decrease pollution emissions.

3.9.2 CERAMICS FOR POWER GENERATION

Our electricity in the past has come almost completely from large-scale central power plants (coal, nuclear, natural gas burning gas turbine engines, and hydroelectric) and from nonrenewable fuels. Ceramics are already used extensively in these plants in the form of concrete, wear and corrosion resistant parts, and electrical components. In the case of nuclear, the fuel pellets, control rods, and high reliability seals and valves all are ceramic. Reactors of the future, such as pebble bed reactors, have the fuel encapsulated in multiple layers of SiC and graphite and are much safer because they are virtually immune to meltdown with this type fuel.

Large gas turbine engines burning natural gas have made major strides in the past 10 years to increase efficiency and reduce pollution emissions. Some large engines are now operating at greater than 60% efficiency. This has been done partially through increasing temperature and pressure, both of which have required major innovations in advanced materials, design technology (such as cooling), and fabrication. Ceramics have played an important role in the fabrication of advanced metal superalloys and also as thermal barrier coatings on the surface of superalloy turbine engine components. Ceramics and ceramic matrix composites have potential to provide further improvements in efficiency, although extensive development is still necessary.

Although large central power stations have been the standard in the past, there is increasing interest in *distributed generation* and in greater use of renewable energy sources. Distributed refers to generation of the electricity at or close to the point of use. This potentially has some benefits: reduced transmission losses, better matching of the amount generated to the need, and greater opportunity to effectively use any waste heat generated. This latter benefit is especially important because a "combined heat and power" system can use over 80% of the chemical energy in the fuel, which is more than double the current average efficiency (about 33% minus transmission losses) for central power stations.

Gas turbine engines, fuel cells, diesel engines, photovoltaic cells, wind turbines, and geothermal systems all have considerable potential for distributed generation. All of these have at least one thing in common: they need advanced materials and engineering to meet the goals of efficiency, pollution emission standards, and cost to be able to compete with central stations.

3.9.3 Ceramics in the Transportation Sector

Burning coal for power generation is one of the largest consumers of nonrenewable fuels and sources of carbon dioxide and other emissions such as sulfur oxides and nitrogen oxides. But the power plants are usually remote so that we do not see the pollution. Another major consumer of nonrenewable fossil fuel and emitter of pollutants is the automobile (along with other petroleum-burning vehicles). In this case, though, the pollution is produced right in the middle of our urban areas where we can really see (and smell) the effects.

Ceramics have played a very important role in our efforts to reduce the pollutants emitted from automobiles. By 1970 motor vehicles were producing 200 million tons of air pollution per year in the United States. This prompted passage by Congress of the Clean Air Act that mandated that automobile companies reduce emissions by 90%. This was achieved by the mid-1970s through development of the *catalytic converter*. The catalytic converter is comprised of a ceramic honeycomb structure coated inside exhaust-gas-carrying channels with a thin layer of catalysts such as platinum and palladium and mounted inside a metal canister. A cutaway of a catalytic converter is shown in Figure 3.57. The ceramic honeycomb and a close-up view of the narrow channels are illustrated in Figures 3.58 and Figure 3.59. As the hot exhaust gases from the engine pass through the long narrow channels, different chemical reactions take place as molecules contact the catalyst particles. Hydrocarbons are catalyzed to react with oxygen to form water and carbon dioxide; carbon monoxide is oxidized to carbon dioxide; and nitrogen oxides are reduced to nitrogen gas. Since catalytic converters were introduced, they have saved us from close to two billion tons of automotive pollution emissions.

The catalytic converter does not remove all of the pollutants and certainly does not remove the carbon dioxide. With the increase in population and the increase in the number of automobiles, vehicular pollution emissions is still a major issue. For example, each gallon of gasoline consumed produces 10.8 kg of carbon dioxide. The solutions to this are (1) reduce the number of miles driven by using mass transit or reversing population growth, (2) increase the gas mileage of the vehicles, (3) improve pollution control devices on vehicles, and (4) transition to a propulsion system that produces less pollutants. Materials are playing an enabling role in items (2) through (4). Hybrid propulsion

FIGURE 3.57 Cutaway of a catalytic converter showing the ceramic honeycomb substrate. (Photo courtesy of Corning, Inc., Corning, NY.)

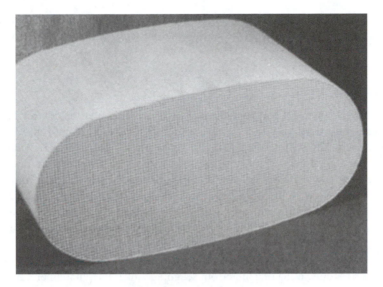

FIGURE 3.58   Catalytic converter ceramic honeycomb manufactured by Corning, Inc. by extrusion of a low thermal expansion cordierite-base ceramic composition. (Photo by D.W. Richerson.)

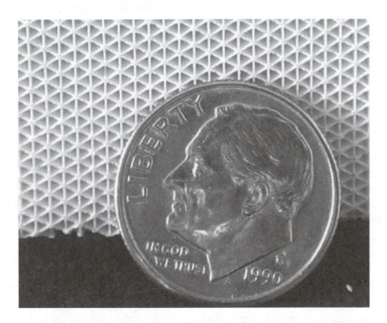

FIGURE 3.59   Close-up view of the ceramic honeycomb showing the geometric uniformity and the thin walls and small size of each cell. Size compared to a U.S. dime, which is 1.3 cm across. (Photo by D.W. Richerson.)

systems are a good first step by significantly increasing miles per gallon of fuel burned. Transitioning to fuel cell propulsion systems that burn hydrogen and only emit water vapor can be the next major step. And perhaps someday we will develop solar cells that are efficient enough and low enough in cost so that we will not have to use fuel in our cars. To accomplish these solutions will require many material and engineering innovations.

Another area of concern is diesel engines. We have all followed a vehicle with a diesel engine and complained about the smelly smoke emitted. This smoke is made up largely of unburned hydrocarbon particles. Technologies are now being introduced that use ceramic materials to trap or filter out the particles and then decompose or burn them during intermittent high temperature cycles.

3.9.4 OTHER USES OF CERAMICS FOR ENERGY EFFICIENCY AND POLLUTION CONTROL

Ceramics are used in many other ways for energy efficiency and pollution control. Dams for hydro-electric power are constructed of concrete, but ceramics are also important for electrical insulators in safely handling and transporting the electricity generated. High technology wind turbine blades are fabricated from high-strength light-weight composites containing ceramic fibers in a polymer matrix. Oxide ceramic fibers are woven into filter bags to remove ash particles from burning of coal and particles from a variety of industrial processes. Porous ceramic filters are important in water purification. Ceramics are even being developed to encapsulate nuclear waste to keep it in a stable nonsoluble form to safely go through the many years of radioactive decay without being released into the environment.

Successful applications can be wherever your imagination takes you. A surprising example is the use of ceramics for containing oil spills from seagoing tankers. High temperature ceramic fibers, a floatable porous ceramic core, stainless steel, and polyvinyl chloride are designed into a floating structure called a *fireboom*. The fireboom consists of a string of floating logs each about 7 feet long and 12 inches in diameter. This string can be towed by a boat to encircle the oil spill, which can then be safely ignited and allowed to burn. During the famous Exxon Valdez oil spill in Prince William Sound, Alaska in 1989, approximately 15,000 to 30,000 gallons of oil were encircled and towed away from the main oil slick and 98% of it was successfully burned in 45 minutes.

3.10 MILITARY

Ceramics are used extensively in military devices and systems. Some examples include electro-magnetic windows, armor, composite structures, stealth technology, lasers, ring laser gyroscopes, rocket nozzles, afterburner seals, a wide variety of sensors, and components in fuel cells, gas turbine engines, communication systems, smart weapons, night vision devices, and flash-blindness prevention goggles.

Electromagnetic windows were mentioned earlier. Metals are opaque and reflective to the electromagnetic waves needed for communications with an aircraft or a missile. Many ceramics are transparent to these wavelengths, even if they are not transparent to visible light. Figure 3.60 shows two ceramic radomes for missile applications. Besides the requirement to be transparent to the desired wavelengths (radar, radio waves, microwaves, infrared, etc.), the ceramic must have special thermal and mechanical properties to withstand conditions such as thermal shock and particle impact (especially rain drops when impacted while the missile is traveling at well over 1000 miles per hour).

Another ceramic application that seems to surprise people is armor. There is continuous competition between development of armor and projectiles that can defeat the armor. One projectile innovation was to insert a tungsten carbide core into the bullet. Because the tungsten carbide was much harder than metal armor and also had a very high specific gravity, it was able to penetrate the metal better than a conventional bullet. To defeat the WC armor-piercing projectile, high-hardness ceramic armor was developed. This consisted of an outer layer of boron carbide, aluminum oxide, or silicon carbide backed with multiple layers of fiberglass composite. The ceramic was hard enough to pulverize the WC core, but also resulted in extensive localized fracture of the ceramic. The fiberglass backing then acted like a catcher's mitt to distribute the stress and catch the debris. Figure 3.61 shows an example of boron carbide personnel armor and the bullets it can defeat. About one-third inch thickness of this armor can stop an armor-piercing projectile that can penetrate 2 inches of steel.

FIGURE 3.60 Ceramic radomes to perform as electromagnetic windows to allow electromagnetic waves of specific frequencies and wavelengths to pass through to the electronic equipment inside. (From Richerson, D.W., *The Magic of Ceramics*, The American Ceramic Society, Westerville, OH, 2000, p. 69; photo courtesy of Ceradyne, Inc., Costa Mesa, CA. Reprinted with permission of the American Ceramic Society, www.ceramics.org)

3.11 RECREATION

Let's finish this chapter on a happy note and discuss some of the many ways we use ceramics in recreation. The most obvious is ceramic art. Millions of people enjoy pottery, ceramic decorating, enameling, glass blowing and casting, and other forms of ceramic art as a hobby. They use some of the same basic processes that we will discuss in later chapters for making high technology ceramics.

A second category of use that is less obvious but adds immensely to our quality of life is music. Piezoelectric ceramics are important in many speakers and microphones, and magnetic ceramics are required in other speakers, tape recording heads, audio tape, and other memory storage devices. Completely electronic musical instruments with incredible purity of sound can be made with piezoelectric ceramics. Ceramics are also involved in the electronics of radios, CD players, computers, and televisions, essentially any device for playing recorded music. Music is enhanced by visual images, and ceramics certainly are involved in producing visual images. The use of ceramic phosphors in the screen of a television was discussed earlier. Any projection equipment requires high-quality glass lenses.

Another recreation category is sports. Polymer matrix composites with high strength ceramic fibers are showing up in nearly every sport: shafts for golf clubs, kayaks and paddles, boat hulls,

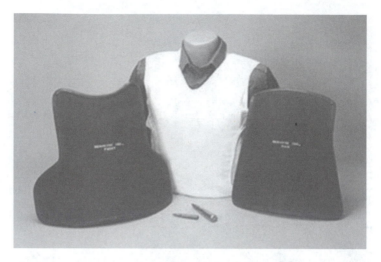

FIGURE 3.61   Boron carbide personnel armor with the projectiles that it can defeat. (From Richerson, D.W., *The Magic of Ceramics*, The American Ceramic Society, Westerville, OH, 2000, p. 259; photo courtesy of Ceradyne, Inc., Costa Mesa, CA. Reprinted with permission of the American Ceramic Society, www.ceramics.org)

FIGURE 3.62   Ceramic putter and spikeless golf cleats manufactured by Coorstek, Golden, CO. (Photo by D.W. Richerson.)

hockey sticks, skis, and tennis rackets. In skis, for example, different ceramic fibers with different elastic modulus (such as glass versus carbon) can be laid up in different proportions in different directions to control longitudinal and torsional stiffness. By this approach, skis can be designed for racing, moguls, powder, beginners, and other specializations.

Ceramics function in other ways for sports. Golf putters have been constructed with a ceramic head or insert. Some drivers also have a ceramic insert. Some golf cleats that do not tear up greens also are made of ceramics. The preferred ceramic for each of these is transformation-toughened zirconia or transformation-toughened alumina, each of which is amazingly strong and resistant to breakage. Figure 3.62 shows a putter and cleats. Some bowling balls contain a ceramic core. Some race cars have been equipped with ceramic valves and composite ceramic brakes.

3.12 SUMMARY

Hopefully you now have a better feeling for the diverse ways ceramics are important in modern applications and products. Even more important, though, hopefully you are beginning to understand that successful use of a ceramic or other material in a product requires a broad understanding of its structure, microstructure, processing, property, design, and performance relationships. The remainder of this book focuses on providing you with an introductory understanding of these factors individually and combined.

REFERENCES

1. Richerson, D.W., *The Magic of Ceramics*, The American Ceramic Society, Westerville, OH, 2000.
2. Freitag, D.W. and Richerson, D.W., *Opportunities for Advanced Ceramics to Meet the Needs of the Industries of the Future*, Final Report, DOE/ORO 2076, Oak Ridge National Laboratory, Oak Ridge, TN, 1999.
3. World Almanac 2002, World Almanac Books, New York, 2002.
4. ASM Handbook, Vol. 15: Casting, ASM Int., Materials Park, Ohio, 1988.
5. Tooley, F.V., *The Handbook of Glass Manufacture*, 3rd ed., Vols. 1–2, Books of the Glass Industry Div., Ashlee Publishing Co., New York, 1984.
6. Geiger, G., Recent advances in glass manufacturing, *Am. Ceram. Soc. Bull.*, 72(2), 27–33, 1993.
7. Foster, B.D. and Patton, J.B., Eds., *Ceramics in Heat Exchangers, Advances in Ceramics* Vol. 14, The American Ceramic Society, Westerville, OH, 1985.
8. Pietsch, A. and Styhr, K., Ceramic heat exchanger applications and developments, in *Ceramics for High Performance Applications-II*, Burke, J.J., Lenoe, E.N., and Katz, R.N., Eds., Brook Hill Publishing Company, Chestnut Hill, MA, 1978, pp. 385–395.
9. Ullmann, F., *Ullmann's Encyclopedia of Industrial Chemistry*, 5th ed., VCH Publishers, New York, 1985.
10. Strigle, R.F. Jr., *Packed Tower Design and Applications: Random and Structural Packings*, 2nd edition, Gulf Publishing, Houston, TX, 1994.
11. van Roode, M., Ferber, M.K., and Richerson, D.W., Eds., *Ceramic Gas Turbine Design and Test Experience*, ASME Press, New York, 2002.
12. van Roode, M., Ferber, M.K., and Richerson, D.W., Eds., *Ceramic Gas Turbine Component Development and Characterization*, ASME Press, New York, 2003.
13. Woods, M.E., Mandler, W.F. Jr., and Scofield, T.L., Designing ceramic insulated components for the adiabatic engine, *Am. Ceram. Soc. Bull.*, 64(2), 287–293, 1985.
14. Svec, J.J., Orbiter has ceramic skin, *Ceram. Ind.*, 107(4), 20–24, 1976.
15. Korb, L.J., Morant, C.A., Calland, R.M., and Thatcher, C.S., The shuttle orbiter thermal protection system, *Am. Ceram. Soc. Bull.*, 60(11), 1188–1193, 1981.
16. Paxton, R.R., Carbon and graphite materials for seals, bearings, and brushes, *Electrochem. Technol.*, 5(5–6), 174–182, 1967.
17. Boyd, G.L., Moy, J., and Fuller, F., Hybrid-ceramic circumferential carbon ring seal, SAE Paper 2002-01-2956, Society of Automotive Engineers, Warrendale, PA, 2002.
18. Buckley, D.H. and Miyoshi, K., Tribological properties of structural ceramics, in *Structural Ceramics, Treatise on Materials Science and Technology*, Vol. 29, Wachtman, J.B. Jr., Ed., Academic Press, San Diego, California, 1989, pp. 293–365.
19. Baumgartner, H.R., Ceramic bearings for turbine applications, in *Ceramics for High Performance Applications-II*, Burke, J.J., Lenoe, E.N., and Katz, R.N., Eds., Brook Hill Publishing, Chestnut Hill, MA, 1978, pp. 423–443.
20. Katz, R.N. and Hannoosh, J.G., Ceramics for high performance rolling element bearings, *Int. J. High Tech. Ceram.*, 1(1), 68–79, 1985.
21. Dow Whitney, E., *New Advances in Ceramic Tooling*, SME Tech Rept. MRR76-15, Society of Manufacturing Engineers, Dearborn, MI., 1976.
22. *Fuel Cell Handbook*, 6th ed., U.S. Department of Energy, Washington, D.C.
23. Pfaender, H.G., *Schott Guide to Glass*, 2nd ed., Chapman & Hall, London, 1996.

24. Mallick, P.K., Introduction: definitions, classifications, and applications, in *Composite Engineering Handbook*, Mallick, P.K. Ed., Marcel Dekker, Inc., New York, 1997, pp. 1–50.
25. *Annual Energy Outlook 1999*, DOE/EIA-0383(99), Dec. 1998, published by the Department of Energy (DOE) Energy Information Administration, Washington, D.C.
26. *National Energy Policy*, Report of the National Energy Policy Development Group, May 2001, U.S. Government Printing Office, Washington, D.C.

STUDY GUIDE

1. Ceramics have been important throughout history for their high temperature resistance. Is high melting temperature the only key property required for these applications, or must the ceramics also have other key properties? Explain with a couple examples.
2. Increasing the operation temperature of a gas turbine engine increases power output, increases efficiency, reduces fuel consumption, and reduces pollution emissions. How have ceramics helped to achieve higher turbine engine operation temperatures for metal components?
3. Describe the function and value of a heat exchanger.
4. What combination of reasons makes zirconium oxide a good choice as a thermal barrier coating for some metal alloys?
5. What characteristics must a ceramic material have to survive in the interior of a gas turbine engine?
6. What combination of properties must a ceramic material have to meet the needs of an automotive spark plug?
7. List some important applications of ceramics in an automobile.
8. The Space Shuttle is covered with about 33,000 ultra-light-weight ceramic tiles. What combination of characteristics was engineered into these tiles to provide an effective thermal protection for the Shuttle?
9. What are some of the engineering requirements for a ceramic seal?
10. List some ceramic parts that can increase the life of a pump.
11. What are some of the attributes of ceramics that should make them good candidates for bearings?
12. Until silicon nitride was invented, why were prior ceramics not successful for high-speed, high-load bearings?
13. What factors delayed commercialization of silicon nitride bearings? Are these factors as important as properties in selecting a material for an engineering application? Explain.
14. List several ways that ceramics are important in cutting and grinding.
15. What is the largest economic sector for ceramics? Does this surprise you? Why or why not?
16. Describe the complexity of an integrated circuit and discuss the challenge of engineering and fabricating such a device.
17. What is a "hybrid package" and why is it important?
18. To what does "dielectric" refer, and why are ceramics in this category important?
19. Are metals the only electrically conductive materials? Explain.
20. What is the potential importance of the solid-oxide fuel cell?
21. What are some challenges for solid-oxide fuel cells? Which are materials issues, design issues, or a combination?
22. What are the differences between "hard ferrites" and "soft ferrites" and their applications?
23. What are some important optical behaviors of ceramics?
24. What are some ways to strengthen glass?
25. Describe the refinement in quality of optical fibers compared to window glass.
26. List some important applications of phosphorescent and fluorescent ceramics.

27. List different ways that ceramics and glass are used to produce light.
28. Describe a couple examples of the use of ceramics in composites.
29. What are some engineering advantages and challenges of composite structures?
30. What do you think are the most remarkable medical applications of ceramics? Why?
31. What is a catalytic converter and why is it important?
32. List some ways that ceramics help us to solve energy and pollution challenges.
33. What uses of ceramics in recreation are most important to you?

Part II

Structures and Properties

The nature of a material is largely controlled by the atoms present and their bond mechanism. Chapter 4 discusses the types of atomic bonding and the atomic elements that are most likely to combine to form ceramic, metallic, and organic materials. Chapter 5 discusses crystal structures and the crystal chemistry relationships that govern which atoms are most likely to form which structures. Chapter 6 reviews the concepts of phase equilibria and phase equilibrium diagrams. Chapter 7 discusses physical and thermal behavior in terms of the atomic elements present, the bond mechanism, and the crystal structure. Chapters 8, 10, and 11 do the same for mechanical, electrical, dielectric, magnetic, and optical behavior. Chapter 9 discusses time-dependent and environment-dependent behavior, especially the effects on mechanical properties and chemical stability. Chapters 7 to 11 all identify specific ceramic materials that exhibit the behavior being discussed.

4 Atomic Bonding and Crystal Structure

Chapter 4 reviews the configuration of electrons in atoms of different elements and discusses how these configurations control the nature of bonding between different atoms.[1–8] Metallic, ionic, covalent, and van der Waals bonding are described. Some concepts of crystal structure, polymorphism, and noncrystalline structures are introduced. The chapter concludes with a brief review of organic structures. A discussion of organic materials is included for the purpose of comparison, but also because organic materials are used extensively in ceramic fabrication processes.[9–12]

4.1 ELECTRONIC CONFIGURATION OF ATOMS

An atom can be visualized in a simplified manner as a positively charged nucleus surrounded by negatively charged electrons. The energy of the electrons varies such that specific electrons are located in specific shells around the nucleus. These are called *quantum shells*. Each shell is referred to by a *principal quantum number n*, where $n = 1, 2, 3, \ldots$. The total number of electrons in a shell is $2n^2$. Thus the lowest-energy quantum shell ($n = 1$) has only 2 electrons and successively higher energy shells have 8 ($n = 2$), 18 ($n = 3$), 32 ($n = 4$), and so on, electrons, respectively.

Although electrons within a quantum shell have similar energy, no two are identical. To distinguish among these electrons, shells are divided into subshells called *orbitals*, which describe the probability of where pairs of electrons will be within the shell with respect to the nucleus. The first quantum shell has only two electrons, both in the *s* orbital, with a spherical probability distribution around the nucleus at a radius of approximately 0.5 Å. These two electrons have identical energy, but opposite magnetic behavior or spin.

The second shell has eight electrons, two in *s* orbitals and six in *p* orbitals. All have higher energy than the two electrons in the first shell and are in orbitals farther from the nucleus. (For instance, the *s* orbitals of the second shell of lithium have a spherical probability distribution at a radius of about 3 Å.) The *p* orbitals are not spherical, but have dumbbell-shaped probability distributions along the orthogonal axes, as shown in Figure 4.1. These *p* electrons have slightly higher energy than *s* electrons of the same shell and are in pairs with opposite spins along each axis when the shell is full.

The third quantum shell has *d* orbitals in addition to *s* and *p* orbitals. A full *d* orbital contains 10 electrons. The fourth and fifth shells contain *f* orbitals in addition to *s*, *p*, and *d* orbitals. A full *f* orbital contains 14 electrons.

A simple notation is used to show the electron configurations within shells, to show the relative energy of the electrons, and thus to show the order in which the electrons can be added to or removed from an atom during bonding. This "*electron notation*" can best be illustrated by a few examples.

Example 4.1 Oxygen has eight electrons and has the electron notation $1s^2 2s^2 2p^4$. The 1 and 2 preceding the *s* and *p* designate the quantum shell, the *s* and *p* designate the subshell within each quantum shell, and the superscripts designate the total number of electrons in each subshell. For oxygen the $1s$ and $2s$ subshells are both full, but the $2p$ subshell is two electrons short of being full.

Example 4.2 As the atomic number and the number of electrons increase, the energy difference between electrons and between shells decreases and overlap between quantum groups occurs.

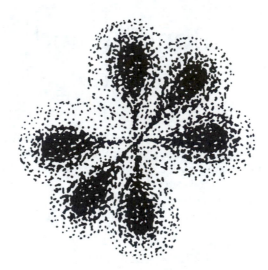

FIGURE 4.1 Electron probability distributions for *p* orbitals. The highest probability electron positions are along the orthogonal axes. Two electrons, each with opposite spin, are associated with each axis, resulting in a total of six *p* electrons if all the *p* orbitals in the shell are filled.

For example, the 4*s* subshell of iron fills before the 3*d* subshell is full. This is shown in the electron notation by listing the order of fill of energy levels in sequence from the left of the notation to the right:

$$Fe = 1s^22s^22p^63s^23p^63d^64s^2$$

Example 4.3 Electronic notation helps a person visualize which electrons are available for bonding and to estimate the type of bond that is likely to result. Unfilled shells contribute to bonding. Electron notation is often abbreviated to include only the unfilled and outer shells. The iron electron notation is thus abbreviated to $3d^64s^2$, which tells the reader that all the subshells up to and including 3*s* are filled. Yttrium is abbreviated from $1s^22s^22p^63d^{10}4s^24p^64d^15s^2$ to $4d^15s^2$ or even more simply, $4d5s^2$ Figure 4.2 lists the abbreviated electron configurations of the elements arranged according to the periodic table. Another form of abbreviation lists the nearest inert gas of lower atomic number and also identifies the electrons in outer shells. For example, Mg consists of the neon inner structure $1s^22s^22p^6$ plus $3s^2$ and is abbreviated $[Ne]3s^2$. Similarly, Ti can be abbreviated $[Ar]3d^24s^2$ and I abbreviated $[Kr]4d^{10}5s^25p^5$.

Based on the examples above, one would predict the electron notation for carbon to be $1s^22s^22p^2$. However, a more stable lower energy state results if the electrons are arranged as $1s^22s^12p^3$. This is referred to as "hybridization" and is sometimes denoted simply as sp^3. A similar sp^3 electron structure occurs for silicon and germanium, as shown in Figure 4.2. As you will see later in this chapter and the next chapter, sp^3 bonds result in some very important covalently bonded ceramic materials.

4.2   BONDING

The unfilled outermost electron shells are involved in bonding. The elements He, Ne, Ar, Kr, Xe, and Rn have full outer electron shells and thus are very stable and do not easily form bonds with other elements. Elements with unfilled electron shells are not as stable and interact with other atoms in a controlled fashion such that electrons are shared or exchanged between these atoms to achieve stable full outer shells.

The three primary interatomic bonds are referred to as metallic, ionic, and covalent. These provide the bond mechanism for nearly all the solid ceramic and metallic materials discussed in later

Periodic table (Figure 4.2)

1	2	3	4	5	6	7	8	9	10	11	12	13	14	15	16	17	18
1 H $1s$																	2 He $1s^2$
3 Li $2s$	4 Be $2s^2$											5 B $2s^22p$	6 C $2s^12p^3$	7 N $2s^22p^3$	8 O $2s^22p^4$	9 F $2s^22p^5$	10 Ne $2s^22p^6$
11 Na $2p^63s$	12 Mg $2p^63s^2$											13 Al $3s^23p^1$	14 Si $3s^13p^3$	15 P $3s^23p^3$	16 S $3s^23p^4$	17 Cl $3s^23p^5$	18 Ar $3s^23p^6$
19 K $3p^64s$	20 Ca $3p^64s^2$	21 Sc $3d4s^2$	22 Ti $3d^24s^2$	23 V $3d^34s^2$	24 Cr $3d^44s^2$	25 Mn $3d^54s^2$	26 Fe $3d^64s^2$	27 Co $3d^74s^2$	28 Ni $3d^84s^2$	29 Cu $3d^104s$	30 Zn $3d^104s^2$	31 Ga $4s^24p$	32 Ge $4s^14p^3$	33 As $4s^24p^3$	34 Se $4s^24p^4$	35 Br $4s^24p^5$	36 Kr $4s^24p^6$
37 Rb $4p^65s$	38 Sr $4p^65s^2$	39 Y $4d5s^2$	40 Zr $4d^25s^2$	41 Nb $4d^45s$	42 Mo $4d^55s$	43 Tc $4d^65s$	44 Ru $4d^75s$	45 Rh $4d^85s$	46 Pd $4d^{10}$	47 Ag $4d^105s$	48 Cd $4d^105s^2$	49 In $5s^25p$	50 Sn $5s^25p^2$	51 Sb $5s^25p^3$	52 Te $5s^25p^4$	53 I $5s^25p^5$	54 Xe $5s^25p^6$
55 Cs $5p^66s$	56 Ba $5p^66s^2$	57-71 La $5d6s^2$	72 Hf $5d^26s^2$	73 Ta $5d^36s^2$	74 W $5d^46s^2$	75 Re $5d^56s^2$	76 Os $5d^66s^2$	77 Ir $5d^9$	78 Pt $5d^96s$	79 Au $5d^106s$	80 Hg $5d^106s^2$	81 Tl $6s^26p$	82 Pb $6s^26p^2$	83 Bi $6s^26p^3$	84 Po $6s^26p^4$	85 At $6s^26p^5$	86 Rn $6s^26p^6$
87 Fr $6p^67s$	88 Ra $6p^67s^2$	89 Ac $6d7s^2$															

58 Ce $4f^26s^2$	59 Pr $4f^36s^2$	60 Nd $4f^46s^2$	61 Pm $4f^56s^2$	62 Sm $4f^66s^2$	63 Eu $4f^76s^2$	64 Gd $4f^75d6s^2$	65 Tb $4f^85d6s^2$	66 Dy $4f^106s^2$	67 Ho $4f^116s^2$	68 Er $4f^126s^2$	69 Tm $4f^136s^2$	70 Yb $4f^146s^2$	71 Lu $4f^145d6s^2$
90 Th $6d^27s^2$	91 Pa $5f^26d7s^2$	92 U $5f^36d7s^2$	93 Np $5f^46d7s^2$	94 Pu $5f^67s^2$	95 Am $5f^77s^2$	96 Cm $5f^76d7s^2$	97 Bk $5f^96d7s^2$	98 Cf $5f^107s^2$	99 Es $5f^117s^2$	100 Fm $5f^127s^2$	101 Md $5f^137s^2$	102 No $5f^147s^2$	103 Lw $5f^146d7s^2$

FIGURE 4.2 Periodic table, including abbreviated electron configuration notation for each element. Bold numbers 1 to 18 designate the group number for each column.

chapters. Other secondary mechanisms referred to as van der Waals bonds also occur, but are discussed only briefly.

4.2.1 METALLIC BONDING

As the name implies, *metallic bonding* is the predominant bond mechanism for metals. It is also referred to as *electronic bonding*, from the fact that the valence electrons (electrons from unfilled shells) are freely shared by all the atoms in the structure. Mutual electrostatic repulsion of the negative charges of the electrons keeps their distribution statistically uniform throughout the structure. At any given time, each atom has enough electrons grouped around it to satisfy its need for a full outer shell. It is the mutual attraction of all the nuclei in the structure to this same cloud of shared electrons that results in the metallic bond.

Because the valence electrons in a metal distribute themselves uniformly and because all the atoms in a pure metal are of the same size, close-packed structures result. Such close-packed structures contain many slip planes along which movement can occur during mechanical loading, producing the ductility that we are so accustomed to for metals. Pure metals typically have very high ductility and can undergo 40 to 60% elongation prior to rupturing. Highly alloyed metals such as the superalloys also have close-packed structures, but the different-size alloying atoms disrupt slip along planes and movement of dislocations (discussed in Chapter 8) and decrease the ductility. Superalloys typically have 5 to 20% elongation.

The free movement of electrons through the structure of a metal results in high electrical conductivity under the influence of an electrical field and high thermal conductivity when exposed to a heat source. These properties are discussed in more detail in Chapters 7 and 10.

Metallic bonding occurs for elements to the left and in the interior of the periodic table (see Figure 4.2 and the complete periodic table, at end of book). Alkali metals such as sodium (Na) and potassium (K) are bonded by outer s electrons and have low bond energy. These metals have low strength and low melting temperatures and are not overly stable. Transition metals such as chromium (Cr), iron (Fe), and tungsten (W) are bonded by inner electrons and have much higher bond strengths. Transition metals thus have higher strength and higher melting temperatures and are more stable.

4.2.2 IONIC BONDING

Ionic bonding occurs when one atom gives up one or more electrons and another atom or atoms accept these electrons such that electrical neutrality is maintained and each atom achieves a stable, filled electron shell. This is best illustrated by a few examples.

Example 4.4   Sodium chloride (NaCl) is largely ionically bonded. The Na atom has the electronic structure $1s^22s^22p^63s$. If the Na atom could get rid of the $3s$ electron, it would have the stable neon (Ne) structure. The chlorine atom has the electronic structure $1s^22s^22p^63s^23p^5$. If the Cl atom could obtain one more electron, it would have the stable argon (Ar) structure. During bonding, one electron from the Na is transferred to the Cl, producing a sodium ion (Na^+) with a net positive charge and a chlorine ion (Cl^-) with an equal negative charge, resulting in a more stable electronic structure for each. This is illustrated in Figure 4.3. The opposite charges provide a Coulombic attraction that is the source of ionic bonding. To maintain overall electrical neutrality, one Na atom is required for each Cl atom and the formula becomes NaCl.

Example 4.5   The aluminum (Al) atom has the electronic structure $1s^22s^22p^63s^23p^1$. To achieve a stable Ne structure three electrons would have to be given up, producing an ion with a net positive charge of 3 (Al^{3+}). The oxygen (O) atom has the electronic structure $1s^22s^22p^4$ and needs two electrons to achieve the stable Ne structure. Bonding occurs when two Al atoms transfer three electrons each to provide the six electrons required by three O atoms to produce the electrically neutral Al_2O_3 compound.

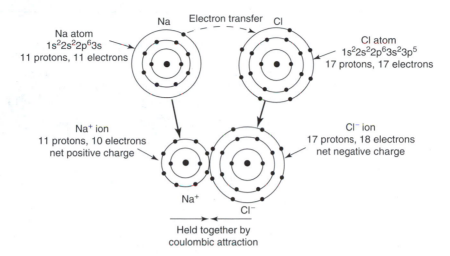

FIGURE 4.3 Schematic illustration of ionic bonding between Na and Cl to produce NaCl. (© ASM International. Drawing courtesy of ASM International from D.W. Richerson, Lesson 2: Atomic Bonding and Crystal Structure, *Introduction to Modern Ceramics*, ASM Materials Engineering Institute Course #56.)

The crystal structure of an ionically bonded material is determined by the number of atoms of each element required for electrical neutrality and the optimum packing based on the relative sizes of the ions. The size of the ions is usually stated in terms of ionic radius. The ionic radius can vary slightly depending on the number and type of oppositely charged ions surrounding an ion. Table 4.1 shows ionic radii for ions with CNs of 4 and 6.[1,2] A more extensive table of ionic radii is included at the end of the book. The *coordination number* (CN or [n]) is defined as the number of nearest neighbor atoms or ions surrounding an atom or ion.

The relative size of the ions determines the coordination number of the ions and the crystal structure. For most ionically bonded structures the anions are positioned in a close-packed arrangement. The cations fit between the anions in the interstitial positions that are closest to the size of the cations. By using the table of ionic radii one can predict the CN of a cation by comparing its size to the size of the anion and then comparing this ratio to the relative size of the possible interstitial positions.

Figure 4.4 shows stable and unstable configurations. For cation-to-anion ratios between 0.155 and 0.225, a CN of 3 is most probable with the anions at the corners of a triangle around the cation. For ratios in the range 0.225 to 0.414, a CN of 4 is most probable with the anions at the corners of a tetrahedron around the cation. Similarly, ratios of 0.414 to 0.732 and greater than 0.732 are most likely to result in CNs of 6 and 8, respectively.

Example 4.6 What is the most likely CN for the cation in a structure made up of Mg^{2+} and O^{2-}? Si^{4+} and O^{2-}? Cr^{3+} and O^{2-}?

From Table 4.1,

$$\frac{Mg^{2+}}{O^{2-}} = \frac{0.72}{1.40} = 0.51 \quad \text{CN of } Mg^{2+} = 6$$

$$\frac{Si^{4+}}{O^{2-}} = \frac{0.40}{1.40} = 0.29 \quad \text{CN of } Si^{4+} = 4$$

$$\frac{Cr^{3+}}{O^{2-}} = \frac{0.62}{1.40} = 0.44 \quad \text{CN of } Cr^{3+} = 6$$

As predicted, Mg^{2+} has a CN of 6 in MgO. The O^{2-} ions also have a CN of 6 and are arranged in a cubic close-packed structure with the Mg^{2+} ions filling the octahedral interstitial positions. This

TABLE 4.1

Ionic Radii for 6 and 4 Coordination (4 Coordination in Parentheses)

Ag$^+$	Al^{3+}	As^{5+}	Au$^+$	B^{3+}	Ba^{2+}	Be^{2+}	Bi^{5+}	Br$^-$	C^{4+}	Ca^{2+}	Cd^{2+}	Ce^{4+}
1.15	0.53	0.50	1.37	0.23	1.36	0.35	0.74	1.96	0.16	1.00	0.95	0.80
(1.02)	(0.39)	(0.34)	—	(0.12)	—	(0.27)	—	—	(0.15)	—	(0.84)	—

Cl$^-$	Co^{2+}	Co^{3+}	Cr^{2+}	Cr^{3+}	Cr^{4+}	Cs$^+$	Cu$^+$	Cu^{2+}	Dy^{3+}	Er^{3+}	Eu^{3+}	F$^-$
1.81	0.74	0.61	0.73	0.62	0.55	1.70	0.96	0.73	0.91	0.88	0.95	1.33
—	—	—	—	—	(0.44)	—	—	(0.63)	—	—	—	(1.31)

Fe^{2+}	Fe^{3+}	Ga^{3+}	Gd^{3+}	Ge^{4+}	Hf^{4+}	Hg^{2+}	Ho^{3+}	I$^-$	In^{3+}	K$^+$	La^{3+}	Li$^+$
0.77	0.65	0.62	0.94	0.54	0.71	1.02	0.89	2.20	0.79	1.38	1.06	0.74
(0.63)	(0.49)	(0.47)	—	(0.40)	(0.96)	—	—	—	—	—	—	(0.59)

Mg^{2+}	Mn^{2+}	Mn^{4+}	Mo^{3+}	Mo^{4+}	Na$^+$	Nb^{5+}	Nd^{3+}	Ni^{2+}	O^{2+}	P^{5+}	Pb^{2+}	Pb^{4+}
0.72	0.67	0.54	0.67	0.65	1.02	0.64	1.00	0.69	1.40	0.35	1.18	0.78
(0.49)	—	—	—	—	(0.99)	(0.32)	—	—	(1.38)	(0.33)	(0.94)	—

Rb$^+$	S^{2-}	S^{6+}	Sb^{5+}	Sc^{3+}	Se^{2-}	Se^{6+}	Si^{4+}	Sm^{2+}	Sn^{2+}	Sn^{4+}	Sr^{2+}	Ta^{5+}
1.49	1.84	0.30	0.61	0.73	1.98	0.42	0.40	0.96	0.93	0.69	1.16	0.64
—	—	(0.12)			(0.29)	(0.26)						—

Te^{2-}	Te^{6+}	Th^{4+}	Ti^{2+}	Ti^{4+}	Tl$^+$	Tl^{3+}	U^{4+}	U^{5+}	V^{2+}	V^{5+}	W^{4+}	W^{6+}
2.21	0.56	1.00	0.86	0.61	1.50	0.88	0.97	0.76	0.79	0.54	0.65	0.58
—	—	—	—	—	—	—	—	—	—	(0.36)	—	(0.41)

Y^{3+}	Yb^{3+}	Zn^{2+}	Zr^{4+}
0.89	0.86	0.75	0.72
—	—	(0.60)	—

Source: Compiled from Shannon, R.D. and Prewitt, C.T., *Acta Crystallogr.*, B25, 925, 1969 and Kingery, W.D., Bowen, H.K., and Uhlmann, D.R., *Introduction to Ceramics*, 2nd ed., Wiley, New York, 1976, Chap. 2.

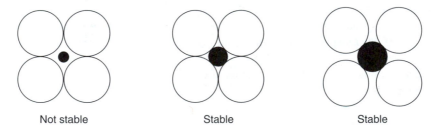

Not stable   Stable   Stable

FIGURE 4.4 Stable and unstable configurations that determine atomic CN within a structure. (Adapted from Shannon, R.D. and Prewitt, C.T., *Acta Crystallogr.*, B25, 925, 1969.)

is the same structure that Na$^+$ and Cl$^-$ bond together to form. This structure is called the rock salt structure after NaCl. Other important ceramic structures are listed in Table 4.2 and are discussed in Chapter 5.

Most of the ionic structures are close packed. Bonding is associated with the *s* electron shells (which have a spherical probability distribution) and would be nondirectional if purely ionic. However, there is a tendency for increased electron concentration between atom centers, which provides a degree of nonionic character. The degree of ionic character of a compound can be estimated using the electronegativity scale (Figure 4.5) derived by Pauling.[3] *Electronegativity* is a measure of

TABLE 4.2
Ionic Crystal Structures

Name of Structure	Packing of Anions	Coordination of Anions	Coordination of Cations	Examples
Rock salt	Cubic close-packed	6	6	NaCl, MgO, CaO, LiF, CoO, NiO
Zinc blende	Cubic close-packed	4	4	ZnS, BeO, SiC
Perovskite	Cubic close-packed	6	12, 6	$BaTiO_3$, $CoTiO_3$, $SrZrO_3$
Spinel	Cubic close-packed	4	4, 6	$FeAl_2O_4$, $MgAl_2O_4$, $ZnAl_2O_4$
Inverse spinel	Cubic close-packed	4	4 (6, 4)[a]	$FeMgFeO_4$, $MgTiMgO_4$
CsCl	Simple cubic	8	8	CsCl, CsBr, CsI
Fluorite	Simple cubic	4	8	CaF_2, ThO_2, CeO_2, UO_2, ZrO_2, HfO_2
Antifluorite	Cubic close-packed	8	4	Li_2O, Na_2O, K_2O, Rb_2O
Rutile	Distorted cubic close-packed	3	6	TiO_2, GeO_2, SnO_2, PbO_2, VO_2
Wurtzite	Hexagonal close-packed	4	4	ZnS, ZnO, SiC
Nickel arsenide	Hexagonal close-packed	6	6	NiAs, FeS, CoSe
Corundum	Hexagonal close-packed	4	6	Al_2O_3, Fe_2O_3, Cr_2O_3, V_2O_3
Ilmenite	Hexagonal close-packed	4	6, 6	$FeTiO_3$, $CoTiO_3$, $NiTiO_3$
Olivine	Hexagonal close-packed	4	6, 4	Mg_2SiO_4, Fe_2SiO_4

[a] First Fe in tetrahedral coordination, second Fe in octahedral coordination.

Source: Adapted from Kingery, W.D., Bowen, H.K., and Uhlmann, D.R., *Introduction to Ceramics*, 2nd ed., Wiley, New York, 1976, Chap. 2.

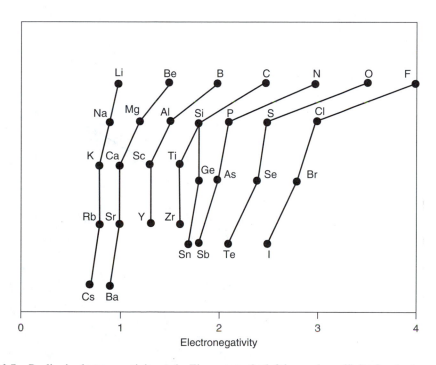

FIGURE 4.5 Pauling's electronegativity scale. Elements to the left have a low affinity for electrons and those to the right have a high affinity. (Reprinted with slight modifications from Linus Pauling, *The Nature of the Chemical Bond*, 3rd ed., © 1960 by Cornell University. Used by permission of the publisher, Cornell University Press.)

an atom's ability to attract electrons and is roughly proportional to the sum of the energy needed to add an electron (electron affinity) and to remove an electron (ionization potential). The larger the electronegativity difference between atoms in a compound, the larger the degree of ionic character. The semiempirical curve derived by Pauling is shown in Figure 4.6.

> **Example 4.7**   What is the degree of ionic character of MgO? of SiO_2? of SiC?
>   From Figures 4.5 and 4.6,
>
> $$E_{Mg} - E_o = 2.3 \text{ fraction ionic MgO} \cong 0.75$$
> $$E_{Si} - E_o = 1.7 \text{ fraction ionic SiO}_2 \cong 0.5$$
> $$E_{Si} - E_c = 0.3 \text{ fraction ionic SiC} < 0.1$$

The monovalent ions in groups 1 (Li, Na, K, etc.) and 17 (F, Cl, Br, etc.) produce compounds that are highly ionic, but relatively low strength, low melting temperature, and low hardness. Ionic compounds with more highly charged ions such as Mg^{2+}, Al^{3+}, and Zr^{4+} have stronger bonds and thus have higher strength, higher melting temperature, and higher hardness. Specific properties for specific materials are discussed in Chapters 7 through 11.

In summary, the following properties are characteristic of ionic bonding and the resulting ceramic materials:

1. There is an electron donor plus an electron acceptor needed to achieve electrical neutrality.
2. Structure is determined by atom (ion) size and charge with a tendency to achieve as close packing as sizes will permit.
3. Bonding is nondirectional.
4. Materials can be transparent to visible wavelengths of light and some other electromagnetic wave lengths.
5. There is low electrical conductivity at low temperature.

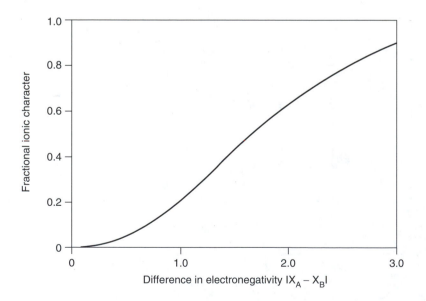

FIGURE 4.6   Semiempirical curve derived by Pauling for using the electronegativity difference between two elements to estimate the degree of ionic character. (Reprinted with slight modifications from Linus Pauling, *The Nature of the Chemical Bond*, 3rd ed., © 1960 by Cornell University. Used by permission of the publisher, Cornell University Press.)

6. Sometimes there is ionic conductivity at high temperature.
7. Metal ions with group 17 anions are strongly ionic (NaCl, LiF).
8. Compounds with higher-atomic-weight elements of group 16 (S, Se, Te) are increasingly less ionic.
9. Strength of ionic bonds increases as charge increases; many oxides composed of multiple-charged ions are hard and melt at high temperatures (Al_2O_3, ZrO_2, Y_2O_3).

4.2.3 COVALENT BONDING

Covalent bonding occurs when two or more atoms share electrons such that each achieves a stable, filled electron shell. Unlike metallic and ionic bonds, covalent bonds are directional. Each covalent bond consists of an electron shared between two atoms such that the probability distribution for the electron resembles a dumbbell. This produces the directionality of the bond. The bonding of carbon atoms to produce diamond is a good example.

Carbon has an atomic number of 6 and an electronic structure of $1s^2 2s^1 2p^3$ and thus has four valence electrons available for bonding. Each $2s$ and $2p$ electron shares an orbital with an equivalent electron from another carbon atom, resulting in a structure in which each carbon atom is covalently bonded to four other carbon atoms in a tetrahedral orientation. This is shown schematically in Figure 4.7a for one tetrahedral structural unit. The central carbon atom has its initial six electrons plus one shared electron from each of the adjacent four carbon atoms, resulting in a total of 10 electrons. This is equivalent to the filled outer shell of a neon atom and is a very stable condition. Each of the four outer carbon atoms of the tetrahedron is bonded directionally to three additional carbon atoms to produce a periodic tetrahedral structure with all the atoms in the structure (except the final outer layer at the surface of the crystal) sharing four electrons to achieve the stable electronic structure of neon.

The continuous periodic covalent bonding of carbon atoms in diamond results in high hardness, high melting temperature, and low electrical conductivity at low temperature. Silicon carbide has similar covalent bonding and thus high hardness, high melting temperature, and low electrical conductivity at low temperature.*

Covalently bonded ceramics typically are hard and strong and have high melting temperatures. However, these are not inherent traits of covalent bonding. For instance, most organic materials have

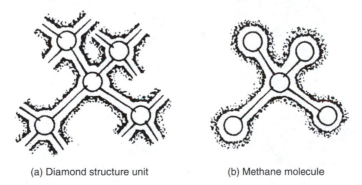

(a) Diamond structure unit   (b) Methane molecule

FIGURE 4.7 Schematic example of covalently bonded materials. (a) Diamond with periodic three-dimensional structure. (b) Methane with single-molecular structure. Shaded regions show directional electron probability distributions for pairs of electrons.

---

* Silicon carbide doped with appropriate impurities has significantly increased electrical conductivity and is an important semiconductor material.

covalent bonds but do not have high hardness or high melting temperatures. The deciding factor is the strength of the bond and the nature of the structure. For instance, methane (CH_4) forms a tetrahedral structural unit like diamond, but the valence electrons of both the carbon atom and the four hydrogen atoms are satisfied within a single tetrahedron and no periodic structure results. Methane is a gas under normal ambient conditions. A methane molecule is shown schematically in Figure 4.7b. Organic bonding and structures are discussed in more detail later in the chapter.

Diatomic gases (H_2, O_2, N_2, etc.) are another example of covalent bonding where molecules rather than interconnected solid structures are formed. Two hydrogen atoms each share their $1s$ electron to form H_2. Two oxygen atoms share two electrons to form O_2. Similarly, two nitrogen atoms share three electrons to form N_2. Multiple sharing leads to a particularly strong bond and a stable molecule. N_2 is often used as a substitute for the inert gases He or Ar. Figure 4.8 illustrates covalent bonding in various diatomic gases.

The directional bonding of covalent materials results in structures that are not close packed. This has a pronounced effect on the properties, in particular density and thermal expansion. Close-packed materials such as the metals and ionic-bonded ceramics have relatively high thermal expansion coefficients. The thermal expansion of each atom is accumulated for each close-packed adjacent atom throughout the structure to yield a large thermal expansion of the whole mass. Covalently bonded ceramics typically have a much lower thermal expansion because some of the thermal growth of the individual atoms is absorbed by the open space in the structure.

Covalent bonding occurs between atoms of similar electronegativity that are not close in electronic structure to the inert gas configuration. (Refer to the electronegativity scale in Figure 4.5.) Atoms such as C, N, Si, Ge, and Te are of intermediate electronegativity and form highly covalent structures. Atoms with a greater difference in electronegativity form compounds having a less covalent bond nature. Figures 4.5 and 4.6 can be used to estimate the relative covalent bond nature. However, the curve in Figure 4.6 is empirical and can be used only as an approximation, especially in intermediate cases.

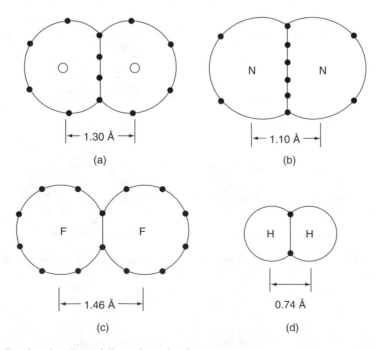

FIGURE 4.8 Covalent bonding of diatomic molecules.

Example 4.8 What is the approximate degree of covalent character of diamond? of Si_3N_4? of SiO_2?
From Figure 4.5,

$$-E_C - E_C = 0$$
$$E_{Si} - E_N = 1.2$$
$$E_{Si} - E_o = 1.7$$

From Figure 4.6,

Fraction covalent C = 1.0 − fraction ionic = 1.0 − 0 = 1.0
Fraction covalent Si_3N_4 = 1.0 − fraction ionic = 1.0 − 0.3 = 0.7
Fraction covalent SiO_2 = 1.0 − fraction ionic = 1.0 − 0.5 = 0.5

In summary, the following properties are characteristic of covalent bonding and of the resulting ceramic materials:

1. Electron are shared to fill outer electron shells and achieve electrical neutrality.
2. Atoms having similar electronegativity form covalent bonds.
3. Bonding is highly directional.
4. Structures are not close packed, but typically three-dimensional frameworks containing cavities and channels.
5. Compounds often have high strength, hardness, and melting temperature.
6. Structures often have relatively low thermal expansion.

4.2.4 IONIC AND COVALENT BOND COMBINATIONS

Many ceramic materials have a combination of ionic and covalent bonding. An example is gypsum ($CaSO_4 \cdot 2H_2O$), from which plaster is manufactured. The sulfur is covalently bonded to the oxygen to produce SO_4^{2-}, which is two electrons short of having full outer electron shells for each of the five atoms. The calcium donates its two valence electrons and is thus bonded ionically to the SO_4^{2-}:

$$Ca + \cdot \ddot{O}:\ddot{S}:\ddot{O}\cdot \longrightarrow Ca^{2+} + :\ddot{O}:\ddot{S}:\ddot{O}:$$

A similar type of combined bonding results in the many silicate compositions that are so important to ceramic technology. These silicate structures are based on the SiO_4^{4-} tetrahedron and the various ways these tetrahedra can be linked. As illustrated in Table 4.3, the tetrahedra are linked by sharing corners, resulting in framework, chain, ring, and layer structures. These silicate materials make up much of the Earth's crust and are the raw materials for most of our traditional ceramic industries.

4.2.5 VAN DER WAALS BONDS

Ionic and covalent bonds are referred to as primary bonds and account for the atomic bonding of most ceramic materials. However, other weaker secondary bond mechanisms also occur that have major effects on the properties of some ceramic materials. These secondary bonds are grouped together under the name *van der Waals forces*.

One type of van der Waals force is referred to as dispersion. In all molecules (including monatomic molecules) there is a fluctuating electrical dipole that varies with the instantaneous positions of the

TABLE 4.3
Bonding of Silicate Structures

Bonding of Tetrahedra	Structure Classification	Schematic	Examples
Independent tetrahedra	Orthosilicates	SiO_4^{4-}	Zircon ($ZrSiO_4$), mullite ($Al_6Si_2O_{13}$), forsterite (Mg_2SiO_4), kyanite (Al_2SiO_5)
Two tetrahedra with one corner shared	Pyrosilicates	$Si_2O_7^{6-}$	Ackermanite ($Ca_2MgSi_2O_7$)
Two corners shared to form ring or chain structures	Metasilicates	$Si_6O_{18}^{12-}$ $(SiO_3)_n^{2n-}$	Spodumene [$LiAl(SiO_3)_2$], wollastonite ($CaSiO_3$), beryl ($Be_3Al_2Si_6O_{18}$), asbestos [$Mg_3Si_2O_5(OH)_4$]
Three corners shared	Layer silicates	See Reference 2, pp. 75–79	Kaolinite clay [$Al_2(Si_2O_5)(OH)_4$], mica (muscovite var.) [$KAl_2(OH)_2(AlSi_3O_{10})$], talc [$Mg_3(Si_2O_5)_2(OH)_2$]
Four corners shared	Framework silicates	See Reference 2, pp. 71–74, for SiO_2 and derivatives	Quartz, cristobalite, tridymite (SiO_2), feldspars (albite var.) ($NaAlSi_3O_8$), zeolites (mordenite var.) [$(Ca, Na_2)Al_2Si_9O_{22} \cdot 6H_2O$]

electrons. Interaction of these fluctuating dipoles between molecules leads to very weak forces of attraction, which can result in bonding when other forces are absent. The dispersion effect is the mechanism of condensation of the noble gases at very low temperature.

Another type of van der Waals force is *molecular polarization*, in which an electrical dipole forms in asymmetrical molecules. Hydrogen fluoride can be used as a simple example. When the single shared electron is orbiting the fluorine nucleus, a dipole moment results where the hydrogen side of the molecule has a net positive charge and the fluorine side has a net negative charge. Because the fluorine nucleus is larger than the hydrogen nucleus, the shared electron spends more time around the fluorine nucleus. Consequently, the center of positive charge and the center of negative charge do not coincide and a weak electric dipole results that can contribute to weak bonding of one molecule to another.

A third type of van der Waals force is the *hydrogen bridge* or *hydrogen bonding*. It is a special case of molecular polarization in which the small hydrogen nucleus is attracted to the unshared electrons in a neighboring molecule. The most common case is water (H_2O), although this type of bonding is also found with other hydrogen-containing molecules, such as ammonia (NH_3).

van der Waals forces are very important in layer structures such as clays, micas, graphite, and hexagonal boron nitride. All of these ceramic materials have strong primary bonding within their layers but depend on van der Waals-type bonds to hold their layers together. Highly anisotropic properties result. All of these layer structures have easy slip between layers. In the clay minerals this property makes possible plasticity with the addition of water and was the basis of the early use of clay for pottery. In fact, it was the basis of almost all ceramic fabrication technology prior to the 20th century and is still an important factor in the fabrication of pottery, porcelain, whiteware, brick, and many other items.

The easy slip between layers in graphite and hexagonal boron nitride has also resulted in many applications of these materials. Both can be easily machined with conventional cutting tools and provide low-friction, self-lubricating surfaces for a wide variety of seals. Both are also used as solid lubricants and as boundary-layer surface coatings.

The weak bond between layers of mica and the resulting easy slip has recently led to a new application for these materials. Small synthetic mica crystals are dispersed in glass to form a non-porous composite having excellent electrical resistance properties. The presence of the mica permits machining with conventional low-cost machine tools of the composite to close tolerances with no chipping or breakage.

Although van der Waals forces are weak, they are adequate to cause adsorption of molecules at the surface of a particle. For particles of colloid dimensions ($100\,\text{Å}$ to $3\,\mu m$), adsorbed ions provide enough charge at the surface of a particle to attract particles of opposite charge and to repel particles of like charge. This has a major effect on the rheology (flow characteristics of particles suspended in a fluid) of particle suspensions used for slip casting and mixes used for extrusion, injection molding, and other plastic-forming techniques (see Chapter 13).

The discussions in this chapter of electronic structure, bonding, and crystal structure have been brief and simplified. More detailed discussions are available in References 1 through 8 and in Chapter 5.

4.3 POLYMORPHIC FORMS AND TRANSFORMATIONS

As described in the sections on bonding, the stable crystal structure for a composition is dependent on the following:

1. Balance of electrical charge
2. Densest packing of atoms consistent with atom size, number of bonds per atom, and bond direction
3. Minimization of the electrostatic repulsion forces

As the temperature of or the pressure on a material change, interatomic distance and the level of atomic vibration change such that the initial structure may not be the most stable structure under the new conditions. Materials having the same chemical composition but a different crystal structure are called polymorphs and the change from one structure to another is referred to as a *polymorphic transformation*.

Polymorphism is common in ceramic materials and in many cases has a strong impact on useful limits of application. For instance, the stable form of zirconium oxide (ZrO_2) at room temperature is monoclinic, but it transforms to a tetragonal form at about $1100°C$. This transformation is accompanied by a large volume change that results in internal stresses in the ZrO_2 body large enough to cause fracture or substantial weakening. In attempts to avoid this problem, it was discovered that appropriate additions of MgO, CaO, or Y_2O_3 to ZrO_2 produced a cubic form that did not undergo a transformation and was thus useful over a broader temperature range.

Before selecting a material for an application, it is necessary for an engineer to verify that the material does not have an unacceptable transformation. A good first step is to check the phase equilibrium diagram for the composition. Even if more than one polymorph is present within the intended temperature range of the application, the material may be acceptable. The important criterion is that no large or abrupt volume changes occur. This can be determined by looking at the thermal expansion curve for the material. For example, Figure 4.9 compares the thermal expansion curves for unstabilized ZrO_2 and stabilized ZrO_2. The large volume change associated with the monoclinic–tetragonal transformation is readily visible for the unstabilized ZrO_2.

Many ceramic materials exist in different polymorphic forms. Among these materials are SiO_2, SiC, C, Si_3N_4, BN, TiO_2, ZnS, $CaTiO_3$, Al_2SiO_5, FeS_2, and As_3O_5. The properties of some of these are discussed in later chapters.

Two types of polymorphic transformations occur. The first, *displacive transformation*, involves distortion of the structure, such as a change in bond angles, but does not include breaking of bonds. It typically occurs rapidly at a well-defined temperature and is reversible. The martensite transformation

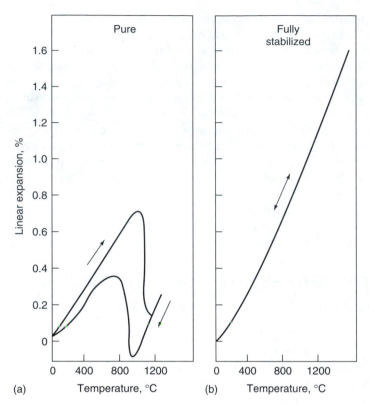

FIGURE 4.9 Thermal expansion curves for (a) unstabilized ZrO_2 and (b) stabilized ZrO_2 showing the abrupt volume change in the unstabilized ZrO_2 due to the monoclinic–tetragonal, polymorphic phase transformation. (From Ryshkewitch, E. and Richerson, D.W., *Oxide Ceramics*, 2nd ed., General Ceramics/Academic Press, 1985. With permission.)

in metals is a displacive transformation. So also are the cubic–tetragonal $BaTiO_3$ and tetragonal–monoclinic ZrO_2 transformations.

Displacive transformations are common in the silicate ceramics. In general, the high-temperature form has higher symmetry, larger specific volume, and larger heat capacity and is always the more open structure. The low-temperature form typically has a collapsed structure achieved by rotating the bond angle of alternating rows of SiO_4 tetrahedra in opposite directions.

The second type of transformation is the *reconstructive transformation*. Bonds are broken and a new structure formed. Much greater energy is required for this type of transformation than for a displacive transformation. The rate of reconstructive transformation is sluggish, so the high-temperature structure can usually be retained at low temperature by rapid cooling through the transformation temperature.

The activation energy for a reconstructive transformation is so high that transformation frequently will not occur unless aided by external factors. For example, the presence of a liquid phase can allow the unstable form to dissolve, followed by precipitation of the new stable form. Mechanical energy can be another means of overcoming the high activation energy.

Silica (SiO_2) is a good example for illustrating transformations. Both displacive and reconstructive transformations occur in SiO_2 and play an important role in silicate technology. Figure 4.10 shows the temperature-initiated transformations for SiO_2. The stable polymorph of SiO_2 at room temperature is quartz. However, tridymite and cristobalite are also commonly found at room temperature in ceramic components as metastable forms because the reconstructive transformations in SiO_2 are very sluggish and do not normally occur. Quartz, tridymite, and cristobalite all have displacive

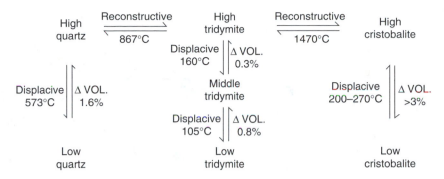

FIGURE 4.10 Transformations and volume changes for SiO_2 polymorphs. (Adapted from Shannon, R.D. and Prewitt, C.T., *Acta Crystallogr.*, B25, 925, 1969.)

transformations in which the high-temperature structures are distorted by changes in bond angle between SiO_4 tetrahedra to form the low-temperature structures. These displacive transformations are rapid and cannot be restrained from occurring.

It is important to note the size of the volume changes associated with displacive transformations in SiO_2. These limit the applications, especially of cristobalite and quartz. Ceramic bodies containing moderate to large amounts of quartz or cristobalite either fracture during thermal cycling through the transformation temperature or are weakened. In the fabrication of silica brick for high-temperature applications, a small amount of $CaCO_3$, or CaO is added to act as a flux at the firing temperature to dissolve the quartz and precipitate the SiO_2 as tridymite. The tridymite has much less shrinkage during transformation and is thus less likely to result in fracture or weakening of the refractory brick.

4.4 NONCRYSTALLINE STRUCTURES

The structures described so far have all had units of atomic arrangement that were repeated uniformly throughout the solid. For example, in a silicate the crystal structure is made up of an ordered repetition of SiO_4 tetrahedra. Each atom of a given type has the same neighboring atoms at the same bond angles and the same interatomic distances. This type of structure, in which both short-range and long-range order occur, is called a *crystalline structure*.

Structures that have short-range order but no long-range periodicity are referred to as noncrystalline. Figure 4.11 illustrates the difference between a crystalline and a noncrystalline material. Noncrystalline solids such as glass, gels, and vapor-deposited coatings have many applications and are very important to a broad range of engineering disciplines.[9-12]

4.4.1 GLASSES

Glasses are the most widely used noncrystalline ceramic. A *glass* is formed when a molten ceramic composition is cooled so rapidly that the atoms do not have time to arrange themselves in a periodic structure. At temperatures below the solidification temperature, glasses are not stable thermodynamically and the atoms would rearrange into a crystalline structure if they had the mobility. Over long periods of time glass can crystallize, as evidenced by the presence of cristobalite in some volcanic glass (obsidian). The crystallization can be speeded up by increasing the temperature to a level at which atomic mobility is increased. Most engineers who have used fused silica in high-temperature applications have encountered this. At use-temperatures well below the melting temperature of 1713°C, cristobalite crystals form in the fused silica (slowly at 1200°C and relatively rapidly at 1400°C). Fused silica has a very low, nearly linear thermal expansion curve and is one of the best thermal-shock-resistant ceramic materials for applications where rapid thermal cycling occurs.

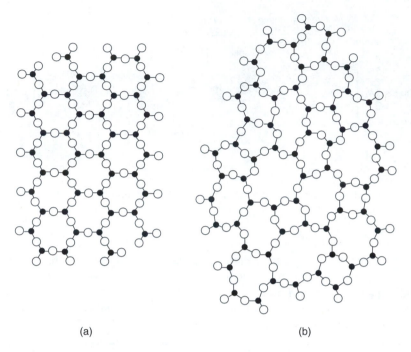

(a)   (b)

FIGURE 4.11   (a) Crystalline material, characterized by both short-range and long-range periodicity of the atomic structure. (b) Noncrystalline material, characterized by short-range order but no long-range periodicity. (From Shannon, R.D. and Prewitt, C.T., *Acta Crystallogr.*, B25, 925, 1969. With permission.)

Crystallization to cristobalite is undesirable because the large volume change of the polymorphic transformation from high cristobalite to low cristobalite in the range 200 to 270°C results in cracking of the fused silica component.

Some of the lesser known but important engineering applications of glasses are discussed in Chapters 3 and 11. Further information about the structure and thermodynamics of glass can be found in References 8 and 9.

Although a wide variety of properties can be achieved with glasses, the following are general characteristics:

1.   Short-range atomic order but no long-range order
2.   Structure isotropic, so the properties are uniform in all directions
3.   Typically transparent to optical wavelengths, but can be formulated to absorb or transmit a wide variety of wavelengths
4.   Typically good electrical and thermal insulators
5.   Soften before melting, so they can be formed by blowing into intricate hollow shapes

4.4.2   GELS

Gels are noncrystalline solids that are formed by chemical reaction rather than melting. Silica gel, which is highly useful as a bonding agent in the ceramic and metal industries, is produced by a reaction of ethyl silicate with water in the presence of a catalyst. $Si(OH)_4$ results, which is then dehydrated to form SiO_2. A silica gel can also be formed by the reaction of sodium silicate with acid.

Another noncrystalline inorganic gel, $Al(H_2PO_4)_3$, can be produced by reacting aluminum oxide (Al_2O_3) with phosphoric acid (H_3PO_4). Like the silica gels, this aluminum phosphate gel is produced at room temperature and is an excellent inorganic cement. The technology and important applications of ceramic cements are discussed in Chapter 14.

4.4.3 VAPOR DEPOSITION

An important class of noncrystalline materials is produced by rapid condensation of a vapor on a cold substrate or by reaction of a gas on a hot substrate. The vapor can be produced by sputtering, electron-beam evaporation, or thermal evaporation. Vapor contacting a cold substrate solidifies so rapidly that the atoms do not have time to rearrange into a crystalline structure.

Condensation from a vapor has been used to produce noncrystalline coatings of materials that are difficult or impossible to produce as noncrystalline solids by other approaches. These coatings are usually nonporous and very fine grained and have unique properties.

4.5 MOLECULAR STRUCTURES

So far we have discussed the bonding and structures of metals and ceramics, but have ignored organic materials. Organic materials are extremely important in modern engineering and their general characteristics should be understood just as well as those of metals and ceramics.

The majority of organic materials are made up of distinct molecules. The atoms of each molecule are held together strongly by covalent bonds with the outer electron shells filled. Because all the shells are filled, the individual molecules are stable and do not have a drive to bond with other molecules (as mentioned earlier for methane).

Organic molecular structures are usually formed from the nonmetallic elements and hydrogen. The most common are the hydrocarbons, which consist primarily of carbon and hydrogen but may also have halogens (especially Cl and F), hydroxide (OH), acetate ($C_2H_3O_2$), or other groups replacing one or more of the hydrogens. Other molecular structures include ammonia, which is made up of N and H, and the silicones, which contain Si in the place of carbon.

4.5.1 HYDROCARBONS

The hydrocarbons and modified hydrocarbons are perhaps the most frequently encountered engineering organic materials. Some of the simple compositions and molecular structures are illustrated in Figure 4.12. The straight lines between the atoms represent individual covalent bonds between pairs of electrons. The bond between two carbon atoms has an energy of about 83 kcal/g-mol. The bond energy between a carbon and a hydrogen is about 99 kcal/g-mol and that between a carbon and chlorine is about 81 kcal/g-mol. Some pairs of carbon atoms in Figure 4.12 have two covalent bonds between them. This double bond has an approximate energy of 146 kcal/g-mol.[13]

Hydrocarbons with only single bonds have no open structural positions where additional atoms can bond and are thus referred to as *saturated*. The paraffins are good examples. They have a general formula of C_nH_{2n+2}. Methane is $n = 1$ and ethane is $n = 2$. These, as well as compositions with n up to 15, are either liquid or gas at room temperature and are used as fuels. As the size of the molecules increase, the melting temperature increases; thus, paraffins with about 30 carbon atoms per molecule are relatively rigid at room temperature. The increase in melting temperature with molecular size is partially due to decreased mobility, but mostly to increased van der Waals bonding between molecules. The larger molecules have more sites available for van der Waals bonds.

Hydrocarbons with double or triple bonds between a pair of carbon atoms are referred to as unsaturated. Under the appropriate conditions, these bonds can be broken and replaced by single bonds that can link small molecules together to form large molecules. This is referred to as *polymerization*.[13–16]

4.5.2 ADDITION POLYMERIZATION

When a double bond is broken, it provides two sites at which new bonds may form, and the molecule is referred to as bifunctional. Ethylene, vinyl chloride, tetrafluoroethylene, styrene, and methyl methacrylate are all bifunctional. Addition polymerization can be achieved with bifunctional molecules by applying enough energy to break the double carbon bond. This energy can be in the form

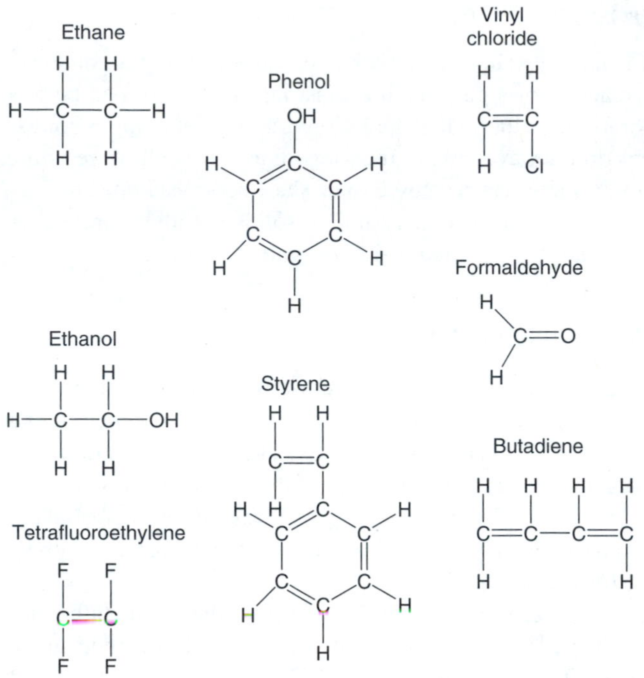

FIGURE 4.12 Examples of hydrocarbon structural units that are important building blocks for polymers.

of heat, pressure, light, or a catalyst. Once the double bonds of a group of molecules are broken, an unstable electron structure is present and the separate molecular units, called mers, bond together to form a long chain (or polymer). The energy released during addition polymerization is greater than the energy that was required to start the reaction. The following illustrates addition polymerization of vinyl chloride to form polyvinyl chloride:

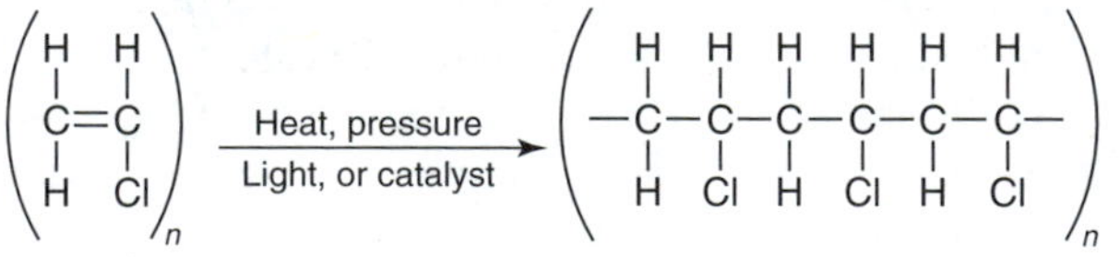

$$
\begin{pmatrix} \text{H} & \text{H} \\ | & | \\ \text{C} = \text{C} \\ | & | \\ \text{H} & \text{Cl} \end{pmatrix}_n \xrightarrow[\text{Light, or catalyst}]{\text{Heat, pressure}} \begin{pmatrix} \text{H} & \text{H} & \text{H} & \text{H} & \text{H} & \text{H} \\ | & | & | & | & | & | \\ -\text{C} - \text{C} - \text{C} - \text{C} - \text{C} - \text{C} - \\ | & | & | & | & | & | \\ \text{H} & \text{Cl} & \text{H} & \text{Cl} & \text{H} & \text{Cl} \end{pmatrix}_n
\tag{4.1}
$$

In more general terms, addition polymerization can be represented by

$$
n\text{A} \longrightarrow (\text{---A---})_n
\tag{4.2}
$$

Addition polymerization can also occur if more than one double bond is present (as in the polymerization of butadiene to make unvulcanized rubber), but only one of the bonds is broken and the resulting molecule is still linear.

Addition polymerization can be achieved with mixtures of two or more different monomers to achieve modified properties. This is called *copolymerization* and the resulting structure is referred to as a copolymer. This is analogous to solid solution in metals and ceramics and can be represented by

$$
n\text{A} + m\text{B} \xrightarrow{\text{heat, pressure, etc.}} (\text{---A}_n\text{B}_m\text{---})
\tag{4.3}
$$

The polymers produced by addition polymerization are typically thermoplastic; that is, they soften when heated and can be plastically worked to produce a shape and then return to their initial properties upon cooling. Thermoplastic polymers with complex shapes can be produced in large

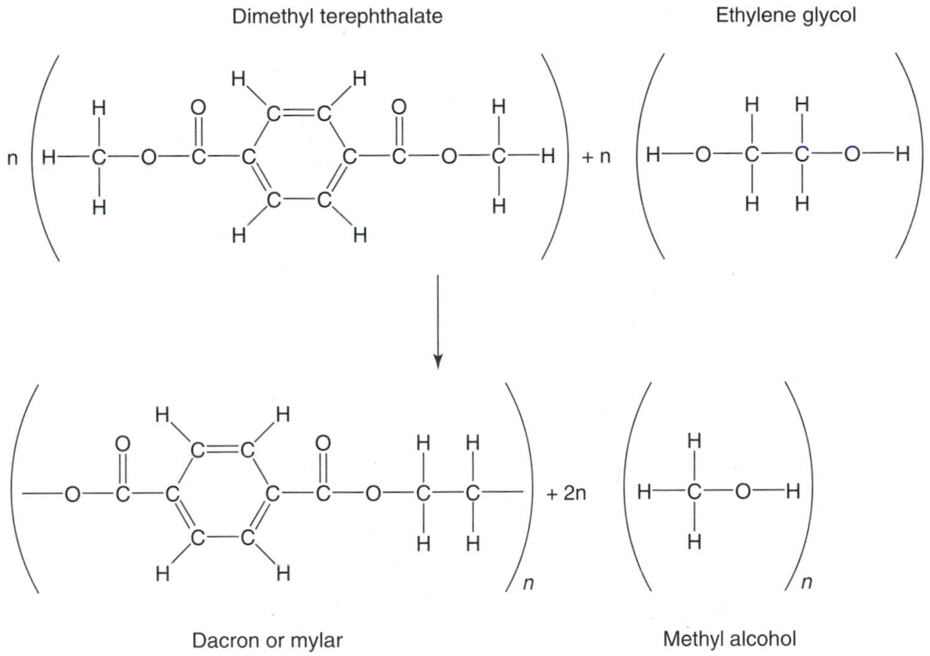

Dimethyl terephthalate

Ethylene glycol

Dacron or mylar

Methyl alcohol

FIGURE 4.13 Formation of dacron by condensation polymerization.

quantities by such low-cost methods as injection molding, extrusion, and sheet forming. Because of the reversible nature of plasticity, many of the thermoplastic polymers can be recycled.

4.5.3 CONDENSATION POLYMERIZATION

Condensation polymerization involves reaction of two different organic molecules to form a new molecule, accompanied by release of a by-product:

$$pC + pD \xrightarrow{\text{heat, pressure, etc.}} (\text{—E—}) + pF \tag{4.4}$$

Either a linear or framework polymer can result from condensation polymerization, depending on whether one double bond or more than one double bond is broken. The by-product is often water, but can also be other simple molecules, such as an alcohol or an acid.

Dacron is a linear polymer produced by condensation polymerization. It is synthesized from dimethyl terephthalate and ethylene glycol and forms methyl alcohol (CH_3OH) as the by-product. The reaction is shown in Figure 4.13. Note that no carbon double bonds were broken in this case, just two C—O bonds on each dimethyl terephthalate and two C—OH bonds on each ethylene glycol.

Phenol (C_6H_5OH) and formaldehyde (CH_2O) combine by condensation polymerization to form a network structure as shown in Figure 4.14. The C=O bond in the formaldehyde is broken and a C—H bond in two adjacent phenol molecules is broken. The remaining CH_2 of the formaldehyde then has two unsatisfied carbon bonds and acts as a bridge between the two phenol molecules. The O from the formaldehyde and the two H from the phenols combine to form water as a by-product. This reaction occurs at several C—H bonds in each phenol and results in the network structure. The phenol–formaldehyde polymer is known by several commercial names, Bakelite and Texalite being two. Other condensation polymers include nylon (hexamethylamineadipic acid) and Melmac (melamine–formaldehyde).

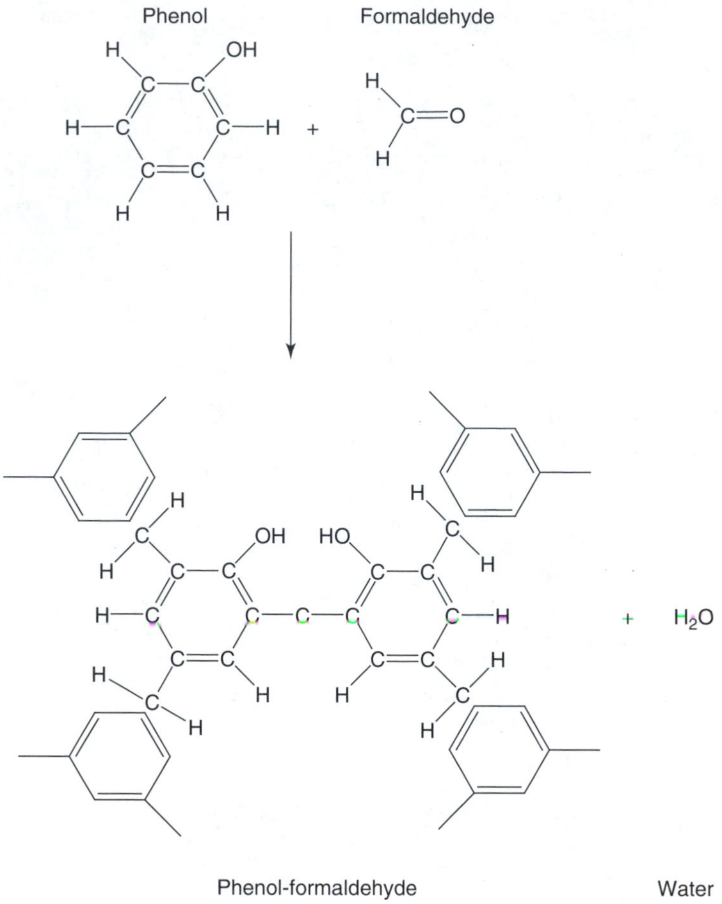

FIGURE 4.14   Formation of a network structure by condensation polymerization of phenol and formaldehyde.

Most of the condensation polymers are thermosetting resins. Once polymerization has occurred, especially for the framework structures, the material is relatively rigid and does not increase in plasticity with increase in temperature. In general, the thermosetting resins have higher strength and higher-temperature capability than the thermoplastic resins, but are not as economical to fabricate.

4.5.4   POLYMER CRYSTALLIZATION

The large molecules in a polymer can be oriented to produce a degree of crystallinity, usually resulting in modified properties. For instance, if the linear molecules in a polymer are random, van der Waals bonding between molecules will occur only in the limited number of positions where the appropriate atoms are adjacent. However, as alignment of the molecules increases, more atoms are in suitable positions for van der Waals bonds to form. Therefore, as the crystallinity increases, the strength tends to increase and the rate of creep decreases.

The shape of the polymer molecules affects the ease and degree of crystallization and also the properties. Crystallization occurs most easily if the individual monomers all have identical ordering.

4.5.5   CROSS-LINKING AND BRANCHING

Crystallization causes moderate changes in properties. Major changes can occur in linear polymers by cross-linking or branching. In cross-linking, adjacent chains are bonded together, usually by bridges

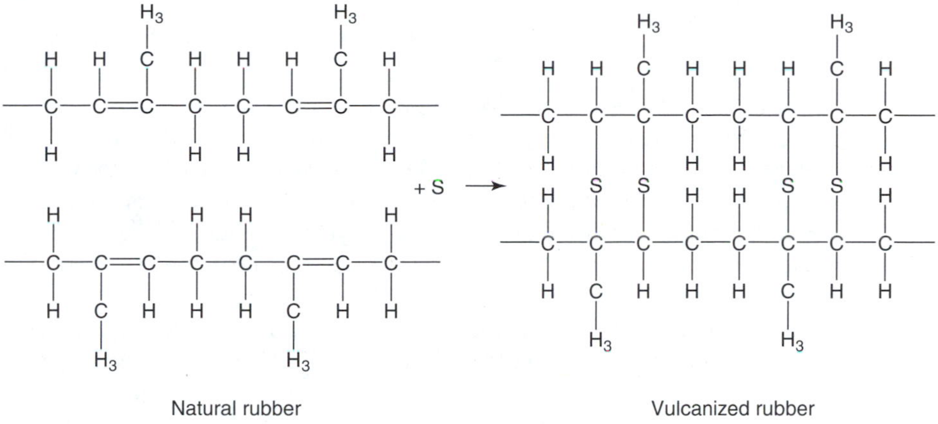

FIGURE 4.15 Cross-linking with sulfur in the vulcanization process for natural rubber.

between unsaturated carbon atoms. The vulcanization of natural rubber with sulfur is a classic example. The reaction is shown in Figure 4.15. The degree of cross-linking can be controlled by the amount of sulfur added. Both the hardness and strength increase as the amount of cross-linking increases.

REFERENCES

1. Shannon, R.D. and Prewitt, C.T., Effective ionic radii in oxides and fluorides. *Acta Crystallogr.*, B25, 925, 1969.
2. Kingery, W.D., Bowen, H.K., and Uhlmann, D.R., *Introduction to Ceramics*, 2nd ed., Wiley, New York, 1976, Chap. 2.
3. Pauling, L., *The Nature of the Chemical Bond*, 3rd ed., Cornell University Press, Ithaca, New York, 1960.
4. West, A.R., *Solid State Chemistry and Its Application*, Wiley, New York, 1984.
5. Oxtoby D.W. and Nachtrieb, N.H., *Principles of Modern Chemistry*, Saunders, Philadelphia, 1986.
6. Shackelford, J.F., *Introduction to Materials Science for Engineers*, 2nd ed., Macmillan, New York, 1988.
7. Callister, W.D., Jr., *Materials Science and Engineering: An Introduction*, 3rd ed., Wiley, New York, 1994.
8. Mangonon, P.L., *The Principles of Materials Selection for Engineering Design*, Prentice Hall, Upper Saddle River, NJ, 1999.
9. Rawson, H., *Inorganic Glass-Forming Systems*, Academic Press, New York, 1967.
10. Doremus, R.H., *Glass Science*, Wiley, New York, 1973.
11. Pfaender, H.G., *Schott Guide to Glass*, 2nd ed., Chapman and Hall, London, 1996.
12. Zallen, R., *The Physics of Amorphous Solids*, Wiley, New York, 1983.
13. Billmeyer, F.W., Jr., *Textbook of Polymer Science*, Interscience, New York, 1962.
14. Rudin, A., *The Elements of Polymer Science and Engineering*, Academic Press, San Diego, 1999.
15. Czekaj, C., Ed., *Encyclopedia of Polymer Science and Technology*, Volumes 1–12, Wiley, New York, 2002.
16. Fried, J.R., *Polymer Science and Technology*, 2nd ed., Prentice Hall, Upper Saddle River, NJ, 2003.

PROBLEMS

1. Explain why hydrogen and oxygen are present in the atmosphere as H_2 and O_2 (diatomic molecules) and helium and argon are present as He and Ar (monatomic molecules).
2. Show the complete electron notation for Co. Why does the $4s$ shell fill before the $3d$ shell?
3. How many $3d$ electrons are in each of the following?
   (a) Fe
   (b) Fe^{3+}

    (c) Fe^{2+}

    (d) Cu

    (e) Cu^{2+}

4. What chemical formula would result when yttrium and oxygen combine? What is the most likely CN for Y^{3+}? Once you know the chemical formula and the CN for the cation, can you figure out how to calculate the CN of the anion? What is the CN of the O^{2-} when it combines with Y^{3+}? What is the relative percent ionic character? What relative hardness and melting temperature would be expected for yttrium oxide?

5. Titanium has an atomic number of 22. How many electrons does the Ti^{4+} ion contain? How many electrons does Mn^{2+} have? Be^{2+}? Y^{3+}?

6. What chemical formula would result when potassium and oxygen combine? When calcium and fluorine combine? When tin and oxygen combine? When magnesium and oxygen? Titanium and oxygen? Zirconium and oxygen? Silicon and oxygen? Uranium and oxygen? Lanthanum and oxygen?

STUDY GUIDE

1. Describe the role of electrons in atomic bonding.

2. What are the key differences between metallic, ionic, and covalent bonding?

3. Explain why metallic bonding typically results in a close-packed structure that has high ductility and electrical conductivity?

4. Explain what happens to the behavior of a metal when other elements are added to form an alloy.

5. What factors determine the crystal structure for an ionically bonded material?

6. Explain the meaning and significance of CN?

7. Explain how to estimate the CN for the cations from the relative ionic radii of the ions.

8. Since you need the coordination for both the cations and anions in order to visualize the structure, how might you calculate the CN for the anions once you know the CN for the cations?

9. What factors determine the strength of an ionic bond or a covalent bond?

10. Based on the periodic table, which groups of elements are likely to bond ionically?

11. Unlike ionic bonding (which is nondirectional), covalent bonding is directional. Does this place constraints on the crystal structure? Explain.

12. Diamond and methane both are covalently bonded. Explain why they have such different behavior.

13. Since electrons are tied up between adjacent atoms in ionic and covalent bonding, how might this affect properties such as electrical conduction and response to light?

14. Based on the periodic table, which types of elements are most likely to bond covalently? Give some specific examples.

15. Are bonds purely ionic or covalent, or can a material result with mixed-bond characteristics? Give examples.

16. Are covalently bonded structures typically close packed? Explain.

17. List several ceramic materials where van der Waals bonding plays an important role in the material's behavior.

18. Explain polymorphism and what causes one polymorph to transform to another polymorph.

19. Identify two types of polymorphic transformation.

20. What is a noncrystalline structure compared to a crystalline structure?

21. What are some ways of achieving noncrystalline ceramic materials?

5 Crystal Chemistry and Specific Crystal Structures

Chapter 4 introduced the basic concepts of atomic bonding and crystal structure. Chapter 5 provides a more-detailed discussion of crystal structure. An extended discussion of crystal structure is necessary to help the reader understand the wide range of ceramic materials and their properties. We start by defining the notation that is commonly used to describe different aspects of crystals and crystal structure. We then discuss crystal chemistry and finish by reviewing specific metallic and ceramic crystal structures.

5.1 CRYSTAL STRUCTURE NOTATIONS

5.1.1 Crystal Systems and Bravais Lattices

Atoms bond together in metals and ceramics in distinct geometric arrangements that repeat throughout the material to form a crystal structure. The crystal structure that results depends upon the type of atomic bonding, the size of the atoms (or ions), and the electrical charge of the ions.

The smallest grouping of atoms that shows the geometry of the structure (and can be stacked as repeating units to form a crystal of the structure) is referred to as the unit cell. Seven unit cell geometries are possible. These are referred to as crystal systems and are illustrated in Figure 5.1. Note that the crystal systems are distinguished from each other by the length of the unit cell edges (called lattice constants or lattice parameters) and the angles between the cell edges.

Different options are available for stacking atoms in unit cells. For example, three options exist for a cubic cell: (1) at the eight corners, (2) at the eight corners plus at the center of each cube, and (3) at the eight corners plus at the center of each face of the cube. These atom positions are referred to as lattice points. Fourteen options are possible for lattice points in the seven crystal systems. These are illustrated in Figure 5.2 and are referred to as the Bravais lattices.

5.1.2 Crystal Directions and Planes

The closeness of packing of atoms varies in different planes within the crystal structure. For example, the planes of the atoms parallel to the diagonal of the cube in a face-centered cubic structure are close packed. Planes in other directions are not close packed. This variation in packing of atoms results in variations in properties of the crystal along different directions. It is therefore important to have a simple and constant method of identifying the directions and planes within a crystal structure.

Since the unit cell is the simplest representation of a crystal structure, it is used as the basis for defining directions and planes within the overall crystal. Figure 5.3 shows a unit cell with three sides lying along the axes of a rectangular coordinate system. This unit cell is generalized and could represent a cubic unit cell (where $a = b = c$), a tetragonal unit cell (where $a = b \neq c$), or an orthorhombic unit cell (where $a \neq b \neq c$).

Crystal directions are identified by the square-bracketed coordinates of a ray that extends from the origin to a corner, edge, or face of the unit cell. For example, a ray along the x axis intersects the corner of the unit cell at $x = 1$, $y = 0$, and $z = 0$. The direction [hkl] is thus defined as [100]. If the

System	Axial lengths and angles	Unit cell geometry
Cubic	$a = b = c,\ \alpha = \beta = \gamma = 90°$	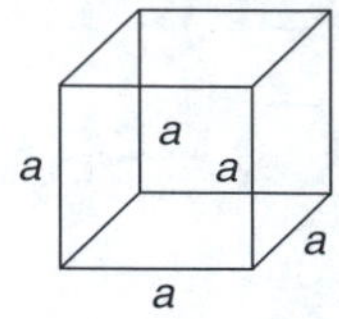
Tetragonal	$a = b \neq c,\ \alpha = \beta = \gamma = 90°$	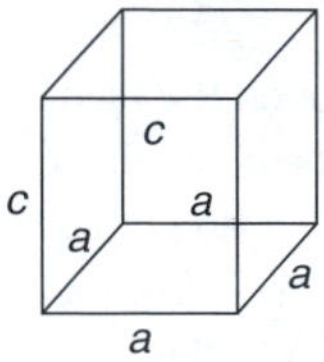
Orthorhombic	$a \neq b \neq c,\ \alpha = \beta = \gamma = 90°$	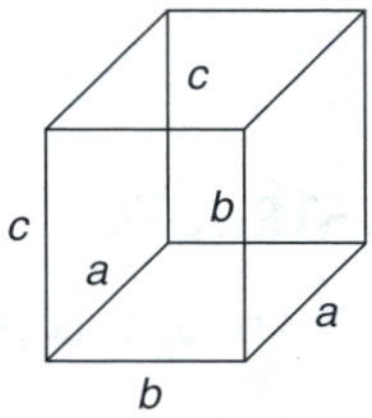
Rhombohedral	$a = b = c,\ \alpha = \beta = \gamma \neq 90°$	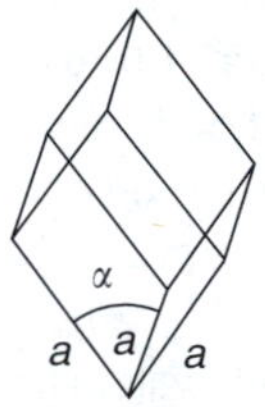
Hexagonal	$a = b \neq c,\ \alpha = \beta = 90°,\ \gamma = 120°$	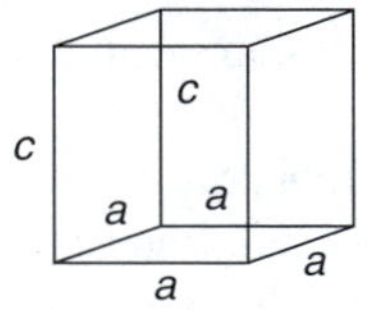
Monoclinic	$a \neq b \neq c,\ \alpha = \gamma = 90°, \neq \beta$	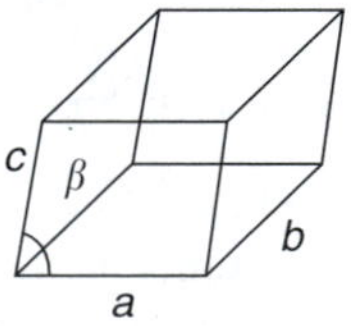
Triclinic	$a \neq b \neq c,\ \alpha \neq \beta \neq \gamma \neq 90°$	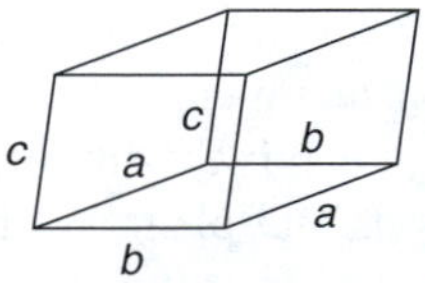

FIGURE 5.1 Geometrical characteristics of the seven crystal systems. (From Shackelford, J.F., *Introduction to Materials Science for Engineers*, Macmillan, New York, 1985. With permission.)

ray is drawn on the *x* axis in the negative direction, the crystal direction is [$\bar{1}$00]. Similarly, a ray along the *y* axis is the crystal direction [010] and along the *z* axis is [001]. A ray along the diagonal of the unit cell face bounded by the *x* and *z* axes intersects the corner of the unit cell at *x* = 1, *y* = 0, and *z* = 1, so the crystal direction is [101]. A ray across the diagonal of the whole unit cell intersects at

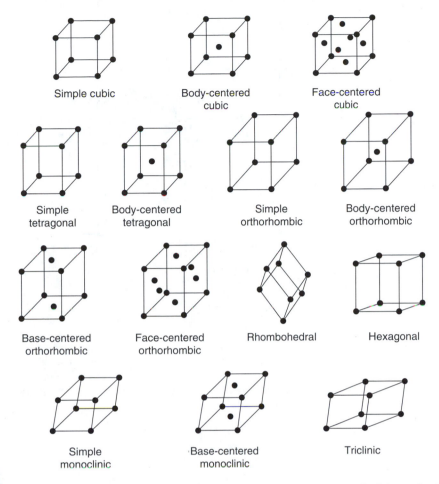

FIGURE 5.2 Bravais lattices. (From Shackelford, J.F., *Introduction to Materials Science for Engineers*, Macmillan, New York, 1985. With permission.)

$x = 1$, $y = 1$, and $z = 1$, so that the crystal direction is [111]. Several of these examples are illustrated in Figure 5.3.

The crystal direction notation is made up of the lowest combination of integers and represents unit distances rather than actual distances. A [222] direction is identical to a [111], so [111] is used. Fractions are not used. For example, a ray that intersects the center of the top face of the unit cell has coordinates $x = 1/2$, $y = 1/2$, and $z = 1$. All have to be multiplied by 2 to convert to the lowest combination of integers [112]. Finally, all parallel rays have the same crystal direction. For instance, the four vertical edges of a unit cell all have the direction $[hkl] = [001]$.

Crystal planes are designated by symbols referred to as *Miller indices*. The Miller indices are indicated by the notation (hkl) where h, k, and l are the reciprocals of the intercepts of the plane with the x, y, and z axes. This is illustrated in Figure 5.4 for a general unit cell. A plane that forms the right side of the unit cell intercepts the y axis at 1, but does not intercept the x or z axes. Thus $h = 1/\infty$, $k = 1/1$, and $l = 1/\infty$, which gives (010). To distinguish from the notation for directions, parentheses are used for planes. Parallel planes have the same Miller indices, so the left side of the unit cell in Figure 5.4 also is (010). Similarly, the top and bottom are (001) and the front and back are (100). Determination of Miller indices for several other planes is also illustrated in Figure 5.4.

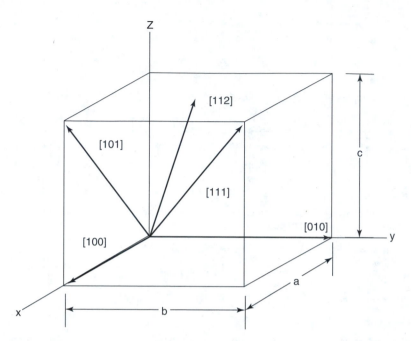

FIGURE 5.3 Definitions of crystal directions for a cubic, tetragonal or orthorhombic crystal. *a*, *b*, and *c* are lattice dimensions of the unit cell, *x*, *y*, and *z* are the orthogonal axes of a rectangular coordinate system. The rays represent crystal directions with their [*hkl*] designations. (From Richerson, D.W., Lesson 2: Atomic bonding and crystal structure, in *Introduction to Modern Ceramics*, ASM Materials Engineering Institute Course #56, 1990. With permission.)

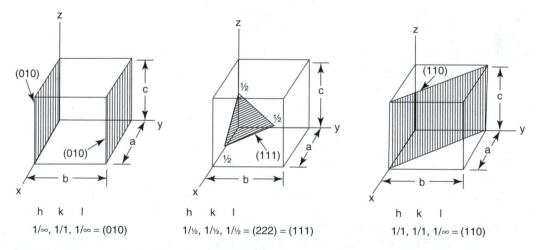

FIGURE 5.4 Examples of Miller indices notation for crystal planes. (From Richerson, D.W., Lesson 2: Atomic bonding and crystal structure, in *Introduction to Modern Ceramics*, ASM Materials Engineering Institute Course #56, 1990. With permission.)

5.1.3 STRUCTURE, COMPOSITION, AND COORDINATION NOTATIONS

Many crystal structures are named after the first chemical composition determined to have the structure. We need a simple notation to distinguish between the structure and composition. When we are referring to the structure we will use brackets. For example, [NaCl], [fluorite], [spinel], and [CsCl] all refer to crystal structures, whereas NaCl, CaF_2, $MgAl_2O_4$, and CsCl refer to compositions that have these structures.

Second, it is convenient to have a simple notation for a general composition that fits into a structural class. The letters A, B, and C will be used throughout this book to designate cations in a structure and X, Y, and Z will designate anions. These are combined in the same fashion as for a chemical formula. Examples are AX (for NaCl, BeO, etc.), AX_2 (for CaF_2, ZrO_2, etc.), and A_2BX_4 (for Mg_2SiO_4, etc.).

Finally, it is useful to have a notation that shows the coordination of each ion in a composition or structure. This is done by the use of the coordination number (CN) [n] as a superscript for each element of structural position. For example, both Na^+ and Cl^- were shown in Chapter 4 to have CN of 6 in [NaCl]. Thus, the coordination formula for NaCl is $Na^{[6]}Cl^{[6]}$. Muller and Roy[3] use an alternate notation. They use Roman numerals rather than a bracketed integer. For example, the coordination formula for NaCl using their notation would be $Na^{VI}Cl^{VI}$.

The following example further demonstrates the use of coordination formulas.

Example 5.1 In $FeAl_2O_4$ the Fe^{2+} cation is in a tetrahedral interstitial site with a CN of 4, the Al^{3+} is in an octahedral interstitial site with a CN of 6, and the O^{2-} anions are in a cubic close-packed arrangement having coordination with adjacent cations of 4. The coordination formula for $FeAl_2O_4$ is $Fe^{[4]}Al_2^{[6]}O_4^{[4]}$ and the general coordination formula for all [spinel] structure compositions analogous to $FeAl_2O_4$ is $A^{[4]}B_2^{[6]}O_4^{[4]}$. The coordination formula must balance the same as a chemical formula. The sum of the products of the subscripts and superscripts for the cations must equal the sum of the products for the anions:

$$A_1^{[4]} = (1) \times (4) = 4 \qquad B_2^{[6]} = (2) \times (6) = 12$$

$$X_4^{[4]} = (4) \times (4) = 16$$

$$A_1^{[4]} + B_2^{[6]} \text{ must equal } X_4^{[4]}, \text{ i.e., } 4 + 12 = 16$$

5.2 CRYSTAL CHEMISTRY OF CERAMICS

As we have discussed, pure metals form crystal structures that are relatively close packed and consist of atoms all of the same size and electrical charge distribution. Such is not the case with ceramic crystal structures. In general, the ceramic is made up of more than one type of atom and a combination of bond types. Thus, additional factors must be considered:

1. Size of the different atoms
2. Balance of charge to maintain overall neutrality
3. Degree of directionality of the bonds

Much knowledge has been gained during the past 60 years regarding these factors and is organized into the technology category referred to as *crystal chemistry.*

5.2.1 CRYSTAL CHEMISTRY CONCEPTS

What can an understanding of crystal chemistry do for us, and why is it relevant? The following are some important reasons:

1. It provides a simple means of understanding how atoms join together to form a ceramic crystal structure.
2. It provides a basis for understanding how atoms can substitute into a structure by solid solution and alter the behavior of the material and will help us to understand the electrical, magnetic, optical, thermal, and mechanical properties of ceramics.

3. It explains how distorted and defect structures occur or can be produced.
4. It explains why ceramics behave differently from each other and from metals and organic materials.

The following sections describe the concepts and techniques of crystal chemistry.

5.2.1.1 Ionic Radius

The size of an ion is the most critical parameter in the science of crystal chemistry. The term *ionic radius* was introduced in Chapter 4. It is an approximation of the radius an ion will have in a particular structure. Note the last few words in the description: "in a particular structure." The volume occupied by an ion in a structure is strongly affected by the nature of the ion, by the nature of the surrounding ions, and by the relative amount of ionic or covalent bonding. It is also affected by the spin state. (Transition metal ions can exist in a high-spin or a low-spin state.) Over the past 60 years, the interatomic separation has been determined for each element in many different compositions and structures, and values of ionic radii have been estimated for both cations and anions. These ionic radii are presented in a table at the back of the book.

We can use this listing to see how factors such as valence, CN, and spin state affect the ionic radius. Let us look at the element iron (Fe). The ferrous ion Fe^{2+} with a valence of $+2$ has an ionic radius of 0.75 Å. The ferric ion Fe^{3+} with a valence of $+3$ has a smaller ionic radius of 0.69 Å. This is not surprising, since the Fe^{3+} has one less electron orbiting the nucleus. Fe^{2+} in the high-spin state has an ionic radius of 0.92 Å compared to 0.75 Å for Fe^{2+} in the low-spin state. Finally, Fe^{2+} surrounded by six anions (CN of 6) has a larger ionic radius (0.92 Å) then Fe^{2+} surrounded by only four anions (CN = 4, r = 0.77).

5.2.1.2 Ionic Packing

Most of the ionic ceramic crystal structures consist of a three-dimensional stacking of the anions with the smaller cations fitting into interstitial positions. The size of the interstitial position varies depending on the mode of stacking. Eight spheres stacked to form a simple cube have a relatively large interstitial position in the center. For example, in a body-centered cubic metal this position is filled by an atom of the same size as the corner atoms, resulting in a structure that is not quite close packed. The atom or ion in the center has a CN of 8. Six spheres stacked to form an octahedron have a smaller interstitial position (CN of 6), and four spheres stacked to form a tetrahedron have a still smaller interstitial position (CN of 4). The cubic close-packed and hexagonal close-packed structures (discussed later in this chapter) both have interstitial positions of these two types. They are clearly visible in Figure 5.5, which shows a cutaway of a face-centered cubic (FCC) structure. The larger octahedral sites are visible from the side view and the smaller tetrahedral sites are visible from the diagonal (cutaway) view. There is one octahedral site at the center of the FCC unit cell and one at the midspan of each of the 12 edges.

Only ions of an appropriate size range are stable in each interstitial position. An ion that is too small to fill the site completely is not stable. One that closely fills the interstitial site or one that is slightly larger than the site is stable. This was illustrated previously in Figure 4.4. Note that the ion larger than the interstitial site causes adjacent ions to deviate from close packing.

One can calculate using simple geometry the size of a sphere that fits exactly into each interstitial position. This represents the minimum size that is stable. The maximum size is approximately the size of the next larger interstitial position. These size ranges based on the radius ratio of the interstitial cation to the host anion polyhedron are listed in Table 5.1. Ratios smaller than 0.225 typically involve independent tetrahedra or octahedra bonded together by sharing of a corner or edge and result in CNs of 2 or 3.

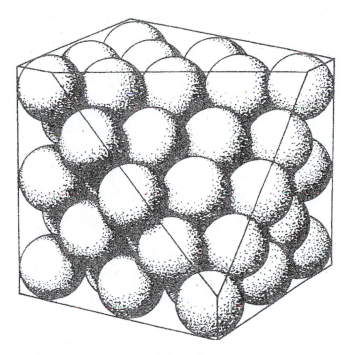

FIGURE 5.5 Cutaway of a FCC structure showing the larger octahedral interstitial holes visible from the side view and the smaller tetrahedral interstitial holes visible from the diagonal view. (From Kingery, W.D., Bowen, H.K., and Unlmann, D.K., *Introduction to Ceramics*, 2nd ed., Wiley, New York, 1976, p. 49. With permission.)

TABLE 5.1
Critical Radius Ratios for Various CNs for Cations in Interstitial Positions in Anion Polyhedra

Coordination Number	Position of Anions	Minimum Ratio of Ionic Radii
4	Corners of tetrahedron	0.225
6	Corners of octahedron	0.414
8	Corners of cube	0.732

The critical radius ratios in Table 5.1 are useful in predicting the types of structures that will be formed by various ion combinations. However, as we discussed in Chapter 4, atoms and ions are not really hard spheres. Anions with high atomic number are large and can be easily deformed, especially by a cation with a high charge. In addition, most materials do not have pure ionic bonding, but have some covalent contribution that may affect the CN. Thus, ions fit into a greater range of CN than is indicated by predictions based on the critical radius ratios of Table 5.1. Table 5.2 identifies deviations that have been observed in ionic bonding between various cations and oxygen. Some cations (Al^{3+}, Na^+, Ca^{2+}, and K^+) have been experimentally observed in several different coordinations within ceramic structures.

5.2.1.3 Effect of Charge

So far we have only considered the effect of the relative size of the cations and anions on their stacking into a structure. The charge on each ion is equally important. Electrical neutrality is required at

TABLE 5.2
CN and Bond Strength of Various Cations with Oxygen

Ion	Radius (CN = 4)	Predicted Coordination Number	Observed Coordination Number	Strength of Electrostatic Bond
B^{3+}	0.16	3	3, 4	1 or 3/4
Be^{2+}	0.25	4	4	1/2
Li^+	0.53	6	4	1/4
Si^{4+}	0.29	4	4, 6	1
Al^{3+}	0.38	4	4, 5, 6	3/4 or 1/2
Ge^{4+}	0.39	4	4, 6	1 or 2/3
Mg^{2+}	0.51	6	6	1/3
Na^+	0.99	6	4, 6, 8	1/6
Ti^{4+}	0.44	6	6	2/3
Sc^{3+}	0.52	6	6	1/2
Cr^{4+}	0.51	6	6, 8	2/3 or 1/2
Ca^{2+}	0.71	6, 8	6, 7, 8, 9	1/4
Ce^{4+}	0.57	6	8	1/2
K^+	0.99	8, 12	6, 7, 8, 9, 10, 12	1/9
Cs^+	1.21	12	12	1/12

Source: From Pauling, L., *Nature of the Chemical Bond*, 3rd ed., Cornell University Press, Ithaca, NY, 1960.

the unit-cell level as well as throughout the crystal structure. The charge of each cation and anion must be mutually balanced by the combined charge of the surrounding ions of the opposite charge. This places a limitation on the positions of the ions in the structure.

The portion of the charge of each cation is equal to the valence of the cation V_C divided by the CN of the cation $(CN)_C$. Likewise, the portion of the charge of each anion is equal to the valence of the anion V_A divided by the CN of the anion $(CN)_A$. To meet the charge-balance criteria discussed above, these two ratios must be equal in a stable crystal structure; that is,

$$|V_C|/(CN)_C = |V_A|/(CN)_A \qquad (5.1)$$

Each of these ratios is referred to as the *bond strength*. Bond strengths for oxide ceramics are included in Table 5.2.

Let us look at an example to illustrate the way that ionic radius, CN, and bond strength can be used to estimate the structure of an ionically bonded ceramic.

Example 5.2   What are the CNs of Na^+ and Cl^- in NaCl and what type of structural arrangement results? From ionic radius table at the back of the book, the ionic radius of Na^+ is 1.16 Å and of Cl^- is 1.67 Å. The radius ratio = $Na^+/Cl^- = 1.16$ Å/1.67 Å = 0.69. Now we compare this radius ratio with the ranges in Table 5.1 for each CN. The value 0.69 lies in the range for sixfold coordination. Thus, the simple calculation provides a strong indication that each Na^+ cation is in an octahedral interstitial position surrounded by six Cl^- anions at the corners of an octahedron. Now, to estimate the coordination of the anion, we insert the known information into Equation (5.1) and solve for the unknown $(CN)_A$:

$$(CN)_A = |V_A|(CN)_C/|V_C| = (1)(6)/1 = 6$$

This indicates that the CN of the anion Cl^- is also 6 and that each Cl^- ion is surrounded by 6 Na^+ ions positioned at the six corners of an octahedron. We now have all the information needed to visualize the arrangement of the ions in the crystal structure. Try sketching the NaCl structure.

TABLE 5.3
Ions of Similar Size that Substitute for each other in Crystal Structures

Valence	Size	Substitutions
+1	Large	Na, K, Rb, Cs
+2	Small	Be, Zn
+2	Medium	Mg, Fe, Ni, Co
+2	Large	Ca, Ba, Sr, Pb
+3	Small	Al, B, Ga
+3	Medium	Al, Fe, Cr, Ti
+3	Large	Y, rare earths (La-Lu)
+4	Small	Si, Ge
+4	Medium	Ti, Zr, V, W, Sn, Hf, Nb, Mo
+4	Large	Th, U
+5	Medium	Ta, Nb, As, V, Sb
−1	Large	Cl, Br, I
−2	Large	O, S, Se

Source: From Richerson, D.W., Lesson 2: Atomic bonding and crystal structure, in *Introduction to Modern Ceramics*, ASM Materials Engineering Institute Course #56, 1990.

5.2.2 Crystal Chemical Substitutions

To review, the atomic arrangement or crystal structure that results when two or more ions bond together is largely determined by the relative size and charge of the ions. If two ions have similar size, they can substitute for one another in the same structure. This is referred to as *solid solution*. Table 5.3 identifies ions with similar sizes that frequently substitute for each other.

The degree of solid solution varies. If the ionic charge of one cation is substantially different from that of the cation in the host structure, only a small amount of substitution may occur. For example, B^{3+} with an ionic radius of 0.26 Å can only replace less than 1% of the Si^{4+} ions ($r = 0.40$ Å) in SiC. In contrast, Fe^{2+} with an ionic radius of 0.92 Å can replace all the Mg^{2+} ($r = 0.86$ Å) in Mg_2SiO_4. This is referred to as *complete* or *continuous solid solution*. Both Mg_2SiO_4 and Fe_2SiO_4 have the same structure and an orthorhombic unit cell, as do all combinations of $(Mg, Fe)_2SiO_4$ in between. However, they have different unit cell dimensions and other properties. The Fe^{2+} ion is slightly larger than the Mg^{2+} ion, so the unit cell of Fe_2SiO_4 has slightly larger dimensions:

$$Mg_2SiO_4: \quad a = 4.755\,\text{Å}, \quad b = 10.198\,\text{Å}, \quad c = 5.982\,\text{Å}$$

$$Fe_2SiO_4: \quad a = 4.820\,\text{Å}, \quad b = 10.485\,\text{Å}, \quad c = 6.093\,\text{Å}$$

Many compositions have the same structures and continuous solid solution between them. Such information is available on phase equilibrium diagrams, as is discussed in Chapter 6.

5.2.3 Derivative Structures

The crystal chemistry concepts identified so far represent only a small portion of the options that can be used to modify a crystal structure and concurrently alter the properties of the ceramic. Other important options include ordering, nonstoichiometry, stuffing, and distortion. These result in structures referred to as *derivative structures*.

5.2.3.1 Ordering

As the name implies, ordering involves positioning of the host and substitution ions in an ordered, repetitious pattern rather than in a random pattern. This results in a difference between the atom sites of the host and substitution ions, which leads to distortions in the structure or to a change in the dimensions of the unit cell. These result in a change in the behavior of the material. Ordering often results when the size of the substitute ion is significantly different from the size of the host ion. The following are examples.

Example 5.3 Mn^{2+} and Fe^{2+} have ionic radii of 0.97 and 0.92 Å such that they distribute randomly in $(Mn, Fe)CO_3$ and result in a complete solid solution between $MnCO_3$ and $FeCO_3$. $CaCO_3$ and $MgCO_3$ have the same structure as $MnCO_3$, $FeCO_3$, and $(Mn, Fe)CO_3$. However, when Ca^{2+} substitutes in $MgCO_3$, or Mg^{2+} substitutes in $CaCO_3$, an ordered structure of composition $CaMg(CO_3)_2$ results because of the difference in ionic radii of Ca^{2+} ($r = 1.14$ Å) and Mg^{2+} ($r = 0.86$ Å). The new structure has alternating layers with Ca^{2+} and Mg^{2+} ions. In the Ca layer the CaO interatomic distance is 2.390 Å and in the Mg layer the Mg—O distance is 2.095 Å.[3]

Example 5.4   The ordering in $CaMg(CO_3)_2$ is only in the octahedral lattice site and results in two distinct octahedral crystallographic positions. In some structures, ordering can occur on tetrahedral lattice sites or on both tetrahedral and octahedral sites. For example, in γ-$Li_2ZnMn_3O_8$ one Li^+ ion orders with the Zn^{2+} on tetrahedral sites while the other Li^+ ion orders with the three Mn^{2+} ions on octahedral sites.[3] In this and many other cases the ordering is induced by the charge difference between the ions.

5.2.3.2 Nonstoichiometry

The second type of derivative structure involves stoichiometry and the presence of either lattice vacancies or excess interstitial ions. *Stoichiometry* refers to the composition of a material and the positioning of the atoms within the crystallographic structure. A stoichiometric ceramic has all lattice positions filled according to the ideal structure and composition. A nonstoichiometric ceramic has a deficiency of either cations or anions accommodated by vacancies in adjacent positions of oppositely charged ions to allow for charge balance. Wüstite has the nonstoichiometric composition $Fe_{0.95}O$.[4] It contains vacancies in some of the cation positions. Similarly, Ca-doped ZrO_2 ($Zr_{1-x}Ca_xO_{2-x}$) contains oxygen anion vacancies. $Zn_{1-x}O$ is another important nonstoichiometric structure. All of these are types of defect structures and have interesting electrical properties.

5.2.3.3 Stuffing

The third type of derivative structure is the stuffed derivative. It involves substitution of a lower-valence ion for a higher-valence ion and "stuffing" of an additional ion into the structure to balance the charge. Many of the silicate compositions are stuffed derivatives of forms of SiO_2. SiO_2 consists of SiO_4 tetrahedra linked into a three-dimensional network structure by sharing of corners. The structure has relatively large open spaces between the tetrahedra. Some of the Si^{4+} ions can be replaced by Al^{3+} ions, which are of similar ionic radius. For each of the Si^{4+} replaced, the equivalent of an ion with $+1$ charge is stuffed into open spaces in the structure to obtain charge balance. Typical ions that are stuffed into the structure include Na^+, K^+, NH_4^+, Ba^{2+}, Ca^{2+}, and Sr^{2+}. In addition to SiO_4, a variety of other tetrahedral coordinations can be involved: GeO_4, GaO_4, AlO_4, ZnO_4, MgO_4, LiO_4, SO_4, BeO_4, BeF_4, FeO_4, and LiF_4. Frequently two different tetrahedra occur in a single structure. In some cases this results in ordering as in $KAlSi_3O_8$ or NH_4LiSO_4. In other cases the substitution positions are random (disordered), such as $BaMgSiO_4$.[3]

The following are additional examples of stuffed derivatives: $BaAl_2O_4$, $BaFe_2O_4$, $CaAl_2O_4$, $NaAlSi_3O_8$, $BaZnGeO_4$, $KLiBeF_4$, $PbGa_2O_4$, $CsBePO_4$, and $BaSrFe_4O_8$.

5.2.3.4 Distortion

The final derivative structure involves distortion of the original structure. This typically results when ions are substituted that have a significant difference in ionic radius or valence compared to the host ions. The sizes of the tetrahedral or octahedral structural units for the host ion and the substituted ion are different and a more-complex, less-symmetrical structure results. Since separate polyhedral structural units (coordination polyhedron) are bonded together to form the overall structure, distortions frequently occur in each polyhedron; that is, cation–anion interatomic distances are different for the different ions in the polyhedron. Ordering often accompanies the distortions.

5.3 METALLIC AND CERAMIC CRYSTAL STRUCTURES

We now have enough background information on crystal chemistry and on crystal structure notation to begin our review of crystal structures. We will start with the simple metallic structures and then proceed to the ceramic structures. Simple metallic structures are based on single-sized atoms in a close-packed arrangement. Many of the ceramic structures consist of close-packed arrangements of anions with one or more types of cations positioned in octahedral or tetrahedral sites. These structures tend to be dominated by ionic bonding. Other ceramic structures consist of isolated tetrahedra and octahedra that are bonded together by sharing of corners or edges. These are not close packed and have a higher degree of directional covalent characteristics.

5.3.1 METALLIC CRYSTAL STRUCTURES

As we discussed earlier, metallic bonding involves atoms of a single element immersed in a cloud of free electrons. This bonding forces the atoms into a uniform three-dimensional array with each atom exposed to identical surroundings. Most pure metals have a structure that is face-centered cubic (FCC), body-centered cubic (BCC), or hexagonal close-packed (HCP).

The *FCC* structure is common among metals (copper, nickel, aluminum, lead, silver). As shown in Figure 5.6, it consists of atoms at each corner of a cube and at the center of each cube face. Each unit cell contains four atoms. Each atom is surrounded by 12 identical atoms and thus has a CN of 12. A CN of 12 is the tightest packing possible for atoms all of a single size and results in a close-packed structure with a packing factor of 0.74. The packing factor (PF) is determined by using the hard-ball model of a unit cell in Figure 5.6a. The PF equals the volume of the balls divided by the volume of the total unit cell. As we know from prior discussions, an atom is not a hard ball but mostly open space. However, the electrons orbiting around the nucleus do form a sphere of influence that for the purposes of crystal structure discussions can be approximated by a solid sphere. Thus, another name for FCC is cubic close packed.

The BCC structure is less common than FCC, but is extremely important commercially because it is the structure of iron (Fe) at room temperature. Chromium (Cr), vanadium (V), tantalum (Ta), and tungsten (W) also have the BCC structure. The BCC structure consists of eight atoms at the corners of a cube and a single atom at the center of the cube. This is illustrated in Figure 5.7. Each atom is in an equivalent geometric position and is surrounded by eight identical atoms. It thus has a CN of 8. The stacking of atoms is not quite as tight as for FCC, so the structure is not close packed. It has a PF of 0.68.

The third major crystal structure for pure metals is HCP. Magnesium (Mg), zinc (Zn), cobalt (Co), zirconium (Zr), titanium (Ti), and beryllium (Be) have the HCP structure. The HCP structure is illustrated in Figure 5.8. Each atom is surrounded by 12 identical atoms and the CN is 12. The packing factor is 0.74, the same as FCC. In both HCP and FCC, each atom touches six atoms in its own close-packed plane, three in the preceding close-packed plane, and three in the following close-packed plane. The difference between the two structures is the order of stacking of these planes.

$$PF = \frac{4(4\pi r^3/3)}{a^3} = \frac{16\pi r^3(2\sqrt{2})}{(3)(64r^3)} = 0.74$$

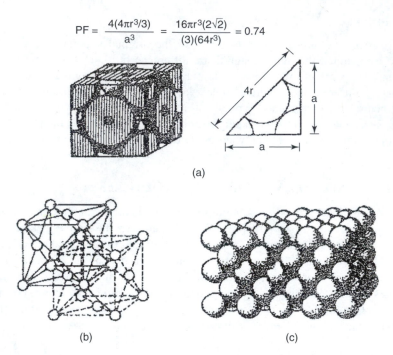

(a)

(b)   (c)

FIGURE 5.6 Representations of the FCC structure. (a) Hard-ball model showing that each unit cell contains a total of four atoms in the closest possible packing (each atom with 12 adjacent atoms), (b) schematic showing the location of atom centers and that each atom is in an equivalent geometric position, and (c) hard-ball model showing the repeating three-dimensional structure. (From Van Vlack, L.H., *Elements of Materials Science*, 2nd ed., Addison-Wesley, Reading, MA, 1964, pp. 59–60. With permission.)

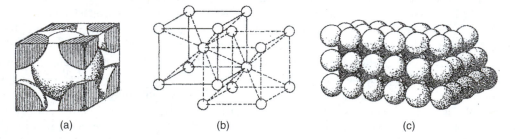

(a)   (b)   (c)

FIGURE 5.7 Representations of the BCC structure. (a) Hard-ball model showing that each unit cell contains a total of two atoms and that the packing is slightly less than for FCC (PF of 0.68 for BCC compared to 0.74 for FCC), (b) schematic showing the location of the atom centers and that each atom is in an equivalent geometric position, and (c) hard-ball model showing the repeating three-dimensional structure. (From Van Vlack, L.H., *Elements of Materials Science*, 2nd ed., Addison-Wesley, Reading, MA, 1964, pp. 59–60. With permission.)

This is illustrated in Figure 5.9. The atoms in one close-packed plane are identified as b. Note that the atoms in the plane above and the plane below can position on either site a or c. For the HCP structure, both planes adjacent to the b plane are positioned on the same site so that the sequence is ababab. The planes alternate positions for FCC, so the sequence is abcabc.

5.3.2 CERAMIC STRUCTURES WITH A SINGLE ELEMENT

Ceramic structures with a single element are rare. The major one is the [diamond] structure. The structure consists of C atoms with each C covalently sharing one electron with each of four

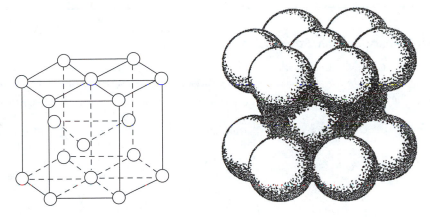

FIGURE 5.8   HCP structure. (From Van Vlack, L.H., *Elements of Materials Science*, 2nd ed., Addison-Wesley, Reading, MA, 1964, p. 62. With permission.)

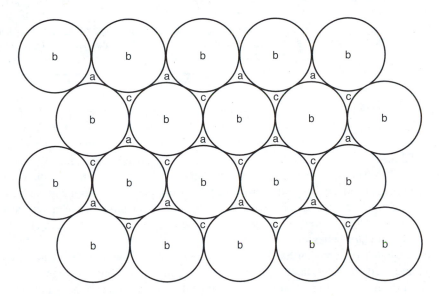

FIGURE 5.9   Comparison of the atom stacking sequence for FCC and HCP structures in a pure metal. Positions b are for a close-packed plane of atoms. Positions a and c are possible sites for atoms in adjacent close-packed planes. For FCC, atoms on one side stack on a and on the other side on c, giving the sequence abcabc. For HCP, atoms above and below both stack on the a position, giving the sequence ababab.

surrounding C atoms. Thus, each C atom is either at the center of a tetrahedron or at the corner of a tetrahedron. The lattice positions are equivalent. The CN is 4 and the coordination formula is $C^{[4]}$. Adjacent C atoms are bonded together by very strong covalent forces. This strong bonding results in a high elastic modulus, the highest hardness of any naturally occurring material, and extremely high temperature stability (over 3700°C [6700°F] in a nonoxidizing atmosphere).

Diamond is only one of the possible structures of carbon. Another alternative is graphite, which consists of the carbon atoms arranged in sheets. The bonding within the sheets is very strong, but the sheets are held together only by weak van der Waals forces. Due to this structure difference, graphite has dramatically different properties than diamond.

A third variant (polymorph) discovered in 1985 is more like an organic structure. It involves 60 carbon atoms arranged into alternating six-member and five-member rings to form a molecule shaped

TABLE 5.4
Summary of Binary Ceramic Structures

Structure Name	General Formula	Coordination Formula	Anion Packing	Fraction Cation Sites Occupied	Examples
[Rock salt]	AX	$A^{[6]}X^{[6]}$	Cubic close packed	All octahedral	NaCl, KCl, LiF, MgO, VO, NiO
[Cesium chloride]	AX	$A^{[8]}X^{[8]}$	Simple cubic	All cubic	CsCl, CsBr, CsI
[Zinc blende]	AX	$A^{[4]}X^{[4]}$	Cubic close packed	1/2 Tetrahedral	ZnS, BeO, β-SiC
[Würtzite]	AX	$A^{[4]}X^{[4]}$	Hexagonal close packed	1/2 Tetrahedral	ZnS, ZnO, α-SiC, BeO, CdS
[Nickel arsenide]	AX	$A^{[6]}X^{[6]}$	Hexagonal close packed	All octahedral	NiAs, FeS, FeSe, CoSe
[Fluorite]	AX_2	$A^{[8]}X_2^{[4]}$	Simple cubic	1/2 Cubic	CaF_2, ThO_2, CeO_2, UO_2, ZrO_2, HfO_2
[Rutile]	AX_2	$A^{[6]}X_2^{[3]}$	Distorted close packed	1/2 Octahedral	TiO_2, GeO_2, SnO_2, PbO_2, VO_2, NbO_2
[Silica types]	AX_2	$A^{[4]}X_2^{[2]}$	Connected tetrahedra	–	SiO_2, GeO_2
[Antifluorite]	A_2X	$A_2^{[4]}X^{[8]}$	Cubic close packed	All tetrahedral	Li_2O, Na_2O, sulfides
[Corundum]	A_2X_3	$A_2^{[6]}X_3^{[4]}$	Hexagonal close packed	2/3 Octahedral	Al_2O_3, Fe_2O_3, Cr_2O_3, V_2O_3, Ga_2O_3, Rh_2O_3

like a soccer ball. This has been named buckminsterfullerene (also referred to as buckyball) after R. Buckminster Fuller who invented the geodesic dome that a buckyball closely resembles. Buckyballs tend to arrange into a weakly bonded FCC array with a C_{60} molecule at each corner and cube face and a unit-cell length of 1.41 nm. This material is relatively weak and soft and is an electrical insulator. However, ions such as K^{1+} can be substituted into interstitial positions to produce materials with unusual and unprecedented properties. For example, K_3C_{60} has many characteristics of a metal including high electrical conductivity. It also becomes superconductive at very low temperature (below 18°K). Atoms can also be inserted into the center of the C_{60} molecules and result in interesting semiconductor behavior.

A fourth carbon structure that has been discovered more recently is the "carbon nanotube." Imagine a single graphite sheet that has rolled so the carbon atoms along the edges bond to form a tube. Now imagine half a buckyball on each end of the tube. Nanotubes and the whole field of nanomaterials are exciting areas for future research and development.

5.3.3 BINARY CERAMIC STRUCTURES

Binary refers to a structure with two distinct atom sites, one typically for an anion and one for a cation. As discussed previously under crystal chemistry concepts, a variety of elements can substitute in solid solution on these sites without a change in structure. Thus the term binary identifies the number of sites rather than chemical elements. Table 5.4 summarizes important binary structures and some of their characteristics.

5.3.3.1 [Rock Salt] Structure $A^{[6]}X^{[6]}$

The [rock salt] structure is named for the mineral NaCl. It is also referred to as the [NaCl] structure. The arrangement of ions is illustrated in Figure 5.10. The structure is cubic with the anions arranged in cubic close packing with all the interstitial octahedral sites occupied by the cations. As is easily seen in Figure 5.10, the structure consists of alternating cations and anions along each of the three unit-cell axes ([100], [010], and [001]) crystal directions. KCl, LiF, KBr, MgO, CaO, SrO,

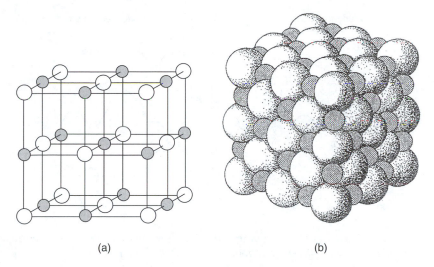

(a)   (b)

FIGURE 5.10 [Rock salt] crystal structure. (From Kingery, W.D., Bowen, H.K., and Unlmann, D.K., *Introduction to Ceramics*, 2nd ed., Wiley, New York, 1976, p. 42. With permission.)

BaO, CdO, VO, MnO, FeO, CoO, NiO, and the alkaline earth sulfides all have the [rock salt] structure. The atomic bonding is largely ionic, especially for monovalent ion compositions.

5.3.3.2   [Nickel Arsenide] Structure $A^{[6]}X^{[6]}$

The [nickel arsenide] structure involves the same size range of cations as the [NaCl] structure, except that the anions are in an HCP stacking arrangement rather than an FCC arrangement. Both the anions and the cations are in sixfold coordination, so the general coordination formula is $A^{[6]}X^{[6]}$, NiAs, FeS, FeSe, and CoSe have the [NiAs] structure.

5.3.3.3   [Cesium Chloride] Structure $A^{[8]}X^{[8]}$

The [CsCl] structure involves cations that are too large to fit into the octahedral interstitial site. They fit into the larger cavity at the center of a simple cube with the anions at each corner. This is similar to the BCC structure of a metal. Compositions with the [CsCl] structure include CsCl, CsBr, and CsI.

5.3.3.4   [Zinc Blende] and [Wurtzite] Structures $A^{[4]}X^{[4]}$

Cations too small to be stable in octahedral sites fit into the smaller tetrahedral interstitial position and form either the [zinc blende] structure or the [wurtzite] structure. The anions in the [zinc blende] structure are in an FCC arrangement. Those in the [wurtzite] structure are HCP. Note in Table 5.4 that some substances (ZnS and SiC) are listed as having both structures. These are referred to as polymorphic forms. The cubic structure for SiC and ZnS is stable at low temperatures and the hexagonal structure is stable at high temperatures.

The [zinc blende] and [wurtzite] structures are illustrated schematically in Figure 5.11. Note that the [zinc blende] structure is similar to that of diamond, with the cations and anions alternating in the C atom positions. The bonding is largely ionic for BeO. The degree of covalent bonding is high for SiC, and SiC has comparatively higher strength and hardness than BeO.

5.3.3.5   [Fluorite] Structure $A^{[8]}X_2^{[4]}$

The [fluorite] structure is named after the mineral fluorite, CaF_2. The [fluorite] structure is similar to the [CsCl] structure. The cations are large and fit into the large body-centered position formed

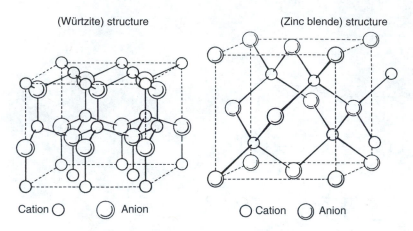

FIGURE 5.11   [Wurtzite] and [zinc blende] structures. (From Kingery, W.D., Bowen, H.K., and Unlmann, D.K., *Introduction to Ceramics*, 2nd ed., Wiley, New York, 1976, p. 63. With permission.)

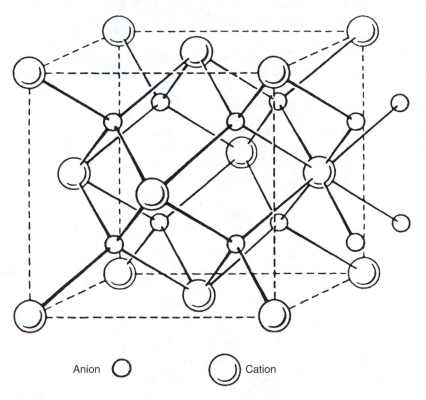

FIGURE 5.12   [Fluorite] structure. (From Kingery, W.D., Bowen, H.K., and Unlmann, D.K., *Introduction to Ceramics*, 2nd ed., Wiley, New York, 1976, p. 68. With permission.)

by the simple cubic arrangement of the anions. Because the cations have twice the charge of the anions, only half of the cation sites are occupied. This results in a structure with relatively large holes. The [fluorite] structure is illustrated in Figure 5.12. The unit cell is based on a FCC packing of the cations.

CaF_2, ThO_2, UO_2, CeO_2, and high-temperature forms of ZrO_2 and HfO_2 have the [fluorite] structure. The room-temperature forms of ZrO_2 and HfO_2 are of a distorted [fluorite] structure with monoclinic symmetry. The crystal structure of ZrO_2 at different temperatures is of great interest currently. Techniques have been developed to use the change from one structure to another to produce a high-toughness ceramic. The mechanism is similar to the martensitic toughening of steel and is discussed in later chapters.

UO_2 also has the [fluorite] structure and is used as a nuclear fuel. Nuclear fission results in fission products that take up a larger volume than the starting material. The large structural holes in UO_2 accommodate much of these fission products and minimize external volume changes.

5.3.3.6 [Antifluorite] Structure $A^{[4]}X^{[8]}$

Li^+, Na^+, K^+, and Rb^+ combine with O^{2-} or S^{2-} to form A_2X compositions with the anions and cations in the reverse lattice positions of the fluorite structure; that is, the anions are arranged in a cubic close-packed structure (FCC) with the cations filling all the tetrahedral interstitial sites. This is called the [antifluorite] structure.

5.3.3.7 [Rutile] Structure $A^{[6]}X^{[3]}$

The [rutile] structure involves medium-sized cations with a charge of 4+. Compositions with the [rutile] structure include TiO_2, SnO_2, GeO_2, PbO_2, VO_2, NbO_2, TeO_2, MnO_2, RuO_2, OsO_2, and IrO_2. The cation has a CN of 6. The CN of the anion equals $|V_A|(CN)_C/|V_C| = (2)(6)/4 = 3$. A CN of 3 cannot be achieved in the cubic or hexagonal close-packed arrangements. Instead, what results is a distorted close-packed structure. The cations fill only half the octahedral positions.

Figure 5.13 shows two representations of the [rutile] structure using the tetragonal form of TiO_2 as the example. Strings of edge-shared distorted octahedra (two Ti—O spacings at 1.988 Å and four at 1.944 Å) extend in the C-direction of the structure. These share corners with adjacent strings of octahedra to form a three-dimensional framework structure.

5.3.3.8 The Silica Structures $A^{[4]}X_2^{[2]}$

The silica structures involve small cations with a charge of 4+. SiO_2 is the model system. The radius ratio is approximately 0.33, which indicates that tetrahedral coordination is stable for the Si^{4+} cation. The coordination of the anion is determined by:

$$(CN)_A = |V_A|(CN)_C/V_C = (2)(4)/4 = 2$$

To accommodate this, the Si^{4+} is at the center of a tetrahedron of O^{2-} anions. Each O^{2-} at the corner of each tetrahedron is shared with an adjacent tetrahedron. This results in a directional structure that is not close packed and that has a combination of ionic and covalent character.

SiO_2 has a wide variety of structures at various temperatures and pressures. The high-temperature polymorphs consist of different arrangements of undistorted SiO_4 tetrahedra linked together by a sharing of corners. The lower-temperature polymorphs have similar structures but are distorted. Figure 5.14 compares unit cells of the high-temperature cristobalite and tridymite polymorphs of SiO_2. A third polymorph of SiO_2 is quartz, which constitutes a significant percentage of the earth's crust. A schematic illustrating the uniformity in the high-temperature form of quartz and the distortion in the low-temperature form is shown in Figure 5.15.

The SiO_4 tetrahedron is the building block of an extensive variety of derivative structures. Many of these structures are discussed later in this chapter under the section on ternary structures. However, at this time it is important that we visualize the relationship of these more complex structures to the silica structures. Figure 5.16 shows some of the ways that SiO_4 tetrahedra can be linked to form derivative structures. Note that in SiO_4^{4-}, the tetrahedra are independent and have four bond positions available to link up with cations or other coordination polyhedra. Note also that $Si_2O_7^{6-}$ involves the sharing of one corner and that $Si_3O_9^{6-}$, $Si_3O_{18}^{12-}$, $(SiO_3)_n^{2n-}$ and $(Si_4O_{11})_n^{6n-}$ involve the

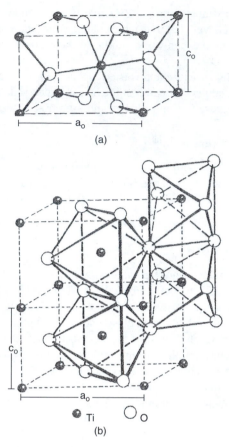

FIGURE 5.13 Illustrations of the [rutile] structure. (a) The tetragonal unit cell of TiO_2 showing the two different Ti—O bond lengths, and (b) the edge and corner sharing of octahedra to produce a three-dimensional structure. (From Muller, O. and Roy, R., *The Major Ternary Structural Families*, Springer-Verlag, Berlin, 1974, p. 112. With permission.)

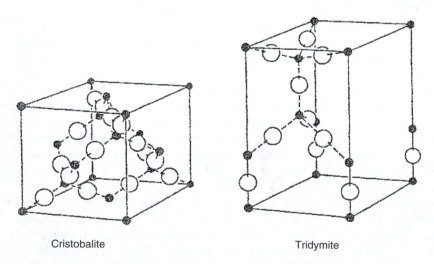

FIGURE 5.14 Comparison of the high-temperature structures of the cristobalite and tridymite structures of SiO_2. (From Muller, O. and Roy, R., *The Major Ternary Structural Families*, Springer-Verlag, Berlin, 1974, p. 218. With permission.)

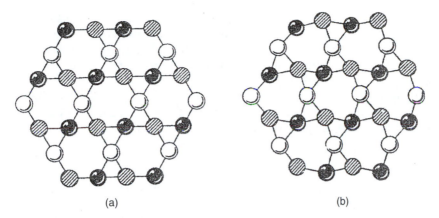

FIGURE 5.15 Schematic comparing the structure of quartz at (a) high temperature, and (b) low temperature. (From Muller, O. and Roy, R., *The Major Ternary Structural Families*, Springer-Verlag, Berlin, 1974, p. 74. With permission.)

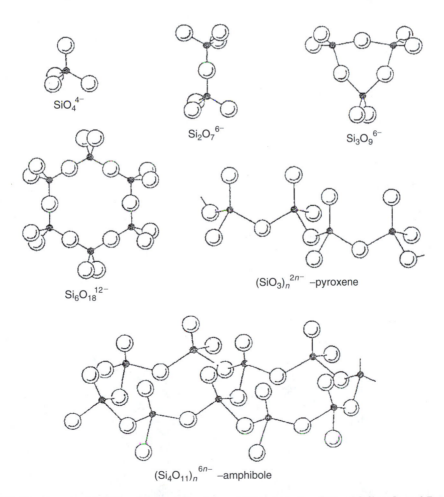

FIGURE 5.16 Structural building blocks derived from SiO_4 tetrahedra. (From Muller, O. and Roy, R., *The Major Ternary Structural Families*, Springer-Verlag, Berlin, 1974, p. 71. With permission.)

sharing of two corners to yield ring or chain structures. Sheet structures are also possible by linking of $(Si_2O_5)_n$ layers of SiO_4 tetrahedra to $AlO(OH)_2$ octahedral layers. This occurs, for example, for the important clay mineral kaolinite, $Al_2(Si_2O_5)(OH)_4$. Finally, additional three-dimensional framework structures are possible by substitution of Al^{3+} or other small cations for Si^{4+} and stuffing with other cations to achieve charge neutrality.

It should now be apparent to the reader that a variety of derivative structures and hundreds of compositions, each with different properties, are possible; the important point is that each of these is based upon relatively few simple baseline structures.

5.3.3.9 [Corundum] Structure $A_2^{[6]}X_3^{[4]}$

A final binary structure of major importance is the [corundum] structure, $A_2^{[6]}X_3^{[4]}$. Aluminum oxide (Al_2O_3) is the most important material with this structure. Others are Fe_2O_3, Cr_2O_3, Ti_2O_3, V_2O_3, Ga_2O_3, and Rh_2O_3. The O^{2-} anions are arranged in nearly hexagonal close packing. The cations fill two thirds of the octahedral interstitial sites. The CN of the cations is 6 and of the anions 4. To achieve a uniform distribution of cations and anions, each Al—O octahedron shares one face and three edges with adjacent octahedra.

5.3.4 TERNARY CERAMIC STRUCTURES

We are now ready to move on to a discussion of structures containing more than one cation position and more than two elements. We will concentrate on ternary structures. A ternary structure generally consists of a close-packed or nearly close-packed arrangement of ions with two different sizes or charges of cations fitting into the appropriate interstitial positions. The guidelines are exactly the same as those we just reviewed for the binary structures. Many commercially important ceramic materials with ternary structures have compositions with more than three elements where more than one element of comparable size occupies a single type of structural position. Thus, as we explore the ternary structures, we shall emphasize the structural positions rather than the chemical composition. The ternary structures are extremely important to advanced ceramics technology. Many of the compositions of materials used in advanced dielectric, magnetic, refractory, structural, and optical applications have ternary structures. In addition, most of the mantle of the earth is made up of ceramic materials with ternary structures.

Some of the ternary structures are identified in Table 5.5. We will review only a few of the ternary structures in this book. These include the A_2BX_4 structures ([spinel], [phenacite] and [olivine]), and ABX_4 structures ([zircon], ordered SiO_2 derivatives, and ternary derivatives of the [rutile] and

TABLE 5.5
Summary of Some Ternary Structures

Structure Name	General Name	Coordination Formula	Examples
[Spinel]	AB_2X_4	$A^{[4]}B_2^{[6]}X_4^{[4]}$	$FeAl_2O_4$, $ZnAl_2O_4$, $MgAl_2O_4$
[Inverse spinel]	AB_2X_4	$B^{[4]}A^{[6]}B^{[6]}X^{[4]}$	$FeMgFeO_4$, Fe_3O_4, $MgTiMgO_4$
[Phenacite]	A_2BX_4	$A_2^{[4]}B^{[4]}X^{[3]}$	Be_2SiO_4, Zn_2SiO_4, β-Si_3N_4, Li_2MoO_4
[β-K_2SO_4]	A_2BX_4	$A^{[10]}A^{[9]}B^{[4]}X_4^{[6]}X^{[5]}$	Rb_2So_4, K_2WS_4, Ba_2TiS_4, $NaYSiO_4$
[Olivine]	A_2BX_4	$A_2^{[6]}B^{[4]}X_4^{[4]}$	Mg_2SiO_4, Fe_2SiO_4, Al_2BeO_4, Mg_2SnSe_4
[Barite]	ABX_4	$A^{[12]}B^{[4]}X^{[4]}$	$BaSO_4$, $KMnO_4$, $CsBeF_4$, $PbCrO_4$, $BaFeO_4$
[Zircon]	ABX_4	$A^{[8]}B^{[4]}X_4^{[3]}$	$ZrSiO_4$, YVO_4, $TaBO_4$, $CaBeF_4$, $BiVO_4$
[Ordered SiO_2]	ABX_4	$A^{[4]}B^{[4]}X_4^{[2]}$	$AlPO_4$, $AlAsO_4$, $FePO_4$
[Calcite]	ABX_3	$A^{[6]}B^{[3]}X_3^{[3]}$	$CaCO_3$, $MgCO_3$, $FeCO_3$, $MnCO_3$
[Ilmenite]	ABX_3	$A^{[6]}B^{[6]}X_3^{[4]}$	$FeTiO_3$, $NiTO_3$, $CoTiO_3$
[Perovskite]	ABX_3	$A^{[12]}B^{[6]}X_3^{[6]}$	$BaTiO_3$, $CaTiO_3$, $SrTiO_3$, $SrZrO_3$, $SrSnO_3$, $SrHfO_3$

[fluorite] binary structures), the ABX_3 structures ([perovskites] in particular), and a brief listing of some of the other ternary structures. The objective is to select examples illustrating the factors that control atom arrangements and influence the properties of specific compositions we discuss in later chapters.

5.3.4.1 A_2BX_4 Structures

5.3.4.1.1 [Spinel] Structures $A^{[4]} B_2^{[6]} X_4^{[4]}$ or $B^{[4]}A^{[6]}B^{[6]}X_4^{[4]}$

The [spinel] structures are cubic with a unit cell containing 32 oxygen ions, 16 octahedral cations, and 8 tetrahedral cations. In the normal [spinel], the A^{2+} occupy one eighth of the tetrahedral sites and the B^{3+} one half of the octahedral sites. In the [inverse spinel] structure, the A^{2+} cations and half the B^{3+} cations occupy octahedral sites while the remaining B^{3+} cations are on tetrahedral sites. Many of the important ceramic materials with magnetic properties have the [inverse spinel] structure. The [spinel] structure is illustrated in Figure 5.17.

A wide variety of ion sizes and charges can fit into the [spinel] structure. In [spinels] where the anion is O^{2+}, the tetrahedral cation can be as large as Cd^{2+} (0.94 Å) or as small as Si^{4+} (0.40 Å) and the octahedral cation can be as large as Ag^+ (1.29 Å) or as small as Ge^{4+} (0.68 Å). This allows for a wide range of compositions. Charge combinations for oxide spinels include $A_2^{2+}B^{4+}O_4$, $A_2^+ B^{6+}O_4$, $A_2^{3+}B^{2+}O_4$ plus others. [Spinels] also exist with other anions (e.g., F^-, S^{2-}, Se^{2-} and Te^{2-}). Tellurium has even been found as a cation in the [spinel] structure. Table 5.6 lists some of the many [spinel] compositions.

5.3.4.1.2 [Phenacite] Structure $A_2^{[4]}B^{[4]}X_4^{[3]}$

The [phenacite] structure is named after the naturally occurring mineral phenacite, Be_2SiO_4. Both the Be^{3+} and Si^{4+} ions are small and fit into fourfold coordination with an oxygen ion at each corner of a tetrahedron. The tetrahedra are linked together into a three-dimensional network structure by sharing of each corner. The resulting structure is not close packed and has a significant degree of directional covalent character. The unit cell has a rhombohedral symmetry with a cylindrical

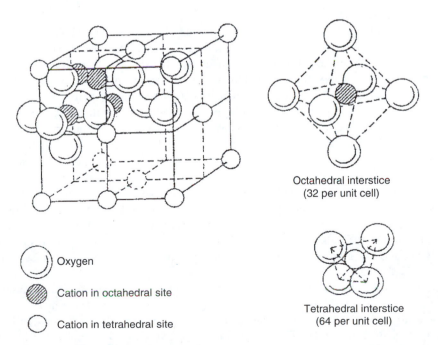

Octahedral interstice
(32 per unit cell)

○) Oxygen

▨ Cation in octahedral site

○ Cation in tetrahedral site

Tetrahedral interstice
(64 per unit cell)

FIGURE 5.17   Relative atom positions in the normal [spinel] structure. (From von Hippel, A.R., *Dielectrics and Waves*, Wiley, New York, 1954. With permission.)

TABLE 5.6

Examples of Compositions with the [Spinel] or Closely Related Structure

$MgAl_2O_4$	$MnAl_2O_4$	$ZnAl_2O_4$	$FeAl_2O_4$	Fe_3O_4	$CoFe_2O_4$
$MnFe_2O_4$	$MgFe_2O_4$	Mo_2GeO_4	Fe_2GeO_4	Zn_2GeO_4	Ni_2GeO_4
Fe_2VO_4	Zn_2VO_4	Co_2TiO_4	Mg_2TiO_4	Ag_2MoO_4	Na_2WO_4
Zn_2SnO_4	Li_2MoO_4	$MgCr_2O_4$	$LiAl_5O_8$	$Li_4Ti_5O_{12}$	$Zn_7Sb_2O_{12}$
$LiMgVO_4$	$LiCrGeO_4$	$ZnTiO_3$	$Li_2CoTi_3O_8$	Li_2NiF_4	$Fe_3O_{3.5}F_{0.5}$
Cu_2FeO_3F	Al_3O_3N	$ZnCr_2S_4$	$MnAl_2S_4$	Y_2MgSe_4	$CuCr_2Te_4$

channel approximately 2 Å in diameter aligned parallel to the c axis. Some compositions with the phenacite structure include Zn_2SiO_4, Li_2MoO_4, Li_2SeO_4, Zn_2GeO_4, and Li_2BeF_4.

Be_2SiO_4 doped with Mn^{2+} ions was one of the early phosphor materials used for home lighting. Zn_2SiO_4 (willenite) doped with Mn^{2+} ions also has strong phosphorescence and was once widely used as a cathodoluminescent material.

5.3.4.1.3 [β-Silicon Nitride] Structure $A_2^{[4]}B^{[4]}X_4^{[3]}$

The [β-Si_3N_4] structure is essentially the same as the [phenacite] structure, except the unit cell is more compact and of a hexagonal symmetry. The cell dimensions of β-Si_3N_4 are a = 7.607 Å and c = 2.911 Å compared to Be_2SiO_4 with a = 12.472 Å and c = 8.252 Å. This is because all the Si^{4+} cations are in equivalent crystallographic positions, whereas there are three distinct positions in Be_2SiO_4 (two for the Be^{2+} and one for the Si^{4+}). The β-Si_3N_4 has strong covalent bonding.

β-Si_3N_4 is an important material for advanced structural applications such as bearings, heat engine components, and metal-cutting tools. It has been determined that considerable Al^{3+} and O^{2-} can substitute into β-Si_3N_4 (via solid solution of Al_3O_3N) and still retain the [β-Si_3N_4] structure. This has led to a series of compositions called *sialons* with a wide range of properties.

5.3.4.1.4 [Olivine] Structure $A_2^{[6]}B^{[4]}X_4^{[4]}$

The [olivine] structure is named after the mineral olivine, $(Mg, Fe)_2SiO_4$, which is a solid solution between the minerals forsterite (Mg_2SiO_4) and fayalite (Fe_2SiO_4). The structure consists of a slightly distorted, hexagonal, close-packed anion arrangement with the smaller "B" cations positioned in one eighth of the tetrahedral interstitial sites and the larger "A" cations in half of the octahedral sites.

Fe_2SiO_4 has been carefully studied and can be used as an example to help visualize the [olivine] structure. Independent SiO_4^{4-} tetrahedra share corners and edges with FeO_6^{10-} octahedra. This results in distortions of both polyhedra such that the Fe^{2+} have two distinct positions within the structure. One position has two oxygen ions at an interatomic distance of 2.122 Å, two at 2.127 Å, and two at 2.226 Å. The other position has two at 2.088 Å, one at 2.126 Å, one at 2.236 Å, and two at 2.289 Å. The tetrahedra have Si—O spacings of one at 1.634 Å, two at 1.630 Å, and one at 1.649 Å. The arrangement of the atoms in the [olivine] structure is illustrated in Figure 5.18. Many ternary structures are distorted. Such distortions are necessary to accommodate the varieties of ion sizes and charges involved. Each size and charge combination results in a slightly different degree or type of distortion. Further modifications result when additional ions are substituted by solid solution. These distortions and modifications of the crystal structure result in modifications in the behavior of the material.

A variety of ionic charge combinations fit into the [olivine] structure. Oxide examples include $A_2^{2+}B^{4+}O_4$, $A_2^{3+}B^{2+}O_4$, $A^{2+}A^{3+}B^{3+}O_4$, $A^+A^{2+}B^{5+}O_4$, and $A^+A^{3+}B^{4+}O_4$. There are also a few with F, S, or Se as the anions. A^{2+} positions can be occupied by Ca, Mn, Mg, Fe, Co, Ni, or Cd cations; A^+ by Li or Na; A^{3+} by Al, Y, Ho, Er, Tm, Yb, Lu; B^{4+} by Si, Ge or Sn; B^{2+} by Be; B^{3+} by B; B^{5+} by V; X^- by F; X^{2-} by O, S or Se. Several of the more unusual examples include Al_2BeO_4, $LiMgVO_4$, γ-Na_2BeF_4, Mn_2SiS_4, and Mg_2SnSe_4.

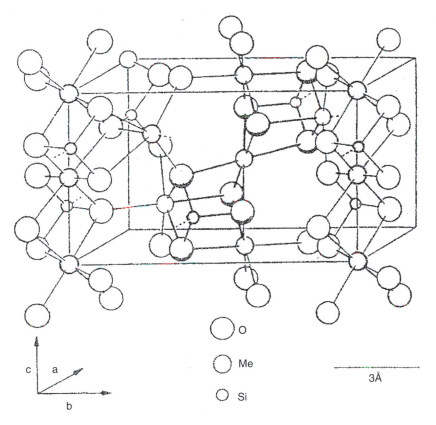

FIGURE 5.18 The [olivine] structure showing the edge sharing of distorted octahedra. Me represents the variety of metal ions that can fit in the structure. (From Muller, O. and Roy, R., *The Major Ternary Structural Families*, Springer-Verlag, Berlin, 1974, p. 36. With permission.)

Olivine and [olivine] structure compositions, along with spinels, constitute a major portion of the earth's mantle. They are also contained in many refractories used in steel plants.

5.3.4.2 ABX₄ Structures

The materials with ABX_4 structures are less important to advanced ceramic technology than some of the A_2BX_4 materials we have discussed. Therefore, we provide only the brief summary in Table 5.7 of selected structures and compositions.

The major importance of the ABX_4 compositions is as ores for Ba, W, Zr, Th, Y, and the rare earths (lanthanide series: La, Ce, Pr, Nd, Sm, Eu, Gd, Tb, Dy, Ho, Er, Tm, Yb, and Lu). Barite $(BaSO_4)$ is a major source of barium, monazite $(LnPO_4)$* for the rare-earth elements, zircon $(ZrSiO_4)$ for the metal zirconium and the ceramic ZrO_2, scheelite $(CaWO_4)$ and wolframite $(Fe_{0.5}Mn_{0.5}WO_4)$ for tungsten metal, and thorite $(ThSiO_4)$ for thorium. Zircon is also used for some ceramic-technology applications because it has a low coefficient of thermal expansion. In addition, some [zircon] and [scheelite] compositions are fluorescent. The [zircon] composition YVO_4 doped with Eu is a red phosphor once used for color television, and the [scheelite] composition, $CaWO_4$, doped with Nd, is a laser host.

---

* Ln refers to a variety of substitutions of the lanthanide series rare-earth elements.

TABLE 5.7

Summary of Some ABX$_4$ Structures

Structure Name	Simplified Coordination Formula	Special Structural Features	Compositional Occurrence
[Barite]	Ba$^{[12]}$S$^{[4]}$O$_4$	Isolated SO$_4$ tetrahedra	Alkali permanganates, perchlorates and fluoroborates with large alkali ions (K$^+$ or bigger). Divalent sulfates with large cations
[Scheelite]	Ca$^{[8]}$W$^{[4]}$O$_4$	Isolated WO$_4$ tetrahedra	Zr, Hf and actinide germinates. Ba, Sr, Ca, Pb tungstates, and molybdates. Alkali periodates, perrhenates and pertechnetates. LnLiF$_4$ where Ln = lanthanide or Y
[Zircon]	Zr$^{[8]}$Si$^{[4]}$O$_4$	Isolated SO$_4$ tetrahedra; less dense than [scheelite]	Zr, Hf, and actinide silicates. Lanthanide phosphates, arsenates, chromates V, and vanadates
Ordered SiO$_2$ structures (several structures)	Al$^{[4]}$P$^{[4]}$O$_4$	Three-dimensional linking of AlO$_4$ and PO$_4$ tetrahedra; ordered variants of the [quartz], [tridymite] [cristobalite] structures	Phosphates and arsenates of Al, Ga, Fe, B; also: BeSO$_4$
[Rutile]	(Cr$_{0.5}$Nb$_{0.5}$)$^{[6]}$O$_2$	Edge-shared strings of randomly mixed CrO$_6$ and NbO$_6$ octahedra, held together in three dimensions by corner sharing	Antimonates and tantalates of Al, Ga, Rh, Cr, Fe. Niobates of Rh, Cr, Fe
[Fluorite]	(Nd$_{0.5}$U$_{0.5}$)$^{[8]}$O$_4$	Disordered NdO$_8$ and UO$_8$ cubes	LnPaO$_4$, LnUO$_4$ LnNpO$_4$ and NaLnF$_4$ where Ln = lanthanide ion

Source: From Muller, O. and Roy, R., *The Major Ternary Structural Families*, Springer-Verlag, Berlin, 1974.

5.3.4.3 ABX$_3$ Structures

A variety of materials of extreme importance to modern technology have ABX$_3$ structures. These are summarized in Table 5.8. The most important are the [perovskite] structure compositions with ferroelectric properties and with high dielectric constant. Compositions such as BaTiO$_3$ and PbZr$_{0.65}$Ti$_{0.35}$O$_3$ are used for capacitors, ferroelectrics, and piezoelectric transducers. Compositions have been altered by crystal chemical substitutions to provide a wide range of properties optimized for specific applications. Other [perovskite] structure compositions of importance include solid solutions between KTaO$_3$ and KNbO$_3$, which are used as electro-optic modulators for lasers. Other laser modulator materials are LiNbO$_3$ and LiTaO$_3$, which have [ilmenite]-related structures. These high-temperature ferroelectric materials are also used as piezoelectric substrates, as optical waveguides, and as a holographic-storage medium.

Other important ABX$_3$ compositions include CaCO$_3$ of the [calcite] structure and the rare earth LnFeO$_3$ ferrites. Transparent single-crystal calcite is used for Nicol prisms in the polarizing microscope. Compositions in the LnFeO$_3$ family are used for magnetic bubble domain devices.

5.3.4.3.1 [Calcite] Structure A$^{[6]}$B$^{[3]}$X$_3^{[3]}$

The [calcite] structure involves a large cation such as Ca, Mg, Fe, or Mn in the "A" position and a very small cation limited to C^{4+}, B^{3+}, or N^{5+} in the "B" position. The CaCO$_3$ [calcite] composition

TABLE 5.8
Summary of Some ABX$_3$ Structures

Structure Name	Simplified Coordination Formula	Special Structural Features	Compositional Occurrence
[Calcite]	$Ca^{[6]}C^{[3]}O_3$	Highly anisotropic, high birefringence	Trivalent borates, divalent carbonates and alkali nitrates
[Aragonite]	$Ca^{[9]}C^{[3]}O_3$	Denser packing of CO_3^{2-} and Ca^{2+} than in [calcite]	Divalent carbonates and lanthanide borates with larger cations
[Ilmenite]	$Fe^{[6]}Ti^{[6]}O_3$	Example of face-shared octahedra	$A^{2+}B^{4+}O_3$ with both A^{2+} and B^{4+} smaller or intermediate-sized cations
[Perovskite]	$Sr^{[12]}Ti^{[6]}O_3$	Close-packed structure with corner-shared octahedra	$A^{2+}B^{4+}O_3$ and $A^+B^{2+}F_3$ with large A and medium-sized B cation
Hexagonal structures (several types)	$Ba^{[12]}Mn^{[6]}O_3$	Close-packed structure with face-shared octahedra	As for [Perovskite] but with slightly smaller B cation
[Pyroxenes] and related structures	$Mg^{[6]}Si^{[4]}O_3$	Sharing of two edges of tetrahedra to form single chains	Small B cations, medium A cations

Source: After Muller, O. and Roy, R., *The Major Ternary Structural Families*, Springer-Verlag, Berlin, 1974, p. 153.

has been studied in detail and is a good example. The structure is illustrated in Figure 5.19. Each C^{4+} is surrounded by three O^{2-} anions, all in the same plane at a C—O interatomic distance of 1.283 Å. Each of these CO_3 groups has six Ca^{2+} neighbors with a Ca—O interatomic distance of 2.36 Å. This results in a rhombohedral unit cell that is much longer in one direction than in other directions, resulting in highly anisotropic properties (different properties in different directions). For example, when calcite is heated, it has very high thermal expansion ($25 \times 10^{-6}/°C$) parallel to the c axis but a negative thermal expansion ($-6 \times 10^{-6}/°C$) perpendicular to the c axis. The anisotropy is so high that even light is affected when passing through a transparent crystal of calcite. When looking through a calcite crystal held in the proper orientation, a double image is seen.

A variety of compositions have the [calcite] structure. Some examples include $MgCO_3$, $CuCO_3$, $FeCO_3$, $MnCO_3$, $FeBO_3$, VBO_3, $TiBO_3$, $CrBO_3$, $LiNO_3$, and $NaNO_3$.

5.3.4.3.2 [Ilmenite] Structure $A^{[6]}B^{[6]}X_3^{[4]}$
The [ilmenite] structure involves intermediate-size cations that fit into octahedral interstitial positions to produce an ordered derivative of the [corundum] structure. One layer of the structure contains A-position cations and the adjacent layer contains B-position cations. Composition examples include $MgTiO_3$, $NiTiO_3$, $CoTiO_3$, $MnTiO_3$, $NaSbO_3$, $MgSnO_3$, $ZnGeO_3$, and $NiMnO_3$. Compositions with closely related ordered structures include $LiNbO_3$ and $LiTaO_3$.

5.3.4.3.3 [Perovskite] Structure $A^{[12]}B^{[6]}X_3^{[6]}$
As mentioned before, the [perovskite] structure and related structures are the most important of the ABX$_3$ structures to advanced-technology applications of ceramics. Many of the [perovskites] are cubic and referred to as [ideal perovskites]. Others are ordered or distorted and have other crystal forms such as tetragonal, orthorhombic, or rhombohedral.

The [ideal perovskite] structure involves a large cation that is similar in size to the anion plus a smaller, second cation. The large cation joins the anions in a cubic close-packed arrangement and thus has a CN of 12. The smaller cation fills one fourth of the octahedra interstitial sites. The structure can be visualized as strings of corner-shared AX octahedra extending in three dimensions at

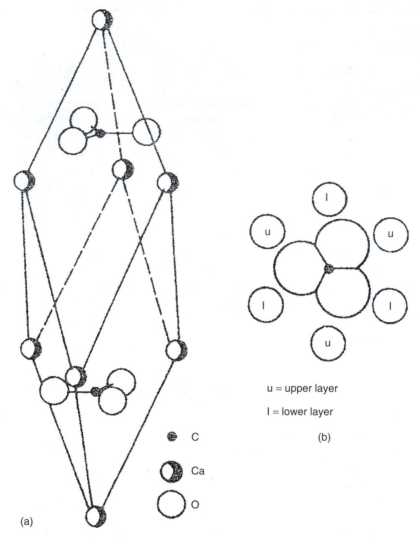

FIGURE 5.19 Illustrations of the [calcite] structure. (a) Side view of elongated rhombohedral cell, and (b) projected top view. (© ASM International. With permission.)

right angles to each other along the [100], [010], and [001] crystal directions. This is illustrated in Figure 5.20 along with an alternative view.

Compositions with the [ideal perovskite] structure include the following $SrTiO_3$, $KNbO_3$, $NaTaO_3$, $CsIO_3$, $LaAlO_3$, $KMgF_3$, $BaLiF_3$, $SrLiH_3$, $Ba(Zn_{0.33}Nb_{0.67})O_3$, and $Pb(Fe_{0.67}W_{0.33})O$.

The [perovskite] structure compositions with important electrical properties are distorted. The B-position cation is displaced slightly off center resulting in distorted eccentric octahedra. The effect of this on the dielectric properties is discussed in Chapter 11.

Many compositions have distorted, ordered, or other nonideal [perovskite] structures. Some are summarized in Table 5.9.

5.3.4.3.4   [Pyroxene] Structure $A^{[6]}B^{[4]}X_3$

Pyroxene structures involve small cations in the B position such that BX_4 tetrahedra share two corners to form chains. The tetrahedra can consist of SiO_4, GeO_4, PO_4, AsO_4, BO_4, or BeF_4 groups. Some pyroxene silicate compositions include $MgSiO_3$, $MgCa(SiO_3)_2$, and $LiAl(SiO_3)_2$.

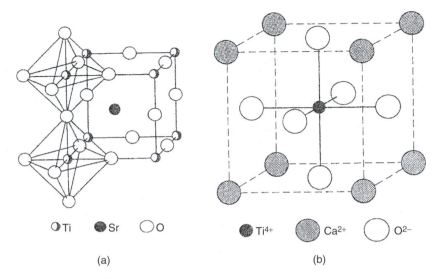

Ti   Sr   O Ti^{4+} Ca^{2+} O^{2-}

(a)   (b)

FIGURE 5.20 Two schematic views of the [ideal perovskite] structure. (a) View with the A cation at the center of a cube, the B cation at the corners of the cube and the X anions at the center of each cube edge, showing the strings of corner-shared BX_6 octahedra: and (b) alternate view with the anions in FCC positions, the A cation at cube corners and the B cation at the center of the cube. (From Muller O. and Roy, R., *The Major Ternary Structural Families*, Springer-Verlag, Berlin, 1974. With permission.)

TABLE 5.9

Examples of Compositions with Nonideal [Perovskite]-Type Structures

[ReO$_3$] structure distortions and metal-stuffed "bronze" compositions with high metallic electrical conductivity:	WO_3, $Na_{0.5}WO_3$, $Li_{0.3}WO_3$ and similar compositions with Mo, Re, Nb, Ta, Ti or V in place of W
Ordered compositions:	$SrSnO_3$, $CsNaInF_6$, K_3MoF_6, Ca_2LaNbO_6
Tetragonal distortions:	$BaTiO_3$, $PbTiO_3$, $PbHfO_3$, $CsSrCl_3$, $KCrF_3$
Rhombohedral distortions:	$LaAlO_3$, $LaCoO_3$, $BiFeO_3$, $CeAlO_3$
Orthorhombic distortions:	$GdFeO_3$, $YFeO_3$, $CaTiO_3$, $CaSnO_3$

5.3.4.4 Other Structures

A few additional structure types need to be mentioned but will not be discussed at this time. These include the [garnet] structure $A_3^{[8]}B_2^{[6]}C_3^{[4]}X_{12}^{[4]}$, the [pyrochlore] structure $A_2^{[8]}B_2^{[6]}X_6^{[4]}X^{[4]}$, and the [pseudobrookite] structure $A_2^{[6]}B^{[6]}X_5$.

5.3.4.5 Carbide and Nitride Structures

The structures discussed so far have involved ionic, covalent, or a combination of ionic and covalent bonding. Some compositions form structures where the bonding is intermediate between covalent and metallic. This is the case for the transition metal carbides. The large metal atoms form a close-packed structure with the small C ions present in interstitial positions. These materials have some characteristics typical of ceramics and some more typical of metals. Combinations of elements such as Si and C with similar electronegativity have covalent bonds.

Nitride structures are similar to carbide structures, except the metal–nitrogen bond is usually less metallic than the metal–carbon bond.

REFERENCES

1. Shackelford, J.F., *Introduction to Materials Science for Engineers*, Macmillan, New York, 1985.
2. Richerson, D.W., Lesson 2: Atomic bonding and crystal structure, in *Introduction to Modern Ceramics*, ASM Materials Engineering Institute Course #56, 1990.
3. Muller, O. and Roy, R., *The Major Ternary Structural Families*, Springer-Verlag, Berlin, 1974.
4. Kingery, W.D., Bowen, H.K., and Unlmann, D.K. *Introduction to Ceramics*, 2nd ed., Wiley, New York, 1976.
5. Pauling, L., *Nature of the Chemical Bond*, 3rd ed., Cornell University Press, Ithaca, New York, 1960.
6. Van Vlack, L.H., *Elements of Materials Science*, 2nd ed., Addison-Wesley, Reading, Mass., 1964.

ADDITIONAL RECOMMENDED READING

1. Evans, R.C., *An Introduction of Crystal Chemistry*, 2nd ed., Cambridge University Press, 1964.
2. Wyckoff, R.W.G., Ed., *Crystal Structures*, 2nd ed., Vols. 1–5, Wiley, New York, 1963–1971.
3. Rogers, B.A., *The Nature of Metals*, ASM International, Ohio, 1951.

PROBLEMS

1. The anion in a binary structure has a valence of -2 and a CN of 4. The cation has a valence of $+4$. What is the CN of the cation?
2. Whats is the notation for the direction along the c axis of a tetragonal crystal?
3. The close-packed plane in an FCC structure is along the cube diagonal. What are the Miller indices for this plane?
4. Which of the following does not have the [perovskite] structure?
   $SrTiO_3$
   $BaLiF_3$
   $MgTiO_3$
   $LaAlO_3$
5. Which composition does *not* have the [spinel] or closely related structure?
   Na_2WO_4
   Fe_3O_4
   $ZrSiO_4$
   $MgAl_2O_4$
6. $NaZr_2(PO_4)_3$ is the type material for the [NZP] structure. Name several ions that are likely substitutes for Na in the structure. Name the most likely ion to substitute for P in the tetrahedral position in the structure.
7. Copper, nickel, and gold have an FCC close-packed structure (i.e., unit cubic structure with atoms at the corners and center of the faces of a cube). Assuming that each atom can be modeled by a hard sphere, how many nearest neighbors does each atom have in the structure? Calculate the percent open space present in such a close-packed structure. The remaining space is filled by the spheres and is referred to as the *atomic packing factor*.
8. Draw the following directions for a cubic crystal: [010], [110], [221], [021].
9. Draw the following planes for a cubic crystal: (001), (111), (120), (101).

STUDY GUIDE

1. Why is it important to have a simple and constant method of identifying directions and planes within a crystal structure?
2. Draw the following directions for a cubic crystal: [001], [111], [121], and [021].

3. Draw the following planes within a cubic crystal structure: (100), (121), (021), and (112).

4. What is a coordination formula and why is it useful?

5. What is "ionic radius" and how does it vary depending upon ion valence, CN, and the nature of the surrounding atoms and ions?

6. Identify or describe the key interstitial positions that can occur in a crystal structure. Feel free to supplement with drawings. Be sure to include the coordination number (CN) for each and the general size of atom (ion) that can fit in (large, medium, or small).

7. What effect does the electrical charge of ions have on the structure they combine to form? Explain.

8. Define bond strength in terms of CN and valence.

9. What equation of valence and CN results in a material structure with electrical neutrality?

10. Explain the meaning of "solid solution"?

11. What are the dominant factors in determining whether one atom can substitute for another to result in complete solid solution?

12. a. List two divalent ions most likely to fit into a tetrahedral site. _____ _____

    b. Two +4 ions. _____ _____

13. Is a monovalent ion likely to fit into a tetrahedral site? Yes_____; No_____

14. Identify some options for achieving derivative structures, which make possible a wide variety of ceramic materials and properties.

15. What is meant by nonstoichiometry?

16. What is meant by stuffing? Give an example.

17. Explain the meaning of "binary structure"?

18. What is the difference between the [CsCl] and [NaCl] structures?

19. What size of cations fit into a [zinc blende] or [wurtzite] structure? Name some examples.

20. a. List three divalent cations that most likely fit into an octahedral site. _____ _____ _____

    b. Three trivalent cations. _____ _____ _____

    c. Three tetra valent cations. _____ _____ _____

21. What are the CNs for Si^{4+} and for O^{2-} when they bond together to form SiO_2?

22. How do SO_4^{2-} tetrahedra bond together, and what types of structures can result? How do these structures differ from ionically bonded structures?

23. Describe at least two structural similarities of diamond, silicon carbide, and silicon nitride.

24. Explain the meaning of "ternary structure."

25. Note the enormous range of compositions possible for each ternary structure by making crystal chemical substitutions in each structural position, especially for [spinel] and [olivine] structures. Where are we likely to encounter [spinel] or [olivine] structure compositions?

26. What are some important applications of ABX_3 structure ceramics?

6 Phase Equilibria and Phase Equilibrium Diagrams*

Ceramic materials are generally not pure. They contain impurities or additions that result in solid solution, noncrystalline phases, multiple crystalline phases, or mixed crystalline and noncrystalline phases. The nature and distribution of these phases have a strong influence on the properties of the ceramic material, as well as on the fabrication parameters necessary to produce the ceramic. Therefore, before we can progress to a discussion of properties or processing in later chapter, we need to explore ways to estimate the nature and distribution of phases. An understanding of equilibrium and nonequilibrium and phase equilibrium diagrams is a good starting point.

6.1 PHASE EQUILIBRIUM DIAGRAMS

A phase equilibrium diagram is a graphical presentation of data that gives considerable information about a single compound (such as SiO_2, Al_2O_3, or MgO) or the nature of interactions between more than one compound.[1-3] The following list indicates some of the information that can be read directly from the diagram:

1. Melting temperature of each pure compound
2. The degree of reduction in melting temperature as two or more compounds are mixed
3. The interaction of two compounds (such as SiO_2 plus Al_2O_3) to form a third compound ($3Al_2O_3 \cdot 2SiO_2$, mullite)
4. The presence and degree of solid solution
5. The effect of temperature on the degree of solid solution
6. The temperature at which a compound goes from one crystal structure to another (polymorphic phase transformation)
7. The amount and composition of liquid and solid phases at a specific temperature and bulk composition
8. The presence at high temperature of immiscible liquids (liquids that are not soluble in each other).

6.1.1 CONCEPT OF PHASE EQUILIBRIA[4-9]

The state referred to as *equilibrium* in a materials system has been defined by the laws of thermodynamics as the condition when the free energy in the system is a minimum for a particular set of external conditions. As long as the external conditions do not change, the system does not change. However, when an external condition such as composition, temperature, or pressure changes, the state of the materials within the system changes until a new state of equilibrium is reached. The relationship of free energy to the material components present in a system and the external conditions follows the mathematical relationship of Equation (6.1):

$$G = PV - TS + \mu_1 X_1 + \cdots \mu_i X_i \qquad (6.1)$$

---

* Unreferenced figures in this chapter are adapted from Ref. 1 (Hummel) and redrawn by ASM International as part of Lesson 3 of "Introduction to Modern Ceramics," ASM Materials Engineering Institute Course #56 by David W. Richerson, 1990.

where

    G = Gibbs free energy
    P = pressure of the system
    V = volume of the system
    T = absolute temperature
    S = entropy
    μ_i = chemical potential of ith component
    X_i = mole fraction of ith component

As stated, equilibrium is reached when G is a minimum for a given combination of system parameters. For most practical cases, these parameters include temperature, pressure, and concentration of the components. In some cases, the presence of an electric, magnetic, or gravitational fields or surface forces can also influence the state of equilibrium.

6.1.2 THE PHASE RULE

The construction of phase equilibrium diagrams is based upon the *phase rule*, which was presented in 1874 by J. Willard Gibbs and is stated in Equation (6.2):

$$P + V = C + 2 \qquad \text{or} \qquad P + F = C + 2 \tag{6.2}$$

where

       P = number of phases present at equilibrium
  $V \text{ or } F$ = the variance or number of degrees of freedom
       C = the number of components

The phase rule was derived from thermodynamics and defines the conditions under which equilibrium can exist.

Before we can proceed, we need to understand the definition of the various terms in the phase rule. A *phase* is defined as any part of the system that is physically homogeneous and bounded by a surface so that it is mechanically separable from the rest of the system. The states of matter (gas, liquid, and solid) represent phases. Normally, only one gas phase is present in a system since all gases are miscible under typical pressure. However, more than one liquid phase can be present if the liquids are not miscible (water, oil, and mercury, for example). Many solid phases can coexist.

The definition of *component* is "the smallest number of independently variable chemical constituents necessary and sufficient to express the composition of each phase present." This definition will become clearer as we discuss one-, two-, and three-component systems later in this chapter.

The *variance* is "the number of intensive variables such as temperature, pressure, and concentration of components in a phase that can be altered independently and arbitrarily without causing the disappearance of one phase or the appearance of a new phase." This will also become clear as we discuss single- and multiple-component systems.

6.1.3 ONE-COMPONENT PHASE DIAGRAMS

Examples of single-component systems include H_2O, SiO_2, C, and TiO_2. It is useful to have a simple diagram identifying which phases and polymorphic forms of the single component are stable as a function of temperature and pressure. A generalized schematic of such a diagram is shown in Figure 6.1. We will use this diagram to illustrate the typical effects of temperature and pressure on a material and to help understand the meaning of the terms phase, component, and variance.

Let us look at specific points, lines, and regions of the diagram in terms of the phase rule. First, only one component is present at all positions on the diagram so $C = 1$ and $P + F = 1 + 2 = 3$.

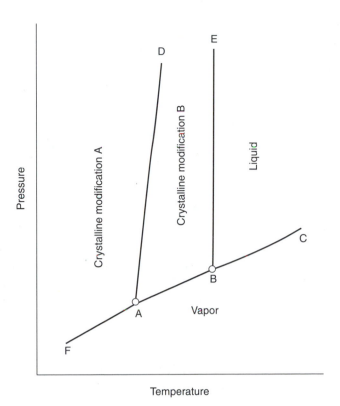

FIGURE 6.1 Schematic illustrating the phase relations in a one-component system. (From Levin, E.M., Robbins, C.R., and McMurdie, H.F., *Phase Diagrams for Ceramists*, M.K. Reser, Ed. American Ceramic Society, Ohio, 1964. With permission.)

At point *A* three phases are in equilibrium (polymorph A, polymorph B plus vapor). Three phases are also in equilibrium at point *B* (polymorph B, liquid plus vapor). Thus, at both *A* and *B*, $F = 3 - P = 3 - 3 = 0$. Because $F = 0$, *A* and *B* are referred to as *invariant* or triple points. Any change in pressure or temperature will cause the disappearance of one phase.

Curves *F–A* and *A–B* are sublimation curves for polymorphs A and B and represent the equilibrium between a solid and vapor. Curve *B–C* is the vapor pressure curve for the liquid and represents the equilibrium between the liquid and vapor phases. *A–D* is the phase transformation (transition) curve between the A and B polymorphs. *B–E* is the fusion or melting curve where polymorph B and liquid are in equilibrium. For all points along these curves, two phases coexist. Thus, $F = 3 - 2 = 1$ for these boundary lines. This is referred to as *univariant equilibrium* and means that pressure and temperature are dependent on each other. The specification of one will automatically fix the other.

The regions between boundary lines contain a single phase and are known as fields of stability. $F = 3 - 1 = 2$ in these regions. This is referred to as *bivariant equilibrium*. Within a specific region, the temperature and pressure can be varied independently without a phase change.

Other information is available from a single-component diagram. Let us refer again to Figure 6.1. Point *C* is called the *critical point*, the temperature above which the gas cannot be liquified no matter how high the pressure. The slope of line *B–E* provides information about the relative density of the liquid and the solid. For most materials, the slope of the curve is positive such that the density of the solid is higher than that of the liquid. Exceptions are H_2O, Bi, and Sb. Important exceptions in ceramic systems are the glasses of lithium aluminosilicate (beta-spodumene) and magnesium aluminosilicate (cordierite), which are more dense than the crystalline phases.

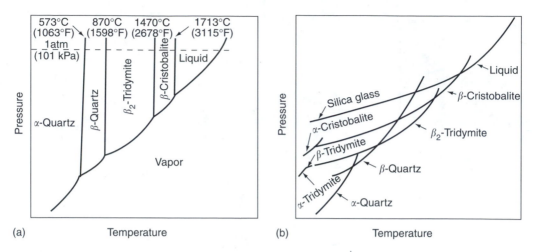

(a) Temperature   (b) Temperature

FIGURE 6.2 Comparison of the estimated equilibrium and metastable phases of SiO_2. (a) Equilibrium diagram, and (b) diagram including metastable phases. (From Kingery, W.D. et al., *Introduction to Ceramics*, 2nd ed., pp. 274, 275, Wiley, New York, 1976. With permission.)

Figure 6.2a shows the estimated equilibrium phases for SiO_2. However, because the rates of change between the SiO_2 phases are very slow, metastable phases are generally present in real materials and often dominate the behavior. The retention of metastable phases is illustrated in Figure 6.2b.

6.1.4 Two-Component Systems

A two-component system is referred to as a *binary system*. Addition of the second component results in a change in the equilibrium of the system and a change in the properties of the resulting materials. It also results in greater complexity and the need for a three-dimensional (p–T–x) diagram to illustrate the equilibrium characteristics for the three variables pressure (p), temperature (T), and composition (x). Figure 6.3 shows a portion of such a diagram with pressure, temperature, and composition as the three orthogonal axes. To the left is the single-component p–T equilibrium diagram for component A and to the right is the similar diagram for component B. In a complete diagram a similar set of "pseudo-one-component" sublimation, melting, and vaporization curves would be drawn for each composition of A + B. The loci of these are curved surfaces that require a three-dimensional model. The general shape of the curved surfaces can be visualized by drawing a plane through the diagram and looking at the intersection of the curved surfaces with the plane. This is illustrated for a simple binary eutectic system[10,11] (discussed later in this chapter) in Figure 6.4. The plane is drawn perpendicular to the pressure axis at a constant pressure of 1 atm and parallel to the temperature and composition axes. Only the intersection of the plane with the melting curve is shown. Note that addition of A to B or B to A results in a reduction in the melting temperature.

For many ceramic systems the effect of pressure is negligible because the vapor pressure of the liquid and solid phases in these systems is very low. Thus, the two-dimensional diagram produced by the intersection of the one-atmosphere-pressure plane with the three-dimensional surfaces in the p–T–x diagram in Figure 6.4 closely approximates the true equilibrium condition. Under this condition where pressure is held constant, the system is referred to as a *condensed* system; that is, the vapor phase is ignored and only the liquid and solid phases are considered. As stated, the system can now be illustrated by a single two-dimensional temperature–composition (T–x) diagram. The phase rule now becomes $F = C - P + 1$ (1 rather than 2 because only temperature is considered as a variable). For two components, $F = 2 - P + 1 = 3 - P$.

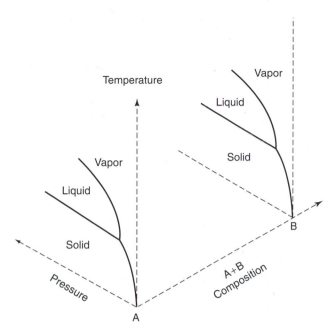

FIGURE 6.3 Schematic illustrating the need for a three-dimensional model to diagram the equilibrium of a two-component system as a function of pressure, temperature, and composition.

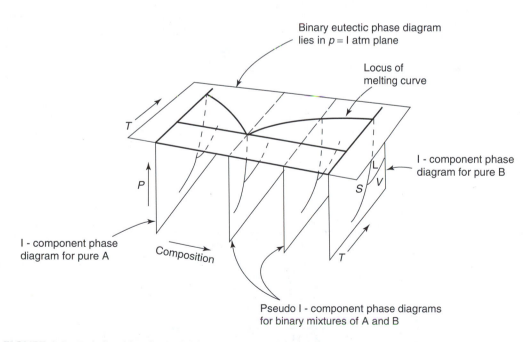

FIGURE 6.4 Relationship of two single-components and pseudo-one-component mixtures of the two single components to a condensed binary diagram. (After Shackelford, J.F., *Introduction to Materials Science for Engineers*, 2nd ed., p. 199, Macmillan, New York, 1988; Schneider, S.J., in *Mechanical and Thermal Properties of Ceramics*, Proc. Symp., NBS Spec. Publ. 303, 1969. With permission. Artwork courtesy of ASM International.)

A variety of behavior can occur for condensed, two-component ceramic systems. In subsequent sections we review the following variations:

1. Simple eutectic (no intermediate compounds between A and B)
2. Intermediate compounds
   a. Congruent melting
   b. Incongruent melting
   c. Dissociation
3. Solid solution
   a. Complete
   b. Partial
4. Liquid immiscibility
5. Exsolution (unmixing of solid solutions)
6. Polymorphism

The emphasis of these subsequent discussions is to identify how the information is shown on the diagram and how it relates to material behavior.

6.1.4.1 Binary Eutectic Systems

The generation of a condensed binary eutectic diagram was illustrated in Figure 6.4. Figure 6.5 shows the generalized regions within a condensed binary eutectic diagram. The vertical lines to the left and right of the diagram identify the pure phases A and B. The melting temperature of A is T_A and of B is T_B. Both A and B melt congruently, which means that each has a well-defined melting temperature at which the pure compound goes directly from a solid to a liquid of the same composition. For pure A or pure B, no liquid is present until temperatures T_A or T_B are reached.

The curves $T_A - T_e$ and $T_B - T_e$ are referred to as *liquidus* curves. Above these curves only liquid is present at equilibrium, and the variance is $F = 3 - 1 = 2$. In this liquid region the system is

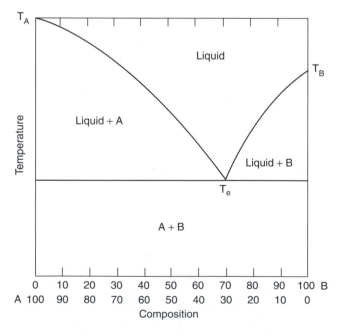

FIGURE 6.5 Schematic of a simple condensed binary eutectic diagram.

bivariant and temperature and composition can both be varied independently without a change in phases. Along the liquids lines $T_A - T_e$ and $T_B - T_e$, two phases coexist, so $F = 3 - 2 = 1$. The system at each point along these lines is univariant so that temperature and composition are dependent upon each other. A specification of one automatically fixes the other.

Three phases (liquid, A, and B) coexist in equilibrium at point T_e. This point is called the *binary eutectic*. Below the eutectic temperature, no liquid can exist in the system under equilibrium. The horizontal line on which T_e lies is called the *solidus*. $F = 3 - 3 = 0$ at the binary eutectic, so T_e is an invariant point. Any change in temperature or composition results in the loss of one or more phases. For example, an increase in temperature without a change in composition results in the melting of all A and B to form a homogeneous liquid. Conversely, a decrease in temperature results in freezing of the liquid by crystallization of a mixture of solid A and B. Similarly, changes in composition to the right of T_e results in complete melting of compound A and crystallization of a portion of the compound B. Changes in composition to the left of T_e result in complete melting of B and crystallization of a portion of A. These changes in phase resulting from changes in temperature or composition are clearly illustrated on the binary eutectic diagram. Even the precise percentages of liquid and solid present at a specific temperature and composition can be calculated. The methods of calculation are discussed later in this chapter.

Now let us take a closer look at the binary eutectic system (and two-component systems in general) and explore the significance of the various points and lines on the diagram. First, as we have observed before, addition of a second component causes a reduction in the melting temperature. The steeper the slope of the liquidus line, the greater the effect of the second component on reducing the melting temperature of the first. In Figure 6.5 the addition of A to B has a greater effect than B to A. A 30% addition of A to B reduces the melting temperature by over 50%, whereas a 30% addition of B to A results in only a 15% reduction. More important though, even a fraction of a percent addition of a second component leads to the presence of liquid at a temperature (the eutectic temperature T_e), well below the temperature at which liquid would occur for a pure single component (T_A or T_B). In the case of component A in Figure 6.5 a small amount of B results in the presence of a liquid at approximately 35% of the melting temperature of pure A. Although the amount of liquid is small for small additions of the second component, the effects on the densification characteristics during fabrication and on the properties of the resulting material can be significant. These effects are discussed in detail in later chapters, especially Chapters 7–11 and 14.

Figure 6.6 gives an example of an actual binary eutectic diagram. This diagram is for the system NaCl–NaF. Note from the diagram that NaCl melts congruently at 800.5°C (1472.9°F), NaF melts congruently at 994.5°C (1822.1°F), and the binary eutectic is at 680°C (1256°F) at a composition of 33.5 mol% NaF and 66.5 mol% NaCl. Note the large drop in temperature at which the first liquid forms when a small amount of NaCl is added to NaF. No liquid occurs for pure NaF until 994.5°C. It is apparent from this example that impurities or controlled additives can have large effects on the melting behavior.

6.1.5 INTERMEDIATE COMPOUNDS

Frequently the two components A and B react to form one or more intermediate compounds such as AB, A_2B, or AB_4. The intermediate compounds can melt congruently (liquid phase and solid phase both of the same composition coexist in equilibrium at the melting temperature), melt incongruently (solid phase changes to a liquid plus a solid phase, both with compositions different from the original phase), or dissociate. Schematic binary diagrams with congruently melting and incongruently melting intermediate compounds are compared in Figure 6.7 and Figure 6.8. Figure 6.7 consists of a two-binary eutectic diagram, A–A_2B and A_2B–B joined together. Two eutectics are now present, with e_1 as the eutectic for the A–A_2B portion of the diagram and e_2 as the eutectic for the A_2B–B portion of the diagram.

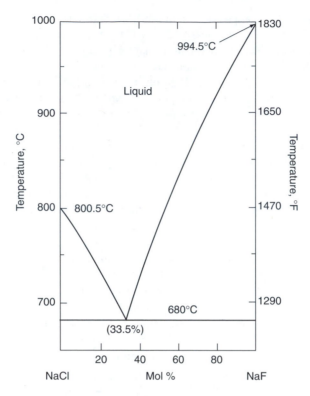

FIGURE 6.6 Simple binary eutectic relationship for NaCl–NaF system. (From Grjotheim, K., Halvorsen, T., and Holm, J.H., *Acta. Chem. Scand.*, 21(8), 2300, 1967. With permission.)

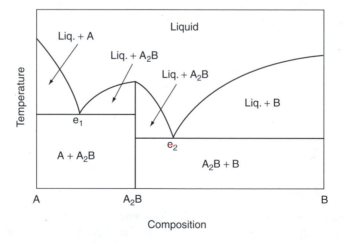

FIGURE 6.7 Schematic of a binary system with a congruently melting intermediate compound.

The binary diagram with an incongruently melting intermediate compound shown schematically in Figure 6.8 requires further discussion. The temperature at which the incongruently melting compound goes from a solid to a liquid plus a different solid is called the *peritectic temperature* (T_p). The point of intersection between this peritectic temperature and the liquidus is the *peritectic point*, P_1. Three phases are in equilibrium at this point (B, AB_2, and liquid), so it is an invariant point

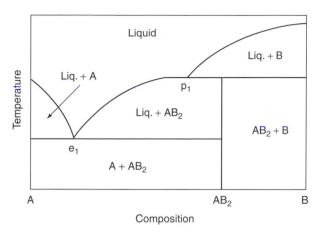

FIGURE 6.8 Schematic of a binary system with an incongruently melting intermediate compound.

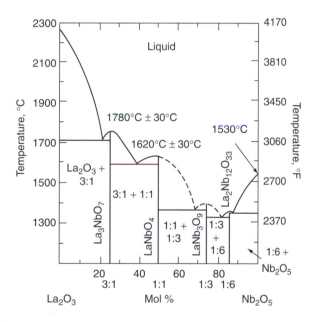

FIGURE 6.9 The binary system La_2O_3–Nb_2O_5 containing four congruently melting intermediate compounds. (From Savchenko, E.P., Godina, N.A., and Keler, E.K., *Chemistry of High-Temperature Materials*, p. 111, N.A. Toropov, Ed., Consultants Bureau, New York, 1969. With permission.)

($F = 3 − 3 = 0$). An additional invariant point e_1 is present in the diagram. In Figure 6.8, the phases A, AB_2, and liquid are in equilibrium at e_1. Note that the eutectic is at a lower temperature than the peritectic. Compositions to the right of the intermediate compound in Figure 6.8 have higher temperature capability (refractoriness) than compositions to the left of the intermediate compound, since they do not contain any liquid until T_p is reached.

Figure 6.9 to Figure 6.13 provide examples of condensed binary phase equilibrium diagrams with congruently and incongruently melting compounds. Figure 6.9 shows the high-temperature region of the diagram for the system La_2O_3–Nb_2O_5. It has four congruently melting compounds and consists essentially of five simple eutectic diagrams side by side. The compounds are identified both by chemical formula (such as La_3NbO_7) and by an abbreviated molar ratio (such as 3:1). The 3:1

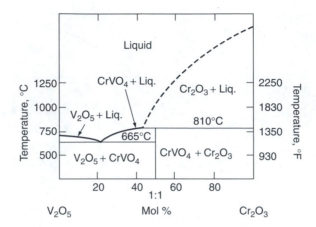

FIGURE 6.10 The binary system Cr_2O_3–V_2O_5 containing one incongruently melting intermediate compound. (From Burdese, A., *Ann. Chim. (Rome)*, 47(7,8), 801, 1957. With permission.)

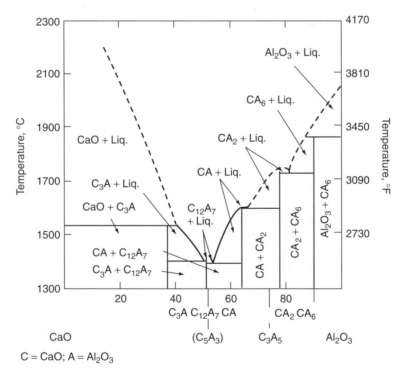

FIGURE 6.11 The binary system CaO–Al_2O_3. The dashed lines indicate regions of the diagram with some uncertainty in the precise position of the boundary lines. (From Lea, F.M. and Desch, C.H., *The Chemistry of Cement and Concrete*, 2nd ed., p. 52, Edward Arnold, London, 1956. With permission.)

molar ratio is shorthand for $3La_2O_3 \cdot 1Nb_2O_5$, which equals $La_6Nb_2O_{14}$, which reduces to La_3NbO_7. Figure 6.10 is the diagram for the Cr_2O_3–V_2O_5 system illustrating an incongruently melting compound $CrVO_4$ having a 1:1 molar composition $Cr_2O_3 \cdot V_2O_5$ ($Cr_2V_2O_8 = CrVO_4$). Note that the lowest temperature for liquid formation of V_2O_5-rich compositions is controlled by the eutectic temperature 665°C (1229°F), and the lowest temperature for liquid formation of Cr_2O_3-rich composition is controlled by the peritectic temperature 810°C (1490°F).

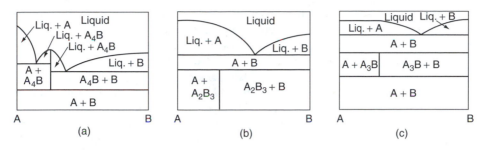

FIGURE 6.12 Representation of dissociating compounds on a phase equilibrium diagram. (a) Lower limit of stability, (b) upper limit of stability, and (c) lower and upper limit.

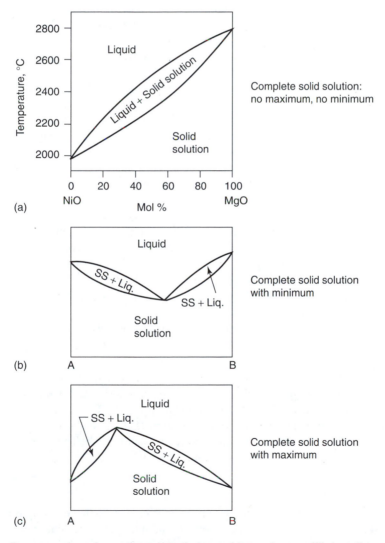

FIGURE 6.13 Representation of complete solid solution on binary phase equilibrium diagrams.

Some binary systems can contain many intermediate compounds and appear quite complex. An example is the $CaO-Al_2O_3$ system, which contains important cementitious and refractory (high-temperature) compositions. Although the diagram looks complex, close observation reveals that it is merely a combination of simple eutectic and peritectic diagrams.

The final category of intermediate compounds involves compounds that are stable only over a limited temperature range. As illustrated in Figure 6.12, some have a lower limit of stability, some have an upper limit of stability, and some have both upper and lower limits of stability. Compounds that demonstrate these types of solid–solid dissociation are uncommon. Examples can be found in the $Al_2O_3-Y_2O_3$ and $SiO_2-Y_2O_3$ systems. As an exercise, locate these diagram in Reference 5.

6.1.5.1 Solid Solution

Solid solution was discussed under crystal chemistry in Chapter 5. Solid solution involves the ability of one atom or group of atoms to substitute into the crystal structure of another atom or group of atoms without resulting in a change in structure. Solid solution must be distinguished from mixtures. In a mixture, two or more components are present, but they retain their own identity and crystal structure. Examples of mixtures are component A plus component B in Figure 6.5 and component A_2B plus component B in Figure 6.7. Similarly, La_2O_3 plus La_3NbO_7 in Figure 6.9 is a mixture, as is $CrVO_4$ plus Cr_2O_3 in Figure 6.10.

In a solid solution one component is "dissolved" in the other component such that only one continuous crystallographic structure is detectable.

Crystallogrpahic substitutions take place most easily if two atoms are similar in size and valence. For instance, Mg^{2+}, Co^{2+}, and Ni^{2+} are all similar in size and can readily replace each other in the cubic [rock salt] structure. In fact, each can replace the other up to 100% in the oxide, resulting in continuous solid solution. Figure 6.13a is the phase equilibrium diagram for the system $MgO-NiO$, showing complete solid solution between the MgO and NiO.[12] This is the most common type of continuous solid solution. Figure 6.13b and c show less common types in which either a maximum or minimum is present. These maxima and minima are neither compounds nor eutectics, just limits in melting temperature for the solid solution.

Solid solution does not have to be complete between two different components and generally is not. Usually, one chemical component will have limited solid solubility in the other. The limits are determined by the similarity in the crystal structures and the size of ions or atoms. Figure 6.14 illustrates partial solid solution for a binary eutectic system and a binary peritectic system. For the eutectic system in Figure 6.14a, component A can contain up to about 40% B in solid solution (at the eutectic temperature) and B can contain up to about 17% A. For the peritectic system shown, A can contain up to 20% B and B up to 60% A.

Figure 6.15 and Figure 6.16 present examples of actual binary systems that have varying degrees of solid solution. Figure 6.15 illustrates complete solid solution between Al_2O_3 and Cr_2O_3. Both of these have the same crystal structure and the Al^{3+} and Cr^{3+} ions are similar in size (0.67 Å vs. 0.76 Å). In contrast, MgO and CaO show limited solid solution (Figure 6.16). The ionic radii of Mg^{2+} and Ca^{2+} are not close (0.86 Å for Mg^{2+} and 1.26 Å for Ca^{2+}). Other important ceramic systems with solid solution include $MgO-Al_2O_3$, $TiO_2-Nb_2O_5$, $MgO-Cr_2O_3$, and $SiO_2-Al_2O_3$.

Solid solution has a strong effect on the densification characteristics and properties of materials. Solid solution reduces the amount of liquid present at equilibrium. The implications of this are discussed later in this chapter and in Chapter 9 and Chapter 14.

6.1.5.2 Liquid Immiscibility

Most of the phase equilibrium diagrams we have examined so far have a single liquid that is a homogeneous solution for all compositions above the liquidus. In some binary systems, composition regions exist above the liquidus where two distinct liquids coexist. This behavior is referred to as *immiscibility* and is illustrated schematically in Figure 6.17. Note that each liquid is a distinct

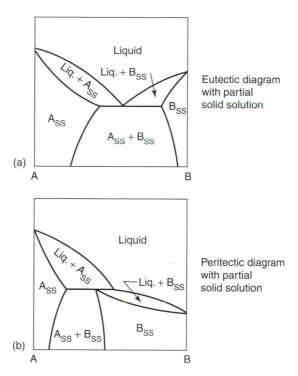

FIGURE 6.14 Binary phase equilibrium diagrams showing partial solid solution.

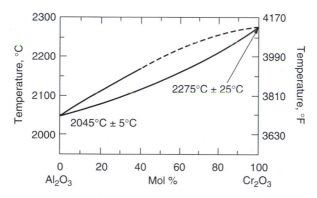

FIGURE 6.15 The system Al_2O_3–Cr_2O_3. (From Bunting, E.N., *J. Res. Natl. Bur. Stand.*, 6(6), 948, 1931. With permission.)

separate phase. When a composition in the two-liquid region is melted and rapidly quenched to a noncrystalline state, the two phases can be observed by transmission electron microscopy. An example is illustrated in Figure 6.18. The two liquid phases typically have a difference in viscosity, density or surface tension that allows them to maintain an interface between them. One liquid is usually rich in A and one in B, and one is often dispersed as very tiny, nearly spherical droplets in the other.

Many SiO_2 systems with divalent oxides (MgO, CaO, SrO, MnO, ZnO, FeO, NiO, and CoO) exhibit liquid immiscibility. Liquid immiscibility also occurs in many systems consisting of B_2O_3 plus another oxide. An example illustrating liquid immiscibility in a real system is presented in Figure 6.19.

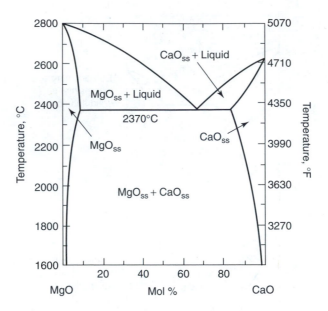

FIGURE 6.16 The CaO–MgO binary system showing limited solid solution. (From Doman, R.C., Barr, J.B., McNally, R.N., and Alper, A.M., *J. Am. Ceram. Soc.*, 46(7), 314, 1963. With permission.)

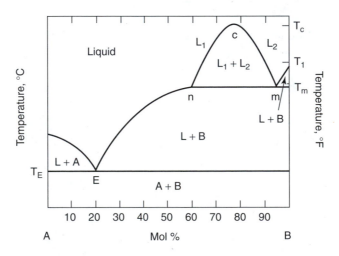

FIGURE 6.17 Schematic representation of liquid immiscibility on a binary phase equilibrium diagram.

6.1.5.3 Exsolution

We can look at liquid immiscibility in another way. At very high temperature only one liquid is present. As the temperature is reduced, two liquids with different characteristics become more stable and "unmixing" occurs. A similar behavior can occur in the solid state in continuous solid solutions. At high temperature, continuous solid solutions are stable. But as the temperature decreases in some systems, two solid solutions become more stable and unmixing or *exsolution* occurs. Typically, one solid solution is rich in A and one in B. Exsolution occurs in the system SnO_2–TiO_2.

6.1.5.4 Polymorphism

Polymorphic transformations are also shown on phase equilibrium diagrams. Figure 6.20a is a schematic of a binary eutectic diagram with no solid solution and with three different polymorphs

FIGURE 6.18 Transmission electron micrograph showing an example of liquid immiscibility. (Courtesy of D. Uhlmann, University of Arizona.)

of the A composition. The different polymorphs are usually designated by letters of the Greek alphabet. Figure 6.20b is a schematic of a binary eutectic diagram with three A polymorphs, each with partial solid solution of B.

Figure 6.21 illustrates a real binary system with polymorphs. Polymorphic transformations are also present in Figure 6.19.

6.1.6 THREE-COMPONENT SYSTEMS

A three-component system is referred to as a *ternary system*. The addition of a third component increases the complexity of the system and of the phase equilibrium diagram. The phase rule becomes $F = 3 - P + 2 = 5 - P$. As with binary ceramic systems, diagrams are usually drawn with pressure as a constant (condensed system). The phase rule for the condensed ternary system is $F = 3 - P + 1 = 4 - P$. Thus, an invariant point involves equilibrium between four phases, which requires a three-dimensional model.

6.1.6.1 Simple Eutectic Ternary Diagram

We shall use the simple eutectic ternary system as an example to describe the general characteristics of a condensed ternary system. A schematic three-dimensional model is illustrated in Figure 6.22. The base is an equilateral triangle with the three pure compositions A, B, and C at the corners. Points on the line A–B only contain A and B, points on B–C only contain B and C, and points on A–C only contain A and C. Points in the interior of the equilateral triangle contain A, B, and C. We discuss the techniques for determining the composition for each point on the diagram later in this chapter.

The vertical axis in Figure 6.22 is temperature. Points T_A, T_B, and T_C, which lie directly above A, B, and C, respectively, are the melting temperatures of the pure components A, B, and C. Note that each melts congruently. Note also that the three sides of the diagram are simply the condensed binary eutectic diagrams for AB, AC, and BC. Point e_1 is the binary eutectic for the binary system AC, e_2 for AB, and e_3 for BC.

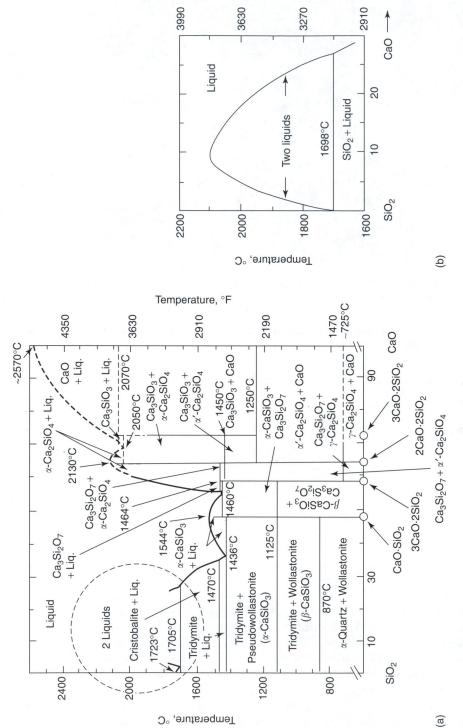

FIGURE 6.19 Liquid immiscibility in the CaO–SiO₂ binary system. (a) Complete diagram. (From Phillips, B. and Muan, A., *J. Am. Ceram. Soc.*, Figure 237, 42(9), 414, 1959. With permission.) (b) Two-liquid region showing complete dome. (From Ya. I. Ol'shanskii, *Dokl. Akad. Nauk SSSR* 76(1), 94, 1951. With permission.)

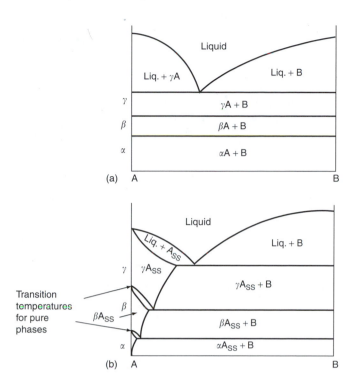

FIGURE 6.20 (a) Simple binary eutectic showing polymorphic transformations. (b) Binary eutectic with solid solution and polymorphic transformation.

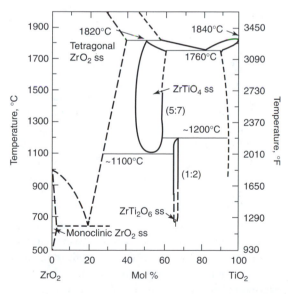

FIGURE 6.21 The system ZrO_2–TiO_2 showing polymorphic transformations with various degrees of solid solution. (After McHale, A.E. and Roth, R.S., *J. Am. Ceram. Soc.*, 69(11), 827, 1986. With permission.)

The top surface of the diagram in Figure 6.22 is the liquidus. The curved lines on this surface represents *isotherms* (lines of constant temperature). All points that lie above the liquidus surface are homogeneous liquid, so $F = 4 - 1 = 3$. All points on the liquidus surface have a solid phase in equilibrium with the liquid phase such that $F = 4 - 2 = 2$. The liquidus surface slopes downward

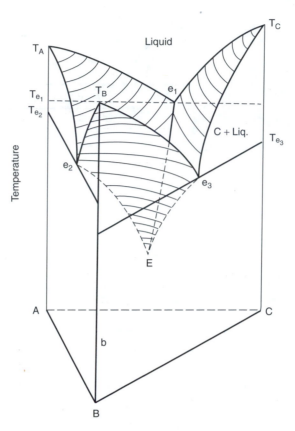

FIGURE 6.22   Schematic of the three-dimensional model for a condensed simple eutectic ternary system.

to point E. This is the ternary eutectic where a homogeneous ternary liquid is in equilibrium with three solid phases. This is an important invariant point ($F = 4 - 4 = 0$), as we discuss later in this chapter. Note that the ternary eutectic is at a lower temperature than the binary eutectics. The lines connecting E with e_1, e_2, and e_3 are univariant ($F = 4 - 3 = 1$) and are referred to as *boundary lines*. On boundary line $E–e_2$ the two solid phases A and B are in equilibrium with a liquid phase. On $E–e_1$, A and C are in equilibrium with liquid. On $E–e_3$, B and C are in equilibrium with liquid.

The three-dimensional diagram in Figure 6.22 is helpful for showing the relationship of the ternary system to the three binary systems, but is a little too complex for general usage. Three techniques are used to transfer information from the three-dimensional model to a two-dimensional diagram. One is to project the boundary lines and contours of the liquidus onto a plane, as shown in Figure 6.23 for a ternary eutectic system. The second is to draw an "isothermal section," that is, the intersection of a horizontal plane of constant temperature with the three-dimensional diagram. Two isothermal sections for the system in Figure 6.23 are shown in Figure 6.24. The third technique is to draw a "vertical section," which is the intersection of a vertical plane with each region of the condensed ternary diagram. An example is illustrated in Figure 6.25 for a system having an intermediate compound BC.

Several features on the liquidus projection in Figure 6.23 require explanation. The arrows on the boundary lines indicate directions of decreasing temperature. The points at the apices of the triangle labeled A, B, and C represent the melting temperature of the three pure components. The regions labeled a, b, and c are the " primary fields" of components A, B, and C. They are important for defining the sequence of crystallization of solid phases from a homogeneous liquid melt. For example, all compositions within the primary field of A initially crystallize A during cooling. Final crystallization occurs at the ternary eutectic. Crystallization paths are discussed later in this chapter.

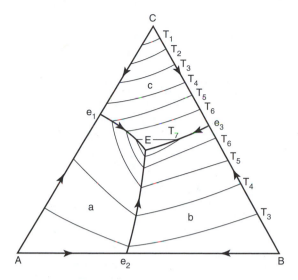

FIGURE 6.23 Two-dimensional projection of the boundary lines and the temperature contours of the liquidus surface for the simple eutectic diagram shown in Figure 6.22.

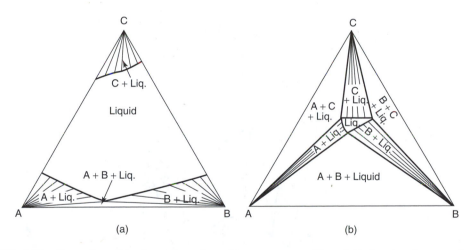

FIGURE 6.24 Examples of isothermal sections for temperatures T_3 and T_7 for the diagram shown in Figure 6.23.

6.1.6.2 Ternary System with Congruently Melting Binary Compound AB

Figure 6.26 illustrates the three-dimensional model for a ternary system with a congruently melting binary compound AB. The two-dimensional projection is shown on the base triangle. Note that the congruently melting compound AB lies within the primary phase field (*ab*) for AB. A line drawn from AB to C separates the diagram into two single ternary eutectic diagrams, one with the ternary eutectic E_1 and one with the ternary eutectic E_2. This line is an *Alkamade line** and represents the intersection of a vertical plane with the liquidus surface. It intersects the boundary line between E_1 and E_2 at point *m* and helps us to quickly recognize temperature relationships. Specifically, temperature decreases along the Alkamade line from C and AB to reach a minimum at *m*. The

---

* An Alkamade line is defined as a straight line in a ternary diagram connecting the composition points of two primary phases whose areas are adjacent and the intersection of which forms a boundary curve.

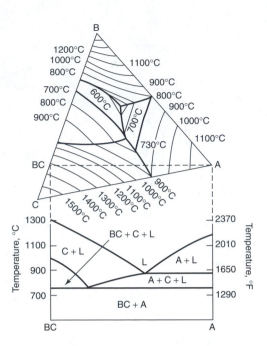

FIGURE 6.25 Example of a vertical section through the join between component A and an incongruently melting intermediate compound BC. (From Bergeron, C. and Risbud, S., *Introduction to Phase Equilibria in Ceramics*, American Ceramic Society, Ohio, 1984, Figure 6.36. With permission.)

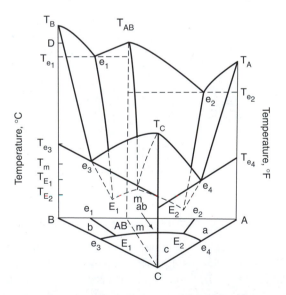

FIGURE 6.26 Ternary system with an intermediate congruently melting binary compound AB; liquidus surface projected on the base triangle to form a two-dimensional diagram. (From Hummel, F.A., *Phase Equilibria in Ceramic Systems*, Marcel Dekker, New York, 1984, Figure 5.52. With permission.)

temperature then decreases from m to E_1 and E_2; that is, m is the highest temperature point along the boundary line E_1–E_2. The lines A–C, B–C, A–AB and B–AB are all Alkamade lines. A–B is not an Alkamade line because the primary fields of A and B are not adjacent and thus do not have a common boundary line.

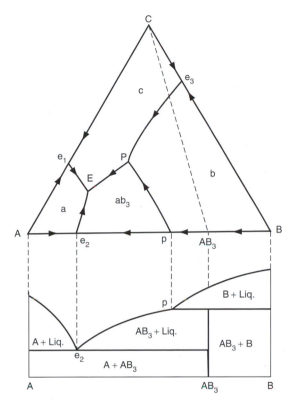

FIGURE 6.27 Ternary system with an intermediate incongruently melting binary compound AB_3 showing the relationship to the binary system AB. (© ASM International.)

The Alkamade line C–AB separates the ternary diagram into two "compatibility triangles." All compositions in triangle A–AB–C terminate their crystallization paths at E_1, and compositions in triangle B–AB–C finally crystallize at E_2. The significance of this will become apparent later in this chapter when we discuss crystallization paths.

6.1.6.3 Ternary System with Incongruently Melting Binary Compound AB

Figure 6.27 illustrates the two-dimensional projection of a ternary system with the incongruently melting binary compound AB_3. The binary for AB is included to show the relationship of the features on the ternary diagram to those on the binary. Note that the composition of AB_3 lies outside the primary phase field for AB_3. This allows us to easily spot an incongruently melting binary compound on a ternary diagram. Note also the presence of a ternary eutectic E and a ternary peritectic P. All compositions in composition triangle B–AB_3–C follow a crystallization path that terminates at the ternary peritectic P, whereas those compositions in A–AB_3–C finally crystallize at E. E is a lower temperature than P, so compositions in A–AB_3–C contain a liquid phase to a lower temperature. Thus, compositions in triangle B–AB_3–C are more refractory.

6.1.6.4 Ternary Compounds

So far we have only discussed intermediate compounds of binary composition. Intermediate ternary compounds are also common. Figure 6.28 compares simple ternary diagrams with a congruently melting ternary compound and an incongruently melting compound. The two are easy to distinguish. The congruently melting composition (ABC) lies within its primary field (*abc*) whereas the incongruently melting composition lies outside its primary field.

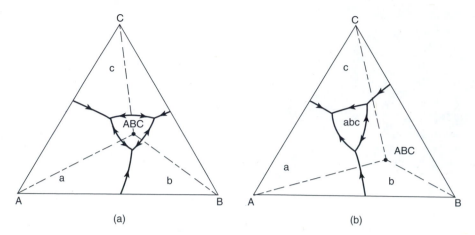

FIGURE 6.28 Schematics of condensed ternary diagrams. (a) Congruently melting ternary compound, and (b) incongruently melting ternary compound. (© ASM International.)

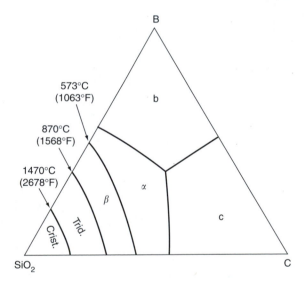

FIGURE 6.29 Schematic ternary diagram showing the method for illustrating polymorphic phase transformations. (From Bergeron, C. and Risbud, S., *Introduction to Phase Equilibria in Ceramics*, American Ceramic Society, Ohio, 1984, p. 101. With permission.)

6.1.6.5 Polymorphic Transformations

Figure 6.29 uses a system containing SiO_2 to illustrate the representation of polymorphic forms of a compound on a ternary phase equilibrium diagram. Alpha quartz is stable up to 573°C (1063°F), beta quartz from 573 to 870°C (1063 to 1598°F), tridymite between 870 and 1470°C (1598 and 2678°F), and cristobalite above 1470°C (2678°F). The stable temperature range of each polymorph is therefore indicated by curved lines that coincide with isotherms. However, remember that equilibrium is not always reached in real systems and that metastable polymorphs are commonly encountered, especially in system containing SiO_2 and ZrO_2.

6.1.6.6 Immiscible Liquids in Ternary Systems

Liquid immiscibility occurs in ternary systems as well as in binary systems. Figure 6.30 shows a three-dimensional model and the two-dimensional projection of the liquidus surface for a ternary system where the liquid immiscibility extends from the binary CB into the primary field of C.

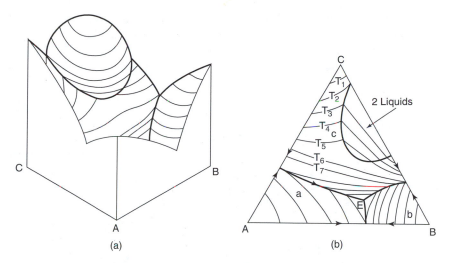

FIGURE 6.30 Schematics illustrating liquid immiscibility for ternary diagrams. (a) Three-dimensional model, and (b) plane projection of liquidus surface.

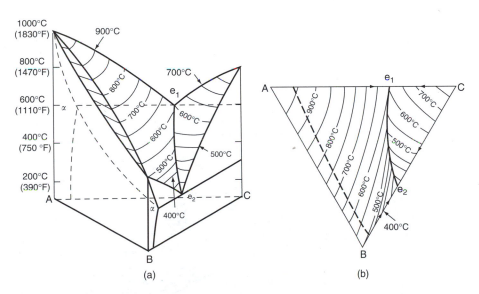

FIGURE 6.31 Schematic of ternary system containing binary compounds with several levels of solid solution. (a) Three-dimensional model, and (b) planar projection of liquidus surface and of α ternary solid-solution range. (From Hummel, F.A., *Phase Equilibria in Ceramic Systems*, Marcel Dekker, New York, 1984, pp. 321, 322. With permission.)

6.1.6.7 Solid Solution in Ternary Systems

Figure 6.31 identifies how solid solution is designated on a ternary diagram. The three-dimensional model in Figure 6.31a contains one binary (A–B) that has complete solid solution between A and B plus two binaries (B–C and A–C) that have limited solid solution of C in A and C in B. B and A have no solid solubility in C. The resulting ternary contains a volume of ternary-composition solid solution identified as α. No ternary eutectic is present. The lowest temperature point on the liquidus is the binary eutectic e_2 on the B–C side of the diagram.

The planar projection of the liquidus surface is illustrated in Figure 6.31b. The solid-solution region is indicated by the dashed line to the left of the diagram.

It is a little difficult to visualize the phases present in different regions of the diagram shown in Figure 6.31. Sometimes, drawing an isothermal plane helps. Figure 6.32a shows the intersection of a 550°C (1020°F) isothermal plane with the three-dimensional model. Figure 6.32b identifies the phases present in each region. Note that the presence of solid solution in the A corner of the

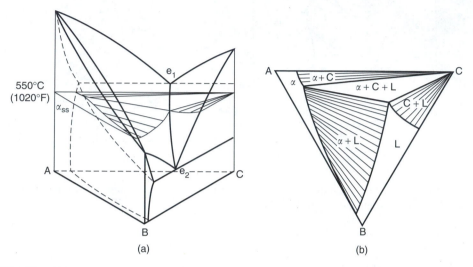

FIGURE 6.32 (a) Intersection of 550°C (1020°F) isothermal section with the system illustrated in Figure 6.31a. (b) Planar view of 550°C (1020°F) isotherm identifying phases present for different compositions. (From Hummel, F.A., *Phase Equilibria in Ceramic Systems*, Marcel Dekker, New York, 1984, p. 323. With permission.)

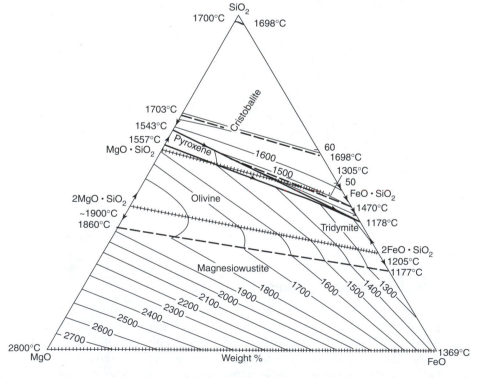

FIGURE 6.33 Extensive solid solution in the ternary system MgO–FeO–SiO₂. (From Bergeron, C. and Risbud, S., *Introduction to Phase Equilibria in Ceramics*, American Ceramic Society, Ohio, 1984, p. 108. With permission.)

diagram results in compositions with much higher temperature capability than compositions in other regions of the diagram or of compositions in a similar ternary system without solid solution.

Solid solution is also identified on ternary diagrams by cross-hatching along a line (called a *join*) connecting two compounds. This is illustrated in Figure 6.33 for the system $MgO–FeO–SiO_2$. Complete solid solution exists for MgO and FeO and $2MgO \cdot SiO_2$ and $2FeO \cdot SiO_2$. Nearly complete solid solution exists for $MgO \cdot SiO_2$ and $FeO \cdot SiO_2$.

6.1.6.8 Real Ternary Systems

Figure 6.34 shows the $MgO–Al_2O_3–SiO_2$ ternary system. A variety of the features we have discussed are illustrated by this diagram:

1. Congruently melting binary compounds ($2MgO \cdot SiO_2$ and $MgO \cdot Al_2O_3$)
2. Incongruently melting binary compound ($MgO \cdot SiO_2$)
3. Ternary eutectics and peritectics
4. Incongruently melting ternary compound ($2MgO \cdot 2Al_2O_3 \cdot 5SiO_2$)
5. Liquid immiscibility in the SiO_2-rich corner
6. Phase transformations in the SiO_2 primary-phase field
7. Solid solution of Al_2O_3 in $MgO \cdot Al_2O_3$ (spinel) and of $3Al_2O_3 \cdot 2SiO_2$ (mullite) in $2MgO \cdot 2Al_2O_3 \cdot 5SiO_2$ (cordierite).

The $MgO–Al_2O_3–SiO_2$ system is of considerable importance to the ceramic industry. Many of the compositions including SiO_2, Al_2O_3, forsterite, mullite, and spinel are used in refractories for high-temperature furnace linings. Forsterite, Al_2O_3, $MgO \cdot SiO_2$, and cordierite are good electrical

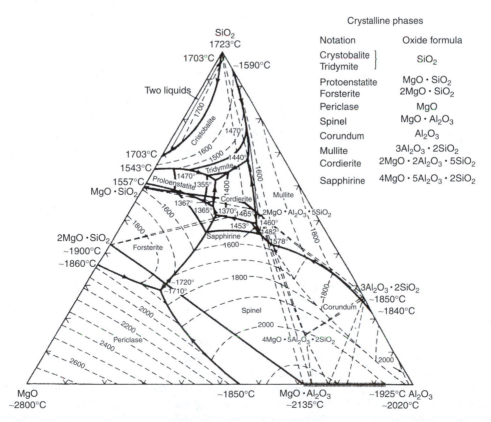

FIGURE 6.34 Ternary phase equilibrium diagram for the system $MgO–Al_2O_3–SiO_2$. (From Ref. 5, Fig. 712.)

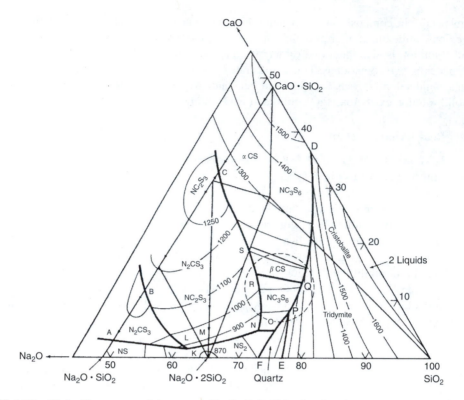

FIGURE 6.35   High-silica corner of the system $Na_2O–CaO–SiO_2$ showing the region (circled) of commercial soda-lime glass compositions. (Adapted from Morey, G.W. and Bowen, N.L., *J. Soc. Glass. Technol.*, 9, 232–233, 1925.)

insulators and are used in the electronics industry. Cordierite has low thermal expansion and has excellent thermal shock resistance.

Figure 6.35 illustrates the silica-rich region of the $Na_2O–CaO–SiO_2$ ternary diagram. This diagram is of particular importance to glass technology. Most of the compositions for window, plate, and container glass are located in the circled region of the diagram modified with small additions of other components. For example, one window glass composition contains approximately 71% SiO_2, 10% CaO, 15% Na_2O, 3% MgO, 1% K_2O, and 1% Al_2O_3.

6.2   PHASE EQUILIBRIUM DIAGRAM COMPOSITION CALCULATIONS

A phase equilibrium diagram contains considerable information regarding percentage of liquids and solids, specific compositions, and change in composition of the phases during temperature change. This information can be obtained by relatively simple calculations and geometric manipulations.

This section describes the types of calculations that are useful and outlines the procedures.

6.2.1   COMPOSITION CONVERSIONS

Compositions on a phase equilibrium diagram are generally in mole%. A *mole* is defined as an amount of substance that contains as many elementary units as there are atoms of carbon in exactly 12 g of pure carbon-12. The elementary units can be atoms, molecules, ions, or any other chemical entity so long as the same elementary unit is specified consistently within a calculation. The abbreviation for mole is mol. It is necessary to be able to convert back and forth between mol% and weight (wt)%. The following are illustrative examples.

Example 6.1. Pure zirconium dioxide (ZrO_2) undergoes a destructive phase transformation between a monoclinic and tetragonal polymorph. An addition of approximately 7 mol% of yttrium oxide (Y_2O_3) stabilizes the ZrO_2 in a cubic structure that can be successfully fabricated and utilized. Calculate the wt% of ZrO_2 and Y_2O_3 to prepare a 1000-g batch of powder of 93 mol% ZrO_2 and 7 mol% Y_2O_3.

Step 1: Use the atomic weight of Zr, Y, and O from the periodic table or the Handbook of Chemistry to determine the molecular weight of ZrO_2 and Y_2O_3:

Atomic weight Zr = 91.22
Atomic weight Y = 88.91
Atomic weight O = 16.00
Molecular weight ZrO_2 = 91.22 + (2)(16.00) = 123.22
Molecular weight Y_2O_3 = (2)(88.91) + (3)(16.00) = 285.82

Step 2: Convert the mol% to wt%.

(0.93 mol)(123.22 ZrO_2/mol) = 114.59 ZrO_2
(0.07 mol)(225.82 Y_2O_3/mol) = 15.81 Y_2O_3

$$\text{Weight fraction } ZrO_2 = \frac{114.59}{114.59 + 15.81} = 0.879 = 87.9\%$$

$$\text{Weight fraction } Y_2O_3 = \frac{15.81}{114.59 + 15.81} = 0.121 = 12.1\%$$

$$\text{Weight fraction } Y_2O_3 = \frac{15.81}{114.59 + 15.81} = 0.121 = 12.1\%$$

Step 3: Calculate the number of grams of ZrO_2 and Y_2O_3 to yield the 1000-gram batch.

Grams ZrO_2 = 0.879 × 1000 g = 879 g
Grams Y_2O_3 = 0.121 × 1000 g = 121 g

Example 6.2. The system La_2O_3–Nb_2O_5 has several intermediate compounds, one of which is $La_2Nb_{12}O_{33}$. What is the molar ratio of La_2O_3 to Nb_2O_5? What is the mol%? To obtain the molar ratio, we separate $La_2Nb_{12}O_{33}$ into individual oxides and determine by simple division the number of moles of each present:

$$La_2Nb_{12}O_{33} = La_2O_3 \cdot Nb_{12}O_{30}; \quad \frac{Nb_{12}O_{30}}{6} = 6Nb_2O_5$$

Therefore,

$$La_2Nb_{12}O_{33} = La_2O_3 \cdot 6Nb_2O_5 \text{ and the molar ratio is 1:6.}$$

$$\text{Mol\% } La_2O_3 = \frac{1}{1+6} = \frac{1}{7} = 14.3 \text{ mol\%}$$

$$\text{Mol\% } Nb_2O_5 = \frac{6}{1+6} = \frac{6}{7} = 85.7 \text{ mol\%}$$

Example 6.3. The ternary system CaO–Al_2O_3–SiO_2 contains a ternary compound $CaO \cdot Al_2O_3 \cdot 2SiO_2$. What are the weight percents of each component to yield this ternary compound?

Solution: Total unit weight equals the sum of the molecular weights of each component, so we first need to calculate the molecular weights for CaO, Al_2O_3, and SiO_2. From the periodic table, *Handbook of Chemistry*, or other suitable source:

Atomic weight Ca = 40.08
Atomic weight Al = 26.98

Atomic weight Si = 28.09
Atomic weight O = 16.00
Molecular weight CaO = 40.08 + 16.00 = 56.08
Molecular weight Al_2O_3 = (2)(26.98) + (3)(16.00) = 101.96
Molecular weight SiO_2 = 28.09 + (2)(16.00) = 60.09
Weight fraction can now be determined for each component:

$$\text{Weight fraction CaO} = (56.08)/[(1)(56.08) + (1)(101.96) \\ + (2)(60.09)] \\ = (56.08)/(278.22) = 0.202$$

$$\text{Weight fraction } Al_2O_3 = (101.96)/(278.22) = 0.366$$

$$\text{Weight fraction } SiO_2 = (2)(60.09)/(278.22) = 0.432$$

Multiplying each by 100 to convert to percent, the weight % CaO, Al_2O_3, and SiO_2 to yield the compound $CaO \cdot Al_2O_3 \cdot 2SiO_2$ equals:

$$20.2\% \text{ CaO, } 36.6\% \text{ } Al_2O_3, 43.2\% \text{ } SiO_2.$$

6.2.2 BINARY COMPOSITION CALCULATIONS

The compositions of all points on a binary diagram are easy to determine. Let us use the simple eutectic diagram in Figure 6.36 to review the techniques. The vertical axis is temperature. The horizontal axis is composition with 100 mol% A to the left and 100 mol% B to the right and with mixed compositions between. The composition of the eutectic point T_e is 70 mol% B and 30 mol% A.

The composition at Y_1, Y_2, Y_3, and C_1 is 90 mol% A and 10 mol% B. Although the total composition is the same at each of these four points (and at all points along the 90–10 vertical C_1 composition line), the percent liquid and solid varies for each point below the liquidus and above the

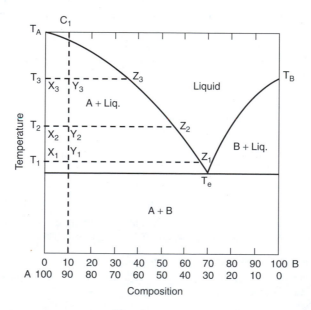

FIGURE 6.36 Binary eutectic diagram used in examples to describe the methods of calculating composition for liquids and solids.

solidus. The percent liquid and solid for any point can be calculated using the "lever rule." For example, at Y_1:

$$\% \text{ Solid} = \frac{X_1 Z_1 - X_1 Y_1}{X_1 Z_1} = \frac{67 - 10}{67} = 85\%$$

$$\% \text{ Liquid} = \frac{X_1 Z_1 - Y_1 Z_1}{X_1 Z_1} = \frac{67 - 57}{67} = 15\%$$

The composition of the liquid at Y_1 is defined by the intersection of the horizontal T_1 line with the liquidus curve, that is, point Z_1. Thus, the composition of the liquid at Y_1 is 67% B and 33% A. We can now check to see if our calculations are correct by adding up the A and B in the liquid and solid to see if they equal the original C_1 composition of 90% A and 10% B:

	Percent in Solid	Percent in Liquid	Total
A	85	$33 \times 0.15 = 5$	90
B	0	$67 \times 0.15 = 10$	10

Example 6.4. What percentages of solid and liquid are present for composition C_1 at temperature T_2? What is the composition of the liquid?

$$\% \text{ Solid at } T_2 = \frac{X_2 Z_2 - X_2 Y_2}{X_2 Z_2} = \frac{55 - 10}{55} = 82\%$$

$$\% \text{ Liquid at } T_2 = \frac{X_2 Z_2 - Y_2 Z_2}{X_2 Z_2} = \frac{55 - 45}{55} = 18\%$$

Composition of liquid at $T_2 = Z_2 = 55\%$ B and 45% A. To check:

	Percent in Solid	Percent in Liquid	Total
A	82	$45 \times 0.18 = 8$	90
B	0	$55 \times 0.18 = 10$	10

Example 6.5. What percentage of solid and liquid are present for composition C_1 at temperature T_3? What is the composition of the liquid?

$$\% \text{ Solid at } T_3 = \frac{X_3 Z_3 - X_3 Y_3}{X_3 Z_3} = \frac{35 - 10}{35} = 71\%$$

$$\% \text{ Liquid at } T_3 = \frac{X_3 Z_3 - Y_3 Z_3}{X_3 Z_3} = \frac{35 - 25}{35} = 29\%$$

Composition of liquid at $T_3 = Z_3 = 35\%$ B and 65% A.

6.2.3 TERNARY COMPOSITION CALCULATIONS

Compositions in a condensed ternary diagram are easy to determine using the planar projection. Two techniques may be used, as illustrated in Figure 6.37. In the first technique, as shown in Figure 6.37a, the percent of each component is proportional to the length of the perpendicular drawn from each edge of the diagram to the composition point. The composition at point X is 20% A, 40% B,

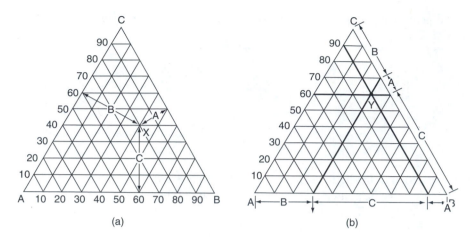

FIGURE 6.37 Techniques used to determine compositions on a ternary diagram. (From Hummel, F.A., *Phase Equilibria in Ceramic Systems*, Marcel Dekker, New York, 1984, p. 165. With permission.)

and 40% C. In the second technique, as shown in Figure 6.37b, lines are drawn through the composition point parallel to the three sides of the triangle. The intersections of these lines with any side of the triangle separate that side into fractions equivalent to A, B, and C in the composition. This technique can be used easily with any composition triangle having any angles and side lengths. The composition at point Y is 10% A, 30% B, and 60% C. For practice, try the first technique to determine the composition of Y and the second technique to determine the composition of X.

6.3 ISOPLETHAL CRYSTALLIZATION PATHS

The final mixture of phases in a ceramic has a strong influence on the properties. The equilibrium phases present as a function of temperature can be estimated from the phase equilibrium diagram. One technique to do this is an isoplethal analysis. *Isoplethal* refers to constant composition. Isoplethal analysis thus involves examining the changes that occur for a single composition as a function of temperature under equilibrium conditions.

6.3.1 BINARY ISOPLETHAL ANALYSIS

Portions of an isoplethal analysis were conducted in Examples 6.4 and 6.5 for the composition 90 mol% A and 10 mol% B. Let us now conduct an isoplethal analysis starting with a liquid at temperature C_1 in Figure 6.36 and cooling slowly to room temperature. From C_1 to the liquidus we have a homogeneous liquid of composition 90% A and 10% B. As we move just below the liquidus, we start to crystallize a very small amount of component A because we are now in the A + liquid region. The segment of the lever for the percent solid is very short (compared to the segment for the percent liquid) just below the liquidus. However, as the temperature cools, the lever segment for percent solid A gets longer while the segment for liquid remains constant. Thus, we calculate a steady increase in the amount of solid A as the temperature continues to decrease. At the same time, the composition of the liquid is changing and becoming richer in B as A crystallizes.

By the time the temperature has reached T_3, as we calculated in Example 6.5, the material is 71% solid and 29% liquid and the composition of the liquid has changed to 35% B and 65% A (compared to the original 90% A and 10% B). By T_2 the solid content has increased to 82% and by T_1 to 85%.

When T_e is reached, the temperature remains constant until all the liquid has crystallized to a combination of A plus B. Below T_e only solid A and B are present.

What is the significance of our isoplethal analysis and of the quantitative calculations we have conducted? It is of greatest significance for the specific example that the material goes from completely solid just below the eutectic temperature to nearly 15% liquid just above the eutectic temperature.

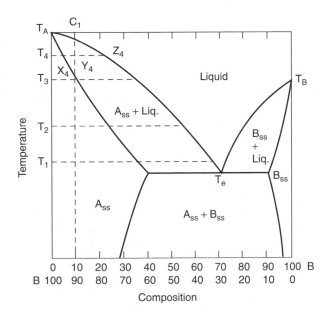

FIGURE 6.38 Binary eutectic diagram with partial solid solution used in examples to describe isoplethal analysis of composition C_1.

Reducing the quantity of B reduces the amount of liquid and increasing B increases the quantity of liquid.

In comparison, let us conduct an isoplethal analysis for the 90A–10B composition for a binary system containing greater than 10% solid solubility of B in A. This is illustrated in Figure 6.38. At C_1 a homogeneous liquid of composition 90A–10B is present. Solid begins to crystallize at the liquidus. This solid is the solid solution of B in A (A_{ss}). By temperature T_4, using the lever rule:

$$\% \text{ Solid} = \frac{X_4 Z_4 - X_4 Y_4}{X_4 Z_4} = \frac{(21 - 6) - (10 - 6)}{21 - 6} = 73\%$$

$$\% \text{ Liquid} = \frac{X_4 Z_4 - Y_4 Z_4}{X_4 Z_4} = \frac{15 - 11}{15} = 27\%$$

The composition of the liquid at T_4 is determined by the intersection of the lever with the liquidus at Z_4 and is 79% A and 21% B. The composition of the solid in the solid solution is defined by the intersection at point X_4 and is 94% A and 6% B. As before, we should check to see if these numbers correspond to the known composition of 90A–10B:

	Percent in Solid	Percent in Liquid	Total
A	$0.73 \times 94 = 69$	$0.27 \times 79 = 21$	90
B	$0.73 \times 6 = 4$	$0.27 \times 21 = 6$	10

As we continue to reduce the temperature, we reach the solid-solution curve just above T_3. The remaining liquid crystallizes. By the time T_3 is reached, the material is completely solid.

Now we can compare the 90A–10B isoplethal analysis for the two systems shown in Figure 6.36 and Figure 6.38. The example with solid solution has a much higher temperature capability. Specifically, at T_3 the system with no solid solution contains 29% liquid, whereas the system with solid solution contains no liquid. In fact, enough liquid is present in the former system at temperatures as low as the eutectic to adversely affect the properties of the material at high temperature. Conversely, the system with no solid solution can be densified during the fabrication process at a much lower temperature than the solid-solution composition. For a material that is intended only for room-temperature or moderate-temperature use, the lower-temperature-densifying system may have perfectly acceptable properties and substantially lower fabrication cost. When selecting a material, these types of trade-offs need to be considered. An understanding of phase equilibrium diagrams can be very valuable in assessing the trade-offs.

6.3.2 TERNARY SYSTEM ISOPLETHAL ANALYSIS

As with binary systems, it is often useful in ternary systems to estimate the sequence of crystallization. This can be done conveniently using the two-dimensional projection of the liquidus surface. A simple example is illustrated in Figure 6.39. The point X has the composition 20% A, 10% B, and 70% C and lies in the primary field of C. As a uniform liquid of composition X is cooled, it remains liquid until temperature T_3 is reached. T_3 is the intersection of composition X with the liquidus surface. As the temperature is cooled slightly below the liquidus, a small amount of solid begins to crystallize. This first solid is C because composition X lies in the primary field of C. As the temperature further decreases, more C crystallizes and the composition of the liquid changes. The composition of the liquid at T_4 is equivalent to the composition of point 4', which is 23.5% A, 11% B, and 65.5% C. The percent liquid and solid at T_4 is determined by the lever rule:

$$\text{Fraction liquid at } T_4 = \frac{\text{CX}}{\text{C4}'} = \frac{32}{37} = 0.865$$

$$\text{Fraction solid (C) at } T_4 = \frac{\text{X4}'}{\text{C4}'} = \frac{5}{37} = 0.135$$

As the temperature is further decreased, the liquid composition moves away from C along the C6' and the percent solid and liquid continue to be determined by the lever rule. At temperature T_6 the

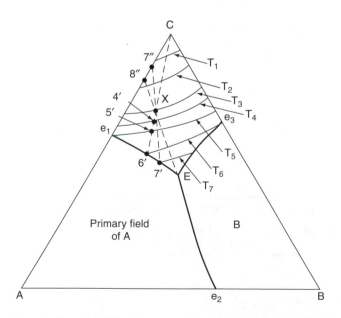

FIGURE 6.39 Simple eutectic ternary diagram used with examples in text to illustrate the crystallization path for composition X and the use of the lever rule to calculate percent liquid and solid at each temperature. (From Ref. 1, p. 169.)

liquid composition is 32% A, 15% B, and 53% C and the percent liquid and solid are 62.7% and 37.3% based on the calculations:

$$\text{Fraction liquid at } T_6 = \frac{CX}{C6'} = \frac{32}{51} = 0.627$$

$$\text{Fraction solid at } T_6 = \frac{X6'}{C6'} = \frac{19}{51} = 0.373$$

The solid is all C. At T_6 the boundary line between the primary phase fields of C and A is intersected. This is the boundary line e_1–E. We now need to decide which direction to follow along the boundary line. We do this by locating the Alkamade line that intersects this boundary line. In Figure 6.39 the Alkamade line is A–C. The Alkamade theorem states that the highest temperature along the boundary line is the point of intersection of the Alkamade line with the boundary. Thus, e_1 is the highest temperature point, 6′ must be a lower temperature, and E must be still lower. Therefore, the crystallization path follows the boundary line from 6′ to E.

As the temperature is reduced below T_6, C and A simultaneously crystallize and the crystallization path follows along the boundary line e_1–E toward the ternary eutectic E. At T_7 the liquid has the composition 29% A, 21% B, and 50% C. The percent liquid and solid are still determined by the lever rule, but the tie lines no longer intersect point C. Instead, they extend from the point on the boundary line through the composition point X and intersect the A–C side of the triangle. The percent liquid and solid now become 45% and 55% based on the calculations:

$$\text{Fraction liquid at } T_7 = \frac{7''X}{7''7'} = \frac{18}{40} = 0.45$$

$$\text{Fraction solid at } T_7 = \frac{X7'}{7''7'} = \frac{22}{40} = 0.55$$

The fraction solid at T_7 contains C plus A:

$$\text{Fraction A} = \frac{C7''}{CA} = 0.13 \text{ or } 13\%$$

$$\text{Fraction C} = \frac{7''A}{CA} = 0.87 \text{ or } 87\%$$

As cooling continues below T_7, A and C continue to crystallize from the melt until the eutectic temperature is reached at E. At E, B begins to crystallize so that the solid A, B, and C are in equilibrium with liquid. Below E the material is completely solid. At a temperature slightly above E, the liquid composition is approximately 25% A, 30% B, and 45% C. The fraction liquid and solid are determined by the lever rule, where

$$\text{Fraction liquid at } E^+ = \frac{8''X}{8''E} = 0.333 = 33.3\%$$

$$\text{Fraction solid at } E^+ = \frac{XE}{8''E} = 0.667 = 66.7\%$$

As we can see from this example, determination of the crystallization path is easy for a simple ternary eutectic system. Techniques for quantitative evaluation of crystallization paths in ternary systems with intermediate compounds are very similar but will not be presented in this introductory text. References 1 and 2 provide a variety of examples of ternary isoplethal analysis and are recommended if the reader wishes further study.

6.4 NONEQUILIBRIUM BEHAVIOR

Phase equilibrium diagram can help us estimate the composition versus temperature behavior of ceramic systems. However, we must be careful and recognize that equilibrium is not always reached quickly in real systems. Sometimes this causes difficulties; other times we can use it to our benefit.

Nonequilibrium can result for a number of reasons:

1. Sluggish kinetics
2. Rapid heating or cooling
3. Nucleation difficulty
4. Elastic constraint of polymorphic transformation

Each of these will be discussed and some practical examples identified.

6.4.1 SLUGGISH KINETICS

Kinetics refers to the rate at which an event occurs. For a ceramic compound to form, atoms must diffuse from one place to another. Diffusion through a liquid is fast but is generally slow through a solid. Crystallization of solids from a melt requires diffusion of atoms through solids. Similarly, densification of compacted ceramic particles involves diffusion.

Most ceramic products are produced by high-temperature reaction of a compact of particles. Let us look at a few examples where kinetics dominate and make equilibrium difficult to attain. One example is phenacite (Be_2SiO_4). The phase equilibrium diagram suggests that heating a mixture of the correct proportion of BeO and SiO_2 in the temperature range 1200 to 1500°C (2190 to 2730°F) should yield Be_2SiO_4. However, the rate of reaction between BeO and SiO_2 is so slow that very little reaction occurs within a reasonable time. However, addition of a little Zn_2SiO_4 allows the reaction to occur. The Zn_2SiO_4 is referred to as *mineralizer* or *fluxing agent*. Other ceramics that are difficult to form by heating the component oxides are mullite ($3Al_2O_3 \cdot 2SiO_2$), cordierite ($2MgO \cdot 2Al_2O_3 \, 5SiO_2$), and $CaSiO_3$.

Another case of sluggish kinetics is reconstructive polymorphic transformation such as occurs in the SiO_2 system. SiO_2 glass, cristobalite, and tridymite are all high-temperature forms of SiO_2 that can be cooled readily to room temperature where the thermodynamically stable equilibrium phase should be quartz. They can then remain at room temperature in the metastable form for many years without converting to the stable quartz form. SiO_2 glass has a very low thermal expansion coefficient and is useful in many furnace tube applications where resistance to thermal shock is important. However, if used at too high a temperature for too long, cristobalite will crystallize and lead to fracture during thermal cycling.

ZrO_2 is another important system in which a metastable polymorphic phase is formed. Pure ZrO_2 is cubic above approximately 2400°C (4350°F). Addition of MgO to ZrO_2 decreases the range of stability of the cubic phase to about 1400°C (2550°F). This ZrO_2 composition with MgO addition can then be cooled to room temperature and the cubic phase retained metastable indefinitely. The equilibrium phase at room temperature is a monoclinic form.

6.4.2 RAPID HEATING OR COOLING

Rapid cooling (quenching) is a well-known technique for forming a glass, which is a noncrystalline metastable phase. Glass that can remain indefinitely in the metastable noncrystalline form at room temperature can be produced readily from a number of systems, including SiO_2, SiO_2–B_2O_3, Na_2O–CaO–SiO_2 and Na_2O–B_2O_3–SiO_2.

Rapid or controlled-rate cooling can be used to obtain a microstructure different from that predicted by an isoplethal analysis from a phase equilibrium diagram. It can alter the percentage of crystalline phase or can result in a combination of glass plus one or more crystalline phases. This can have a dramatic effect on the properties, as is discussed in subsequent chapters.

Rapid heating can also result in nonequilibrium conditions. This can cause problems such as incomplete reaction, crystalline nuclei in a glass, and inhomogeneity. It can also be used to advantage through techniques such as reactive liquid sintering. Look at Figure 6.36 and Figure 6.38 again. The compositions in the solid-solution region contained no liquid until very high temperature and would be difficult to densify. Once densified, though, they would have excellent high-temperature properties. Conversely, the compositions with no solid solution contained substantial liquid and would be easy to densify, but would have poor high-temperature properties. However, by rapid heating of the proper reactants, one can form a nonequilibrium liquid before solid solutions have time to form. This liquid can then permit rapid liquid-phase-aided densification, but subsequently be absorbed into solid solution and yield superior high-temperature properties.

6.4.3 NUCLEATION DIFFICULTY

A crystal cannot form in a liquid until a critical number of atoms come together in a crystal structure to form a nucleus. The formation of a critical-size nucleus is referred to as *nucleation*. Nucleation is difficult in many mixed oxide- and silica-containing melts. These melts are very viscous and the mobility of the atoms is low. In addition, the crystal structures are complex. As a result, the melts tend to supercool and take a long time to reach equilibrium. This is used to advantage in forming glasses and glass ceramics. In the case of a glass ceramic, the liquid is supercooled and solidified and then held for extended time at an elevated temperature until nucleation occurs. The conditions are controlled such that many nuclei are formed simultaneously to achieve a uniform, fine microstructure. The nucleation and growth of crystalline grains within a glass is called *devitrification*.

6.4.4 ELASTIC CONSTRAINT OF A POLYMORPHIC TRANSFORMATION

As we discussed in Chapter 4, polymorphic transformations can be displacive or reconstructive. The reconstructive transformations often are very sluggish and can lead to metastable phases, but the displacive transformations generally are rapid and reversible over a small temperature range. The displacive transformations generally involve a significant volume change, such as occurs for the alpha-beta cristobalite and tetragonal–monoclinic ZrO_2 transformations. It is possible to prevent a polymorphic transformation from occurring by restraining the volume change and thus achieving a metastable phase. An example of current commercial importance is ZrO_2 with selected additives. Pure ZrO_2 goes through a transformation between tetragonal and monoclinic crystal forms in the temperature range 800 to 1100°C (1470 to 2010°F). During cooling, this transformation from tetragonal to monoclinic involves several percent volume increase. If the ZrO_2 particles are small enough and if a controlled amount of Y_2O_3 (or one of several other specific oxides) is added as a stabilizer and if the surrounding microstructure is strong enough, the volume increase can be restrained so that the tetragonal phase is retained metastably to room temperature. The resulting material has remarkable mechanical properties, including unusually high toughness and strength for a ceramic material. The mechanism of the toughening involves controlled transformation of the metastable tetragonal phase to the stable monoclinic phase and is referred to as *transformation toughening*. This important mechanism is discussed in detail in subsequent chapters.

6.4.5 ADDITIONAL INFORMATION ON NONEQUILIBRIUM

The discussion of nonequilibrium in ceramic systems in this chapter has been brief. It is an important topic worthy of further study. Chapter 6 in Reference 1 and Chapter 7 in Reference 2 are recommended.

The concepts we have learned in this chapter about equilibrium and nonequilibrium plus the concepts we learned in Chapters 4 and 5 about atomic bonding and crystal structure now prepare us to discuss in subsequent chapters the factors that control properties and fabrication of advanced ceramics.

REFERENCES

1. Hummel, F.A., *Phase Equilibria in Ceramic Systems*, Marcel Dekker, New York, 1984.
2. Bergeron, C. and Risbud, S., *Introduction to Phase Equilibria in Ceramics*, American Ceramic Society, Ohio, 1984.

3. Masing, G. and Rogers, B.A., *Ternary Systems*, Reinhold, New York, 1944, Paperback, Dover, New York, 1980.

4. Gordon, P., *Principles of Phase Diagrams in Materials Systems*, McGraw-Hill, New York, 1968.

5. Levin, E.M., Robbins, C.R., and McMurdie, H.F., *Phase Diagrams for Ceramists*, M. K. Reser, Ed. American Ceramic Society, Ohio, 1964.

6. Rhines, F.N., *Phase Diagrams in Metallurgy*, McGraw-Hill, New York, 1956.

7. Bowden, S.T., *The Phase Rule and Phase Reactions*, Macmillan, New York, 1938.

8. Findlay, A., Campbell, A.N., and Smith, N.O., *The Phase Rule and Its Applications*, 9th ed., Dover, New York, 1951.

9. Wetmore F.E.W. and LeRoy, D.J., *Principles of Phase Equilibrium*, McGraw-Hill, New York, 1951.

10. Shackelford, J.F., *Introduction to Materials Science for Engineers*, 2nd ed., p. 199, Macmillan, New York, 1988.

11. Schneider, S.J., in *Mechanical and Thermal Properties of Ceramics*, Proc. Symp. NBS Spec. Publ., 303, 1969.

12. Wartenberg H.V. and Prophet, E., Schmelzdiagramme Hochstfeuerfester Oxyde, *Z. Anorg. Allg. Chem.*, 208, 379, 1932.

PROBLEMS

1. Figure Q6.1 contains the compounds A, B, AB, AB_2, AB_3, and A_3B.
   (a) Label each compound on the diagram. Which of these melt congruently? At what temperatures? Are there any compounds that do not melt congruently? If so, describe their behavior.
   (b) Label each region of the diagram. Which material remains completely solid to the highest temperature? Which solid material decomposes and disappears at the lowest temperature? Would you expect metastable behavior in either case during rapid cooldown?
   (c) Are there any eutectics or peritectics? If so, label them and estimate the temperature of each. Are there any other points that need to be labeled? If so, label them and estimate their temperature.

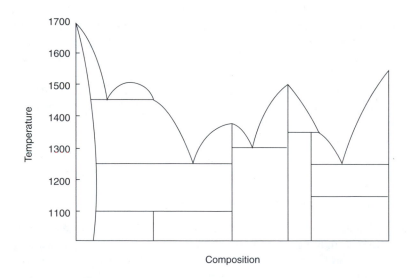

FIGURE Q6.1

2. Construct a phase equilibrium diagram from the following information. Label key points and regions.
   (a) Compounds A and B melt congruently at 1400°C and 1100°C.
   (b) At the composition 70B–30A, homogeneous liquid solidifies directly to a solid containing A plus a high temperature polymorph of B.

    (c) B exhibits a displacive transformation at 300°C.

    (d) No solid solution or intermediate compounds were detected.

3. Construct a phase diagram from the following information:

    (a) Component A melts congruently at 1600°C.

    (b) Component B melts congruently at 1400°C.

    (c) An intermediate compound has composition 50A–50B and melts congruently at 1400°C.

    (d) An intermediate compound has composition 80B–20A and melts incongruently at 1000°C.

    (e) The congruently melting compound has maximum solubility of 50% in a high temperature polymorph of A at 1200°C, maximum solubility of 25% in an intermediate temperature polymorph at 800°C, and maximum solubility of 10% in the lowest temperature polymorph at 500°C.

    (f) The lowest composition at which B is in equilibrium is 75B–25A.

    (g) Invariant points occur at 30% B and 65% B.

4. The binary diagram for TiO_2–Nb_2O_3 contains an intermediate phase $TiO_2 \cdot Nb_2O_5$. What is the wt% TiO_2?

5. Yttrium, aluminum, and oxygen combine to form the intermediate compound $Y_3Al_5O_{12}$. What is the molar ratio Y_2O_3 to Al_2O_3?

6. The ternary system MgO–Al_2O_3–SiO_2 contains the important low thermal expansion compound cordierite ($Mg_2Al_4Si_5O_{18}$). How many grams of Al_2O_3 are required to prepare 1000 g of cordierite?

7. You wish to prepare a ZrO_2 material stabilized with 12 mol% CeO_2. How many grams of CeO_2 are required for a 100-g batch?

8. Use Figure Q6.8 to answer the following:

    (a) The peritectic is located at which point?

    (b) What is the maximum solubility of A in B?

    (c) What is the eutectic composition?

    (d) What is the composition of the liquid at point X?

    (e) What phases are present in region K?

    (f) What is the approximate percent of each phase at point Y?

    (g) The total amount of A at point Z is 12%. What percent of the A is in the liquid and what percent is in the solid?

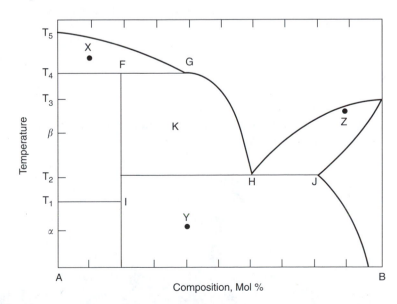

FIGURE Q6.8

STUDY GUIDE

1. Why is an understanding of phase equilibrium and phase equilibrium diagrams useful in understanding and/or controlling the behavior of materials?
2. What are some of the key pieces of information that are shown on a phase equilibrium diagram?
3. A 3D model is required to show the temperature/pressure/composition relationships for a two-component system. How do we simplify to a 2D diagram that still contains much of the important information present in a 3D model?
4. What do the vertical lines on the left and right of a binary diagram represent?
5. Explain the physical significance of the liquidus curve and illustrate with a simple sketch.
6. What is the binary eutectic, and what is its significance?
7. What is an intermediate compound and why does it occur in some systems?
8. What is the difference between congruent and incongruent melting?
9. Explain the distinction between the peritectic temperature and the peritectic point.
10. Explain the difference between a solid solution and a mixture?
11. What factors determine the degree of solubility?
12. Explain the meaning of "liquid immiscibility"? What might be the practical value?
13. How are polymorphic transformations shown on a binary diagram? Sketch an example.
14. Describe how a three-component system is depicted in 2-D.
15. What is a boundary line and why is it useful?
16. What is an isotherm?
17. In what ways is solid solution shown on a ternary diagram?
18. Draw (1) a simple ternary, (2) a ternary with an intermediate congruently melting binary compound BC, (3) a ternary with an intermediate incongruently melting binary compound A_3B, (4) a ternary with a congruently melting ternary compound, and (5) a ternary with an incongruently melting ternary compound.
19. Why might it be useful to look at a phase diagram and visualize a crystallization path?
20. Phase equilibrium diagrams depict our best estimate of the phases present under conditions of thermodynamic equilibrium. What conditions might result in non-equilibrium in a real material?
21. Review the calculations for the amount and composition of solid and liquid phases in Figures 6.36 and 6.38. Explain the difference in liquid content and the lowest temperature where liquid is present for the case of solid solution versus no solid solution.
22. How might you use the information in question 21 to design a ceramic material with superior properties at high temperature?
23. How might you use the information in question 21 to design a ceramic material that can be sintered (densified) at a minimum temperature?

7 Physical and Thermal Behavior

Many applications of ceramics are based upon unique physical and thermal properties. Space shuttle thermal protection tiles, for example, require ultralight weight, high temperature resistance, high thermal shock resistance, and low conduction of heat. Density and melting temperature are important physical properties. Heat conduction is a thermal property. Thermal shock resistance is a combination of thermal and mechanical properties. Chapter 7 identifies key physical and thermal properties and describes the source of the behavior based on the concepts of atomic bonding, crystal structure, and phase equilibrium learned in Chapters 4 to 6.

7.1 PHYSICAL PROPERTIES

7.1.1 DENSITY

Density (ρ) is a measure of the mass (m) per unit volume (V) of a material and is reported in units such as grams per cubic centimeter or pounds per cubic inch. Several factors influence density: the size and atomic weight of the elements, the tightness of packing of the atoms in the crystal structure, and the amount of porosity in the microstructure. The term density can be used in various ways, each with a different meaning. To be sure of the correct meaning, we need to use modifying words:

1. Crystallographic density: the ideal density of a specific crystal structure calculated from chemical composition data and from inter-atomic spacing data obtained by x-ray diffraction
2. Theoretical density: the density of a material that contains zero microstructural porosity (takes into account multiple phases, defect structures, and solid solution)
3. Bulk density: the measured density of a bulk ceramic body (includes all porosity, lattice defects, and phases)
4. Specific gravity: the density of a material relative to the density of an equal volume of water at 4°C (usually based upon crystallographic or theoretical density).

7.1.1.1 Crystallographic Density

The crystallographic density is calculated by dividing the mass (weight) of a unit cell of the material by the volume of the unit cell, as is illustrated in the following examples.

Example 7.1 Estimate the crystallographic density of copper. Copper has a face-centered cubic (FCC) structure. Each unit cell contains four atoms (one eighth each of eight corner atoms plus one half each of six face-centered atoms). The weight per unit cell is:

$$\frac{\text{(Number of atoms per unit cell)(atomic weight)}}{\text{(Avogadro's number*)}}$$

$$= \frac{(4 \text{ atoms})(63.54 \text{ g/mol})}{(6.022 \times 10^{23} \text{ atoms/mol})}$$

$$= 4.22 \times 10^{-22} \text{ g}$$

---

* Avogadro's number $= 6.022 \times 10^{23}$ atoms/mol.

The atomic radius (r) for Cu is $1.278\,\text{Å}$ (Å = angstrom = $10^{-8}\,\text{cm}$) so the length of the edge (a) of the unit cell is:

$$a = \frac{4r}{\sqrt{2}} = \frac{(4)(1.278\,\text{Å})}{\sqrt{2}} = 3.61\,\text{Å}$$

$$3.61\,\text{Å} = 3.61 \times 10^{-8}\,\text{cm}$$

thus

$$V_{\text{unitcell}} = (3.61 \times 10^{-8}\,\text{cm})^3$$
$$= 0.470 \times 10^{-22}\,\text{cm}^3$$

The crystallographic density ρ_c then is:

$$\frac{m}{V} = \frac{4.22 \times 10^{-22}\,\text{g}}{0.470 \times 10^{-22}\,\text{cm}^3}$$
$$= 8.98\,\text{g/cm}^3$$

where m = mass and V = volume. The density of copper is listed as $8.92\,\text{g/cm}^3$ in the *Handbook of Chemistry*.

The atomic weights of the elements that comprise the unit cell have the major effect on the crystallographic density of the material. Elements with low atomic weight such as H, Be, C, N, O, Si, and B result in materials with low crystallographic density. Examples of densities include $2.27\,\text{g/cm}^3$ for graphite (C), $2.51\,\text{g/cm}^3$ for B_4C, $2.65\,\text{g/cm}^3$ for α-quartz (SiO_2 polymorph), $3.22\,\text{g/cm}^3$ for SiC, $2.7\,\text{g/cm}^3$ for Al metal, $1.74\,\text{g/cm}^3$ for Mg metal, and $0.9\,\text{g/cm}^3$ for polyethylene. Elements with high atomic weight such as W, Hf, and U result in materials with high crystallographic density. Examples include $19.4\,\text{g/cm}^3$ for W metal, $15.70\,\text{g/cm}^3$ for WC, and $10.11\,\text{g/cm}^3$ for HfO_2. Elements with intermediate atomic weight result in compounds with intermediate crystallographic density: $6.02\,\text{g/cm}^3$ for $BaTiO_3$, $3.99\,\text{g/cm}^3$ for α-Al_2O_3, and $4.25\,\text{g/cm}^3$ for TiO_2.

The relative effect of atomic weight can be illustrated by comparing the density of two materials that have identical structure and continuous solid solution. Let us use ZrO_2 and HfO_2 as an example. As shown in Chapter 4 and Chapter 5, Zr and Hf have similar ionic radii and outer-electron configurations and can substitute for each other in the same structure. ZrO_2 and HfO_2 have the same structure and form continuous solid solutions. The atomic-weight of Zr is 91.22 and of Hf is 178.49. The crystallographic density of monoclinic ZrO_2 is $5.83\,\text{g/cm}^3$ and of monoclinic HfO_2 is $10.11\,\text{g/cm}^3$.

The stacking of ions or atoms in a structure also affects the crystallographic density. Close packing in metals and ionic ceramics results in higher crystallographic density than in the non-close-packed covalent structures. For example ZrO_2 has a close-packed structure, but zircon ($ZrSiO_4$) has a more open structure resulting from the linking of SiO_4 tetrahedra. Comparison of the atomic weights of the constituent atoms and the molecular weights of the compounds suggests that zircon would have a higher density. However, the open structure of zircon dominates. Thus, zircon has a lower crystallographic density ($4.67\,\text{g/cm}^3$) than monoclinic ZrO_2 ($5.83\,\text{g/cm}^3$).

A more graphic example is the comparison between the diamond and graphite structures of carbon. Diamond has strong bonding, a very short bond length, and close packing. The crystallographic density of diamond is $3.52\,\text{g/cm}^3$, which is surprisingly high for such a low-atomic-weight element. Graphite has strong bonding and relatively close packing in one plane, but weak van der Waals' bonds in the perpendicular direction. The crystallographic density of graphite is $2.27\,\text{g/cm}^3$.

The large difference in crystallographic density of the diamond and graphite polymorphs of carbon is an extreme case. Polymorphs of other materials have a density difference, but not quite as large.

For example, the quartz, tridymite, and cristobalite polymorphs of SiO_2 have crystallographic densities, respectively, of 2.65, 2.19, and 2.33 g/cm^3. As one would expect, the higher-temperature polymorphs typically have a lower density because the structures are more open. An exception is ZrO_2. The crystallographic density for room-temperature monoclinic ZrO_2 has been reported as 5.83 g/cm^3, compared to 6.10 for the higher-temperature tetragonal form and 6.09 for the still-higher-temperature cubic structure.

Crystallographic density values are not always easy to recognize in the literature. Part of the problem is interpretation of the terminology. For example "specific weight," "x-ray density," and "true density" all appear in the literature and, from the evidence present, were used to refer to crystallographic density.

7.1.1.2 Bulk Density

Most commercial ceramic materials contain more than one crystalline phase and often a noncrystalline phase. Each of these has a different density based upon the atoms present and the packing arrangement of the atoms. In addition, there is porosity present in the microstructure. The crystallographic density does not adequately characterize such a mixed-phase porous material. In this case we use the term bulk density B, where

$$B = \frac{\text{mass}}{\text{bulk volume}}$$

$$= \frac{\text{mass}}{\text{volume solids} + \text{volume porosity}}$$

Bulk density can be measured by several techniques. The simplest involves calculating the volume for measured dimensions and dividing this volume into the measured weight. This can readily be done for simple geometrical shapes such as rectangular bars and cylindrical rods as in Equations (7.1) and (7.2):

$$\text{Solid cylinder } B_c = \frac{D}{V_c} = \frac{D}{\pi r^2 h} \tag{7.1}$$

$$\text{Rectangular bar or plate } B_r = \frac{D}{V_r} = \frac{D}{lwh} \tag{7.2}$$

where B_c and B_r and V_c and V_r are, respectively, the bulk density and volume of the solid cylinder and the rectangular bar, D is the measured weight of the shape, r is the radius, h is the height, l is the length, w is the width, and π is 3.14.

The bulk density of a complex shape for which no simple geometric equation is available is measured using Archimedes' principle, where the difference in the weight of the shape in air compared to the weight suspended in water permits calculation of the volume. Parts with no surface-connected porosity can be immersed directly in water. Parts containing surface-connected porosity must be either coated with a wax or other impervious material of known density or boiled as defined in American Society for Testing and Materials Specification ASTM C373.[1] The latter technique permits direct measurement of bulk density, open porosity, water absorption, and apparent specific gravity and indirect assessment of closed porosity. The procedure involves first measuring the dry weight D. The part is then boiled in water for 5 h and allowed to cool in the water for 24 h. The wet weight in air W and the wet weight suspended in water S are then measured. The following can then be calculated:

$$\text{Exterior volume } V = W \cdot S \tag{7.3}$$

$$\text{Bulk density } B = \frac{D}{V} \tag{7.4}$$

$$\text{Apparent porosity } P = \frac{W - D}{V} \tag{7.5}$$

$$\text{Volume of impervious material} = D - S \tag{7.6}$$

$$\text{Apparent specific gravity } T = \frac{D}{D - S} \tag{7.7}$$

$$\text{Water absorption } A = \frac{W - D}{D} \tag{7.8}$$

The apparent specific gravity can then be compared with the true specific gravity or theoretical density to estimate the amount of closed porosity.

Another technique for measuring the bulk density is to use a series of calibrated heavy liquids. If the density of the liquid is identical to that of the ceramic, the ceramic will neither sink nor float. If the density of the liquid is lower, the ceramic will sink. If it is higher, the ceramic will float. This technique is useful for determining the bulk density of very small ceramic samples which contain no open porosity.

7.1.1.3 Theoretical Density

For many application it is desirable to produce a ceramic material that contains minimum open and closed porosity. If this ceramic could be densified completely to contain no open or closed porosity, it would consist only of a mixture of solid phases. This pore-free condition would represent the maximum bulk density achievable for the specific composition and is referred to as the theoretical density. Theoretical density is often used as a standard against which to compare the actual bulk density achieved for a material. For example, if a material contained 10% porosity, it would be defined as 90% of theoretical density.

Theoretical density can be calculated if the crystallographic density and volume fraction of each solid phase in the microstructure are known.

This is illustrated by the following examples.

Example 7.2 A ceramic composite material consists of 30 volume% SiC whiskers in an Al_2O_3 matrix. Estimate the theoretical density (TD) if SiC has a crystallographic density of $3.22 \, g/cm^3$ and Al_2O_3 has a crystallographic density of $3.95 \, g/cm^3$.

$$\begin{aligned} TD &= (\text{Volume fraction SiC})(\rho_c \text{SiC}) + (\text{Volume fraction } Al_2O_3)(\rho_c Al_2O_3) \\ &= (0.3)(3.22 \, g/cm^3) + (0.7)(3.95 \, g/cm^3) \\ &= 0.966 + 2.765 = 3.731 \, g/cm^3 \end{aligned}$$

Example 7.3 A mixture of 30 volume% SiC whiskers and 70 volume% Al_2O_3 powder is hot pressed. The measured bulk density is $3.65 \, g/cm^3$. What is the percent theoretical density? How much porosity is present?

$$\% \, TD = \frac{(\text{bulk density})}{(\text{theoretical density})} \, (100)$$

$$= \frac{(3.65 \text{ g/cm}^3)}{(3.731 \text{ g/cm}^3)} \ (100)$$

$$= 97.8\% \text{ TD}$$

$$\% \text{ Porosity} = 100\% - \text{solids}\%$$
$$= 100 - 97.8 = 2.2\% \text{ porosity}$$

Is this likely to be open porosity or closed porosity? How would you determine this?

The theoretical density can be measured directly by the pycnometer method. The test sample is pulverized until the particles are so small that no closed pores are present. The powder is placed in a known-volume pycnometer bottle and weighed. A liquid is added and the pycnometer bottle is heated to remove air bubbles trapped between particles or air adsorbed on the surface of particles. This assures that only liquid and solid of a known total volume are present. The bottle is weighed again. The difference in weight gives the volume of the liquid. This volume, subtracted from the total known volume of the pycnometer bottle, gives the volume of the solids. The theoretical density is now calculated as the weight of the solids divided by the volume of the solids.

7.1.1.4 Specific Gravity

A final density-related term is specific gravity. Specific gravity (SG) is simply the ratio of the density of a solid or liquid to the density of water at 4°C (39.2°F); that is,

$$SG = \frac{\text{density of materials}}{\text{density of water at 4°C}}$$

$$= \frac{\text{g/cm}^3}{\text{g/cm}^3} = \text{a unitless number}$$

Thus, a material with a specific gravity of 4.5 has a density 4.5 times the density of water at 4°C. Similarly, an equal volume weighs 4.5 times as much as water.

7.1.1.5 Open Porosity

Open porosity is an important density parameter and is often crucial to measure. It can have a strong effect on the properties of a ceramic material. Open porosity can reduce the strength, allow permeability to gases or liquids, alter the electrical characteristics, or compromise the optical behavior. Therefore, it is often important to determine the nature of the porosity in addition to determining the bulk density.

One useful technique that provides information about the distribution as well as the quantity of porosity is mercury porosimetry. The sample is placed in an impermeable container. The container is evacuated and back-filled with a measured volume of mercury. Pressure is then applied to the mercury in increasing increments. At each pressure increment the apparent volume of the mercury plus the sample is measured. At low pressure the mercury can only be forced into large-diameter pore channels in the samples, so the volume change is small. As the pressure is increased, smaller pore channels are permeated, and a greater volume reduction results. The degree of mercury intrusion depends upon the applied pressure, the pore diameter, the surface tension, and the contact wetting angle according to Equation (7.9):

$$p = \frac{4\gamma \cos \Theta}{d} \tag{7.9}$$

where

p = pressure in kg/cm^2
d = pore diameter in μm
Θ = contact angle
γ = surface tension of Hg at test temperature

For most oxide ceramics, the contact angle is 140°, so Equation (7.9) simplifies to Equation (7.10):

$$p \approx \frac{14}{d} \tag{7.10}$$

Modern mercury porosimeters interface with a personal computer and have software to calculate and plot pore size vs. pressure, pore surface area versus pressure, percent porosity versus pressure, volume and surface area distributions versus pore radius, and percent pores of different size ranges. From these plots, an individual can gain insight regarding the porosity distribution.

Figure 7.1 shows an example of the use of a mercury porosimeter to study the fabrication of reaction-bonded silicon nitride. Reaction-bonded silicon nitride is fabricated by first forming a compact of silicon particles, presintering* the compact in vacuum at about 1200°C (2190°F) to achieve adequate strength for green machining,* and then slowly converting the silicon particles to silicon nitride by reaction in nitrogen at temperatures up to 1400°C (2550°F). The material must remain permeable to nitrogen gas to allow complete conversion of the silicon to silicon nitride, yet the objective is to minimize the size and quantity of final porosity to maximize strength and oxidation resistance. Mercury porosimetry was used at each step in the fabrication process to help understand the changes that occurred in the amount and size of the porosity. As can be clearly seen in Figure 7.1, porosity was high in the as-cast silicon compact and was largely just under 0.1 μm in pore channel diameter. Presintering slightly increased the pore channel diameter. Nitriding resulted in a large decrease in the pore quantity and pore channel diameter.

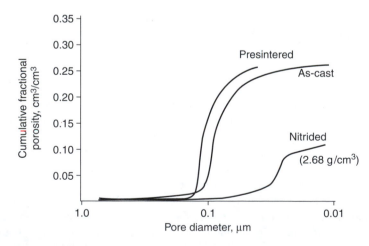

FIGURE 7.1 Mercury porosimetry curves for various stages in the fabrication of reaction-bonded silicon nitride. (From Johansen, K., Richerson, D., and Schuldies, J., *Ceramic Components for Turbine Engines*, Phase II Technical Report, February 29, 1980, U.S. Air Force Contract F33615-77-C-5171. With permission.)

---

* The terms *sintering* and *green machining* are discussed in detail in later chapters.

7.1.1.6 Density Comparisons

The prior sections have described different terms for density and identified the factors that determine the density for a material. Table 7.1 lists density values for a variety of ceramics, metals, and polymers. The values have been assembled from a wide range of literature. Most of the values represent crystallographic density determined for pure phases by calculations from x-ray diffraction data. Other values were reported from experimental measurements on samples that were not verified to be single-phase and pore-free; these values should not be considered accurate for quantitative calculations, but only for general comparisons.

7.1.2 Melting Behavior

Chapter 6 pointed out that different materials have different melting characteristics. Some melt congruently and some melt incongruently. Some sublime and others decompose. The melting or sublimation behavior is largely determined by the strength of the atomic bond. Materials with strong primary atomic and structural bonds tend to have high melting temperatures. Examples are diamond, tungsten metal, and HfC. Materials with weak primary bonds or with van der Waal's bonds as the major structural link have low melting temperatures. Examples include Na metal, polymers, B_2O_3, and NaCl. Table 7.2 compares the approximate melting temperatures of a variety of ceramic, metallic, and organic materials.

Careful examination of Table 7.2 and other melting-temperature data reveals the following trends: (1) weakly bonded alkali metals and monovalent ionic ceramics have low melting temperatures; (2) more-strongly-bonded transition metals (Fe, Ni, Co, etc.) have much higher melting temperatures; (3) multivalent ionic ceramics with increasing covalent bond behavior have increasing melting temperatures; (4) strongly bonded covalent ceramics have very high melting or dissociation temperatures; and (5) strongly bonded metals such as W, Ta, and Mo have very high melting temperatures.

Organic materials have low melting or decomposition temperatures because of the weak van der Waals bonding between molecules. Linear structures such as thermoplastics melt, whereas network structures such as thermosetting resins tend to decompose or degrade. Cross-linking or chain branching tends to increase the melting temperature of thermoplastic compositions.

7.2 THERMAL PROPERTIES

7.2.1 Heat Capacity

The heat capacity c is the energy required to raise the temperature of a material, or more specifically, the quantity of heat required to increase the temperature of a substance one degree. The units are cal/°C or cal/g · °C in the metric system and Btu/°F or Btu/lb · °F in the English system.

Several different terms are used to refer to heat capacity. One is thermal capacity, which is synonymous with heat capacity. A second is molar heat capacity, which is the quantity of heat necessary to increase the temperature of one molecular weight of a material by one degree. A third term is specific heat. Specific heat is defined as the ratio of the heat capacity of a material to that of water at 15°C (59°F). Since it is a ratio, it is dimensionless. The heat capacity of water at 15°C is 0.99976 cal/g · °C (~1.0). Thus, the numerical values of heat capacity in cal/g · °C and specific heat are essentially equal. As a result, the term "specific heat" is often incorrectly used interchangeably with heat capacity and attributed the units cal/g · °C.

The heat capacity of a material is largely determined by the effects of temperature on: (1) the vibrational and rotational energies of the atoms within the materials; (2) the change in energy level of electrons in the structure; and (3) changes in atomic positions during formation of lattice defects (vacancies or interstitials), order–disorder transitions, magnetic orientation or polymorphic

TABLE 7.1
Density of Ceramic, Metallic, and Organic Materials

Material	Composition	Reported Density,[a] g/cm^3 (lb/in.^{3})	X-Ray Density,[b] g/cm^3 (lb/in.^{3})
Ceramic Materials			
α-Aluminum oxide	α-Al_2O_3	3.95 (0.14)	3.987 (0.14)
Aluminum nitride	AlN	3.26 (0.12)	–
Mullite	$Al_6Si_2O_{13}$	–	3.166 (0.11)
Boron carbide	B_4C	2.51 (0.09)	–
Boron nitride	BN	2.20 (0.08)	–
Beryllium oxide	BeO		3.010 (0.10)
Barium titanate	$BaTiO_3$	5.80 (0.21)	–
Diamond	C	–	3.516 (0.13)
Graphite	C	–	2.267 (0.08)
Fluorite	CaF_2	–	3.179 (0.11)
Cerium oxide	CeO_2	–	7.216 (0.33)
Chromium oxide	Cr_2O_3	–	5.225 (0.19)
Spinel	$MgAl_2O_4$	–	3.583 (0.13)
Iron aluminum spinel	$FeAl_2O_4$	–	4.265 (0.15)
Magnetite	$FeFe_2O_4$	–	5.202 (0.13)
Hafnium oxide	HfO_2	9.68 (0.35)	10.108 (0.36)
β-Spondumene	$LiAlSi_2O_6$	–	2.379 (0.08)
Cordierite	$Mg_2Al_4Si_5O_{18}$	–	2.513 (0.09)
Magnesium oxide	MgO	–	3.584 (0.13)
Forsterite	Mg_2SiO_4	–	3.214 (0.12)
Quartz	SiO_2	–	2.648 (0.09)
Tridymite	SiO_2	–	2.192 (0.08)
Cristobalite	SiO_2	–	2.334 (0.08)
Silicon carbide	SiC	3.17 (0.11)	3.22 (0.12)
Silicon nitride	Si_3N_4	3.19 (0.12)	–
Titanium dioxide (rutile)	TiO_2		4.245 (0.15)
Tungsten carbide	WC	15.70 (0.57)	–
Zirconium oxide (monoclinic)	ZrO_2	5.56 (0.20)	5.827 (0.21)
Zircon	$ZrSiO_4$	–	4.669 (0.17)
Metals			
Aluminum	Al	2.7 (0.09)	–
α-Iron	Fe	–	7.875 (0.28)
Magnesium	Mg	1.74 (0.06)	–
1040 Steel	Fe-base alloy	7.85 (0.28)	–
Hastelloy X	Ni-base alloy	8.23 (0.29)	–
HS-25 (L605)	Co-base alloy	9.13 (0.33)	–
Brass	70 Cu-30 Zn	8.5 (0.30)	–
Bronze	95 Cu-5 Sn	8.8 (0.31)	–
Silver	Ag	10.4 (0.37)	10.501 (0.38)
Tungsten	W	19.4 (0.70)	–
Platinum	Pt	–	21.460 (0.77)
Organic Materials			
Polystyrene	Styrene polymer	1.05 (0.03)	–
Teflon	Polytetrafluoroethylene	2.2 (0.08)	–
Plexiglass	Polymethyl methacrylate	1.2 (0.04)	–

[a] Values reported from a variety of literature sources, but not specified whether bulk, crystallographic, or otherwise.

[b] X-ray crystallographic values, mostly from R. Robie, P. Bethke, and K. Beardsley, U.S. Geological Survey Bulletin 1248.

TABLE 7.2
Melting Temperatures of Ceramic, Metallic, and Organic Materials

Material	Approximate Melting Temperature	
	°C	°F
Polystyrene (GP grade)	65–75	150–170[a]
Polymethyl methacrylate	60–90	140–200[a]
Na metal	98	208
Polyethylene	120	250a
Nylon 6[b]	135–150	275–300[a]
Polyamides	260	500[a]
Teflon	290	550[a]
B_2O_3	460	860
Al metal	660	1220
NaCl	801	1474
Ni-base superalloy (Hastelloy X)	1300	~2370
Co-base superalloy (Haynes 25)	1330–1410	2425–2570
Stainless steel (304)	1400–1450	2550–2650
CaF_2	1423	2593
Fused SiO_2	~1650	3000
Si_3N_4[b]	~1750–1900	3180–3450
Mullite	1850	3360
Al_2O_3	2050	3720
Spinel	2135	3875
B_4C	2425	4220
SiC[b]	2300–2500	4170–4530
BeO	2570	4660
ZrO_2 (stabilized)	2500–2600	4530–4710
MgO	2620	4750
WC	2775	5030
UO_2	2800	5070
TiC	3100	5520
ThO_2	3300	5880
W metal	3370	6010
Ca	3500	6240
HfC	3890	6940

[a] Maximum temperature for continuous use.
[b] Sublimes or decomposes.

transformations. For most crystalline ceramics, the heat capacity increases from zero at 0 K to a value near 6.0 cal/g atom · °C around 1000°C (1830°F). Figure 7.2 illustrates the change in heat capacity for several polycrystalline ceramic materials as a function of temperature.

It is apparent from Figure 7.2 that heat capacity is relatively insensitive to the crystal structure or composition of the material. For practical applications, other factors need to be considered. The major one is porosity. A piece of ceramic containing high porosity has less solid material per unit volume than a piece with no porosity. The ceramic with no porosity requires more heat energy than a porous ceramic to heat to a specific temperature. For example, less heat energy is required to raise the temperature of porous firebrick or a fibrous furnace lining than is required for dense firebrick. As a result, a furnace lined with porous firebrick or fibrous insulation can be heated or cooled much more quickly and efficiently.

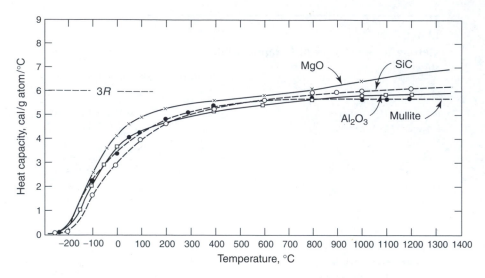

FIGURE 7.2 Comparison for several polycrystalline ceramics of the heat capacity vs. temperature. (From Kingery, W.D., Bowen, H.K., and Uhlmann, D.R., *Introduction to Ceramics*, 2nd ed., Wiley, New York, 1976, Chapter 12. With permission.)

7.2.2 THERMAL CONDUCTIVITY

Thermal conductivity k is the rate of heat flow through a material[2,3] and is usually reported in units of cal/sec · cm² · °C · cm, where calories are the amount of heat, cm² is the cross section through which the heat is traveling, and cm is the distance the heat is traveling. The SI units are W/m · K. Figure 7.3 shows the thermal conductivity of a variety of ceramic, metallic, and organic materials as a function of temperature.[4,5] In the following paragraphs we explore the parameters that affect thermal conductivity.

The amount of heat transfer is controlled by the amount of heat energy present, the nature of the heat carrier in the material, and the amount of dissipation. The heat energy present is a function of the volumetric heat capacity c. The carriers are electrons or phonons, where phonons can be thought of simply as quantized lattice vibrations. The amount of dissipation is a function of scattering effects and can be thought of in terms of attenuation distance for the lattice waves (sometimes referred to as the mean free path).

The thermal conductivity k is directly proportional to the heat capacity c, the quantity and velocity of the carrier v, and the mean free path λ:[6]

$$k \propto cv\lambda \qquad (7.11)$$

Increasing the heat capacity, increasing the number of carriers and their velocity, and increasing the mean free path (i.e., decreasing attenuation or scattering) results in increased thermal conductivity.

In metals the carriers are electrons. Because of the nature of the metallic bond, these electrons are relatively free to move throughout the structure. This large number of carriers plus the large mean free path results in the high thermal conductivity of pure metals. Alloying reduces the mean free path, resulting in decreased thermal conductivity. For example, the thermal conductivity for a mild steel is 51 W/m · K (0.12 cal/sec · cm² · °C · cm), compared to about 400 W/m · K for pure copper or silver.

Most organic materials have low thermal conductivity due to the covalent bonding, the large molecule size, and the lack of crystallinity. Rubber, polystyrene, polyethylene, nylon, polymethyl methacrylate, Teflon, and most other common commercial organic materials have thermal conductivity values at room temperature in the range 0.08 to 0.33 W/m · K (0.0002 to 0.0008 cal/sec · cm² · °C · cm).

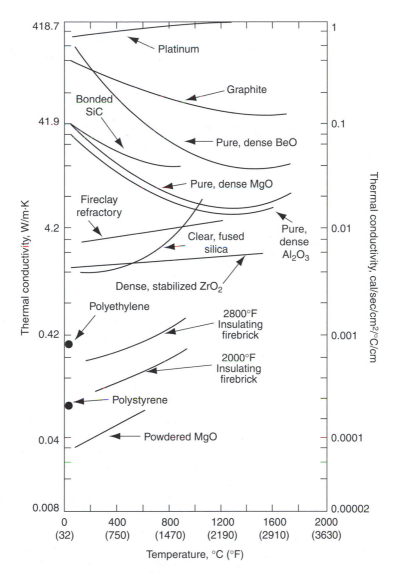

FIGURE 7.3 Thermal conductivity vs. temperature for a variety of ceramic, metallic, and organic materials. (Adapted from Kingery, W.D., Bowen, H.K., and Uhlmann, D.R., *Introduction to Ceramics*, 2nd ed., Wiley, New York, 1976, Chapter 12, p. 643, plus data from Van Vlack, L.H., *Elements of Materials Science*, 2nd ed., Addison-Wesley, Reading, MA, 1964, p. 420.)

Foamed polymers provide very good thermal insulation. The thermal conductivity of solid polymers can be increased by adding conductive fillers such as metals or graphite, but the other properties are also substantially modified.

Ceramic materials exhibit a wide range of thermal conductivity behavior as shown in Figure 7.4. The primary carriers of thermal energy are phonons and radiation. The highest conductivities are achieved in the least cluttered structures, that is, structures consisting of a single element, structures made up of elements of similar atomic weight, and structures with no extraneous atoms in solid solution.

Diamond and graphite are good examples of single-element ceramic structures having high thermal conductivity. Diamond has a thermal conductivity at room temperature of 900 W/m · K, which is more than double that of copper. Because of its layer structure, graphite is anisotropic.

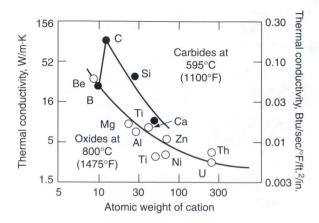

FIGURE 7.4   Effect of cation atomic weight on the thermal conductivity of some ceramic oxides and carbides. (From Kingery, W.D., Bowen, H.K., and Uhlmann, D.R., *Introduction to Ceramics*, 2nd ed., Wiley, New York, 1976, Chapter 12, p. 619. With permission.)

Within the layers the bonding is strong and periodic and does not result in severe scattering of thermally induced lattice vibrations, resulting in high thermal conductivity in this direction (2000 W/m · K for dense, strongly aligned pyrolytic graphite[7]). Only weak van der Waals bonding occurs between layers, and lattice vibrations are quickly attenuated, resulting in much lower thermal conductivity in this direction (10 W/m · K for the same pyrolytic graphite). The reader should note that different fabrication techniques result in different degrees of orientation of the graphite crystals in the microstructure. As a result, the actual thermal conductivity of a bulk sample of graphite depends largely upon the fabrication procedure. Graphite prepared by pressing has nearly random orientation of the graphite crystals and thus has more-uniform bulk thermal conductivity, that is, about 100 W/m · K (0.24 cal/sec · cm² · °C · cm) parallel to the pressing direction and 130 W/m · K (0.31 cal/sec · cm² · °C · cm) perpendicular to the pressing direction. Extruded graphite has greater alignment, but less than for the pyrolytic graphite mentioned previously. The thermal conductivity parallel to the extrusion direction for one example is about 230 W/m · K (0.55 cal/sec · cm² · °C · cm) compared to 140 W/m · K (0.33 cal/sec · cm² · °C · cm) perpendicular to the extrusion direction.[7]

BeO, SiC, and B_4C are good examples of ceramic materials composed of elements of similar atomic weight and size and that have high thermal conductivity. Lattice vibrations can move relatively easily through these structures because the lattice scattering is small. In materials such as UO_2 and ThO_2, where there is a large difference in the size and atomic weight of the anions and cations, much greater lattice scattering occurs and the thermal conductivity is low. UO_2 and ThO_2 have less than one tenth of the thermal conductivity of BeO and SiC. Materials such as MgO, Al_2O_3, and TiO_2 have intermediate values. Figure 7.4 illustrates the correlation of atomic weight and thermal conductivity for various cations in carbide and oxide structures.

Solid solution decreases the thermal conductivity of ceramics. A good example is MgO. Ni^{2+} ions have an ionic radius (for six coordination) of 0.69 Å. This is very close to Mg^{2+} (0.72 Å). NiO and MgO exhibit complete solid solution. Even a slight difference in size and electron distribution results in enough lattice distortion to increase lattice wave scattering and decrease thermal conductivity. The effects of 1 to 35% additions of NiO to MgO on the mean free path are illustrated in Figure 7.5. Another example is Cr_2O_3 addition to Al_2O_3. Al_2O_3 and Cr_2O_3 also exhibit complete solid solution. The effects on thermal diffusivity* are shown in Figure 7.6.

---

* Thermal diffusivity is the thermal conductivity (k) divided by the heat capacity at constant pressure (C_p) and the density of the material (ρ), that is, $k/C_p\rho$.

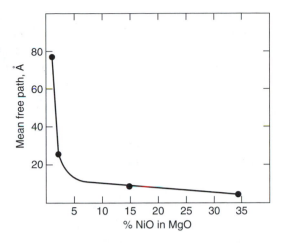

FIGURE 7.5 Reduction of the mean free path in MgO by solid solution with NiO. (Adapted from Kingery, W.D., Bowen, H.K., and Uhlmann, D.R., *Introduction to Ceramics*, 2nd ed., Wiley, New York, 1976, Chapter 12.)

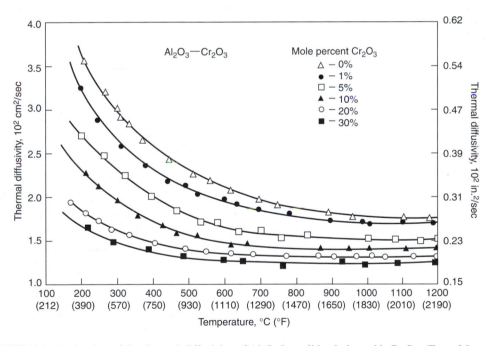

FIGURE 7.6 Reduction of the thermal diffusivity of Al_2O_3 by solid solution with Cr_2O_3. (From Matta, J.E. and Hasselman, D.P.H., *J. Am. Ceram. Soc.*, 58(9–10), 458, 1975. With permission.)

Another example of the effects of solid solution is aluminum nitride (AlN). AlN powder contains Al_2O_3 in solution. Samples fabricated by hot pressing with no additives have thermal conductivity of about 50 W/m · K. Samples fabricated with additions of rare earth or alkaline earth oxides (especially Y_2O_3 and CaO) and held at high temperature for about 1000 min have thermal conductivity over 200 W/m · K.[8] The additive reacts with the Al_2O_3 on the surface of the powder and in solid solution within particles to form a liquid phase at the grain boundaries. This aids in densification, but also acts as a getter to remove Al_2O_3 from solid solution. A long time at high temperature is required for the Al_2O_3 to diffuse from the center of particles to the grain boundaries. The resulting aluminate phase has low thermal conductivity. However, it concentrates at three-grain junctions and along some two-grain faces, but does not continuously coat the AlN grains. As a result, AlN–AlN grain contacts form a continuous network through the sample and the aluminate phase has a relatively small effect on the thermal conductivity.[8] AlN in being developed as a high-thermal-conductivity alternative to Al_2O_3 in advanced electronics applications. New Si and GaAs devices have such a high concentration of circuits that excessive heat is produced during normal operation. AlN can remove this heat by conduction much more rapidly than Al_2O_3.

Temperature has a strong effect on the thermal conductivity of ceramic materials. To understand the mechanisms we need to examine the relationship $k \propto cv\lambda$ as a function of temperature. The heat capacity c initially increases, but approaches a constant. The velocity v remains relatively constant. The mean free path λ is inversely proportional to the temperature T (i.e., $\lambda \propto 1/T$). For crystalline ceramics, where lattice vibrations are the primary mode of heat conduction, the effect of λ dominates and thermal conductivity decreases as temperature increases. For glasses the structure is disordered even at room temperature and λ is low to start with and does not change significantly as the temperature increases. Thus, for glasses the heat capacity c has a major effect and the thermal conductivity typically increases with temperature, as shown for fused silica (SiO_2 glass) in Figure 7.3.

Another temperature-dependent factor is radiant heat transfer. Radiant heat transfer is proportional to an exponential function of T and is usually in the range $T^{3.5}$ to T^5. This can dominate the behavior of glass, transparent crystalline ceramics, and highly porous ceramics and result in an increase in thermal conductivity at elevated temperature. For example, the increase in conductivity with temperature for powdered MgO and insulating firebrick in Figure 7.3 is due to radiation across the pores. In spite of this increase, the thermal conductivity of these porous materials is still very low and is an important factor in their application as high-temperature thermal insulation.

Additional factors that affect the thermal conductivity of ceramics are second-phase dispersions, impurities that do not go into solid solution, porosity, microcracks and noncrystallinity (such as glass), grain boundaries, and grain size. An example of the effect of impurities is illustrated in Figure 7.7 for BeO. The impurities result in a reduction in the thermal conductivity. However, the degree of reduction is not as great as was shown previously for solid solution.

The effect of dispersed phases on conductivity varies depending upon the conductivity of each phase and the distribution of the phases.[4] Figure 7.8 shows schematically various phase distributions and lists examples of each. Equations have been derived to estimate the thermal conductivity of the bulk material based upon the thermal conductivity of each phase. For the parallel slab, the thermal conduction is directional. Parallel to the plane of the slabs the conduction is dominated by the better conductor, as in a parallel electrical circuit, and is expressed in Equation (7.12):

$$k_m = V_1 k_1 + V_2 k_2 \tag{7.12}$$

where
$\quad k_m$ = bulk thermal conductivity
$\quad k_1, k_2$ = thermal conductivity of the two materials and,
$\quad V_1, V_2$ = volume fraction of the two materials (based upon the relative cross-sectional areas).

Perpendicular to the plane of the slabs the conduction is dominated by the lower conductivity material, as in a series electrical circuit, and is expressed in Equation (7.13):

$$\frac{1}{k_m} = \frac{V_1}{k_1} + \frac{V_2}{k_2} \tag{7.13}$$

or Equation (7.14):

$$k_m = \frac{k_1 k_2}{V_1 k_2 + V_2 k_1} \tag{7.14}$$

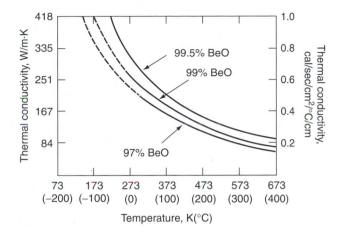

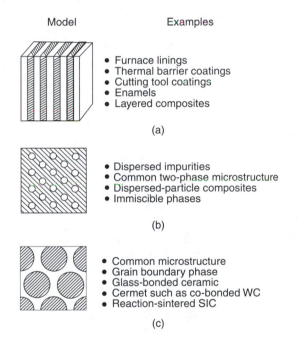

FIGURE 7.7   Thermal conductivity of beryllium oxide (BeO) as a function of temperature for three different purities. (From National Beryllia Division of General Ceramics, Haskell, NJ. With permission.)

FIGURE 7.8   Several models for distribution of two phases in a material: (a) Parallel slabs, (b) continuous matrix phase, discontinuous particulate dispersion, and (c) large isolated grains separated by a continuous minor phase. (From Richerson, D.W., Lesson 4: *Physical and Thermal Behavior, Introduction to Modern Ceramics*, ASM Materials Engineering Institute Course #56, 1990. With permission.)

For dispersed phases, the bulk conductivity k_m is approximated by Equation (7.15):

$$k_m = k_c \left(\frac{1 + 2V_d(1 - k_c/k_d)/(2k_c/k_d + 1)}{1 - V_d(1 - k_c/k_d)/(k_c/k_d + 1)} \right) \tag{7.15}$$

where
k_c = thermal conductivity of the continuous phase
k_d = thermal conductivity of the dispersed phase
V_d = volume fraction of the dispersed phase

The slab configuration aligned perpendicular to the direction of heat flow is extensively used to achieve a reduction in thermal conduction by selecting a low-conductivity material for at least one of the layers. One example is the refractory lining of a furnace. The lining is typically in layers. The internal lining is selected to be stable to the temperature and environment inside the furnace. Ultralow thermal conductivity is not a requirement. The inner lining is surrounded by a low-conductivity material that acts as a barrier to heat transfer and keeps the heat in the furnace. A solid outer layer, such as sheet metal, is then used to hold all the layers together. Another example is the use of a layer of low-thermal-conductivity zirconium dioxide (usually ZrO_2 stabilized with yttrium oxide) as a thermal barrier over the surface of a superalloy gas turbine engine component. If the metal is cooled from the side opposite the ceramic, a temperature drop of about 300°C (570°F) can be achieved across a ZrO_2 layer only about 0.6 mm (0.024 in.) thick. This temperature drop allows the engine to be operated at higher temperature than the uncoated metal could tolerate and results in improved efficiency.

The slab configuration previously was not common. However, recently there has been a rapid increase in the application of coatings applied by a wide variety of techniques: plasma spray and other molten-particle-deposition approaches, chemical vapor deposition, physical vapor deposition, electrophoresis, ion implantation, glazing, enameling, and others. Understanding the thermal conductivity behavior based on the slab model is likely to be important for many applications involving ceramic coatings. However, the other configurations illustrated in Figure 7.8 are more useful for understanding or predicting the behavior of bulk ceramics.

The "continuous matrix, discontinuous dispersion" configuration in Figure 7.8b is very common in ceramics. It occurs in a very dilute fashion when impurities are present that do not form a solid solution with the matrix. An example was shown previously for BeO in Figure 7.7. The thermal conductivity of the major matrix phase dominates until the volume fraction of the dispersant becomes high enough to form continuous paths through the matrix. This results in an S-shaped curve when the thermal conductivity is plotted on one axis and the volume fraction is plotted on the other. Such behavior is typical for many two-phase polycrystalline ceramics where crystallization occurred under approximately equilibrium conditions to form an A + B region, as discussed in Chapter 6. Similar behavior occurs for particulate or whisker dispersion in ceramic matrix composites. Examples are the Al_2O_3–TiC and Al_2O_3–SiC whisker composites used for cutting tool inserts for metal turning applications.

The final configuration, "large isolated grains separated by a continuous minor phase," shown schematically in Figure 7.8c, is also common in ceramic microstructures. Examples include glass-bonded ceramics, many porcelain compositions, Si_3N_4-bonded SiC refractories, reaction-sintered SiC, and many fireclay compositions. Included are ceramics where the cooling cycle after densification is too rapid for equilibrium to occur, so that a portion of the liquid is quenched in as a glassy grain boundary coating on the polycrystalline phases. Because the minor phase is continuous, it tends to dominate the thermal conductivity behavior, similar to the effect of coatings for the slab model.

A special case of second-phase dispersion is porosity. Dead air space is a poor conductor of heat, so porosity typically results in a reduction in thermal conductivity. Figure 7.9 shows a rapid decrease in thermal conductivity of BeO as porosity increases.

7.3 THERMAL EXPANSION

Thermal expansion is the general term used to describe the change in dimensions that occurs with most materials as the temperature is increased or decreased. Thermal expansion data are typically reported in terms of the linear thermal expansion coefficient α as in Equation (7.16):

$$\alpha = \frac{\Delta l / l_0}{\Delta T} \tag{7.16}$$

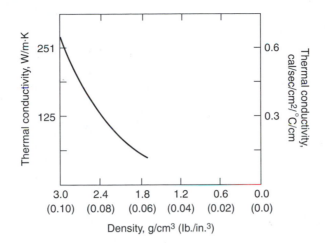

FIGURE 7.9 Effect of porosity on the thermal conductivity of beryllium oxide (BeO). (From National Beryllia Division of General Ceramics, Haskell, NJ. With permission.)

where

l_0 = length at temperature T_1
Δl = change in length ($l - l_0$) between T_1 and T_2
ΔT = change in temperature $T_2 - T_1$

Units are in./in./°F, cm/cm/°C, or simply 1/°F or 1/°C. Frequently, data are plotted in percent expansion vs. temperature or in parts per million expansion vs. temperature. Percent expansion vs. temperature is plotted in Figure 7.10 and Figure 7.11 for a variety of metallic, ceramic, and organic materials.

Since thermal expansion is reported in different ways such as percent, parts per million, and coefficient of thermal expansion, it is useful to be able to calculate back and forth. This is illustrated in the following examples.

Example 7.4   From the % expansion versus temperature curves in Figure 7.10, estimate the average linear coefficient of expansion of Al_2O_3 between 0 and 1000°C.

From Figure 7.10, the percent expansion of Al_2O_3 between 0°C and 1000°C is approximately 0.86%. Thus, the change in length per unit length $\Delta l/l_0$ is 0.86% = 0.0086. ΔT is 1000 − 0 = 1000. Thus:

$$\alpha = \frac{\Delta l/l_0}{\Delta T} = \frac{0.0086}{1000}$$

$$= 8.6 \times 10^{-6}/°C$$

This is very close to values reported in the literature.

Example 7.5   A sample of Si_3N_4 is reported to have an average thermal expansion of 3840 ppm between room temperature (25°C) and 1225°C. What is the average coefficient of thermal expansion?

$$3840\,ppm = \frac{3840}{1,000,000} = 3.84 \times 10^{-3} = \Delta l/l_0$$

$$\alpha = \frac{\Delta l/l_0}{\Delta T} = \frac{3.84 \times 10^{-3}}{(1225 - 25)} = 3.2 \times 10^{-6}/°C$$

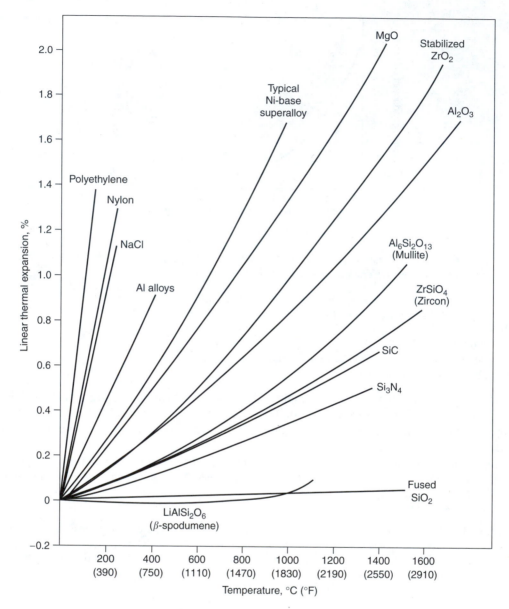

FIGURE 7.10 Thermal expansion characteristics of typical metals, polymers, and polycrystalline ceramics. (Data from numerous sources.)

7.3.1 FACTORS INFLUENCING THERMAL EXPANSION

The change in dimensions results from an increase in the amplitude of vibration between the atoms in the material as the temperature is increased. The amount of dimensional change is determined by the strength of bonding between the atoms and by the arrangement of the atoms in the material.[9,10] Let us examine these factors for various metallic, ceramic, and organic materials.

7.3.1.1 Thermal Expansion of Metals

As discussed in Chapter 5, metallic structures are close packed. When the temperature of a metal is increased, the amplitude of vibration for each atom increases. This results in an increase in the interatomic spacing between adjacent atoms. Since the atoms are close-packed, the increase is cumulative

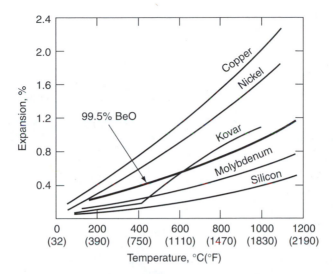

FIGURE 7.11   Percent thermal expansion versus temperature for materials commonly used for electronics devices. (From Fleischner, P.L., Beryllia ceramics in microelectronic applications, *Solid State Technol.*, 20(1), 25–30, 1977. With permission.)

across the whole sample of metal; that is, the total expansion in one direction is equal to the sum of the expansions for each atom.

Metals with weak atomic bonding and a close-packed structure have very high thermal expansion. Examples include the alkali metals Na and K. As the bond strength increases, the thermal expansion decreases. Thus, Al, Fe Cu, and Ni sequentially decrease in thermal expansion, as shown in Figures 7.10 and 7.11. W and Mo have still lower thermal expansion.

7.3.1.2   Thermal Expansion of Ceramics

Materials with a high degree of ionic bonding have thermal expansion behavior similar to metals. The structures are close-packed so that the total expansion is the sum of the thermal expansion of the individual atoms. Weakly bonded ionic substances such as NaCl and KCl have very high expansion. As the bond strength increases and as the percent covalent bond increases, the thermal expansion decreases.

An interesting correlation has been observed between the linear thermal expansion coefficient (α_L) and the melting temperature (M_p) of relatively close-packed metal and ceramic structures.[11] For cubic close packing, $(\alpha_L)(M_p) \approx 0.016$. For rectilinear structures $(\alpha_L)(M_p) \approx 0.027$. This is illustrated in Figure 7.12 where melting point versus linear thermal expansion is plotted for a variety of ceramics. The solid lines are constant product curves for $(\alpha_L)(M_p) \approx 0.027$ (top curve extending from LiI to ThO_2) and $(\alpha_L)(M_p) \approx 0.016$ (bottom curve extending from AgBr to HfO_2). A similar plot is shown in Figure 7.13, which illustrates that data points for many metals and close-packed carbides and borides are close to the $(\alpha_L)(M_p) \approx 0.016$ curve.

It is not surprising that a correlation exists between melting temperature and thermal expansion for close-packed structures. Both melting temperature and expansion coefficient are controlled primarily by bond strength and the magnitude of thermal vibrations. As the bond strength increases, the melting temperature increases and the thermal expansion coefficient decreases.

Close-packed structures with cubic symmetry have uniform thermal expansion along the three crystallographic axes and are referred to as isotropic. Single crystals of noncubic structures have different thermal expansion along different crystallographic directions and are referred to as anisotropic or nonisotropic. Table 7.3 lists the average coefficient of expansion in different crystallographic

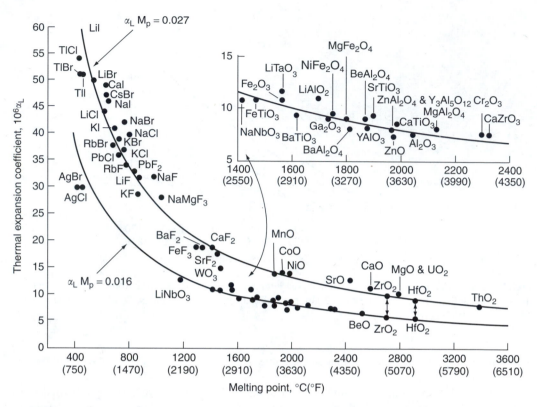

FIGURE 7.12   Plot of average thermal expansion coefficient versus melting temperature for ceramics with close-packed structures. (From Van Uitert, L.G., et al., Thermal expansion—an empirical correlation, *Mater. Res. Bull.*, 12, 261–268, 1977. With permission.)

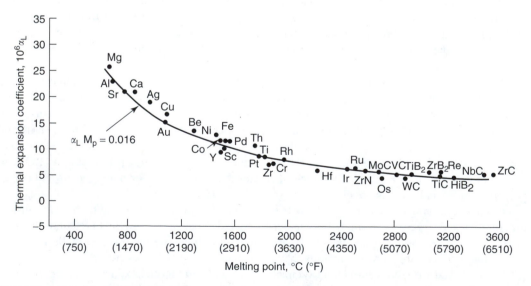

FIGURE 7.13   Average linear thermal expansion vs. melting point for metals, carbides, and borides with close-packed structures. (From Van Uitert, L.G., et al., Thermal expansion—an empirical correlation, *Mater. Res. Bull.*, 12, 261–268, 1977. With permission.)

TABLE 7.3

Ceramic Materials Having Anisotropic Thermal Expansion Behavior

Material	Linear Thermal Expansion Coefficient	
	Normal to c Axis	Parallel to c Axis
Al_2O_3	$8.3 \times 10^{-6}/°C$	$9.0 \times 10^{-6}/°C$
$3Al_2O_3 \cdot 2SiO_2$ (mullite)	4.5	5.7
TiO_2	6.8	8.3
$ZrSiO_4$	3.7	6.2
SiO_2 (quartz)	14	9.0
Graphite	1.0	27
Al_2TiO_5 (aluminum titanate)	-2.6	11.5
$CaCO_3$ (calcite)	-3.7	25.1
KNO_3	-8.5	243.2
$LiAlSi_2O_6$ (β-spondumene)	6.5	-2.0
$LiAlSiO_4$ (β-eucryptite)	8.2	-17.6
$NaZr_2P_3O_{12}$ (NZP)	-4.8	1.9
$BaZr_4P_6O_{24}$ (BZP)	3.8	-1.0

directions for a variety of anisotropic ceramics. These values only apply to a single crystal of each material or a single grain in a polycrystalline ceramic. If the grains in a polycrystalline ceramic have random orientation, the bulk thermal expansion of the ceramic body will be isotropic. The bulk value will be intermediate between the single crystal values for the different crystallographic directions. For example, polycrystalline Al_2O_3 has an average coefficient of expansion of about 8.6 to $8.8 \times 10^{-6}/°C$, compared to a single crystal which has $8.3 \times 10^{-6}/°C$ perpendicular to the c axis and $9.0 \times 10^{-6}/°C$ parallel to the c axis.

Let us look more closely at the ceramics in Table 7.3 to see how the expansion characteristics correlate with the nature of the structures and atomic bonding. Al_2O_3, TiO_2, and mullite have moderately close-packed noncubic structures with intermediate atomic bond strength. As a result, they have moderate thermal expansion with only slight differences in the different crystallographic directions. In contrast, $CaCO_3$ (calcite) and KNO_3 have highly anisometric crystal structures and very large differences in thermal expansion along the different crystallographic directions. The expansion for calcite is so large along the c axis that the crystal must physically contract along the a axis to accommodate the motion.

The anisotropic crystallographic characteristics of graphite were discussed previously in this chapter under the section on thermal conductivity. Within the graphite layers, the atomic bonding is very strong (C—C covalent bonds) and the coefficient of expansion is low ($1 \times 10^{-6}/°C$). The atomic bonding between layers is very weak (van der Waal's bonds) and the thermal expansion coefficient is high ($27 \times 10^{-6}/°C$).

Quartz, Al_2TiO_5, β-spondumene, β-eucryptite, and NZP are all three-dimensional network structures with directional covalent bonds and nonclose-packed atom arrangements containing open cavities or channels. The structures consist of polyhedra of atoms linked together by sharing of corners, as shown for NZP in Figure 7.14. Quartz, β-spondumene and β-eucryptite consist of rings of SiO_4 tetrahedra. β-eucryptite resembles a spiral arrangement along the c axis and has been likened to the coil of a spring. Such structures are not rigid like close-packed structures or structures made up of edge- or face-shared tetrahedrons or octahedrons. During temperature changes, corner-shared bonds have the flexibility to change angle by tilting or rotating. This can absorb thermal vibration motion of atoms that would normally (in a close-packed structure) cause bulk expansion. It can also result in highly anisotropic expansion behavior where the expansion is negative (contraction) in one direction and positive in the perpendicular direction. This is illustrated schematically in Figure 7.15

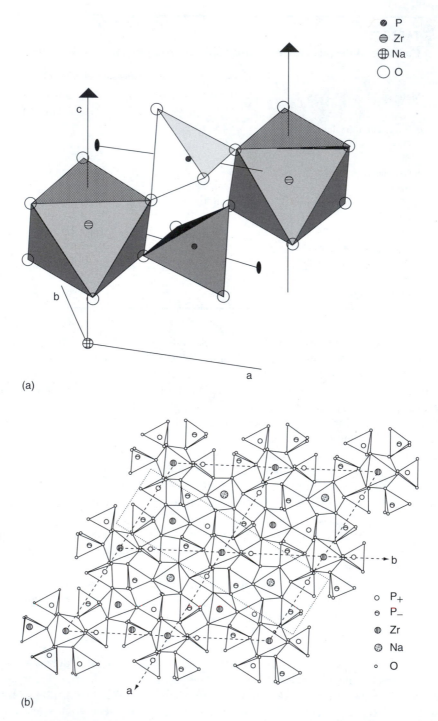

(a)

(b)

FIGURE 7.14   (a) Corner sharing of ZrO_6 octahedra and PO_4 tetrahedra in the NZP structure. (From Chi-Yuen Huang, Thermal Expansion Behavior of Sodium Zirconium Phosphate Structure Type Materials, Ph.D. thesis, Pennsylvania State University, May, 1990. With permission.) (b) Projection of the NZP structure perpendicular to the c axis. (From Chi-Yuen Huang, Thermal Expansion Behavior of Sodium Zirconium Phosphate Structure Type Materials, Ph.D. thesis, Pennsylvania State University, May, 1990. With permission.)

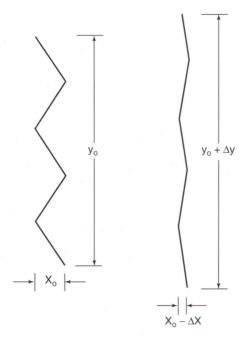

FIGURE 7.15 Schematic illustrating one option for change in dimensions of a corner-shared network structure as the temperature increases and the polyhedra rotate or tilt to straighten the bond angle.

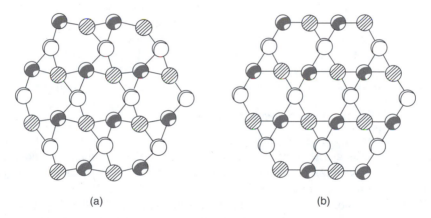

(a) (b)

FIGURE 7.16 Schematic comparing (a) the slightly collapsed structure of quartz at low temperature, and (b) the fully extended structure of quartz at high temperature. (From Kingery, W.D., Bowen, H.K., and Uhlmann, D.R., *Introduction to Ceramics*, 2nd ed., Wiley, New York, 1976, Chapter 12, p. 74. With permission.)

and for the quartz structure in Figure 7.16. At the lower temperature the structure is slightly collapsed so that alternating bonds form a zigzag pattern through the structure. As the temperature increases, the angle of the SiO_4 tetrahedra corner bonds rotates or tilts so that the zigzag pattern begins to straighten. This results in greater expansion parallel to the zigzags and lesser expansion or even contraction perpendicular to the zigzags. In a ring structure such as quartz, this results in expansion of the *a* axis ($14 \times 10^{-6}/°C$) by a hoop growth of the ring and a lesser expansion in the length of the *c* axis ($9 \times 10^{-6}/°C$). For β-spondumene and β-eucryptite, this results in expansion along the *a* axis and contraction along the *c* axis, as shown in Table 7.3.

When anisotropic materials are fabricated into a polycrystalline ceramic body, the net thermal expansion can be very low, as shown for lithium aluminum silicate, $LiAlSi_2O_6$ (spondumene; also

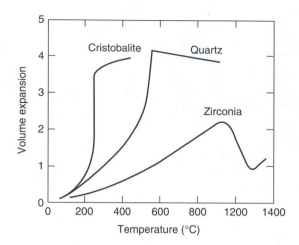

FIGURE 7.17   Volume expansion curves for cristobalite, quartz, and ZrO_2 showing the effects of displacive polymorphic transformations. (From Kingery, W.D., Bowen, H.K., and Uhlmann, D.R., *Introduction to Ceramics*, 2nd ed., Wiley, New York, 1976, Chapter 12. With permission.)

frequently called LAS) in Figure 7.10. Such a material has very little dimensional change as a function of temperature and can therefore withstand extreme thermal cycling or thermal shock without fracturing. Thermal shock resistance is discussed in detail in Chapter 9.

Prior paragraphs have emphasized the large influence of crystal structure on thermal expansion behavior. As discussed in Chapter 5 and Chapter 6, the crystal structure changes for many ceramics at specific temperatures through polymorphic transformation. These transformations can result in an abrupt change in the rate of expansion or in an abrupt volume change, both of which show up on a thermal expansion curve. Displacive transformations in SiO_2 and ZrO_2 are illustrated in Figure 7.17. Order–disorder transformations also are usually visible on a thermal expansion curve.

7.3.1.3   Thermal Expansion of Noncrystalline Solids

Glasses and other noncrystalline solids range in thermal expansion behavior much like crystalline solids. However, in glasses the thermal expansion characteristics are controlled not only by the composition and the resulting structural aspects, but also by the thermal history, that is, the temperature of the initial melt, rate of cooling, and subsequent heat treatments.

Low-thermal-expansion glasses are important in many applications that require a transparent material resistant to thermal shock. Some borosilicate glasses are in this category. Fused silica is also, and it is one of the best thermal-shock-resistant materials available. Fused silica has also been fabricated in a porous foam used for lining critical surfaces of the space shuttle that are exposed to high temperatures during reentry. The low thermal expansion prevents thermal shock damage and the very low thermal conductivity protects underlying structures that are less thermal resistant.

7.3.1.4   Thermal Expansion of Organic Solids

Organic materials, especially the linear structures, have very high thermal expansion, as shown in Figure 7.10. Polyethylene has a thermal expansion at room temperature of 180×10^{-6} cm/cm · °C. Vulcanized rubber, Teflon, and polystyrene have values of 81×10^{-6}, 99×10^{-6}, and 63×10^{-6} cm/cm · °C, respectively. The weak bonding between molecules results in the high thermal expansion. Polymers with network structures typically have lower thermal expansion because the bonding is stronger. The values for melamine-formaldehyde, urea-formaldehyde, and phenol-formaldehyde are 27×10^{-6}, 27×10^{-6} and 72×10^{-6} cm/cm · °C, respectively.[5] These values are still substantially higher than the values for most ceramics and metals.

7.3.1.5 Importance of Thermal Expansion

Many materials applications expose the material to a range of temperature. A large temperature gradient in a material or a mismatch in thermal expansion behavior between two adjacent materials can result in high enough stresses to fracture or distort the material. Thus, for many applications the thermal expansion behavior of the materials is crucial.

Some applications require as close a match in thermal expansion as possible. For example, as shown previously in Figure 7.11, the Kovar metal alloy was developed to have thermal expansion behavior as close as possible to Al_2O_3, BeO, and molybdenum to allow assembly of hermetically sealed electrical devices with these materials. Another example is the use of stabilized ZrO_2 as a thermal barrier coating on superalloys. The coating and the alloy are exposed to large, rapid temperature fluctuations in gas turbine engines. A moderately close thermal expansion match is necessary to minimize thermal stresses and prevent the ceramic coating from spalling off the surface of the metal.

Some applications require very low thermal expansion. A well-known domestic application is the use of LAS-based polycrystalline ceramics for heat-resistant cooking ware (Corningware) and stove tops. Because of the low expansion of the LAS, the cooking ware can be removed directly from and oven and immersed in cold water without breaking. Similarly, the stove top can withstand the very high temperature gradients between the position of the heating element and adjacent areas or when a pan of frozen vegetables is placed directly on the preheated, red-hot burner. The newer Visionware is also a low thermal expansion ceramic material.

Another important consumer application of a low thermal expansion ceramic is the catalyst substrate for a catalytic convertor for pollution control in automobiles. All U.S. automobiles are required to have such a pollution control device. One design consists of a honeycomb structure of cordierite (magnesium aluminum silicate). The exhaust gases from the automobile are at a high temperature and cause large temperature fluctuations and gradients in the honeycomb. A material is required that is stable at high temperature and has a low enough thermal expansion to avoid failure by thermal shock. Thermal shock is discussed in Chapter 9.

Advanced optical and guidance devices also require materials with near-zero thermal expansion. Examples include laser mirrors and the frames of ring laser gyroscopes.

REFERENCES

1. ASTM Specification C373, ASTM Standards, Part 13, American Society for Testing and Materials, Philadelphia, 1969.
2. Kittel, C., *Introduction to Solid State Physics*, 3rd ed., Wiley, New York, 1968.
3. M.I.T., Thermal conductivity, *J. Am. Ceram. Soc.*, 37(2), 67–110, 1954.
4. Kingery, W.D., Bowen, H.K., and Uhlmann, D.R., *Introduction to Ceramics*, 2nd ed., Wiley, New York, 1976, Chapter 12.
5. Van Vlack, L.H., *Elements of Materials Science*, 2nd ed., Addison-Wesley, Reading, Mass., 1964, p. 420.
6. Van Vlack, L.H., *Physical Ceramics for Engineers*, Addison-Wesley, Reading, Mass., 1964, Chapter 9.
7. *Handbook of Chemistry and Physics*, 63rd ed., p. E-11,CRC Press, Boca Raton, FL, 1982.
8. Virkar, A.V., Jackson, T.B., and Cutler, R.A., Thermodynamics and kinetic effects of oxygen removal on the thermal conductivity of aluminum nitride, *J. Am. Ceram. Soc.*, 72(11), 2031–2042, 1989.
9. Megaw, H.D., Crystal structure and thermal expansion, *Mater. Res. Bull.*, 6, 1007–1018, 1971.
10. Megaw, H.D., Thermal expansion of crystals in relation to their structure, *Z. Krist.*, 100, 58–76, 1969.
11. Van Uitert, L.G., et al., Thermal expansion—an empirical correlation, *Mater. Res. Bull.*, 12, 261–268, 1977.

PROBLEMS

1. An engineer is designing a furnace for hot pressing. He wishes to apply the heat circumferentially and the pressure axially. He would like to obtain as uniform as possible a temperature in the interior of the furnace, especially across the diameter, yet needs to avoid

excessive heat loss axially, which might damage the hydraulic press. He could line the top and bottom of the furnace with porous refractory, but available porous refractories are not strong enough to withstand the hot pressing pressures. What material could he select that would help him achieve a uniform temperature in the furnace, yet minimize axial heat loss? Why?

2. Porosity has a strong influence on the properties of ceramic materials. Therefore, bulk density is routinely measured and used in the material specification to determine if the material is acceptable or should be rejected. What effect would you expect porosity to have on thermal conductivity? Explain.

3. Fiber refractories have become important for lining furnaces. They are used in the form of loose blankets (like fiberglass insulation), cottonlike bundles, and fiber boards. What functions do they fulfill? What advantages do they have over conventional brick refractories?

4. Compare the mechanisms of thermal conductivity at room temperature and high temperature for a glass, a polycrystalline ceramic, and a porous refractory brick.

5. Figure 7.10 shows the percent linear thermal expansion as a function of temperature. The average coefficient of thermal expansion can be calculated from the curves for each material over any selected temperature range. Calculate the average thermal expansion coefficient of Al_2O_3 from room temperature to 1000°C. How does it compare with the values listed in Table 7.3? Does this comparison give you any insight regarding the microstructure of the alumina sample depicted in Figure 7.10? Explain.

6. Note that the curves in Figure 7.10 are not all straight lines. This indicates that the rate of expansion is different at different temperatures. For example, the curve for $LiAlSi_2O_6$ appears to have three distinct regions: (1) a negative slope from 0°C to about 500°C, (2) a positive but very low slope from about 500°C to 950°C, and (3) a steeper positive slope from 950°C to 1100°C. Estimate the coefficient of expansion in each region.

7. β-SiC has the cubic [zinc blende] structure, as shown in Figure 5.11. According to the *Handbook of Chemistry*, the lattice parameter a equals 4.359 Å. What is the crystallographic density of β-SiC?

8. What is the crystallographic density of the ferroelectric tetragonal polymorph of $BaTiO_3$? X-ray diffraction data identify that the lattice parameters for $BaTiO_3$ are $a = 3.994$ Å and $c = 4.038$ Å. As shown in Figure 5.20 each $BaTiO_3$ unit cell contains one Ti, one Ba, and three O.

9. Which one of the following would you estimate to have the highest theoretical density?
   (a) ZrO_2
   (b) Zr
   (c) $ZrSiO_4$
   (d) ZrB_2

10. Aluminum, gold, silver chloride, and ZrO_2 all have close-packed structures and roughly follow the relationship $(\alpha_L)(M_p) \approx 0.016$. Which of these materials would you guess to have the highest coefficient of thermal expansion?

11. A sample of SiC has an average thermal expansion of 6300 ppm between 0°C and 1400°C. What is the average coefficient of thermal expansion of this SiC?

12. A cutting tool insert contains 35 vol% of a uniform dispersion of TiC particles. If the thermal conductivity of the Al_2O_3 is 25 W/m · K and that of the TiC is 100 *W/m · K, estimate the thermal conductivity of the cutting tool insert. (*Hypothetical value; actual value generally much lower.)

13. What is the approximate theoretical density of the Al_2O_3–35 vol% TiC cutting tool insert? Assume that TiC has a theoretical density of 4.93 g/cm³ and Al_2O_3 has a theoretical density of 3.99 g/cm³.

14. A sample of BeO is fabricated by pressureless sintering and only has a bulk density of 2.84 g/cm^3. Using density data presented in Chapter 7, calculate the percent theoretical density of this BeO sample.

15. A test bar of reaction-bonded Si_3N_4 is 3 mm thick, 4 mm wide, and 40 mm long and weighs 1.34 g. Calculate the bulk density of this bar.

16. The atomic packing factor can also be estimate knowing the specific gravity of the material. Copper and diamond both have cubic structures. Calculate the packing factor for each. Explain the difference in terms of the nature of the interatomic bonding. Assume spherical atoms and the following data: specific gravity of copper and diamond, 8.96 and 3.51; atomic radii, 1.278 Å and 0.77 Å; and Avogadro's number, 6.02×10^{23} atoms per gram atom.

17. A block of hot-pressed boron carbide (B_4C) weighs 27.36 g and is 6.0 cm long, 4.0 cm wide, and 5.0 mm thick. Assuming that the theoretical density of B_4C is 2.51 g/cm^3, what is the percent porosity in the B_4C block?

STUDY GUIDE

1. What factors influence the density of a material?
2. The term "density" is ambiguous. Usually when we say "density" we mean "bulk density." Explain "bulk density."
3. Identify three methods of measuring bulk density.
4. Identify the groups of materials that fit into each category (examples: multivalent ionic, transition metals, thermoset polymers) and a couple specific materials in each group.

	Ceramics	Metals	Polymers
Relatively low melting temperature	Under 1000°C	Under 1000°C	Under 100°C
Relatively higher melting temperature	1000–1500°C	1000–1500°C	100–200°C
Relatively highest melting temperature	Above 1500°C	Above 1500°C	200–300°C
Decomposes rather than melts			

5. Explain the meaning of thermal conductivity?
6. What factors determine the thermal conductivity of a pure material?
7. How does heat get transmitted through a metal (What are the carriers)?
8. How does heat get transmitted through a ceramic?
9. Identify five ceramic materials with very high thermal conductivity.
10. In Figure 7.3, why does thermal conductivity of dense crystalline ceramics go down with temperature but goes up for transparent fused silica (glass)?
11. Explain why the thermal conductivity of polymers is so low?
12. Explain the effect of solid solution on thermal conductivity.
13. Real ceramics are generally made up of a mixture of crystalline phases, glass, and porosity. How does this affect thermal conductivity in comparison to a pure material?
14. How can we use layering to engineer the rate of heat flow through a material? Identify a couple practical examples.
15. NaCl and MgO both have the same crystal structure. Explain their difference in thermal expansion.
16. Explain why organic materials have high thermal expansion.
17. Explain why some covalently bonded ceramics have the lowest expansion of all materials.
18. Why do materials have anisotropic expansion? Give a couple extreme examples.
19. Name three ceramic materials with near-zero thermal expansion.
20. What are some reasons that knowledge of thermal expansion and manipulation of thermal expansion is important?

8 Mechanical Behavior and Measurement

The mechanical properties of a material determine its limitations for structural applications where the material is required to sustain a load. To make a judicious material selection for such applications, it is useful to understand mechanical properties terminology, theory, and test approaches as well as to obtain the specific property data for the candidate materials.

The objective of this chapter is to review the basic principles of elasticity and strength[1-4] and to develop an understanding of why the actual strength of a ceramic component is far below its theoretical strength. This information is applied concurrently to an evaluation of the property data measurement techniques and the limitations that must be considered when using these data for material selection and component design.[5]

8.1 ELASTICITY

When a load is applied to a material, deformation occurs because of a slight change in the atomic spacing. The load is defined in terms of *stress* σ, which is typically in units of pounds per square inch (psi) or megapascals (MPa). The deformation is defined in terms of *strain* ε, which is typically in units of inches (or centimeters) of deformation per inches (or centimeters) of the initial length or in percent.

The amount and type of strain is dependent on the atomic bond strength of the material, the stress, and the temperature. Up to a certain stress limit for each material the strain is reversible; that is, when the stress is removed, the atomic spacing returns to its original state and the strain disappears. This is referred to as *elastic deformation*, and the stress and strain are related by a simple proportionality constant. For tensile stress

$$\sigma = E\varepsilon \tag{8.1}$$

and the proportionality constant E is called the *modulus of elasticity* or *Young's modulus*.

For shear loading

$$\tau = G\gamma \tag{8.2}$$

where τ is the shear stress, γ the shear strain, and G the proportionality constant, referred to as the *shear modulus* or the *modulus of rigidity*.

At ambient and intermediate temperatures for short-term loading, most ceramics behave elastically with no plastic deformation up to fracture, as illustrated in Figure 8.1a. This is known as *brittle fracture* and is one of the most critical characteristics of a ceramic that must be considered in design for structural applications.

Metals also behave elastically up to a certain stress, but rather than fracture in a brittle manner like ceramics, most metals deform in a ductile manner as the stress is further increased. This is referred to as *plastic deformation* or *plastic strain* and is not reversible. Some metals, aluminum for instance, have a smooth transition from elastic strain to plastic strain, as shown in Figure 8.1b. Others, low-carbon steels for instance, have a discontinuity at the outset of plastic strain. This is called the yield point and is illustrated in Figure 8.1c.

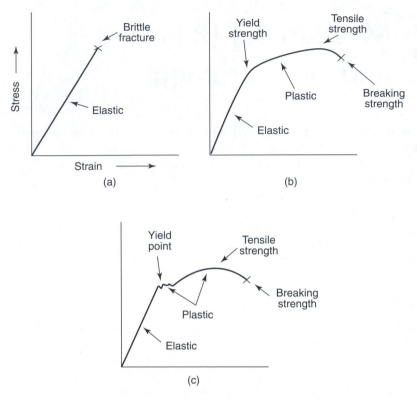

FIGURE 8.1   Types of stress–strain behavior. (a) Brittle fracture typical of ceramics. (b) Plastic deformation with no distinct yield point. (c) Plastic deformation with yield point.

Not all ceramics behave in a brittle fashion and not all metals behave in a ductile fashion. Most ceramic materials undergo plastic deformation at high temperature. Even at room temperature ceramics such as LiF, NaCl, and MgO undergo plastic deformation, especially under sustained loading. These ceramics all have the rock salt structure, which has cubic symmetry and thus has many slip systems available for plastic deformation by dislocation movement.

Pure metals have the greatest degree of ductile behavior. Addition of alloying elements reduces ductility to the point where some metals are brittle at room temperature. Cast iron is a good example. Intermetallic compounds also have little or no ductility at room temperature.

8.1.1 MODULUS OF ELASTICITY

As shown in Equation (8.1), the modulus of elasticity E is the proportionality constant between elastic stress and elastic strain and can be thought of simply as the amount of stress σ required to produce unit elastic strain ε.

$$E = \frac{\sigma}{\varepsilon} \tag{8.3}$$

The magnitude of the elastic modulus is determined by the strength of the atomic bonds in the material. The stronger the atomic bonding, the greater the stress required to increase the interatomic spacing, and thus the greater the value of the modulus of elasticity. Ceramics with weak ionic bonding have low E values. For example, NaCl has a modulus of elasticity of 44.2 GPa (6409 ksi). Ceramics with strong covalent bonding have high E values. The elastic modulus of diamond is 1035 GPa (150,075 ksi).

Metals show a similar trend. Metals with weaker bonding, such as aluminum or magnesium, have low elastic modulus. The elastic modulus of aluminum is 69 GPa (10,005 ksi). Metals with intermediate bond strength have intermediate E values. For example, most iron alloys and Ni-based or Co-based superalloys have an elastic modulus of around 200 GPa (30,000 ksi). Metals like tungsten and molybdenum have strong atomic bonds and high values of elastic modulus.

Organic materials have relatively low values of modulus of elasticity. Their behavior is dominated by the weak van der Waals bonds between molecules, rather than the strong covalent bonds between atoms within each molecule. The elastic modulus of organic materials depends upon the size of the molecules and the degree of polymerization. Rubber can vary from roughly 0.0035 to 3.5 GPa (0.50 to 508 ksi). Urea-formaldehyde, which has a high degree of cross-linking, has an elastic modulus of 10.4 GPa (1508 ksi).

Bond strength, and thus E, varies in different crystallographic directions. This anisotropy must be considered when dealing with single crystals and the crystallographic orientation identified when reporting data. Elastic anisotropy occurs in metals as well as ceramics and in cubic as well as less-symmetrical crystal structures. A single crystal of iron, which has the body-centered cubic structure, has an elastic modulus of 283 GPa (41,035 ksi) in the [111] direction and 124 GPa (17,980 ksi) in the [100] direction. The [111] direction involves atoms that are most closely packed and have the highest net bond strength. This results in a higher E value. The packing density is lower in the [100] direction, resulting in a lower E value.

Most ceramics are polycrystalline and are made up of many crystals in random orientation. If the orientation is truly random, the overall elastic modulus is an average of the elastic moduli for various crystallographic directions. The elastic modulus values most commonly reported for ceramic materials are average values for polycrystalline bodies. Even though these polycrystalline ceramics have an apparent single elastic modulus, the reader must be aware that the individual crystals within the microstructure are anisotropic and that internal stresses may be present that can affect the application of the material.

Table 8.1 compares the average room temperature elastic modulus values of a variety of ceramic, metallic, and organic materials.

Figure 8.2 shows the effect of temperature on the elastic moduli of various ceramics. In each case E decreases slightly as the temperature increases. This results from the increase in interatomic spacing due to thermal expansion. As the interatomic spacing increases, less force is necessary for further separation.

Many materials encountered by an engineer are made up of more than one composition or phase and have an elastic modulus intermediate between the moduli of the two constituent phases. Examples are cermets (Co-bonded WC, for instance), glass- or carbon-reinforced organics, oxide-dispersion-strengthened metals, and glass-bonded ceramics (machinable glass bonded mica, for instance). In cases where the elastic modulus value is not available, it can be estimated using the law of mixtures:

$$E = E_a V_a + E_b V_b \tag{8.4}$$

where E_a and E_b are the elastic moduli of the constituents, V_a and V_b the volume fractions, and E the estimated elastic modulus of the mixture. This is a simplified relationship and is suitable only for rough estimates.

Porosity also affects the elastic modulus, always resulting in a decrease. MacKenzie[9] has derived a relationship for estimating the elastic modulus of porous materials:

$$E = E_0(1 - 1.9P + 0.9P^2) \tag{8.5}$$

where E_0 is the elastic modulus of nonporous material and P the volume fraction of pores. This relationship is valid for materials containing up to 50% porosity and having a Poisson's ratio of 0.3.

TABLE 8.1

Typical Room-Temperature Elastic Modulus Values for Important Engineering Materials

Material	Average Elastic Modulus, E	
	GPa	psi
Rubber	0.0035–3.5	5×10^2–5×10^5
Nylon	2.8	0.4×10^6
Polymethyl methacrylate	3.5	0.5×10^6
Urea-formaldehyde	10.4	1.5×10^6
Bulk graphite	6.9	1×10^6
Concrete	13.8	2×10^6
NaCl	44.2	6.4×10^6
Aluminum alloys	69	10×10^6
Fused SiO_2	69	10×10^6
Typical glass	69	10×10^6
ZrO_2	138	20×10^6
Mullite ($Al_6Si_2O_{13}$)	145	21×10^6
UO_2	173	25×10^6
Iron	197	28.5×10^6
MgO	207	30×10^6
Ni-base superalloy (IN-100)	210	30.4×10^6
Spinel ($MgAl_2O_4$)	284	36×10^6
Si_3N_4	304	44×10^6
BeO	311	45×10^6
Al_2O_3	380	55×10^6
SiC	414	60×10^6
TiC	462	67×10^6
Diamond	1035	150×10^6

Source: Data from Kingery, W.D., Bowen, H.K., and Uhlmann, D.R., *Introduction to Ceramics*, 2nd ed., Wiley, New York, 1976, Chap. 15; Van Vlack, L.H., *Elements of Materials Science*, Addison-Wesley, Reading, MA, 1964, pp. 418–420; Van Vlack, L.H., *Physical Ceramics for Engineers*, Addison-Wesley, Reading, MA, 1964, p. 118.

8.1.2 ELASTIC MODULUS MEASUREMENT

Two techniques are commonly used to measure the elastic modulus. The first involves direct measurement of strain as a function of stress, plotting the data graphically and measuring the slope of the elastic portion of the curve. This technique can be conducted accurately at room temperature using strain gauges, but is limited at temperatures above which strain gauges can be reliably attached. Some individuals have calculated the elastic modulus from load-deflection curves obtained from tensile strength testing. Such data may provide a rough approximation, but it is not likely to be accurate because the deflection curve typically contains other components in addition to the elastic strain of the material, such as deflection of fixtures and looseness in both the load train and data-recording systems.

A second method for determining elastic modulus is based on measurement of the resonant frequency of the material and calculation of E from the equation

$$E = CMf^2$$

$$(8.6)$$

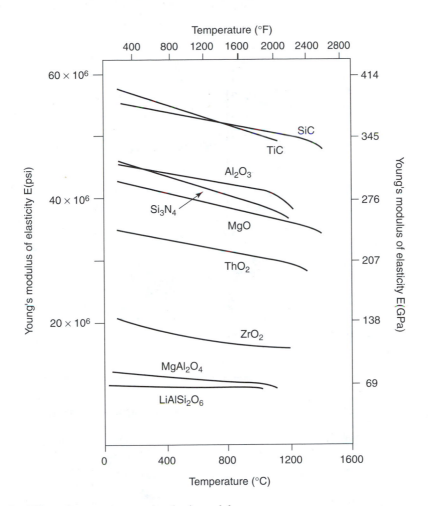

FIGURE 8.2   Effect of temperature on the elastic modulus.

where C is a constant depending on the specimen size and shape and on Poisson's ratio, M is the mass of the specimen, and f is the frequency of the fundamental transverse (flexural) mode of vibration. E can also be determined using the longitudinal or torsional vibration modes, but the equations will be different. These equations plus tables that give values for C can be found in ASTM Specification C747.[10]

This technique can be used accurately over the complete temperature range and for the various crystallographic directions of single crystals as well as for the average elastic modulus of a polycrystalline material.

8.1.3   POISSON'S RATIO

When a tensile load is applied to a material, the length of the sample increases slightly and the thickness decreases slightly. The ratio of the thickness decrease to the length increase is referred to as Poisson's ratio v:

$$v = -\frac{\Delta d/d}{\Delta l/l} \tag{8.7}$$

This is shown schematically in Figure 8.3.

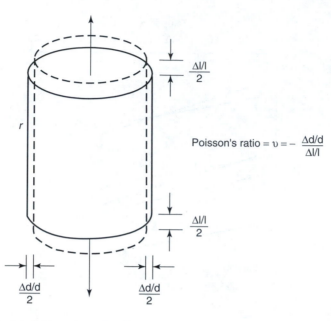

FIGURE 8.3   Physical definition of Poisson's ratio.

TABLE 8.2
Poisson's Ratio at Room Temperature for Various Materials

Material	Approximate Poisson's Ratio
SiC	0.14
$MoSi_2$	0.17
HfC	0.17
Concrete	0.20
B_4C	0.21
Si_3N_4	0.24
SiO_2	0.25
Al_2O_3	0.26
Steels	0.25–0.30
Most metals	0.33
BeO	0.34
MgO	0.36

Poisson's ratio typically varies from 0.1 to 0.5. Values for various materials at room temperature are listed in Table 8.2.

For isotropic and polycrystalline ceramics, Poisson's ratio, Young's modulus, and the shear modulus are related by

$$E = 2G(1 + v) \tag{8.8}$$

8.2 STRENGTH

The term *strength* is ambiguous for both metals and ceramics. One must use modifiers to be specific, that is, yield strength, tensile strength, compressive strength, flexural strength, ultimate strength,

---

TABLE 8.3

Comparison of Theoretical Strength and Actual Strength

Material	E [GPa (psi)]	Estimated Theoretical Strength [GPa (psi)]	Measured Strength of Fibers [GPa (psi)]	Measured Strength of Polycrystalline Specimen [GPa (psi)]
Al_2O_3[a]	380 (55×10^6)	38 (5.5×10^6)	16 (2.3×10^6)	0.4 (60×10^6)
SiC	440 (64×10^6)	44 (6.4×10^6)	21 (3.0×10^6)	0.7 (100×10^6)

[a] From Stokes, R.J., *The Science of Ceramic Machining and Surface Finishing*, NBS Special Publication 348, U.S. Government Printing Office, Washington, DC, 1972, p. 347.

---

fracture (or breaking) strength, and theoretical strength. The following sections discuss the types of strength and attempt to provide the reader with an understanding of the strength characteristics of ceramics and the criteria that must be considered when selecting a ceramic material for a structural application.

8.2.1 THEORETICAL STRENGTH

Theoretical strength can be defined as the tensile stress required to break atomic bonds and pull a structure apart. The equation

$$\sigma_{th} = \left(\frac{E\gamma}{a_0} \right)^{1/2} \tag{8.9}$$

has been derived for estimating the theoretical strength under tensile loading, where σ_{th} is the theoretical strength, E the elastic modulus, a_0 the interatomic spacing, and γ the fracture surface energy. The theoretical strength for ceramic materials typically ranges from one-tenth to one-fifth of the elastic modulus.

Aluminum oxide (Al_2O_3), for instance, has an average elastic modulus of 380 GPa (55×10^6 psi) and would thus have a theoretical strength in the range 38 GPa (5.5×10^6 psi) to 76 GPa (11×10^6 psi). However, the theoretical strength of a ceramic material has not been achieved. This is due to the presence of fabrication flaws and structural flaws in the material, which result in stress concentration and fracture at a load well below the theoretical strength.

Table 8.3 compares the theoretical strengths of Al_2O_3 and silicon carbide (SiC) with typical tensile strengths reported for specimens fabricated by different approaches. Most ceramic products are fabricated from the polycrystalline approach. The fracture strengths of the polycrystalline versions of SiC and Al_2O_3 are only about 1/100 of the theoretical strength.

8.2.2 EFFECTS OF FLAW SIZE

The presence of a flaw such as a crack, pore, or inclusion in a ceramic material results in stress concentration. Inglis[11] showed that the stress concentration at the tip of an elliptical crack in a nonductile material is

$$\frac{\sigma_m}{\sigma_a} = 2 \left(\frac{c}{\rho} \right)^{1/2} \tag{8.10}$$

where σ_m is the maximum stress at the crack tip, σ_a the applied stress, $2c$ the length of the major axis of the crack, and ρ the radius of the crack tip.

To obtain an idea of the effect of flaws on stress concentration, one can assume that the crack tip radius is approximately equal to the atomic spacing a_0 (~2 Å) and use some recent flaw size and strength data for reaction-bonded silicon nitride (Si_3N_4). For a flaw size c of 170 μm (~0.007 in.) fracture occurred at 21.7×10^3 psi (150 MPa). Substituting in Equation (8.10), the stress concentration factor is 1840. It is apparent that even a small flaw in ceramics is extremely critical and leads to substantial stress concentration.

Griffith[12] proposed an equation of the form

$$\sigma_f = A\left(\frac{E\gamma}{c}\right)^{1/2}$$

(8.11)

for relating the fracture stress to the material properties and the flaw size, where σ_f is the fracture stress, E the elastic modulus, γ the fracture energy, c the flaw size, and A a constant that depends on the specimen and flaw geometries.

Evans and Tappin[13] have presented a more general relationship:

$$\sigma_f = \frac{Z}{Y}\left(\frac{2E\gamma}{c}\right)^{1/2}$$

(8.12)

where Y is a dimensionless term that depends on the flaw depth and the test geometry, Z is another dimensionless term that depends on the flaw configuration, c is the depth of a surface flaw (or half the flaw size for an internal flaw), and E and γ are defined as above. For an internal flaw that is less than one tenth the size of the cross section under tensile loading, $Y = 1.77$. For a surface flaw that is much less than one tenth the thickness of a cross section under bend loading, Y approaches 2.0. Z varies according to the flaw shape, but is usually between 1.0 and 2.0.

The effect of a planar elliptical crack at the surface of a ceramic specimen is the easiest to analyze. This type of crack results commonly from machining but can also be due to impact, thermal shock, glaze crazing, or a number of other causes. The Z value depends on the ratio of the depth of the flaw (c) to the length of the flaw (l), as illustrated in Figure 8.4.

The effects of three-dimensional flaws such as pores and inclusions have not been analyzed as rigorously. However, it is evident that the severity of strength reduction is affected by a combination of factors:

1. The shape of a pore
2. The presence of cracks or grain boundary cusps adjacent to a pore
3. The distance between pores and between a pore and the surface
4. The size and shape of an inclusion
5. The differences in elastic moduli and coefficients of thermal expansion between the inclusion and the matrix

These factors are discussed individually in the following paragraphs.

8.2.2.1 Pore Shape

A simple spherical pore theoretically would have less stress concentration effect than a sharp crack. However, pores in ceramics are not perfectly spherical. Some are roughly spherical but most are highly irregular. Roughly spherical pores can result from air entrapment during processing, as shown in Figure 8.5a for a specimen of reaction-bonded silicon nitride (Si_3N_4) fabricated by slip casting. This specimen fractured at 150 MPa (21,800 psi) during four-point bend testing. Several models were evaluated to relate the flaw size to the fracture stress.[14] The best correlation between measured and

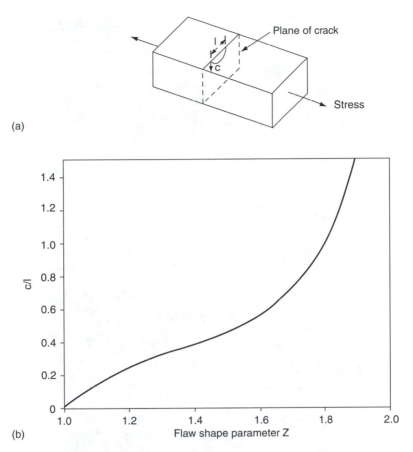

(a)

(b)

FIGURE 8.4 Flaw shape parameter Z values for elliptical surface crack morphologies. (a) Geometry. (b) Z versus c/l. (Reprinted with permission from *Progress in Materials Science*, Vol. 21, Pts. 3/4: *Structural Ceramics*. A.G. Evans and T.G. Langdon, © 1976, Pergamon Press, Ltd. With permission.)

calculated fracture stress was obtained by assuming that the pore was equivalent to an elliptical crack of the cross section outlined by the arrows in Figure 8.5a.

For this calculation, $Y = 2$, $Z = 1.58$ (from Figure 8.4), $E = 219 \times 10^9 \, \text{N/m}^2$ ($32 \times 10^6 \, \text{psi}$), and $\gamma = 11.9 \, \text{J/m}^2$, resulting in a calculated σ_f of 157 MPa (22,800 psi). This is very close to the value measured by four-point bend testing.

8.2.2.2 Pore-Crack Combinations

The simplest and most common combination between a pore and a crack involves the intersection of the pore with grain boundaries of the ceramic material. If the pore is much larger than the grain size of the material (as in the previous example), the extremities of the pore provide a good approximation of the critical flaw size. If the size of the pore approaches the size of the grains, the effect of cracks along the grain boundaries probably will predominate, and the effective flaw size will be larger than that of the pore. Evans and Langdon[14] provide a more detailed treatment of pores, cracks, and pore crack combinations and provide an extensive list of references.

8.2.2.3 Internal Pores

The effect on strength of an internal pore depends on the shape of the pore and the position of the pore with respect to the surface. If the pore is close to the surface, the bridge of material separating it from

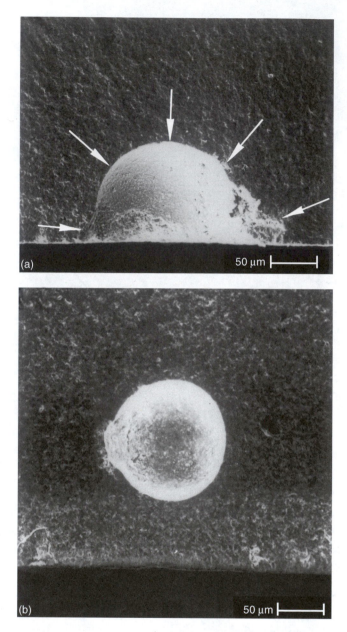

FIGURE 8.5   Scanning electron photomicrographs of fracture surfaces of reaction-bonded silicon nitride containing nearly spherical pores resulting from air entrapment during processing. Arrows outline flaw dimensions used to calculate fracture stress.

the surface may break first and result in a critical flaw whose dimensions would be the size of the pore plus the bridge. In this case, the measured strength is likely to be less than would be predicted simply by the dimensions of the pore and less than if the pore intersected the surface.[13] Experimental data by the author for internal spherical pores that were within about half a radius of the surface correlated well with this ligament theory, resulting in calculated stresses that were within 10% of the measured values.[15] However, for pores progressively further from the surface, the ligament theory did not apply. In fact, other theories proposed in the literature also did not give acceptable correlation. It appears that further study is necessary if we are to understand more fully the stress concentration effects of internal pores and to derive mathematical relationships that provide accurate predictions of fracture stress.

Figure 8.5b shows an internal nearly spherical pore in reaction-bonded Si_3N_4 that was approximately one radius from the surface. The measured bend strength was 215 MPa (31,200 psi). This represents the peak tensile stress at the specimen surface at the time of fracture. Since the tensile stress in a bend specimen decreases linearly inward from the surface (reaching zero at the midplane of the bar), the stress at the plane of the flaw can easily be calculated. At the plane intersecting the deepest portion of the pore, the corrected stress was 182 MPa (26,400 psi). At the plane intersecting the centerline of the pore, the corrected stress was 193 MPa (28,000 psi). Using the ligament theory, the calculated fracture stress was 126 MPa (18,300 psi), well below the measured fracture stress. Assuming a flaw size equal to the pore diameter and $Y = 1.77$ (for an internal flaw) and the same E and γ values as used earlier, the calculated stress was 182 MPa (26,400 psi). This appears to be excellent correlation. However, of the eight specimens evaluated in this way, the average difference between measurement and calculation was 16%, with the highest being 37%. This further illustrates that the effect of internal flaws is not well enough understood for accurate quantitative analysis and that additional experimental and analytical effort is required.

8.2.2.4   Pore Clusters

If a group of pores are close together, the material bridges between them can crack first, linking the pores together and producing a much larger flaw that results in much lower strength. The probability is high that pores separated by less than one pore radius will link.

8.2.2.5   Inclusions

Inclusions typically occur in ceramic materials through contamination of the ceramic powders during processing. Sources of contamination are discussed in detail in Part III. The degree of strength reduction due to an inclusion depends on the thermal and elastic properties of the inclusion compared to the matrix material. Thermal expansion differences can result in cracks forming adjacent to the inclusion during cooling from the fabrication temperature. Elastic modulus difference can result in the formation of cracks when a stress is applied. The worst decrease in strength occurs when the inclusion has a low coefficient of thermal expansion and a low elastic modulus compared to the matrix material. In this case, the effective flaw size is larger than the visible inclusion size. It is equivalent to the inclusion size plus the length of the adjacent cracks.

Inclusions with high thermal expansion coefficient or a high elastic modulus have less effect on strength. These conditions produce circumferential cracks rather than radial cracks, and the effective flaw size approaches that of a flat elliptical crack equivalent in elliptical dimensions to the inclusion.

8.2.3   STRENGTH MEASUREMENT

8.2.3.1   Tensile Strength

Strength can be measured in a number of different ways, as illustrated in Figure 8.6. Tensile strength testing is typically used for characterizing ductile metals. A metal tensile test specimen is attached to threaded fixtures of any universal test machine that can provide a calibrated pull load at controlled rate. Yield strength, breaking strength, and elongation are measured in a single test. The *tensile strength* is defined as the maximum load P (the stress at fracture for a ceramic) divided by the original cross-sectional area A:

$$\sigma_t = \frac{P}{A} \tag{8.13}$$

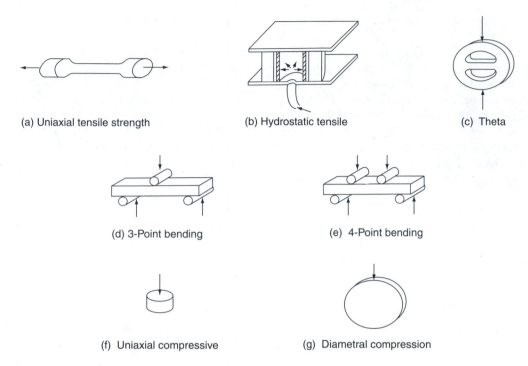

FIGURE 8.6   Schematics of strength tests.

Ceramics are not normally characterized by tensile testing because of the high cost of test specimen fabrication and the requirement for extremely good alignment of the load train during testing. Any misalignment introduces bending and thus stress concentration at surface flaws, which results in uncertainty in the tensile strength measurement. For accurate tensile strength measurement with ceramics, strain gauges must be used to determine the amount of bending and stress analysis must be conducted to determine the stress distribution within the test specimen. A ceramic tensile test specimen is shown in Figure 8.7. This specimen was designed for testing in a very sophisticated test facility at Southern Research Institute, Birmingham, Alabama, in which close alignment was achieved through the use of gas bearings.[16] Other sophisticated tensile test facilities have been established at Oak Ridge National Laboratory, National Institute for Standards and Testing, and the University of Dayton Research Institute. Substantial tensile testing has been conducted at these facilities since the mid-1980s, especially on silicon nitride and silicon carbide ceramics.

Tensile strength can also be measured by applying a hydrostatic load to the inside of a thin-walled, hollow-cylinder specimen configuration (Figure 8.6b).[17] This has been used for room-temperature strength measurement but has not been adapted for elevated temperatures. Achieving adequate seal of the pressurizing fluid at high temperature would be difficult. Another limitation of this test is the likelihood for fracture to occur at flaws on the corners at the ends of the hollow cylinder. It is difficult to machine these corners without producing chips or cracks. The chips or cracks can be removed by radiusing the edges, but this results in reduced specimen thickness, which adversely modifies the stress distribution during testing.

Another method of obtaining tensile strength of a ceramic material is known as the theta test.[18] The configuration is shown in Figure 8.6c. Application of a compressive load to the two arches produces a uniaxial tensile stress in the crossbeam. Very little testing has been conducted with this configuration owing largely to difficulty in specimen fabrication.

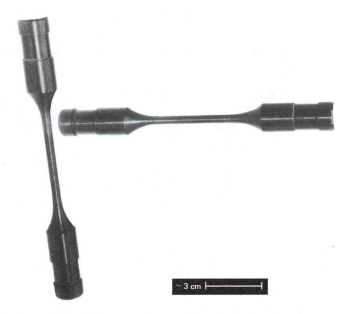

FIGURE 8.7   Typical ceramic tensile test specimen configuration.

8.2.3.2   Compressive Strength

Compressive strength is the crushing strength of a material, as shown in Figure 8.6f. It is rarely measured for metals, but is commonly measured for ceramics, especially those that must support structural loads, such as refractory brick or building brick. Because the compressive strength of a ceramic material is usually much higher than the tensile strength, it is often beneficial to design a ceramic component so that it supports heavy loads in compression rather than tension. In fact, in some applications the ceramic material is prestressed in a state of compression to give it increased resistance to tensile loads that will be imposed during service. The residual compressive stresses must first be overcome by tensile stresses before additional tensile stress can build up to break the ceramic. Concrete prestressed with steel bars is one example. Safety glass is another example.

Rice[19] has conducted an extensive review of the literature and has proposed that the upper limit of compressive strength is the stress at which microplastic yielding (deformation involving slip along crystallographic planes) occurs. He suggests that the microplastic yield stress can be estimated by dividing the measured microhardness, either Vickers or Knoop, by three, and that the compressive strength of current well-fabricated ceramic materials ranges from about one half to three fourths of the yield stress. Table 8.4 lists some of the hardness and compressive strength data reported by Rice. There does appear to be some correlation between the microhardness and compressive strength.

The following factors probably contribute to reduction of the measured compressive strength below the microplastic yield stress for the material: flaws in the ceramic, such as cracks, voids, and impurities; twinning; elastic and thermal expansion anisotropy; and misalignment during testing. Grain size also appears to have a large effect. In general, the compressive strength increases as the grain size decreases.

8.2.3.3   Bend Strength

The strength of ceramic materials is generally characterized by bend testing (also referred to as flexure testing), as illustrated in Figure 8.6d and e. The test specimen can have a circular, square, or rectangular cross section and is uniform along the complete length. Such a specimen is much less expensive to fabricate than a tensile specimen.

TABLE 8.4

Comparison of Hardness and Compressive Strength for Polycrystalline Ceramic Materials

Material	Vickers Hardness		Calculated Stress Hv/3 Yield		Measured Compressive Strength	
	kg/mm²	kpsi	kg/mm²	kpsi	kg/mm²	kpsi
Al_2O_3	2370	3360	790	1120	650	924
BeO	1140	1620	380	540	360	512
MgO	660	930	220	310	200	284
$MgAl_2O_4$	1650	2340	550	780	400	569
Fused SiO_2	540	780	180	260	190	270
ZrO_2 (CaO)	1410	1980	470	660	290	412
$ZrSiO_4$	710	1140	270	380	210	299
SiC	3300	4680	1100	1560	–	–
Diamond	9000	13,780	3000	4260	910	1294
NaCl	21	30	7	10	6	8
B_4C	4980	7080	1660	2360	414	589

Source: From Rice, R.W., The compressive strength of ceramics, in *Materials Science Research, Vol. 5: Ceramics in Severe Environments* (Kriegel, W.W. and Palmour III, H., Eds.), Plenum, New York, 1971, pp. 195–229.

Bend testing is conducted with the same kind of universal test machine used for tensile and compressive strength measurements. As shown in Figure 8.6d and e, the test specimen is supported at the ends and the load is applied either at the center (three-point loading) or at two positions (four-point loading). The bend strength is defined as the maximum tensile stress at failure and is often referred to as the modulus of rupture (MOR). The bend strength for a rectangular test specimen can be calculated using the general flexure stress formula:

$$S = \frac{Mc}{I} \tag{8.14}$$

where M is the moment, c the distance from the neutral axis to the tensile surface, and I the moment of inertia. For a rectangular test specimen $I = bd^3/12$ and $c = d/2$, where d is the thickness of the specimen and b is the width. Figure 8.8 illustrates the derivation of the three-point and four-point flexure formulas for rectangular bars.

The strength characterization data for ceramics are reported in terms of MOR or bend strength. Specimens are relatively inexpensive and testing is straightforward and quick. However, there is a severe limitation on the usability of MOR data for ceramics: the measured strength will vary significantly depending on the size of the specimen tested and whether it is loaded in three-point or four-point. To understand the magnitude and reason for this variation, data generated for hot-pressed $Si_3N_4^*$ during the late 1970s may be used as an example.

For specimens having a rectangular cross section of 0.32 × 0.64 cm (0.125 × 0.250 in.), three-point bend testing over a 3.8-cm (1.5-in.) span resulted in an average MOR of about 930 MPa (135 kpsi). Four-point bend testing of bars from the same batch resulted in an average MOR of only 724 MPa (105 kpsi). Uniaxial tensile testing of a comparable cross section of the same hot-pressed Si_3N_4 yielded a strength of only 552 MPa (80 kpsi). Which of these strengths should an engineer

---

* Norton Company NC-132 hot-pressed Si_3N_4.

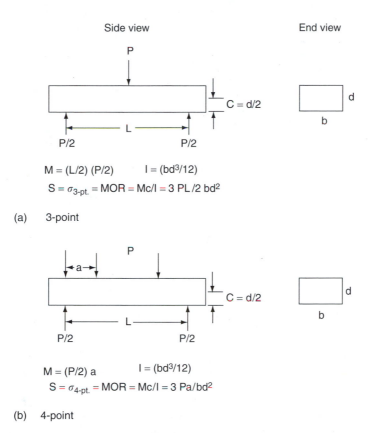

FIGURE 8.8 Derivation of the modulus of rupture equations. (a) For three-point bending, and (b) four-point bending.

use? Why are they different? The engineer can answer the first question only if he or she under-stands the answer to the second question. The answer to the second question can best be visualized by referring to Figure 8.9.

The stress distribution for three-point bending is shown in Figure 8.9a. The peak stress occurs only along a single line on the surface of the test bar opposite the point of loading. The stress decreases linearly along the length of the bar and into the thickness of the bar, reaching zero at the bottom supports and at the neutral axis, respectively. The probability of the largest flaw in the specimen being at the surface along the line of peak stress is very low. Therefore, the specimen will fracture at either a flaw smaller than the largest flaw or a region of lower stress, whichever one satisfies Equation (8.11) first. For the case of hot-pressed Si_3N_4, where $\sigma_{3\text{-pt.}} = 930$ MPa (135,000 psi) with the assumptions that $E = 303 \times 10^9$ N/m^2, $\gamma = 30$ J/m^2, $Z = 1.5$, and $Y = 2$, c from Equation (8.11) equals 10 μm (0.0004 in.). In other words, a flaw of 10-μm depth on the surface at midspan would result in fracture at 930 MPa. Halfway between midspan and the bottom support, the load at fracture would be $(1/2)(930) = 465$ MPa (67,400 psi). The critical flaw size to cause fracture at this point would be 41 μm. If the 41-μm flaw had been at midspan, the bar would have fractured at 326 MPa (47,200 psi) rather than 930 MPa. The point of this example is that the MOR or bend strength measured in a three-point test does not reveal the strength limit of the material or even the local stress and flaw size that caused fracture. All it tells is the peak stress on the tensile surface at the time of fracture, as provided by Equation (8.14).

Three-point bend testing results in synthetically high strength values. These values can be used for design only if treated statistically or probabilistically. This point is discussed in Chapter 18.

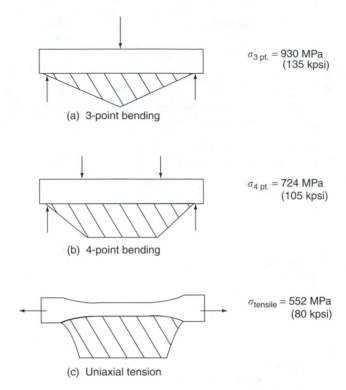

$\sigma_{3\ pt.} = 930$ MPa
(135 kpsi)

(a) 3-point bending

$\sigma_{4\ pt.} = 724$ MPa
(105 kpsi)

(b) 4-point bending

$\sigma_{tensile} = 552$ MPa
(80 kpsi)

(c) Uniaxial tension

FIGURE 8.9   Comparison of the tensile stress distributions for three-point, four-point, and uniaxial tensile test specimens along with typical average strengths as measured by each technique for Norton NC-132 hot-pressed Si_3N_4. Shaded area represents the tensile stress, ranging from zero at the supports of the bend specimens to maximum at midspan, and being uniformly maximum along the whole gauge length of the tensile specimen.

Four-point bend testing results in lower strength values for a given ceramic material than does three-point bending. Figure 8.9b illustrates the approximate stress distribution in a four-point bend specimen. The peak stress is present over the area of the tensile face between the load points. The tensile stress decreases linearly from the surface to zero at the neutral axis and from the load points to zero at the bottom supports. The area and volume under peak tensile stress or near peak tensile stress is much greater for four-point bending than for three-point bending, and thus the probability of a larger flaw being exposed to high stress is increased. As a result, the MOR or bend strength measured in four-point is lower than that measured in three-point. For the Si_3N_4, the average four-point bend strength was 724 MPa (105,000 psi), which would correspond to a critical flaw size of 17 µm (0.0007 in.) at the tensile surface in the region of peak stress. As was shown for three-point bending, fracture could also occur at lower stress regions from larger flaws.

Uniaxial tensile strength testing results in lower strength values for a given ceramic than does bend testing. Figure 8.9c illustrates that the complete volume of the gauge section of a tensile test specimen is exposed to the peak stress. Therefore, the largest flaw in this volume will be the critical flaw and will result in fracture when the critical stress as defined by Equation (8.11) is reached. To continue the Si_3N_4 example, the average fracture stress in uniaxial tensile loading was 552 MPa (80,000 psi) and the critical flaw size was 29 µm (0.0011 in.).

In summary, the observed strength value is dependent on the type of test conducted. More specifically, it is dependent simultaneously on the flaw size distribution of the material and the stress distribution in the test specimen. As the uniformity of the flaws within a material increases, the strength values measured by bend and tensile testing approach each other. Such is the case for

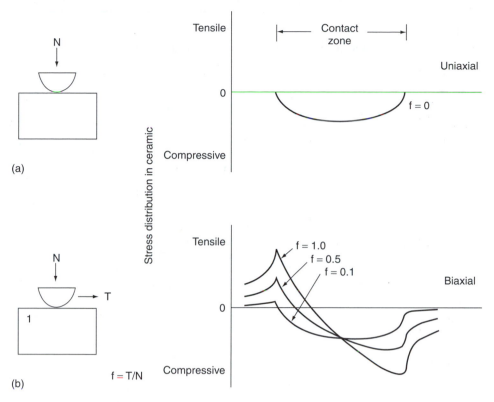

FIGURE 8.10 Contact loading showing uniaxial and biaxial effects on stress distribution. (Adapted from Finger, D.G., *Contact Stress Analysis of Ceramic-to-Metal Interfaces*, Final Report, Contract N00014-78-C-0547, Sept. 1979.)

most metals. For most ceramic materials, the apparent strength will decrease when going from three-point to four-point to tensile testing and as specimen size increases.

8.2.3.4 Biaxial Strength

The tensile, compressive, and bend tests discussed in the previous paragraphs all involve loading in a single direction and thus produce uniaxial stress fields. Many applications for materials impose multiaxial stress fields. Very few data are available for the response of ceramics to multiaxial stress fields.

The specimen test shown in Figure 8.6g provides strength data under a biaxial stress condition. A flaw in the material is exposed simultaneously to both tensile and shear stresses.[20,21]

Biaxial loading frequently occurs at the contact zone between two ceramic parts or between a ceramic and a metal part, especially during relative motion due to mechanical sliding or thermal cycling. Under certain conditions, very localized surface tensile stresses much higher than the applied load can result.[22,23] This is illustrated in Figure 8.10.

Application of only a normal force N results in compressive stresses. Simultaneous application of a tangential force T results in localized tensile stress at the edge of the contact zone opposite the direction of the tangential force. This tensile stress is a maximum at the surface of the ceramic and rapidly decreases inward from the surface. The magnitude of the tensile stress increases as the coefficient of friction increases. It reaches a peak when the static friction is highest, but is immediately reduced once sliding begins because the dynamic coefficient of friction is lower.

Most engineers are not aware of this mechanism of tensile stress generation, yet it is a common cause of chipping, spalling, cracking, and fracture of ceramic components.

TABLE 8.5
Typical Room-Temperature Strengths of Ceramic Materials

Material	MOR		Tensile	
	MPa	kpsi	MPa	kpsi
Sapphire (single-crystal Al_2O_3)	620	90	–	–
Al_2O_3 (0–2% porosity)	350–580	50–80	200–310	30–45
Sintered Al_2O_3 (<5% porosity)	200–350	30–50	–	–
Alumina porcelain (90–95% Al_2O_3)	275–350	40–50	172–240	25–35
Sintered BeO (3.5% porosity)	172–275	25–40	90–133	13–20
Sintered MgO (<5% porosity)	100	15	–	–
Sintered stabilized ZrO_2 (<5% porosity)	138–240	20–35	138	20
Sintered Mullite (<5% porosity)	175	25	100	15
Sintered spinel (<5% porosity)	80–220	12–32	–	19
Hot-pressed Si_3N_4 (<1% porosity)	620–965	90–140	350–580	50–80
Sintered Si_3N_4 (~5% porosity)	414–580	60–80	–	–
Reaction-bonded Si_3N_4 (15–25% porosity)	200–350	30–50	100–200	15–30
Hot-pressed SiC (<1% porosity)	621–825	90–120	–	–
Sintered SiC (~2% porosity)	450–520	65–75	–	–
Reaction-sintered SiC (10–15% free Si)	240–450	35–65	–	–
Bonded SiC (~20% porosity)	14	2	–	–
Fused SiO_2	110	16	69	10
Vycor or Pyrex glass	69	10	–	–
Glass-ceramic	245	10–35	–	–
Machinable glass-ceramic	100	15	–	–
Hot-pressed BN (<5% porosity)	48–100	7–15	–	–
Hot-pressed B_4C (<5% porosity)	310–350	45–50	–	–
Hot-pressed TiC (<2% porosity)	275–450	40–65	240–275	35–40
Sintered WC (2% porosity)	790–825	115–120	–	–
Mullite porcelain	69	10	–	–
Steatite porcelain	138	20	–	–
Fire-clay brick	5.2	0.75	–	–
Magnesite brick	28	4	–	–
Insulating firebrick (80–85% porosity)	0.28	0.04	–	–
2600°F insulating firebrick (75% porosity)	1.4	0.2	–	–
3000°F insulating firebrick (60% porosity)	2	0.3	–	–
Graphite (grade ATJ)	28	4	12	1.8

8.2.4 STRENGTH DATA FOR CERAMIC MATERIALS

Table 8.5 summarizes strength data for a variety of ceramic materials. The purpose of this table is to provide the reader with a general indication of the strengths typically measured for the different types of ceramic materials. The data are adequate for making an initial material selection for an application, but are not suitable for analytical design or life prediction calculations. The best procedure is to select candidate materials from Table 8.5 and then contact current suppliers and obtain their most up-to-date data. As discussed earlier, to understand the data adequately, one must know the porosity content of the material, the size of the test specimen, and the geometry of the strength test (i.e., three-point versus four-point and span of fixture). The data in Table 8.5 were obtained from many sources and were undoubtedly measured on a wide variety of specimen sizes and fixture configurations.

For many applications the high-temperature strength is more important than the room-temperature strength. The strength of nearly all ceramic materials decreases as the temperature is increased (graphite is an exception). Figures 8.11 and 8.12 compare strength versus temperature for some ceramic and metallic materials.

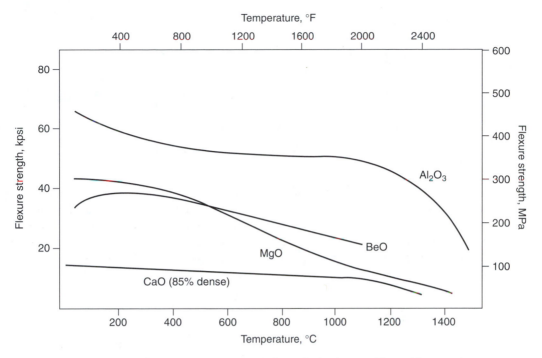

FIGURE 8.11 Examples of strength vs. temperature for typical polycrystalline oxide ceramics.

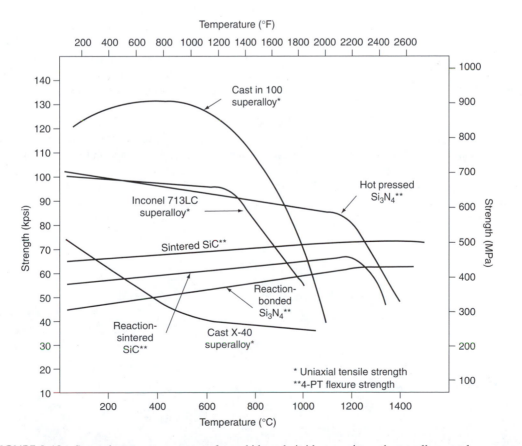

FIGURE 8.12 Strength versus temperature for carbide and nitride ceramics and superalloy metals.

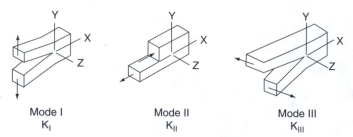

Mode I   Mode II   Mode III
K_I K_{II} K_{III}

FIGURE 8.13   Stress intensity factor notations for various displacement modes.

The strength would be expected to decrease with increasing temperature in proportion to the decrease in elastic modulus. This appears to occur for most ceramic materials at intermediate temperatures. At still higher temperatures, the rate of strength decrease is more rapid, generally because of nonelastic effects. For instance, most ceramics have secondary chemical compositions concentrated at grain boundaries that soften at high temperature and decrease the load-bearing capability of the bulk ceramic.

8.3   FRACTURE TOUGHNESS

Discussions so far have considered strength and fracture in terms of critical flaw size. An alternative approach considers fracture in terms of crack surface displacement and the stresses at the tip of the crack. This is the fracture mechanics approach.

The stress concentration at a crack tip is denoted in terms of the stress intensity factors K_I, K_{II}, and K_{III}. The subscripts refer to the direction of load application with respect to the position of the crack. If the load is perpendicular to the crack, as is typically the case in a tensile or bend test, the displacement is referred to as mode I and is represented by K_I. This is also called the opening mode and is most frequently operational for ceramics. Similarly, shear loading is referred to as mode II and mode III and is represented by K_{II} and K_{III}.[24] The load directions for the three modes are illustrated in Figure 8.13.

Mode I is most frequently encountered for ceramic materials. Stress analysis solutions for K_I for a variety of crack locations within simple geometries have been reported by Paris and Sih.[25] Experimental data for a variety of materials have been obtained for the critical stress intensity K_{Ic}. This is the stress intensity factor at which the crack will propagate and lead to fracture. It is also referred to as *fracture toughness*. The higher the fracture toughness, the more difficult it is to initiate and propagate a crack.

The mode I stress intensity factor is related to other parameters by the following equations:[25]
For plane strain,

$$K_I = \left(\frac{2\gamma E}{1 - v^2} \right)^{1/2} \tag{8.15}$$

For plane stress,

$$K_I = (2\gamma E)^{1/2} \tag{8.16}$$

and for applied stress σ_a and crack length $2c$,

$$K_I = \sigma_a Y c^{1/2} \tag{8.17}$$

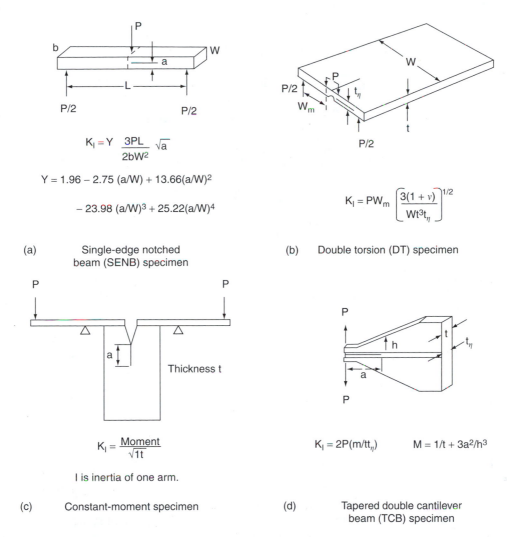

FIGURE 8.14 Techniques for experimental determination of mode I stress intensity factor. (Compiled from Evans, A.G., in *Fracture Mechanics of Ceramics*, Vol. 1 (Bradt, R.C., Hasselman, D.P.H., and Lange, F.F., Eds.), Plenum, New York, 1974, pp. 25–26.)

where γ is the fracture energy, E the elastic modulus, ν the Poisson's ratio, and Y a dimensionless term determined by the crack configuration and loading geometry.

A variety of experimental techniques[26] for measuring fracture toughness are available and are discussed by Evans.[27] Schematics of the specimen geometries and equations for the calculation of K_I are included in Figure 8.14.

8.4 DUCTILE VERSUS BRITTLE BEHAVIOR

Typical stress–strain curves were compared for metals and ceramics in Figure 8.1. The typical ceramic fractures in a brittle mode with only elastic deformation prior to fracture. The typical metal fractures in a ductile mode with initial elastic deformation followed by a yield point and plastic deformation. The metal elongates and decreases in cross section during plastic deformation. Fracture occurs when the cross section decreases to the degree that the applied stress can no longer be sustained.

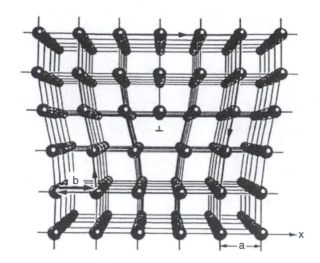

FIGURE 8.15   Simple schematic illustrating an edge dislocation and showing that the displacement *b* (Burgers vector) is equal to one unit cell edge *a*. (From Guy, A.G., *Elements of Physical Metallurgy*, Addison-Wesley, Reading, MA, 1959, p. 110. With permission.)

Not all metals fail in a ductile mode and not all ceramics fail in a brittle mode, so it is not adequate for us to automatically define metals as ductile and ceramics as brittle. We need to understand the factors that control these behaviors. We have already discussed some of the factors that cause brittle behavior, such as the stress concentration at microstructural defects. Now we need to review the mechanisms of plastic deformation.

8.4.1   Mechanism of Plastic Deformation

Plastic deformation at room temperature involves slip between planes of atoms under the influences of an applied stress.[28,29] The applied stress can be tensile or compressive but acts on the crystal structure as a shear stress. Early theories assumed that one complete plane of atoms would be displaced with respect to the adjacent plane. Theoretical calculations estimated that the stress needed to cause this type of slip would be about $E/20$. Actual measurements for many metals determined that plastic deformation occurred at much lower stress.

The current theory that is compatible with experimental measurements and theoretical calculations involves movement of dislocations. A *dislocation* is a defect in the way planes of atoms are stacked in a crystal structure. There are two primary types of dislocations. One is called an *edge dislocation*. It consists of a partial plane of atoms that terminates within the crystal structure, as shown schematically in Figure 8.15. The other is called a *screw dislocation* and is illustrated in Figure 8.16. The dislocation produces a line of discontinuity in the crystal structure. Along this line, the structure is distorted and under localized stress even when the overall material is not under an applied stress. This residual stress state can be visualized by examining Figure 8.17. The dislocation line extends into the structure perpendicular to the surface of the page. Note that the structure is distorted so as to fill in the space of the missing half-plane of atoms. This results in a state of residual tensile stress just below the extra plane of atoms balanced by compressive stress in the region above the dislocation.

The presence of the dislocations and the associated residual stress allows slip to occur along atom planes at a fraction of the $E/20$ value that would be required in the absence of dislocations. For example, 99.97% purity nickel begins to plastically deform at room temperature at about 110 MPa (16 ksi).

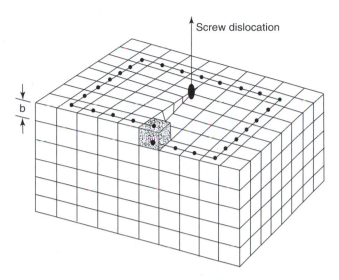

FIGURE 8.16 Simple schematic illustrating a screw dislocation. (From Van Vlack, L.H., *Elements of Materials Science*, Addison-Wesley, Reading, MA, 1964, pp. 418–420, p. 92. With permission.)

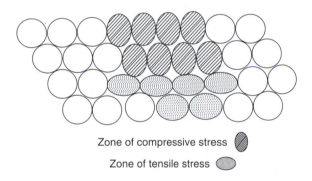

Zone of compressive stress

Zone of tensile stress

FIGURE 8.17 Schematic of the residual stress state showing compressive stress above the dislocation and tensile stress below the dislocation.

Deformation by dislocation motion is a little difficult to visualize. Nicholson[30] offers a helpful analogy as illustrated in Figure 8.18. The caterpillar forms a hump to move its end segment ahead a small distance. The hump then moves forward. When it reaches the head, the whole caterpillar has moved forward one hump distance. Less energy was required for the caterpillar to move this way than to move its whole body in a single motion. Another analogy involves laying carpet. Invariably when you first unroll the carpet into the room, it is one or two inches short on one side and climbs the wall the same amount on the other side. If you now grab hold of the carpet on the short side and try to pull it, an enormous amount of force is required to get the carpet to move. This is analogous to the force required to slide one plane of atoms over another plane in the absence of dislocations. However, if you go to the other side of the room and make a hump in the carpet, you can easily advance the hump across the room similarly to the way the caterpillar's hump moved. When the hump reaches the end of the carpet, the carpet has moved a distance equal to the height of the hump. This is analogous to the deformation of a material by one atom position by the motion of a single dislocation through the material.

A single dislocation does not result in significant plastic deformation within a material. However, under an applied stress, dislocations can form and multiply. Typical deformed metals contain millions

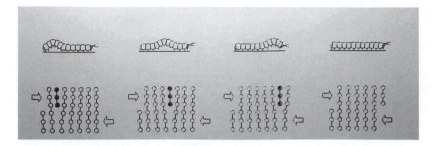

FIGURE 8.18 Analogy of the movement of a caterpillar to the movement of a dislocation through a material. (From Nicholson, P.S., *A Disastrous Approach to Materials Science*, Mosaic Press, Oakville, Ontario, 2004, p. 138. With permission.)

of dislocations per cubic centimeter. The actual slip occurs in bands along preferred crystal planes. The preferred planes are those that require the least applied stress to initiate dislocation movement. The preferred plane for hexagonal close-packed structures is (0001). Slip readily occurs along several families of planes in cubic close-packed structures: {111}, {100}, and {110}.

8.4.2 DEFORMATION BEHAVIOR OF METALS

The deformation behavior of metals at all temperatures up to the melting point is dominated by the effects of dislocations. Pure metals have very uniform structures consisting of close-packed atoms all of one size. Each atom position in the structure is equivalent to every other nonsurface atom position. With dislocations present, slip is initiated at very low stress in such a simple, homogeneous structure. Pure aluminum, for example, at room temperature, has a yield stress of 15 to 20 MPa (2175 to 2900 psi), a tensile strength of 40 to 50 MPa (5800 to 7250 psi), and elongation of 50 to 70%.

The yield stress of a metal can be increased by inhibiting the motion of dislocations. This can be accomplished in several ways: (1) solid solution, (2) inclusions or precipitates, (3) grain boundaries, and (4) work hardening.

Solid solution, as we discussed earlier, involves substitution of foreign atoms into lattice positions or into interstitial positions. These foreign atoms are a different size than the metal atoms and tend to fit into the crystal structure in a position of minimum energy. An atom of smaller size than the base metal will preferentially substitute in the region of residual compressive stress adjacent to a dislocation. A larger atom will substitute in the region of residual tensile stress. In each case, the residual stress is reduced and a larger applied stress is necessary to initiate slip. The same effect occurs when a moving dislocation encounters a foreign atom. The effect of substitution of foreign atoms is illustrated for nickel in Table 8.6.

Inclusions or precipitates can also decrease the residual stress, but their primary mechanism is to "pin" the dislocation. Think of the dislocation as a stiff wire sliding sideways across a table. It can be moved relatively easily. However, if it encounters a nail sticking out of the table, its motion is restricted. Superalloys contain precipitates as well as solid-solution substitutions. Alloy B-1900, which contains 64% nickel, has yield strength of 825 MPa (120 ksi), tensile strength of 970 MPa (140 ksi), and only 8% ductility at room temperature. Gray cast iron, which contains relatively large graphite inclusions or precipitates, has only 0.6% ductility and fractures in essentially a brittle mode. Other cast iron compositions have ductility ranging from about 1 to 25%.

Grain boundaries also inhibit dislocation motion. The slip within a grain is along a specific crystallographic plane. Adjacent grains have different crystal orientations. When slip reaches a grain boundary (the edge of the initial grain), it cannot continue directly into the next grain. Figure 8.19 shows the effect of grain size on the ductility and tensile strength of annealed 70-30 brass (70 weight% copper, 30% percent zinc). The smaller the grain size, the greater the number of grain boundaries and the lower the ductility.

TABLE 8.6
Change in Strength and Ductility of Nickel as Impurities Are Added

Nickel Alloy	% Nickel	Yield Strength MPa	(ksi)	Tensile Strength MPa	(ksi)	Elongation (%)
Commercial purity	99.97	110	(15.9)	345	(50.0)	50
Nickel 200	99	148	(21.5)	462	(67.0)	47
Duranickel	93	910	(131.9)	1276	(185.0)	28

Source: Data compiled from *Metals Handbook*. ASM International, Materials Park, OH.

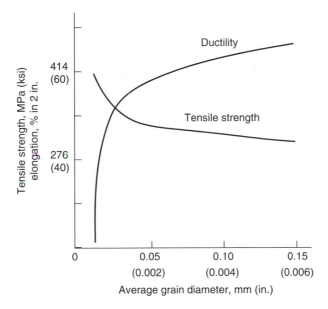

FIGURE 8.19   Effect of grain size on the strength and ductility of annealed 70-30 brass. (From Van Vlack, L.H., *Elements of Materials Science*, 2nd ed., Addison-Wesley, Reading, MA, 1964, p. 92. With permission.)

Dislocations can interfere with each other, with the surface of the material, with grain boundaries, or with an inclusion. The dislocations pile up at the barriers and require increased stress for further slip to occur. This increase in the stress requirement for further plastic deformation is referred to as *work hardening*, *strain hardening*, or *cold working*. Cold working reduces the ductility and increases the tensile strength. Metals are often shot-peened to preferentially cold work the surface to increase surface strength and hardness. Strain hardening can be removed by annealing at high temperature. For example, several stages of annealing may be required during wire drawing to achieve the desired reduction in diameter. Table 8.7 further illustrates the effects of work hardening and annealing.

8.4.3   DEFORMATION BEHAVIOR IN CERAMICS

8.4.3.1   Single Crystals

Ductile behavior has been observed in some single-crystal ceramics at room temperature and in some polycrystalline ceramics at elevated temperature. The criteria for plastic deformation are the same as for metals: (1) the presence of dislocations, (2) a mechanism of generation of new

TABLE 8.7
Effects of Work Hardening of Aluminum

Amount of Cold Work (%)	Yield Strength		Tensile Strength		Elongation (%)
	MPa	(ksi)	MPa	(ksi)	
Annealed	15–20	(2.2–2.9)	40–50	(5.8–7.2)	50–70
40	50–60	(7.2–8.7)	80–90	(11.6–13.0)	15–20
70	65–75	(9.4–10.8)	90–100	(13.0–14.5)	10–15
90	100–120	(14.5–17.4)	120–140	(17.4–20.3)	8–12

Source: Data compiled from *Metals Handbook*, ASM International, Materials Park, OH.

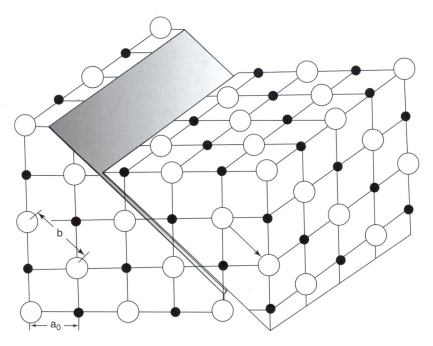

FIGURE 8.20 Primary mode of slip for single crystals of the [rock salt] structure. (From Kingery, W.D., et al., *Introduction to Ceramics*, 2nd ed., Wiley, New York, 1976, p. 713. With permission.)

dislocations under an applied stress, and (3) a path along which the dislocations can move at a stress lower than the fracture strength of the materials.

Plastic deformation occurs in carefully prepared crystals that have the [rock salt] (NaCl, KCl, KBr, LiF, MgO) and [fluorite] (CaF$_2$, UO$_2$) cubic structures. Slip occurs along {110} planes in the [110] direction in the [rock salt] structure. This motion involves the minimum distance of slip to restore the original structure arrangement, as illustrated in Figure 8.20. It also allows ions of the same polarity to remain at maximum distance from each other, thus avoiding having to overcome a large repulsion energy barrier. Stated in a different way, plastic deformation will not occur in an ionic ceramic unless: (1) the electrostatic balance between positive and negative ions is retained around each ion, (2) the structural geometry is not altered, and (3) the cation–anion ratio is maintained.

Figure 8.21 shows stress–strain curves for KBr and MgO single crystals. The yield stress is about 2 MPa (300 psi) for the KBr and 160 MPa (23 ksi) for the MgO. As with metals, the stress to initiate dislocation movement can be increased through solid solution. Figure 8.22 shows yield

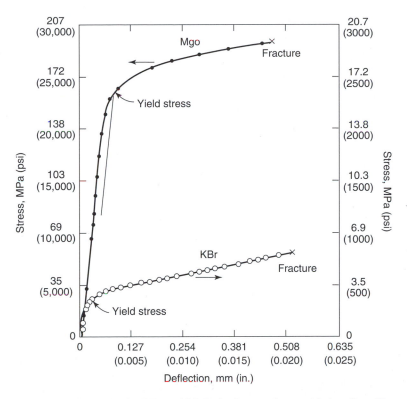

FIGURE 8.21   Stress–strain curves for KBr and MgO single crystals tested in bending. (From Gorum, A.E., Parker, E.R., and Pask, J.A., *J. Am. Ceram. Soc.*, 41, 161, 1958. With permission.)

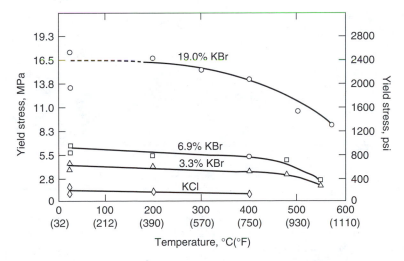

FIGURE 8.22   Effects of solid solution on the yield stress of KBr–KCl single crystal loaded in compression. (From Stoloff, N.S., et al., *J. Appl. Phys.*, 34, 3315, 1963. With permission.)

stress versus temperature for KCl and several KCl–KBr solid solutions. Figure 8.23 shows the effect of additions of NiO on the yield stress of MgO and on the shape of the stress–strain curve.

Limited plastic deformation through dislocation motion can also occur in more-complex, less-symmetrical single crystals of ceramics, but the yield stress is much higher and the mechanism of

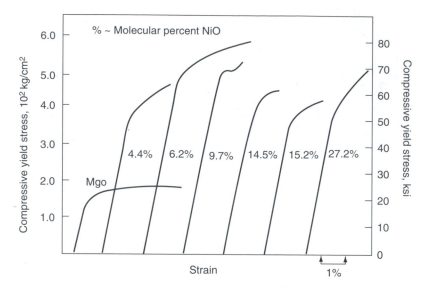

FIGURE 8.23 Effects of solid solution on the yield stress for the MgO–NiO system. Curves are displaced horizontally for clarity. (From Liu, T.S., Stokes, R.J., and Li, C.H., *J. Am. Ceram Soc.*, 47, 276–279, 1964. With permission.)

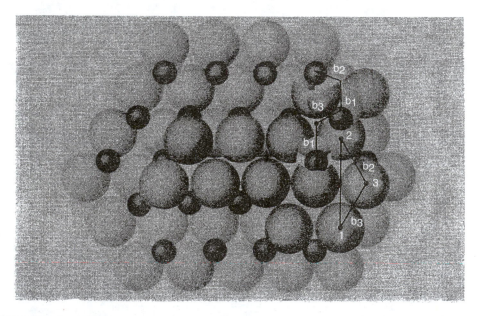

FIGURE 8.24 Crystal structure of Al_2O_3 showing complex paths O^{2-} and Al^{3+} ions must follow to allow slip to occur under an applied stress. (From Kingery, W.D., et al., *Introduction to Ceramics*, 2nd ed., Wiley, New York, 1976, p. 732. With permission.)

slip more restrictive. Slip does not occur at room temperature, so these materials fracture at room temperature and intermediate temperature in a brittle mode with no plastic deformation. Slip only occurs at high temperature where additional dislocation motions become energetically possible. Al_2O_3 is a good example.

Al_2O_3 has a hexagonal structure. As shown in Figure 8.24, ions or pairs of ions cannot move in a straight path through this structure to get to the next equivalent structural position. For example,

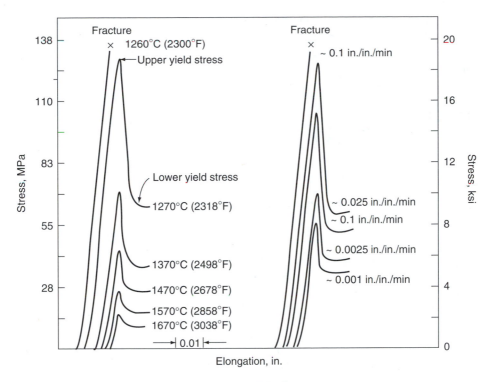

FIGURE 8.25 Deformation behavior of single crystals of Al_2O_3 at various temperatures and strain rates. (From Kingery, W.D., et al., *Introduction to Ceramics*, 2nd ed., Wiley, New York, 1976, p. 730. With permission.)

movement of oxygen ion 1 directly to the site of oxygen ion 2 would require a large amount of energy. Less energy would be required for a move from site 1 to site 3 to site 2, but this would require the formation of two partial dislocations involving stacking faults within the structure. Similarly, the aluminum ions must follow an even more circuitous route through troughs in the structure as shown by the path b_1' to b_2' to b_1'' to b_2''. Such movements become possible for Al_2O_3 single crystals only at temperatures in excess of 900°C (1650°F). Data for various temperatures and strain rates are shown in Figure 8.25. Note the yield behavior. High stress is required for initial yield, but less stress is required to sustain plastic deformation. Once deformation is initiated, additional dislocations are generated that then allow plastic deformation to occur at lower stress. Note also that the yield stress is less at higher temperatures and at low strain rates.

8.4.3.2 Polycrystalline Ceramics

Polycrystalline ceramics fracture in a brittle mode at room temperature. The random orientation of the individual crystals (grains) severely inhibits dislocation motion. The dislocations terminate at grain boundaries. For this to be overcome and for plastic deformation to be possible, it has been determined that five independent slip systems must be present in the crystal structure. At room temperature, most ceramic structures have three or less slip systems. A few ceramics have five at elevated temperature. These include NaCl, LiF, MgO, NaF, TiC, UC, diamond, CaF_2, UO_2, and $MgAl_2O_4$.

8.4.4 Ceramics Deformation Summary

Some single-crystal ceramics can have substantial plastic deformation prior to fracture. However, most single-crystal and polycrystalline ceramics fracture in a brittle mode with no plastic deforma- tion. The behavior is determined by a competition between creation and movement of dislocations and stress concentration at microstructural defects. If the applied stress first reaches the yield stress for dislocation motion, plastic deformation will occur. If the local stress first reaches the critical fracture stress at a microstructural defect (such as a pore, crack, or inclusion) in the material, brittle fracture will occur.

REFERENCES

1. McClintock, F.A. and Argon, A.S., *Mechanical Behavior of Materials*, Addison-Wesley, Reading, Mass., 1966.
2. Wachtman, J.B., Ed., *Mechanical and Thermal Properties of Ceramics*, NBS Special Publication 303, U.S. Government Printing Office, Washington, D.C., 1969.
3. Bradt, R.C., Hasselman, D.P.H., and Lange, F.F., Eds., *Fracture Mechanics of Ceramics*, Vols. 1 and 2, Plenum, New York, 1974.
4. Bradt, R.C., Hasselman, D.P.H., and Lange, F.F., Eds., *Fracture Mechanics of Ceramics*, Vols. 3 and 4, Plenum, New York, 1978.
5. Cranmer D.C. and Richerson, D.W., Eds., *Mechanical Testing Methodology for Ceramic Design and Reliability*, Marcel Dekker, Inc., New York, 1998.
6. Kingery, W.D., Bowen, H.K., and Uhlmann, D.R., *Introduction to Ceramics*, 2nd ed., Wiley, New York, 1976, Chap. 15.
7. Van Vlack, L.H., *Elements of Materials Science*, Addison-Wesley, Reading, Mass., 1964, pp. 418–420.
8. Van Vlack, L.H., *Physical Ceramics for Engineers*, Addison-Wesley, Reading, Mass., 1964, p. 118.
9. MacKenzie, J.K., *Proc. Phys. Soc., (Lond.)* B63, 2, 1950.
10. ASTM Specification C747, *Annual Book of ASTM Standards*, American Society for Testing and Materials, Philadelphia, pp. 1064–1074.
11. Inglis, C.E., Stresses in a plate due to the presence of cracks and sharp corners, *Trans. Inst. Nav. Arch.*, 55, 219, 1913.
12. Griffith, A.A., The phenomenon of rupture and flow in solids, *Philos. Trans. R. Soc. Lond. Ser.*, A 221(4), 163, 1920.
13. Evans, A.G. and Tappin, G., *Proc. Br. Ceram. Soc.*, 20, 275–297, 1972.
14. Evans, A.G. and Langdon, T.G., Structural ceramics, in *Progress in Materials Science*, Vol. 21, Pts. 3/4: *Structural Ceramics*, Pergamon, Elmsford, N.Y., 1976.
15. Johansen, K.M., Richerson, D.W., and Schuldies, J.J., *Ceramic Components for Turbine Engines*, Phase II Technical Report under Air Force Contract F33615-77-C-5171, Feb. 29, 1980.
16. Pears, C.D. and Starrett, H.W., *An Experimental Study of the Weibull Volume Theory*, AFML-TR-66-228, Mar. 1967.
17. Sedlacek, R. and Halden, F.A., Method of tensile testing of brittle materials, *Rev. Sci. Instrum.*, 33(3), 298–300, 1962.
18. Shook, W.B., *Critical Survey of Mechanical Property Test Methods for Brittle Materials*, ADS-TDR-63-491, July 1963.
19. Rice, R.W., The compressive strength of ceramics, in *Materials Science Research, Vol. 5: Ceramics in Severe Environments* (Kriegel W.W. and Palmour III, H., Eds.), Plenum, New York, 1971, pp. 195–229.
20. Rudnick, A., Hunter, A.R., and Holden, F.C., An analysis of the diametral compression test, *Mater. Res. Std.*, 3(4), 283–289, 1963.
21. Rudnick, A., Marschall, C.W., Duckworth, W.H., and Emrich, B.R., *The Evaluation and Interpretation of Mechanical Properties of Brittle Materials*, AFML-TR-67-316, DCIC 68-3, Mar. 1968.
22. Finger, D.G., *Contact Stress Analysis of Ceramic-to-Metal Interfaces*, Final Report, Contract N00014-78-C-0547, Sept. 1979.
23. Richerson, D.W., Carruthers, W.D., and Lindberg, L.J., Contact stress and coefficient of friction effects on ceramic interfaces, in *Surfaces and Interfaces in Ceramic and Ceramic-Metal Systems, Materials Science Research*, Vol. 14 (Pask, J. and Evans, A., Eds.), Plenum, New York, 1981, pp. 661–676.
24. Braiden, P.M., in *Mechanical Properties of Ceramics for High Temperature Applications*, AGARD Report No. 651, NTIS No. ADA034 262, Dec. 1976.
25. Paris, P.C. and Sih, G.C., *Stress analysis of cracks, ASTM STP No.* 381, 1965, p. 30.

26. Wachtman, Jr., J.B., Highlights of progress in the science of fracture of ceramics and glass, *J. Am. Ceram. Soc.*, 57(12), 509–519, 1974.

27. Evans, A.G., in *Fracture Mechanics of Ceramics*, Vol. 1 (Bradt, R.C., Hasselman, D.P.H., and Lange, F.F., Eds.), Plenum, New York, 1974, pp. 25–26.

28. Cottrell, A.H., *Dislocations and Plastic Flow in Crystals*. Clarendon Press, Oxford, 1953.

29. Tressler, R.E. and Bradt, R.C., Eds., *Plastic Deformation of Ceramic Materials*, Plenum Press, New York, 1975.

30. Nicholson, P.S., *A Disastrous Approach to Materials Science*, Mosaic Press, Oakville, Ontario, 2004.

ADDITIONAL RECOMMENDED READING

1. Rice, R.W., *Mechanical Properties of Ceramics and Composites*, Marcel Dekker, Inc., New York, 2000.
2. Rice, R.W., *Porosity of Ceramics*, Marcel Dekker, Inc., New York, 1998.
3. Wachtman, Jr., J.B., Ed., *Structural Ceramics*, Treatise on Materials Science and Technology, Vol. 29, Academic Press, Boston, 1989.
4. Jahanmir, S., Ed., *Friction and Wear of Ceramics*, Marcel Dekker, Inc., New York, 1994.
5. *Engineered Materials Handbook*, Vol. 4: *Ceramics and Glass*, ASM Int., Materials Park, Ohio, 1991.

PROBLEMS

1. The average thermal expansion coefficient of MgO is 5×10^{-6} in./in.$\cdot$°F. What temperature change is required to produce the same linear change as a stress of 15,000 psi?
2. Fused silica has a thermal expansion coefficient of about 2.8×10^{-7} in./in.$\cdot$°F. What temperature change is required to produce the same linear change as a stress of 15,000 psi? How does this compare with the answer to Problem 1? What degree of thermal shock resistance would you expect for each material?
3. What would have the greater effect on the elastic modulus of an Al_2O_3 material, 5 vol% intergranular glass or 5 vol% porosity? Verify with calculations.
4. Ten rectangular bars of BeO 0.5 in. wide by 0.25 in. thick were tested in three-point bending over a 2.0-in. span. The failure loads for each (in ascending order) were 280, 292, 296, 299, 308, 317, 319, 330, 338, and 360 lb. Calculate the modulus of rupture (MOR) for each and the mean MOR for the group.
5. A ceramic material is tested in three-point and four-point bending and in uniaxial tension. The resulting MOR and strength values are 80,000, 60,000, and 25,000 psi. What can we deduce from these data about the flaw distribution and uniformity of this material?
6. A Si_3N_4 material has a tensile strength of 650 MPa. What does this equal in psi?
7. A transformation-toughened ZrO_2 sample fractures at a stress of 900 MPa at a flaw estimated to be about 50 μm by observation of the fracture surface. Assuming that the geometrical constant Y is 1.3. What is the approximate value of the fracture toughness of the ZrO_2 sample?
8. A ceramic test bar with a bulk elastic modulus of 57×10^6 psi deforms elastically up to the point of fracture at 45×10^3 psi. What is the strain to failure for this test bar?
9. A rectangular bar of ceramic 3 mm thick, 4 mm wide, and 50 mm long fractures in a three-point flexure test at a load of 270 N. If the span of the three-point fixture is 40 mm, what is the modulus of rupture of the test bar?

STUDY GUIDE

1. Explain the difference between elastic and plastic deformation. Include sketches of the stress-strain behavior.
2. For a pure, dense material the magnitude of the modulus of elasticity (E) correlates well with the strength of the atomic bonds in a material. What other factors will influence the value of E that you will measure for a real engineering material?

3. Compare the average elastic modulus for the following engineering materials.

Rubber	Nylon
Bulk graphite	Concrete
Typical glass	Zirconium Oxide
Iron	Silicon Nitride
Aluminum Oxide	Silicon Carbide
Diamond	

4. Explain the meaning of Poisson's ratio and give a typical value for the following materials:

Most metals	Magnesium Oxide
Aluminum Oxide	Silicon Carbide
Silicon Nitride	

5. Compare the theoretical strength of SiC with a typical measured strength of polycrystalline SiC. Explain why there is such a large difference.

6. What are some types of defects in ceramics that can dramatically reduce strength?

7. What is the distinction between tensile, compressive and bend strength?

8. Explain the pros and cons for use of modulus of rupture (MOR) as a means of characterizing the strength of ceramics.

9. Explain why tensile testing results in a lower strength value than bend testing?

10. What is the difference between strength and fracture toughness?

11. What is the meaning of "critical stress intensity"?

12. What is a dislocation?

13. Explain how the presence of dislocations affects the stress required for ductile deformation?

14. How can the motion of dislocations be reduced in a metal?

15. Compare the ductility and strength of pure Al and pure Ni with highly alloyed Ni.

16. Can metals have brittle fracture? Give an example.

17. What are the criteria for plastic deformation in a metal or a ceramic?

18. Under what circumstances has plastic deformation been demonstrated for a ceramic?

19. Explain why plastic deformation occurs in aluminum oxide only at high temperature.

20. What are the competing factors that determine if a ceramic deforms or fractures? Explain.

9 Time, Temperature, and Environmental Effects on Properties

The elastic properties of ceramics and relationships between flaw size and strength as determined by flexure and tensile testing were discussed in Chapter 8. In flexure and tensile strength tests, the load is applied rapidly and fracture occurs from single flaws or groups of flaws initially in the material. However, in real life engineering applications stresses are applied for long duration and often at high temperature. Material degradation mechanisms can occur under these conditions that result in failure of the material at a stress lower than was measured during "fast fracture" strength testing. These "time-dependent" degradation mechanisms plus other environmentally induced sources of failure are discussed in this chapter:

- Creep — plastic deformation at elevated temperature
- Slow crack growth (static fatigue, stress rupture) — flaws initially in the material grow with time and result in fracture at a lower stress than the fast fracture stress
- Chemical attack — chemical reactions with gases, liquids, or solids that come into contact with the material and either cause recession of the material or a change in the flaw size
- Mechanically induced effects — recession due to erosion and wear and new surface flaws due to stress concentrations at localized points of contact
- Thermal shock — new material flaws caused by thermally induced stress

9.1 CREEP

The term *creep* is normally used to refer to deformation at a constant stress as a function of time and temperature. Creep is plastic deformation rather than elastic deformation and thus is not recovered after the stress is released. A typical creep curve has four distinct regions, as shown in Figure 9.1. The secondary creep region is the most useful for predicting the life of a ceramic component. It is typified by a constant rate of deformation and is often referred to as *steady-state creep*.

Steady-state creep can be represented by the equation

$$\dot{\varepsilon} = A\sigma^n e^{-Q_c/RT} \tag{9.1}$$

where σ is the stress, T the absolute temperature, Q_c the activation energy for creep, and A and n are constants for the specific material. The constant n is usually referred to as the *stress exponent*.

The activation energy for creep Q_c can be obtained by measuring the slope of a plot of $\dot{\varepsilon}$ vs. $1/T$. The stress exponent and the activation energy provide information about the mechanism of creep, that is, whether it is controlled by viscoelastic effects, diffusion, porosity, or some other mechanism. Understanding the mechanism provides information about the kinetics of flaw growth. This flaw-growth information plus a knowledge of the initial flaw-size distribution (as determined by statistical evaluation of room-temperature strength tests) can be used in conjunction with fracture mechanics

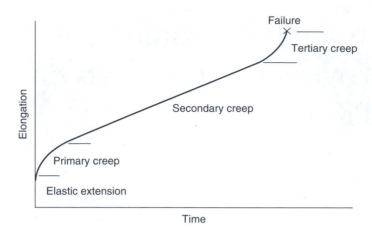

FIGURE 9.1 Typical constant-temperature, constant-stress creep curve.

theory and stress rupture testing to estimate the life of the ceramic component under similar stress and temperature conditions.

The creep rate of ceramic materials is affected by temperature, stress, crystal structure of single crystals, microstructure (grain size, porosity, grain boundary chemistry) of polycrystalline ceramics, viscosity of noncrystalline ceramics, composition, stoichiometry, and environment.[1]

9.1.1 Effects of Temperature and Stress on Creep

Temperature and stress both have strong effects on creep, as would be expected from examination of Equation (9.1). In general, as the temperature or stress increases, the creep rate increases and the duration of steady-state creep decreases. This is shown schematically in Figure 9.2. When comparing creep data for different materials under different conditions, it is often useful to plot the data as steady-state creep vs. either $1/T$ or log σ to produce the type of curves shown in Figure 9.3.

9.1.2 Effects of Single-Crystal Structure on Creep

The mechanism of creep in single crystals is the movement of dislocations through the crystal structure. Such movement is accommodated by slip along preferential crystal planes or by homogeneous shear (twinning). Highly symmetrical cubic structures such as NaCl and MgO have many planes available for slip.[1] At low temperature slip occurs along (110) planes in the [110] direction. At high temperature slip also occurs along (001) and (111) planes in the [110] direction, resulting in five independent slip systems. Diamond, silicon, CaF_2, UO_2, TiC, UC, and $MgAl_2O_4$ (spinel) all have five independent slip systems at high temperature. However, the temperatures at which these systems become operative and the stresses required for slip vary depending on the bond strength of the material. Slip occurs at low temperature and low stress in weakly ionic bonded NaCl, but requires much higher temperature and stress for strongly covalent bonded diamond or TiC.

Single crystals with less symmetry than cubic crystals have fewer slip systems available. Graphite, Al_2O_3, and BeO are hexagonal and have only two independent slip systems. Slip of (0001) planes occurs in the [1120] direction.

For measurable creep of a single crystal to occur, dislocations must be present or created and then start moving. Energy is required to form dislocations, to initiate their movement, and to keep them moving. Increasing the stress or the temperature increases the energy available for forming and moving dislocations.

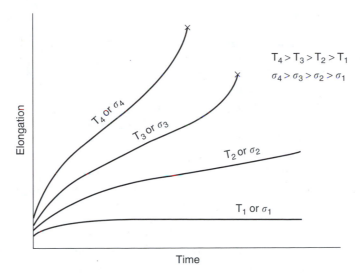

FIGURE 9.2 Effects of temperature and stress on creep rate.

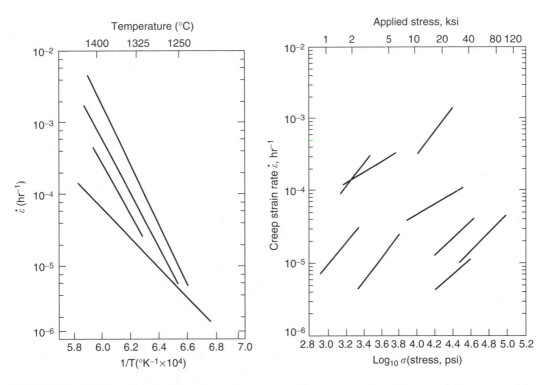

FIGURE 9.3 Additional methods for summarizing creep data and comparing differences between materials. Each line represents a different material.

Defects in the lattice structure decrease creep by pinning dislocations. Dislocations oriented across the direction of slip and large precipitates have the largest effect on blocking slip motion. Solid solution and point defects have less effect. However, even their effect can be substantial. Chin et al.[2] reported that additions of 840 ppm of Sr^{2+} ions in solid solution in KCl increased the compressive stress required to produce yield from about 2 MPa to over 20 MPa.

9.1.3 EFFECTS OF MICROSTRUCTURE OF POLYCRYSTALLINE CERAMICS ON CREEP

Creep of polycrystalline ceramics is usually controlled by different mechanisms than those that control single crystals. Dislocation movement is generally not a significant factor because of the random orientation of the individual crystals (more typically referred to as grains when in a polycrystalline structure) and the difficulty for a dislocation to travel from one grain through the grain boundary into the adjacent grain.

Creep in polycrystalline ceramics is usually controlled by the rate of diffusion or by the rate of grain boundary sliding. Diffusion involves the motion of ions, atoms, or vacancies through the crystal structure (bulk diffusion) or along the grain boundary (grain boundary diffusion). Grain boundary sliding often involves porosity or a different chemical composition at the grain boundary. Grain boundary sliding is an important (and undesirable) contribution to fracture in many ceramic materials densified by hot pressing or sintering. Additives are required to achieve densification. These additives concentrate at the grain boundaries together with impurities initially present in the material. If a glass is formed, it may soften at a temperature well below the temperature at which the matrix material would normally creep, allowing slip along grain boundaries. Grain boundary sliding is normally accompanied by the formation of pores at the grain boundaries, especially by cavitation at triple points (points at which three grains meet). The combination of reduced load-bearing capability due to softening of the grain boundary glass phase and formation of new flaws usually results in fracture before appreciable plastic deformation can occur.

The creep mechanism varies for materials at different temperatures and stresses. Ashby[3] has constructed a "deformation mechanism map" in which he plots shear stress vs. temperature. His map for MgO is shown in Figure 9.4. It indicates that at low temperature and high stress MgO creeps by dislocation motion and that at lower stress and increased temperature MgO creeps by diffusion mechanisms.

The creep mechanism for Al_2O_3 also changes according to stress and temperature, as described by Kingery et al.[1] In single crystals Al_2O_3 and polycrystalline Al_2O_3 with grains larger than 60 µm, slip can occur along (0001) planes at 2000°C at about 7 MPa (1000 psi) and at 1200°C at about 70 MPa (10,000 psi). Other planes require stresses greater than 140 MPa (20,000 psi) to cause slip

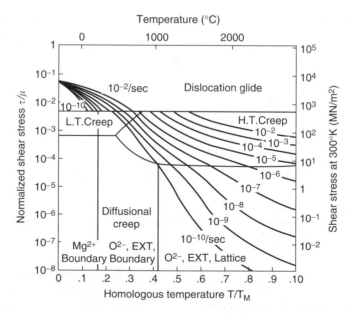

FIGURE 9.4 Deformation mechanism map for MgO. (Reprinted with permission from *Acta Metall.*, 20, M. Ashby, "Deformation mechanism maps," © 1972, Pergamon Press Ltd.)

even at 2000°C and thus are usually not a factor. In polycrystalline Al_2O_3 with grain size in the range 5 to 60 µm, the creep rate is controlled by Al^{3+} ion diffusion through the lattice. Below 1400°C and for finer grain sizes, Al^{3+} ion diffusion along the grain boundaries appears to be rate-controlling.

The examples for MgO and Al_2O_3 show that creep mechanism and rate can vary according to temperature, stress, and grain size. Porosity also has a substantial effect. Figure 9.5 shows the effect of porosity on the creep rate of a polycrystalline Al_2O_3 ceramic.[1] In this case, the creep rate was substantially increased by increased porosity, perhaps because of the decreased cross-sectional area available to resist creep. Similar results were reported by Kingery et al.[1] for MgO, where material with 12% porosity had six times the creep rate of comparable material with only 2% porosity.

9.1.4 CREEP IN NONCRYSTALLINE CERAMICS

Creep in glasses is controlled by viscous flow and is a function of the viscosity of the glass at the temperature of interest. The viscosity of glass varies over a wide range with temperature. Soda–lime–silica glass (common container or window glass) has a viscosity of 10^{15} P at 400°C and 10^2 P at 1300°C. The softening point of glass is defined as $10^{7.6}$ P, which occurs for soda–lime–silica glass at about 700°C. Soda–lime–silica glass melts in the range 1300 to 1500°C. This glass creeps at low temperatures.

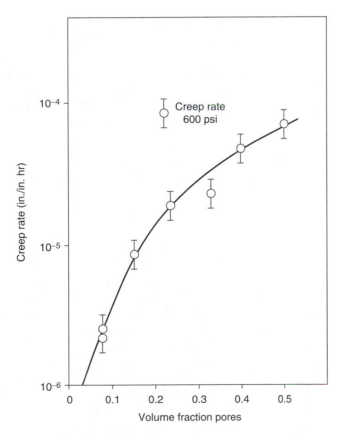

FIGURE 9.5 Effect of porosity on the creep rate of a polycrystalline Al_2O_3 ceramic. (From Kingery, W.D., Bowen, H.K., and Uhlmann, D.R., *Introduction to Ceramics*, 2nd ed., Wiley, New York, 1976, Chap. 14. With permission.)

The viscous flow of glasses is an important mechanism of creep in many commercial poly-crystalline ceramics. These polycrystalline ceramics contain secondary glass phases at the grain boundaries. The glass phases result from additions made during processing to achieve densification and from impurities initially present in the ceramic powders or picked up during processing. The viscosity characteristics of the glass are highly dependent on composition. For silicate glasses, the viscosity decreases with increased concentration of modifying cations. For example, the viscosity of fused silica is reduced by $10^4 P$ at 1700°C by the addition of 2.5 mol% $K_2O \cdot F^-$, Ba^{2+}, and Pb^{2+} are particularly effective at reducing the viscosity of glasses. Additions of SiO_2 or Al_2O_3 typ-ically increase viscosity.

9.1.5 EFFECTS OF COMPOSITION, STOICHIOMETRY, AND ENVIRONMENT

Composition governs the bonding and structure of single-crystal, polycrystalline, and noncrys-talline ceramics and thus determines the baseline creep tendencies. Grain size, porosity, grain boundary glass phases, and second-phase dispersions all modify the baseline creep properties; so does stoichiometry. A *stoichiometric ceramic* is one that has all crystallographic lattice positions filled according to the normal chemical formula. A *nonstoichiometric ceramic* is one that has a defi-ciency of one type of atom. For instance, $TiC_{0.75}$ is a nonstoichiometric form of TiC in which insuf-ficient carbon atoms were available during formation. The structure is basically a TiC structure with vacancies to make up for the missing carbon atoms. Nonstoichiometric materials have differ-ent diffusion characteristics than their stoichiometric equivalent and thus have different creep characteristics.

Environment apparently can also affect creep, although very little work has been done to study or quantify the effects. Joffe et al.[4] observed that NaCl had brittle behavior in air but was ductile when immersed in water. Evidently, the NaCl tested in air fractured from surface defects before creep had a chance to occur. The surface of NaCl in water was dissolved, removing the surface defects and allowing plastic deformation by creep. Although it has not been studied in detail, it is likely that high-temperature oxidation and corrosion mechanisms can also alter the surface flaw size or provide crack blunting to change the creep characteristics of ceramic materials.

9.1.6 MEASUREMENT OF CREEP

Creep has been measured in tension, compression, torsion, and bending. The data from these dif-ferent approaches may not be comparable. Therefore, it is important to know as many details as pos-sible about how the data were generated, that is, test configuration, specimen size, sensitivity of deflection measurement, and so on.

Creep testing consists of measurement of deflection at a constant load and constant tempera-ture. The deflection is usually measured with a transducer. Corrections should be made for thermal expansion within the test specimen and deflection measurement system and for elastic and plastic deflection in the test fixture and deflection measurement system. Corrections can often be checked by comparing the cumulative creep measured by the deflection system with an actual physical measurement of the specimen dimensions before and after testing.

Table 9.1 summarizes torsional creep data for some ceramic materials. The polycrystalline oxides have much lower creep rates than the glasses or the insulating firebrick. The high creep rates of the firebrick can be attributed largely to the presence of glass phases and porosity.

During the 1970s substantial study was conducted to measure and understand creep in nonox-ide ceramics. Figure 9.6 shows the creep characteristics of several Si_3N_4 and SiC materials. Some of the data were measured in compression by Seltzer[5] and others were measured in tension by Larsen and Walther.[6] In general, the creep resistance of pure nitrides and carbides is very high because of their strong covalent bonding. However, other factors can decrease the creep resistance. Let us look at each curve in Figure 9.6, starting at the upper left.

TABLE 9.1
Torsional Creep of Some Ceramic Materials

Material	Creep Rate at 1300°C (in./in. · h)
	At 1800 psi (12.4 Mpa)
Polycrystalline Al_2O_3	0.13×10^{-5}
Polycrystalline BeO	$(30 \times 10^{-5})^a$
Polycrystalline MgO (slip cast)	33×10^{-5}
Polycrystalline MgO (hydrostatic pressed)	3.3×10^{-5}
Polycrystalline $MgAl_2O_4$ (2–5 μm)	26.3×10^{-5}
Polycrystalline $MgAl_2O_4$ (1–3 mm)	0.1×10^{-5}
Polycrystalline ThO_2	$(100 \times 10^{-5})^a$
Polycrystalline ZrO_2 (stabilized)	3×10^{-5}
Quartz glass	$20,000 \times 10^{-5}$
Soft glass	$1.9 \times 10^9 \times 10^{-5}$
Insulating firebrick	$100,000 \times 10^{-5}$
	At 10 psi
Quartz glass	0.001
Soft glass	8
Insulating firebrick	0.005
Chromium magnesite brick	0.0005
Magnesite brick	0.00002

[a] Extrapolated.
Source: From Kingery, W.D., Bowen, H.K., and Uhlmann, D.R., *Introduction to Ceramics*, 2nd ed., Wiley, New York, 1976, Chap. 14.

NC-132* was Si_3N_4 hot-pressed with MgO additives. During the high-temperature hot-pressing operation, a complex glassy phase formed at the grain boundaries. This glass phase was primarily a magnesium silicate modified by Ca, Fe, Al, and other impurities initially present in the Si_3N_4 powder. At temperatures roughly above 1100°C, grain boundary sliding occurred under loading, resulting in a mechanism for creep. Figure 9.7 shows the large amount of creep that resulted for an initially straight bar of NC-132 hot-pressed Si_3N_4 exposed in bending to 276 MPa (40,000 psi) for 50 h at 1100°C (~2000°F).

NCX-34* was Si_3N_4 hot-pressed with Y_2O_3 additives. A grain boundary phase was also present in this material, but it was more refractory than the complex magnesium silicate phase in NC-132 and required a higher temperature to initiate grain boundary sliding. Thus, NCX-34 had somewhat improved creep resistance.

NC-350 was Si_3N_4 prepared by the reaction-bonding process (described in Chapter 14). No additives were required to densify this material, so no grain boundary phase formed. The creep resistance was greatly improved.

NC-435* was a SiC-Si material prepared by a reaction sintering process (described in Chapter 14). Direct bonding occurred between SiC particles and SiC produced by the process. No glassy grain boundary phases were present and the creep resistance was relatively good.

The sintered α-SiC[†] was prepared by pressureless sintering at very high temperatures (>1900°C) with low levels of additives. A grain boundary glass phase did not form. The creep rate was very low.

The Si_3N_4 of Greskovich and Palm[7] was densified with additions of $BeSiN_2$ and SiO_2 and had a creep rate only slightly higher than Carborundum's α-silicon carbide. The low creep resulted from

* Manufactured by the Norton Company, Worcester, MA.
[†] Manufactured by the Carborundum Co., Niagara Falls, NY.

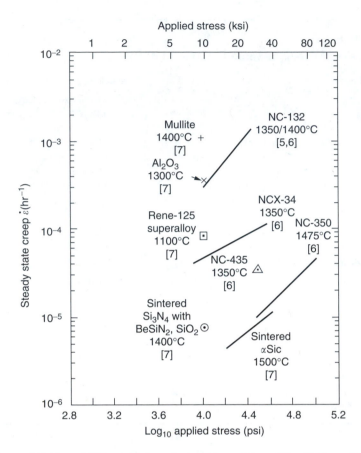

FIGURE 9.6 Creep of Si_3N_4 and SiC materials and comparison with mullite, Al_2O_3, and Rene-125 superalloy. (Data from Seltzer, M.S., *Bull. Am. Ceram. Soc.*, 56(4), 418, 1977; Larsen D.C. and Walther, G.C., *Property Screening and Evaluation of Ceramic Turbine Engine Materials*, Interim Rept. 6, AFML Contract F33615-75-C-5196, 1978; Greskovich, C.D. and Palm, J.A., Development of high performance sintered Si_3N_4, *Highway Vehicle Systems Contractors Coordination Meeting 17th Summary Report*, NTIS Conf. 791082, Dist. Category UC-96, 1979, pp. 254–262.)

the thinness and refractoriness of the grain boundary phase. Transmission electron microscopy using high-resolution lattice imaging techniques showed that the grain boundary phase was only about 10 Å thick.

As the mechanisms of creep in Si_3N_4 materials became better understood, new materials evolved with improved creep resistance.[8–10] Much of this work was directed towards qualifying silicon nitride ceramics for long-term, high-temperature (>1300°C) service in gas turbine engines. Since 1980 greater than six orders of magnitude increase has been achieved in creep life as illustrated in Figure 9.8.[10] These improvements were achieved through a combination of composition, processing, and microstructure manipulation.

- The composition modifications involved (1) high purity starting powders and (2) sintering aids that resulted in higher temperature grain boundary phases. MgO sintering aid was first replaced by yttrium oxide or yttrium oxide plus aluminum oxide and later by lanthanum oxide or lutecium oxide.
- Hot pressing and reaction bonding were replaced by hot isostatic pressing and later by higher temperature over-pressure sintering (these processes are described later in Chapter 14). In addition, heat treatments were implemented that converted most of the grain

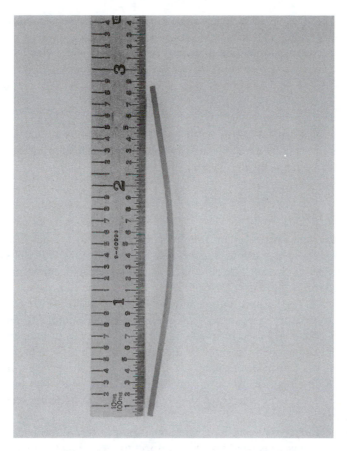

FIGURE 9.7 Hot-pressed Si_3N_4 specimen deformed by creep under a load of 276 MPa (40,000 psi) at 1100°C (~2200°F) for 50 h.

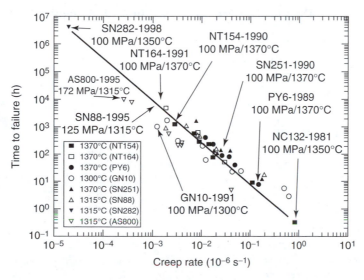

FIGURE 9.8 Monkman-Grant plot comparing the time to failure vs. creep rate for various silicon nitride materials and showing the improvements that were achieved between 1981 and 1998. (Data from Ferber, M.K., Lin, H-T., Jenkins, M.G., and Ohji, T., *Mechanical Characterization of Monolithic Ceramics for Gas Turbine Applications*, ASME Press, New York, 2003, pp. 353–395.)

boundary phase from a glass to a crystalline ceramic and substantially decreased the grain boundary sliding mechanism of creep.
- The microstructure was modified to consist of elongated, intertwined grains that inhibited deformation by creep and also increased fracture toughness.

9.1.7 CREEP CONSIDERATION FOR COMPONENT DESIGN

Creep is not normally a critical consideration for low-temperature applications. However, for intermediate- and high-temperature applications where structural loading is present, creep may be life-limiting and must be carefully considered.

9.2 STATIC FATIGUE

Static fatigue, also known as stress rupture, involves subcritical (slow) crack growth at a stress that is lower than required for instantaneous fracture. Flaws initially in the material slowly grow under the effects of the stress to a size such that instantaneous fracture can occur according to the Griffith relationship [Equation (8.11)] discussed in Chapter 8. Static fatigue is measured by applying a static tensile load at a constant temperature and recording the time to fracture. The experimental setup is similar to that used to measure creep, except that deformation measurements are not required.

Stress rupture is commonly used for predicting the life of metals at temperatures and stresses that simulate the application conditions. Stress rupture testing can also be useful in predicting the life of ceramic materials and in determining life-limiting characteristics that might be improved with further material development. However, stress rupture data for ceramics are usually not easy to interpret because of the large scatter, as illustrated in Table 9.2 for hot-pressed and reaction-bonded Si_3N_4 materials.

Polycrystalline ceramics typically have stress and temperature thresholds below which slow crack growth does not occur or is negligible with respect to the life requirement of the component.

TABLE 9.2
Stress Rupture Data for Hot-Pressed Si_3N_4,[a] Illustrating Typical Scatter

Temperature		Stress		Time of Test (h)		
°C	°F	MPa	kpsi	No Failure	Failure	Source of Data
1066	1950	310	45	50		
1066	1950	310	45		0.46	
1066	1950	310	45		0.18	[b]
1066	1950	310	45	50		
1066	1950	310	45	50		
1200	2192	262	38		380	
1200	2192	276	40	480		
1200	2192	262	38		180	
1200	2192	262	38		105	[c]
1200	2192	262	38		52	
1200	2192	276	40		32	

[a] NC-132 from Norton Co., Worcester, MA.
[b] AiResearch Rept. 76-212188(10), *Ceramic Gas Turbine Engine Demonstration Program*, Interim Rept. 10, Aug. 1978, prepared under contract N00024-76-C-5352.
[c] G.D. Quinn, *Characterization of Turbine Ceramics After Long-Term Environmental Exposure*, AMMRC TR 80-15, April 1980, p. 18.

This is not surprising. Energy is required to cause a flaw or crack to increase in size. The amount of energy is dependent on the bond strength of the material and on the mechanisms available for crack growth. Crack growth is relatively easy if the grain boundaries of the material are coated with a glass phase. At high temperature, localized creep of this glass can occur, resulting in grain boundary sliding. Figure 9.9a shows the fracture surface of an NC-132 hot-pressed Si_3N_4 specimen

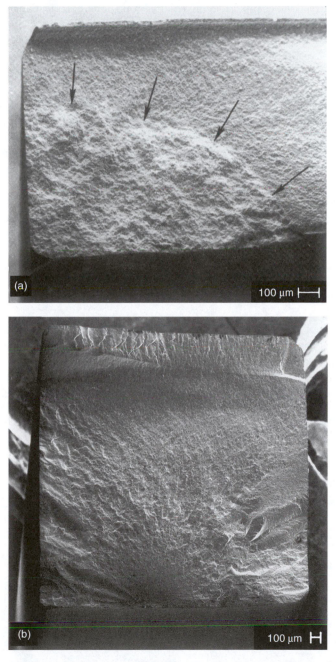

FIGURE 9.9 Comparison of a slow crack growth fracture vs. a normal bend fracture for hot-pressed Si_3N_4. (From Richerson, D.W. and Yonushonis, T.M., Environmental effects on the strength of silicon nitride materials, *DARPA-NAVSEA Ceramic Gas Turbine Demonstration Engine Program Review*, MCIC Rep. MCIC-78-36, 1978, pp. 247–271. With permission.)

that fractured after 2.2 min under a static bending load of 276 MPa (40,000 psi) at ~1100°C (~2000°F). The initial flaw was probably a shallow (20 to 40 μm) machining crack. It linked up with cracks formed by grain boundary sliding and separation and pores formed by triple-point cavitation to produce the new flaw or structurally weakened region seen in Figure 9.9a as the large semicircular area extending inward from the tensile surface. This was the effective flaw size at fracture. Figure 9.9b shows the fracture surface of an NC-132 hot-pressed Si_3N_4 specimen from the same batch, but fractured under rapid loading (normal four-point bend test at a cross-head speed of 0.02 in./min) at room temperature, where slow crack growth did not occur. This specimen fractured at 876 MPa (127,000 psi).

The example above illustrates the importance of knowing the static fatigue properties of the ceramic material. To use the material successfully in an application requires an understanding of the time-dependent properties in addition to the short-term properties.

The factors discussed previously that influence creep also influence static fatigue. Slow crack growth tends to increase with increasing temperature or stress and with the presence of glassy phases. It is affected by grain size and by the presence of secondary phases. For instance, particulate dispersion tends to interrupt slow crack growth and increase the stress rupture life.

To characterize completely the stress rupture life of a ceramic requires many test repetitions at a variety of temperatures and stresses. When selecting candidate materials for an application, it is often desirable to obtain a quick determination of the susceptibility of the material to static fatigue. This can be done by variable-stressing rate experiments. Samples are loaded at three or four rates (as controlled by the cross-head speed of the test equipment) to fracture at the temperature of interest and the data are plotted as strength vs. load rate, as shown in Figure 9.10. If the strength changes as a function of load rate, the ceramic is susceptible to slow crack growth and can probably be removed from consideration if a long stress rupture life is required. If little or no strength change is detected as a function of load rate, the material is suitable for further evaluation by longer-term static fatigue tests.

Static fatigue is not restricted to high temperatures. It also occurs at room temperature for silicate glasses and for many other ceramics. The mechanism involves localized corrosion at the crack tip or where stress is concentrated at a flaw. The corrosion is accelerated by the presence of water or, in some cases, other chemicals.[12] This mechanism is referred to as *stress corrosion* and should be considered during the design of glass components and polycrystalline ceramics known to contain a glass phase.

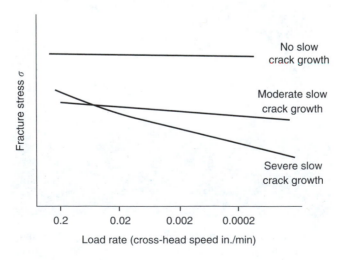

FIGURE 9.10 Use of variable stressing rate experiments for estimating the susceptibility of a ceramic to slow crack growth.

9.3 CHEMICAL EFFECTS

The resistance of a ceramic to chemical attack is largely a function of the strength of atomic bonding and the kinetics of thermochemical equilibrium for the ceramic and the surrounding environment at the temperature of exposure. Weakly ionic bonded ceramics have relatively low resistance to chemical attack. For instance, NaCl is soluble in water and $CaCO_3$ is dissolved by weak acids. Strongly ionic- and covalent-bonded ceramics are much more stable. Al_2O_3 and $ZrSiO_4$ are stable enough that they can be used for crucible and mold linings for melting and casting metal alloys at temperatures above 1200°C (2200°F). Si_3N_4 and Si_2ON_2 (silicon oxynitride) have also been used in contact with molten metals and molten ceramics without appreciable reaction, especially in the processing of aluminum. Alumina-, magnesia-, chromia-, and zirconia*-based ceramic refractories are indispensable for lining the high-temperature furnaces used for metals refining and glass manufacturing.

Ceramics are often selected instead of metals for applications because of their superior chemical stability over a broad temperature range. Al_2O_3, porcelain, and many other oxides, silicates, borides, carbides, and nitrides are resistant to acids and bases and are used in a variety of corrosive applications where metals do not survive. Al_2O_3 and porcelain are particularly important to consider for these applications because they are currently in high-volume production for liners, seals, laboratory ware, and a variety of specialty items and can be obtained for prototype evaluation at moderate cost with reasonable delivery.

Selection of ceramics for high-temperature corrosive applications requires knowledge of chemical resistance plus other properties, such as thermal conductivity, thermal expansion coefficient, strength, and creep resistance. The following paragraphs review the response of some ceramics to high-temperature environments and define the combinations of properties required for some current and projected applications.

9.3.1 GAS-SOLID REACTIONS

9.3.1.1 Oxidation

Many engineering operations are conducted at high temperature in an air or oxygen environment. Oxides and stoichiometric silicates are typically stable at high temperature in an oxygen or mixed oxygen–nitrogen atmosphere. Most of the carbides, nitrides, and borides are not. Si_3N_4 and SiC, which are becoming important commercial ceramics, are used as examples.

At elevated temperature, oxygen interacts with any exposed surface of SiC or Si_3N_4. If the oxygen partial pressure is roughly 1-mm Hg or higher, SiO_2 will form a protective layer at the surface. This is referred to as *passive oxidation*. Formation of SiO_2 will initially be rapid, but will decrease as the thickness of the layer increases. Under these conditions, oxidation will be controlled by oxygen diffusion through the SiO_2 layer and will be parabolic. The oxidation rate increases as the temperature increases, as shown by data from Singhal summarized in Figure 9.11.

At low partial pressure (1-mm Hg or below) of oxygen, inadequate oxygen is present to form a protective SiO_2 layer. Instead, SiO (silicon monoxide) gas forms. This is called *active oxidation*. It is roughly linear and continuous and the component can be completely consumed.

Surface recession of SiC and Si_3N_4 can also occur at high temperature when substantial water vapor is present in the combustion environment. A silica layer forms by oxidation but reacts with the H_2O to form gaseous $Si(OH)_4$.[13,14] The resulting rate of recession is determined by temperature, partial pressure of water, and the flow rate of the gaseous atmosphere over the surface. Water vapor can also increase the oxidation rate by increasing diffusion rates and solubilities of impurities that enhance oxidation.[15]

---

* It is common practice to refer to the oxides by a contracted form ending in "ia" (e.g., Al_2O_3, alumina; MgO, magnesia; Cr_2O_3, chromia; ZrO_2, zirconia).

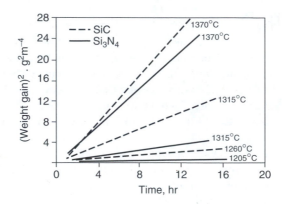

FIGURE 9.11 Oxidation rate vs. temperature for Si$_3$N$_4$ and SiC materials. (From Singhal, S.C., in *Ceramics for High Performance Applications*, Burke, J.J., Gorum, A.E., and Katz, R.N., Eds., Brook Hill Publishing Co., Chestnut Hill, MA, 1974, pp. 533–548. With permission.)

Silica volatilization was identified as a major problem in the late 1990s following a 948 h field test by Solar Turbines Inc. of a SiC–SiC ceramic composite combustor liners in an industrial (~5000 kW) gas turbine engine.[16] The liners were degraded in some regions to a depth of about 0.5 mm, but were still functional. Testing was continued to 5028 h, at which time the combustor liners were removed and cut up to study the depth of degradation. In some regions of the liners, especially those regions exposed to the highest temperatures, a major portion of the thickness has been either eaten away or severely degraded. Development of an oxide surface coating was pursued in an effort to protect the SiC–SiC composite from the moist combustion environment. So far this has been quite successful. Combustor liners with oxide "environmental barrier coatings" (EBCs) have now survived more than 15,000 hours of engine testing. Figure 9.12 shows polished cross sections of portions of SiC–SiC combustor liners tested with and without EBCs in the Solar Turbines engine.

Surface recession is not the only concern for gas-solid reactions, especially if the ceramic material is exposed to stress. Another concern is whether the reaction will alter the size of strength-controlling flaws. Here again we can use early research on silicon nitride as an illustrative example. Figure 9.13 shows the strength of NC-132 hot-pressed silicon nitride samples vs. the direction of surface grinding and subsequent high temperature oxidation exposures.

Specimens with grinding grooves parallel to the length (longitudinal) and thus parallel to the stressing direction were much stronger than those with grinding grooves perpendicular to both the length (transverse) and stressing direction. Low-temperature oxidation (1000°C) significantly increased the strength for the transverse machined specimens by blunting or reducing the severity of the surface flaws associated with the grinding damage. High-temperature oxidation completely removed the grinding grooves, but resulted in formation of surface pits, which had an equivalent or worse effect on the strength. The surfaces before and after oxidation are compared in Figure 9.14. Based on these and other studies, a static oxidation exposure became routine for surface ground dense silicon nitride samples and engine components to minimize strength reduction due to grinding damage.

A static oxidation exposure in a furnace does not simulate the conditions that a material will experience in a real combustion environment such as in a gas turbine engine. In fact, the conditions are generally impossible to duplicate without putting samples inside the actual operating engine. This generally is not feasible, so an engineer tries to design a test rig that at least approximates key conditions in the engine. Figure 9.15 shows the schematic of a test rig that was designed to evaluate ceramic materials for potential use in a gas turbine engine. Extensive testing was conducted in this rig for silicon nitride and silicon carbide materials developed between the late 1970s and the mid 1990s.[17,18] Four-point flexure test bars were mounted in a rotating holder and cycled in and out

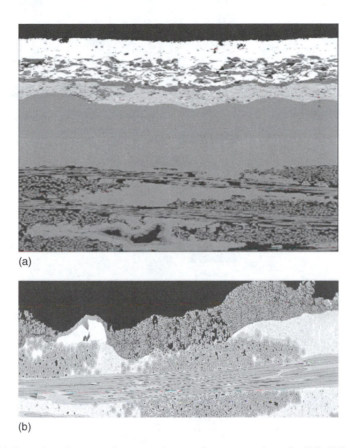

(a)

(b)

FIGURE 9.12   (a) Scanning electron microscope image of a cross section of a SiC–SiC composite combustor liner with a oxide environmental coating following engine testing for 14,000 hours. (b) Scanning electron microscope image of a cross section of a SiC–SiC composite combustor liner with no oxide environmental barrier coating following engine testing for only 1400 hours. (Photographs courtesy of Karren More, Oak Ridge National Laboratory, Oak Ridge, TN.)

of the combustion flame of a burner operating on typical gas turbine engine fuel. Test bars could be removed at various time intervals to measure the strength and compare with the strength of unexposed test bars. Some of the data are summarized in Figure 9.16. These tests helped guide development of improved materials and also helped identify mechanisms of material degradation that had not been clearly identified in static oxidation tests.

All carbides, nitrides, and borides are susceptible to oxidation in a high-temperature oxidizing atmosphere. Examples include diamond, graphite, TiB_2, ZrB_2, AlN, AlON, SiAlON, and TiC.

9.3.1.2   Reduction and Other Reactions

Many industrial processes are conducted under atmospheres other than oxygen or air. Examples are hydrogen, ammonia (NH_3), carbon monoxide (CO), argon, nitrogen, and vacuum. H_2 and NH_3 are strongly reducing and can remove oxygen from oxides at high temperature. For example,

$$H_2(g) + SiO_2(s) \rightarrow SiO(g) + H_2O(g) \tag{9.2}$$

The amount of SiO_2 loss can be substantial. Figure 9.17 shows the percent weight loss vs. temperature for various SiO_2-containing refractory brick after exposure to a 100% H_2 atmosphere for

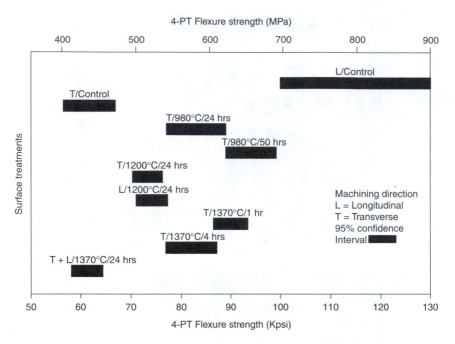

FIGURE 9.13 Oxidation of NC-132 hot-pressed silicon nitride, showing strength increase for low-temperature exposure and strength decrease for high-temperature exposure. (From Richerson D.W. and Yonushonis, T.M., Environmental effects on the strength of silicon nitride materials, *DARPA-NAVSEA Ceramic Gas Turbine Demonstration Engine Program Review*, MCIC Rep. MCIC-78-36, 1978, pp. 247–271. With permission.)

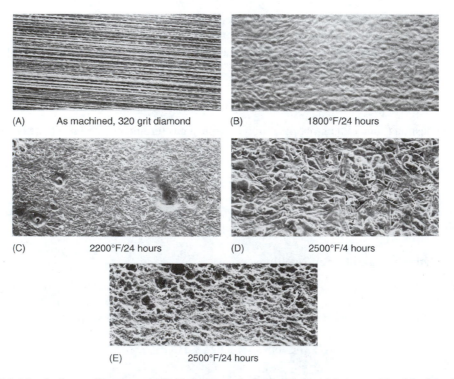

FIGURE 9.14 Surfaces of hot-pressed Si_3N_4 before and after oxidation. (a) As-machined surface, 320-grit diamond; (b) oxidized in air for 50 h at 980°C (1800°F); (c) oxidized in air for 24 h at 1200°C (2200°F); and (d) oxidized in air for 24 h at 1370°C (2500°F).

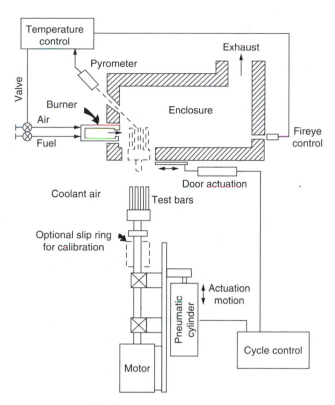

FIGURE 9.15   Schematic of a test rig for long-term cyclic oxidation exposures of ceramic strength test samples of a combustion environment. (From Carruthers, W.D., Richerson, D.W., and Benn, K.W., *3500-Hour Durability Testing of Commercial Ceramic Materials*, NASA CR-159785, July 1980. With permission.)

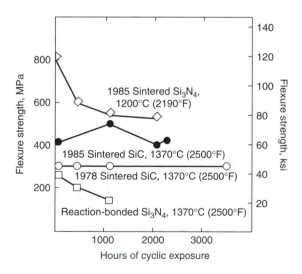

FIGURE 9.16   Comparison of the change in strength of different ceramic materials after cyclic exposure to high-velocity combustion gases. (Data compiled from Carruthers, W.D., Richerson, D.W., and Benn, K.W., *3500-Hour Durability Testing of Commercial Ceramic Materials*, NASA CR-159785, July 1980, and Lindberg, L.J., Durability testing of ceramic materials for turbine engine applications, *Proceedings of the 24th Automotive Technology Development Contractors Coordination Meeting*, SAE Publication P-197, April 1987, p. 149.)

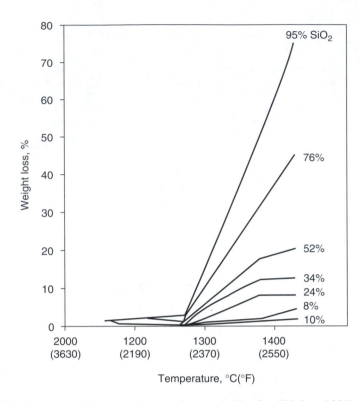

FIGURE 9.17 Weight loss of silicon-containing refractory brick after 50 h in a 100% reducing hydrogen atmosphere. (From Crowley, M.S., Hydrogen-silica reactions in refractories, *Bull. Am. Ceram. Soc.*, 46, 679, 1967. With permission.)

50 h. The losses are low below 1250°C (2280°F) but increase rapidly above 1250°C, especially for high-SiO$_2$ compositions.[19]

The degree of reduction depends on the stability of the oxides. Oxides with a high negative free energy of formation are more difficult to reduce than oxides with a low negative free energy of formation. Figure 9.18 compares the free energy of formation for various oxides. Note that yttrium, for example, has a stronger affinity for oxygen than silicon and thus Y$_2$O$_3$ is more stable than SiO$_2$ in a reducing environment.

9.3.1.3 Thermodynamics

Data are published in sources such as References 20 to 22 that allow an individual to estimate the stability of a material in a specific environment. One simply uses the free energy data to predict which reactions are most likely to occur. This is particularly useful when there are several possible reactions. For example, if a person wanted to synthesize or densify a particular material in a furnace at a selected temperature, he or she would write potential chemical equations for each possible reaction involving the materials, the furnace lining, heating elements, known impurities, and the furnace atmosphere. The free energy of formation for each chemical equation would then be compared to determine the most likely reaction or reactions.

The reduction of SiO$_2$ by H$_2$ represents one type of detrimental reduction reaction. Another type involves a change in oxidation state. An example is an alumina refractory containing chromium oxide (Cr$_2$O$_3$) plus silica. At high oxygen partial pressure ($\sim 10^0$ to 10^{-9} atm), Cr$_2$O$_3$ with trivalent Cr is stable. The Cr$_2$O$_3$–SiO$_2$ eutectic is above 1700°C (3090°F). Reduction of the Cr to the 2+ state at oxygen partial pressure less than about 10^{-9} atm results in a drop in the eutectic temperature to

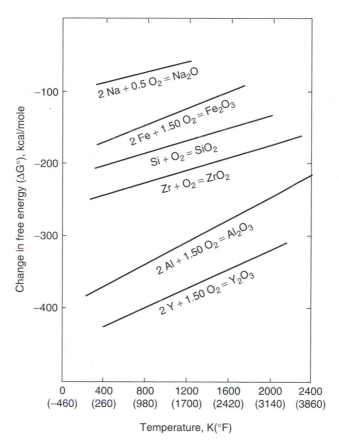

FIGURE 9.18 Comparison of the free energy of formation of various oxides. (Compiled with data from Pankratz, L.B., *Thermodynamic Properties of Elements and Oxides*, Bull., 672, U.S. Bureau of Mines, Washington, DC.)

as low as about 1400°C (2550°F). The resulting liquid phase would be detrimental to the function of the refractory. Oxygen pressure in this low range can occur in a coal gasification process.

9.3.1.4 Interactions with Water Vapor

Water vapor can have a strong effect on many ceramic materials, even at room temperature. A particularly important room-temperature example is *stress corrosion* as discussed earlier. Water vapor catalyzes corrosion at the tip of a surface crack in glass and some polycrystalline ceramics and allows the crack to slowly grow. This results in a cumulative reduction of the strength of the material. Another example is hydrolysis of Si_3N_4. Fine particles of Si_3N_4 powder react at room temperature with water vapor to form a surface film of SiO_2, as in Equation (9.3).

$$Si_3N_4 + 6H_2O \rightarrow 3SiO_2 + 4NH_3 \tag{9.3}$$

A distinct ammonia odor is present when most containers of Si_3N_4 powder (that have not been stored under dry nitrogen or argon) are opened.

Many ceramic materials are hygroscopic (sensitive to water). The incomplete atomic bonds at the surface attract water molecules and form weak bonds equivalent to van der Waals bonds. Some salts can actually absorb enough water to dissolve themselves. An example is NaCl.

9.3.1.5 Vaporization and Dissociation

The single-component phase equilibrium diagrams studied in Chapter 6 all contained a region where vapor was in equilibrium with the solid phase. The amount of the vapor at any given temperature is identified in terms of the "vapor pressure." The vapor pressure for most ceramics is negligible at room temperature and becomes appreciable only at very high temperature. If the material is held in an open container at a temperature at which the vapor pressure is positive, the material will slowly vaporize (evaporate). If the material is in a sealed container, it will only evaporate until the container is filled with vapor at the equilibrium vapor pressure for the material at the applied temperature. Knowledge of the vapor pressure characteristics of a ceramic is important for optimizing high-temperature densification conditions during fabrication and for evaluating suitability of application in high-temperature environments.

Some ceramics *dissociate* rather than vaporize. This means that they decompose to a different composition. SiO_2 placed in a vacuum or inert gas at high temperature dissociates to $SiO + O_2$:

$$2SiO_2(s) \rightarrow 2SiO(g) + O_2(g) \tag{9.4}$$

Si_3N_4 dissociates to $Si + N_2$:

$$Si_3N_4(s) \rightarrow 3Si(l) + 2N_2(g) \tag{9.5}$$

The dissociation reaction is similar to vaporization. It occurs according to which phases are in equilibrium at a given temperature and environment. The degree of dissociation is negligible or low at low or intermediate temperatures and increases as the temperature is increased. It also is accelerated if the material is in an open container or vacuum.

Dissociation must be considered during the fabrication of ceramic materials. Si_3N_4, for example, undergoes some dissociation at the temperature required to achieve densification. Use of vacuum or an inert atmosphere would favor the dissociation reaction. Use of a positive nitrogen over-pressure, packing the parts in a bed of Si_3N_4 powder, or enclosing the parts in a sealed container depresses the dissociation reaction and allows densification to occur.

Decomposition reactions are very important in the processing of oxide ceramics and refractories. The oxides are usually derived from metal salts such as carbonates, hydroxides, nitrates, sulfates, acetates, oxalates, or alkoxides. These salts are either naturally occurring raw materials or the results of chemical refining operations. They are used as a portion of the raw materials in the fabrication of glass, refractories, and a wide variety of ceramic products. They decompose at low to intermediate temperatures to produce a solid oxide plus a gas. Heating rates and decomposition temperatures must be carefully controlled to avoid breaking the ceramic part during processing, as a result of too rapid evolution of the reactant gas. It is also important that complete decomposition occur before melting or densification of the part begins.

9.3.2 Liquid–Solid Reactions

Liquid–solid reactions are generally referred to as corrosion. Corrosion can result from direct contact of a reactive liquid with the ceramic as a simple dissolution reaction or can be more complex, such as interactions with impurities in the ceramic or surrounding gas or liquid environment.

The kinetics of a reaction are often more important for ceramics than whether or not a reaction occurs. A reaction can only occur if fresh reactants can get to the ceramic surface and if reaction products can get away from the surface. Often the reaction products remain on the surface and act as a boundary layer, which limits further reaction. In this case, diffusion through the boundary layer controls the rate of corrosion. For instance, in glass-melting furnaces and in other industrial processes where silicate slags are present, the boundary layer may be on the order of 1 cm thick due to the high viscosities and low fluid velocities involved. This protects the refractory linings and significantly extends their life.

9.3.2.1 Ambient Temperature Corrosion

Ceramics show a broad range of resistance to corrosion at room temperature. Strongly bonded ceramics such as Al_2O_3 and Si_3N_4 are virtually inert to attack by aqueous solutions, including most strong acids and bases. On the other hand, many of the weakly bonded ionic metal salts, including most of the nitrates, oxalates, chlorides, and sulfates, are soluble in water or weak acids.

Ceramic silicates are also very stable. A notable exception is attack by hydrofluoric acid (HF). HF readily dissolves most silicate glass compositions.

Lay[23] has prepared a review of the resistance of many engineering ceramics, including Al_2O_3, BeO, MgO, ZrO_2, spinel, Si_3N_4, mullite, and SiC, to corrosion by acids, alkalis, gases, fused salts, metals, and metal oxides.

9.3.2.2 High-Temperature Corrosion of Oxides

High-temperature corrosion of oxide ceramics is encountered in many cases where the ceramic is in contact either with a molten ceramic or a molten metal. The former case is especially important in the glassmaking industry, where it has been estimated that around 85 million tons of glass were produced worldwide in 1992.[24] This represents a large-volume usage of oxide refractories for glass-melting furnace linings.

Corrosion of Al_2O_3 in a CaO–Al_2O_3–SiO_2 melt provides a good example for comparing some of the rate-controlling factors for the corrosion of an oxide by a glass or slag. The data of Cooper et al.[25] are summarized in Figure 9.19. The corrosion rate increases as the temperature increases (curve B

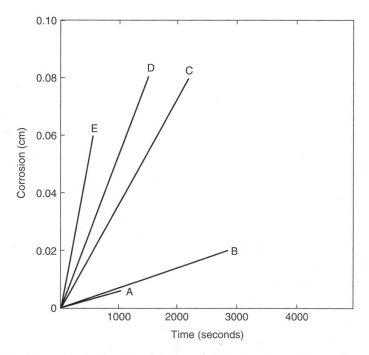

FIGURE 9.19   Corrosion of Al_2O_3 in CaO–Al_2O_3–SiO_2 melt. A, single crystal, 1550°C, natural convection; B, single-crystal disk rotating at 1200 rpm, 1410°C; C, single crystal, 1500°C, forced convection; D, polycrystalline, 1500°C, forced convection; E, single-crystal disk rotating at 1200 rpm, 1550°C. (From Cooper, Jr., A.R., Samaddar, B.N., Oishi, Y., and Kingery, W.D., Dissolution in ceramic systems, *J. Am. Ceram. Soc.*, 47, 37, 1964; 47, 249, 1964; 48, 88, 1965. With permission.)

vs. curve E). The corrosion rate is typically higher for a polycrystalline ceramic than a single crystal (curve D vs. curve C), due to grain boundary effects. The corrosion rate is lower for natural convection than for forced convection or for cases where either the melt is flowing or the ceramic is moving (curves A, C, and E).

Environments for oxide ceramics in metal melting and refining processes are even more severe than the conditions in glass-melting furnaces. An important example is the basic oxygen steelmaking process (BOSP) used in a major portion of iron refining in the United States.[26,27] BOSP is a batch process conducted in a ceramic refractory brick-lined furnace referred to as the *basic oxygen furnace* (BOF). During a typical cycle the BOF is first tilted and charged with up to 100 tons of scrap steel and 250 tons of 1300°C molten crude iron. The BOF is then turned upright and oxygen is introduced through a water-cooled lance. CaO and CaF_2 are added as flux to remove Si, P, S, and other impurities by formation of a slag. The BOF cycle is approximately 1 h long, with the temperature reaching 1600 to 1700°C (~2900 to 3100°F) and the furnace lining bathed in churning molten slag and steel.

The BOF service environment is extremely severe. The refractory lining is exposed to impact, abrasion, and thermal shock during charging of the scrap steel and the 1300°C crude iron and is exposed to both molten slag and molten metal corrosion during operation. Two primary types of refractories are used. One is referred to as *pitch-bonded* or *tar-bonded* and consists mostly of MgO particles coated with pitch and warm-bonded to form bricks. The other is referred to as tar-impregnated and consists of porous MgO ceramic brick formed and then impregnated under vacuum by molten pitch. These two types of bricks are installed in the furnace and then "burned in" under controlled temperature and reducing conditions to pyrolyze the pitch to elemental carbon. The carbon is resistant to wetting by the molten slag and metal and inhibits penetration of the brick. A typical BOF ceramic lining ranges from 18 in. thick at the opening to 36 in. thick in regions of maximum erosion and corrosion. Its life is usually less than 1000 cycles, with a rate of refractory consumption between 0.08 and 0.15 cm (0.03 to 0.06 in.) average recession per cycle.

9.3.2.3 Condensed-Phase Corrosion

Sometimes constituents in a gas will condense at a surface to produce a liquid phase. This condensed phase generally is not present in sufficient volume to eat away large masses of material, but it can result in enhanced corrosion or surface degradation. A classic example is hot corrosion of metals in gas-turbine engines. Sulfur from the fuel combines with NaCl from seaside air or road salt to form sodium sulfate. The Na_2SO_4 condenses on surfaces of the engine that are roughly in the range 650 to 950°C (1200 to 1740°F). Severe corrosion results.

Si_3N_4 and SiC are being developed as gas turbine materials and for other high-temperature applications in which condensed phases may occur. Depending on the nature of the condensed phase, a variety of potential degradation mechanisms are possible:

Change in the chemistry of the SiO_2 layer, increasing the oxygen diffusion rate.
Bubble formation, disrupting the protective SiO_2 layer and allowing increased oxygen access.
Decreased viscosity of the protective surface layer, which is then swept off the surface by the high-velocity gas flow.
Formation of a molten composition at the ceramic surface that is a solvent for the ceramic.
Localized reducing conditions, decreasing the oxygen partial pressure at the ceramic surface to a level at which active oxidation can occur.
Formation of new surface flaws, such as pits or degraded microstructure, which decrease the load-bearing capability of the component.

Singhal[28] evaluated the dynamic corrosion-erosion behavior of Si_3N_4 and SiC in a pressurized turbine test passage operating at 1100°C, 0.9 MPa pressure, and 152 m/sec gas velocity using

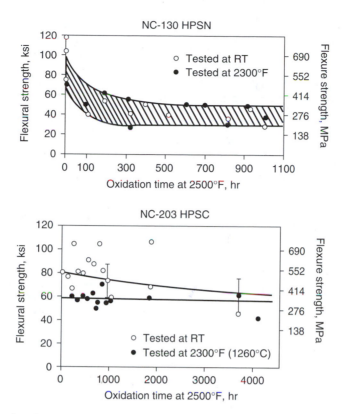

FIGURE 9.20 Results of exposure of hot-pressed Si_3N_4 (HPSN) and SiC (HPSC) in a turbine passage. (From Miller, D.G., Andersson, C.A., Singhal, S.C., Lange, F.F., Diaz, E.S., Kossowsky, E.S., and Bratton, R.J., *Brittle Materials Design High Temperature Gas Turbine—Materials Technology*, AMMRC CTR-76-32, Vol. 4, 1976. With permission.)

Exxon No. 2 diesel fuel. After 250 h of exposure, the surfaces of test specimens were smooth and free of adherent surface deposits. Average surface erosion was only 2.3 μm for the SiC and 3 μm for the Si_3N_4. No strength degradation occurred. Other tests in the same turbine test passage with 4 ppm barium present in the fuel produced drastically different results: massive surface deposits containing barium silicates with iron, magnesium, nickel, chromium, and other trace impurities. This example illustrates that small quantities of some impurities can have a pronounced effect on the corrosion behavior. This is especially important in applications where heavy residual fuels or coal-derived fuels are being considered.

Continuation of turbine passage testing without barium fuel additions was reported by Miller et al.[29] for longer time and higher temperature. The results are summarized in Figure 9.20. The strength of the hot-pressed Si_3N_4 was significantly degraded due to the formation of surface pits. The hot-pressed SiC showed no degradation in high-temperature strength after 4000 h of exposure. The slight decrease in room-temperature strength was attributed to cracking of the SiO_2 surface layer during cooling, probably due to the displacive phase transformation of crystalline cristobalite present in the layer.

McKee and Chatterji[30] evaluated the stability of 97 to 99% dense sintered SiC at 900°C in a gas stream flowing at 250 m/min at 1 atm total pressure. Gases evaluated were N_2, O_2, air, O_2–N_2 mixture, 0.2% SO_2–O_2, 2% SO_2–N_2, 2% SO_2–5% CH_4–balance N_2, 10% H_2S–H_2, and H_2. Gas–molten–salt environments were also evaluated using constant gas compositions and the following salt mixtures: Na_2SO_4, $Na_2SO_4 + C$ (graphite), $Na_2SO_4 + Na_2CO_3$, $NaSO_4 + Na_2O$, $Na_2SO_4 + NaNO_3$, and

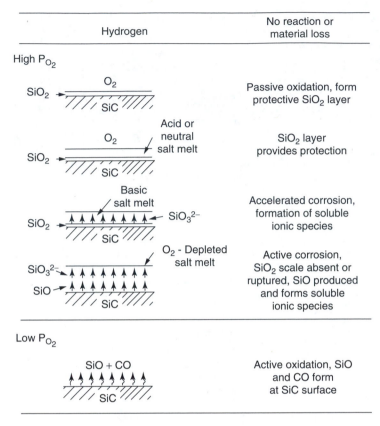

FIGURE 9.21 Behavior of SiC in gas-molten salt environments. (From McKee D.W. and Chatterji, D., Corrosion of silicon carbide in gases and alkaline melts, *J. Am. Ceram. Soc.*, 59(9–10), 441–444, 1976. With permission.)

$Na_2SO_4 + Na_2S$. Corrosion behavior was studied by continuous monitoring of the specimen weight in the test environment using standard thermo-balance techniques, followed by examination of polished sections of the specimen cross section. The results are summarized in Figure 9.21.

McKnee and Chatterji concluded that SiC at 900°C is inert in H_2, H_2S, and high-purity N_2; that passive oxidation provides protection under normal gas-turbine operating conditions or when thin condensed layers of Na_2SO_4 are present in an oxidizing atmosphere; and that corrosion occurs in the presence of Na_2O, a carbonaceous condensed phase, or a thick Na_2SO_4 surface layer.

Richerson and Yonushonis[11] evaluated the effect on the strength of Si_3N_4 materials of 50 h of cyclic oxidation-corrosion at specimen temperatures up to 1200°C using a combustor rig burning typical aircraft fuels with 5 ppm sea salt additions. Under conditions where Na_2SO_4 was present in the condensed form, corrosion resulting in slight material recession occurred. This was accompanied by buildup of a glassy surface layer containing many bubbles that appeared to be nucleating at the Si_3N_4 surface (Figure 9.22). The strength of Si_3N_4 hot-pressed with MgO was degraded by 30%. The strength of 2.5-g/cm³ density reaction-bonded Si_3N_4 was degraded by as much as 45%.

9.3.2.4 Corrosion in Coal Combustion Environments

Coal-burning steam plants accounted in 2000 for 52% of the electricity generated in the United States and also for an enormous amount to air pollution. As illustrated by Tables 9.3 and 9.4, coal contains many impurities that contribute to the air pollution but that also result in combustion environments that are corrosive. On the average, each ton of coal burned produces 2.3 tons of carbon

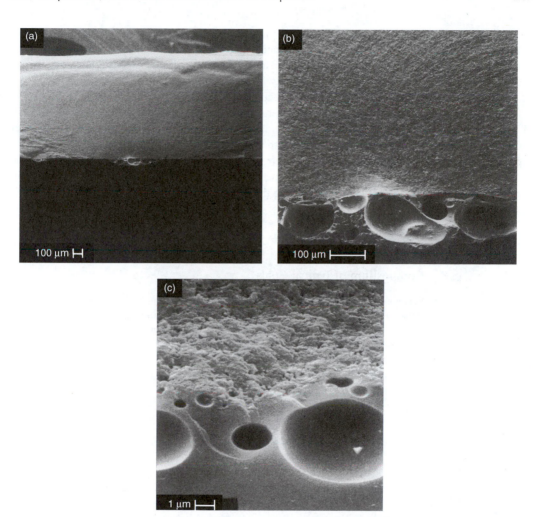

FIGURE 9.22   Reaction-bonded Si_3N_4 after exposure in a combustion rig with 5 ppm sea salt addition for 25 cycles of 1.5 h at 900°C, 0.5 h at 1120°C, and a 5-min air-blast quench. (a), (b), and (c) show the fracture surface at increasing magnification and illustrate the glassy buildup in the region of combustion gas impingement. (From Richerson D.W. and Yonushonis, T.M., Environmental effects on the strength of silicon nitride materials, *DARPA-NAVSEA Ceramic Gas Turbine Demonstration Engine Program Review*, MCIC Rep. MCIC-78-36, 1978, pp. 247–271. With permission.)

dioxide, 10 kg of sulfur dioxide (SO_2), and 5 kg of nitrogen oxides (NO_x). Since about one billion tons of coal are burned in the United States each year, this represents a large amount of pollution emissions and corrosive species in the combustion environment.

Conventional coal steam plants are relatively inefficient (only converting about 30 to 33% of the energy in the coal to electrical energy) and require expensive postcombustion processes to remove some of the sulfur dioxide, nitrogen oxides, and ash: flue gas desulfurization, selective catalytic reduction, and electrostatic precipitation. These are expensive and can only remove about 95% of the SO_2, 90% of the NO_x, and 0% of the CO_2. Alternate coal burning power generation processes have been under development that have the potential for higher efficiency and decreased emissions: fluidized bed combustion, gasification systems, and combined cycle systems. For example, the Wabash River gasification facility demonstrated 38% efficiency and was projected to achieve greater than 60% efficiency when combined in the future with an advanced gas turbine engine or fuel cells.

Another study evaluated ceramic linings for a coal gasification system. The evaluations were conducted at the Conoco Coal Development Company's Lignite Gasification Pilot Plant in Rapid City, South Dakota. The reactor vessel was 21.3 m (70 ft) high with an inside diameter of 168 cm (66 in.) lined with 45.7 cm (18 in.) of low-density insulating castable refractory plus 15.2 cm (6 in.) of dense abrasion-resistant castable refractory. The dense castable consisted of approximately 37% SiO_2, 57% Al_2O_3, and 6% CaO. The gasifier operated for 5 years at 843°C (1550°F) and 1.04 MPa (150 psi) without major problems with the refractory lining. No reactions attributed to the coal were reported. The major chemical reaction involved the refractory and the steam used in the gasification process. Calcium aluminates in the ceramic refractory lining reacted hydrothermally with silica under the influence of the steam to yield calcium aluminum silicate. This acted as a strong bonding phase for the refractory, significantly increasing the compressive strength and abrasion resistance, both of which were beneficial.

TABLE 9.3
Composition of Typical Coals

Location: Rank:	Illinois High-Volatile bituminous B	Wyoming Subbituminous B
Analysis (% by weight)		
Moisture	5.8	15.3
Volatile matter	36.2	33.5
Fixed carbon	46.3	45.2
Ash	11.7	5.7
Sulfur	2.7	2.3

Source: Compiled from *Steam*, Babcock and Wilcox, 38th ed., 1972, pp. 5–15 and *Combustion Engineering*, Combustion Engineering, Inc., 1966, pp. 13–15.

TABLE 9.4
Variations in Slag Composition for Typical Coals

Location: Seam: Rank: Ash Composition (% wt)	Zap, N.D. Zap Lignite	Ehrenfield, Pa. L. Freeport Medium-Volatile bituminous	Victoria, Ill. Illinois 6 High-Volatile bituminous	Hanna, Wy. 80 Subbituminous
SiO_2	20–23	37	50	29
Al_2O_3	9–14	23	22	19
Fe_2O_3	6–7	34	11	10
TiO_2	0.5	0.8	0.9	0.8
CaO	18–20	0.8	9.0	18.7
MgO	6–7	0.4	1.1	2.9
Na_2O	8–11	0.2	0.35	0.2
K_2O	0.3	1.3	2.2	0.7
SO_3	21	1.6	1.6	17.78
% Ash	7–11	15.1	14.3	6.6
% Sulfur	0.6–0.8	4.04	2.6	1.2

9.3.3 SOLID–SOLID REACTIONS

Solid–solid reactions occur by solid-state diffusion of atoms, that is, by the motion of individual atoms along the surface or through the bulk of a material. The rate of solid-state diffusion is extremely low at room temperature, but can be substantial at high temperature. The following can occur:

1. Localized change in composition, resulting in a change in properties or even the formation of a liquid phase
2. Bonding at an interface

9.4 MECHANICALLY INDUCED EFFECTS

Mechanical effects can reduce the life of a ceramic by producing surface flaws that are larger than those defects initially in the material or by removal of material from the surface (wear).

9.4.1 SURFACE FLAW FORMATION

Mechanically induced surface flaws are invariably small cracks caused by localized stress concentration. The stress concentration can result from a machining operation, from impact, from point or line contact loading at an interface or attachment, or from a combination of normal plus tangential loading. These different types of loading and the resultant mechanisms of crack formation are described in the following paragraphs and figures.

Most ceramics undergo a final machining operation to achieve the required shape and dimensional tolerances. The machining is typically an abrasive grinding approach, usually involving diamond particles imbedded into the surface of a metal, resin, or rubber wheel. The diamond is harder than all other materials. When the rotating grinding wheel is pressed against the surface of the ceramic, each diamond particle plows a furrow into the ceramic. A schematic of a single furrow (grinding groove) is illustrated in Figure 9.23. Very high localized compressive stress and temperature occur where the diamond particle is pressed against the ceramic. The stress is so high that the ceramic is locally plastically deformed, as indicated by the criss-crossed area. A high tensile stress results at the base of the deformed area and produces a crack that extends into the ceramic perpendicular to the surface. This is called the *median crack* and is the deepest crack produced. Because it extends to the greatest depth, it has the potential to cause the greatest degree of strength reduction compared to the other types of machining cracks. As the abrasive particle passes, the plastically deformed region rebounds. This results in tensile stresses along the edges of the plastically deformed area and produces subsurface cracks roughly parallel to the surface. These cracks generally curve toward the surface and result in tiny chips of material spalling off, which accounts for much of the ceramic removed during machining. These cracks are called *lateral cracks*. Because they do not

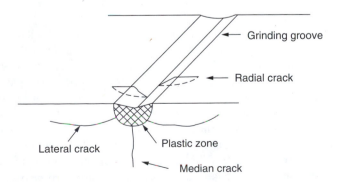

FIGURE 9.23 Schematic showing the cracks and material deformation that occur during grinding with a single abrasive particle.

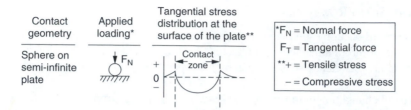

FIGURE 9.24 Hertzian loading involving a sphere pressed with a normal (perpendicular) force against a flat plate. (From Richerson, D.W., Contact stress at ceramic interfaces, *Progress in Nitrogen Ceramics*, Riley, F.L., Ed. Martinas Nijhoff Publishers, The Hague, 1983. With permission.)

extend into the ceramic, they have little effect on the strength. As the abrasive particle cuts through the ceramic, large friction forces result. These lead to tensile stress and intermittent cracks in the ceramic perpendicular to the direction of the grinding groove. These are called *radial cracks*. The radial cracks are not as deep as the median crack and cause less strength reduction.

A second important mode of surface flaw formation is *impact*. Impact is the sudden application of a localized stress (usually by point loading). It can result from a separate body or particle striking the material or from the material striking or colliding with another surface. Metals are relatively tolerant to impact. They can deform in a ductile fashion to distribute the stress. Metals may dent or deform, but generally do not fracture. Conventional ceramics have no mechanism for redistributing the stress and generally fracture. The new transformation-toughened ceramics and ceramic-matrix composites are more resistant to impact than conventional ceramics, but still not as good as metals.

Impact is essentially point contact stress. This can be modeled by a sphere pressed against a plate as shown in Figure 9.24 where the amount of the load applied at a 90° angle to the surface is referred to as the *normal force*, F_N. This type of loading is called *Hertzian* loading and results in compressive stress directly beneath the sphere and tensile stress surrounding the circular contact between the sphere and plate. Because the tensile stress is distributed in a circular pattern, it is referred to as *hoop* tensile stress. The peak tensile stress is on the surface of the plate at the edge of the contact (as shown by the cross section of the contact zone in Figure 9.24) and decreases radially away from the contact and beneath the surface. A crack initiates in the ceramic plate when the tensile stress exceeds the critical value for a material flaw as defined by the Griffith equation in Chapter 8. No tensile stress is present in the bulk of the plate, so the crack that forms is very shallow. However, tensile stress is present around the circumference of the contact zone, so the shallow crack follows this path. What results is a ring crack surrounding the contact zone. If the applied stress is high enough to initiate cracks at more than one material flaw, a series of concentric ring cracks will result. This is illustrated schematically in Figure 9.25 and can be easily demonstrated experimentally. Tape a piece of window glass on a flat surface. Drop a marble or ball bearing onto the glass from progressively greater heights until the force is high enough to cause Hertzian ring cracks.

Impact and point loading are common sources of mechanically induced damage that reduce the life of a ceramic part and are major reasons why ceramics have not been used broadly for structural applications. Another source is sliding contact. This can be modeled as shown in Figure 9.26 by a sphere on a flat plate with both a normal force F_N and tangential force F_T applied to the sphere. The addition of the tangential force changes the stress distribution. Rather than a uniform hoop tensile stress surrounding the contact zone, the tensile stress is now all concentrated on the side of the contact opposite the applied tangential force, that is, on the trailing edge of the contact. This can be easily demonstrated experimentally. Hold a segment of toilet paper on the surface of a table. Press your finger tightly against the toilet paper and slide your finger in one direction. Note where the rip occurs in the paper.

The tensile stress resulting from the tangential loading can vary from zero to very high depending on the coefficient of friction. The coefficient of friction f is the tangential force F_T divided by the normal force F_N. If the coefficient of friction is zero, the tangential stress is zero and the degree

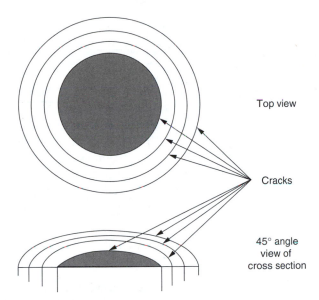

Top view

Cracks

45° angle
view of
cross section

FIGURE 9.25   Schematic of a top view and angled cross-sectional view of concentric cracks formed in a ceramic by impact or point contact. The contact zone is shaded.

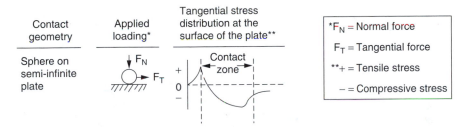

FIGURE 9.26   Model of the stress distribution for sliding contact where both normal and tangential forces are present. (From Richerson, D.W., Contact stress at ceramic interfaces, *Progress in Nitrogen Ceramics*, Riley, F.L., Ed., Martinas Nijhoff Publishers, The Hague, 1983. With permission.)

of the tensile stress is determined only by the normal force. If the coefficient of friction is high, the magnification of the tensile stress by the tangential force is high. This is illustrated in Figure 9.27 for an aluminum cylinder in contact with a hot-pressed silicon nitride (HPSN) plate. The tensile stress is at a maximum at the surface, but decreases rapidly below the surface as shown in Figure 9.28. This explains why surface cracks form, but do not propagate through the material.

An equation for estimating the stress magnification for biaxial (normal load plus tangential load) vs. normal contact loading has been derived:[31]

$$\frac{P_N}{P_S} = \left(1 + \frac{3\pi(4 + v)}{8(1 - 2v)} f \right)^3 \tag{9.6}$$

where

P_N = load causing Hertzian damage with normal loading of a spherical indenter
P_S = load causing damage with sliding spherical indenter
v = Poisson's ratio of the material
f = coefficient of friction
π = 3.14

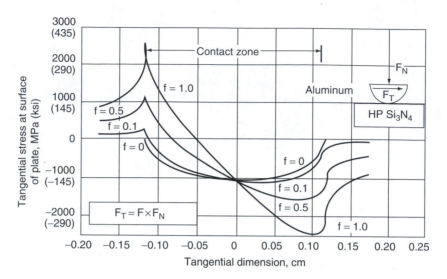

FIGURE 9.27 Effect of coefficient of friction (f) on magnification of the tensile stress at the trailing edge of the contact zone during sliding contact. (From Finger, D.G., ONR contract N00014-78-C-0547, 1979. With permission.)

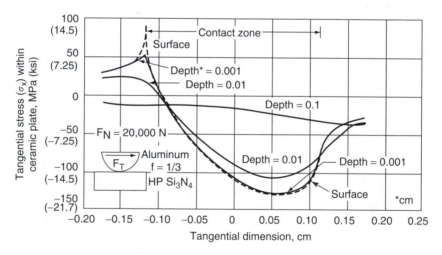

FIGURE 9.28 Stress distribution vs. depth from the surface for sliding contact. (From Finger, D.G., ONR contract N00014-78-C-0547, 1979. With permission.)

Let us look at two examples.

Example 9.1 Reaction-bonded Si_3N_4 (RBSN) has a Posson's ratio of 0.18 and in one circumstances has a coefficient of friction of 0.5. Under this circumstance, how much less load will cause surface damage during sliding contact than during only normal contact? When we use Equation (9.6), $P_S = P_N/114$, so a sliding load 114 times less than a stationary load will cause surface damage.

Example 9.2 A sintered SiC material has a Poisson's ratio of 0.17 and a coefficient of friction of 0.3 in a specific application. Compare the load to cause surface damage under static and sliding conditions. When we use Equation (9.6), $P_N/P_S = 34$. The reduction in coefficient of friction (compared to Example 9.1) has significantly reduced the stress magnification.

Figure 9.29 shows some actual data comparing the strength of RBSN with no contact load ($\square$), with a stationary contact load applied at 90°C ($\circ$, $\bullet$), and with a sliding contact load ($\blacklozenge$). Stationary

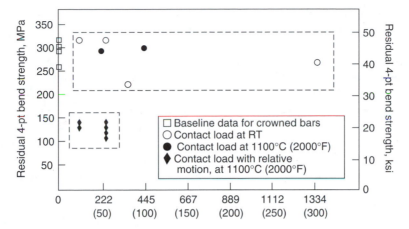

FIGURE 9.29 Comparison of the strength of reaction-bonded Si_3N_4 (RBSN) as fabricated and after static and sliding contact. (From Richerson, D.W., Carruthers, W.D., and Lindberg, L.J., Contact stress and coefficient of friction effects on ceramic interfaces, *Materials Science Research, Vol. 14, Surfaces and Interfaces in Ceramic and Ceramic–Metal Systems*, Pask J. and Evans, A., Eds., Plenum Press, New York, 1981, pp. 661–676. With permission.)

contact loads at least up to 140 kg (308 lb.) did not result in strength reduction, which indicates that no surface damage occurred. Sliding contact resulted in greater than 50% drop in strength at a load of only 10 kg (22 lb). Subsequent examination of the surface showed a series of concentric semicircular surface cracks that had resulted from the sliding contact.

Surface damage and strength reduction are major concerns in the design and application of ceramics. Ceramics are always susceptible to impact during manufacture, assembly, and handling and are exposed to sliding contact or biaxial stress distribution more commonly than most people realize. Examples include (1) at or adjacent to points of attachment; (2) surface irregularities at a shrink fit and the region at the margin of a shrink fit; (3) surface being machined; (4) contact surfaces of bearing and race; (5) contact region of a cutting tool; (6) contact surfaces of heated parts that have different thermal expansion; and (7) contact surfaces of seals.

9.4.2 REMOVAL OF SURFACE MATERIAL

The second major category of mechanically induced effects is removal of surface material. This is generally classified as *wear*.[32,33,34] Wear occurs by several modes. Two important modes are adhesion and microfracture.

Adhesion involves localized bonding of two surfaces, followed by transfer by pullout of material from one surface to the other as the surfaces are moved relative to each other. The mechanism involves atomic bonding. As we learned in Chapter 5, atoms in the interior of a material are completely surrounded by other atoms such that their bond forces are satisfied. Atoms at the surface have unsatisfied bonds and a driving energy to bond to a surface that comes in contact. The tendency to bond is especially high if the material has mobile electrons and a simple close-packed, nondirectional structure. Metals fit this description and have a strong tendency for surface bonding and adhesive wear. Inertial welding is an example of the case of adhesive bonding of metal surfaces. Lubricants are used to minimize adhesive bonding as well as minimize other wear mechanisms of metals.

Ceramics are much less susceptible to adhesive wear than metals, especially at room temperature and moderate temperatures. The atoms and electrons in ceramics have low mobility and have minimal tendency to bridge the interface between the surfaces. However, diffusion (motion) of atoms increases at high temperatures so that adhesive bonding and adhesive wear of ceramics can occur at elevated temperature. The temperature at which adhesion occurs can sometimes result from frictional heating of sliding surfaces.

Wear of ceramics generally occurs by microfracture mechanisms. These mechanisms include gouging, interference of asperities, impact, and contact stress. All four involve the type of contact stress distributions described in the prior section. *Gouging* occurs when one material is harder than the other, such as when a diamond particle mounted in a grinding wheel is raked across the surface of the workpiece. The larger the difference in hardness, the easier wear by gouging is. The hardness values of various materials are listed in Table 9.5.

A second mechanism is interference of surface asperities. *Asperities* are protrusions and depressions on the surface that wedge against each other during sliding contact and cause stress concentrations so high that microfracture occurs. Microfracture can involve the top of a protrusion breaking off or can involve cracks that extend beneath the surface and weaken the material.

A third mechanism of wear is *impact*. It is also referred to as *erosion* and generally involves small particles of an abrasive material entrained in a flowing fluid.

A fourth mechanism of wear is *contact stress*. This can act independently as well as being a source of tensile stress magnification to aid gouging, asperity breakage, and erosion. As discussed earlier, reduction in the coefficient of friction at the interface will help to reduce wear by contact stress microfracture.

Wear is a major cause of equipment failure and down time in the military, in industry, and in commercial products. Associated friction is a major consumer of energy. It has been estimated that over four quadrillion Btu's of energy were lost in the United States in 1978 due to friction and wear.[35] Based on a cost of oil of $15/barrel, this translated into a loss of $10 billion.

The market for materials for wear resistance is dominated by cermets such as cobalt-bonded tungsten carbide and by specialty steels. The Industry Analysis Division of the U. S. Department of Commerce estimated a total market for wear-resistant materials of $3.3 billion in 1980, $4.5 billion in 1985, $6.0 billion in 1990, $7.5 billion in 1995, and $9.0 billion in 2000.

Ceramics have been targeted as a means of reducing wear and friction and avoiding the chemical and galvanic attack to which bonded tungsten carbide and steels are vulnerable. The potential advantages of ceramics compared to metals and polymers include high hardness, chemical stability, ability to be ground with a very smooth surface to high tolerances, strength retention over a broad temperature range, and low cost. Based on these characteristics, ceramics are currently in production or being evaluated for a variety of wear applications such as seals, sandblast nozzles, wear pads, and liners. The barriers to more extensive use include brittleness and low toughness (leading to the tendency to fracture during handling or under contact loading), inadequate understanding of the design requirements necessary to overcome the mechanical limitations, and insufficient design database.

The use of ceramic wear parts is currently very low compared to the overall market. The Industry Analysis Division of the U.S. Department of Commerce estimated that in 1980 only

TABLE 9.5
Erosion Resistance vs. Hardness

Material, in Increasing Order of Erosion Resistance	Knoop Hardness (kg/mm^2)
MgO	370
SiO$_2$	820
ZrO$_2$	1160
Al$_2$O$_3$	2000
Si$_3$N$_4$	2200
SiC	2700
B$_4$C	3500
Diamond	7000–8000

$20 million in ceramic wear parts were supplied in the United States. This represents less than 1% of the total wear parts market. Predictions for 1985 through 2000 estimated a slow steady increase in the use of ceramics: $45 million (1%) in 1985, $180 million (3%) in 1990, $375 million (5%) in 1995, and $540 million (6%) in 2000. Ceramics could be used much more extensively.

The "barriers" to the use of ceramics as wear parts may be more a matter of perception than technology. Existing ceramics can perform effectively in many wear applications if they are properly designed into the system. However, in many cases the system manufacturer is either not aware of this or not willing to risk the design development funds. As case histories of successful ceramic uses become more widely publicized, system manufacturers will be more willing to explore ceramic designs.

Besides improvements in design, major improvements in the properties of ceramics have occurred during the past 30 years. The area of improvement that has provided the greatest benefit in resistance to contact stress and wear has been increase in fracture toughness. This is summarized in Table 9.6. Conventional ceramics have fracture toughness values less than $4 \, \text{MPa} \cdot \text{m}^{1/2}$

TABLE 9.6

Typical Toughness Values for Some Single Crystals, Polycrystalline Ceramics, Glass, Composites, and Metals

Material	K_{Ic}	
	$\text{MPa} \cdot \text{m}^{1/2}$	$\text{ksi} \cdot \text{in.}^{1/2}$
Glass	0.7	0.64
Single Crystals		
NaCl	0.3	0.27
Si	0.6	0.55
MgO	1	0.91
ZnS	1	0.91
SiC	1.5	1.37
Al_2O_3	2	1.82
WC	2	1.82
Polycrystalline Ceramics		
Al_2O_3	3.5–4.0	3.19–3.64
SiC	3.0–3.5	2.73–3.19
Stabilized ZrO_2	2	1.82
RBSN	2.5	2.28
Sintered Si_3N_4	4–6	3.64–5.46
Transformation-Toughened Ceramics		
Mg-PSZ	9–12	8.19–10.92
Y-TZP	6–9	5.46–8.19
Ce-TZP	10–16	9.10–14.56
Al_2O_3-ZrO_2	6.5–13.0	5.92–11.83
Dispersed-Particle Ceramics		
Al_2O_3-TiC	4.2–4.5	3.82–4.09
Si_3N_4-TiC	4.5	4.09
Dispersed-Whisker Ceramics		
Al_2O_3-SiC whiskers	6–9	5.46–8.19
Metals		
Ductile cast iron	25–35	22.75–31.85
Aluminum alloys	33–44	30.03–40.04
Medium-toughness steel	44–66	40.04–60.06

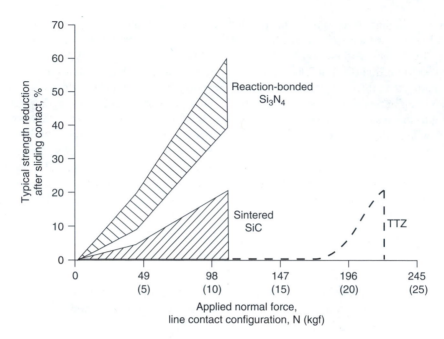

FIGURE 9.30 Comparison of the relative contact stress resistance of a transformation-toughened ZrO_2 (TTZ) material with sintered SiC and reaction-bonded Si_3N_4. (From Lindberg, L.J. and Richerson, D.W., Comparison of the contact stress and friction behavior of SiC and ZrO_2 materials, *Ceram. Eng. and Sci. Proc.*, 6(7–8), 1985. Reprinted with permission of the American Ceramic Society, www.ceramics.org)

($3.64\,ksi \cdot in.^{1/2}$). For example, most glass compositions are less than $1\,MPa \cdot m^{1/2}$ ($0.91\,ksi \cdot in.^{1/2}$), Al_2O_3 is about $3.5\,MPa \cdot m^{1/2}$ ($3.18\,ksi \cdot in.^{1/2}$), SiC is about 2.5 to $3.0\,MPa \cdot m^{1/2}$ (2.27 to $2.73\,ksi \cdot in.^{1/2}$), and reaction-bonded Si_3N_4 is about 2.0 to $2.5\,MPa \cdot m^{1/2}$ (1.82 to $2.27\,ksi \cdot in.^{1/2}$). Transformation-toughened ZrO_2 ceramics range in toughness from 5 to $15\,MPa \cdot m^{1/2}$ (4.55 to $13.65\,ksi \cdot in.^{1/2}$). Sintered Si_3N_4 with a fibrous microstructure has toughness in the range of 5 to $7\,MPa \cdot m^{1/2}$ (4.55 to $6.37\,ksi \cdot in.^{1/2}$). Al_2O_3 reinforced with SiC whiskers has toughness of about 5 to $8\,Mpa \cdot m^{1/2}$ (4.55 to $7.28\,ksi \cdot in.^{1/2}$). The benefit of increased toughness to resisting contact stress and microfracture wear damage is illustrated in Figure 9.30. The toughened zirconia was substantially more resistant to formation of surface damage.

Methods of toughening ceramics are discussed in Chapter 20.

9.5 THERMAL SHOCK

Thermal shock refers to the thermal stresses that occur in a component as a result of exposure to a temperature difference between the surface and interior or between various regions of the component. For shapes such as an infinite slab, a long cylinder (solid or hollow), and a sphere (solid or hollow), the peak stress typically occurs at the surface during cooling according to the equation

$$\sigma_{th} = \frac{E\alpha\,\Delta T}{1 - v} \tag{9.7}$$

where σ_{th} is the thermal stress, E the elastic modulus, α the coefficient of thermal expansion, v Poisson's ratio, and ΔT the temperature difference.

Equation (9.7) indicates that the thermal stress increases as the elastic modulus and thermal expansion coefficient of the material increases and as the imposed ΔT increases. From a materials point of view, the ΔT can be decreased by increasing the thermal conductivity of the material. From

TABLE 9.7

Thermal Shock Resistance Parameters

Parameter Designation	Parameter Type	Parameter[a]	Physical Interpretation/ Heat Transfer Conditions	Typical Units
R	Resistance to fracture initiation	$\dfrac{\sigma(1 - v)}{\alpha E}$	Maximum ΔT allowable for steady heat flow	°C
R'	Resistance to fracture initiation	$\dfrac{\sigma(1 - v)k}{\alpha E}$	Maximum heat flux for steady flow	cal/cm · sec
R''	Resistance to fracture initiation	$\dfrac{\sigma(1 - v)\alpha_{TH}}{\alpha E}$	Maximum allowable rate of surface heating	cm² · °C/sec
R'''	Resistance to propagation damage	$\dfrac{E}{\sigma^2(1 - v)}$	Minimum in elastic energy at fracture available for crack propagation	(psi)$^{-1}$
R''''	Resistance to propagation damage	$\dfrac{\gamma E}{\sigma^2(1 - v)}$	Minimum in extent of crack propagation on initiation of thermal stress fracture	cm
R_{st}	Resistance to further crack propagation	$\left(\dfrac{\gamma}{\alpha^2 E}\right)^{1/2}$	Minimum ΔT allowed for propagating long cracks	°C/m$^{1/2}$

[a] σ, tensile strength; v, Poisson's ratio; α, coefficient of thermal expansion; E, Young's modulus of elasticity; k, thermal conductivity; α_{TH}, thermal diffusivity; γ, fracture surface energy.

Source: From Hasselman, D.P.H., Thermal stress resistance parameters for brittle refractory ceramics: a compendium, *Bull. Am. Ceram. Soc.*, 49, 1033–1037, 1970.

a design point of view, the ΔT in the material can be decreased by configuration modification and possibly by modification of the heat transfer conditions.

Hasselman[36] has defined thermal stress resistance parameters based on Equation (9.7) and other equations for various heat transfer conditions and conditions of crack initiation vs. crack growth. These are summarized in Table 9.7. Note that the effects of E, σ, and v are opposite for crack initiation vs. crack propagation. Low E and v with high σ provide high resistance to propagation of existing cracks. This provides both a paradox and a pronounced alternative to designing for thermal shock conditions.

When selecting a ceramic material for an application where thermal shock is expected to be a problem, calculation of the appropriate thermal shock parameter for the various candidate materials can sometimes be useful. Table 9.8 shows an example of calculations of the parameter R for several materials. Of this group, the LAS is by far the most thermal-shock-resistant. Its thermal shock resistance is entirely due to its low coefficient of thermal expansion. Fused silica glass also has a low coefficient of thermal expansion and, similarly, excellent thermal shock resistance. Silicon nitride has a much higher strength than either lithium aluminum silicate (LAS) or fused silica, but much lower thermal shock resistance because of its higher thermal expansion coefficient and elastic modulus. Silicon carbide has still higher α and E and correspondingly less calculated thermal shock resistance. Al_2O_3, other oxides, and strongly ionic bonded ceramics have even higher α and even lower thermal shock resistance.

The R parameter has only been used as an illustrative example. Other factors, such as thermal conductivity and fracture toughness, have substantial effects. For instance, for some configurations and under some heat transfer conditions, SiC appears to be more thermal-shock-resistant than Si_3N_4. The higher thermal conductivity of SiC redistributes the heat and decreases stress-causing gradients. Another example is the partially stabilized transformation-toughened ZrO_2. Even though

TABLE 9.8

Calculated Values of the Thermal Shock Parameter R for Various Ceramic Materials Using Typical Property Data

Material	Strength,[a] σ (psi)	Poisson's Ratio, v	Thermal Expansion, α (in./in. °C)	Elastic Modulus, E (psi)	$R = \sigma(1 - v)/\alpha E$ E (°C)
Al_2O_3	50,000	0.22	7.4×10^{-6}	55×10^6	96
SiC	60,000	0.17	3.8×10^{-6}	58×10^6	230
RSSN[b]	45,000	0.24	2.4×10^{-6}	25×10^6	570
HPSN[b]	100,000	0.27	2.5×10^{-6}	45×10^6	650
LAS[b]	20,000	0.27	-0.3×10^{-6}	10×10^6	4860

[a] Flexure strength used rather than tensile strength.

[b] RSSN, reaction-sintered silicon nitride; HPSN, hot-pressed silicon nitride; LAS, lithium aluminum silicate (β-spodumene).

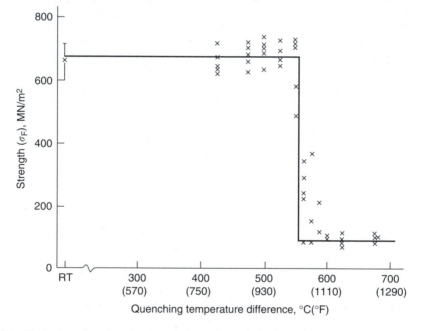

FIGURE 9.31 Typical results of retained strength vs. thermal shock ΔT for quench test. Example is for hot-pressed Si_3N_4 material containing 3% MgO as a densification aid. (From Ziegler, G., in *Progress in Nitrogen Ceramics*, Riley, F.L., Ed., Martinas Nijhoff Publishers, The Hague, 1983. With permission.)

it has a very high coefficient of thermal expansion ($\sim 12 \times 10^{-6}$ per °C) and only moderate strength, it is extremely thermal-shock-resistant due to the high fracture toughness. Polycrystalline aluminum titanate has shown similar thermal shock resistance but due to a slightly different mechanism. Aluminum titanate has extremely anisotropic thermal expansion properties in single crystals, and polycrystalline material spontaneously microcracks during cooling after fabrication. These fine intergranular cracks limit the strength of the material, but provide an effective mechanism for absorbing strain energy during thermal shock and preventing catastrophic crack propagation.

A simple quench test has been devised to gather data experimentally to compare with the R parameter calculations. Flexure strength test bars are heated to a selected temperature and dropped into a water bath of known temperature. The bars are subsequently strength-tested. If no thermal shock damage has occurred, the strength will be equivalent to the strength of control bars that have

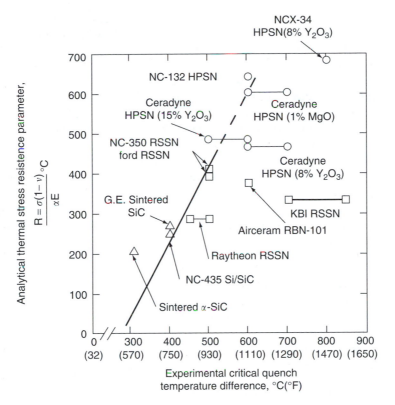

FIGURE 9.32 Comparison of analytical and experimental thermal shock ΔT results for various Si_3N_4 and SiC materials. (From Larsen, D.C., AFML-TR-79-4188, Oct. 1979. With permission.)

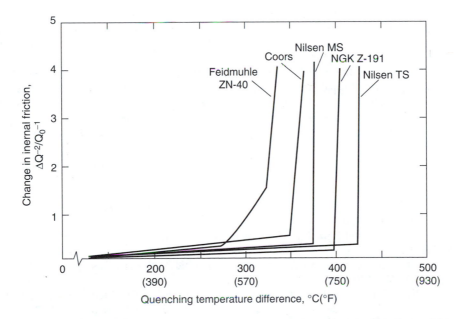

FIGURE 9.33 Water-quench thermal shock results for various transformation-toughened materials using the internal friction technique to determine the critical ΔT. (From Larsen D.C. and Adams, J., IIT Research Institute, Ill. From presentation at DOE/NASA/DOD Task Group Meeting, AMMRC, Watertown, MA, 25–26 September 1984. With permission.)

not been heated and quenched. The procedure is continued at successively higher temperature difference between the heat soak and the water bath. At the critical ΔT where thermal shock cracks initiate, the strength of the bars will be lower than the controls. Figure 9.31 shows a plot of strength data vs. ΔT for a typical quench test. Figure 9.32 shows ΔT measured by the quench test for a variety of Si_3N_4 and SiC ceramics and compared with calculations.

The critical ΔT can be determined by another approach. The quench test is conducted as before, but the ΔT is not determined by strength measurement. Instead, internal friction measurements are made on the test bar. An internal friction measurement involves the use of a piezoelectric transducer to send acoustic waves through the length of the test bar.

If no thermal shock cracks are present, the waves go through the bar easily with relatively little scattering or back reflections. If cracks are present, the waves are scattered and reflected and have difficulty getting through the bar. This is easily detected with the acoustic wave measurement equipment. Data for several materials are illustrated in Figure 9.33. This technique is good for initial screening to obtain a rough idea of the critical ΔT. It can often be accomplished by repeated quenches with a single bar.

REFERENCES

1. Kingery, W.D., Bowen, H.K., and Uhlmann, D.R., *Introduction to Ceramics*, 2nd ed., Wiley, New York, 1976, Chap. 14.
2. Chin, G.Y., Van Uitert, L.G., Green, M.L., Zydzik, G.J., and Komentani, T.Y., Strengthening of alkali halides by bivalent ion additions, *J. Am. Ceram. Soc.*, 56, 369, 1973.
3. Ashby, M., A first report on deformation mechanism maps, *Acta Met.*, 20, 887, 1972.
4. Joffe, A., Kupitschewa, M.W., and Lewitzky, M.A., Deformation and strength of crystals, *Z. Phys.*, 22, 286, 1924.
5. Seltzer, M.S., High temperature creep of silicon-base compounds, *Bull. Am. Ceram. Soc.*, 56(4), 418, 1977.
6. Larsen, D.C. and Walther, G.C., *Property Screening and Evaluation of Ceramic Turbine Engine Materials*, Interim Rept. 6, AFML Contract F33615-75-C-5196, 1978.
7. Greskovich, C.D. and Palm, J.A., Development of high performance sintered Si_3N_4, *Highway Vehicle Systems Contractors Coordination Meeting 17th Summary Report*, NTIS Conf. 791082, Dist. Category UC-96, 1979, pp. 254–262.
8. Richerson, D.W., Ferber, M.K., and van Roode, M., The ceramic gas turbine — retrospective, current status and prognosis, Chapter 29 in *Ceramic Gas Turbine Component Development and Characterization*, van Roode, M., Ferber, M.K., and Richerson, D.W., Eds., ASME Press, New York, 2003, pp. 693–741.
9. Carruthers, W.D., Pollinger, J., Becher, P.F., Ferber, M.K., and von Roode, M., *Future Needs and Developments in Structural Ceramics Materials for Gas Turbines*, ASME Press, New York, 2003, pp. 743–766.
10. Ferber, M.K., Lin, H-T., Jenkins, M.G., and Ohji, T., *Mechanical Characterization of Monolithic Ceramics for Gas Turbine Applications*, ASME Press, New York, 2003, pp. 353–395.
11. Richerson, D.W. and Yonushonis, T.M., Environmental effects on the strength of silicon nitride materials, *DARPA-NAVSEA Ceramic Gas Turbine Demonstration Engine Program Review*, MCIC Rep. MCIC-78-36, 1978, pp. 247–271.
12. Wiederhorn, S.M. and Bolz, L.H., Stress corrosion and static fatigue of glass, *J. Am. Ceram. Soc.*, 53(10), 543–548, 1970.
13. Opila, E.J., Variation of the oxidation rate of silicon carbide with water-vapor pressure, *J. Am. Ceram. Soc.*, 82(3), 625–636, 1999.
14. More, K.L., Tortorelli, P.F., Ferber, M.K., and Keiser, J.R., Observations of accelerated silicon carbide recession by oxidation at high water-vapor pressures, *J. Am. Ceram. Soc.*, 83(1), 211–213, 2000.
15. Deal, B.E. and Grove, A.S., General relationship for the thermal oxidation of silicon, *J. Appl. Phys.*, 36(12), 3770–3778, 1965.
16. Brentnall, W.D., van Roode, M., Norton, P.F., Gates, S., Price, J.R., Jimenez, O., and Miriyala, N., Ceramic gas turbine development at Solar Turbines Incorporated, Chapter 7 in *Ceramic Gas Turbine*

Design and Test Experience, van Roode, M., Ferber, M.K., and Richerson, D.W., Eds., ASME Press, New York, 2002.

17. Carruthers, W.D., Richerson, D.W., and Benn, K.W., Combustion rig durability testing of turbine ceramics, in *Ceramics for High Performance Applications III: Reliability*, Lenoe, E.M., Katz, R.N., and Burke, J.J., Eds., Plenum Press, New York, 1983, pp. 571–595.

18. Schienle, J.L., *Durability Testing of Commercial Ceramic Materials*, Final Report DOE/NASA/0027-1, NASA CR-198497, Jan. 1996.

19. Crowley, M.S., Hydrogen-silica reactions in refractories, *Bull. Am. Ceram. Soc.*, 46, 679, 1967.

20. Pankratz, L.B., *Thermodynamic Properties of Elements and Oxides*, Bull., 672, U.S. Bureau of Mines, Washington, D.C.

21. JANAF *Thermochemical Tables*, 2nd ed., NSRDS-NBS 37, National Institute of Standards and Technology (formerly, National Bureau of Standards), Washington, D.C., 1971.

22. Pankratz, L.B., Stuve, J.M., and Gokcen, N.A., *Thermodynamic Data for Mineral Technology*, Bull. 677, U.S. Bureau of Mines, Washington, D.C.

23. Lay, L.A., *The Resistance of Ceramics to Chemical Attack*, NPL Rept. CHEM 96, Nat. Physical Lab., Jeddington, Middlesex, U.K., Jan. 1979.

24. Pfaender, H.G., *Schott Guide to Glass*, Chapman & Hall, London, 1996.

25. Cooper, Jr., A.R., Samaddar, B.N., Oishi, Y., and Kingery, W.D., Dissolution in ceramic systems, *J. Am. Ceram. Soc.*, 47, 37, 1964; 47, 249, 1964; 48, 88, 1965.

26. Carniglia, S.C., Boyer, W.H., and Neely, J.E., MgO refractories in the basic oxygen steelmaking furnace, *Ceramics in Severe Environments*. Wurth Kriegel W. and Palmour III, H., Eds., Plenum, New York, 1971, pp. 57–88.

27. Freitag, D.W. and Richerson, D.W., Opportunities for Advanced Ceramics to meet the Needs of the Industries of the Future, DOE/ORO 2076, Oak Ridge, National Laboratory, Dec. 1998.

28. Singhal, S.C., Corrosion behavior of silicon nitride and silicon carbide in turbine atmospheres, *Proceedings of the 1972 Tri-Service Conference on Corrosion*, MCIC 73-19, 1973, pp. 245–250.

29. Miller, D.G., Andersson, C.A., Singhal, S.C., Lange, F.F., Diaz, E.S., Kossowsky, E.S., and Bratton, R.J., *Brittle Materials Design High Temperature Gas Turbine—Materials Technology*, AMMRC CTR-76-32, Vol. 4, 1976.

30. McKee, D.W. and Chatterji, D., Corrosion of silicon carbide in gases and alkaline melts, *J. Am. Ceram. Soc.*, 59(9–10), 441–444, 1976.

31. Gilroy, D.R. and Hirst, W., *J. Phys. D: Appl. Phys.*, 2, 1784–1787, 1969.

32. Rabinowicz, E., *Friction and Wear of Materials*, Wiley, New York, 1965.

33. Bowden, F.P. and Tabor, D., *Friction and Lubrication*, Mehuen and Co., Ltd., London, 1956.

34. Jahanmir, S., Ed., *Friction and Wear of Ceramics*, Marcel Dekker Inc., New York, 1984.

35. Fehrenbacher, L.L. and Levinson, T.M., *Identification of Tribological Research and Development Needs for Lubrication of Advanced Heat Engines*. Battelle Pacific Northwest Lab. Rept. PNL-5537. Sept. 1985.

36. Hasselman, D.P.H., Thermal stress resistance parameters for brittle refractory ceramics: a compendium, *Bull. Am. Ceram. Soc.*, 49, 1033–1037, 1970.

PROBLEMS

1. A sample of SiC is exposed to a rapid temperature change of 250°C (450°F). If the elastic modulus of the SiC is 60×10^6 psi, the thermal expansion coefficient is 4.5×10^{-6}/°C, and Poisson's ratio is 0.17, estimate the thermal stress that could result in the SiC due to the rapid temperature change.

2. A Si_3N_4 block is resting on a horizontal steel surface such that the vertical (normal) force against the steel is 25 kg (55 lb). A horizontal force of 7.5 kg (16.5 lb) is required to cause the Si_3N_4 to slide against the steel surface. What is the coefficient of friction between the two surfaces?

3. If hardness was the dominant criterion, which of the following materials would be the most resistant to erosion?

   (a) boron carbide

   (b) magnesium oxide

(c) zirconium oxide

(d) silicon carbide

4. If impact damage was the primary cause of erosion in a specific application such that erosion resistance was dominated by fracture toughness of the ceramic, which ceramic would most likely have superior erosion resistance?

   (a) boron carbide

   (b) magnesium oxide

   (c) Mg-PSZ

   (d) silicon carbide

5. A ceramic is strength-tested in four-point flexure at 1200°C (2190°F) at a high loading rate and a low loading rate. The high loading rate results in an average modulus of rupture (MOR) of 450 MPa (65 ksi). The low loading rate results in an average MOR of only 320 MPa (46 ksi). What is likely to be the strength-limiting factor for this material?

6. A specific ceramic has a microstructure consisting of crystalline grains separated by a thin, noncrystalline grain boundary phase. The material exhibits relatively high creep at elevated temperature. What is the likely mechanism of creep at the elevated temperature?

   (a) dislocation movement

   (b) bulk diffusion

   (c) grain boundary diffusion

   (d) viscous flow

7. A normal contact load of 200 Newtons (N) is required to cause Hertzian damage in the surface of a ceramic that has a Poisson's ratio of 0.24. If the coefficient of friction between the two surfaces is 0.15, what sliding load would cause Hertzian damage?

8. A sample of reaction-bonded Si_3N_4 is exposed at 1100°C (2010°F) to an atmosphere containing a mixture of 5% oxygen and 95% nitrogen. The sample exhibits an increase in weight of 1 mg/cm^2 after 2 h, 2 mg/cm^2 after 5 h, 3 mg/cm^2 after 15 h, and 3.5 mg/cm^2 after 30 h. What chemical effect would you expect is occurring? Explain.

   (a) active oxidation

   (b) passive oxidation

   (c) corrosion

   (d) dissociation

9. Based on thermodynamic free-energy data, which oxide would you expect to be least susceptible to reduction at 1500°C (2730°F)?

   (a) yttrium oxide

   (b) silicon dioxide

   (c) zirconium oxide

   (d) aluminum oxide

10. Which material would you expect to have the greatest resistance to sliding contact stress?

    (a) reaction-bonded Si_3N_4

    (b) silicon carbide

    (c) transformation-toughened zirconia

    (d) glass

11. Estimate the critical ΔT for thermal shock damage for a sialon cutting tool material with a tensile strength of 650 MPa, an elastic modulus of 320 GPa, a coefficient of thermal expansion of $2.9 \times 10^{-6}/°C$, and a Poisson's ratio of 0.25.

STUDY GUIDE

1. What is "creep"?

2. What factors can affect creep rate?

3. Describe the primary mechanisms of creep in a single crystal.

4. Describe the primary mechanisms of creep in a polycrystalline ceramic.
5. Explain the influence of grain size on creep based on the different creep mechanisms.
6. Explain what controls creep rate in glass.
7. What are the key factors that resulted in a large range in creep behavior of the silicon nitride and silicon carbide materials in Figure 9.6?
8. What is "slow crack growth"?
9. Describe how slow crack growth can result from creep in a polycrystalline ceramic.
10. Explain how slow crack growth can occur independent of creep.
11. What effects do creep and slow crack growth have on mechanical properties under a sustained load?
12. How is "static fatigue" measured?
13. To characterize the stress rupture behavior of a material requires many tests at a variety of stresses and temperatures. This is expensive and time-consuming. How are "variable-stressing-rate" tests used to quickly determine the susceptibility of a material to static fatigue?
14. What is "stress corrosion"?
15. Explain the difference between passive and active oxidation of silicon nitride or silicon carbide.
16. What effects can oxidation have on the strength of silicon nitride? Explain and give examples.
17. What are some other ceramic materials besides SiC and silicon nitride that are susceptible to oxidation at high temperature?
18. What determines the stability of oxide ceramics in a high-temperature reducing environment? Explain and give a couple examples.
19. What are some factors that can affect the corrosion rate of a ceramic furnace lining in contact with molten glass?
20. What type of conditions does a ceramic lining in a basic oxygen furnace have to survive?
21. What corrosion conditions can occur during the burning of coal, such as in an electricity generation steam boiler or in a gasifier?
22. Explain the difference between a Hertzian load and a sliding load on the stress distribution and nature of damage at the surface of a ceramic material. Use drawings to illustrate.
23. Describe briefly how and why each of the following affects the thermal shock resistance of a ceramic material. Consider both crack initiation and crack propagation. Show equations or figures of merit if you choose. For example, high strength results in high thermal shock resistance for a crack to initiate based on the equation $R = \sigma (1 - v)/E\alpha$, while high strength results in decreased thermal shock resistance for crack propagation based on the equation $R''' = E\sigma^2(1 - v)$. The reason is the higher the stress at the time of fracture initiation, the more stored strain energy to drive the crack through the material. Use similar explanations for each of the other properties.

	Crack Initiation	Crack Propagation
Strength		
Elastic modulus		
Poisson's ratio		
Thermal conductivity		
Coefficient of thermal expansion		
Fracture toughness		
Temperature difference imposed by the environment		

10 Electrical Behavior

Ceramic materials have a wide range of electrical properties. Some do not allow passage of an electric current even in a very strong electric field and thus are excellent insulators. Others allow an electric current to pass only under certain conditions or when an energy threshold has been reached and thus are useful semiconductors. Still others do allow passage of an electric current and have application as electrical conductors. Some ceramics do not conduct electricity but undergo internal charge polarization that allows the material to be used for storage of an electrical charge in capacitors. These electrical properties are determined primarily by the nature of the atomic bonding and the crystal structure. Chapter 10 reviews some of the fundamentals of electricity and discusses electronic conduction, ionic conduction, insulators, semiconductors, and superconductors. Chapter 11 reviews polarization (dielectric) behavior along with magnetic and optical behavior.

10.1 FUNDAMENTALS AND DEFINITIONS

What happens when we attach a copper wire from the positive terminal of a battery to one metal contact of a light bulb and another copper wire from the other metal contact of the light bulb to the negative terminal of the battery? The light begins to glow. Why? The difference in electrical charge between the positive and negative terminals of the battery produces a *voltage* (V) that causes free electrons in the copper wire to flow from the negative terminal to the positive terminal. The flow of electrons is referred to as *current* (I). The electrons flowing through the copper wire interact with each other and with the copper atoms. This interaction interferes with the flow of the electrons and is referred to as *electrical resistance* (R). The voltage V, the current I, and the resistance R are related by the equation

$$V = IR \qquad (10.1)$$

where
V = volts
I = amperes
R = ohms

It is apparent from Equation (10.1) that I and R compete with each other. If R is low, then the electrons flow easily through the material and I is high. This is the case for the copper wire. If R is high, the electrons have difficulty flowing through the material and I is low. This is the case for the filament in the light bulb. The filament is made of a very thin cross section of a tungsten alloy that has much higher electrical resistance than the copper. When the electrons reach the thin tungsten filament, they must exert considerable effort to get through. This results in work that is dissipated in the form of heat. So much heat is produced that the tungsten filament becomes white hot and emits bright light.

Different materials have different capabilities to transmit or conduct an electrical charge when an electric field is applied. The degree of electrical conduction is somewhat analogous to thermal conductivity; that is, the *electrical conductivity* (σ) is determined by the number of charge carriers (n), the charge carried by each carrier (q), and the mobility of the carriers (μ) as expressed in Equation (10.2)

$$\sigma = nq\mu \qquad (10.2)$$

285

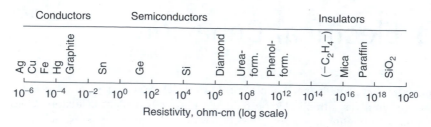

FIGURE 10.1 General range of resistivity for conductors, semiconductors, and insulators. (From van Vlack, L.H., *Elements of Materials Science*, 2nd ed., Addison-Wesley, Reading, MA, 1964, p. 420, p. 109. With permission.)

The units are

$$\frac{1}{\text{ohm-cm}} = \left(\frac{\text{carriers}}{\text{cm}^3}\right)\left(\frac{\text{coulombs}}{\text{carrier}}\right)\left(\frac{\text{cm/sec}}{\text{volts/cm}}\right)$$

As was the case for thermal conductivity, electrical conductivity is controlled largely by the nature of the atomic bonding and the crystal structure. Some materials have high electrical conductivity and are called *electrical conductors*. Most metals, some ceramics, and a few organics fall in this category. Other materials have very low electrical conductivity and are referred to as *electrical insulators*. Most ceramics and organics are electrical insulators. Some materials fall in between and have a moderate level of conductivity under certain conditions. These materials are referred to as *semiconductors*. Most are covalent ceramics.

Figure 10.1 illustrates the range of *electrical resistivity* for some common materials. Resistivity (ρ) is the reciprocal of electrical conductivity and has the units ohm-cm. It is related to the electrical resistance by the equation

$$\rho = \frac{AR}{l} \tag{10.3}$$

where

A = cross-sectional area of a sample
l = gauge length

Table 10.1 compares the electrical resistivity of some ceramic, metal, and organic materials.

10.2 ELECTRONIC CONDUCTIVITY

Metals are the most widely recognized materials having electronic conduction. In fact, the terms *electronic conduction* and *metallic conduction* are often used synonymously. Electrons are the charge carriers for a material with electronic conductivity and the path they follow is referred to as a conduction band.

In order to understand electronic conduction and conduction bands, we need to review some of the concepts studied in Chapter 4. Electrons in atoms have specific energy levels and are located in discrete zones around the nucleus. These zones or energy bands can only contain a limited number of electrons in an individual atom as well as in a structure consisting of a network of atoms bonded together. If all the allowable electron positions are full within a given band, then the electrons are not free to move through the band and no electronic conduction occurs. However, if the band is only partially full, electrons can move freely within the band and result in electronic conduction. If the band is empty, then no carriers will be present and no conduction will occur.

TABLE 10.1
Electrical Resistivity of Some Metals, Polymers, and Ceramics at Room Temperature

Material	Resistivity (ohm-cm)
Metallic Conduction	
Copper	1.7×10^{-6}
Iron	10×10^{-6}
Tungsten	5.5×10^{-6}
ReO_3	2×10^{-6}
CrO_2	3×10^{-5}
Semiconductors	
SiC	10
B_4C	0.5
Ge	40
Fe_3O_4	10^{-2}
Insulators	
SiO_2	$>10^{14}$
Steatite porcelain	$>10^{14}$
Fire-clay brick	10^8
Low-voltage porcelain	$10^{12}-10^{14}$
Al_2O_3	$>10^{14}$
Si_3N_4	$>10^{14}$
MgO	$>10^{14}$
Phenol-formaldehyde	10^{12}
Vulcanized rubber	10^{14}
Teflon	10^{16}
Polystyrene	10^{18}
Nylon	10^{14}

Source: Compiled from Kingery, W.D., Bowen, H.K., and Uhlmann, D.R., *Introduction to Ceramics*, 2nd ed., Wiley, New York, 1976, Chap. 17; van Vlack, L.H., *Elements of Materials Science*, 2nd ed., Addison-Wesley, Reading, MA, 1964, p. 420; van Vlack, L.H., *Physical Ceramics for Engineers*, Addison-Wesley, Reading, MA, 1964, Chap. 9; and *Kirk-Othmer Encyclopedia of Chemical Technology*, Vol. 4, 3rd ed. Mark, H.F., Othmer, D.F., Overberger, C.G., and Seaborg, G.T., Eds., Wiley, New York, 1978.

The atomic bonding in metals results in partially filled energy bands and electronic conduction. The atomic bonding in insulators and semiconductors results in energy bands that are either completely full or completely empty. There is an energy gap between the outermost filled band (valence band) and the adjacent empty band (conduction band) of insulators and semiconductors. This band gap prevents flow of electrons in these materials between the valence and conduction bands and no electronic conduction occurs. These materials will be discussed later in this chapter. Let us return to the topic of electronic conduction in metals.

Figure 10.2 illustrates energy band information for sodium metal. The energy levels for an isolated sodium atom are shown on the left, and for a sodium atom in the metallic sodium structure on the right. Since sodium is monovalent, and since each level in the energy band can contain two electrons, the energy band is half full. When an electric field is applied, electrons can easily move to unfilled levels within the sodium energy band, resulting in a free path from atom to atom for the conduction of electrons. Since the difference in energy is very small between adjacent levels within the energy band, the electric field required to cause conduction is small.

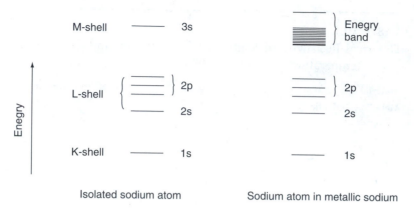

FIGURE 10.2 Schematic of the energy levels of an isolated sodium atom and of a sodium atom bonded into a metallic sodium crystal structure. (Drawing courtesy of ASM International.)

The other alkali metals (Li, K, Rb, Cs, and Fr) all have half-filled energy bands similar to Na, except that the magnitude of the energy is different because $2s$, $4s$, $5s$, $6s$, and $7s$ electrons are involved rather than $3s$.

The alkaline earth metals (Be, Mg, Ca, Sr, Ba, and Ra) behave differently. They each have two valence electrons and thus completely fill the first valence energy band. Since the band is full, one would not expect electronic conduction to occur. However, a second valence energy band is present that overlaps the first band. Under an electric field, electrons from the first band easily move into the second band, providing a mechanism for electronic conductivity.

Aluminum is trivalent and aluminum metal has three valence bands. The first band is filled and the second band is partially filled. Electron conduction can readily occur within the second band when an electric field is applied.

We have now established that electrons are the carriers of electrical charge in metals and that the electrons are available due to the nature of the atomic bonding. We next need to examine the factors that affect the mobility of the electrons.

A major factor that affects electron mobility is temperature. Increasing temperature produces thermal vibrations that have a scattering effect which reduces the mean free path. As a result, the electronic conductivity of metals decreases (and the electrical resistivity, reciprocally, increases) as the temperature increases. Resistivity vs. temperature is shown in Figure 10.3 for some metals.

Other factors that affect electron mobility are impurities, solid solution, and plastic deformation. All three disrupt the short-range uniformity of the structure and decrease the mean free path of the electrons.

Metals are not the only materials that exhibit electronic conductivity. Some transition metal oxides such as ReO_3, CrO_2, ReO_2, TiO, and VO also have high levels of electronic conductivity. This results from overlap of electron orbitals forming wide, unfilled d or f electron energy bands (in contrast to the s electron energy bands formed in metals). Concentrations of quasifree electrons of 10^{22} to 10^{23} per cm^3 result, which is essentially equivalent to metallic conduction.[1] Figure 10.4 illustrates the conductivity vs. temperature for a variety of transition metal oxides. Note that some of the materials have an abrupt increase in conductivity at a specific temperature. This temperature is generally referred to as the *semiconductor-to-metal transition*.

10.3   IONIC CONDUCTIVITY

A second major mechanism of electrical conductivity in materials is *ionic conductivity*.[5,6] In this case, the carrier of electrical charge is an ion, and the most prominent materials that exhibit this behavior are ionically bonded ceramics. Whereas the charge carried by an electron is 1.6×10^{-19} coulomb

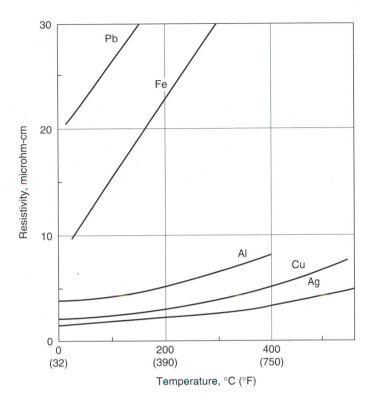

FIGURE 10.3 The change in electrical resistivity vs. temperature for common metals. (From van Vlack, L.H., *Elements of Materials Science*, 2nd ed., Addison-Wesley, Reading, MA, 1964, p. 420, p. 117. With permission.)

(i.e., ampere-seconds), the charge carried by an ion is equal to the valency times the charge of an electron. For example, a monovalent ion carries the same charge as an electron, and a divalent ion carries twice the charge.

10.3.1 Mechanisms of Ionic Conductivity

Electrons are relatively free to move in a metallic conductor under the influence of an electric field, but ions are restricted by their position in a crystal structure. Each positive ion (cation) is surrounded by negative ions (anions) and each negative ion is surrounded by positive ions. The overall charge is balanced and all the ions are in a state of equilibrium. Energy must be added before an ion can move. This is analogous to trying to push an automobile over a hill. We have to push the ion over an energy "ridge" or "barrier." The opposite charge of the nearest neighbor ions tries to hold the ion in the equilibrium position and provides the energy barrier. This is illustrated in Figure 10.5.

The energy barrier to ion movement is high at 0 K ($-460°F$) and for most ionic ceramics at room temperature. However, as the temperature is increased, thermal vibrations move ions further and further away from their equilibrium positions and boost the ions closer to the top of the energy barrier. When the thermal energy is high enough for a specific material for ions to move over the energy barrier under the influence of an electric field, the material becomes ionically conductive. The degree of ionic conductivity then increases as the temperature further increases. This is illustrated for specific materials later in this chapter.

Other factors also affect ionic conductivity. An ion must have a position to move into and must not alter the overall charge balance by the move. The presence of open lattice positions and point defects greatly enhances the ease with which ions move or "diffuse" through a ceramic structure.

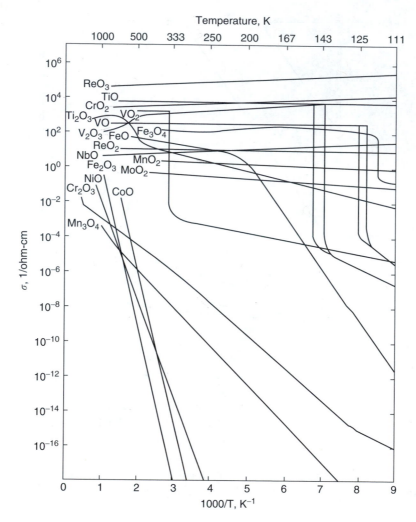

FIGURE 10.4 Conductivity vs. temperature for some transition metal oxides that exhibit electronic conductivity. (From data of D. Adler published in Kingery, W.D., Bowen, H.K., and Uhlmann, D.R., *Introduction to Ceramics*, 2nd ed., Wiley, New York, 1976, Chap. 17, p. 867. With permission.)

An example of an open lattice position is an unfilled cubic, octahedral, or tetrahedral site within the crystal structure. Remember, in Chapter 5 we discussed that these sites are completely filled in some crystal structures and only partially filled in others. When they are partially filled, ions have a lower energy path for movement.

Point defects, which are defects in the crystal structure, can also provide a path for ions to move. There are four common types of point defects: (1) Schottky; (2) Frenkel; (3) vacancy; and (4) interstitial. These types of defects are illustrated in Figure 10.6. A Schottky defect consists of a missing cation plus a missing anion of equal charge. A Frenkel defect consists of an ion that has moved from a lattice position into an interstitial position. A vacancy is simply a missing atom, and an interstitial is simply a foreign atom that has been forced into an interstitial position. All four types of point defects reduce the height of the energy ridge for ion movement. Also, the probability of point defects increases as temperature increases.

Vacancies are particularly important for enabling controlled ionic conduction. The concentration of the vacancies can be controlled during fabrication of the material. This is accomplished

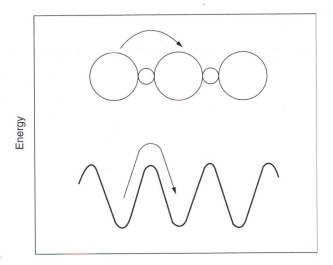

FIGURE 10.5   Energy barrier that must be overcome before an ion can move from its initial position to a new position.

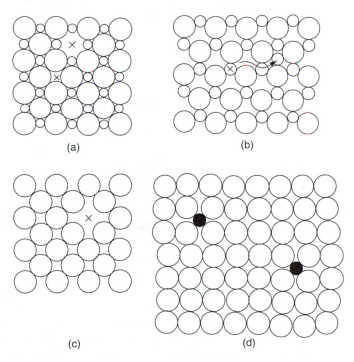

FIGURE 10.6   Point defects that commonly occur in a crystal structure. (a) Schottky defect, (b) Frenkel defect, (c) vacancy, and (d) interstitial.

by substituting ions of similar size but different electrical charge. An example is substitution of CaO into ZrO_2. For each Ca^{2+} that substitutes for a Zr^{4+}, an oxygen vacancy must result so that charge balance is maintained ($Zr_{1-x} Ca_x O_{2-x}$). At elevated temperature the energy barrier is low enough that oxygen ions can diffuse through the structure from vacancy to vacancy when an electric field is applied.

TABLE 10.2

Mode of Conduction and Temperature of Conduction for Various Ceramics

Compound	Temperature, °C (°F)	Transference Number[a]		
		t_+	t_-	$t_{e.h}$
NaCl	400 (750)	1.00	0.00	
	600 (1110)	0.95	0.05	
KCl	435 (815)	0.96	0.04	
	600 (1110)	0.88	0.12	
KCl + 0.02% $CaCl_2$	430 (805)	0.99	0.01	
	600 (1110)	0.99	0.01	
AgCl	20–350 (70–660)	1.00		
AgBr	20–300 (70–660)	1.00		
BaF_2	500 (930)	–	1.00	
PbF_2	200 (390)	–	1.00	
CuCl	20 (70)	0.00	–	1.00
	366 (690)	1.00	–	0.00
ZrO_2 + 7% CaO	>700 (>1290)	0	1.00	10^{-4}
$Na_2O \cdot 11Al_2O_3$	<800 (<1470)	1.00 (Na^+)	–	$<10^{-6}$
FeO	800 (1470)	10^{-4}	–	1.00
ZrO_2 + 18% CeO_2	1500 (2730)	–	0.52	0.48
+ 50% CeO_2	1500 (2730)	–	0.15	0.85
$Na_2O \cdot CaO \cdot SiO_2$ glass	–	1.00 (Na^+)		
15% $(FeO \cdot Fe_2O_3) \cdot CaO$ $\cdot SiO_2 \cdot Al_2O_3$ glass	1500 (2730)	0.1 (Ca^{+2})	–	0.9

[a] Cations t_+, anions t_-, and electrons or holes $t_{e.h}$.

Source: Kingery, W.D., Bowen, H.K., and Uhlmann, D.R., *Introduction to Ceramics*, 2nd ed., Wiley, New York, 1976, Chap. 17, p. 853.

10.3.2 CERAMIC MATERIALS EXHIBITING IONIC CONDUCTIVITY

Several categories of ceramics have high ionic conductivity: (1) halides and chalcogenides of silver and copper where the metal atoms are disordered over several sites (allowing low-energy requirement for diffusion); (2) oxides of the fluorite structure doped to cause a high concentration of oxygen vacancies; and (3) oxides with the beta-alumina structure that contain large structural channels through which monovalent cations can move. The mode of conduction and temperature of conduction for a range of ceramics are shown in Table 10.2.

The term *transference number* is also introduced in Table 10.2. It identifies the fraction of the total conductivity contributed by each type of charge carrier, with the sum equaling 1.0. Cation conduction is designated by t_+, anion conduction by t_-, and electron or electron hole conduction by $t_{e.h}$. Some ceramics have pure cation or anion conduction with $t_+ = 1.0$ or $t_- = 1.0$. Others have mixed ionic and electronic. The type of conduction strongly influences the applications, as is discussed later in the chapter.

Table 10.2 contains much information. The following should be noted:

1. The electrical conduction in the alkali halides (NaCl, KCl, AgCl, etc.) is 100% ionic; the carriers are positive ions.
2. The electrical conduction in the fluorite or defect fluorite structure materials (BaF_2, PbF_2, ZrO_2 with 7 wt% CaO) is essentially 100% ionic; the carriers are negative ions.
3. The electrical conduction of beta-alumina ($Na_2O \cdot 11Al_2O_3$) and beta″-alumina ($Na_2O \cdot 5Al_2O_3$) is essentially 100% ionic; the carriers are sodium ions.

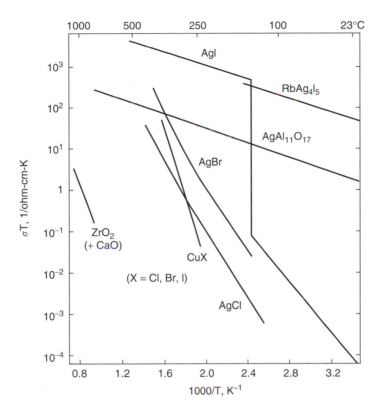

FIGURE 10.7 Ionic conductivity vs. temperature for a variety of ceramic solid-electrolyte materials. (From Kingery, W.D., Bowen, H.K., and Uhlmann, D.R., *Introduction to Ceramics*, 2nd ed., Wiley, New York, 1976, Chap. 17, p. 862. With permission.)

4. The electrical conduction of FeO, ReO_3, CrO_2, VO, TiO, and ReO_2 is essentially 100% electronic; the carriers are electrons or electron holes.
5. Additions of CeO_2 to ZrO_2 result in a combination of ionic and electronic conduction.
6. Some glasses are ionically conductive. Some Na_2O-containing glasses conduct by sodium ion diffusion and have resistivity of about 10^2 ohm-cm at about 350°C (660°F).
7. Some glasses are electronically conductive. These contain multivalent transition metal ions such as iron, cobalt, manganese, or vanadium. The degree of conductivity is low. Resistivity is typically greater than 10^4 ohm-cm.
8. Alkali halides have useful levels of conduction at low to moderate temperatures, that is, room temperature to 600°C (1110°F). Beta″-alumina is adequately conductive above about 300°C (570°F). Zirconia-based compositions require temperatures above 600°C (1110°F). Ionic conductivity vs. temperature is illustrated for a variety of ceramic materials in Figure 10.7.

Table 10.3 identifies some existing and potential applications of ionically conductive ceramics and some of the ceramics that have been explored for these applications. Some of these applications and the characteristics of the specific ceramic materials involved are discussed in subsequent sections.

10.3.3 APPLICATIONS OF ZIRCONIA OXYGEN ION CONDUCTIVE CERAMICS

A variety of applications based on oxygen-ion conduction through doped ZrO_2 are identified in Table 10.3.[7] Oxygen sensors are presently in high volume production for automobiles. Solid oxide fuel cells have great potential for low pollution, efficient generation of electricity.

TABLE 10.3
Applications of Ion-Conducting Ceramics

Application	Material Options
Oxygen sensor	ZrO_2 solid solutions with CaO, MgO, or Y_2O_3
Oxygen pump	Same
Electrolysis	Same
SO_x–No_x decomposition	Same
Solid oxide fuel cell	Same; δ-Bi_2O_3; Bi_2O_3-Y_2O_3; $ZrGd_2O_7$
Na-S battery	Beta-alumina, beta″-alumina; NaZrSiPO compositions; Na_2S-GeS_2-NaI; Na ion-conducting glasses
Sodium heat engine	Beta-alumina, beta″-alumina
Ion pumps	Beta-alumina analogues
Lithium batteries	LiI-Al_2O_3; Li-Al/FeS
Molten carbonate fuel cell	Li_2CO_3-K_2CO_3; $LiAlO_2$
Resistance-heating elements	Cubic ZrO_2 solid solutions; CeO_2-ZrO_2 compositions
Galvanic cells for thermodynamics and kinetics measurements	Doped ZrO_2, CaF_2, ThO_2, and others

---

Pure ZrO_2 has a monoclinic close-packed crystal structure that does not conduct oxygen ions. Addition of CaO, MgO, Y_2O_3, or rare earth oxides results in a cubic defect fluorite structure that contains vacant oxygen lattice sites. As mentioned earlier, when the doped ZrO_2 is heated to temperatures above approximately 600°C (1110°F), relatively low energy is required for the oxygen ions to migrate through the crystal structure.

Most commercial ZrO_2 oxygen-ion conducting devices use either CaO or Y_2O_3 doping. Typical additive levels to achieve the cubic solid solution are 12 to 20 mol% CaO or 5 to 10 mol% Y_2O_3.

10.3.3.1 Oxygen Sensors

Oxygen sensors are based on the relationship expressed in Equation (10.4):

$$E = \frac{RT}{4F} \ln\left(\frac{PO_2}{PO_2'}\right) \tag{10.4}$$

where

E = electromotive force (emf, voltage) across the zirconia membrane
F = the Faraday constant
R = the universal gas constant
T = absolute temperature
ln = natural logarithm
PO_2 and PO_2' = pressures of oxygen on opposite sides of the ZrO_2 membrane

If $PO_2 = PO_2'$, $E = 0$. If PO_2 and PO_2' are different from each other, a measurable emf will be present across the ZrO_2 membrane. The larger the difference between PO_2 and PO_2', the larger the emf. By selecting a reference PO_2 of known value, an unknown PO_2' can be accurately sensed and calculated from Equation (10.4).

The schematic of an oxygen sensor is shown in Figure 10.8a. It consists of a doped ZrO_2 tube with electrically conductive porous electrodes applied to the inside and outside, and connected through an external circuit. A known reference PO_2 (atmospheric oxygen concentration) is maintained on the inside of the tube. The outside of the tube is inserted into the unknown PO_2'. The emf is then measured with a simple voltmeter.

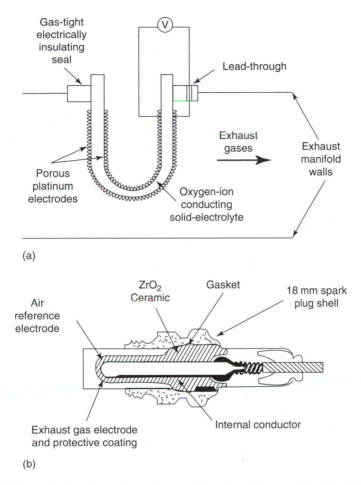

FIGURE 10.8 (a) Schematic showing the key elements of an exhaust gas oxygen sensor. (From McGeehin, P. and Williams, D.E., Ceramics for sensors and monitors, *Ceramics in Advanced Energy Technologies*, D. Reidel, Dordrecht, 1984, p. 461. With permission.) (b) Schematic cross section of a typical automotive ZrO_2 oxygen sensor. (From Logothetis, E.M., ZrO_2 oxygen sensors in automotive applications, *Advances in Ceramics, Vol. 3, Science and Technology of Zirconia*, American Ceramic Society, Ohio, 1981, pp. 388–405. With permission.)

Zirconia oxygen sensors must be operated at approximately 600°C (1110°F) or higher. Below this temperature range, the ionic conductivity is too low for a measurable emf to be generated. The most common sensor applications are for control of combustion conditions in automobile engines and industrial combustion processes and for measurement of oxygen content at various stages of metals melting and refining. All of these applications are above 600°C. For oxygen sensor applications below 600°C, a heater is required to maintain the zirconia membrane at >600°C. Examples where the temperature is below 600°C include measurement of oxygen impurity level in inert-gas cylinders and verification of low oxygen content in some food processing and packaging applications.

The greatest quantity of zirconia oxygen sensors are used for automotive engine-control devices to reduce emissions and fuel consumption. The concentration of oxygen, CO, NO_x, and hydrocarbons in the exhaust gas of the engine are a function of the air-to-fuel ratio (*A/F*) of the mixture introduced into the cylinders of the engine, as well as the efficiency of the catalytic emission-control device. Optimum emission control and fuel economy occur at approximately the stoichiometric *A/F* ratio (~15). A large change in oxygen partial pressure occurs at *A/F* mixtures below and above stoichiometric. This is monitored by the zirconia oxygen sensor that provides the feedback signal to the control system to readjust the *A/F* ratio to an optimum set-point. Figure 10.8b shows the schematic of a typical automotive oxygen sensor.

Industrial uses of oxygen sensors include glass melting, heat treatment, and metals melting and refining. Sensors using CaO-stabilized ZrO_2 tubes have been developed for continuous oxygen monitoring in molten copper, copper alloys, tin, tin alloys, lead, silver, lead-silver alloys, and sodium. Use of ZrO_2 sensors for monitoring oxygen content in steel refining has proven to be more difficult, primarily because of the higher temperature. Disposable one-measurement sensors have been developed. These are economical because oxygen content needs to be determined only once or twice in a heat. The oxygen sensor is immersed directly in the molten steel and provides an oxygen content reading in approximately 20 sec. The former technique required sampling, transfer to a laboratory, and chemical analysis at 2000 to 2500°C (3630 to 4530°F) by vacuum extraction. This technique required 8 to 10 min.

10.3.3.2 Oxygen Pumps

Whereas an oxygen sensor utilizes the natural emf produced by a difference in oxygen pressure on opposite sides of an ionically conducting zirconia membrane, an oxygen pump utilizes an applied voltage to force oxygen ions to move across the membrane. This is illustrated for a simple tubular configuration in Figure 10.9. An oxygen molecule (O_2) contacts the porous electrode. Four electrons are picked up to form two O^{2-} ions. These ions are then forced to diffuse through the zirconia solid electrolyte by the applied electric field. When they reach the porous electrode on the opposite side, they each give up two electrons and combine to form O_2. Only oxygen can be transported in this fashion through the ZrO_2 membrane, thus allowing a means of separating gases. For example, oxygen can be pumped from air to yield 99.999% pure oxygen. This is referred to as an *oxygen concentration cell*. Similarly, oxygen can be removed from another gas, such as final purification of bottled nitrogen or argon gas. This is referred to as an *oxygen extraction cell*. Carefully controlled partial pressures can be achieved by this technique. A zirconia oxygen extraction pump has been used to remove oxygen from flowing nitrogen wherein an O_2 partial pressure of 10^{-38} atm at 530°C (985°F) was achieved.

10.3.3.3 Electrolysis and Thermolysis

High-temperature zirconia solid electrolyte cells can be used for production of hydrogen by *electrolysis* of water. The cell is operated at 1000°C (1830°F) with an applied voltage of approximately 0.95 V. Zirconia stabilized with CaO or Y_2O_3 and having only ionic conduction is used as the solid electrolyte. H_2 is released at the cathode and O_2 at the anode.

Hydrogen can also be produced by the direct thermal decomposition of water. This process, called *thermolysis*, requires a very high temperature (>2000°C [>3630°F]) and an oxygen-permeable

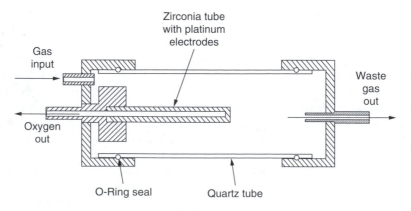

FIGURE 10.9   Schematic of a tubular oxygen pump. (Courtesy Ceramatec, Inc.)

membrane. In this case, a zirconia solid electrolyte with a combination of ionic and electronic conduction is required. ZrO_2–CeO_2–Y_2O_3 and ZrO_2–Cr_2O_3–Y_2O_3 compositions have been demonstrated.

10.3.3.4 SO_x–NO_x Decomposition

Sulfur dioxide and nitrogen oxides are dangerous pollutants that result from combustion processes. These pollutants are major contributors to the acid rain problems. Bench-scale experiments have demonstrated that SO_2 and NO_x can be decomposed to harmless substances with a zirconia solid electrolyte cell. At the appropriate temperature and voltage, SO_2 and NO_x molecules contact the electrode and dissociate into S, N, and O ions. The oxygen ions are ionically conducted through the ZrO_2 membrane and are therefore not available to recombine with the S and N ions. The N ions combine to form harmless N_2 gas. The S ions polymerize during cooling to form liquid sulfur that can be easily separated from the gas stream.

10.3.3.5   Solid Oxide Fuel Cells

A schematic of a simple solid oxide fuel cell (SOFC) is shown in Figure 10.10. The fuel cell is operated at about 900 to 1000°C. A high oxygen pressure exists on the air side and a low oxygen pressure exists on the fuel side. This gradient causes oxygen molecules to accept electrons from the air electrode to form oxygen ions and for the oxygen ions to diffuse across the doped zirconia membrane toward the fuel electrode. Oxygen ions reaching the fuel electrode give up their extra electrons, form O_2 molecules, and electrochemically combine with the hydrogen in the fuel. This results in the generation of an electric current in an external circuit between the electrodes and in high efficiency, noise-free combustion with the only combustion products being water and heat (and CO_2 for some fuels). The waste heat can be utilized in a cogeneration cycle. A combined efficiency of about 80% is predicted. This is much greater than can be achieved by diesel engines and gas turbine engines, which typically have efficiency in the 25 to 35% range.

Major efforts are underway internationally to develop viable solid oxide fuel cells of a variety of designs varying from a couple below to over 200 kW.[8–11]

Perhaps the most mature SOFC is the tubular design (shown earlier in Chapter 3) of Siemens-Westinghouse. A 100 kW 50 Hz unit was operated for 16,667 h at a site in Westervoort, Netherlands between 1998 and March 2001.[10] It demonstrated 46% electrical efficiency in producing 109 kW of electricity and also provided 64 kW of hot water, so it thus had an impressive combined heat and power efficiency. In addition the SOFC emitted less than one part per million NO_x and volatile hydrocarbons. This SOFC was subsequently moved to Essen, Germany where it accumulated another 3700 h by January 2002 and is still running in May of 2004.

A potential advantage of SOFCs is multi-fuel capability. Two 25 kW Siemens-Westinghouse units were tested successfully on natural gas, diesel, and jet fuels at the National Fuel Cell Research Center (NFCRC) at the University of California at Irvine. Another potential advantage is linking to a gas turbine engine in a combined cycle. A 200 kW SOFC linked to a 20 kW gas turbine engine demonstrated at NFCRC 53% electrical efficiency with potential to reach 60 to 70%.[10] A third projected advantage of SOFC systems is isolation and recovery of CO_2 emissions to combat global warming.

Besides high reliability electricity generation and combined heat and power, the SOFC is being developed for auxiliary power systems for vehicles and later for residential total energy systems. Compact plan designs consisting of stacked plates rather than tubes are under development for these applications.

10.3.3.6   Resistance Heating Elements

Doped zirconia is one of the few materials that can be operated as a resistance heating element at temperatures greater than 1800°C (3270°F) in an oxidizing environment. An important application

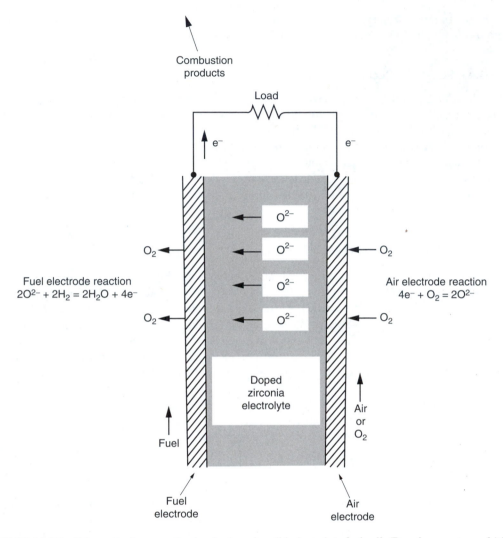

FIGURE 10.10 Schematic diagram of a simple zirconia solid-electrolyte fuel cell. (Drawing courtesy of ASM International.)

that has utilized this capability is the production of optical fibers. Figure 10.11 shows the schematic of an induction furnace with an ionically conductive zirconia susceptor. The zirconia must first be heated to ~1000°C (~1830°F) before it has high enough conductivity to perform as a susceptor for the induction coil. This preheating is achieved by inserting a carbon rod into the internal diameter of the zirconia tube and inductively heating the carbon. When the zirconia becomes adequately heated by the carbon rod, the rod is removed and the remainder of the heating is achieved by direct induction to the ZrO_2. For this application, the conductivity of the ZrO_2 does not have to be purely ionic, but can be a combination of ionic and electronic.

10.3.3.7 Galvanic Cells for Thermodynamic and Kinetic Measurements

Doped zirconia solid electrolytes have been used extensively for obtaining thermodynamic and kinetic measurements for metals, alloys, intermetallics, and oxides. Galvanic cells have been constructed for measurements of the standard free energy of formation of oxides and intermetallics, chemical activities in molten and solid metals and alloys, and diffusivity of oxygen in molten and solid metals.

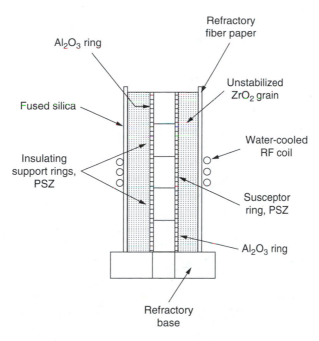

FIGURE 10.11   Schematic of an induction-heated 2400°C (4350°F) furnace with a ZrO_2 susceptor. (From Keguang, S., Pike, R., and Chapman, A., *Am. Ceram. Soc. Bull.*, 65(12), 1604–1605, 1986. With permission.)

10.3.4   ALTERNATE OXYGEN ION CONDUCTORS

Although most of the present applications for oxygen ion conductors utilize CaO-doped or Y_2O_3-doped ZrO_2, other ceramic materials are also known to have high oxygen ion conductivity. Examples are shown in Figure 10.12.

10.3.5   SODIUM ION CONDUCTORS AND APPLICATIONS

Oxygen ion-conducting ceramics require relatively high temperature before the electrical conductivity is high enough for commercial applications. Sodium ion-conducting ceramics operate at lower temperature, as shown in Figure 10.13. Sodium ion conduction is the basis for a high-power-density (~200 W/kg) battery that has received intermittent development for automotive, utility load leveling, and satellite applications. The battery is referred to as the sodium/sulfur (Na/S) battery and was first announced in 1967 by Ford Motor Company. A simple schematic is shown in Figure 10.14. A thin-walled tube of an impervious sodium ion conductor separates molten sodium metal from molten sulfur. The most successful sodium ion conductor developed so far for the Na/S battery has been beta″-alumina. At 300 to 350°C (570 to 660°F) sodium ions rapidly migrate through the beta″-alumina solid electrolyte from the sodium side to the sulfur side. Each sodium atom that ionizes releases an electron that travels through an external circuit and provides an overall open circuit voltage of 2.08 V for a typical cell. Sodium ions that reach the sulfur side of the beta″-alumina membrane combine with sulfur to form sodium polysulfide compounds. This represents the discharge cycle of the battery. An external voltage can be applied to the battery to force the polysulfides to dissociate and the Na ions to return to the molten sodium side of the electrolyte, resulting in recharge of the battery.

Beta-alumina was originally thought to be a polymorph of Al_2O_3, but was subsequently shown to be a separate compound that contains sodium as part of the crystal structure. The nominal composition $Na_2O \cdot 11Al_2O_3$ is referred to as beta-alumina. A closely related compound of composition

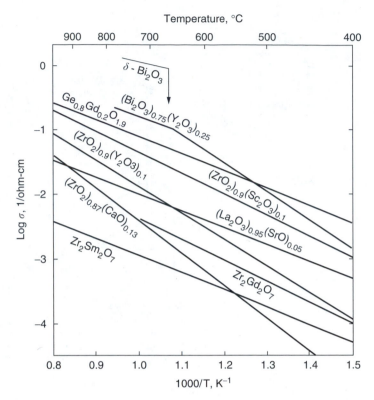

FIGURE 10.12   Examples of ceramics that have high oxygen ion conductivity. (After Steele, B.C.H., Ceramic materials for electrochemical energy conversion devices, *Ceramics in Advanced Energy Technologies*, D. Reidel, Dordrecht, 1984, pp. 386–412.)

$Na_2O \cdot 5Al_2O_3$ is referred to as beta″-alumina. The structures of beta alumina and beta″-alumina are illustrated in Figure 10.15. Both are layered structures consisting of blocks similar to spinel that are linked together by oxygen and sodium ions. Each spinel block consists of four close-packed oxygen layers with aluminum ions residing in both octahedral and tetrahedral positions. The sodium and oxygen layers are not close-packed and allow relatively easy movement of the sodium ions perpendicular to the c axis under the influence of an electric field. Movement of the sodium ions parallel to the c axis is restricted by the close-packed oxygen layers. The ionic conductivity of beta-alumina and beta″-alumina is therefore highly anisotropic and usually referred to as two-dimensional.

The oxygen planes in the beta″-alumina are staggered and result in slightly larger spacings through which the sodium ions can migrate. Thus, beta″-alumina has greater ionic conductivity than beta-alumina. Beta″-alumina is not stable above approximately 1500°C (2730°F) and is difficult to obtain by conventional sintering because the sintering temperature is typically above 1545°C (2815°F). The beta″-alumina structure can be stabilized at higher temperatures by the addition of cations that are capable of occupying either the octahedral or tetrahedral sites in the lattice of the spinel blocks. Li^+ and Mg^{2+} have been identified as the most effective.

Beta″-alumina is an interesting material. Because of the large size of the Na^+-containing channels and the stability of the surrounding structure, other mobile ions besides Na^+ can be ion-exchanged into the structure. For example, Ag^+ ions can replace the Na^+ ions. This modified material can then be used as an electrolyte across which Ag^+ ions can be pumped when a voltage is applied. Similarly, the Na^+ can be replaced by Li^+, K^+, and other ions.

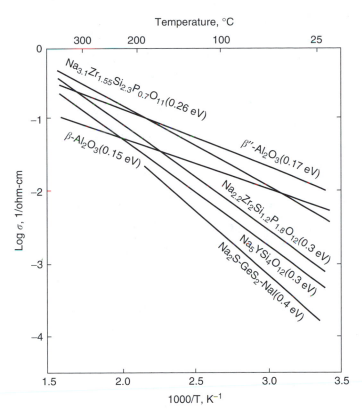

FIGURE 10.13   Examples of ceramics that have high sodium ion conductivity. (From Steele, B.C.H., Ceramic materials for electrochemical energy conversion devices, *Ceramics in Advanced Energy Technologies*, D. Reidel, Dordrecht, 1984, pp. 386–412. With permission.)

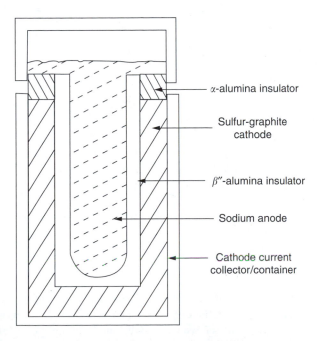

FIGURE 10.14   Schematic of a single cell of a sodium/sulfur (Na/S) battery.

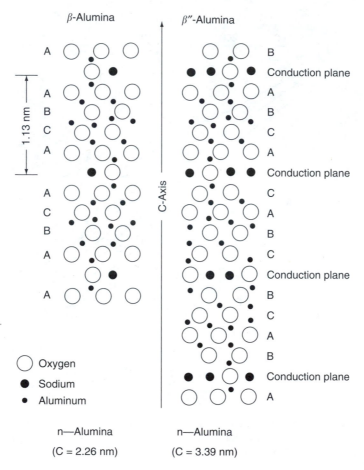

FIGURE 10.15 Comparison of the crystal structure of beta-alumina and beta″-alumina; projection of the unit cells on (1120). (From Stevens, R. and Binner, J.G.P., *J. Matter. Sci.*, 19, 695–715, 1984. With permission.)

10.3.6 LITHIUM ION CONDUCTION AND APPLICATIONS

The Na/S battery discussed earlier is a "secondary" battery, which means that it is rechargeable. Another category of batteries that are not recharged are referred to as "primary" batteries. The alkaline cells and a variety of Li batteries fall in this category. These batteries operate at room temperature and have a relatively low output and a slow discharge rate. Applications include batteries for watches, calculators, medical products (especially pacemakers), photographic equipment, radios, and toys. Figure 10.16 identifies some solid electrolyte lithium ion conductors. Some current commercial systems utilize a semisolid electrolyte that consists partially of a polymer (poly-2-vinyl-pyridine). Another secondary Li ion-conducting battery has LiAl and FeS as the electrodes separated by molten LiCl-KCl eutectic as the electrolyte.

10.4 CONDUCTIVE POLYMERS

Polymers are covalently bonded and have electron configurations similar to insulators and semiconductors, that is, their valence band is full, their conduction band is empty, and the two are separated by a large energy gap. Therefore, polymers have been traditionally thought of as electrical insulators.

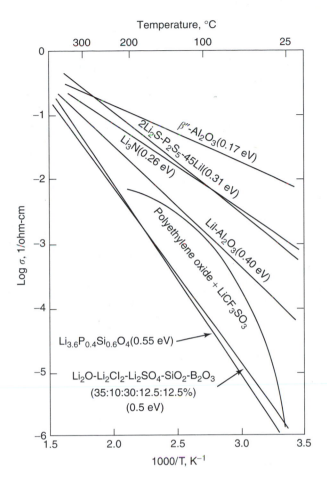

FIGURE 10.16 Lithium ion conductivity of some solid electrolytes. (From Steele, B.C.H., *Ceramic materials for electrochemical energy conversion devices, Ceramics in Advanced Energy Technologies*, D. Reidel, Dordrecht, 1984, pp. 386–412. With permission.)

Work at Tokyo Institute of Technology and University of Pennsylvania in the 1970s determined that certain polymers could become electron conductors with relatively heavy levels of doping (approximately one dopant atom for every 15 carbon atoms in the polymer).[12] Conductivities at room temperature equivalent to copper have been achieved. Conductivity has been achieved in polyacetylene, polyparaphenylene, polypyrrole, polythiophene, and polyaniline. All of these have alternation of double and single carbon bonds.

The initial studies were conducted with polyacetylene doped with iodine. The mechanism is more complex than the doping in ceramics to achieve ionic conduction or semiconduction. The dopant does not substitute into the structure, but instead causes a slight change in the position of atoms along the carbon chain of the polymer. This prompts the formation of one of three types of charge "islands" called solitons, polarons, and bipolarons. With a high enough dopant level, the islands overlap and result in new energy levels that bridge the valence and conduction bands. The polymer becomes conductive and even has a metallic appearance.

Polymers have also been developed that are hosts for ionic conduction.[5] An example is a polyphosphazene polymer containing Li cations and $F_3CSO_3^-$ anions. The Li^+ ions are mobile at room temperature and result in low levels of conductivity of about 10^{-4} $(ohm-cm)^{-1}$.

10.5 ELECTRICAL INSULATORS

We have discussed in detail mechanisms by which electrons or ions can move through a ceramic (or other material) under an applied electric field. Specifically, electrons can move when the material has electron energy levels that overlap or are not filled. Ions can move when point defects are present in the crystal structure (such as vacancies) or if the structure contains weakly bonded ions in large structural channels (such as in beta-alumina). Most ceramics do not have mobile electrons or ions and do not permit passage of an electrical current when placed in an electric field. These nonconductive ceramics are called *electrical insulators*. Most polymers are also electrical insulators. Check back in this chapter to Table 10.1 to see typical resistivity values for ceramic and polymer insulator materials. Note that the values for SiO_2, Al_2O_3, MgO, Si_3N_4, teflon, polystyrene, and nylon are all $>10^{14}$ ohm-cm.

The high value of electrical resistivity for these materials results from the way that the electrons are tied up during atomic bonding. In each case, valence electrons are either shared (covalently) or transferred (ionically) such that each atom achieves a full outer shell of electrons. This leaves no overlap of electron energy bands and no low-energy mechanisms for electron conduction. This is illustrated in Figure 10.17.

Temperature and impurities have strong effects on the conductivity of insulator materials. The effect of temperature is shown for a variety of materials in Figure 10.18. The effect of temperature and impurities is shown for BeO in Figure 10.19. The resistivity decreases as the temperature increases due to the increased thermal energy of the electrons. The resistivity also decreases as the amount of impurities increases. The impurities increase the concentration of structural defects, which either increase the number of electrons or electron holes available for conduction or provide a lower-energy path for migration of ions.

Table 10.4 lists examples of ceramic materials that are normally electrical insulators. Most ceramics and glasses are electrical insulators at room temperature. Most oxides and silicates are electrical insulators also up to high temperatures.

10.5.1 APPLICATIONS OF ELECTRICAL INSULATORS

Electrical insulators are required to isolate all electrical circuits. This duty is shared by polymers and ceramics. Ceramics are used where high strength, elevated temperature, heat dissipation, or long-term

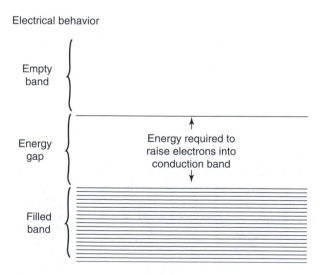

FIGURE 10.17   Schematic of the energy bands in an insulator such as MgO or Al_2O_3 showing the larger energy gap between the filled band and the next available empty bands. (Drawing courtesy of ASM International.)

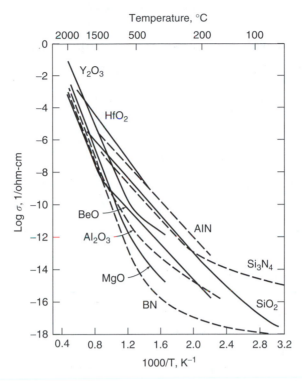

FIGURE 10.18 Effect of temperature on the electrical conductivity of ceramic insulator materials. (From Bauer, A.A. and Bates, J.L., Battelle Memorial Institute Rept. 1930, July 31, 1974. With permission.)

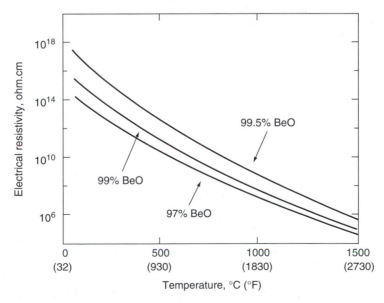

FIGURE 10.19 The effect of temperature and impurities on the electrical resistivity of beryllium oxide. (From Ryshkewitch, E. and Richerson, D., *Oxide Ceramics*, 2nd ed., General Ceramics, Inc., N.J., 1985, p. 566. With permission.)

TABLE 10.4
Examples of Ceramics That Are Electrical Insulators

Silicon dioxide	(SiO_2)
Aluminum oxide	(Al_2O_3)
Beryllium oxide	(BeO)
Magnesium oxide	(MgO)
Mullite	$(Al_6Si_2O_{13})$
Cordierite	$(Mg_2Al_4Si_5O_{18})$
Forsterite	(Mg_2SiO_4)
Silicon nitride	(Si_3N_4)
Aluminum nitride	(AlN)
Some spinels	$(MgAl_2O_4)$
Porcelains	
Steatites	
Most silicate glasses	
Most glass ceramics	
Most ceramics at room temperature	

TABLE 10.5
Some Applications of Ceramic Electrical Insulators

Integrated circuit chip carriers
Substrates for printed circuits
Spark plug insulators
Power line insulators
Electron-tube cathode insulator
High-frequency tube envelope
Microwave tube components
Thermocouple protection tubes
Filament support for incandescent lights
X-ray tube components
Television and oscilloscope components
Computer components
Household appliance components
Automotive electrical system components
Aerospace components

---

hermeticity (sealed from exposure to the atmosphere) are required. Table 10.5 lists some applications of ceramic electrical insulators.

10.5.1.1 Integrated Circuit Substrates and Packages

The largest application for ceramic electrical insulators is for substrates and "packages" for silicon chips for integrated circuits. The ceramic provides three major benefits: (1) miniaturization, (2) hermetic sealing, and (3) heat conduction. Miniaturization is permitted because of the high degree of electrical insulation of the ceramic. Thin-film, narrow-line conductive circuits can be printed on the ceramic into a tight pattern with very little separation between lines. This results in very short

---

TABLE 10.6

Improvements in Alumina Ceramic Substrates Resulting from Improvements in Fabrication and Starting Materials

Parameter	1960	1962	1966	1970	1972	1973
Percent Al_2O_3	96	96 glazed	99.5	99.5	99.7	99.95
Density (g/cm³)	3.70	3.70	3.88	3.83	3.89	3.96
Crystal size (μm)	6	6	3	1.5	1.2	1.0
Surface finish (μin.)	18	1	8	5	2.5	1
Laser reflectometer				0.400	1.300	1.800
Intrinsic pore size (μm)	4	4	1.5	1.5	0.5	0.3
Flexural strength (ksi)	60		70	85	85	100
Film application	Thick	1st thin	Thin	Thin	Thin	Thin

Source: Williams, J.C., *Am. Ceram. Soc. Bull.*, 56(6), 580, 1977.

---

circuit paths with low resistance loss and high rate of response. Silicon chip semiconductor devices are highly susceptible to degradation by moisture in the atmosphere and by heat buildup during their normal operation. To avoid this degradation, the chip is sealed into a ceramic carrier. Let us review some of the history of this technology and describe how chip carriers have evolved.

The rapid growth driven by the need for miniaturization of the electronics industry in the early 1960s led to the need for thin electrical insulator substrates for conductive, resistive, and dielectric* thick- and thin-film circuits. The substrates had to be thin, smooth, flat, and inexpensive as well as have acceptable mechanical and electrical properties. Substrates in the early 1960s were fabricated by compacting with a press a ceramic powder containing about 95% Al_2O_3 and firing the thin plate of pressed powder at high temperature to achieve densification. Thick films of conductors were then applied to the Al_2O_3 substrate by silk screening, followed by another firing operation to bond the thick film to the ceramic. Substrates with printed circuits were fabricated in this fashion for automobile instrument circuits and for microelectronic solid-state logic devices. Thin films could not be applied directly to initial ceramic substrates because the surfaces were too rough. Adequate continuity could not be achieved with thin films. Use of a glass substrate or a glazed (glass-coated) ceramic substrate was required to achieve the desired smoothness.

Improved ceramic powders and processing techniques were developed to achieve improved smoothness of ceramic substrates. The results are summarized in Table 10.6. Thinner lines could be printed on the improved substrates, resulting in advances in miniaturization. In addition, multilayer ceramic chip carriers were now possible that consisted of complex metallized circuit paths fired directly into the interior of Al_2O_3-based ceramics. Figure 10.20 illustrates schematically the cross section of one variety of integrated circuit multilayer package. This variety is referred to as a *pin grid array*. The array of metal pins provides the interconnection with the overall system and the power source. Each pin is connected to thin-film metallization that forms a continuous circuit path through the ceramic to the silicon chip (which is mounted on the top surface of the multilayer ceramic, or in most cases, into a recessed cavity in the ceramic). The techniques for fabrication of multilayer ceramic packages are described in Chapter 13.

Early silicon semiconductor devices could only perform a single circuit function and therefore did not require a complex or sophisticated substrate or package. As time passed, however, increased numbers of devices were integrated into individual silicon chips and increased numbers of silicon

---

* The term "dielectric" is discussed in Chapter 11.

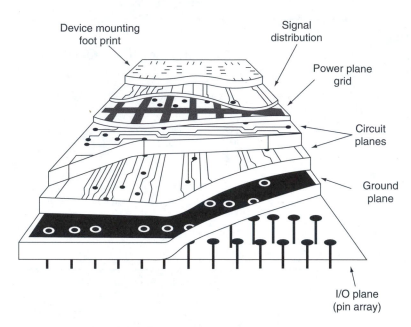

FIGURE 10.20 Schematic cross section showing the complexity of the metallized circuit patterns in a ceramic multilayer integrated circuit package. (From Ryshkewitch, E. and Richerson, D., *Oxide Ceramics*, 2nd ed., General Ceramics, Inc., N.J. 1985, p. 466. With permission.)

chips were interconnected into larger ceramic substrates. By the mid-1980s, individual modules could contain 100,000 circuits.[13] This resulted in tremendous improvements in miniaturization. An example of a state-of-the-art multilayer ceramic package for a mainframe computer is 90 mm by 90 mm by 5.5 mm and contains up to 33 layers of molybdenum-metallized Al_2O_3. It provides all the power distribution and interconnections for up to 133 logic and array chips.[14] A typical package of this type contains 350,000 vias for layer-to-layer interconnection, 130 m of wiring, and 1800 brazed-on pins for external connection. Cooling is achieved with water-cooled metal pistons in contact with the back of each chip.

Alumina continues to be the major ceramic package material for silicon chips. The metallization for Al_2O_3 is tungsten or molybdenum with additions to obtain a closer match in thermal expansion between the metal and ceramic. The resulting metal does not have particularly high electrical conductivity. This results in resistance heat generation and slow circuit time. Efforts are in progress to develop a ceramic composition that has the electrical characteristics of Al_2O_3, but which can be processed into multilayer packages containing a more highly conductive metallization such as copper. This requires a ceramic composition that will densify at around 1000°C (1830°F) rather than the 1500 to 1600°C (2730 to 2910°F) normally required for Al_2O_3. Some success has been reported with compositions containing a high glass content and with cordierite-based compositions.

Heat buildup is another concern in advanced integrated circuit applications. The thermal conductivity of Al_2O_3 was adequate to dissipate heat of simple modules, but is not adequate for the newer modules with many circuits. Cooled designs and the use of high thermal conductivity insulator ceramics (BeO and AlN) have been developed.

10.5.1.2 SPARK PLUG INSULATORS

A second widespread application of ceramic electrical insulators is in spark plugs. Spark plug insulators for automobiles are fabricated from high-Al_2O_3 compositions. The spark plug insulator is exposed to severe conditions. It must withstand several thousand volts at each spark discharge, a pressure pulse of about 10.4 MPa (1.5 ksi), and the thermal radiation from 2400°C (4350°F) combustion temperature.

10.5.1.3 Power Line Insulators

Another demanding application for high-resistivity ceramics is insulators for high-voltage power lines. Most people are aware that ceramics are used for these insulators, but do not realize how demanding the application really is. First, the insulators must be very strong because they support the weight of the power lines. Next time you drive through the countryside, notice how far apart the power line towers are and imagine the load the insulators must endure during a heavy wind or snow storm. Then consider how long these insulators are designed to last. They must be highly reliable. Second, the insulator must be resistant to weather damage and to absorption of water. Internal adsorption can result in arcing at the high voltages involved.

10.6 SEMICONDUCTORS

By definition, *semiconductors* are materials that have electrical resistivity between that of conductors and insulators, as was illustrated earlier in Figure 10.1. Semiconductors have an energy gap between the filled and empty electron bands such that conduction will occur only when sufficient external energy is supplied to overcome the energy gap. Whereas the band gap is large for insulators and difficult to overcome, it is small enough for semiconductors that a sufficient increase in temperature or in electric field will allow electrons to bridge the gap. The following section describes the mechanisms of semiconduction and some of the key materials.

10.6.1 Mechanisms of Semiconduction

Semiconduction can be *intrinsic* or *extrinsic*. *Intrinsic* refers to natural semiconduction within a pure material. The degree of intrinsic conduction is largely determined by the width of the energy gap within the electron structure of the material. Examples of the energy gap at room temperature for a variety of covalent and ionic materials are shown in Table 10.7. Note that the ionic ceramics tend to have a large energy gap and thus have very little likelihood of electron conduction at room temperature. In comparison, some of the covalently bonded materials have small energy gaps, especially elements from Group 14 (Si, Ge, Sn) and compounds from Groups 13 and 15 (GaAs, PbTe, PbS) of the periodic table. Values of resistivity correlate with the relative value of the energy gap. For example, diamond, Si, Ge, and gray Sn all have the same structure but different energy gaps and resistivities, that is, energy gaps of >6.0, 1.0, 0.7, and $0.08\,eV$ respectively, (electron volts) and resistivities of $>10^6$, 6×10^4, 50, and <1 ohm-cm, respectively.

Most intrinsic semiconductors do not have low enough resistivity to be useful at room temperature in an electronic circuit. Resistivity in the range 10^{-2} to 10^2 ohm-cm is required. This can be achieved by adding a controlled impurity to produce electronic imperfections. The resulting material is referred to as an *extrinsic* semiconductor. Let us use silicon as an example. Aluminum atoms are similar in size to silicon atoms and can replace a portion of the silicon atoms in the crystal structure. Each Al atom contains one less valence electron than the Si atom it replaces, resulting in the formation of an electron vacancy or electron hole in the valence band. Under an applied electric field, electrons can use the electron holes to move through the material. The electron hole conduction is equivalent to motion of a positive charge carrier. This mode is thus called *p-type extrinsic semiconduction*.

Other additives to Si can result in extrinsic semiconduction. Addition of phosphorus with five valence electrons results in one more electron than is required to fulfill the covalent bonds in the silicon structure. This leaves one electron free for conduction for each phosphorus atom added. Since the carrier in this case is an electron, the material is referred to as an *n-type extrinsic semiconductor*.

Extrinsic semiconduction can also be achieved through nonstoichiometry, especially in oxide ceramics. For example, $Fe_{0.95}O$ contains some Fe^{3+} in a structure of Fe^{2+} plus O^{2-}. This provides a mechanism for electron holes to move from one iron atom to the next under an applied electric field,

TABLE 10.7
Value of the Energy Gap at Room Temperature for Intrinsic Semiconduction

Crystal	E_g (eV)	Crystal	E_g (eV)
$BaTiO_3$	2.5–3.2	TiO_2	3.05–3.8
C (diamond)	5.2–5.6	CaF_2	12
Si	1.1	BN	4.8
Ge	0.7	Sn (gray)	0.08
α-SiC	2.8–3	CdO	2.1
PbS	0.35	LiF	12
PbSe	0.27–0.5	Ga_2O_3	4.6
PbTe	0.25–0.30	CoO	4
Cu_2O	2.1	GaP	2.25
Fe_2O_3	3.1	Cu_2O	2.1
AgI	<2.8	CdS	2.42
KCl	7	GaAs	1.4
MgO	>7.8	ZnSe	2.6
Al_2O_3	>8	CdTe	1.45

Source: Kingery, W.D., Bowen, H.K., and Uhlmann, D.R., *Introduction to Ceramics*, 2nd ed., Wiley, New York, 1976, Chap. 17, p. 868.

TABLE 10.8
Partial List of Materials That Can Receive Additives or Annealing Treatments to Achieve Extrinsic Semiconduction

n-Type

TiO_2	Nb_2O_5	CdS	Cs_2Se	$BaTiO_3$	Hg_2S
V_2O_5	MoO_2	CdSe	BaO	$PbCrO_4$	ZnF_2
U_3O_8	CdO	SnO_2	Ta_2O_5	Fe_3O_4	
ZnO	Ag_2S	Cs_2S	WO_3		

p-Type

Ag_2O	CoO	Cu_2O	SnS	Bi_2Te_3	MoO_2
Cr_2O_3	SnO	Cu_2S	Sb_2S_3	Te	Hg_2O
MnO	NiO	Pr_2O_3	CuI	Se	

Amphoteric

Al_2O_3	SiC	PbTe	Si	Ti_2S
Mn_3O_4	PbS	UO_2	Ge	
Co_3O_4	PbSe	IrO_2	Sn	

Source: Kingery, W.D., Bowen, H.K., and Uhlmann, D.R., *Introduction to Ceramics*, 2nd ed., Wiley, New York, 1976, Chap. 17, p. 890.

resulting in a p-type semiconductor. Another example where a deficit in the metal content results in p-type conduction is $Cu_{2-x}O$. Excess metal (or oxygen shortage) results in an n-type semiconductor. An example is $Zn_{1+x}O$. This type of nonstoichiometric defect structure is usually achieved by annealing the stoichiometric ceramic in a reducing atmosphere.

Table 10.8 lists examples of materials that can receive additives or annealing treatments to achieve extrinsic semiconduction. The term *amphoteric* means that either the n-type or p-type can be achieved, depending on the addition or treatment.

TABLE 10.9
Some Semiconductor Applications of Ceramics

Device or Application	Base Material with Suitable Doping
Rectifier	Cu_2O
Thermistor	$BaTiO_3$, $Fe_3O_4/MgAl_2O_4$ solid solution
Thermal switch	VO_2, V_2O_3
Solar cell	Si, CdS, InP, TiO_2, $SrTiO_3$
Varistor	ZnO, SiC
Electrode	Ti_4O_7, $LaCrO_3$, $LaMnO_3$, $La_{0.5}FeO_3$
Heating element	SiC, $MoSi_2$, graphite
Sensors	ZnO, SnO_2, Fe_2O_3

One other category of material has especially interesting electrical behavior that can range from semiconductor to metallic. This category includes forms of carbon with the graphene structure. While diamond has sp^3 bonding to form a three-dimensional framework structure with a relatively wide band gap, graphite has sp^2 bonds to form a planar (sheet) graphene structure with zero band gap within the plane. Polycrystalline graphite can range from semiconductor behavior to near-metallic conduction behavior based on the orientation of the microstructure, grain size, and other factors. This allows grades of graphite to be used for the conductive electrodes in aluminum electrolytic cells and many other applications and also for semiconducting heating elements. Another exciting option is carbon nanotubes. Carbon nanotubes consist of a graphine sheet rolled into a tube or a series of concentric tubes. This can occur with various geometries, some with zero band gap and others with semiconductor band gaps.

10.6.2 Applications of Ceramic Semiconductors

Ceramics are used for a variety of semiconductor devices that have numerous applications. Examples are listed in Table 10.9. A *rectifier* is a device that allows current to flow only in one direction and can thus convert alternating current to direct current. Rectifiers have many applications. A *thermistor* is a device that has a controlled variation in electrical resistance as a function of temperature.[15] Thermistors are used in temperature sensors, temperature compensators, switches, infrared sensors, and heater systems. A *varistor* is a device whose resistance varies, depending on the applied voltage.[16] It has high resistance at low voltage and low resistance at high voltage. One use for varistors is for voltage-surge protection. The varistor is wired in the circuit between line and ground. If a high-voltage surge occurs, the high current selectively passes through the varistor to ground rather than through the circuit of the electrical apparatus.

Electrodes are the positive and negative terminals of a galvanic (electrochemical) cell. They must have adequate conductivity to transfer electrical charge, but be corrosion-resistant in the chemical and thermal environment of the cell and often must have catalytic capabilities. Semiconductive ceramic electrodes are used for fuel cells, some batteries, photoelectrolysis, chloralkali cells, and electrochlorination.

Electrochlorination is used for water purification, chlorination of swimming pools, and disinfecting sea water used for cooling at power stations or on large ships. A small electric current applied through a ceramic electrode causes decomposition of water and the dissolved sodium chloride adjacent to the electrode to yield hypochlorous and/or hypochlorite ions. These ions have strong disinfecting capabilities. The *chloralkali process* involves the electrolysis of NaCl to produce chlorine, hydrogen, hypochlorite, and NaOH (caustic soda). Currently, electrodes for the chloralkali process are graphite or titanium-coated, ruthenium-doped TiO_2.

A major industrial application of semiconducting ceramics is for *heating elements*. Electricity must work to pass through a semiconductor. This work results in heat. By control of the resistivity of the material, the cross section of the heating element, and the applied voltage, a wide range of operating conditions can be achieved. Ceramic heating elements are used for industrial heating, glass melting, a wide range of ceramics processing, stove- and oven-heating elements, and natural gas igniters.

Igniters for home appliances were developed in the late 1960s to help meet new energy conservation laws that were passed in California. The ceramic igniters were developed for use in home appliances such as gas clothes dryers and gas stoves to replace the pilot light. The igniters are safer and save on fuel consumption. Prior to development of these igniters for home appliances, approximately 35 to 40% of the gas used was wasted by the pilot light. It has been estimated that this wastage for appliances in the United States was over 6 million cubic feet of natural gas per day prior to widespread installation of igniters. Initial igniters were developed for gas clothes dryers and became available on a limited basis in 1968. In 1980 about 750,000 per year were being manufactured. Marketing of igniters for gas ranges started in 1974 and reached a volume of about 1,700,000 per year by about 1980.

10.7 SUPERCONDUCTIVITY

Materials normally have some resistance to the motion of charge carriers when an electric field is applied. This is even true of the best metallic conductors such as copper, aluminum, and silver. The charge carriers interact with other charge carriers and with the atoms in the structure. As we discussed at the beginning of this chapter, this is referred to as electrical resistance (R) and is related to the current (I) and the voltage (V) by the equation $V = IR$. Electrical resistance has three adverse effects on electrical devices and on transmission of electricity: (1) it consumes a portion of the electrical energy and results in decreased efficiency; (2) it results in heat generation that limits some applications or causes extra design sophistication to dissipate the heat; and (3) it slows down the response time of an electrical circuit, which is a critical limitation in the advancement of high-speed computers and other electrical devices.

An exciting phenomenon was discovered in 1911 by a Dutch scientist, Heike Kamerlingh-Onnes. He believed that the major cause of electrical resistance was thermal vibrations and that a decrease in temperature to absolute zero (0 K or $-460°F$) where thermal vibrations are zero would result in zero electrical resistance. He cooled high-purity mercury with liquid helium. The electrical resistivity decreased slowly as the temperature was decreased and then abruptly dropped to zero at about 4.2 K ($-452°F$). This was the first time that zero electrical resistance was observed. Kamerlingh-Onnes called the phenomena supraconductivity, which later evolved into the present term *superconductivity*. The transition temperature at which the resistance becomes zero is referred to as the *critical temperature* (T_c).

The broad implications of superconductivity were immediately recognized. Onnes received the Nobel prize in 1913 for his discovery. A zero-resistance material would make possible a large increase in the efficiency of transmission of electricity, permit development of extremely powerful electromagnets, allow easy storage of electricity, and would open up a wide variety of directions for new electrical applications that were previously not possible. However, the extremely low T_c required cooling with liquid helium. Liquid helium was generally not available in 1911. It had to be prepared in the laboratory by a very tedious procedure that was only known to a few individuals. As a result, commercial application of superconductivity was not feasible. Even the conduct of research was highly restricted because of the difficulty of liquid helium preparation and because several individuals had a monopoly on the liquid helium supply. Liquid helium did not become generally available until the 1950s.

Progress has been slow in the development of superconductivity. Research has been directed toward achieving an understanding of the mechanism of superconductivity, identifying materials that have superconductive behavior, and attempting to increase the temperature of superconductivity.

10.7.1 MECHANISM OF SUPERCONDUCTIVITY

Many theories of the mechanism of superconductivity were hypothesized and explored. Finally, in 1957 a theory was presented that adequately accounted for the major aspects of superconductivity of materials studied up to that time. The theory was devised by John Bardeen, Leon Cooper, and J. Robert Schrieffer and is called the *BCS theory*. They won the Nobel Prize for Physics in 1972 for this theory. Simply stated, when the thermal vibrations become low enough, electrons team together in pairs that are able to travel through the material under an electric field without interacting or colliding with atoms or electrons. The pairs all move in phase with other pairs. This is illustrated in Figure 10.21 in comparison with the scattering that occurs for normal conduction. The difference between normal conduction and superconduction can be likened to a large crowd of people. If there is no organization and if each person in the crowd is moving individually and at random, the overall crowd moves very slowly, even if everyone is moving roughly in the same direction. This is analogous to the case of normal conduction. However, if the crowd is organized and all individuals are in step moving in a single direction, the overall crowd moves smoothly and swiftly. This is analogous to superconductivity.

10.7.2 CHARACTERISTICS OF SUPERCONDUCTIVITY

Zero electrical resistance is only one of the important characteristics of a superconductor. A second important characteristic is *diamagnetism*. Diamagnetism involves the way a magnetic field interacts with a material. A nondiamagnetic material will not be affected by a magnetic field. The lines of magnetic force will penetrate the material as if it were not there. This is illustrated in Figure 10.22a. In a diamagnetic material, the magnetic field does not penetrate, but is repelled. This is illustrated in Figure 10.22b. Diamagnetism tied to zero electrical resistance in a superconductor results in special behavior referred to as the *Meissner effect*. A magnet placed over the surface of the superconductor levitates, that is, is suspended in midair. The magnetic field of the magnet cannot penetrate the diamagnetic superconductor. Instead, it induces a current in the zero resistance superconductor that produces a mirror image magnetic field of the magnet. Because the two fields are coupled, stable levitation results. As long as the material stays cool enough to be superconductive, the magnet remains suspended. Figure 10.23 shows a magnet levitated over a ceramic superconductor material.

The Meissner effect is a quick and decisive means to determine if a material or a portion of a material is superconductive. It is easier and more reliable than only conducting an electrical resistance measurement. Sophisticated equipment and careful test procedures are required to accurately detect at cryogenic temperatures whether a sample has zero resistance or only low resistance.

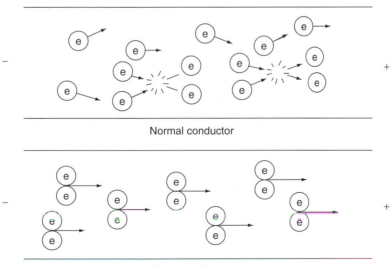

FIGURE 10.21 Schematic illustrating the difference, according to the BCS theory, between normal conduction and zero-resistance superconduction.

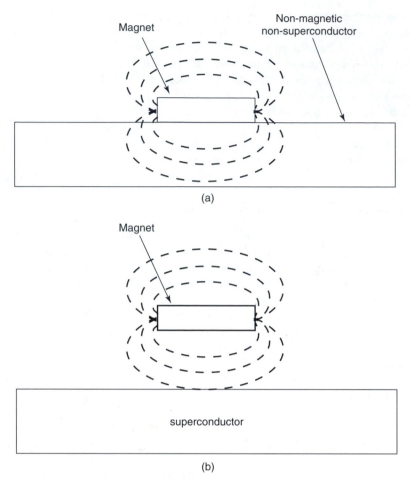

(a)

(b)

FIGURE 10.22 Comparison between the interaction of a magnet with a nondiamagnetic material (a) and a diamagnetic superconductor (b).

Other characteristics of superconductive materials are also important. The transition temperature T_c is particularly important. The higher the T_c, the easier it is to cool the material to the superconductive state and to maintain it at that temperature. Until 1987 liquid helium was required. In 1987 a ceramic composition was discovered that had T_c of about 90 K ($-298°F$) which is above the temperature of liquid nitrogen (77 K or $-321°F$). The significance of this is discussed later in this chapter.

The response of the superconductive material to the amount of current being carried or to an applied magnetic field is also very important. Too high a current density or magnetic field can destroy the superconductive behavior. Each material has a different response.

10.7.3 EVOLUTION OF SUPERCONDUCTOR MATERIALS

Figure 10.24 shows the historical progression in discovery of superconductive materials with higher T_c. Progress was extremely slow up to 1986, averaging about 4 K per decade. Initial materials identified to be superconductive were elemental metals (Hg, Pb, Nb), followed primarily by solid solutions (NbTi) and intermetallics (Nb_3Sn, V_3Si, Nb_3Ge). Until the early 1960s relatively few materials had been identified with superconductive behavior. Superconductivity was thought to be an anomalous property. Since 1960, techniques have been available to achieve temperatures closer to absolute zero (on the order of 0.0002 K) and to simultaneously apply high pressure. Under these conditions many

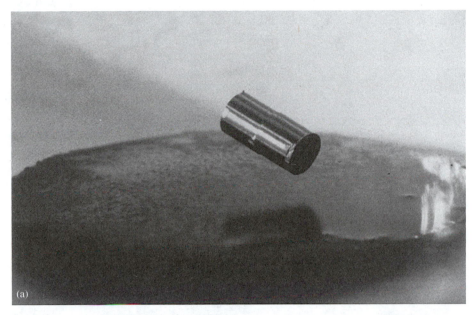

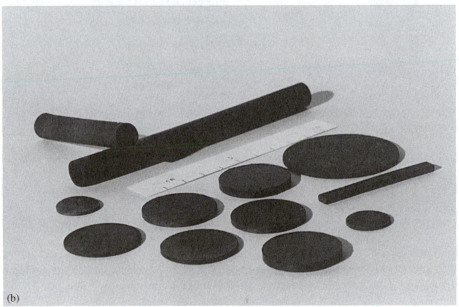

FIGURE 10.23 Example of the Meissner effect showing the levitation of a magnet at liquid nitrogen temperature by $YBa_2Cu_3O_{7-x}$ ceramic superconductor. (Courtesy of Ceramatec, Inc.)

more elements, solid solutions, intermetallics, and ceramics have been demonstrated to have superconductivity.

Several ceramic compositions were identified to be superconductive. These included tungsten, molybdenum, and rhenium "bronze" compositions A_xWO_3, A_xMoO_3, and A_xRhO_3, where A was Na, K, Rb, Cs, NH_4, Ca, Sr, Ba, etc.; oxygen-deficient $SrTiO_3$ and $LiTiO_3$; and $BaPb_{1-x}Bi_xO_3$. The bronzes had T_c up to 6 K ($-449°F$). All of these ceramic compositions are derivatives of the perovskite structure [ABO_3].

Until 1986 the highest T_c achieved was 23 K ($-419°F$). Liquid helium was still required for cooling. Then a major breakthrough was achieved by Bednorz and Mueller[17] of IBM in Zurich. They

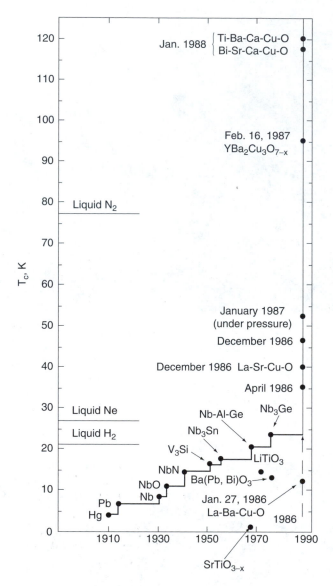

FIGURE 10.24   Progress in the discovery of superconductive materials with increased T_c. (From Clarke, D.R., *Special Issue of Advanced Ceramic Materials*, 2, 3B, 288, July 1987. With permission.)

reported in April 1986 that a La–Ba–Cu–O composition was superconductive at 35 K ($-397°$F). By December 1986 this was verified by researchers in the U.S., Japan, and China and a composition with strontium substitution for barium was reported to be superconductive at 40 K ($-388°$F). Chu and coworkers at the University of Houston applied high pressure to the La–Sr–Cu–O composition and found that the T_c increased to 52 K ($-366°$F). Shortly thereafter, in February 1987, they announced that substitution of yttrium for lanthanum in the La–Ba–Cu–O material resulted in transition to superconductivity at 92 K ($-294°$F). Similar results were independently reported by the Institute of Physics in Beijing and the Indian Institute of Sciences in Bangalore. This was a major and dramatic breakthrough. For the first time (after over 70 years of searching) a material exhibited superconductivity at a temperature where cooling could be achieved without the requirement for liquid helium. This new material could be cooled at liquid nitrogen temperature (77 K or $-321°$F).

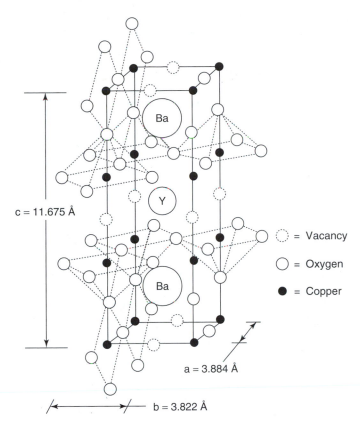

FIGURE 10.25 Unit cell of orthorhombic ceramic superconductor $YBa_2Cu_3O_{7-x}$. (From Clarke, D.R., *Special Issue of Advanced Ceramic Materials*, 2, 3B, 289, July 1987. With permission.)

Since 1987 extensive work has been conducted on a wide range of aspects of high-temperature superconductors: new compositions with higher T_c, increased stability in a magnetic field, thin film and oriented structures, fabrication (including long length flexible wire and tape), and potential applications.[18-20] Some compositions in the families Bi–Sr–Ca–Cu–O and Tl–Ca–Ba–Cu–O have exhibited T_c around 120 K ($-244°F$).

10.7.4 STRUCTURE OF HIGH T_c CERAMIC SUPERCONDUCTORS

The initial ceramics samples that exhibited superconductivity around 90 K ($-298°F$) were determined to be multiphase. The superconductive phase was determined to be of an orthorhombic defect perovskite structure with the composition $Y_1Ba_2Cu_3O_{7-x}$. Because of the Y:Ba:Cu ratio of 1:2:3, this material is typically referred to as the 1:2:3 phase. The structure of a unit cell is illustrated in Figure 10.25. Note that each copper ion is at the center of an octahedron of oxygen ions, except that one oxygen position is vacant. The oxygen vacancies are charge-compensated by the copper ions such that Cu^2 and Cu^3 states are both thought to be present. The mixed valence of the copper ions, the ordering of the oxygen vacancies, and the combination of the copper and oxygen ions are all believed to be important in determining the superconductive behavior.

10.7.5 CHARACTERISTICS OF THE 1:2:3 CERAMIC SUPERCONDUCTOR

The most beneficial characteristic of the 1:2:3 ceramic phase is the temperature vs. electrical resistance curve. This is illustrated in Figure 10.26. The resistance drops abruptly to zero at the T_c of about

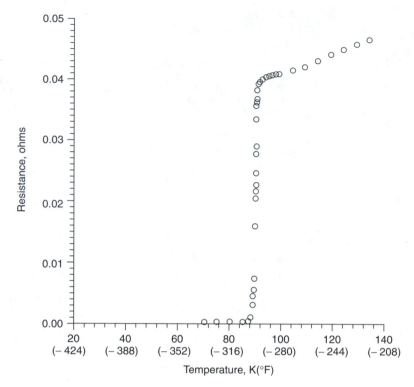

FIGURE 10.26   Change in electrical resistance vs. temperature for $YBa_2Cu_3O_{7/x}$ shows T_c at about 90 K. (From Robinson, Q., et al., *Special Issue of Advanced Ceramic Materials*, 2, 3B, 384, July 1987. With permission.)

90 K ($-298°F$). This allows the material to be cooled below T_c with liquid nitrogen. Liquid nitrogen is abundant, inexpensive, easy to store, and easy to use. The ability to cool with liquid nitrogen represents a tremendous advantage over prior superconductors and raises hopes for achieving applications not previously considered viable economically. However, the 1:2:3 material has many deficiencies that must be resolved before widespread applications are achieved:

1. Low current-carrying capability compared to Nb_3Sn and NbTi superconductors. A current density of 10^7 amps/cm² has been demonstrated for Nb_3Sn at 4.2 K ($-452°F$). Bulk, randomly oriented 1987-vintage 1:2:3 material had a current-carrying capacity of only about 10^3 amps/cm². Some improvement has been achieved for thin films and single crystals. A thin film applied to a polycrystalline substrate had a critical current density of 3×10^4 amps/cm² and one grown epitaxially on a single crystal was $>10^5$ amps/cm².
2. The 1:2:3 material progressively loses its diamagnetic characteristics as the applied magnetic field increases above a critical value, that is, the magnetic flux can penetrate the bulk material at high levels of magnetic field. This will restrict the use of 1:2:3 material in high-magnetic-field applications.
3. Low strength and brittle fracture characteristics.
4. Difficulty of synthesis and fabrication due to the requirement of careful control of oxygen stoichiometry to achieve optimum T_c.
5. Sensitivity to moisture.
6. Difficulty of fabrication of bulk shapes and specialized configurations such as wire.

Extensive research and development continues in an effort to resolve the deficiencies of the 1:2:3 material and alternate materials and to achieve still higher-temperature T_c. The field of study is wide open, and it is likely that applications for high-temperature ceramic superconductors will result.

10.7.6 APPLICATIONS OF SUPERCONDUCTORS[21]

The total sale of superconductive materials in 1986 was estimated to be about $20 million. Most of this was NbTi or Nb_3Sn wire and tape used for magnets. The major magnet application was medical magnetic resonance imaging. Minor magnet applications included laboratory magnets and high-energy physics. Other applications included SQUIDS (superconductive quantum interference devices) and shielding. Predicted future applications of helium-cooled low-temperature superconduction include Josephson junction devices, magnetometers, instrumentation, levitated trains, the superconducting supercollider, magnetic separation, magnetic confinement fusion, magnetic launchers, and magneto-hydrodynamics. The new, high-temperature ceramic superconductors raise hopes for reduced cost of the prior mentioned applications and for additional applications such as antenna, microwave devices, motors, magnetic shielding, power transmission, energy storage, and interconnects for computers.

REFERENCES

1. Kingery, W.D., Bowen, H.K., and Uhlmann, D.R., *Introduction to Ceramics*, 2nd ed., Wiley, New York, 1976, Chap. 17.
2. van Vlack, L.H., *Elements of Materials Science*, 2nd ed., Addison-Wesley, Reading, Mass., 1964, p. 420.
3. van Vlack, L.H., *Physical Ceramics for Engineers*, Addison-Wesley, Reading, Mass., 1964, Chap. 9.
4. Kirk-Othmer, *Encyclopedia of Chemical Technology*, Vol. 4, 3rd ed. Mark, H.F., Othmer, D.F., Overberger, C.G., and Seaborg, G.T., Eds., Wiley, New York, 1978.
5. Shriver, D.F. and Farrington, G.C., *Solid Ionic Conductors*, C&EN, 42–44, 50–57, May 20, 1985.
6. Farrington, G.C. and Briant, J.L., A general review of solid electrolytes and beta-alumina, *Science* 204, 1371, 1979.
7. Heuer, A.H. and Hobbs, L.W., Eds., *Advances in Ceramics, Vol. 3, Science and Technology of Zirconia*, American Ceramic Society, Ohio, 1981.
8. Subbarao, E.C., Ed., *Solid Electrolytes and Their Applications*, Plenum, 1980.
9. Appleby, A.J. and Foulkes, F.R., *Fuel Cell Handbook*, Van Nostrand Reinhold, New York, 1989.
10. Fuel Cell Handbook, 6th Edition, U.S. Department of Energy National Energy Technology Laboratory, Pittsburgh and Morgantown, November 2002.
11. Steele, B.C.H., Ed., *Ceramic Oxygen Ion Conductors and the Technological Applications*, British Ceramic Proc. No. 56, The Institute of Materials, London, 1996.
12. Kaner, R.B. and MacDiarmid, A.G., Plastics that conduct electricity, *Sci. Am.*, 258(2), 106–111, Feb. 1988.
13. Schwartz, B., Microelectronics packaging: II, *Bull. Am. Ceram. Soc.*, 63(4), 577–581, 1984.
14. Tummala, R.R., Ceramics in microelectronic packaging, *Am. Ceram. Soc. Bull.*, 67(4), 752–758, 1988.
15. Ketron, L., Ceramic sensors, *Am. Ceram. Soc. Bull.*, 68(4), 860–865, 1989.
16. Levinson, L.M., Advances in varistor technology, *Am. Ceram. Soc. Bull.*, 68(4), 866–868, 1989.
17. Bednorz, J.G. and Mueller, K.A., Perovskite-type oxides—the new approach to high T_c superconductivity, *Reviews of Modern Physics*, 60(3), 585–600.
18. Nair, K.M., Balachandran, U., Chiang, Y.M., and Bhalla, A.S., Eds., *Superconductivity and Ceramic Superconductors II, Ceramic Transactions* Vol. 18, The American Ceramic society, Westerville, Ohio, 1991.
19. Goyal, A., Wong-Ng, W., Murakami, M., and Driscoll, J., Eds., *Processing of High-Temperature Superconductors, Ceramic Transactions* Vol. 140, American Ceramic Society, Westerville, Ohio, 2003.
20. Goyal, A., Wong-Ng, W., Freybardt, H.C., and Matsumoto, K., Eds., *Fabrication of Long-Length and Bulk High Temperature Superconductors, Ceramic Transactions* Vol. 149, The American Ceramic Society, Westerville, Ohio, 2004.
21. Malozemoff, A.P., Gallagher, W.J., and Schwall, R.E., Applications of high temperature superconductors, ACS Symposium Series 351, *Chemistry of High Temperature Superconductors*, American Chemical Society, Washington, D.C., 1987.

ADDITIONAL RECOMMENDED READING

1. Nair, K.M. and Bhalla, A.S., Eds., *Electronic Ceramic Materials and Devices, Ceramic Transactions Vol. 106*, The American Ceramic Society, Westerville, Ohio, 2000.
2. Hummel, R.E., *Electronic Properties of Materials*, Springer, New York, 2001.
3. Knauth, P., Tarascon, J-M., Traversa, E., and Tuller, H.L., Eds., *Solid-State Ionics 2002*, MRS Vol. 756, Materials Research Society, 2003.
4. Buchanan, R.C., *Ceramic Materials for Electronics*, Marcel Dekker, New York, 2003.

PROBLEMS

1. A furnace with a ceramic heating element operates on 110 V and draws a current of 8 A. What is the resistance?
2. Which of the following is *not* an electrical insulator?
   (a) Al_2O_3
   (b) graphite
   (c) nylon
   (d) Si_3N_4
3. A material exhibits the Meissner effect. What is this material?
   (a) insulator
   (b) conductor
   (c) semiconductor
   (d) superconductor
4. Which of the following materials has the largest band gap at room temperature?
   (a) diamond
   (b) Si
   (c) Ge
   (d) Al_2O_3
5. Which of the following can be either an *n*-type or *p*-type extrinsic semiconductor depending on the dopant?
   (a) ZnO
   (b) SiC
   (c) Cu_2O
   (d) Fe_3O_4
6. A doped SiC igniter with a radius of 4 mm (0.16 in.) and a length of 3 cm (1.2 in.) is heated from room temperature to 1000° (1830°F) in 7 sec at 110 V and 3 A. What is the resistivity of this material?
7. Which of the following ceramics does *not* exhibit oxygen ion conduction?
   (a) Ca-doped ZrO_2
   (b) beta-alumina
   (c) Gd-doped CeO_2
   (d) δ-Bi_2O_3
8. Which material will decrease in resistivity as the temperature is increased?
   (a) Cu
   (b) superconductor
   (c) Al_2O_3
   (d) TiO
9. Which of the following applications is *not* based on ionic conduction?
   (a) solid oxide fuel cell
   (b) sodium-sulfur battery
   (c) zirconia oxygen sensor
   (d) thermistor

STUDY GUIDE

1.   What factors determine the electrical conductivity of a material?
2.   Explain the role of energy bands in the conduction of electrons in a metal.
3.   Explain how the energy bands are different for most ceramics compared to metals.
4.   Explain how conduction by electrons can occur in some ceramic materials. How many electrons can be available for conduction? Give some examples of these ceramic materials.
5.   What is the "semiconductor-to-metal transition" and how might it be used for a practical product?
6.   What conditions must be met for an ion to move through a ceramic and provide ionic conductivity?
7.   Identify and define four types of point defects that can be present in a crystal structure.
8.   Explain how the degree of ionic conductivity can be engineered in zirconium oxide.
9.   Explain how a zirconia oxygen sensor works.
10. Describe the role of ionic conduction in a solid oxide fuel cell.
11. Polymers are normally very effective electrical insulators. Explain how electrical conductivity has been achieved in polymers.
12. Explain why most ceramics and polymers are electrical insulators.
13. Identify three key reasons ceramics are important for "packages" for silicon chips for integrated circuits.
14. Describe the complexity of a hybrid ceramic-metal package for a mainframe computer application.
15. How does a semiconductor differ from a conductor and an insulator?
16. Explain the difference between "intrinsic" and "extrinsic" semiconductors and identify examples of each.
17. Describe the function of each of the following and list one or more ceramic examples for each: Rectifier, Varistor, Thermistor, Thermal Switch, Heating Element, Solar Cell.
18. What have been the benefits of replacing pilot lights with ceramic igniters?
19. What is superconductivity, and what benefits could its application make possible?
20. Explain why the ceramic superconductor materials discovered since 1986 are so exciting and have opened up new opportunities.

11 Dielectric, Magnetic, and Optical Behavior*

Chapter 10 discussed the difference between electrical conduction, semiconduction, superconduction, and insulation. This chapter continues the discussion of electrical properties, followed by a review of magnetic and optical properties. Specific topics include dielectric polarization, dielectric constant, dielectric loss, dielectric strength, capacitors, ferroelectricity, pyroelectricity, piezoelectricity, ferromagnetism, ferrimagnetism, paramagnetism, electromagnetic spectrum, absorption, transparency, index of refraction, color, phosphorescence, and lasers. Although some of these terms sound difficult and mysterious, they are really quite simple to understand.

11.1 DIELECTRIC PROPERTIES

Ceramic materials that are good electrical insulators are referred to as *dielectric materials*. Although these materials do not conduct electric current when an electric field is applied, they are not inert to the electric field. The field causes a slight shift in the balance of charge within the material to form an electrical dipole, thus the source of the term "dielectric." This is illustrated in Figure 11.1.

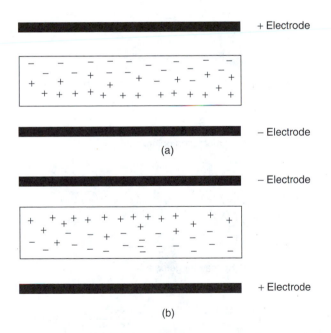

FIGURE 11.1   Shift in the distribution of charge in a ceramic insulator when it is placed in an electric field between two electrically conductive electrodes. (Drawing courtesy of ASM International.)

---

* Much of this chapter is from D.W. Richerson, *Introduction to Modern Ceramics*, Lesson 6, ASM Materials Engineering Institute Home Study Course 56, ASM International, Materials Park, OH, 1990, which was prepared in collaboration with preparation of the 2nd edition of *Modern Ceramic Engineering*.

11.1.1 POLARIZATION

The formation of an electrical dipole is called *polarization*. There are several contributors to polarization: (1) electronic polarization; (2) orientation (dipolar) polarization; (3) space charge polarization; and (4) atomic or ionic polarization. These are illustrated schematically in Figure 11.2.[1]

Electronic polarization occurs in all dielectric materials. The electrons surrounding each nucleus are shifted very slightly in the direction of the positive electrode and the nucleus is very slightly shifted in the direction of the negative electrode. As soon as the electric field is removed, the electrons and nuclei return to their original distributions and the polarization disappears. The effect is analogous to elastic stress and strain that we studied in Chapter 8. The displacement of charge is very small for electronic polarization, so the total amount of polarization is small compared to the other mechanisms of polarization.

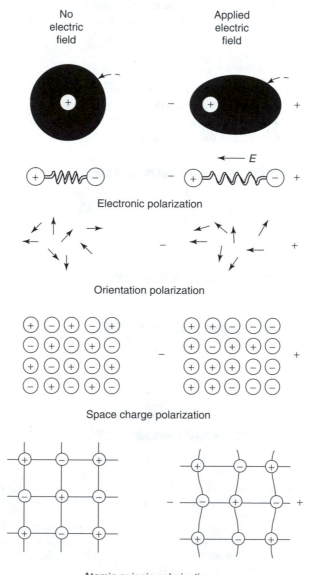

FIGURE 11.2 Schematic representation of different mechanisms of polarization. (From von Hippel, A.R., *Dielectrics and Waves*, Wiley, New York, 1954. With permission.)

Orientation polarization involves nonsymmetrical molecules that contain permanent electric dipoles. An example is H_2O. The covalent bonds between the hydrogen and oxygen atoms are directional such that the two hydrogens are on one side of the oxygen. The hydrogen side of the molecule has a net positive charge and the oxygen side has a net negative charge. Under an electric field, the molecules will align with the positive side facing the negative electrode and the negative side facing the positive electrode. Other examples of asymmetrical molecules that result in orientation polarization included HCl, CH_3Br, HF, and $C_2H_5(NO_2)$. Orientation polarization results in a much higher degree of polarization than electronic polarization. This is because larger charge displacement is possible in the relatively large molecules compared to the spacing between the electrons and nucleus in individual atoms.

The third source of polarization is *space charge polarization*. Space charges are random charges caused by cosmic radiation, thermal deterioration, or are trapped in the material during the fabrication process.

The fourth type of polarization is *atomic* or *ionic polarization*. It involves displacement of atoms or ions within a crystal structure when an electric field is applied. A wide range of polarization effects is possible through this mechanism, depending on the crystal structure, the presence of solid solution, and other factors. Much of this chapter addresses atomic and ionic polarization phenomena. Examples include pyroelectricity, piezoelectricity, and ferroelectricity. Before we study these topics, though, let us explore some of the other basic characteristics of dielectrics and define terms such as the dielectric constant, dielectric strength, dielectric loss, and capacitance.

11.1.2 DIELECTRIC CONSTANT

The degree of polarizability or charge storage capability of a material is identified by the term *relative dielectric constant (K')*. This is also referred to as relative permittivity. The concept of the relative dielectric constant is illustrated in Figure 11.3. An electric field is applied to two flat plates of a metal such that one plate becomes positive and the other negative. The electric field causes polarization in the material in the space between the conductive plates. The relative dielectric constant compares the polarizability or charge storage capability of the material with that of a vacuum between the plates:

$$K' = \frac{K_{\text{material}}}{K_{\text{vacuum}}} = \frac{E^*}{E_0} \tag{11.1}$$

E^* equals the permittivity of the material and E_0 equals the permittivity of vacuum (8.85×10^{-14} farads/cm).

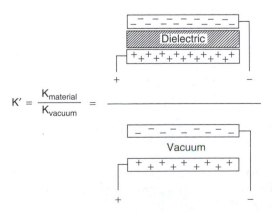

FIGURE 11.3 Schematic illustration of the definition of the relative dielectric constant K'. (Drawing courtesy of ASM International.)

K' for a vacuum is 1.0. K' for dry air at 0°C and standard atmospheric pressure is 1.0006. A material that has three times the polarizability of a vacuum has $K' = 3$. Similarly, a material with 100 times the polarizability of a vacuum has $K' = 100$. Table 11.1 lists the relative dielectric constant for a variety of materials.

Note the wide range in values of dielectric constant for different materials. Materials with low dielectric constant are used for electrical insulator applications. For example, Al_2O_3, BeO, AlN, and some polymers are used for insulation in integrated circuits (substrates and packages). Rubber and other flexible polymers are used for insulation on electrical wires. Electrical porcelain, forsterite, and steatite are used as electrical insulators. Materials with high dielectric constant are used in capacitors for charge storage and other functions. Capacitors and their functions are discussed later in this chapter.

The dielectric constant is affected by temperature. The nature of the effect is dependent on the source of polarization. Electronic polarization is relatively insensitive to temperature, so temperature has little effect on the dielectric constant. Molecular orientation polarization is opposed by thermal agitation, so the dielectric constant goes down as the temperature increases. Atomic and ionic

TABLE 11.1
Relative Dielectric Constant for a Variety of Materials at Room Temperature

Material	K'
NaCl	5.9
LiF	9.0
KBr	4.9
Mica	2.5–7.3
MgO	9.6
BaO	34
BeO	6.5
Diamond	5.7
Al_2O_3	8.6–10.6
Mullite	6.6
TiO_2	15–170
Cordierite	4.5–5.4
Porcelain	6.0–8.0
Forsterite (Mg_2SiO_4)	6.2
Fused SiO_2	3.8
Steatite	5.5–7.5
High-lead glass	19.0
Soda-lime-silica glass	6.9
Zircon	8.8
$BaTiO_3$	1600
$BaTiO_3$ + 10% $CaZrO_3$ + 1% $MgZrO_3$	5000
$BaTiO_3$ + 10% $CaZrO_3$ + 10% $SrTiO_3$	9500
Paraffin	2.0–2.5
Beeswax	2.7–3.0
Rubber, polystyrene, polyacrylates, polyethylene	2.0c3.5
Phenolic	7.5

Source: Compiled from *Handbook of Chemistry and Physics*, 63rd ed., CRC Press, Boca Raton, Fla., 1982; Kingery, W.D., Bowen, H.K., and Uhlmann, D.R., *Introduction to Ceramics*, 2nd ed., Wiley, New York, 1976, Chaps. 17–19; Thomas, D.G., in *Physics of Electronic Ceramics*, Part B, Hench, L.L. and Dove, D.B., Eds., Marcel Dekker, New York, 1972, pp. 1057–1090; Henry, E.C., *Electronic Ceramics*, Doubleday, New York, 1969.

polarization tends to increase with temperature due to an increase in charge carriers and ion mobility. The dielectric constants of BeO and some Al_2O_3-based compositions are shown as a function of temperature in Figure 11.4. The increase in dielectric constant suggests that atomic and ionic polarization is the dominant mechanism.

The dielectric constant is also affected by the frequency of the applied electric field or the frequency of other electromagnetic fields impinging on the material. The polarization requires time to respond to an applied field. Electronic polarization occurs very rapidly and is present even at high frequencies. For example, visible light is of relatively high frequency (10^{15} cycles per second) and has an electric interaction with the electronic polarization of a dielectric. This is discussed later in this chapter under optical properties. Molecular polarization is only affected by low frequencies. In a high-frequency field, the molecules do not have time to realign with each cycle. Examples of the effect of frequency and the combined effect of frequency and temperature for several materials are illustrated in Figure 11.5 through 11.7.

11.1.3 DIELECTRIC STRENGTH

A second important dielectric property is dielectric strength. Dielectric strength is the capability of the material to withstand an electric field without breaking down and allowing electrical current to pass. It has units of volts per unit of thickness of the dielectric material. Volts per mil (thousandth of an inch) or volts per centimeter are often used. Table 11.2 lists values of dielectric strength for a range of materials. Note that some organic materials have very high dielectric strength. A phenolic resin has a value of about 2000 V/mil, which means that a thin film 0.001 in. thick can block the

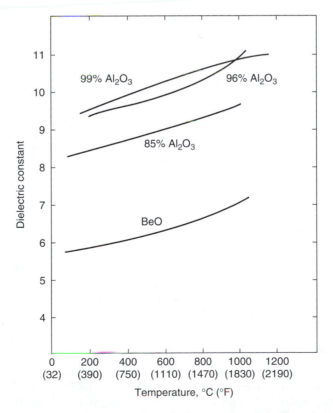

FIGURE 11.4 Relative dielectric constant of BeO- and Al_2O_3-based compositions measured at 4 GHz. (After Comeforo, J.E., *The Electrical Engineer*, 26(4), 82–87, 1967.)

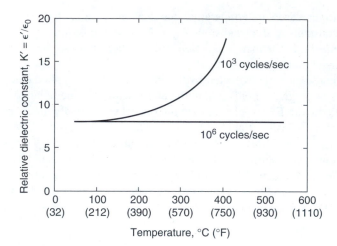

FIGURE 11.5 Effect of frequency and temperature on the dielectric constant of an Al_2O_3 crystal with the electric field aligned normal to the *c* axis. (From Kingery, W.D., Bowen, H.K., and Uhlmann, D.R., *Introduction to Ceramics*, 2nd ed., Wiley, New York, 1976, Chaps. 17–19, p. 936. With permission.)

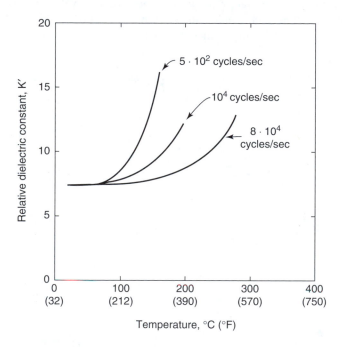

FIGURE 11.6 Effect of frequency and temperature on the dielectric constant of a soda-lime-silica glass. (From Strutt, M.J.O., *Arch. Elektrotech.*, 25, 715, 1931. With permission.)

passage of an electric current under an applied voltage gradient of 2000 V. This is one of the reasons organic materials such as rubber and phenolic resin are in widespread use as electrical insulation. Another interesting high-dielectric-strength material identified in Table 11.2 is mica. Mica is a naturally occurring silicate composition with a layered structure with pronounced cleavage between the layers. Thin flexible sheets can be peeled off of single crystals of mica. Dielectric strength as high as 500 V/mil has been measured for a mica material.

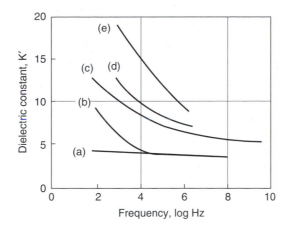

FIGURE 11.7   Dielectric constant vs. frequency for various ceramic materials. (a) Fused SiO_2 at 100°C (212°F), (b) fused SiO_2 at 400°C (750°F), (c) AlSiMag A-35™ at 150°C (300°F), (d) ZrO_2 porcelain, and (e) Al_2O_3. (From van Vlack, L.H., *Elements of Materials Science*, 2nd ed., Addison-Wesley, Reading, MA, 1964, p. 221. With permission.)

TABLE 11.2
Dielectric Strength of Various Materials

Material	Dielectric Strength, V/mil
Glass	
Fused silica	400, at 50°C
Borosilicate	
Soda-lime	
Soda-lead	
Ceramics	
Alumina	40–400
Beryllia	250–300
Rutile	50–300
Forsterite	Over 200
Cordierite	Over 200
Steatite	Over 200
Electrical porcelain	100–400
Zircon	Over 200
Titanates	100–300
Mica	125–5500
Organics	
Rubber	400–1200
Phenolic resin	2000

Source: Adapted from Henry, E.C., *Electronic Ceramics*, Doubleday, New York, 1969, p. 93.

11.1.4   DIELECTRIC LOSS

An ideal dielectric would allow no flow of electric charge, only a displacement of charge via polarization. If a thin plate of such an ideal material were placed between parallel plate electrodes to form a capacitor, and if an alternating (sine wave) electric field was applied, the current would lead

TABLE 11.3

Dielectric Loss for Some Ceramic Crystals and Glasses at 25°C (75°F) and 10^6 Hz

Material	Tan $\delta = \dfrac{K''}{K'}$
LiF	0.0002
MgO	0.0003
KBr	0.0002
NaCl	0.0002
TiO_2 (∥ c axis)	0.0016
TiO_2 (⊥ c axis)	0.0002
Al_2O_3 (∥ c axis)	0.0010
Al_2O_3 (⊥ c axis)	0.0010
BaO	0.001
KCl	0.0001
Diamond	0.0002
Mullite	–
Mg_2SiO_4 (forsterite)	0.0003
Fused silica glass	0.0001
Vycor ($96SiO_4$–$4B_2O_3$) glass	0.0008
Soda-lime-silica glass	0.01
High-lead glass	0.0057

Source: Adapted from Kingery, W.D., Bowen, H.K., and Uhlmann, D.R., *Introduction to Ceramics*, 2nd ed., Wiley, New York, 1976, Chaps. 17–19, p. 933.

the voltage by a phase angle of $\pi/2$ (i.e., 90°). Under this circumstance, no power would be absorbed by the dielectric and the capacitor would have zero loss. Real materials always have some loss. The phase angle between the current and voltage is not exactly 90°; the current lags slightly behind what it would be in an ideal dielectric. The angle of lag is defined as δ and the amount of lag becomes tan δ. Tan δ is referred to as the *loss tangent* and is equal to K''/K', where K' is the relative dielectric constant and K'' is defined as the relative loss factor. Table 11.3 lists values of loss tangent and dielectric constant for various examples of crystals and glasses. The dielectric loss and dielectric constant are often higher for polycrystalline materials.

Dielectric loss results from several mechanisms: (1) ion migration; (2) ion vibration and deformation; and (3) electronic polarization. The most important mechanism to most ceramics is ion migration. Ion migration is strongly affected by temperature and frequency. The losses due to ion migration increase at low frequencies and as the temperature increases. Several examples are given in Figure 11.8.

11.1.5 CAPACITANCE

Now that some dielectric terms have been defined, let us look back at Figure 11.3 and examine the characteristics of an important application of dielectrics. The configuration in Figure 11.3, which consists of a dielectric material between two electrically conductive electrodes, is called a *capacitor*. When placed in an electric circuit, the capacitor is able to store electrical charge. The higher the degree of polarizability of the dielectric material, the higher the relative dielectric constant and the more charge that can be stored.

The amount of charge (Q) that can be stored is equal to the applied voltage (V) times the capacitance (C):

$$Q = CV \qquad (11.2)$$

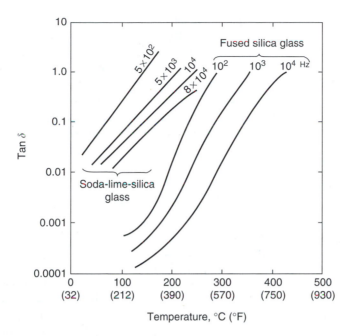

FIGURE 11.8   Effect of frequency and temperature on dielectric loss of fused silica and on soda-lime-silica glass composition. (From Kingery, W.D., Bowen, H.K., and Uhlmann, D.R., *Introduction to Ceramics*, 2nd ed., Wiley, New York, 1976, Chaps. 17–19, p. 939. With permission.)

The unit of charge Q is the *coulomb*, which is an ampere-second. The unit of capacitance C is the *farad*. A capacitor has a capacitance of one farad when one coulomb charges it to an electric potential of one volt. One farad is a very large level of capacitance. Most capacitors have capacitance of a much lower level such as a microfarad (10^{-6} farad), a nanofarad (10^{-9} farad), or a picofarad (10^{-12} farad).

The capacitance is dependent on the relative dielectric constant K' and the geometry of the capacitor. For a parallel-plate capacitor such as shown in Figure 7.3, this relationship is expressed as

$$C = \frac{E_0 K' A}{t} \tag{11.3}$$

where

A = total area of the electrodes
t = thickness of the dielectric
E_0 = permittivity of vacuum

Note that the capacitance increases as the area and relative dielectric constant increase and as the thickness of the dielectric decreases.

11.1.5.1   Functions of a Capacitor

Capacitors are important elements in an electric circuit and can be used for a variety of different functions: energy storage, blocking, coupling, decoupling, bypassing, filtering, transient voltage suppression, and arc suppression. *Energy storage* involves building up a large charge in the capacitor for release at a later time. A couple of common applications are welding and photoflash bulbs. *Blocking* involves the interaction of the capacitor with direct current (DC) vs. alternating current (AC) currents. Direct current results in polarization in the capacitor and blocks the flow of current. Alternating current results in charge and discharge of the capacitor in opposite directions during each AC cycle, which has the effect of allowing AC current to pass. This characteristic is used to "couple" one circuit to another. *Decoupling* involves the use of a capacitor to isolate specific voltages to different areas of the circuit. *Bypass* involves the simultaneous use of blocking and coupling to separate the DC and AC components of a mixed signal. Bypass is achieved by placing the capacitor in parallel with the circuit device. The AC signal passes through the capacitor and the DC signal passes through the device. *Filtering* involves the use of a capacitor to separate AC signals of different frequencies. The higher the capacitance, the higher the current at any given frequency; conversely, the higher the frequency at a fixed capacitance, the higher the level of current passing through. Thus, filtering provides a means of tuning or frequency discrimination.

11.1.5.2 History of Capacitors

Capacitors in the early 1900s were typically flat plate in configuration and either had an air gap, oiled paper, or mica between the electrodes. One important application was a variable condenser for tuning in the various broadcast stations on a radio receiver. The variable condenser consisted of a capacitor with one stationary metal electrode and one movable electrode. Turning the tuning knob of the radio changed the area of the capacitor and allowed it to function as a frequency filter.

The dielectric materials available in the early 1900s all had dielectric constants below 10. Higher dielectric constant materials were not developed until the late 1930s. At that time it was discovered that some ceramics containing rutile (TiO_2) had dielectric constants of 80 to 100. In 1943 barium titanate ($BaTiO_3$) was discovered to have a dielectric constant of 1200 to 1500. Following this discovery, researchers determined that further increases could be achieved by controlled crystal chemical substitutions into $BaTiO_3$. For example, addition of 10 wt% calcium zirconate plus 1 wt% magnesium zirconate resulted in a dielectric constant above 5000. Addition of 10% calcium zirconate plus 10% strontium zirconate resulted in a room temperature dielectric constant of 9400 to 9500. Dielectric constants up to about 18,000 have been achieved for $BaTiO_3$-based ceramics. Recently, other ceramics have been developed with even higher dielectric constant. Examples include lead magnesium niobate and lead iron tungstate with K' values around 25,000. Internal boundary-layer (IBL) capacitors based on strontium titanate have K' values around 100,000.

The availability of materials with high dielectric constant has been a significant factor in miniaturization of electrical devices. The radio is a good example.[5] A typical transistor radio contains at least 10 capacitors ranging in capacitance from 0.005 to 0.05 microfarad. Each is about 9.5 mm (3/8 in.) across and has a dielectric ceramic with a dielectric constant of about 5000 to 7000. What would be the diameter of an equivalent capacitor made with the highest K' dielectric prior to 1943? Prior to 1935? If we assume the same thickness of the dielectric, a 0.01 microfarad capacitor made with rutile ($K' = 80$) in 1942 would be over 9 cm (3.5 in.) in diameter. A comparable 0.01 microfarad capacitor made in 1930 with mica ($K' = 7$) would be about 40 cm (15.5 in.) in diameter.

11.1.5.3 Mechanism of High Dielectric Constant

The high dielectric constant of $BaTiO_3$-type ceramics results from the crystal structure. $BaTiO_3$ has the perovskite structure, as illustrated in Figure 11.9. Each barium ion is surrounded by 12 oxygen ions. The oxygen ions plus the barium ions form a face-centered cubic lattice. The titanium atoms reside in octahedral interstitial positions surrounded by six oxygen ions. Because of the large size of the Ba ions, the octahedral interstitial position in $BaTiO_3$ is quite large compared to the size of the Ti^{4+} ions. The Ti^{4+} ions are on the margin of being too small to be stable in this octahedral position. It is believed that there are minimum-energy positions off-center in the direction of each of the six oxygen ions surrounding the Ti^{4+} ion. The Ti^{4+} ion positions randomly in one of these six possible minimum-energy sites. This results in spontaneous polarization for each Ti^{4+} ion. Since each Ti^{4+} ion has a +4 charge, the degree of polarization is very high. When an electric field is applied, the titanium ions can shift from random to aligned positions and result in high bulk polarization and high dielectric constant.

Temperature has a strong effect on the crystal structure and polarization characteristics of $BaTiO_3$ as shown in Figure 11.10. Above 120°C (250°F), $BaTiO_3$ is cubic and the behavior described

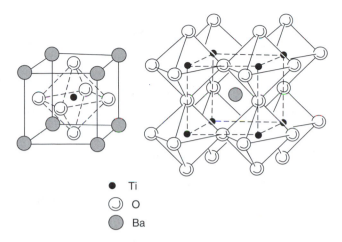

FIGURE 11.9   Schematic of the perovskite structure of $BaTiO_3$ above 120°C (250°F). (From Kingery, W.D., Bowen, H.K., and Uhlmann, D.R., *Introduction to Ceramics*, 2nd ed., Wiley, New York, 1976, Chaps. 17–19, p. 968. With permission.)

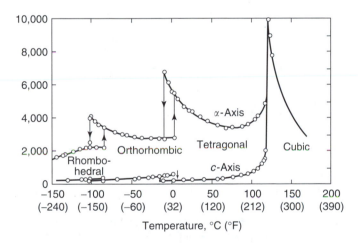

FIGURE 11.10   Changes in dielectric constant of $BaTiO_3$ as a function of temperature and crystallographic form. (After Merz, W.J., *Phys. Rev.*, 76, 1221, 1949.)

above prevails. The thermal vibration is high enough to result in the random orientation of the titanium ions. When the temperature decreases to about 120°C (250°F), the structure changes to tetragonal. The octahedral site is now distorted with the Ti^{4+} ion in an off-center position. This results in a permanent dipole. The temperature of transformation from cubic to tetragonal and from spontaneous random polarization to permanent dipole domains is called the *Curie temperature*.

The polarization characteristics can be modified by crystal chemical alterations of the crystal structure.[6] Ba^{2+} and Pb^{2+} ions are very large and result in a large octahedral site in which the titanium ions can readily move. Substitution of smaller ions for Ba^{2+} or Pb^{2+} reduces the size of the octahedral site and restricts the motion of the Ti^{4+} ion. Substitutions also modify the temperature effect. Additions to $BaTiO_3$ of $CaZrO_3$ and $MgZrO_3$ result in a decrease in sensitivity to temperature by broadening the temperature vs. K' curve. Addition to $BaTiO_3$ of $PbTiO_3$ increases the temperature at which the transformation occurs (Curie temperature) and at which the dielectric constant is a maximum. Additions of $SrTiO_3$, $SrSnO_3$, $CaSnO_3$, or $BaSnO_3$ reduce the Curie temperature. Examples of broadening and shifting are shown in Figure 11.11.

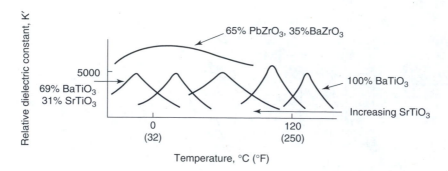

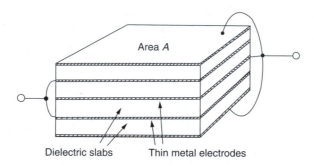

FIGURE 11.11 Broadening and shifting of the dielectric constant vs. temperature behavior achieved through crystal chemical modifications of ferroelastic ceramics. (From Snelling, E.C., *Soft Ferrites*, ILIFFE Books Ltd., London, 1969, p. 157. With permission.)

FIGURE 11.12 Schematic of a multilayer ceramic capacitor. (Drawing courtesy of ASM International.)

11.1.5.4 Types of Capacitors

Ceramic and polymer materials are used as the dielectric for most capacitors currently in use. The ceramic capacitors comprise two types: single-layer and multilayer. The single-layer configuration was depicted in Figure 11.3. Single-layer ceramic capacitors have relatively low capacitance capability because of the relative thickness of the monolithic dielectric layer. A multilayer ceramic capacitor is illustrated schematically in Figure 11.12. Higher capacitance is possible in the multilayer configuration because thinner dielectric layers can be fabricated. The tape casting fabrication technique used to produce multilayer capacitors is described in a later chapter.

Some high-capacitance commercial capacitors are called tantalum and aluminum capacitors. These consist of metal on which a very thin film of oxide has been formed by electrolytically anodizing the surface of the metal. The layer on the aluminum is Al_2O_3 with a dielectric constant of about 8. The layer on the tantalum is tantalum pentoxide (Ta_2O_5) with a dielectric constant of about 27. The high capacitance is achieved based on the thinness of the dielectric layer.

Ceramic and polymer dielectrics are used primarily in small-capacitance applications such as consumer electronics, personal computers, peripherals, and microprocessors. Tantalum and aluminum are used in higher-capacitance applications such as mainframe computers, military systems, and telecommunications. A large increase in the use of multilayer ceramic capacitors occurred in the last 30 years. It is estimated that the sale of multilayer capacitors increased from \$45 million in 1971 to \$340 million in 1981.

11.1.6 Piezoelectricity

Polarization occurs in single crystals of some materials when a stress is applied.[7,8] One side of the crystal derives a net positive charge and the opposite side derives a net negative charge. This effect

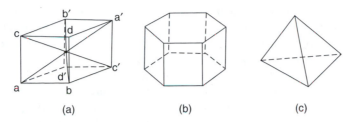

FIGURE 11.13   Examples of a simple cube (a) and a simple hexagonal prism (b) with a center of symmetry, and (c) a tetrahedron with no center of symmetry. (Drawings courtesy of ASM International.)

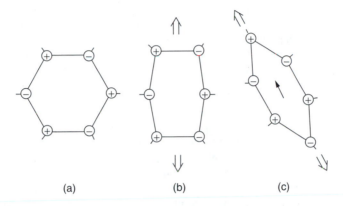

FIGURE 11.14   Directionality of polarization in a piezoelectric material. (a) Unstressed material showing no polarization, (b) stressed material showing deformation but no polarization, and (c) stressed material showing deformation plus polarization. (From Newnham, R.E., *Structure-Property Relations*, Springer-Verlag, New York, 1975, p. 83. With permission.)

is referred to as *piezoelectricity*. The term piezoelectricity translates literally to "pressure electricity." Application of a pressure results in a measurable electrical potential. Conversely, application of an electric field results in a very small amount of mechanical deformation. The piezoelectric phenomenon has led to widespread use of piezoelectric ceramics as transducers in ultrasonic devices, microphones, phonograph pickups, accelerometers, strain gauges, and sonar devices.

The piezoelectric effect was discovered by Pierre and Jacques Curie and first reported in 1880. Piezoelectricity was identified by the Curies in a number of naturally occurring and laboratory grown single crystals. Examples included quartz, zinc blende (sphalerite), boracite, tourmaline, topaz, sugar, and Rochelle salt (potassium sodium tartrate tetrahydrate, $NaKC_4H_4O_6 \cdot 4H_2O$). All of these crystals had one thing in common. None had a center of symmetry. The concept of center of symmetry is defined as follows. A crystal (or a unit cell) has a center of symmetry if an imaginary line is passed from some point on its surface through its center, and a similar point on the opposite surface is intersected when the line is extended an equal distance beyond the center. This is illustrated in Figure 11.13a for a simple cube. The center of symmetry is the point at the center of the cube. A line between a and a' is bisected by this point. Similarly, lines between b and b', c and c', and d and d', and all other opposite points on the surface of the cube intersect at the center point. Thus, the simple cube has a center of symmetry. Draw imaginary lines from opposite corners of the simple hexagonal prism in Figure 11.13b. These also intersect at a center of symmetry. In contrast, no such lines can be drawn for the tetrahedron in Figure 11.14c. Thus, the tetrahedron does not have a center of symmetry.

Therefore, only crystals that are anisotropic with no center of symmetry are piezoelectric. Even these, though, are not piezoelectric in all directions. This is illustrated in Figure 11.14. The unstressed structure is depicted in a. The positive and negative charges are uniformly distributed, so they balance each other in all directions and no polarization is present. The deformed structure under an applied stress is depicted in b and c. Even though the atoms are displaced in b, the positive and negative charges remain uniformly distributed and no polarization occurs. Applying a stress in this direction does not result in piezoelectricity. However, in c the stress is applied in a direction such that polarization does occur, and piezoelectricity results. Quartz (polymorph of SiO_2) is a good example. A compressive stress applied along [100] causes polarization, whereas a stress along [001] does not.

How common is piezoelectricity? In Chapter 5 we discussed seven crystal systems that were categorized according to relative length and angles of the crystallographic axes. We also discussed 14 Bravais lattices categorized according to atom arrangements in unit cells. Neither the crystal systems nor Bravais lattices account for all the symmetry variations possible. If we also consider symmetry, there are 32 classes of crystals possible. Of these, 12 have a center of symmetry and cannot exhibit piezoelectricity. The remaining 20 can. This would lead one to guess that piezoelectricity is very common. Since the initial discovery of piezoelectricity by the Curies, hundreds of dielectric materials have been identified that exhibit piezoelectricity. However, relatively few of these have been optimum for practical application. (Applications are discussed later in this chapter.) Table 11.4 lists a variety of piezoelectric materials.

11.1.7 PYROELECTRICITY

Pyroelectric crystals are a special class of piezoelectric crystals. They contain within their crystal structure a pre-existing spontaneous polarization along at least one crystallographic direction. Heating of the crystal results in mechanical deformation due to thermal expansion, which results in a change in the extent of polarization. The prefix "pyro" is Greek for "fire," so pyroelectric translates to "fire electricity," that is, electricity released by heat. Of the 20 piezoelectric crystal classes, 10 are pyroelectric. Examples of pyroelectric crystals include würtzite (hexagonal ZnS), tourmaline, Rochelle salt, triglycine sulfate, $BaTiO_3$, $Pb(Zr, Ti)O_3$, and lithium sulfate.

TABLE 11.4
Examples of Crystals That Exhibit Piezoelectricity

Quartz
$BaTiO_3$, $Pb(Zr, Ti)O_3$, $KNbO_3$
Many other titanates, niobates, and tantalates with the [perovskite] structure
Rochelle salt ($NaKC_4H_4O_6 \cdot 4H_2O$)
Guanidine aluminum sulfate hydrate [$CN_3H_6Al(SO_4)_2 \cdot 6H_2O$]
$NiSO_3 \cdot 6H_2O$
ZnS, CdS
ZnO
GaAs
Bi_2WO_6, Bi_3TiNbO_9
KH_2PO_4
$LiNbO_3$
SbSI
$BaCoF_4$
HCl
$NaNO_2$
Thiourea
Triglycine sulfate
Lithium sulfate
Tourmaline

An important pyroelectric material is $LiTaO_3$. Most pyroelectric materials lose their pyroelectric behavior as the temperature is increased to a few hundred degrees C. $LiTaO_3$ retains its pyroelectric behavior to 609°C (1128°F). It can therefore be used over a broad temperature range for extremely accurate measurement of temperature changes. Changes on the order of 10^{-6}°C can be detected. As a result, $LiTaO_3$ has been developed into a scanning microcalorimeter capable of sensitivity in the sub-microcalorie range. It has also been used as a high-sensitivity microenthalpimeter for monitoring catalytic processes. Both devices involve a 50 μm (0.002 in.) thick chip of $LiTaO_3$ mounted on a ceramic substrate and coated with a thin-film NiCr heater element. The size of the overall device is about 5 mm × 5 mm × 0.5 mm.

Other applications for pyroelectric ceramics include optical sensors, thermal bolometers for infrared measurement, and devices for gas-flow measurement.

11.1.8 FERROELECTRICITY

Ferroelectric materials are a subclass of pyroelectric crystals. Like the pyroelectric crystals, the ferroelectric crystals contain a spontaneous polarization, that is, they retain a dipole even after an applied voltage has been removed. The key characteristics of a ferroelectric crystal are that the direction of the polarization can be reversed by application of an electric field and that a *hysteresis loop* results. This is illustrated in Figure 11.15, where E is the applied electric field in units such as kV/cm and P is the polarization in units such as μC/cm^2. After we define a few more terms, we will return to this figure.

Ferroelectric behavior is dependent on the crystal structure. The crystal must be noncentric and must contain alternate atom positions or molecular orientations to permit the reversal of the dipole and the retention of polarization after the voltage is removed. Let us look at $BaTiO_3$. Between 120 and 1460°C (250 and 2660°F), $BaTiO_3$ has a cubic structure and is not ferroelectric. In this temperature range the Ti^{4+} ion lies in the center of an octahedron of oxygen ions (as was shown earlier in Figure 11.9). The Ti^{4+} ion does shift position and result in polarization when an electric field is applied, but it returns to its stable central position as soon as the field is removed. Thus, there is no retained polarization, no hysteresis loop, and no ferroelectric behavior.

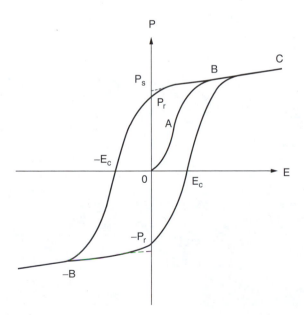

FIGURE 11.15 Ferroelectric hysteresis loop between applied electric field E and polarization P. (Drawing courtesy of ASM International.)

As the temperature of $BaTiO_3$ is lowered slightly below 120°C (250°F), a displacive transformation occurs in which the structure of the $BaTiO_3$ shifts from cubic to tetragonal. One crystallographic axis increases in length (unit cell goes from 4.010 to 4.022 Å) and the other two decrease in length (4.010 to 4.004 Å). The Ti^{4+} ion moves off-center toward one of the two oxygen ions of the long axis, resulting in a spontaneous increase in positive charge in this direction. This is illustrated in Figure 11.16. Application of an electric field opposite to the polarity of this original dipole will cause the Ti^{4+} ion to move through the center of the octahedral site and to an equivalent off-center position. This is shown in Figure 11.16. This results in a reversal in polarization, hysteresis in the E vs. P curve, and ferroelectricity.

During the initial transformation of a crystal from cubic to tetragonal, not all Ti^{4+} ions shift in the same direction. In fact, each Ti^{4+} ion has an equal probability of shifting in six directions toward one of the corners of the octahedron. As a result, the tetragonal crystal contains some dipoles in one portion of the crystal pointing in one direction, whereas others in another portion may point in a direction 90° or 180° away from the first. A region of the crystal in which the dipoles are aligned in a common direction is called a *domain*. An example of $BaTiO_3$ with a ferroelectric domain with aligned dipoles is illustrated in Figure 11.17.

Let us return now to Figure 11.15 and describe what happens in a ferroelectric crystal such as tetragonal $BaTiO_3$ when an electric field is applied. The ferroelectric domains are randomly oriented

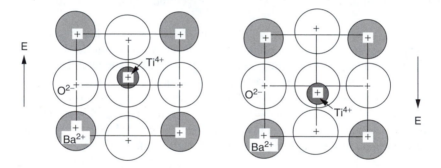

FIGURE 11.16 Reversal in the direction of spontaneous polarization in $BaTiO_3$ by reversal of the direction of the applied electric field. (Drawing courtesy of ASM International.)

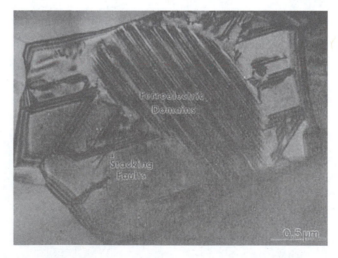

FIGURE 11.17 TEM of a ferroelectric domain in $BaTiO_3$ with aligned dipoles. (Courtesy of W.E. Lee, University of Sheffield, UK.)

prior to application of the electric field, that is, at $E = 0'$, the net polarization equals zero ($P_{net} = 0$). As we apply an electric field and increase the electric field, the domains begin to move in the $BaTiO_3$ and align parallel to the applied field. This results in an increase in net polarization along line OA. The polarization reaches a saturation value (B) when all the domains are aligned in the direction of the field. If we now reduce the electric field to zero, many of the domains will remain aligned such that a *remanent polarization* (P_r) exists. Interpolation of the line BC until it intersects the polarization axis gives a value P_s, which is referred to as the *spontaneous polarization*. If we now reverse the electric field, we force domains to begin to switch direction. When enough domains switch, the domains in one direction balance the domains in the opposite direction and result in zero net polarization. This occurs for an electric field E_c that is called the coercive field. Continued increase in the negative electric field causes net polarization in the opposite direction, reaching the point B where all the available domains are aligned.

Maximum alignment of domains can be achieved by cooling the $BaTiO_3$ crystal through the 120°C (250°F) cubic-to-tetragonal transition while an electric field is applied. This is referred to as *poling*. Poling forces a maximum number of domains to form in one direction (rather than some at 90° or 180° angles) and results in maximum polarization.

As mentioned earlier, the transformation temperature (120°C or 250°F for $BaTiO_3$) at which the crystal becomes ferroelectric is called the Curie temperature. Above this temperature the crystal is referred to as *paraelectric*. Solid-solution additions can change the Curie temperature and also the shape of the hysteresis loop. Some examples were discussed earlier in this chapter and illustrated in Figure 11.11. For $BaTiO_3$, addition of $PbTiO_3$ increases the Curie temperature and additions of $SrTiO_3$, $SrSnO_3$, $CaSnO_3$, or $BaSnO_3$ decrease the Curie temperature.

The shape of the hysteresis loop varies for different temperatures below the Curie temperature. Examples are shown in Figure 11.18 for $BaTiO_3$ at just below 120°C (250°F), at 90°C (195°F), and at 30°C (85°F). Note that the hysteresis loop gets thinner as the temperature increases and becomes a single line above the Curie temperature when the material is no longer ferroelectric.

We have now reviewed most of the terms associated with ferroelectric behavior. All crystals that are ferroelectric are pyroelectric and piezoelectric; all crystals that are pyroelectric are piezoelectric; but not all piezoelectric crystals are pyroelectric or ferroelectric. Let us move on now and look at some ferroelectric crystals other than $BaTiO_3$, discuss polycrystalline ferroelectrics, and review important ferroelectric and piezoelectric applications of ceramics.

11.1.8.1 Types of Ferroelectric Crystals

Ferroelectrics have been categorized in the literature in different ways. One approach classifies them as soft or hard. Soft ferroelectrics are water-soluble, mechanically soft, and have low melting or decomposition temperatures. Soft ferroelectrics include Rochelle salt, some other tartrates, KDP (potassium dihydrogen phosphate, KH_2PO_4), GASH (guanidine aluminum sulfate hydrate, $CN_3H_6Al(SO_4)_2 \cdot 6H_2O$), and some other sulfates, sulfites, nitrates, and nitrites. Most of these involve hydrogen bonding. Above the Curie temperature the hydrogen ions and bonds are distributed randomly in a nonordered

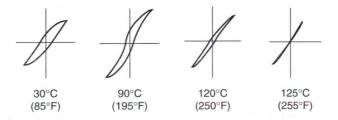

30°C	90°C	120°C	125°C
(85°F)	(195°F)	(250°F)	(255°F)

FIGURE 11.18 Change in the hysteresis loop shape for $BaTiO_3$ at various temperatures. (Drawing courtesy of ASM International.)

fashion. At the Curie temperature the crystal transforms from a disordered paraelectric structure to an ordered ferroelectric structure. The ordered structure has specific pairs of positions that the hydrogen ions can fit into to form the reversible dipoles. Other common sources of dipoles in soft ferroelectric crystals are tetrahedral groups such as PO_4^{3-}.

Hard ferroelectrics include the oxides that are formed at high temperature, are mechanically hard, and are not water-soluble. Examples include $BaTiO_3$, $KNbO_3$, $CdNb_2O_6$, $PbNb_2O_6$, $PbTa_2O_6$, $PbBi_2Nb_2O_9$, and many others. Many of the hard ferroelectrics contain a small, highly charged cation (Ti^{4+}, Zr^{4+}, Nb^{5+}, Ta^{5+}) in an oxygen octahedron and have a similar ferroelectric mechanism to $BaTiO_3$. Others contain asymmetrical ions with a "lone-pair" electron configuration. Examples are Pb^{2+}, Bi^{3+}, Sn^{2+}, Te^{4+}, I^{5+}, and Sn^{3+}. Each of these has two electrons outside a closed d shell. These form a lone-pair orbital on one side of the ion and promote a directional bonding. The resulting structure has dipoles that result in spontaneous polarization when the dipoles do not cancel each other.

Another classification of ferroelectrics is based on the magnitude of atomic displacements and the resulting spontaneous polarization. $BaTiO_3$, $PbTiO_3$, $LiNbO_3$, SbSI, and Bi_2WO_6 all have atomic displacements along a single axis (one-dimensional). Since the complete displacement is concentrated in a single direction, the spontaneous polarization is high ($\approx 25\,\mu C/cm^2$). Atomic displacements in some crystals are along planes (two-dimensional). Examples include $BaCoF_4$, HCl, $NaNO_2$, and thiourea. The spontaneous polarization is only approximately $5\,\mu C/cm^2$. Ferroelectrics with tetrahedral groups or hydrogen bonding have complex structures with three-dimensional effects on polarization. Spontaneous polarization is typically less than $3\,\mu C/cm^2$.

11.1.8.2 Polycrystalline Ferroelectrics

For many years, all ferroelectric devices were prepared from slices of crystals (primarily Rochelle salt) cut in specific directions to optimize the ferroelectric characteristics. After the discovery of the ferroelectric behavior of $BaTiO_3$, a method was optimized to achieve a high level of ferroelectric behavior in polycrystalline $BaTiO_3$. The method involves the following steps:

1. Fabricate the required shape by normal powder compaction and sintering to achieve a dense polycrystalline ceramic.
2. Apply electrically conductive electrodes to the two surfaces perpendicular to the desired polarization direction.
3. Heat the part to above 120°C (250°F) and apply a large enough electric field between the electrodes to force many of the domains to align parallel to the direction of the applied field.
4. Cool the part below the Curie temperature and remove the electric field.

An enormous number of products have been invented that utilize piezoelectric behavior. One reason for this is that piezoelectric behavior can be activated in numerous ways:

1. Apply an electric input to stimulate high frequency (ultrasonic) vibration. Examples: transducers (acoustic wave generators), ultrasonic cleaner, emulsifier, atomizers, ultrasonic machining, microphone, quartz watch, and medical ultrasound physical therapy.
2. Detect and characterize an applied pressure or motion by generating a corresponding electrical output. Examples: hydrophone, loudspeaker, transducer (acoustic wave receivers), accelerometer, wheel balancer, integrated circuit photolithography (vibration detection and elimination), pressure and motion sensors, and atomic force microscope.
3. Combine 1 and 2 in a pulse-echo mode to send vibrations (acoustic waves) through a material and to detect reflected waves. Examples: sonar, fish finders, ocean floor mapping, medical ultrasound imaging, non-destructive inspection, and blood flow monitoring during dialysis.

4. Apply a controlled electrical input to cause a material to vibrate at its resonant frequency; can achieve a wide range of frequencies from ultrasonic to audible. Examples: alarms, musical greeting cards, electronic musical instruments.
5. Apply a mechanical force and obtain a high voltage discharge in the form of a spark. Examples: igniter, lighter (such as for a charcoal grill).
6. Apply an electrical input and obtain a controlled deflection or motion, including very tiny micrometer scale deflections. Examples: actuators, x-y tables, ink jet and dot matrix printers, microswitches, pressure-sensitive valves, precision measurement instruments, and thermal distortion compensation for telescope lenses.
7. Decrease vibration by converting to electrical output (active vibration control). Examples: skis, snowboards.
8. Others: step-up-transformer, piezoelectric motor (such as for autofocus for cameras).

The following paragraphs describe how piezoelectricity is used to achieve some of the applications listed above.

11.1.8.2.1   Phonograph, Microphone, and Loudspeaker

The phonograph, microphone, and loudspeaker were early uses of piezoelectric materials. As shown in Figure 11.19 the stylus of the phonograph was coupled to a piezoelectric ceramic. Contours in the groove of the record caused the stylus to vibrate in a mode unique to the musical content of the record. The piezoelectric material converted the vibrations to electrical signals, again corresponding to the content of the record. These electrical signals then traveled to a speaker where another piezoelectric ceramic converted the electrical signals back to vibrations that duplicated the sound that was initially recorded. A diaphragm attached to the piezoelectric amplified the vibrations so they could be heard. Early commercial phonographs utilized Rochelle salt (which was water soluble). Later, after people learned to pole polycrystalline piezoelectric ceramics, materials such as lead zirconate titanate replaced Rochelle salt.

11.1.8.2.2   Hydrophone

Another early use of piezoelectrics was in hydrophones, which were particularly important during World War II for detecting submarines. A schematic of a hydrophone is shown in Figure 11.20. It consists of a stack of piezoelectric crystal slices (usually lithium sulfate) mounted in parallel that are able to detect very small pressure pulses from sound waves traveling through water. Each tiny sound wave infinitesimally deforms the piezoelectric material and produces an electrical signal that can be observed with an oscilloscope.

11.1.8.2.3   Accelerometers and Other Sensors

While the hydrophone detects tiny sound waves, other piezoelectric devices can be designed to detect virtually any pressure pulse, mechanical force, or movement and thus are the basis of a wide range of sensors. Figure 11.20 illustrates two different types of piezoelectric accelerometers. Applications can be high technology such as the sensors that monitor vibration and shock with high precision on missile and rocket engines, or they can be lower technology such as the wheel balancer that your auto mechanic uses to determine if your automobile tires are in balance.

Perhaps one of the most important applications is to eliminate vibrations during the photolithography steps in the fabrication of silicon chip integrated circuits. During the photolithography steps, images with details down to about 500 nm are projected onto the silicon chip. The slightest vibration during this operation will result in a blurred image that will destroy the function of the integrated circuit. To avoid vibrations, a table is designed with three sets of piezoelectric sensors attached to the base of each leg. Any vibration traveling up the leg is detected and characterized by the triangular placement of the sensors. The electrical output from the sensors goes to a computer that then sends an electrical command to three piezoelectric devices (transducers). mounted further up the leg of the table. The electrical input to these causes them to produce a mechanical force (up

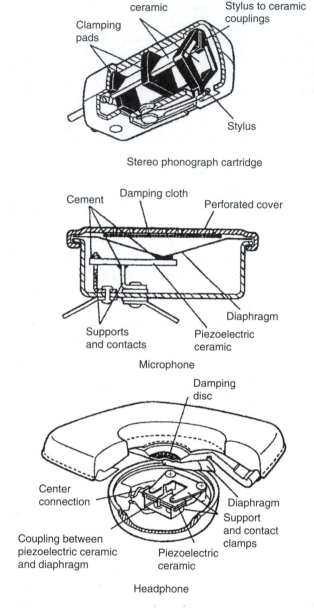

Stereo phonograph cartridge

Microphone

Headphone

FIGURE 11.19   Schematics of piezoelectric applications based on conversion of a mechanical force to an electrical signal. (Reprinted with permission of Morgan Matroc Ceramic Division, Bedford, OH.)

to about 1600 kg against the leg of the table that exactly counteracts the specific vibration and guarantees that no vibrations reach the top of the table where the photolithographic process is conducted.

11.1.8.2.4   Actuators

Measurement instruments and manufacturing processes are continually becoming more sophisticated with greater requirement for a high level of precision. Piezo-actuators can be designed to provide precise control over movements from micrometers to millimeters to control motions during a manufacturing operation, to accurately position a laser beam, and to control many other tasks. As an example, the manufacturing precision of optical instruments such as lasers and cameras and the

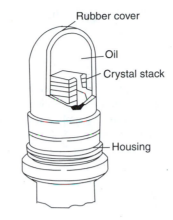

Measurement hydrophone

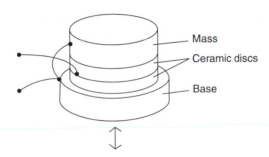

Compression accelerometer

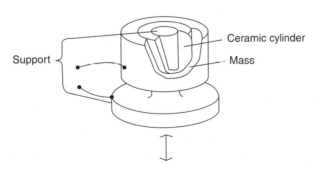

Shear accelerometer

FIGURE 11.20   Schematic of a piezoelectric hydrophone and two types of piezoelectric accelerometers. (Reprinted with permission of Morgan Matroc Ceramic Division, Bedford, OH.)

positioning accuracy for fabricating some semiconductor devices requires position control to less than 1 micrometer. Position errors of this magnitude can result due to a temperature fluctuation of 1°C.

Each year many new patents are granted for piezo-actuator devices. Examples of innovative applications include impact dot matrix printer heads, ink-jet printer heads, precision microscope stages, oil pressure servovalves, cutting error compensation actuators, ultraprecision guide mechanisms, microangle adjusting devices, and deformable optical grids. One interesting application is for telescope mirrors. Temperature changes during the night cause thermal expansion distortions that blur the image. Some telescope mirrors have an array of piezoelectric devices attached on the back that can be computer controlled to slightly change the shape of the mirror to counteract any distortions caused by temperature fluctuations.

11.1.8.2.5 High-Voltage Generator

Whereas an actuator produces a mechanical strain when a voltage is applied, the converse can be applied to produce a high voltage output when a mechanical strain is applied. This is illustrated in Figure 11.21 for a high-voltage "spark pump" that was demonstrated as an igniter for a gasoline engine. An eccentric cam rotates and causes a lever to apply pulses of force to the ceramic cylinders. A pressure pulse of about 48 MPa (7000 psi) to cylinders approximately 1 cm (0.39 in.) diameter by 2 cm (0.78 inch) long yields an electrical output pulse (in the form of a spark) of about 21,000 volts. This same concept is used for charcoal grill lighters and for cigarette lighters.

11.1.8.2.6 DC Step-Up Transformer

The concepts of the actuator and the voltage generator are combined to produce a DC step-up transformer, which is shown schematically in Figure 11.22. A small DC voltage applied to the primary end of the bar causes mechanical distortion that is magnified by the secondary end and results in a stepped-up voltage. The output voltage is controlled by the input voltage, the geometry of the transformer, and the characteristics of the piezoelectric material. For example, one configuration produced 400 V for a 10-V input and 2400 V for a 30-V input. The total size of the transformer was only

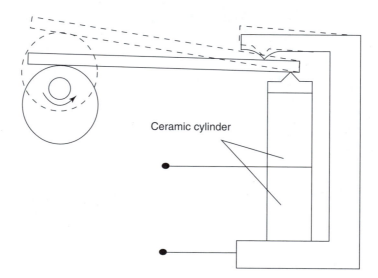

FIGURE 11.21   Schematic of a piezoelectric high-voltage generator "spark pump." (Drawing courtesy of ASM International.)

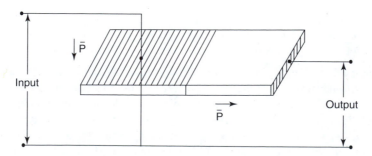

FIGURE 11.22   Schematic of a piezoelectric step-up transformer. (From Henry, E.C., *Electronic Ceramics*, Doubleday, New York, 1969, p. 81. With permission.)

about $7\,cm \times 1\,cm \times 2.5\,mm$. The piezoelectric transformer has advantages over coil-type transformers for many applications, especially if it must perform in a magnetic environment. A coil-type transformer works through induced magnetic fields. Its operation is affected by external magnetic fields, whereas the piezoelectric transformer is not.

11.1.8.2.7 Sonar, Decoys, and Directional Beacons

The applications discussed so far have involved the simple response of a piezoelectric ceramic to either an electrical input or a pressure input. For example, the hydrophone can detect sound waves in the water by converting pressure pulses from acoustic waves into a detectable electrical output. This can tell that another ship or submarine is nearby, but cannot tell exactly where or how far. Sonar stands for "sound navigation and ranging" and was developed to go beyond the capability of the hydrophone. It is a pulse-echo approach analogous to the navigation technique of a bat or the radar gun used by the police to catch speeding motorists. In the case of sonar, a ceramic piezoelectric device (referred to as a transducer) is stimulated by an electrical input to vibrate at a selected frequency that produces acoustic waves in water. The waves travel through the water until they strike a solid object and are reflected back. When these reflected acoustic waves reach the transducer, they are detected by the same pressure to electrical response exhibited by the hydrophone. By knowing the speed of sound in water and the time between sending the pulse and receiving the echo, one can determine the distance of the object.

Sonar was an important military innovation for tracking both underwater and surface ships. However, every military device seems to have a counter-device designed to defeat it. Figure 11.23 shows hemispheres of piezoelectric ceramic that are assembled into spherical "pingers." These send out acoustic waves in all directions and can be released by a submarine or surface ship to act as a decoy. They also can be positioned as a homing or directional beacon.

11.1.8.2.8 Fish Finders, Ocean Floor Mapping, and Boat Speedometers

The same pulse-echo approach as sonar is used for other devices in water. One, referred to as a "fish finder" is effective at locating and tracking schools of fish and has increased the efficiency of the world fishing industry. Another amazing piece of equipment allows mapping the contours of the ocean floor. An array of piezoelectric transducers towed by a ship bounce sound waves off the ocean floor and record the reflected waves to allow calculations of the water depth. As shown in Figure 11.24, the data are combined to construct a three-dimensional image of the ocean floor. A modified procedure allows geologists to gain information about the rock layers underlying the ocean. Instead of transducers to send low amplitude waves, the geologists use air cannons to produce very high intensity waves that can penetrate beneath the ocean bottom. Piezoelectric transducers are still used to record the reflected waves.

Another pulse-echo application is the boat speedometer. Accurately measuring the speed of an ocean-going vessel always has been a challenge. For many years the best system was dragging a rope with knots tied at regular spaces and counting the number of knots that passed a floating object

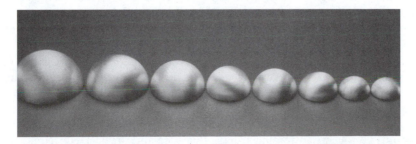

FIGURE 11.23 Hemispheres of piezoelectric ceramic that are assembled into spherical "pingers" to be used as decoys or beacons. (Photo courtesy of Edo Corporation Piezoelectric Products, Salt Lake City, UT.)

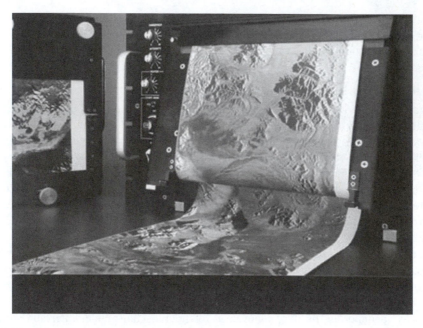

FIGURE 11.24   Ocean floor contour map generated using arrays of piezoelectric transducers in a pulse-echo mode. (Photo courtesy of Edo Corporation Piezoelectric Products, Salt Lake City, UT.)

during a certain time. Now a "dopler sonar velocity system" is used. It is similar to sonar, but consists of two stacks of piezoelectric ceramic slices oriented perpendicular to each other. These send acoustic waves to the bottom of the body of water and use the reflections to calculate the speed of the boat. The speed can be measured at a precision of about 0.5 cm per second and can even correct for the effects of ocean currents.

11.1.8.2.9   Medical Ultrasound Imaging, Nondestructive Inspection, and Sonic Delay Lines

Piezoelectric transducers have been developed that can also send acoustic waves into solid objects and analyze the reflected waves. An important application is medical ultrasound imaging. With this technology (which is a combination of piezoelectric materials, advanced electronics, and advanced computers), a mother and doctor can image the features and movements of a baby in her womb. They can determine the gender and general state of health of the baby. This same technology can allow imaging of the carotid artery in the neck to determine whether plaque is building up on the inside of the artery and to obtain early warning of the progression of cardiovascular disease. A different transducer design can be used to produce higher intensity ultrasonic waves that are effective for physical therapy. The high intensity ultrasonic waves cause an increase in temperature of tissues to stimulate an increase in the rate of healing.

The same pulse-echo approach is used for nondestructive inspection, i.e., locating critical defects in the interior of a material. As we discussed in Chapter 8, defects such as pores and inclusions reduce the strength of a material. These types of defects represent a discontinuity in the material that will result in reflection of an acoustic wave and thus can be detected using piezoelectric transducers and the appropriate electronics. Ultrasonic inspection is performed extensively in the aerospace industry and in many other industries to sort good parts from defective parts.

Another important wave-generation application is the sonic delay line. A delay line consists of a solid bar or rod of a sound-transmitting material (glass, ceramic, metal) with a transducer attached to each end. An electrical signal that is to be delayed is input to the first transducer. The signal is converted to a sonic wave impulse that travels along the sound-transmitting "waveguide." The sonic impulse is then converted back to an electrical impulse by the second transducer. The delay results because a sonic wave travels much more slowly than electrons passing through a wire. The time of delay is controlled by the length of the waveguide. Delay lines are used extensively in military electronics gear and in color television sets. One example is in a radar system to compare information form one echo with the next echo and for range calibration.

11.1.8.2.10 Ultrasonic Cleaner and Emulsifier
High-intensity ultrasonic waves that travel through a container of water or other liquid cause tiny bubbles to form at the surface of any object immersed in the liquid. This results in a vigorous cleaning action. Ultrasonic cleaners are used extensively in industry, in scientific laboratories, and even for cleaning jewelry and false teeth. The bubble formation (cavitation) also provides a strong mixing action that is effective for producing an emulsion. Normally a mixture of water and oil will separate with the oil rising to the surface. However, if the oil droplet size can be reduced to very small size and mixed uniformly with the water, an emulsion will result where the oil and water do not separate. Good examples are hand lotion and some shampoos.

11.1.8.2.11 Quartz Watch
Perhaps the most common piezoelectric application is the quartz watch. Each watch contains a tiny slice of a quartz crystal cut in a specific direction to have piezoelectric behavior. Connected to the battery in the watch, the slice of quartz vibrates at a constant predetermined frequency (thousands of vibrations per second). Each of these vibrations is equal to a tiny unit of time and allows the electronics in the watch to monitor time much more accurately than could be achieved with the old fashioned mechanical watches that depended on springs, levers, gears, and bearings.

11.1.8.2.12 Audible Piezoelectric Devices
For most of the applications discussed so far, the vibrations produced by the piezoelectric transducer are very high frequency and outside the range that our ears can hear. By control of the size of the piezoelectric piece and the electrical input, though, the piezoelectric transducer can be stimulated to vibrate at its resonant frequency. This makes possible production of vibrations at a wide range of frequencies including in the audible range. Such piezoelectric resonators are used for alarms and other noise generators and also to produce very pure musical sounds in special electronic musical instruments.

11.1.8.2.13 Active Vibration Control
An innovative use of ceramic piezoelectric materials is in skis and snowboards. For those of you who ski or snowboard, you know that when you get going very fast (especially in icy snow conditions) your skis vibrate and you begin to lose control. The K2 ski company introduced the "smart skis" with a piezoelectric module mounted just in front of the ski boot. When the skis start to vibrate, the piezoelectric ceramic converts a portion of the vibrations to electrical energy, which in turn is converted to heat energy and dissipated. This is claimed to reduce the vibration by 30 to 40%. This same approach has been explored for reducing vibration in aircraft structures and in reducing shock waves in baseball bats and golf clubs.

11.1.8.2.14 Communications Band-Pass Filters
Ferroelectric materials have a sharp minimum in their electrical impedance at their resonant frequency. Thin wafers can be sized to respond to specific frequencies, that is, to allow these frequencies to pass and to reject all others. These filters are important to separate individual messages in cases where many channels of communications are carried in a narrow range of frequencies.

11.2 MAGNETIC BEHAVIOR

Magnetism in naturally occurring ceramic compositions has been known for centuries, even though the nature and source of the magnetism was not understood. The best known composition is the mineral magnetite (Fe_3O_4), which is also known as lodestone. It was used in compasses beginning in the thirteenth century. The modern age of magnetic ceramics began in 1946 when J.L. Snoeck reported studies conducted at the Philips Laboratories in Holland. The studies demonstrated that oxide ceramics could be synthesized with strong magnetic properties, high electrical resistivity, and low losses. Such properties gave the ceramics a decisive advantage over metals, especially in high-frequency devices. Important applications that have evolved since 1946 include permanent magnets, memory units with rapid switching times in digital computers, recording heads in tape recorders, motors, and circuit elements in radio, television, microwave, and other electronic devices.

The following sections describe the source of magnetism in ceramic materials, define some of the terms used to describe magnetic behavior, and review specific ceramic materials and applications.

11.2.1 Source of Magnetism

The intrinsic magnetic properties of a material are determined by the electronic structure, the crystal structure, and the microstructure (domains).[3] As we discussed in Chapter 4, electrons travel in orbitals with two electrons of identical energy in each orbital. Each electron represents a tiny electrical current rotating around the positively charged nucleus. This produces a magnetic dipole moment. The two electrons in each orbital spin in opposite directions and produce equal but opposite magnetic moments that cancel. Magnetic moments are discussed in greater detail in Reference 3 (see p. 976).

Most crystal structures result in filled electron orbitals where all electrons are in pairs and the magnetic moments cancel. These materials are not magnetic. Some materials contain electron structures with unpaired electrons and have the potential to be magnetic. These include metals with conduction electrons, atoms, and molecules containing an odd number of electrons, and atoms and ions with partially filled inner-electron shells. Elements with partially filled inner-electron shells are especially important to magnetic ceramic structures. These include the transition elements (such as Fe, Co, and Mn that have unfilled shells), the rare earth elements, and the actinide elements.

Table 11.5 compares the electron configurations for ions commonly found in oxide ceramic structures. Ions such as Mg^{2+}, Al^{3+}, Zn^{2+}, and Ti^{4+} have paired electrons and no net magnetic moment.

TABLE 11.5

Comparison of the Electron Configurations and Unpaired Electrons for Various Ions

Ion	Electron Configuration	Number of Unpaired Electrons
Mg^{2+}	$2p^6$	0
Al^{3+}	$2p^6$	0
O^{2-}	$2p^6$	0
Sc^{3+}	$3p^6$	0
Ti^{4+} (Ti^{3+})	$3p^6(3d^1)$	0(1)
V^{3+} (V^{5+})	$3d^2(3p^6)$	2(0)
Cr^{3+} (Cr^{2+})	$3d^3(3d^4)$	3(4)
Mn^{2+} (Mn^{3+})(Mn^{4+})	$3d^5(3d^4)(3d^3)$	5(4)(3)
Fe^{2+}	$3d^6$	4
Fe^{3+}	$3d^5$	5
Co^{2+} (Co^{3+})	$3d^7(3d^6)$	3(4)
Ni^{2+}	$3d^8$	2
Cu^{2+} (Cu^+)	$3d^9(3d^{10})$	1(0)
Zn^{2+}	$3d^{10}$	0
Cd^{2+}	$4d^{10}$	0

Source: From Kingery, W.D., Bowen, H.K., and Uhlmann, D.R., *Introduction to Ceramics*, 2nd ed., Wiley, New York, 1976, Chaps. 17–19, p. 979.

Fe^{3+} and Mn^{2+} have five unpaired electrons and substantial magnetic moments. Fe^{2+}, Co^{3+}, Mn^{3+}, and Cr^{2+} have four unpaired electrons and also have large magnetic moments. However, the presence of a magnetic moment in an ion does not guarantee that the material containing the ion will have a net magnetic moment. Often, the ions will align in a structure such that their moments cancel. FeO is a good example. Fe^{2+} ions in one plane of FeO all contain electrons with spins aligned in one direction, whereas Fe^{2+} ions in adjacent planes have electron spins aligned in the opposite direction. The spins in adjacent planes cancel and the bulk material is not magnetic.

Several ceramic structures contain ions with a magnetic moment and retain a net magnetic moment within the structure. These magnetic ceramics are commonly referred to as *ferrites*.[9–12] Examples of ferrite structures and compositions are listed in Table 11.6.

Lodestone (magnetite, Fe_3O_4) has the [inverse spinel] structure. The structure is illustrated in Figure 11.25. This only shows a portion of a unit cell. A complete unit cell contains 32 oxygen ions

---

TABLE 11.6
Compositions and Structures of Magnetic Ceramics

Cubic Ferrites

Spinel	General structure MFe_2O_4, where Fe is trivalent and M is divalent Ni, Mn, Mg, Zn, Cu, Co, or a mixture.
Garnet	General structure $R_3Fe_5O_{12}$, where Fe is trivalent and R is Y or a trivalent rare earth, typically Gd.

Hexagonal Ferrites

Various structures	$BaFe_{12}O_{19}$, $Ba_2MFe_{12}O_{22}$, $BaM_2Fe_{16}O_{27}$, $Ba_3M_2Fe_{24}O_{41}$, $Ba_2M_2Fe_{28}O_{46}$, $Ba_4M_2Fe_{36}O_{60}$, where M is divalent Ni, Co, Zn, or Mg, and Ba can be replaced by Sr and Pb.

Orthorhombic Ferrites

Perovskite	General structure $RFeO_3$, where R is a trivalent rare earth and Fe is trivalent and can be partially replaced by trivalent Ni, Mn, Cr, Co, Al, Ca, or V^{5+}.

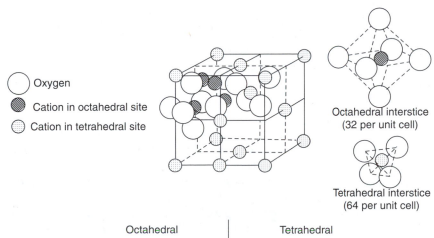

FIGURE 11.25 Schematic illustrating the source of magnetism in the spinel ferrite Fe_3O_4. (Drawing courtesy of ASM International.)

in a close-packed cubic arrangement. This unit cell contains 64 tetrahedral interstitial sites and 32 octahedral interstitial sites. Of these available sites, only 8 tetrahedral and 16 octahedral sites are occupied in the spinel structure. Ions that position on octahedral sites have their magnetic moments all aligned in the same direction. Magnetic moments of ions on tetrahedral sites are aligned in the opposite direction. Now let us return to Figure 11.25 and use Fe_3O_4 as an example. Fe_3O_4 is more properly written as $Fe^{2+} Fe_2^{3+} O_4$, or in terms of a unit cell $Fe_8^{2+} Fe_{16}^{3+} O_{32}$. Eight of the Fe^{3+} occupy the 8 tetrahedral positions. The remaining 8 Fe^{3+} and 8 Fe^{2+} occupy the 16 octahedral sites in the [inverse spinel] structure. As can be seen in Figure 11.25, the moments of the Fe^{3+} in the octahedral sites have opposite direction to the moments of the Fe^{3+} on the tetrahedral sites. These cancel and leave an overall magnetic moment equal to the sum of the moments of the Fe^{2+} ions in the octahedral sites.

Other ions can substitute in solid solution into the [inverse spinel] structure in place of the Fe^{2+} and Fe^{3+} ions. These other ions have different magnetic moments. As a result, by crystal chemical substitution one can engineer the relative magnetic moments on each structural position to obtain a wide range of magnetic behavior. Some options are illustrated in Table 11.7.

A second important cubic ferrite structure is the [garnet] structure. The [garnet] structure is illustrated in Figure 11.26. A unit cell consists of 160 atoms including 96 oxygen, 40 trivalent iron (Fe^{3+}), and 24 trivalent rare earth ions (typically Y or Gd). The trivalent rare earth ions are in dodecahedral (coordination number of 8) interstitial positions. Sixteen of the Fe^{3+} are in octahedral sites and 24 are in tetrahedral sites, as shown in Figure 11.26.

As in the spinel ferrites, the Fe^{3+} ions in octahedral sites have opposite magnetic moment to Fe^{3+} ions in tetrahedral sites. Thus, the 16 in the octahedral sites cancel 16 in the tetrahedral sites. This leaves 8 Fe^{3+} in the tetrahedral sites that provide a net magnetic moment.

The degree of magnetic moment can be varied in the [garnet] structure by crystal chemical substitution. This is also illustrated in Figure 11.26. Substitution of nonmagnetic Al^{3+} ions for Fe^{3+} ions reduces the net magnetic moment.

11.2.2 MAGNETIC TERMINOLOGY

The terminology of magnetic ceramics in many ways is analogous to dielectric ceramics. Analogous terms include magnetization and dielectric polarization, electrical domains and magnetic domains, spontaneous magnetization and spontaneous electric moment, ferromagnetic and ferroelectric, paramagnetic and paraelectric, antiferromagnetism and antiferroelectric, and magnetic field and electric field. The meaning of some of the terms is illustrated in Figure 11.27. A ferromagnetic ceramic has all the magnetic dipoles aligned in one direction. An antiferromagnetic ceramic has equal but opposite magnetic dipoles that cancel each other. A good example is FeO that was described earlier. A ferrimagnetic ceramic has magnetic dipoles in opposite directions, but they are not equal and do not cancel. The most important ferrites including [inverse spinel] and [garnet] structures are ferrimagnetic.

Ferromagnetic and ferrimagnetic ceramics both contain domains that are not necessarily lined up. As shown schematically in Figure 11.28, these domains can be aligned by application of a magnetic field and can be realigned by application of a reverse magnetic field. This results in a hysteresis loop analogous to the one shown previously for ferroelectric materials. A magnetic hysteresis loop is illustrated in Figure 11.29. A magnetic field, H, is applied to the material. Domains begin to align slowly at first. The slope of this initial magnetization curve is referred to as the *initial permeability* (μ_i). As the magnetic field increases, the rate of domain alignment increases and the slope of the magnetization curve increases until a maximum is reached (*maximum permeability*, μ_{max}). The domain boundaries have all been displaced by this maximum. Further increase in magnetic field completes the alignment of the domains, reaching the saturation magnetization B_s (also called the saturation flux density). Application of a reverse magnetic field begins to reverse the direction of the magnetic dipoles. At zero-applied magnetic field, many of the domains are still aligned and a residual or remanent magnetization B_r exists.

TABLE 11.7
Examples of Engineering the Magnetic Behavior of [Spinel] Compositions by Crystal Chemical Substitution

Ferrite	Postulated Ion Distribution		Magnetic Moment of Tetrahedral Ions	Magnetic Moment of Octahedral Ions	Magnetic Moment per Molecule $MeFe_2O_4$	
	Tetrahedral Ions	Octahedral Ions			Theoretical	Experimental
$MnFe_2O_4$	$Fe_{0.2}^{3+} + Mn_{0.8}^{2+}$	$Mn_{0.2}^{2+} + Fe_{1.8}^{3+}$	5	$5 + 5$	5	4.6
Fe_3O_4	Fe^{3+}	$Fe^{2+} + Fe^{3+}$	5	$4 + 5$	4	4.1
$CoFe_2O_4$	Fe^{3+}	$Co^{2+} + Fe^{3+}$	5	$3 + 5$	3	3.7
$NiFe_2O_4$	Fe^{3+}	$Ni^{2+} + Fe^{3+}$	5	$2 + 5$	2	2.3
$CuFe_2O_4$	Fe^{3+}	$Cu^{2+} + Fe^{3+}$	5	$1 + 5$	1	1.3
$MgFe_2O_4$	Fe^{3+}	$Mg^{2+} + Fe^{3+}$	5	$0 + 5$	0	1.1
$Li_{0.5}Fe_{2.5}O_4$	Fe^{3+}	$Li_{0.5}^{+} + Fe_{1.5}^{3+}$	5	$0 + 7.5$	2.5	2.6

Source: Adapted from Smit, J. and Wijn, H.P.J., *Ferrites*, Wiley, New York, 1959.

Dielectric, Magnetic, Optical behaviour

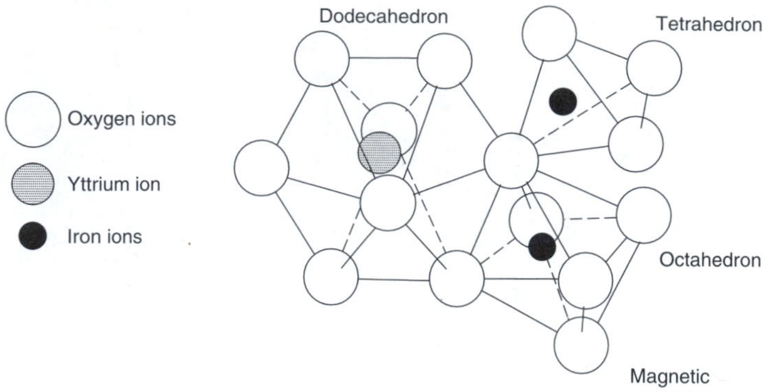

Dodecahedron   Tetrahedron

○ Oxygen ions

◌ Yttrium ion

● Iron ions

Octahedron

	Dodecahedron	Octahedron	Tetrahedron	Oxygen	Magnetic moment (gauss)
Approximate radii	1Å	0.7Å	0.5Å	1.3Å	
YIG	Y^{3+}_{24} Nonmagnetic (dielectric)	Fe^{3+}_{16}	Fe^{3+}_{24}	O^{2-}_{96}	1750
YAIG	Y^{3+}_{24}	Fe^{3+}_{16}	Al^{3+}_{3} Fe^{3+}_{21}	O^{2-}_{96}	1200
YAIG	Y^{3+}_{24}	Al^{3+}_{2} Fe^{3+}_{14}	Al^{3+}_{8} Fe^{3+}_{16}	O^{2-}_{96}	200
YAG	Y^{3+}_{24}	Al^{3+}_{16}	Al^{3+}_{24}	O^{2-}_{96}	Nonmagnetic (dielectric)

FIGURE 11.26 Schematic of the [garnet] structure ferrite and examples of the effect of Al^{3+} substitutions on the magnetic behavior. (After Cohen, H.M., Polycrystalline magnetic garnets. *Bell Labs Record*, Feb. 1968.)

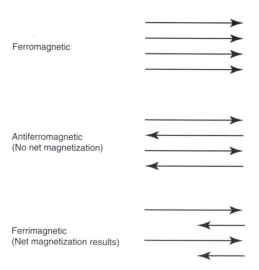

Ferromagnetic

Antiferromagnetic
(No net magnetization)

Ferrimagnetic
(Net magnetization results)

FIGURE 11.27 Definition of several terms that describe magnetic characteristics of a material. (Drawing courtesy of ASM International.)

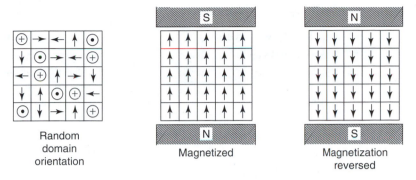

Random domain orientation

Magnetized

Magnetization reversed

FIGURE 11.28 Alignment of magnetic domains by application of a magnetic field. (Drawings courtesy of ASM International.)

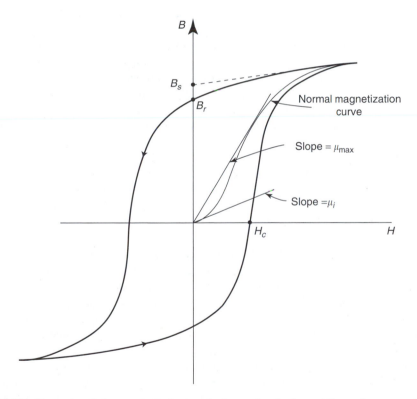

FIGURE 11.29 Example of the magnetic hysteresis loop of a ferrimagnetic or ferromagnetic material. (Drawing courtesy of ASM International.)

11.2.3 APPLICATIONS OF MAGNETIC CERAMICS

Ferrites are often classified according to the nature of their response to a magnetic field. A ferrite that is easily magnetized or demagnetized is referred to as a "soft" ferrite. A ferrite that has permanent magnetization and does not easily demagnetize is called a "hard" ferrite. A ferrite with a sharp square or rectangular hysteresis behavior is called a "square-loop" ferrite. A ferrite that is utilized at microwave frequencies is referred to as a microwave ferrite. Some examples of applications and

TABLE 11.8
Classes of Ferrites and Examples of Applications

Type	Industry where Used	Composition
Soft	Entertainment electronics, radio communication, military electronics	Manganese, zinc, iron oxides Nickel, zinc, iron oxides Nickel, copper, zinc, iron oxides
Square-Loop	Computers	Manganese, magnesium, iron oxides Cobalt, iron oxides
Microwave	Communications, military electronics	Magnesium, manganese, iron oxides Aluminum, nickel, zinc, iron oxides
Hard	Permanent magnet motors	Barium, iron oxides Strontium, iron oxides Lead, iron oxides

Source: From Henry, E.C., *Electronic Ceramics*, Doubleday, New York, 1969, p. 51.

general compositions for each ferrite class are listed in Table 11.8. Shapes of typical magnetization curves are compared in Figure 11.30.

Soft ferrites are generally of a spinel composition. They are used in television, radio, touchtone telephones, communication and radio-interference filters, submarine communications, electronic ignition systems, high-frequency fluorescent lighting, high-frequency welding, transformer cores, and high-speed tape and disk recording heads. The manganese-zinc and nickel-zinc spinel ferrites are soft ferrites that have been widely used.

Hard ferrites are generally of a hexagonal structure such as the magnetoplumbites $BaFe_{12}O_{19}$ and $PbFe_{12}O_{10}$. These have strong permanent magnetization. Some are used in the household as magnets to latch cupboards or hang memos. Others are mixed as a powder with rubber or a plastic and used as a flexible strip on the refrigerator door to act as a latch and to provide sealing. Permanent magnets are also used in an electricity meter to support the weight of the rotor to reduce wear in the bearings.

Another important application of hard ferrites is in small motors that replace copper-wound DC motors. Ceramic ferrite motors are used for electric toothbrushes, electric knives, and automobile windshield wipers, heater blowers, air conditioners, power windows, seat adjustors, and convertible-top raisers. Ceramic permanent magnets are also used in such diverse applications as loudspeakers and linear particle accelerators. For example, the 30 billion electron volt cyclotron particle accelerator at Brookhaven National Laboratory contains about seven tons of ceramic ferrite rings about 35 cm (13.75 in.) in diameter by 2.5 cm (1 in.) thick.

The next category is square-loop ferrites. Square-loop ferrites are generally of the [inverse spinel] or [garnet] structures. One of the most important applications in the past was in digital computers for data storage. Early digital computers were made possible by ceramic square-loop ferrite memory cores. Sales of computers leaped from essentially zero in 1952 to 350 million dollars in 1956 to over 7 billion dollars in 1969. A square-loop magnetic response is required for data storage, that is, the material has to have two stable states of magnetization that can rapidly be switched from one to the other. One state represents the digit "0" and the other the digit "1," allowing storage of data. Data are retrieved by switching. The cycle time for switching is about 5 μsec, which allows for very rapid data handling. Each storage unit or core consists of a ferrite magnet ring or toroid less than 0.15 mm (0.020 in.) in diameter. Each computer of this type had many of these cores. Over 6 billion cores were manufactured in 1969. The use of ferrite memory cores decreased after 1969 because of the development of alternative technologies more amenable to miniaturization and cost reduction.

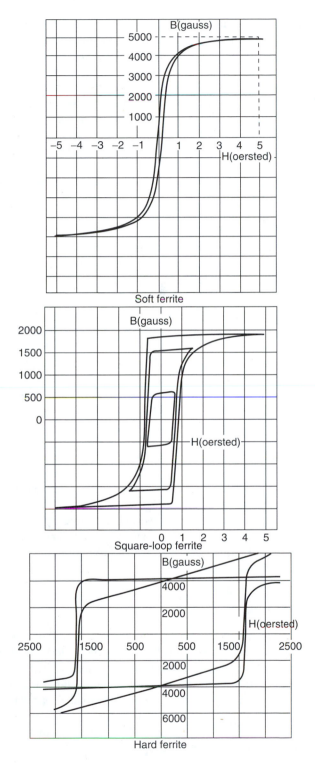

FIGURE 11.30 General shapes of the hysteresis loops for soft ferrites, hard ferrites, and square-loop ferrites. (From Henry, E.C., *Electronic Ceramics*, Doubleday, New York, 1969, pp. 58–59. With permission.)

Rare-earth garnet ferrites have also been used for data storage. One particularly interesting approach involved depositing a thin film ($\approx 5\,\mu m$ or 0.0002 in. thick) of the rare-earth garnet material on a nonmagnetic substrate having a slightly different thermal expansion coefficient.[13] During cooling, the mismatch in thermal contraction induced preferred magnetization perpendicular to the plane of the rare-earth garnet layer and separated into alternating microscopic domains of opposite spin. These domains, which looked like bubbles under a polarizing microscope, provided the binary memory input for a digital computer. Similar bubble domain structures have been achieved in thin single-crystal plates of orthorhombic ferrites.

Other applications of square-loop ferrites include switches and automatic controls for processing and production equipment.

Ceramic ferrites are used extensively for microwave applications. Microwaves are electromagnetic waves that are in the range of a millimeter to tens of centimeters in length. They have longer wavelength and lower frequency than visible light, but shorter wavelength and higher frequency than radio waves. The general range of frequency of microwaves is about 10^8 to 10^{12} cycles per second, compared to about 10^{14} to 10^{15} for visible light and 10^4 to 10^7 for radio waves. Because of their higher frequency, microwaves can carry more information than radio waves and are therefore important for high technology communications. However, because of their shorter wavelengths, microwaves can only travel in a straight line (unless they are channeled through a special coaxial cable) and must be transmitted via repeater stations that must be within line of sight of each other. These are the microwave towers that we see at about 30-mile intervals. Satellites are also utilized extensively to relay microwave signals.

Microwaves are used in a wide range of applications including radar, communications, and heating. Communication applications include transmission of voice, television, computer data, press copy, and electronic facsimiles (e.g., telefax). Ceramic ferrites are required in microwave applications for functions such as wave guides, rectifiers, resonance isolators, junction circulators, and phase shifters.

[Garnet] ferrite compositions are applied extensively to microwave applications. Each composition is engineered by crystal chemical modifications to meet the specific requirements of the specific applications. For example, Gd, Dy, Ho, Tb, and Ce are substituted for Y in the dodecahedral site, In and Mn are substituted for Fe in the octahedral site, and Al and Ga are substituted for Fe in the tetrahedral sites of yttrium iron garnet ($Y_3Fe_5O_{12}$).

A final important application example of ferrites is in recording tape. Individual single crystals of γ-Fe_2O_3 are dispersed in a thin, flexible polymer tape. The crystals are very carefully synthesized to be about $0.75\,\mu m$ long and $0.15\,\mu m$ in diameter, which provides optimum magnetic characteristics. The polymer is filled with about $40\,wt\%$ of these ferrite particles.

11.3 OPTICAL BEHAVIOR

The optical properties of a material include adsorption, transparency, index of refraction, color, and phosphorescence. These properties are determined primarily by the level of interaction between the incident electromagnetic radiation and the electrons within the material. The incident electromagnetic radiation can have a wide range of wavelengths and frequencies. This is illustrated in Figure 11.31. Note that visible light only forms a very narrow portion of the total electromagnetic spectrum, from about 400 to 700 nanometers (nm) (1 nanometer = 10 angstroms; 1000 nm = $1\,\mu m$). We will concentrate our discussions in this section on interaction of materials with visible light and adjacent wavelengths such as ultraviolet, infrared, and microwaves.

11.3.1 ABSORPTION AND TRANSPARENCY

Absorption and transparency are closely related optical properties. If the incident electromagnetic radiation stimulates electrons to move from their initial energy level to a different energy level, the

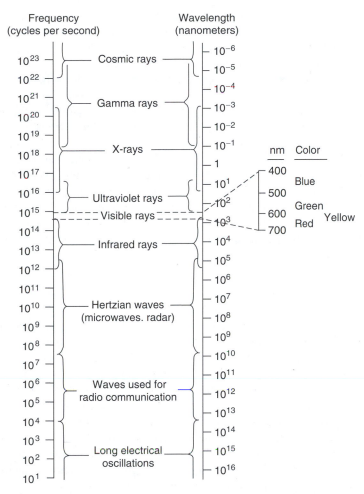

FIGURE 11.31 The electromagnetic spectrum. 1 nm = 10 Å. (Adapted from Kreidl, N.J., in *Physics of Electronic Ceramics*, Part B, Hench, L.L. and Dove, D.B., Eds., Marcel Dekker, New York, 1972, p. 917.)

radiation is absorbed and the material is opaque to this particular wavelength of radiation. Metals have many open energy levels for electron movement and thus are opaque to most wavelengths of electromagnetic radiation.

Ionic ceramics have filled electron shells comparable to inert gas electron configurations and do not have energy levels available for electron movement. Most single crystals of ionic ceramics are transparent to most electromagnetic wavelengths. Covalent ceramics vary in their level of optical transmission. Those that are good insulators and have a large band gap transmit. Those that are semiconductors and have a small band gap can transmit under some conditions, but become opaque as soon as enough energy is present for electrons to enter the conduction band.

Optical absorption can also occur due to *resonance*. This results when the frequency of the electromagnetic radiation is comparable to the natural frequency of the material. The resulting oscillations in the material absorb the radiation and the material is optically opaque.

Absorption due to electron transition and resonance is intrinsic. Absorption can also result from extrinsic effects due to scattering by inclusions, pores, grain boundaries, or other internal flaws. Such absorption is undesirable in most optical applications. For instance, much effort was expended to develop a transparent polycrystalline Al_2O_3 material. The first problem was to control the starting materials and processing to eliminate all porosity, both in the grains and between grains. Until this was achieved, the transparency was severely reduced by scattering from pores. Even when the pores were removed, though, additional scattering resulted from the grain boundaries. This resulted because Al_2O_3 is anisotropic and the light passing through the material was affected differently by grains oriented in different directions. This would not be a problem with polycrystalline Y_2O_3, which is isotropic.

Optical transparency is important in many applications.[14] Glass and a variety of ionic ceramics are transparent in the visible range of the spectrum. This is the range between wavelengths of 400 and 700 nm. There are many applications in this range for windows, lenses, prisms, and filters. Transparency in other wavelength ranges is important for electrooptical and electromagnetic window materials for tactical and strategic missiles, aircraft, remotely piloted vehicles, spacecraft, battlefield optics, and high-energy lasers. MgO, Al_2O_3, and fused SiO_2 are transparent in the ultraviolet (200 to 400 nm), a portion of the infrared (700 to 300 nm), and the radar ($>10^6$ nm) ranges. MgF_2, ZnS, ZnSe, and CdTe are transparent to infrared wavelengths as shown in Table 11.9. Silicon and Ge are transparent in the infrared, but are opaque to visible light. Diamond is transparent to visible and infrared wavelengths.

11.3.2 COLOR

Color is another optical property that leads to many ceramic applications. Color results from the absorption of a relatively narrow wavelength of radiation within the visible region of the spectrum (400 to 700 nm). For this type of absorption, transition of electrons must occur. There are four common types of electron transitions: (1) internal transitions within transition-metal, rare-earth, or other ions with incomplete electron shells; (2) charge-transfer processes in which an electron is transferred from one ion to another; (3) electron transitions associated with crystal imperfections; and (4) band-gap transitions, intrinsic coloration found in many semiconductor compounds. Types (1) through (2) are usually associated with small amounts of impurities or defects, whereas (3) is a bulk property. Color is common in materials containing transition elements having an incomplete d electron shell (V, Cr, Mn, Fe, Co, Ni, Cu) or containing rare-earth elements having an incomplete f electron shell. Color can also occur due to nonstoichiometry within a structure.

The oxidation state and bond field are also important in color formation. For instance, neither S^{2-} nor Cd^{2+} cause visible absorption, but CdS produces strong yellow. Similarly, Fe^{3+} and S^{2-} are responsible for the color of amber glass. Some ions can cause different colors depending on the position of the ion in the structure and the adjacent ions. Cr^{3+} produces a green color in Cr_2O_3 and emerald and a red color in ruby and spinel.

A beam of white light contains the complete spectrum of colors. When one wavelength is absorbed by a material, the remaining wavelengths pass through. The human eye sees the color of the combined wavelengths that pass through. For example, red light is absorbed by Fe^{2+} in olivine [$(Mg,Fe)_2SiO_4$].

TABLE 11.9
Materials Used for Infrared Windows

Material	Transmission Range, μm
MgO	0.2–9.0
MgF_2	0.4–9.0
ZnS	0.3–15
CaF_2	0.2–12
ZnSe	0.5–22
CdTe	0.8–32

The olivine appears to the eye to be a greenish color. The visible color is referred to as the *complementary color* of the *absorbed spectral color*. Table 11.10 lists the wavelength of various spectral colors and their corresponding complementary colors.

Let us return to the CdS example to further explain how coloration occurs due to semiconductor band-gap transitions. CdS has a band gap of 2.45 eV. The blue and violet wavelengths of incident light have enough energy to promote electrons into the conduction band. These wavelengths are absorbed. The longer wavelengths have lower frequency and do not have adequate energy to cause electron transitions. These wavelengths are transmitted through the CdS. Since violet and blue are absorbed, we see the complementary color yellow. In contrast, Si has a band gap of only 1.1 eV and requires less energy for electron transitions. As a result, all visible wavelengths of light are absorbed and Si appears opaque and metallic.

Color can also result from a defect in a crystal structure. The alkali halides such as NaCl, KBr, and KCl are good examples. Pure crystals of these materials are colorless. If the crystals are heated in the presence of the alkali metal vapor or are irradiated with a high-energy source such as x-rays or neutrons, anion vacancies are formed and the crystal changes color. Each vacancy contains a trapped electron. This is called a *color center* or *F-center* (from the German word "Farbzentrum"). Selected visible wavelengths are absorbed by the trapped electrons. F-centers cause NaCl to become yellow, KBr blue, and KCl magenta.

Ceramic colorants are widely used as pigments in paints and other materials produced and used at low temperatures. They are especially important where processing is done at elevated temperature where other types of pigments are destroyed. For instance, porcelain enamels that are fired in the range 750 to 850°C (1380 to 1560°F) require ceramic colorants. Ceramics having the spinel structure AB_2O_4 (such as blue $CoAl_2O_4$) are often used in this temperature range. Doped ZrO_2 and $ZrSiO_4$ are used at higher temperatures (1000 to 1250°C or 1830 to 2280°F) because of their increased resistance to attack by the glass in which they are dispersed. Dopants include vanadium (blue), praseodymium (yellow), and iron (pink).

Complex patterns of colors have been achieved in glass materials using photosensitive glass compositions.[15] These are opal glasses based on $SiO_2-Na_2O-Al_2O_3-ZnO$ compositions containing halides (F^-, Br^-, Cl^-, and/or I^-) and sensitizers (Ag^+, Ce^{3+}, Sn^{2+}, and Sb^{3+}). During cooling, the sensitizers form colloidal precipitates and the halides combine with Na^+ to form supersaturated solutions of NaF, NaBr, etc. The morphology of the precipitates and halide crystallites can be precisely controlled by exposures to ultraviolet light in the 280 to 340 nm band and by subsequent heat treatment. The morphology changes as the intensity and duration of the ultraviolet exposure and heat treatment are varied. The different precipitate and crystallite morphologies have different absorption bands in the ultraviolet and visible regions of the spectrum, resulting in the ability to produce

TABLE 11.10
The Spectral Colors, Their Wavelengths (λ), and Complementary Colors

λ (nm)	Spectral Color	Complementary Color
410	Violet	Lemon-yellow
430	Indigo	Yellow
480	Blue	Orange
500	Blue-green	Red
530	Green	Purple
560	Lemon-yellow	Violet
580	Yellow	Indigo
610	Orange	Blue
680	Red	Blue-green

a broad range of colors. By careful masking, complex combinations of colors can be achieved in a single piece of the glass. This type of glass material is referred to as photochromatic.

11.3.3 PHOSPHORESCENCE

Phosphorescence is another important optical property displayed by some ceramic compositions. *Phosphorescence* is the emission of light resulting from the excitation of the material by the appropriate energy source. Ceramic phosphors are used in fluorescent lights, oscilloscope screens, TV screens, photocopy lamps, and other applications.[16]

The fluorescent light consists of a sealed glass tube coated on the inside with a halogen phosphate [such as $Ca_5(PO_4)_3(Cl,F)$ or $Sr_5(PO_4)_3(Cl,F)$ doped with Sb and Mn] and filled with mercury vapor and argon. A voltage source provides an electric discharge that stimulates radiation of the mercury vapor at a wavelength of 2537 Å. This ultraviolet radiation excites a broad band of radiation in the visible range from the phosphor, producing the light source.

For oscilloscopes and TV sets the phosphor is excited by an electron beam that sweeps across the phosphor-coated screen. The decay time of the light emission of the phosphor is important. For color TV, decay occurs in approximately 1/10 to 1/100 of a second. A phosphor having slower decay is necessary for a radar screen. More than one phosphor is required for a color TV set, each being selected to emit a narrow wavelength of radiation corresponding to one of the primary colors. The most difficult color to achieve was red. Initially, $Zn_3(PO_4)_2$ doped with Mn was used for red, but it had low luminous efficiency and required high electron-beam currents. YVO_4 and Y_2O_2S, both doped with Eu, have been developed more recently and have improved efficiency. The phosphorescence results from transitions of $4f$ electrons in the Eu.[17]

Photocopy lamps also make use of phosphorescent ceramic coatings. $MgGa_2O_4$ doped with Mn is typically used, resulting in a narrow band of light emission in the green wavelength range. Strontium magnesium pyrophosphate doped with Eu^{2+} has also been used. It gives off light in the ultraviolet and blue range.

11.3.4 LASERS

LASER is the acronym for "light amplification by the stimulated emission of radiation."[18–20] Ceramic materials that have been especially important for laser applications include Al_2O_3 doped with Cr^{3+} (the ruby laser), $Y_3Al_5O_{12}$ doped with Nd^{3+} (the yttrium aluminum garnet or YAG laser), and glass doped with Nd^{3+}. All emit radiation of a specific wavelength, which is determined by the dopant: 694 nm* for the Cr^{3+}-doped A_2O_3 and 1006 nm for the Nd^{3+}-doped materials. Ceramics are also important in gas lasers such as the CO_2 laser as a tube with controlled passages through which the gas flows. Besides providing a precise flow path, the ceramic also must be stable at the operating temperature, dissipate heat, and be an electrical insulator. Beryllium oxide has demonstrated the right combination of properties for this application.

A third category of lasers, semiconductor lasers, also utilizes ceramics for the semiconductor junctions and for the substrates upon which the semiconductors are deposited. These lasers have similarities to LEDs and operate with relatively low electrical input. A common application is the laser pointer.

The ceramic component of the first category of lasers (doped ceramic host) consists of a cylindrical rod typically 0.3 to 1.5 cm (0.11 to 0.60 in.) in diameter by 5 to 15 cm (2 to 6 in.) long, with the ends polished to a flatness of $\lambda/10$ (where $\lambda = 590$ nm) and parallelism of $+/-5$ sec of arc. The rod must be as flaw-free as possible to avoid losses due to scattering, and the dopant must be uniformly dispersed. Usually, a tungsten-iodine filament lamp or a rare-gas arc lamp is used to stimulate the rod. A small portion of the lamp output is absorbed by the dopant ions (the rest is dissipated as heat),

---

* 1 nm = 10 Å.

resulting in electron transitions to a higher-energy state. As these electrons drop back into their initial energy state, light of a single wavelength specific to the dopant is emitted. Mirrors placed at the ends of the ceramic rod reflect the stimulated light back into the rod, where further coherent light is emitted and amplification results. The intensity can be built up and then released as a "pulse" by removing the mirror, as in the case of a pulsed laser. Alternatively, the mirror can be only partially reflective and allows emission of a portion of the coherent light as a continuous beam, as in the case of a continuous laser.

The first laser was reported by T.H. Maiman in 1960. It was a ruby laser (Cr^{3+}-doped Al_2O_3) of the design described above and illustrated schematically in Figure 11.32. Since 1960, thousands of applications have been found for lasers including communications, heat treating, cutting, weapons, ceramic powder synthesis, guidance systems, measurement instrumentation, holography, entertainment, medicine, and potentially, nuclear fusion. Each of these applications requires specific laser characteristics made possible by a wide range of dopants and hosts that have been discovered since 1960. Table 11.11 lists a variety of dopants and ceramic hosts along with the laser wavelengths.

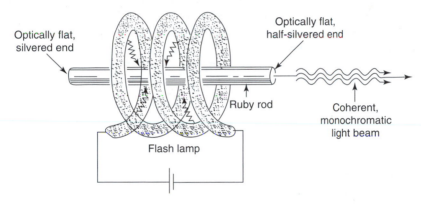

FIGURE 11.32 Schematic of a ruby laser. (From van Vlack, L.H., *Elements of Materials Science*, 2nd ed., Addison-Wesley, Reading, MA, 1964, p. 131. With permission.)

TABLE 11.11
Examples of Ceramic Laser Hosts and Dopants

Dopant	Host	Wavelength, μm	Other Hosts
Cr^{3+}	Al_2O_3	0.694	
Nd^{3+}	$Y_3Al_5O_{12}$(YAG)	1.06	$CaWO_4$, glass, $SrWO_4$, $SrMoO_4$, $CaMoO_4$, $PbMoO_4$, CaF_2, SrF_2, BaF_2, LaF_2, $CaNb_2O_6$, Y_2O_3, Gd_2O_3, $Y_3Ga_5O_{12}$, $Gd_3Ga_5O_{12}$, Na, $Gd(WO_4)_2$, CeF_3
Pr^{3+}	$CaWO_4$	1.047	$CaNb_2O_6$, $SrMoO_4$
Pr^{3+}	LaF_3	0.599	
Eu^{3+}	Y_2O_3	0.611	
Ho^{3+}	$CaWO_4$	2.046	$CaMoO_4$, $CaNb_2O_6$, CaF_2, $Y_3Al_5O_{12}$, glass
Er^{3+}	$CaWO_4$	1.612	
Dy^{2+}	CaF_2	2.36	
U^{3+}	CaF_2	2.24	BaF_2, SrF_2
Ni^{2+}	MgF_2	1.62	
Co^{2+}	MgF_2	1.75	ZnF_2

Source: Adapted from Johnson, L.F., in *Lasers*, Levine, A.K., Ed., Marcel Dekker, New York, 1966, p. 142.

11.3.5 INDEX OF REFRACTION

As discussed earlier in the section on absorption and transparency, light interacts with the electrons in the material and causes polarization for transparent materials, the degree of interaction varies, but results in a decrease in the velocity (V) of the light compared to its velocity in vacuum. The ratio of the velocity of light in vacuum to the velocity in the material is defined as the *index of refraction* (η) of the material:

$$\eta = \frac{V_{vacuum}}{V_{material}} \qquad (11.4)$$

The index of refraction for a material varies according to the wavelength of incident radiation. This is called *dispersion* and normally results in a decrease in η as wavelength increases. Therefore, when comparing the index of refraction of various materials, one should make sure that the wavelength was constant or that dispersion curves are available to make the necessary corrections.

The average index of refraction of various ceramic materials is listed in Table 11.12. The index of refraction is uniform in all directions (isotropic) for cubic crystals and glasses, but varies in different crystallographic directions for noncubic crystals.

The change in velocity when light passes from one material to another causes the light to bend or change direction. This is called *refraction* and is the effect that makes a fish in the water look like it is in a different position than it really is. The angle of refraction r is related to the angle of incidence i and the index of refraction by

$$\sin r = \frac{\sin i}{\eta} \qquad (11.5)$$

for the case where one medium is air or vacuum. This is shown schematically in Figure 11.33. Note that a portion of the light is also reflected.

Equation (11.5) is used extensively in the design of optical devices, sometimes as a correction (as in lenses, to make sure that the image is of the appropriate size and at the desired focal plane) and sometimes to achieve special effects. For instance, it is desirable with gemstones to optimize the color and brilliance of the material. The stones are cut at angles that allow maximum light to enter and be retained to accentuate the color and brilliancy.

Another important example is fiber optics. For communications, the fiber must carry coherent light from a laser for large distances with minimal loss. To minimize losses, a refractive index and angle of refraction are selected such that total internal reflection is achieved. Figure 11.34 shows how total internal reflection is achieved. First, let us imagine that the fiber is being lighted internally from the left. We want to calculate the value of the internal angle that will result in refraction parallel to the length of the fiber. The maximum angle of r occurs when $i = 90°$, and the critical value of r is obtained by

$$\sin r_{crit} = \frac{\sin i}{\eta} = \frac{1}{\eta} \qquad (11.6)$$

The value of r_{crit} for a typical glass (with $\eta = 1.5$) is about 42°. Any angle larger than this will result in total internal reflection. Because the critical angle for total reflection is relatively low, the light can be transmitted around corners without loss.

As discussed earlier, losses in transmission of light through a transparent material can result from scattering and absorption due to internal defects, inclusions, or inhomogeneity. Similar losses can occur in fibers due to surface defects. Scratches, grease, or dust on the surface can cause scattering or change the critical angle for total internal reflection. Interesting engineering approaches have been devised to avoid these losses. One approach is to clad (or encase) the fiber in a thin layer of glass having a lower index of refraction. Another approach is to provide a gradient in the index of refraction from the surface to the interior using ion-exchange techniques. This results in the actual focusing of the light in a sinusoidal path along the fiber axis.[20,22]

TABLE 11.12
Average Index of Refraction Values for Glass, Ceramic, and Organic Compositions

Material	Average Refractive Index
Glass Composition	
From orthoclase ($KAlSi_3O_8$)	1.51
From albite ($NaAlSi_3O_8$)	1.49
From nepheline syenite	1.50
Silica glass, SiO_2	1.458
Vycor glass (96% SiO_2)	1.458
Soda-lime-silica glass	1.51–1.52
Borosilicate (Pyrex) glass	1.47
Dense flint optical glasses	1.6–1.7
Arsenic trisulfide glass, As_2S_3	2.66
Crystals	
Silicon chloride, $SiCl_4$	1.412
Lithium fluoride, LiF	1.392
Sodium fluoride, NaF	1.326
Calcium fluoride, CaF_2	1.434
Corundum, Al_2O_3	1.76
Periclase, MgO	1.74
Quartz, SiO_2	1.55
Spinel, $MgAl_2O_4$	1.72
Zircon, $ZrSiO_4$	1.95
Orthoclase, $KAlSi_3O_8$	1.525
Albite, $NaAlSi_3O_8$	1.529
Anorthite, $CaAl_2Si_2O_8$	1.585
Sillimanite, $Al_2O_3 \cdot SiO_2$	1.65
Mullite, $3Al_2O_3 \cdot 2SiO_2$	1.64
Rutile, TiO_2	2.71
Silicon carbide, SiC	2.68
Litharge, PbO	2.61
Galena, PbS	3.912
Calcite, $CaCO_3$	1.65
Silicon, Si	3.49
Cadmium telluride, CdTe	2.74
Cadmium sulfide, CdS	2.50
Strontium titanate, $SrTiO_3$	2.49
Lithium niobate, $LiNbO_3$	2.31
Yttrium oxide, Y_2O_3	1.92
Zinc selenide, ZnSe	2.62
Barium titanate, $BaTiO_3$	2.40
Organics	
Polyethylene, nylon	1.50–1.54
Epoxy	1.58
Polystyrene	1.59–1.60

Source: Adapted from Kingery, W.D., Bowen, H.K., and Uhlmann, D.R., *Introduction to Ceramics*, 2nd ed., Wiley, New York, 1976, Chaps. 17–19, p. 662.

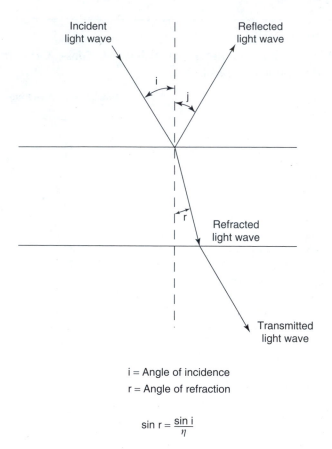

i = Angle of incidence

r = Angle of refraction

$$\sin r = \frac{\sin i}{\eta}$$

FIGURE 11.33 Primary refraction, reflection, and transmission for a plate of transparent material in air. (Drawing courtesy of ASM International.)

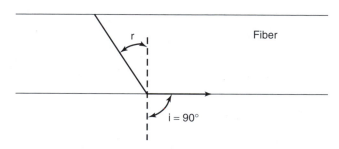

FIGURE 11.34 Minimum angular requirement for *r* to achieve total internal reflection within a fiber (Drawing courtesy of ASM International.)

11.3.6 Electro-Optics and Integrated Optic Devices

Fiber-optic communication systems have rapidly grown in importance and have made possible rapid global communication by internet. It has been estimated that a single fiber-optic system could simultaneously carry 500 million telephone conversations.

Between 1976 and 2000, enough optical fiber was installed worldwide to stretch back and forth between the Earth and the moon 160 times. A fiber-optic system consists of several elements: a transmitter, a transmission line, a repeater, and a receiver. The transmitter consists of a laser or a light-emitting diode to provide the light beam, plus a means of encoding information on the light beam.

The transmission line consists of low-loss glass optical fibers to guide and retain the light beam. The repeaters are amplification stations at regular intervals along the fiber to permit transmission over long distances. The receiver is a photodetector that decodes the information and transforms it back to an electrical signal.

A variety of functions must be conducted to encode information and successfully transfer it between the transmitter and the receiver. These include modulation, coupling, switching, multiplexing, and filtering. Modulation involves a change in frequency, amplitude, or phase of the signal to encode information. Coupling and switching involve transfer of information from the transmitter to the fiber and from one fiber to another. Multiplexing involves combination of several optical signals of different wavelength from separate fibers into one light beam. Filtering involves tuning the system to maintain a signal of a single wavelength. All of these functions in initial systems could only be accomplished by converting back to an electrical signal, which was complex and inefficient. Electro-optical materials have been developed to accomplish these functions directly on the light beams without having to convert back to an electrical signal.

The electro-optical effect consists of a change in the index of refraction of a material when an electric field is applied. $LiNbO_3$ and $LiTaO_3$ single crystals are examples. Both have strong electro-optic effect and requires only a small voltage to produce an effect strong enough for practical devices for switching, modulation, and multiplexing. $Ca_2Nb_2O_7$, $Sr_xBa_{1-x}Nb_2O_6$, KH_2PO_4 $K(Ta_xNb_{1-x})O_3$, and $BaNaNb_5O_{15}$ are other ceramic crystals with strong electro-optic behavior.

The index of refraction of some crystals changes when a pressure or acoustic wave is applied. This is referred to as the *acousto-optic effect*. Crystals of $LiNbO_3$, $LiTaO_3$, $PbMoO_4$, and $PbMoO_5$ exhibit the acousto-optic effect.

This chapter completes our study of electrical, magnetic, and optical properties and applications of ceramics. These technologies are of enormous importance to our modern civilization and represent by far the largest application category of advanced ceramic materials.

REFERENCES

1. von Hippel, A.R., *Dielectrics and Waves*, Wiley, New York, 1954.
2. *Handbook of Chemistry and Physics*, 63rd ed., CRC Press, Boca Raton, Fla., 1982.
3. Kingery, W.D., Bowen, H.K., and Uhlmann, D.R., *Introduction to Ceramics*, 2nd ed., Wiley, New York, 1976, Chaps. 17–19.
4. Thomas, D.G., in *Physics of Electronic Ceramics*, Part B, Hench, L.L. and Dove, D.B., Eds., Marcel Dekker, New York, 1972, pp. 1057–1090.
5. Henry, E.C., *Electronic Ceramics*, Doubleday, New York, 1969.
6. Beeforth, T.M. and Goldsmid, H.J., *Physics of Solid State Devices*, Pion Ltd., London, 1970.
7. Newnham, R.E., *Structure-Property Relations*, Springer-Verlag, New York, 1975.
8. Cady, W.G., *Piezoelectricity*, Vols. 1 and 2, Dover, New York, 1964.
9. Smit, J. and Wijn, H.P.J., *Ferrites*, Wiley, New York, 1959.
10. Riches, E.E., *Ferrites, A Review of Materials and Applications*, Mills & Boon Ltd., London, 1972.
11. Snelling, E.C., *Soft Ferrites*, ILIFFE Books Ltd., London, 1969.
12. Hoshino, Y., Iida, S., and Sugimoto, M., Eds., *Ferrites*, University Park Press, Tokyo, 1971.
13. Bobeck, A.H., Spencer, E.G., Van Uitert, L.G., Abrahams, S.C., Barns, R.L., Grodkiewicz, W.H., Sherwood, R.C., Schmidt, P.H., Smith, D.H., and Walters, E.M., Uniaxial magnetic garnets for domain wall "bubble" devices, *Appl. Phys. Lett.*, 17, 131, 1970.
14. Musikant, S., Tanzilli, R.A., Charles, R.J., Slack, G.A., White, W., and Cannon, R.M., *Advanced Optical Ceramics*, General Electric Co., Document 78SDR2195, prepared as a final report for ONR contract N00014-77-C-0649, 1978.
15. Stookey, S.D., Beall, G.H., and Pierson, J.E., Full-color photosensitive glass, *Cer. Ind.*, 37–39, June 1978.
16. Burrus, H.L., *Lamp Phosphors*, Mills & Boon Ltd., London, 1972.
17. Taylor, M.J., in *Modern Oxide Materials*, Cockayne, B. and Jones, D.W., Eds., Academic Press, London, 1972, pp. 120–146.

18. Harvey, A.F., Ed., *Coherent Light*, Wiley-Interscience, London, 1970.
19. Sona, A., Ed., *Lasers and Their Applications*, Gordon and Breach, New York, 1976.
20. Cockayne, B., in *Modern Oxide Materials*, Cockayne, B. and Jones, D.W., Eds., Academic Press, London, 1972, pp. 1–28.
21. Kingery, W.D., Bowen, H.K., and Uhlmann, D.R., *Introduction to Ceramics*, 2nd ed., Wiley, New York, 1976, Chap. 13.
22. Cole, J.A., Communications through a glass wire, *Ceramic Industry*, April 1977, pp. 28–30.

PROBLEMS

1. Which ion has the potential to produce the highest magnetic moment? Why?
   (a) Co^{2+}
   (b) Fe^{2+}
   (c) Fe^{3+}
   (d) Mn^{3+}
2. A capacitor has a 0.05-cm thick $BaTiO_3$ dielectric with a dielectric constant of 1600 and an electrode area of $0.2\,cm^2$. What is the capacitance?
3. A beam of light is shined on the surface of a material at an angle of incidence of 30°. The light is refracted by the material through an angle of 20°C. What is the refractive index of the material? The material is most likely to be which of the following:
   (a) TiO_2
   (b) Al_2O_3
   (c) epoxy
   (d) glass
4. Which of the following would you expect to have the least magnetic characteristics? Why?
   (a) $MnFe_2O_4$
   (b) $Gd_3Fe_5O_{12}$
   (c) $BaFe_{12}O_{22}$
   (d) $Y_3Al_4FeO_{12}$

STUDY GUIDE

1. What is a "dielectric" material?
2. Draw a diagram that illustrates polarization in a materials when it is placed in an electric field.
3. What are some important examples of atomic or ionic polarization?
4. Explain the meaning of "dielectric constant."
5. Compare the values (or range) of dielectric constant for the following:
   Alumina   Porcelain
   Many polymers TiO_2
   Barium titanate   Mica
6. Explain the meaning of "dielectric strength."
7. Why does dielectric constant often decrease as the frequency of the applied electric field increases?
8. Describe some of the functions of a capacitor in an electric circuit.
9. Explain why the development of high dielectric constant ceramics has been so important to advances in electrical devices.
10. Explain why barium titanate has such a large dielectric constant.
11. How is crystal chemistry used to engineer the dielectric properties of some ceramics? Give specific examples.

12. What is piezoelectricity?
13. What characteristics must exist in a single crystal for it to exhibit piezoelectric behavior?
14. Name some examples of some single crystals that can exhibit piezoelectric behavior in specific crystal directions.
15. What is "pyroelectricity"?
16. What happens in a ferroelectric ceramic when an electric field is applied, then released, then reversed? Use explanation and drawing.
17. What is the Curie temperature, and what is its relationship to crystal structure?
18. How is ferroelectric behavior achieved in a polycrystalline ceramic?
19. Explain how piezoelectricity makes each of the following possible:
    Hydrophone
    Sonar
    Accelerometer
    Step-up transformer
    Actuator
    Medical scanner
    Sound generation
20. What is the source of magnetism in a material, and which elements or ions exhibit this characteristic or behavior?
21. What term is commonly used to refer to a magnetic ceramic?
22. Explain why magnetite is magnetic.
23. Explain how the magnetic behavior of an inverse spinel or a garnet can be engineered?
24. What is the difference between a "soft ferrite" and a "hard ferrite"?
25. List 10 different applications for magnetic ceramics.
26. Identify three different ways that light is absorbed by materials.
27. List some examples of ceramics that are transparent to various wavelengths of the electromagnetic spectrum.
28. What results in color in a ceramic material?
29. What is phosphorescence, and what are some important applications?
30. How are ceramics used in lasers?
31. What are some of the roles of ceramics in fiberoptic communication systems?

Part III

Processing of Ceramics

The relationships among atomic bonding, crystal structure, and properties for ceramics, metals, and polymers were discussed in Part II. It was shown that the theoretical strength is controlled by the strength of bonding, but that in actual ceramic components the theoretical strength is not achieved due to flaws in the fabricated material. The objectives of Part III are to review the fabrication processes used for manufacturing ceramic components, determine where in these processes strength-limiting flaws are likely to occur, and provide the reader with approaches for detecting these flaws and working with the ceramic fabricators to eliminate them.

Most ceramic fabrication processes begin with finely ground powder, Chapter 12 describes the criteria for selection of the starting powder, methods of achieving the proper particle size distribution, and requirements for pretreating the powder before it can be formed into the desired component.

Chapter 13 describes the processes used to form the ceramic powders into the component shapes. Uniaxial and isostatic pressing, slip casting, extrusion, injection molding, tape forming, and green machining are included.

The shapes resulting from the forming processes described in Chapter 13 consist essentially of powder compacts that must be densified by high-temperature processing before they will have adequate strength and other properties. The mechanisms and processes for densification are explored in Chapter 14. Some processes combine forming and densification in a single step. These include hot pressing, chemical vapor deposition, liquid particle spray, and cementitious bonding. These are also discussed in Chapter 14.

Ceramic components requiring close tolerances must be machined after densification. This machining step can be as expensive as all the other process steps together and thus must be thoroughly understood by the engineer. Chapter 15 reviews the mechanisms of material removal during machining, the effects on the strength of the ceramic, and guidelines for selection of a machining method and abrasive material.

Chapter 16 discusses quality control and quality assurance methods for ceramic components. Use of specifications, in-process certification, problem-solving strategies, and some statistical process control (SPC) techniques are described. Nondestructive inspection techniques for both internal and surface flaws are briefly reviewed.

12 Powder Processing

The nature of the raw material has a major effect on the final properties of a ceramic component. Purity, particle size distribution, reactivity, polymorphic form, availability, and cost must all be considered and carefully controlled. In this chapter we discuss the types and sources of raw materials and the processing required to prepare them to the appropriate purity, particle size distribution, and other conditions necessary to achieve optimum results in later processing steps.

12.1 RAW MATERIALS

12.1.1 TRADITIONAL CERAMICS

Ceramics have been produced for centuries. The earliest ceramic articles were made from naturally occurring raw materials. Early civilizations found that clay minerals became plastic when water was added and could be molded into shapes. The shape could then be dried in the sun and hardened in a high-temperature fire.

Many of the raw materials used by the ancient civilizations are still used today and form the basis of a sizable segment of the ceramic industry.[1–3] These ceramic products are often referred to as *traditional ceramics*. Important applications of traditional ceramics are listed in Table 12.1. Some of the naturally occurring minerals and their sources and uses are summarized in the following paragraphs.

The clay minerals are hydrated aluminosilicates that have layer structures. There are a variety of clay minerals, including kaolinite $[Al_2(Si_2O_5)(OH)_4]$, halloysite $[Al_2(Si_2O_5)(OH)_4 \cdot 2H_2O]$, pyrophyllite $[Al_2(Si_2O_5)_2(OH)_2]$, and montmorillonite $[Al_{1.67}(Na,Mg)_{0.33}(Si_2O_5)_2(OH)_2]$. All are secondary in origin, having formed by weathering of igneous rocks under the influence of water, dissolved CO_2, and organic acids. The largest deposits were formed when feldspar $(KAlSi_3O_8)$ was eroded from rocks such as granite and deposited in lake beds and then altered to a clay.

The importance of clays in the evolution of traditional ceramic processing cannot be overemphasized. The plasticity developed when water is added provides the bond and workability so important in the fabrication of pottery, dinnerware, brick, tile, and pipe.

Silica (SiO_2) is a major ingredient in glass, glazes, enamels, refractories, abrasives, and whiteware. Its major sources are in the polymorphic form quartz, which is the primary constituent of sand, sandstone, and quartzite.

Feldspar is also used in glass, pottery, enamel, and other ceramic products. Feldspar minerals range in composition from $KAlSi_3O_8$ to $NaAlSi_3O_8$ to $CaAl_2Si_2O_8$ and act as a flux (reduces the melting temperature) in a composition. Nepheline syenite $(Na_2Al_2Si_2O_8)$ is used in a similar fashion.

TABLE 12.1
Traditional Ceramics

Whitewares	Dishes, plumbing, enamels, tiles
Heavy clay products	Sewer pipe, brick, pottery, sewage treatment, and water purification components
Refractories	Brick, castables, cements, crucibles, molds
Construction	Brick, block, plaster, concrete, tile, glass, fiberglass
Abrasive products	Grinding wheels, abrasives, milling media, sandblast nozzles, sandpaper
Glass	Too numerous to list

Other naturally occurring minerals used directly in ceramic compositions include talc $[(Mg_3(Si_2O_5)_2) (OH)_2]$, asbestos $[(Mg_3Si_2O_5) (OH)_4]$, wollastonite $(CaSiO_3)$, and sillimanite (Al_2SiO_5).

12.1.2 MODERN CERAMICS

During the past 100 years scientists and engineers have acquired a much better understanding of ceramic materials and their processing and have found that naturally occurring minerals could be refined or new compositions synthesized to achieve unique properties.[4,5] These refined or new ceramics are often referred to as *modern ceramics*. They typically are of highly controlled composition and structure and have been engineered to fill the needs of applications too demanding for traditional ceramics. Modern ceramics include the oxide ceramics (such as Al_2O_3, ZrO_2, ThO_2, BeO, MgO, and $MgAl_2O_4$), magnetic ceramics (such as $PbFe_{12}O_{19}$, $ZnFe_2O_4$, and $Y_6Fe_{10}O_{24}$), ferroelectric ceramics (such as $BaTiO_3$), nuclear fuels (such as UO_2 and UN), and nitrides, carbides, and borides (such as Si_3N_4, SiC, B_4C, and TiB_2).[6–8] Table 12.2 summarizes many of the modern ceramic applications. Emphasis in this book will be on the modern ceramics, as they are the ones with which an engineer is most likely to become involved. The following sections describe how some of the modern ceramic starting powders are refined or synthesized.

12.1.2.1 Aluminum Oxide Powder

Aluminum oxide (Al_2O_3) occurs naturally as the mineral corundum, which is better known to most of us when it is in gem-quality crystals called ruby and sapphire. Ruby and sapphire are precious gems because of their chemical inertness and hardness. Al_2O_2 powder is produced in large quantity from the mineral bauxite by the Bayer process. Bauxite is primarily colloidal aluminum hydroxide intimately mixed with iron hydroxide and other impurities. The Bayer process involves the selective leaching of the alumina by caustic soda and precipitation of the purified aluminum hydroxide. The resulting fine-particle-size aluminum hydroxide can then be thermally converted to Al_2O_3 powder, which is used to manufacture polycrystalline Al_2O_3-based ceramics.

Al_2O_3 powder is used in the manufacture of porcelain, alumina laboratory ware, crucibles and metal casting molds, high-temperature cements, wear-resistant parts (sleeves, tiles, seals, etc.), sandblast nozzles, armor, medical components, abrasives, refractories, and a variety of other components. Many hundreds of tons of alumina powder and alumina-based articles are produced each year. It has even been used to make extrusion dies for corn chips and mixing valves for water faucets.

12.1.2.2 Magnesium Oxide Powder

Magnesium oxide occurs naturally as the mineral periclase, but not in adequate quantity or purity for commercial requirements. Most MgO powder is produced from $MgCO_3$ or from seawater. It is

TABLE 12.2
Modern Ceramics

Electronics	Heating elements, dielectrics, substrates, semiconductors, insulators, transducers, lasers, hermetic seals, igniters
Aerospace and automotive	Reentry, radomes, turbine components, heat exchangers, emission control
Medical	Prosthetics, controls, orthodontic brackets
High-temperature structural	Kiln furniture, braze fixtures, advanced refractories
Nuclear	Fuels, controls
Technical	Laboratory ware
Miscellaneous	Cutting tools, wear-resistant components, armor, magnets, glass ceramics, single crystals, fiber optics

extracted from seawater as the hydroxide, which is easily converted to the oxide by heating at the appropriate temperature.

MgO powder is used extensively for high-temperature electrical insulation and refractory brick.

12.1.2.3   Silicon Carbide Powder

SiC has been found occurring naturally only as small green hexagonal plates in meteoric iron. This same hexagonal polymorph (α-SiC) has been synthesized commercially in large quantities by the Acheson process. This fascinating process appears crude, but is cost-effective and simultaneously produces lower-grade SiC for abrasives and high-grade SiC for electrical applications. The Acheson process consists essentially of mixing SiO_2 sand with coke in a large elongated mound and placing large carbon electrodes in opposite ends. An electric current is then passed between the electrodes, resistance-heating the coke in the mound to about 2200°C. In this temperature range the coke reacts exothermically with the SiO_2 to generate enough heat to sustain a chemical reaction to produce SiC plus CO gas. Exothermic heating continues until the reaction is completed in the interior of the mound. After cooling, the mound is broken up and sorted. The core contains intergrown green hexagonal SiC crystals that are low in impurities and suitable for electronic applications. Around the core is a zone of lower purity that is used for abrasives. The outer layer of the mound is a mixture of SiC and unreacted SiO_2 and carbon that is added to silica sand and coke for the next batch.

SiC can be prepared from almost any source of silicon and carbon. For instance, it has been prepared in the laboratory from a mixture of silicon metal powder and sugar. It has also been prepared from rice hulls. It can also be prepared from silicon tetrachloride ($SiCl_4$) and some silanes. In this case, the cubic β-SiC polymorph normally forms.

SiC is used for high-temperature kiln furniture, electrical-resistance heating elements, grinding wheels and abrasives, wear-resistance applications, incinerator linings, varistors, light emitting diodes, and also gem stones cut from single crystals.

12.1.2.4   Silicon Nitride Powder

Silicon nitride does not occur naturally. It has been synthesized by several different processes. Some of the powder available commercially has been made by the reaction of silicon powder with nitrogen at temperatures in the range 1250 to 1400°C and consists of a mixture of α-Si_3N_4 and β-Si_3N_4 polymorphs. This Si_3N_4 is not a ready-to-use powder when it is removed from the furnace. It is loosely bonded and must be crushed and sized first. The resulting powder is not of high purity, but contains impurities such as Fe, Ca, and Al, which were originally present in the silicon, plus impurities picked up during crushing and grinding.

Higher-purity Si_3N_4 powder has been made by reduction of SiO_2 with carbon in the appropriate nitrogen environment and reaction of $SiCl_4$ or silanes with ammonia. Both of these methods produce very fine particle size powder that does not require further grinding for use. In fact, some of these powders are so fine that they require coarsening by heat treating (calcining) before they are suitable for shape-forming operations.

High-purity Si_3N_4 powder has been made in small quantities by laser reaction.[9] A mixture of gaseous silane (SiH_4) and ammonia is exposed to the coherent light output of a CO_2 laser. The silane has high absorption for the wavelengths involved, resulting in the heat required for reaction. The resulting Si_3N_4 is in spherical particles of a uniform size for the given gas flow and laser power conditions. Particles typically in the range 20 to 100 nm can be produced.

12.1.3   RAW MATERIALS SELECTION CRITERIA

The selection criteria for ceramic starting powders are dependent on the properties required in the finished component. Purity, particle size distribution, reactivity, and polymorphic form can all affect the final properties and thus must be considered from the outset.

12.1.3.1 Purity

Purity strongly influences high-temperature properties such as strength, stress rupture life, and oxidation resistance, as discussed in earlier chapters. The effect of the impurity is dependent on the chemistry of both the matrix material and the impurity, the distribution of the impurity, and the service conditions of the component (time, temperature, stress, environment). For instance, Ca severely decreases the creep resistance of Si_3N_4 hot-pressed with MgO as a densification (sintering) aid, but appears to have little effect on Si_3N_4 hot-pressed with Y_2O_3 as the densification aid.[10,11] In the former case, the Ca is concentrated at the grain boundaries and depresses the softening temperature of the grain boundary glass phase. In the latter case, the Ca is apparently absorbed into solid solution by the crystalline structure and does not significantly reduce the refractoriness of the system.

Impurities present as inclusions do not appreciably affect properties such as creep or oxidation, but do act as flaws that can concentrate stress and decrease component tensile strength. The effect on strength is dependent on the size of the inclusion compared to the grain size of the ceramic and on the relative thermal expansion and elastic properties of the matrix and inclusion. Tungsten carbide inclusions in Si_3N_4 have little effect on the strength; Fe and Si have a large effect.

The effects of impurities are important for mechanical properties, but may be even more important for electrical, magnetic, and optical properties. Electrical, magnetic, and optical properties are usually carefully tailored for a specific application, often by closely controlled addition of a dopant. Slight variations in the concentration or distribution of the dopant severely alter the properties. Similarly, the presence of unwanted impurities can poison the effectiveness of the dopant and cause the device to operate improperly.

12.1.3.2 Particle Size and Reactivity

Particle size distribution is important, depending on which consolidation or shaping technique is to be used. In most cases the objective of the consolidation step is to achieve maximum particle packing and uniformity, so that minimum shrinkage and retained porosity will result during densification. A single particle size does not produce good packing. Optimum packing for particles all the same size results in over 30% void space. Adding particles of a size equivalent to the largest voids reduces the void content to 26%. Adding a third, still smaller particle size can reduce the pore volume to 23%. Therefore, to achieve maximum particle packing, a range of particle sizes is required.

Real ceramic particles are generally irregular in shape and do not fit into ideal packing. Porosity after compaction of these powders is generally greater than 35% and sometimes greater than 50%. Large amounts of porosity are difficult to eliminate during densification. Figure 12.1 shows an example of this for Al_2O_3. A high degree of porosity in the compact resulted in high porosity and large grain size after sintering. For example, 40% initial porosity resulted in only 0.25% porosity and 2 μm grain size after sintering, whereas 50% initial porosity resulted in 2% porosity and over 10 μm grain size after sintering and 60% porosity resulted in over 10% porosity and 55 μm grain size after sintering.

Low porosity and fine grain size are beneficial to achieve a ceramic with high strength. However, there are many applications where strength is not the primary criterion. Refractories are a good example. Most refractories contain either large particles or high porosity as an important constituent in achieving the desired properties such as low thermal conductivity and high thermal shock resistance.

Another important aspect of the starting powder is *reactivity*. The primary driving force for densification of a compacted powder at high temperature is the change in surface free energy. Very small particles with high surface area have high surface free energy and thus have a strong thermodynamic drive to decrease their surface area by bonding together. Very small particles, approximately 1 μm or less, can be compacted into a porous shape and sintered at a high temperature to near-theoretical density. Transparent polycrystalline alumina for sodium vapor lamp envelopes is a good example. To achieve transparency, virtually all the pores larger than about 0.4 μm must be removed during sintering. Pores larger than 0.4 μm scatter the visible wavelengths of light (~0.4 to 0.8 μm) and

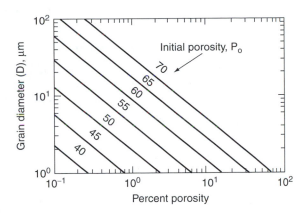

FIGURE 12.1 Effects of porosity in the powder compact on the porosity and grain size of Al_2O_3 after densification. (From Bruch, C.A., Sintering kinetics for the high density alumina process, *Am. Ceram. Soc. Bull.*, 41(12), 799, 1962.)

prevent transmission. Use of a highly reactive starting Al_2O_3 powder with an average particle size of about $0.3\,\mu m$ (plus use of a specific additive discussed in Chapter 14) aids in elimination of pores over $0.4\,\mu m$.

Another example is sintered Si_3N_4. Starting powder of approximately $2\,\mu m$ average particle size only sinters to about 90% of theoretical density. Submicron powder with a surface area roughly greater than $10\,m^2/g$ sinters to greater than 95% of theoretical density.

Particle size distribution and reactivity are also important in determining the temperature and the time at temperature necessary to achieve sintering. Typically, the finer the powder and the greater its surface area, the lower are the temperature and time at temperature for densification. This can have an important effect on strength. Long times at temperature result in increased grain growth and lower strength. To optimize strength, a powder than can be densified quickly with minimal grain growth is desired.

12.2 POWDER PREPARATION AND SIZING

As mentioned earlier, control of particle size and particle size distribution is required to achieve the optimum properties for the intended applications. Each application has specific requirements. High-strength ceramics require very fine particles (typically $<1\,\mu m$) to achieve a fine-grained microstructure with minimum flaw size. Refractories generally require a bimodal or multimodal particle distribution that can contain particles hundreds of microns across. Abrasives must be available in many sizes, each segregated into a narrow size range. Many different powder synthesis and sizing techniques have been developed to achieve the various required distributions. Table 12.3 lists many of these diverse techniques. Note that some are mechanical and involve crushing or separation procedures, some are chemical and yield fine particles directly, and some require a combination.

12.2.1 Mechanical Sizing

12.2.1.1 Screening

Screening is a sorting method of particle sizing. The powder is poured onto a single screen having the selected size openings or on a series of screens, each subsequently with smaller openings. The particles are separated into size ranges; the particles larger than the screen openings remain on the screen and smaller particles pass through until they reach a screen with holes too small to pass through.

TABLE 12.3
Techniques for Powder Preparation and Sizing

Mechanical	Chemical	Miscellaneous
Screening	Precipitation	Calcining
Elutriation	Sol–gel	Fluidized bedRotary kiln
Air classification	Liquid mix	Combustion synthesis
Ball milling	Decomposition	
Attrition milling	Freeze drying	
Vibratory milling	Hot kerosene drying	
Turbomilling	Plasma	
Fluid energy milling	Laser	
Hammer milling	Hydrothermal	
Roll crushing		

TABLE 12.4
ASTM Standard Screen Sizes

"Mesh" Sieve Designation	Sieve Opening	
	mm	in.
4	4.76	0.187
6	3.36	0.132
10	2.00	0.0787
12	1.68	0.0661
16	1.19	0.4969
20	0.84	0.0331
40	0.42	0.0165
80	0.177	0.0070
120	0.125	0.0049
170	0.088	0.0035
200	0.074	0.0029
230	0.063	0.0025
270	0.053	0.0021
325	0.044	0.0017
400	0.037	0.0015

Source: ASTM E11, *Annual Book of ASTM Standards*, American Society for Testing and Materials, Philadelphia, 1970.

Screen sizes are classified according to the number of openings per linear inch and are referred to as *mesh sizes*. A 16-mesh screen has 16 equally spaced openings per linear inch; a 325-mesh screen has 325. Table 12.4 compares the mesh size of standard screens with the actual size of the openings.

Raw and processed ceramic materials are often supplied according to screen size. For instance, a 325-mesh powder has all passed through a 325-mesh screen and should contain no particles larger than 44 μm (0.0018 in.). A powder designated −100 mesh +150 mesh consists of a narrow particle size range that was small enough to pass through a 100-mesh screen but too large to pass through a 150-mesh screen. Powders containing a broad particle size range can also be classified according to screen size, for example

Size Range	Weight (%)
−80 + 100	5
−100 + 150	8
−150 + 220	13
−220 + 280	20
−280 + 325	18
−325	36

Screening can be conducted dry or with the particles suspended in a slurry. Dry screening is used most frequently for larger particles and is a fast and effective approach. It is used in the mining industry and in many phases of the ceramic industry, especially in the sizing of abrasives. For free-flowing particles, dry screening can normally be effective down to about 325 mesh. Below this the particles are so fine that they either tend to agglomerate or clog the screen. Some automatic screen systems use airflow or vibration to aid in screening powders that have a significant portion of particles in the range of 325 mesh or smaller.

Suspending the particles in a dilute water or other liquid suspension (slurry) also aids in screening fine particles. Slurries can normally be screened easily through at least 500 mesh as long as the solids content in the slurry is low and fluidity is high. For very fine powder, this is a useful method of assuring that no particles larger than an acceptable limit (determined by the screen size selected) are left in the powder. Since isolated large particles in a powder of fine particle size distribution often become the strength-limiting flaw in the final component, wet screening can be used as an in-process quality control step.

Screening does have limitations. As with any process, it is accurate only as long as the equipment is properly maintained. Distorted or broken screens pass larger particles than specified. Screens of 325 mesh and finer are frequently damaged because of the fragile nature of the thin filaments required to construct the screen. The user tends to get impatient due to the slow feed rate of some powders and tries to force the powder through by brushing or scraping the screen. Once the screen is damaged, particle size control and knowledge of the particle size distribution are lost.

Another limitation of screening is related to the nature of the powder. If the powder tends to compact or agglomerate, groups of particles will act as a single particle and result in inaccurate screening. Similarly, packing or agglomeration can clog the screen and prevent further screening or decrease efficiency.

12.2.1.2   Air Classification

Air classification (also referred to as air separation) is used to separate coarse and fine fractions of dry ceramic powders. A schematic of an air classifier is shown in Figure 12.2. Separation is achieved by control of horizontal centrifugal force and vertical air currents within the classifier. Particles enter the equipment along the centerline and are centrifugally accelerated outward. As the coarse particles move radially away from the center into the separating zone, they lose velocity and settle into a collection cone. The finer particles are carried upward and radially by the air currents through selector blades. These selector blades impart an additional centrifugal force to the particles and cause additional coarse particles to settle into the coarse collection cone. The fines are then carried by the airflow to a separate cone for collection.

Air classifiers have been used to separate particles in the approximate range of 40 to 400 mesh at rates exceeding 400 tons/h. They are used extensively in the cement industry for sizing fine particles

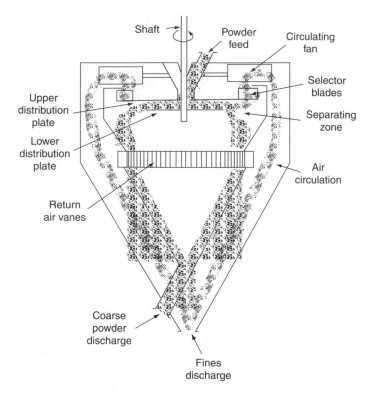

FIGURE 12.2 Drawing of an air classifier, showing the paths of the coarse and fine particles. (Courtesy of Sturtevant Mill Company, Boston, MA.)

to close size and surface area specification (-170 mesh size range). They are also used to remove undesirable fines in other ceramic industries where a coarse aggregate is required.

Special air classifiers are available for isolating fine powders in the range below $20 \mu m$. Separation can be done with reasonable accuracy down to about $10 \mu m$. However, below $10 \mu m$, it is difficult to obtain a cut that is completely free of larger particles. For instance, if a powder were desired with no particles larger than $5 \mu m$, it could not be guaranteed by air classification.

Air classification is frequently linked directly to milling, crushing, grinding, or other comminution equipment in a closed circuit. Particles from the mill are discharged directly into the air classifier. The fines are separated and the coarse is returned to the mill for further grinding. One type of unit combines size reduction and classification into a single piece of equipment. The coarse powder particles are carried by high-velocity air through two opposing nozzles. Where the two air streams meet, particles strike each other and are shattered into smaller particles. The air carrying the particles flows vertically. Large particles pass through a centrifugal-type air classifier at the top of the unit, where additional controlled sizing is accomplished.

Air classification has its advantages and limitations. It is an efficient and high-volume approach for separating coarse particles from fine particles and producing controlled size ranges roughly from 40 to 400 mesh. However, it is limited in its efficiency and accuracy in producing controlled sizing of particles below $10 \mu m$.

Another concern of the engineer is the presence of contamination. Sliding motion and impact of the ceramic particles against equipment surfaces result in some metallic contamination. The amount of contamination is less than for comminution equipment, but may still cause a problem for some applications. Therefore, the engineer should be aware of the air classification process and its potential for affecting material application specification.

During development, an engineer sometimes needs to have an experimental batch of powder air classified. This can be done on a contract basis with many of the equipment manufacturers at their facilities. Again, though, it should be realized by the engineer that contamination can be picked up, especially since research and development equipment is used on a batch basis with a different material in each batch. Even with careful cleaning (which is not automatically done between batches), foreign particles will remain in the equipment. The best procedure is to insist on careful cleaning, run a sacrificial batch of your powder prior to your study batch, and be aware that contamination may occur that would not occur in production.

12.2.1.3 Elutriation

Elutriation is a general term that refers to particle size separation based on settling rate; that is, large or high-specific-gravity particles settle more rapidly from a suspension than do small or low-specific-gravity particles. Air classification is a form of elutriation where the suspending medium is air. For this book the term *elutriation* is used to describe particle sizing by settling from a liquid suspension.

Elutriation is frequently used in the laboratory for obtaining very fine particle distributions free of large particles. The powder is mixed with water or other liquid and usually with a wetting agent and possibly a deflocculant* to yield a dilute suspension. Stirring or mixing is stopped and settling is allowed to occur for a predetermined time. The time is based on the particle size cut desired. The fluid containing the fine particles is then decanted or siphoned and the remaining fluid and residue discarded or used for some other purpose.

A major problem with elutriation is that the fines must be extracted from the fluid before they can be used. This can be done by evaporating the fluid or by filtration. Both tend to leave the fines compacted or crusted rather than as a free-flowing powder, thus requiring additional process steps before the powder can be used. Also, unless the elutriation and liquid removal are conducted in a closed system, chances for contamination are high.

12.2.1.4 Ball Milling

The desired particle size distribution usually cannot be achieved simply by screening, classifying, or elutriating the raw material. More typically a particle size reduction (comminution) step is required. Ball milling is one of the most widely used. Ball milling consists of placing the particles to be ground (the "charge") in a closed cylindrical container with grinding media (balls, short cylinders, or rods) and rotating the cylinder horizontally on its axis so that the media cascade. This is illustrated schematically in Figure 12.3. The ceramic particles move between the much larger media and between the media and the wall of the mill and are effectively broken into successively smaller particles.

The rate of milling is determined by the relative size, specific gravity, and hardness of the media and the particles. High-specific-gravity media can accomplish a specified size reduction much more quickly than low-specific-gravity media. The following media are commonly used and are listed in descending order of specific gravity: WC, steel ZrO_2, Al_2O_3, and SiO_2. Some experimental Si_3N_4 media are shown in Figure 12.4.

Contamination is a problem in milling. While the particle size is being decreased, the mill walls and media are also wearing. Milling Al_2O_3 powder with porcelain or SiO_2 media can result in about 0.1% contamination per hour. Some Si_3N_4 powder milled in a porcelain-lined mill with porcelain cylinders picked up nearly 6% contamination in 72 h of milling time. The contamination in the Si_3N_4 resulted in a decrease in the high-temperature strength by a factor of three and nearly an order-of-magnitude decrease in creep resistance in the final part.

Contamination can be controlled by careful selection of the mill lining and the media. Polyurethane and various types of rubber are excellent wear-resistant linings and have been used successfully with dry milling and with water as a milling fluid. However, some milling is conducted with organic fluids that may attack rubber or polyurethane. Very hard grinding media can reduce contamination because

---

* The wetting agent and deflocculant help to break up agglomerates.

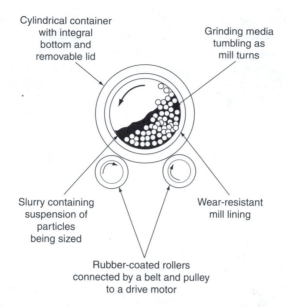

Cylindrical container
with integral
bottom and
removable lid

Grinding media
tumbling as
mill turns

Slurry containing
suspension of
particles
being sized

Wear-resistant
mill lining

Rubber-coated rollers
connected by a belt and pulley
to a drive motor

FIGURE 12.3   Schematic cross section showing the key elements of a typical ball mill. (© ASM International.)

FIGURE 12.4 Si_3N_4 grinding media showing one of the common configurations. Spheres are also commonly used. (Courtesy of KemaNord.)

TABLE 12.5

Comparison of the Advantages of Wet and Dry Ball Milling

Advantages of Wet Milling
Low power required
No dust problems
Higher rotational speeds
Can wet screen through fine screen
Good homogenization
Smaller particle size than dry
Narrower particle size distribution than dry
Compatible with spray drying and casting processes
Advantages of Dry Milling
Avoids drying of the powder
Avoids reaction of the powder/liquid
Less media and lining wear than wet
Can be started/stopped any time
Easier to optimize

Source: Adapted from Lukasiewicz, S., Ceramic powder processing, *NICE Course*, American Ceramic Society, Ohio, 1982.

they wear more slowly. WC is good for some cases because its high hardness reduces wear and its high specific gravity minimizes milling time. If contamination from the media is an especially critical consideration, milling can be conducted with media made of the same composition as the powder being milled. Another approach is to mill with steel media and remove the contamination by acid leaching.

Milling can be conducted either dry or wet. The advantages and disadvantages are listed in Table 12.5. Dry milling has the advantage that the resulting powder does not have to be separated from a liquid. The major concern in dry milling is that the powder does not pack in the corners of the mill and avoid milling. The powder must be kept free-flowing. One method of accomplishing this is to use a dry lubricant such as stearic acid, oleic acid, zinc stearate, ammonium stearate, ethylene glycol, triethanolamine, glycerine, or naphthenic acid. The dry lubricant coats the surface of the powder particles and decreases the affinity for moisture. This, in turn, helps to avoid agglomeration (sticking together of groups of particles) and results in increased grinding rate and greater reduction in particle size. The milling aid also helps improve the compaction characteristics of the powder during a subsequent pressing operation. Small levels of additive provide a large effect. Usually 0.5 wt% or less is all that is required.

An example of the milling and compaction benefits is shown in Figure 12.5 for dry pressing of Al_2O_3 powder prepared by wet milling, dry milling, and dry milling with 0.5% addition of naphthenic acid. The use of the milling and compaction aid dramatically increased the green density and the densification behavior. Green density refers to the density of the powder compact. A solid body of Al_2O_3 containing no porosity would have a density of 3.98 g/cm³ (referred to as the theoretical density). The powder wet-milled for 2 h compacted to 47.7% theoretical density (1.90 ÷ 3.98). The powder dry-milled for 2 h compacted to 52.8% theoretical density. The powder milled with naphthenic acid for 2 h compacted to 57.8% theoretical density.

The example illustrated by Figure 12.5 implies that wet milling is not effective. On the contrary, optimized wet milling is effective and is widely used in the ceramic industry. As is discussed later in this chapter, many ceramic powders are wet-milled and spray dried to achieve a free-flowing powder suitable for automated pressing.

Milling produces a broad particle size distribution rather than a narrow particle size range as achieved by screening. Milling can readily reduce the average particle size to 5 µm or less. Figure 12.6

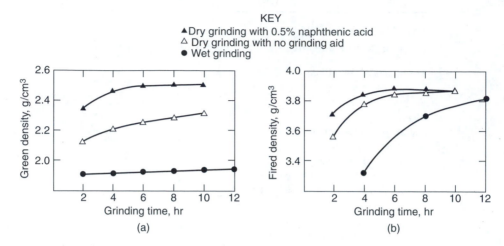

KEY
▲ Dry grinding with 0.5% naphthenic acid
△ Dry grinding with no grinding aid
● Wet grinding

(a)

(b)

FIGURE 12.5 Effects of the use of a milling aid on the pressed density and fired density of a low-soda calcined Al₂O₃. (From Hart, L.D. and Hudson, L.K., *Am. Ceram. Soc. Bull.*, 43(1), 13, 1964. With permission.)

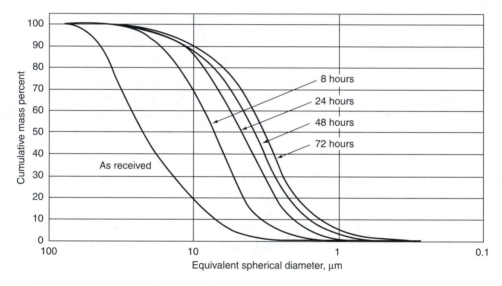

FIGURE 12.6 Particle size distribution of silicon powder as a function of milling time.

shows particle size distribution curves for silicon powder ball-milled for use in the fabrication of reaction-bonded Si_3N_4 parts. Initial particle size reduction was rapid but decreased as the powder became finer.

Besides producing the required particle size distribution, ball milling can also produce a very active powder that is easier to densify in later process steps. In some cases, this is achieved by an active surface condition. In other cases, it appears to be achieved by increased strain energy in the particle.

At this point, the nonceramic engineer may wonder why he or she needs to know the details about particle sizing methods, since he or she will probably only be purchasing the final part and will not be involved in the material processing. This is probably true, but it is still the engineer's responsibility to make sure the part works in the particular application. Current engineering requirements of materials are often far more demanding than they used to be and traditional controls and techniques used by ceramic manufacturers may require upgrading. The engineer knows the application requirements better than the supplier and will thus need extensive liaison with the

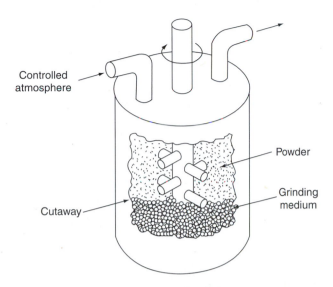

FIGURE 12.7 Schematic of an attrition mill. (Adapted from Herbell, T.P. and Glasgow, T.K., NASA, paper presented at the DOE Highway Vehicle Systems Contractors Coordination Meeting, Dearborn, MI, Oct. 17–20, 1978.)

TABLE 12.6
Changes during Attrition Milling of Silicon Powder

Milling Time (hr)	Surface Area (m²/g)	Iron Content (wt%)	Carbon Content (wt%)	Oxygen Content (wt%)
0	3.0	0.62	0.03	0.60
1	11.5	0.59	0.04	2.29
4	14.5	0.58	0.04	2.57
18	23.3	0.55	0.04	3.67

Source: From Herbell, T.P., Glasgow, T.K., and Yeh, H.C., *Effect of Attrition Milling on the Reaction Sintering of Silicon Nitride*, National Aeronautics and Space Administration, Lewis Res. Ctr., NASA-TM-78965, 1978 (N78-31236).

supplier to achieve the material objectives. It is not unusual for an engineer to request or suggest a process change or modification to the supplier.

12.2.1.5 Attrition Milling

Figure 12.7 shows a schematic of an attrition mill. It is similar to a ball mill since it is cylindrical and contains balls or grinding media, but rather than the cylinder rotating, the very small balls are agitated by a series of stirring arms mounted to an axial shaft. Herbell et al.[12] and Claussen and Jahn[13] report that attrition milling is quicker than ball milling, is more efficient at achieving fine particle size, and results in less contamination. Furthermore, attrition milling can easily be conducted dry, wet, or with a vacuum or inert gas atmosphere.

Data from Herbell et al.[12] for dry attrition milling of silicon powder are summarized in Table 12.6. Although the attrition mill was lined with an iron-base alloy, no significant iron or carbon was picked up, even after 18 h of milling. The oxygen content did increase by interaction with air, as would normally be expected for fine particles of silicon, as the surface area increased.

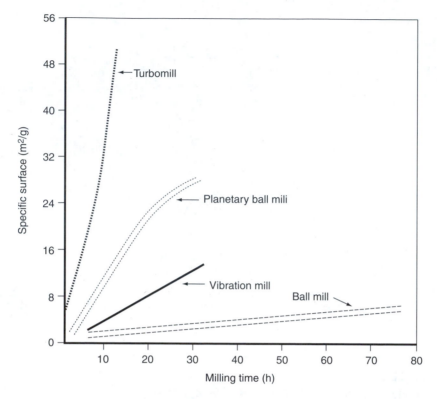

FIGURE 12.8 Turbomilling of barium ferrite powder compared to other milling techniques. Particles 0.1 μm in diameter were produced in 2 h with turbomilling. (From Jesse L. Hoyer, Ultrafine ceramic powders produced by turbomilling, *Am. Ceram. Soc. Bull.*, 67(10), 1663–1668, 1988. With permission.)

Even though the average particle size was substantially reduced by dry attrition milling, some particles in the range 40 μm (0.0018 in.) were still present and ultimately controlled the strength of the final component. Evidently, some particles were trapped or packed in regions of the mill where they did not receive adequate milling. This did not occur for wet milling.

Stanley et al.[14] have described a different configuration of attrition mill that has been used primarily for wet milling. It is referred to as a turbomill and consists of a rotating cylindrical cage of vertical bars surrounded by a stationary cylindrical cage of vertical bars. The material to be ground is mixed with water or other fluid plus sand-size grinding media. Materials such as ZrO_2, Al_2O_3, and SiO_2 have been ground to submicron size in a few hours, compared to 30 h for vibratory milling and much longer times for ball milling. The primary problem with this attrition milling approach is the amount of contamination and the difficulty of separating the powder from the media. For example, in one case 20 to 30% of the media was ground to −325 mesh (<44 μm) and was not successfully separated from the 0.1-μm milled powder.

Figure 12.8 compares the size reduction of a barium ferrite powder for turbomilling versus ball milling and vibratory milling.

12.2.1.6 Vibratory Milling

Vibratory milling is substantially different from ball milling or attrition milling. The energy for comminution is supplied through vibration rather than tumbling or mechanical stirring. The powder is placed in the stationary chamber of the mill together with suitable grinding media and a liquid. When the mill is turned on, vibration is transmitted (usually from the bottom center of the mill) through the chamber and into the media and powder. This results in two types of movement. First, it causes

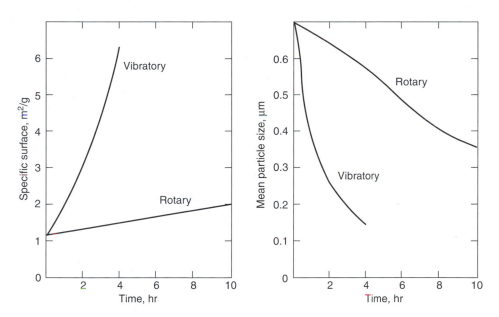

FIGURE 12.9 Comparison of the particle size reduction of a ferrite powder by vibratory milling and ball milling. (Plotted from data of Chol, G.R., *Am. Ceram. Soc. Bull.*, 54(1), 34, 1971.)

a cascading or mixing action of the contents of the milling chamber. Second, it causes local impact and shear fracturing of the powder particles between adjacent grinding media.

Vibratory milling is relatively fast and efficient and yields a finer powder than is usually achieved by ball milling. The vibratory mill chamber is typically lined with polyurethane or rubber and minimizes contamination. Vibratory mills are not used exclusively for powder processing. They are used extensively for deburring and cleaning of metal parts.

Figure 12.9 compares the particle size reduction rate for vibratory versus rotary (ball) milling of a ferrite powder. Fine particle size was achieved rapidly by vibratory milling.

12.2.1.7 Fluid Energy Milling

Fluid energy mills achieve particle size reduction by particle–particle impact in a high-velocity fluid.[15] The fluid can be compressed air, nitrogen, carbon dioxide, superheated steam, water, or any other gas or liquid compatible with the specific equipment design. The powder is added to the fluid and accelerated to sonic or near-sonic velocity through jets leading into the grinding chamber. The grinding chamber is designed to maximize particle–particle impact and minimize particle–wall impact.

Fluid energy milling can achieve controlled particle sizing with minimum contamination. Most jet mills have no moving parts and can be easily lined with polyurethane, rubber, wear-resistant steel, and even ceramics. Examples of particle sizing capabilities are shown in Table 12.7.[16] Output capacity can range from a few-grams per hour to thousands of kilograms per hour, depending on the size of the equipment.

The primary problem with fluid energy milling is collecting the powder. Large volumes of gases must be handled. Cyclones are not efficient for micrometer-sized particles and filters clog rapidly.

12.2.1.8 Hammer Milling

The milling techniques discussed so far are generally used with moderately fine starting powder and accomplish the final, ultrafine powder sizing. Hammer milling typically inputs chunks of ceramic of millimeter size or larger and reduces them by impact to a size manageable for final sizing by techniques such as ball milling. A variety of hammer mill designs have been used over the years. All involve a

TABLE 12.7
Typical Grinding Data for Fluid Energy Milling

Material	Mill Diameter		Grinding Medium	Material Feed Rate		Average Particle Size Obtained	
	cm	in.		kg/hr	lb/hr	μm	in.
Al_2O_3	20.3	8	Air	6.8	15	3	0.00012
TiO_2	76.2	30	Steam	1020	2250	<1	<0.00004
TiO_2	106.7	42	Steam	1820	4000	<1	<0.00004
MgO	20.3	8	Air	6.8	15	5	0.0002
Coal	50.8	20	Air	450	1000	5–6	~0.00025
Cryolite	76.2	30	Steam	450	1000	3	0.00012
DDT 50%	61.0	24	Air	820	1800	2–3	~0.0001
Dolomite	91.4	36	Steam	1090	2400	<44	<0.0018
Sulfur	61.0	24	Air	590	1300	3–4	~0.00014
Fe_2O_3	76.2	30	Steam	450	1000	2–3	~0.0001

Source: From Sturtevant Mill Co., Boston, MA, Bulletin 091, Sturtevant Micronizer Fluid Energy Mills. Courtesy of Sturtevant Mill Company, Boston, MA.

rapidly rotating rigid bar or plate. The particles are dropped in the path of this bar or plate and fragmented by the impact. Further fragmentation occurs as the "batted" particles strike the walls of the mill.

12.2.1.9 Roll Crushing

Roll crushing provides a coarse crushing alternative to hammer milling. The ceramic chunks are directed between two hardfaced rollers that rotate in opposite directions such that the ceramic chunks are pinched between them. Roll crushing is generally done in several stages. The rolls are spaced closer together in each subsequent stage.

12.2.1.10 Miscellaneous Crushing

Naturally occurring ceramic raw materials that are obtained by mining can consist of sizes too large for roll crushing or hammer milling. Techniques commonly used in the mining industry are utilized. These include gyratory, cone, and jaw crushing.

12.2.2 CHEMICAL SIZING

The mechanical powder-sizing techniques discussed in the prior section are widely used, but have disadvantages. One major disadvantage is the pickup of contamination. This has been minimized in some segments of the ceramic industry by lining the mill or other apparatus with dense material of the same composition as the powder and using grinding media of the same composition. A second disadvantage is control of the particle size distribution. To avoid these disadvantages, chemical processes have been developed to produce high-purity powder of controlled size.

12.2.2.1 Precipitation

Precipitation of soluble salts followed by thermal decomposition to the oxide is a widely used method of both particle sizing and purifying of oxide ceramics.[17] Analogous techniques in controlled atmospheres have also been used to produce nonoxide ceramic powders.

The Bayer process for producing Al_2O_3 from bauxite relies on controlled precipitation. The aluminum hydroxide in bauxite is dissolved with caustic soda and separated from the nonsoluble impurities in the bauxite by filtering. Aluminum trihydrate is then precipitated by changing the pH,

TABLE 12.8

Range in Powder Characteristics Available for Al_2O_3 Produced by Precipitation using the Bayer Process

	Type[a]	Percent Na_2O	Average Particle Size (μm)	Surface Area (m²/g)	Compaction Ratio	Final Density
Nonreactive, high purity	A-10	0.06	5	0.2	—	—
	C-75	0.01	2.8	0.49	2.25	2.97[b]
	A-14	0.06	—	0.6	—	—
	RC-122	0.03	2.6	0.35	2.35[c]	3.46[b]
Nonreactive, intermediate purity	A-12	0.24	—	0.5	—	—
	RC-24	0.23	2.9	0.53	2.35[c]	3.50[b]
	C-73	0.2	—	0.34	—	—
Nonreactive, low purity	A-2	0.46	3.25	—	2.21	3.22[b]
	RC-20	0.40	2.7	0.87	2.31[c]	3.37[b]
	C-70	0.50	2.5	0.66	2.28	3.18[b]
	RC-25	0.31	2.6	0.95	2.33[c]	3.52[b]
	A-5	0.35	4.7	—	2.30	3.07[b]
	C-71	0.75	2.2	0.71	2.24	3.33[b]
Reactive, high purity	XA-139	0.008	0.42	6.5	2.00	3.90[d]
	ERC-HP	0.008	0.55	7.4	2.15[e]	3.95[d]
	A-16	0.06	0.6	6.5	—	—
	RC-172	0.04	0.6	4.0	2.21[e]	3.94[d]
Reactive, low purity	A-3	0.36	0.63	9.0	1.79	3.64[b]
	RC-23	0.30	0.6	7.5	1.91[c]	3.86[b]
Reactive, nonreactive mixture	RC-152	0.05	1.5	2.6	2.30[c]	3.46[b]
	A-15	0.07	—	—	—	—

[a] C, Alcan: A, Alcoa; RC, Reynolds.
[b] For 1620°C/1 h sintering.
[c] 20-g pellet pressed at 4000 psi, 4-h grind.
[d] For 1510°C/2 h sintering.
[e] 10-g pellet pressed at 5000 psi, 4-h grind.
Source: Adapted from Sturtevant Mill Co., Boston, MA, Bulletin 091, Sturtevant Micronizer Fluid Energy Mills.

controlling the resulting particle size through addition of seed crystals. Table 12.8 lists the characteristics of commercially available Al_2O_3 powders and illustrates the variations in particle size, surface area, and purity that can be achieved in the precipitation process.[18] The very fine reactive alumina powders first became available in 1966 and resulted in dramatic improvements in the properties of Al_2O_3 components and an expansion in the number and range of applications.

12.2.2.2 Freeze Drying

Freeze drying (also known as cryochemical processing) is a process that was first reported in 1968 by Schnettler et al.[19] It has potential for producing uniform particle and crystallite sizing of very pure, homogeneous powder.

There are four steps in freeze drying:

1. A mixture of soluble salts containing the desired ratio of metal ions is dissolved in distilled water.

2. The solution is formed into droplets usually 0.1 to 0.5 mm in diameter and rapidly frozen so that no compositional segregation can occur and so that the ice crystals that nucleate are very small.
3. The water is removed in vacuum by sublimation, with care to avoid any liquid phase, thus preventing any chance for segregation.
4. The resulting powder is then calcined (heat-treated) at a temperature that decomposes the crystallized salts and converts them to very fine crystallites of the desired oxide or compound.

Most of the reported freeze drying has been conducted with sulfates. Each frozen droplet contains sheaths of sulfate crystals radiating from the center to the surface. During calcining, the oxide crystallites form in an oriented chainlike fashion along these radial sheaths. The size of the crystallites can be controlled by the time and temperature of calcining. Johnson and Schnettler[20] reported that a 10-h 1200°C heat treatment of freeze-dried aluminum sulfate resulted in Al_2O_3 crystallites averaging 1500 Å and ranging from 600 to 2600 Å.

Not all compositions can be achieved by freeze drying. The two primary limitations are (1) the nonavailability of soluble salts containing the required metal ions and (2) the reaction of some salt combinations to form a precipitate. For instance, it has been difficult to find soluble salts that will yield lead zirconate titanate. Similarly, a mixture of barium acetate plus ferrous sulfate results in precipitation of insoluble $BaSO_4$.

Rigterink[21] described a laboratory apparatus and potential production approaches for freeze drying. The laboratory apparatus is quite simple and inexpensive and is adequate for making small batches for material processing development. The apparatus consists of a beaker containing hexane that is mechanically stirred or swirled rapidly. The hexane container is surrounded by a bath of dry ice and acetone. The salt solution is forced through a glass nozzle as uniform drops of the desired size into the swirling hexane. The drops freeze rapidly and settle to the bottom, where they can be removed later by a sieve.

12.2.2.3 Hot Kerosene Drying

Some powders that cannot be prepared by freeze drying can be prepared by hot kerosene drying. Water-soluble compounds are measured in the desired proportions and dissolved in water. The solution is vigorously mixed with kerosene (approximately 50-50 by volume) plus an emulsifying agent to form an emulsion. This emulsion is metered drop by drop in a distillation apparatus into well-stirred kerosene preheated to 170°C (340°F). The water evaporates instantly, leaving the homogeneous dry salt mixture dispersed in the kerosene. The salts are removed by screening and thermally decomposed in air to form the oxide. An example is $MgSO_4$ plus $Al_2(SO_4)_3$ to yield $MgAl_2O_4$. Mullite, cordierite, stabilized ZrO_2, and other pure, homogeneous, fine powders have been prepared by this technique.

This technique is not restricted to kerosene. Many other organic fluids that are liquid above 100°C (212°F) have been successfully used. Examples are benzene, toluene, and mineral oil. Emulsifying agents are available from major chemical suppliers.

12.2.2.4 Sol–Gel

Sol–gel is a generic term that includes a variety of techniques to achieve a high-purity composition with homogeneity at the molecular level.[22] Preparing a powder by a sol–gel approach involves the following steps:

1. Form a stable dispersion (sol) of particles less than 0.1 μm in diameter in a liquid.
2. By change in concentration (evaporation of a portion of the liquid), aging, or addition of a suitable electrolyte, induce polymer-like, three-dimensional bonding to occur throughout the sol to form a gel.

3. Evaporate the remaining liquid from the gel.
4. Increase the temperature to convert the dehydrated gel to the ceramic composition.

The rigidity of the gel prevents migration or segregation of atoms during drying and assures homogeneity at the molecular level. The resulting powder has high surface area and small particle size. Particle size is generally in the range of 20 to 50 nm. Surface areas as high as 500 m²/g have been reported. Due to the high surface area and fine particle size plus the homogeneous distribution of atoms, these sol-gel-derived powders can be densified at lower temperatures than powder prepared by conventional mechanical processes.

The following provide examples of several sol–gel approaches.

Example 12.1.   Silica-based compositions have been prepared using tetraethyl orthosilicate (TEOS) or a colloidal SiO_2. TEOS is soluble in ethyl alcohol. Colloidal SiO_2 is soluble in water and contains a cation stabilizer such as Na^+ or NH_4^+. The water-based procedure is easiest. The colloidal SiO_2 plus nitrates of the required metal cations are mixed into a homogeneous sol. The pH is controlled with nitric acid (HNO_3). Depending on the system and the pH, a gel will form in several minutes to a day.

Example 12.2.   Compositions in the $PbO–TiO_2$ system have been prepared by sol–gel methods starting with $Pb(NO_3)_2$ and $TiCl_4$. These were then mixed with $NH_4(NO_3)$ and NH_4OH to form colloidal complex hydroxide particles, which formed a stable sol after removal of the chloride by washing and filtration. The sol was then gelled by adding it to isopropyl alcohol.

Example 12.3.   Metal alkoxides have been determined to be excellent precursors for formation of oxide powders by sol–gel methods. An alkoxide has the general chemical formula $M(OR)_n$, where M is a metal ion of valence n and R is an alkyl group C_xH_{2x+1}. The alkoxide can be converted to a pure oxide powder either by direct thermal degradation or by hydrolysis plus thermal degradation. These reactions are illustrated for ZrO_2 and Al_2O_3:

$$Zr(OC_4H_9)_4 \xrightarrow[\text{(615 to 930°F)}]{\text{325 to 500°C}} ZrO_2 + 2C_4H_9OH + 2C_4H_8 \qquad (12.1)$$

$$Al(OC_4H_9)_3 + 3H_2O \longrightarrow Al(OH)_3 + 3C_4H_9OH \qquad (12.2)$$

$$2Al(OH)_3 \xrightarrow[\text{(270 to 840°F)}]{\text{150 to 450°C}} Al_2O_3 + 3H_2O \qquad (12.3)$$

12.2.2.5 Liquid Mix Process

This is a generic name for various processes that start with a homogeneous solution containing the desired cations, which use additives and evaporation to convert the homogeneous liquid to a rigid cross-linked polymer, and which utilize heat to convert the polymer into a homogeneous oxide powder. The initial process was pioneered by M. Pechini and is referred to either as the Pechini process or the amorphous citrate process.[23,24] The following steps are used to achieve a powder:

1. An aqueous solution is prepared with metal alkoxides, oxides, hydrated oxides, or carbonates in an alpha-hydroxycarboxylic acid such as citric acid; the ratio of metal ions can be precisely controlled. The acid complexes with the metal ions to form polybasic acid chelates.
2. A polyhydroxy alcohol such as ethylene glycol is added and the liquid is heated to 150 to 250°C to allow the chelates to undergo polyesterification.
3. Heating is continued to remove excess water, resulting in a solid polymeric "resin."
4. The temperature is increased to about 400°C to char or decompose the resin.
5. The temperature is further increased to 500 to 900°C to form crystallites of the mixed oxide composition. The crystallites are typically 20 to 50 nm and clustered into agglomerates.

The Pechini process has been successfully used to produce high-purity powders of high T_c super-conductor compositions, of doped zirconatetitanate dielectric compositions, of doped perovskite [La(Sr)CrO$_3$, La(Sr)MnO$_3$] electrode compositions, and over 100 other oxide compositions.

A variety of modifications of the Pechini process have been developed. One involves the use of metal nitrates. Another replaces citric acid with polyacrylic acid. A process called the glycine–nitrate process adds the amino acid glycine.[25] The glycine performs two functions. First, it forms complexes with the metal cations and increases their solubility. This seems to prevent selective precipitation and segregation during evaporation. Second, the glycine services as a fuel during charring. Specifically, metal nitrates are combined with glycine in water and evaporated until a homogeneous viscous liquid forms. The liquid is further heated to about 180°C and auto-ignites. Temperatures rapidly reach 1100 to 1450°C and nearly instantly convert the material to fine, relatively nonagglomerated crystallites of the mixed oxide composition and structure. An example of powder prepared from the glycine–nitrate process is shown in Figure 12.10.

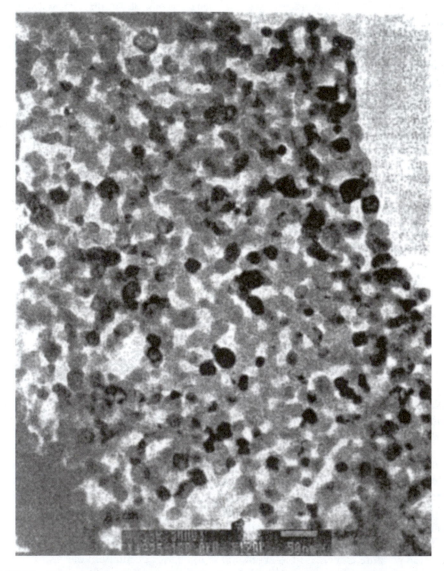

FIGURE 12.10 Transmission electron microscope image of ultrafine La$_{0.76}$Sr$_{0.24}$CrO$_3$ powder prepared by the glycine-nitrate process. (Courtesy of Larry Chick, Battelle Northwest Laboratories, Richland, WA.)

12.2.2.6 Spray Roasting

Spray roasting involves spraying fine atomized droplets of a solution of precursors in water or other fluid into a heated chamber. The temperature in the chamber is selected such that evaporation and chemical reaction occur to yield a high-purity powder containing fine crystallite size.

One variant of spray roasting is the pyrohydrolysis process[26] for synthesis of high-purity metal oxides. A solution of a metal chloride in water is sprayed into a heated ceramic-lined chamber. Depending on the specific metal chloride, a temperature of 300 to 950°C results in reaction of the metal chloride with the water to form the metal oxide plus hydrochloric acid. Figure 12.11 shows the schematic of a spray roaster. Table 12.9 identifies examples of pyrohydrolysis reactions. The resulting oxide powder consists of crystallites approximately 0.2 to 0.4 µm in diameter agglomerated into hollow spheres 100 to 200 µm in diameter.

Other reactions can also be achieved by the spray roasting approach. For example, the glycine nitrate precursor liquid has been successfully converted to homogeneous mixed-oxide powders in a conventional spray drying apparatus.

12.2.2.7 Decomposition

Decomposition reactions are commonly used in ceramic processing. Carbonates, nitrates, sulfates, oxalates, and other compounds containing oxygen ions are commonly used in preparing ceramic

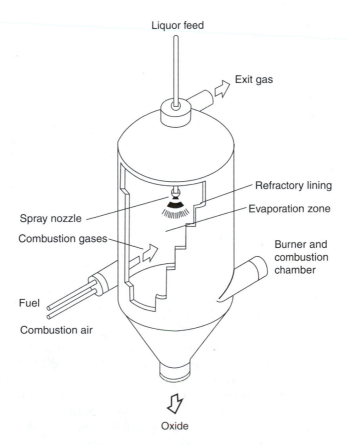

FIGURE 12.11 Schematic drawing illustrating the hydropyrolysis or spray-roasting process. (From Kladnig, W.F. and Karner, W., Pyrohydrolysis for the production of ceramic raw materials, *Am. Ceram. Soc. Bull.*, 69(5), 814–817, 1990. With permission.)

TABLE 12.9

Common Pyrohydrolysis Reactions

$2FeCl_2 + 2H_2O + 1/2O_2 \rightarrow Fe_2O_3 + 4HCl$
$MgCl_2 + H_2O \rightarrow MgO + 2HCl$
$2FeCl_3 + 3H_2O \rightarrow Fe_2O_3 + 6HCl$
$2AlCl_3 + 3H_2O \rightarrow Al_2O_3 + 6HCl$
$3CoCl_2 + 3H_2O + 1/2O_2 \rightarrow Co_3O_4 + 6HCl$
$CoCl_2 + H_2O \rightarrow CoO + 2HCl$
$2MnCl_2 + 2H_2O + 1/2O_2 \rightarrow Mn_2O_3 + 4HCl$
$2FeCl_2 + MnCl_2 + 3H_2O + 1/2O_2 \rightarrow MnFe_2O_4 + 6HCl$
$2FeCl_2 + NiCl_2 + 3H_2O + 1/2O_2 \rightarrow NiFe_2O_4 + 6HCl$
$2AlCl_3 + NiCl_2 + 4H_2O \rightarrow NiAl_2O_4 + 8HCl$
$NiCl_2 + H_2O \rightarrow NiO + 2HCl$
$2CrCl_3 + 3H_2O \rightarrow Cr_2O_3 + 6HCl$
$2(RE)Cl_3 + 3H_2O \rightarrow (RE)_2O_3 + 6HCl$ (RE = rare earths)
$TiOCl_2 + H_2O \rightarrow TiO_2 + 2HCl$

Source: From Kladnig, W.F. and Karner, W., Pyrohydrolysis for the production of ceramic raw materials, *Am. Ceram. Soc. Bull.*, 69(5), 814–817, 1990.

powder batches. These then decompose at elevated temperatures during calcining or sintering to yield the oxide. For example, $MgCO_3$ decomposes to yield MgO. These nonoxide raw materials are used for various reasons such as purity, particle size, availability, or requirement for a soluble composition.

12.2.2.8 Hydrothermal

Hydrothermal synthesis involves crystallization of a composition in hot, pressurized water.[27] Typical temperatures range from 100 to 350°C (212 to 630°F) at pressures up to 15 MPa (2175 psi). Under these conditions, a wide variety of pure, fine-particle ceramic compositions can be synthesized. The feedstock can be oxides, hydroxides, salts, gels, organics, acids, and bases. The conditions can be oxidizing or reducing. The particle size can be controlled by residence time, temperature, and pressure. The resulting powder consists of single crystals of the final composition. No heat treatments or milling operations are required.

Hydrothermal synthesis has been demonstrated on a laboratory scale, but has not yet been scaled up to commercial production.

12.2.2.9 Plasma

A variety of ceramic powders of high purity and very small particle size (10 to 20 nm) have been synthesized in high-temperature plasma* environments.[28] The particles essentially condense in a flowing gas, which accounts for the high purity. Two types of plasma reactors have been successfully used. One is the DC arc jet system. In this system the plasma is in direct contact with the metal electrode that supplies current. This reactor has very high efficiency, but can result in trace impurities from the electrode. The second type of plasma reactor is an rf (radio frequency) induction system. In this system the current is transferred to the plasma through the electromagnetic field of the induction coil. No direct contact occurs, so no contamination results in the powder being synthesized. The efficiency

---

* Plasma is a high-temperature, ionized gas. Because it is electrically conductive, a high degree of electrical heating can be achieved, that is, temperatures in the range 4000 to 10,000°C (7230 to 18,030°F).

of the rf induction system is lower than the DC arc jet system. However, both have produced SiC particles with greater than 70% efficiency using $SiCl_4$, CH_4, and H_2 as the gaseous precursors. Si_3N_4 has also been synthesized by plasma techniques.

12.2.2.10 Laser

The use of laser energy for synthesizing Si_3N_4 powder was discussed earlier in this chapter. It has also been used successfully to produce controlled particle sizes of silicon and SiC. The SiC powder was prepared using a CO_2 laser and a mixture of silane (SiH_4) and methane (CH_4). The CO_2 laser energy is absorbed by this gas mixture. The resulting localized high temperature decomposes the gases and allows reaction to form SiC particles directly in the gas stream. Purity is very high. Particle sizes in the range of from 5 to 200 nm have been achieved. Organosilicon compounds such as 1,1,1,3,3,3-hexamethyldisilazane have also been successfully used.

12.2.3 MISCELLANEOUS POWDER SYNTHESIS/SIZING TECHNIQUES

12.2.3.1 Calcining

Calcining has been mentioned before. It refers to a high-temperature treatment of a powder to modify the characteristics of the powder. Several types of modifications are commonly achieved by calcining: coarsening, decomposition, reaction, and dehydration.

Coarsening involves crystallite growth or fusing or bonding small particles together to produce larger particles. Decomposition, as discussed in the prior section, involves converting compositions such as carbonates and nitrates to oxides. Gases are evolved during these decomposition reactions that could build up enough pressure to crack a powder compact. Therefore, the decomposition is sometimes achieved by calcining the powder prior to compaction. Prereacting powders is sometimes conducted for the same reason.

Dehydration is important in preparation of hydraulic cements (such as Portland cement) and plaster. Plaster powder, for example, consists of partially dehydrated gypsum. Gypsum ($CaSO_4 \cdot 2H_2O$) is dehydrated during calcining to $CaSO_4 \cdot 1/2H_2O$. Addition of water to $CaSO_4 \cdot 1/2H_2O$ powder results in rehydration to $CaSO_4 \cdot 2H_2O$.

12.2.3.2 Rotary Kiln

A kiln is a high-temperature furnace. A rotary kiln is a furnace cylindrical in cross section that can be rotated during a powder calcining or synthesis operation. The rotation keeps the powder free-flowing and minimizes bonding of adjacent particles. The rotation can also provide continuous mixing to help achieve a homogeneous powder.

12.2.3.3 Fluidized Bed

A fluidized bed consists of a powder or granules of material contained in an enclosure and supported from below by a porous plate. A gas is passed through the porous plate with enough pressure to slightly lift the particles off the surface of the porous plate and to allow the complete bed of powder to slowly percolate. A high-temperature fluidized bed can be used for calcining, synthesis, and deposition of coatings by vapor reactions.

12.2.3.4 Self-Propagating Combustion

Self-propagating high-temperature synthesis (SHS) forms ceramic and intermetallic compositions through exothermic reaction.[29] For example, fine particles of Mg, Al, and Ti are highly reactive when heated to moderate temperatures. In air, they "ignite" and "burn" at very high temperature to form oxides. In an inert environment in contact with carbon or boron, they can react exothermically to

---

TABLE 12.10
Examples of Materials Prepared by the SHS Method

Borides	CrB, HfB_2, NbB_2, TaB_2, TiB_2, LaB_6, MoB_2
Carbides	TiC, ZrC, HfC, NbC, SiC, Cr_3C_2, B_4C, WC
Carbonitrides	$TiC-TiN$, $NbC-NbN$, $TaC-TaN$
Cemented carbides	$TiC-Ni$, $TiC-(Ni, Mo)$, $WC-Co$, $Cr_3C_2-(Ni, Mo)$
Chalcogenides	MoS_2, $TaSe_2$, NbS_2, WSe_2
Composites	$TiC-TiB_2$, $TiB_2-Al_2O_3$, $B_4C-Al_2O_3$, $TiN-Al_2O_3$
Hydrides	TiH_2, ZrH_2, NbH_2
Intermetallics	$NiAl$, $FeAl$, $NbGe$, $TiNi$, $CoTi$, $CuAl$
Nitrides	TiN, ZrN, BN, AlN, Si_3N_4, TaN (cubic and hexagonal)
Silicides	$MoSi_2$, $TaSi_2$, Ti_5Si_3, $ZrSi_2$

Source: From Munir, Z.A., Synthesis of high temperature materials by self-propagating combustion methods, *Am. Ceram. Soc. Bull.*, 67(2), 342–349, 1988.

---

produce carbides and borides. The reactions are extremely rapid, often being completed in less than a second. By starting with loosely packed powder, the end product can also be a powder or a loosely compacted mass that is easily comminuted into powder. In many cases the particle size of the reactant can be controlled by the particle size of the starting powder. Table 12.10 identifies some of the materials that have been synthesized by SHS.

12.2.3.5 Gas Condensation

A gas condensation process has been developed in recent years that produces ultra-fine nanometer-sized powders that consist essentially of small clusters of atoms.[30] These powders have successfully been compacted into small pellets and densified at temperatures 400 to 600°C lower than can be achieved with conventional powders. Because of the low densification temperature and other factors, polycrystalline materials with mean grain sizes ranging from 5 to 25 nm have been achieved. These "nanophase" materials have unique properties. TiO_2 with a grain size of 12 nm has shown substantial ductile behavior at room temperature (a strain-rate sensitivity of 0.04, which is one-quarter that of lead).

The apparatus for gas condensation synthesis of nanometer-sized powder consists of an ultra-high vacuum system, resistively heated evaporation sources, a liquid nitrogen-filled condensation tube, and a scraper to remove condensed powder from the surface of the tube.

12.3 PRECONSOLIDATION

The sized powders described in Section 12.2 are compacted into the desired shapes by techniques such as pressings, slip casting, and injection molding (discussed in Chapter 13) and then strongly bonded or densified (discussed in Chapter 14). To achieve a final component having uniform properties and no distortion requires a uniform particle compact. To achieve the required uniformity, the powder usually requires special treatments or processing prior to compaction. Table 12.11 summarizes some of these special preconsolidation considerations for several compaction or consolidation approaches.

The preconsolidation steps are essential to minimize severe fabrication flaws that can occur in later processings steps. For instance, a powder that is not free-flowing can result in poor powder distribution in the pressing die and distortion or density variation in the final part. Similarly, improper viscosity control of a casting slurry can result in incomplete fill of the mold or a variety of other defects during slip casting. Inadequate de-airing of either a slurry or an injection molding mix can result in

TABLE 12.11
Preconsolidation Steps for Several Consolidation Approaches

Pressing	Slip Casting	Injection Molding
Binder addition	Slurry preparation	Thermoplastic addition
Lubricant addition	Binder addition	Plasticizer addition
Sintering aid addition	Deflocculant addition	Wetting agent addition
Preparation of a free-flowing powder	pH control	Lubricant addition
by spray drying or granulation	Viscosity control	Sintering aid addition
	Percent solids control	De-airing
	De-airing	Granulation or pelletizing

a strength-limiting void in the final slip-cast or injection-molded part. Such fabrication flaws can reduce the strength of a material to a fraction of its normal value.

The following sections describe briefly some of the reasons and techniques of preconsolidation steps. Emphasis will be on preparation of a free-flowing powder suitable for pressing. Preconsolidation considerations for slip casting and injection molding are discussed in Chapter 13 as part of those specific processing steps.

12.3.1 ADDITIVES

Additives are required for different reasons, depending on the specific forming process. However, several general comments are relevant to most forming approaches:

1. Binders are added to provide enough strength in the "green" body (unfired compact) to permit handling, "green" machining, or other operations prior to densification.
2. Lubricants are added to decrease particle-particle and particle-tool friction during compaction.
3. Sintering aids are added to activate densification.
4. Deflocculants, plasticizers, wetting agents, and thermoplastics are added to yield the rheological (flow) properties necessary for the specific shape-forming process (discussed in detail in Chapter 13).

Table 12.12 further summarizes the function of additives.[31]

A wide variety of binders are available, as shown by the partial listing in Table 12.13. Selection depends on a number of variables, including green strength needed, ease of machining, compatibility with the ceramic powder, and nature of the consolidation process. Gums, waxes, thermoplastic resins, and thermosetting resins are not soluble in water* and do not provide a benefit for slip casting, but are excellent for the warm mixing used to prepare a powder for injection molding. Organic binders can be burned off at low temperature and result in minimal contamination, whereas inorganic binders become a part of the composition.

12.3.2 SPRAY DRYING

Spray drying is commonly used in ceramic processing to achieve a uniform, free-flowing powder.[32] As shown in Figure 12.12, a spray dryer consists of a conical chamber that has an inlet for hot air. The powder to be spray dried is suspended in a slurry with the appropriate additives. Slurry preparation

---

* Emulsions can be used.

TABLE 12.12

Function of Additives of Ceramics

Additive	Function
Binder	Increase green strength and provide lubrication.
Lubricant	Mold releases, decrease die-wall and interparticle friction.
Compaction aid	Aid in particle rearrangement during pressing.
Plasticizer	Rheological aid, improving flexibility of binder films, allowing plastic deformation of granules.
Deflocculant	pH control, control of surface charge on particles, dispersion, or coagulation.
Wetting agent	Reduction of surface tension.
Water retention agent	Retain water during pressure application.
Antistatic agent	Charge control.
Antifoam agent	Prevent foam or bubble formation.
Foam stabilizer	Enhances foam formation.
Chelating or sequestering agent	Deactivate undesirable ions.
Fungicide and bactericide	Stabilize against degradation with aging.
Sintering aids	Aid in densification.
Dopants	Modify electrical, magnetic, optical properties.
Phase stabilizers	Control the crystalline phases present.

Source: Adapted from Pincus, A.G. and Shipley, L.E., The role of organic binders in ceramic processing, *Ceram. Ind.*, 92, 106–109, 1969.

TABLE 12.13

Examples of Binders Used in Ceramic Processing

Organic	Inorganic
Polyvinyl alcohol (PVA)	Clays
Waxes	Bentonites
Celluloses	Mg-Al silicates
Dextrines	Soluble silicates
Thermoplastic resins	Organic silicates
Thermosetting resins	Colloidal silica
Chlorinated hydrocarbons	Collodial alumina
Alginates	Aluminates
Lignins	Phosphates
Rubbers	Borophosphates
Gums	
Starches	
Flours	
Casein	
Gelatins	
Albumins	
Proteins	
Bitumens	
Acrylics	

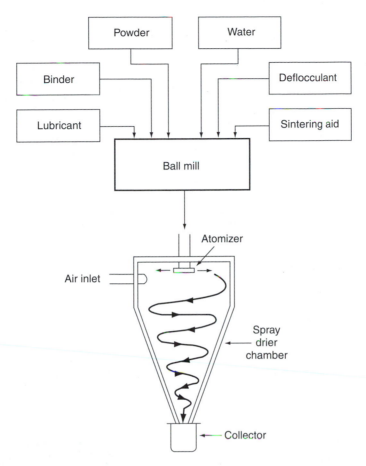

FIGURE 12.12 Schematic of the spray-drying process for achieving free-flowing spherical powder agglomerates containing a uniform level of additives.

is most frequently done in a ball mill. The slurry is fed into the spray dryer through an atomizer and is swirled around by the hot air circulating in the conical spray dryer chamber. The fluid evaporates and the powder forms into roughly spherical, soft agglomerates.

Spray dryers are available in a wide range of sizes and designs. Some atomize the fluid with a pressure nozzle at pressures up to 10,000 psi. The fluid exits the nozzle in a cone-shaped array of droplets that dry to roughly spherical agglomerates. By control of the feed rate and the viscosity, spray-dried powder agglomerates ranging from 30 to 250 µm can be achieved.

Some spray dryers atomize the fluid with a two-fluid nozzle. The fluid is pumped at low pressure through one orifice in the nozzle. A gas (usually air) is passed through another orifice and breaks the fluid stream into fine droplets. The size of spray-dried powder can be varied by control of gas flow, fluid flow, and fluid viscosity. Relatively high viscosities can be accommodated with this type of spray dryer.

Other spray dryers use a rotary atomizer. The slurry is introduced at the center of a rotating wheel or disk. The fluid accelerates radially and breaks into droplets at the periphery of the wheel. A wide range of spray characteristics can be achieved by control of rotor design, rotation speed, feed rate, viscosity, and solids content. Spherical agglomerate sizes are generally in the range of 30 to 120 µm.

Atomization is only one key aspect of a spray dryer. Another important factor is the relationship of the hot air flow to the droplet flow. Some spray dryers flow the air in the same direction as

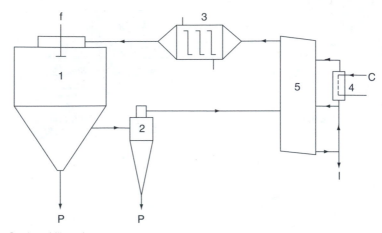

c Coolant (diluent)
f Feed
l Solvent product
p Dried product
1 Spray-dryer chamber
2 Cyclone (or alternative dry collector)
3 Liquid phase indirect heater
4 Heat exchanger
5 Scrubber/condenser

FIGURE 12.13   Schematic of a closed-cycle spray dryer. (From Masters, K., *Spray Drying Handbook*, 4th ed., Longman Scientific & Technical, Essex, England, Wiley, New York, 1985, p. 40. With permission.)

the spray (co-current), some in the opposite direction (counter-current), and some in a mixed-flow mode. The nature of the hot air flow determines how quickly the droplets dry and how much moisture is retained. Droplets that dry quickly result in a hollow, spherical powder agglomerate. Careful control can achieve a solid sphere or a partially collapsed or dimpled sphere.

Most spray drying is conducted with water as the carrier for the ceramic particles. However, some powders (especially nonoxides) react with water. Closed-cycle spray dryers that can utilize alcohol or other nonaqueous fluid and an inert gas have been designed. A schematic of a closed-cycle spray dryer is illustrated in Figure 12.13. Closed-cycle spray drying has been used successfully with AlN, WC, and other powders.

12.3.3 GRANULATION

Granulation is another approach to achieving better flow properties of a powder. In this case, a slurry is not prepared. Instead, only a damp or plastic mix is prepared, usually with equipment such as a mix muller, a sigma mixer, or any of a variety of other commercially available mixer designs. For laboratory-scale batches, this mixing can be accomplished with a mortar and pestle or even by hand. The damp or plastic material is then forced through orifices of the desired size or screened. The resulting particle agglomerates are usually harder and more dense than spray-dried agglomerates and irregular in shape. They do not flow as readily, but do tend to pack to a lower volume.

The major advantage of granulation is that the powder is prepacked and takes up less volume during pressing, extrusion, or injection molding. This is illustrated in Table 12.14 where compaction characteristics are compared for as-received powder, spray-dried powder, and granulated powder. In some cases, powders are precompacted at 703 to 1055 kg/cm^2 (10,000 to 15,000 psi) prior to granulation to assure relatively tight packing of the powder in each agglomerate. However, if this pressure is not equaled or exceeded during the shape-forming-process, the agglomerates will not be crushed

TABLE 12.14
Comparison of Pressing Compaction Characteristics vs. Precompaction Treatments

Powder Preconsolidation Treatment	Density[a] of Granules	Fill Density[a] of Die	Density[a] of Pressed Compact	Pressure to Break Granules, MPa (ksi)
As-received bulk powder	–	5–20	30–60	–
Milled and screened powder	20–40	15–25	40–60	10–20 (1.45–2.90)
Spray dried	30–55	20–35	50–62	1–5 (0.14–0.72)
Pressed and granulated	20–65	15–40	45–62	1–50[b] (1.45–7.25)[b]

[a] Percent of theoretical density.
[b] Depending on prepress pressure and on binder.

or fused together and the finished part will retain the identity of the agglomerates. This will consist of inhomogeneity in the part and for many applications will require that the part be rejected.

12.4   BATCH DETERMINATION

An important aspect of powder preparation and preconsolidation is determination of the proper proportions of each powder and additive. Formulations are sometimes defined in weight percent, sometimes in volume percent, sometimes in mol percent, and sometimes in atomic percent. We need to understand each of these and be able to convert back and forth from one to another. Let us look at some examples of instances where conversions may be necessary.

Example 12.4. Mullite shows up on a phase equilibrium diagram as $3Al_2O_3 \cdot 2SiO_2$. What is the mol% of Al_2O_3? Of SiO_2?

Answer. Mullite consists of 3 mol of Al_2O_3 and 2 mol of SiO_2.

$$\text{mol\% } Al_2O_3 = \frac{3}{3+2} \times 100 = 60 \, \text{mol\%}$$

$$\text{mol\% } SiO_2 = \frac{2}{3+2} \times 100 = 40 \, \text{mol\%}$$

Example 12.5. What is the wt% of Al_2O_3 in Mullite? Of SiO_2? We must first obtain the atomic weight of each element from the periodic table or another source, that is,

$$Al = 27.0, \quad Si = 28.1, \quad O = 16.0$$

We can now calculate the equivalent weight of 3 mol Al_2O_3:

$$3 \, \text{mol } Al_2O_3 = 3[(2)(27.0) + (3)(16.0)] = 306$$

and 2 mol SiO_2:

$$2 \, \text{mol } SiO_2 = 2[(28.1) + (2)(16.0)] = 120.2$$

Thus,

$$\text{wt\% } Al_2O_3 = \frac{\text{wt } Al_2O_3}{\text{total wt}} \times 100$$

$$= \frac{306}{426.2} \times 100 = 71.8 \, \text{wt\%}$$

and

$$\text{wt\% SiO}_2 = \frac{120.2}{426.2} \times 100 = 28.2\,\text{wt\%}$$

Therefore, to end up with 100 g of Mullite, we must mix 71.8 g of Al_2O_3 and 28.2 g of SiO_2.

Example 12.6. We wish to prepare a composite consisting of 80 vol% Al_2O_3 and 20 vol% SiC whiskers. We want a 1000-g batch. How much Al_2O_3 and SiC whiskers do we mix? We first determine the weight of Al_2O_3 and SiC in a selected volume of material. For simplicity, let us select 1 cm³, of which 0.8 cm³ is Al_2O_3 and 0.2 cm³ is SiC. All we need to know is the density of pure Al_2O_3 and pure SiC.

$$Al_2O_3 = 3.98\,\text{g/cm}^3 \text{ and SiC} = 3.19\,\text{g/cm}^3$$

$$\text{wt of } Al_2O_3 \text{ in 1 cm}^3 = 0.8\,\text{cm}^3 \times 3.98\,\text{g/cm}^3 = 3.18\,\text{g}$$

$$\text{wt of SiC in 1 cm}^3 = 0.2\,\text{cm}^3 \times 3.19\,\text{g/cm}^3 = 0.64\,\text{g}$$

$$\text{wt\% } Al_2O_3 = \frac{3.18}{3.18 + 0.64} \times 100 = 83.2\,\text{wt\%}$$

$$\text{wt\% SiC whiskers} = \frac{0.64}{3.18 + 0.64} \times 100$$
$$= 16.8\,\text{wt\%}$$

Thus, our 1000-g batch will consist of 832 g of Al_2O_3 powder and 168 g of SiC whiskers.

Example 12.7. We wish to prepare a barium titanate composition in which we replace 10 atomic % of the Ba ions with Sr ions. What wt% of BaO, SrO, and TiO_2 must we mix to obtain the required composition? Atomic weight of the elements equals 137.34 for Ba, 87.62 for Sr, 16.0 for O, and 47.9 for Ti.

$$0.9\,\text{BaO} = 0.9(137.34 + 16.0) = 138$$
$$0.1\,\text{SrO} = 0.1(87.62 + 16.0) = 10.4$$
$$TiO_2 = 47.9 + (2)(16.0) = 79.9$$

$$\text{wt\% BaO} = \frac{138}{138 + 10.4 + 79.9} = 60.4$$

$$\text{wt\% SrO} = \frac{10.4}{138 + 10.4 + 79.9} = 4.6$$

$$\text{wt\% TiO}_2 = \frac{79.9}{138 + 10.4 + 79.9} = 35.0$$

REFERENCES

1. Norton, F.H., *Elements of Ceramics*, 2nd ed., Addison-Wesley, Reading, Mass., 1974.
2. Tooley, F.V., Ed., *Handbook of Glass Manufacture*, Vols. 1 and 2, Ogden Publishing Company, New York, 1961.
3. Norton, F.H., *Refractories*, 4th ed., McGraw-Hill, New York, 1968.
4. Rhodes, W.H. and Natansohn, S., Powders for advanced structural ceramics, *Am. Ceram. Soc. Bull.*, 68(10), 1804–1812, 1989.
5. Sheppard, L.M., International trends in powder technology, *Am. Ceram. Soc. Bull.*, 68(5), 979–985, 1989.
6. Burke, J.E., Ed., *Progress in Ceramic Science*, Vols. 1–4, Pergamon Press, Elmsford, N. Y., 1962–1966.
7. Alper, A.M., Ed., *High Temperature Oxides*, Parts I–IV, Academic Press, New York, 1970–1971.
8. Ryshkewitch, E., *Oxide Ceramics*, Academic Press, New York, 1960.
9. Haggerty, J.S. and Cannon, W.R., *Sinterable Powders from Laser-Driven Reactions*, MIT/Cambridge, Mass., ONR Contract Rept. AD-A063 064, Oct. 1978.

10. Richerson, D.W. and Washburn, M.E., Hot pressed silicon nitride, U.S. Patent No. 3,836,374, Sept. 17, 1974.
11. Gazza, G.E., Effect of yttria additions on hot-pressed Si_3N_4, *Bull. Am. Ceram. Soc.*, 54, 778–781, 1975.
12. Herbell, T.P., Glasgow, T.K., and Yeh, H.C., *Effect of Attrition Milling on the Reaction Sintering of Silicon Nitride*, National Aeronautics and Space Administration, Lewis Res. Ctr., NASA-TM-78965, 1978 (N78-31236).
13. Claussen, N. and Jahn, J., Mechanical properties of sintered and hot-pressed Si_3N_4–ZrO_2 composites, *J. Am. Ceram. Soc.*, 61, 94–95, 1978.
14. Stanley, D.A., Sadler III, L.Y. and Brooks, D.R., *1st Proceedings of International Conference on Particle Technology*, 1973.
15. Greskovich, C., in *Treatise on Materials Science and Technology, Ceramic Fabrication Processes*, Vol. 9, Wang, F.F.Y., Ed., Academic Press, New York, 1976, pp. 28–33.
16. Sturtevant Mill Co., Boston, Mass., Bulletin 091, Sturtevant Micronizer Fluid Energy Mills.
17. Johnson, D.W. and Gallagher, P.K., Reactive powders from solution, in *Ceramic Processing Before Firing* Onoda, Jr. G.Y. and Hench, L.L., Eds., Wiley, New York, 1978, pp. 125–139.
18. Flock, W.M., Bayer-processed aluminas, in *Ceramic Processing Before Firing*, Onoda, Jr. G.Y. and Hench, L.L., Eds., Wiley, New York, 1978, pp. 85–100.
19. Schnettler, F.J., Monforte, F.R., and Rhodes, W.H., A cryochemical method for preparing ceramic materials, in *Science of Ceramics*, Stewart, G.H., Ed., The British Ceramic Society, Stoke-on-Trent, U.K., 1968, pp. 79–90.
20. Johnson, D.W. and Schnettler, F.J., Characterization of freeze-dried Al_2O_3 and Fe_2O_3, *J. Am. Ceram. Soc.*, 53, 440–444, 1970.
21. Rigterink, M.D., Advances in technology of the cryochemical process, *Am. Ceram. Soc. Bull.*, 51, 158–161, 1972.
22. Johnson, Jr., D.W., Sol-gel processing ceramics and glass, *Am. Ceram. Soc. Bull.*, 64(12), 1597–1602, 1985.
23. Pechini, M., Method of preparing lead and alkaline-earth titanates and niobates and coating method using the same to form a capacitor, U.S. Patent No. 3,330,697, July 11, 1967.
24. Eror, N.G. and Anderson, H.U., Polymeric precursor synthesis of ceramic materials, in *Better Ceramics Through Chemistry, II, MRS Proceedings*, Vol. 73 (C.J. Brinker, D.E. Clark, and D.R. Ulrich, Eds.), MRS Society, Pittsburgh, Pa., 1986, pp. 571–577.
25. Chick, L.A. et al., Synthesis of oxide ceramic powders by the glycine-nitrate process, *Mat. Let.*, 10(1–2), 1990.
26. Kladnig, W.F. and Karner, W., Pyrohydrolysis for the production of ceramic raw materials, *Am. Ceram. Soc. Bull.*, 69(5), 814–817, 1990.
27. Dawson, W.J., Hydrothermal synthesis of advanced ceramic powders, *Am. Ceram. Soc. Bull.*, 67(10), 1673–1678, 1988.
28. Phillips, D.S. and Vogt, G.J., Plasma synthesis of ceramic powders, *Mater. Res. Bull.*, 12(1), 54–58 (1987).
29. Munir, Z.A., Synthesis of high temperature materials by self-propagating combustion methods, *Am. Ceram. Soc. Bull.*, 67(2), 342–349, 1988.
30. Siegel, R.W., *Nanophase Materials Assembled from Atomic Clusters*, MRS Bulletin, Oct. 1990.
31. Pincus, A.G. and Shipley, L.E., The role of organic binders in ceramic processing, *Ceram. Ind.*, 92, 106–109, 1969.
32. Masters, K., *Spray Drying Handbook*, 4th ed., Longman Scientific & Technical, Essex, England, Wiley, New York, 1985.

ADDITIONAL RECOMMENDED READING

1. Messing, G.L., Hirano, S. and Hausner, H., Eds., *Ceramic Powder Science III, Ceramic Transactions*, Vol. 12, American Ceramic Society, Westerville, Ohio, 1990.
2. Messing, G.L., Fuller, Jr, E.R. and Hausner, H., Eds., *Ceramic Powder Science II, A, Ceramic Transactions*, Vol. 1, American Ceramic Society, Westerville, Ohio, 1988.
3. Brinker, C.J., *Sol-Gel Science*, Academic Press, Inc., San Diego, 1990.

4. Shanefield, D.J., *Organic Additives and Ceramic Processing*, 2nd ed., Kluwer Academic Publishers, Boston, 1996.

5. Rahaman, M.N., *Ceramic Processing and Sintering*, Marcel Dekker, Inc., New York, 1995.

PROBLEMS

1. What ceramic material is synthesized by the Acheson process?

2. Seven wt% yttrium oxide has been successfully used as a sintering aid for aluminum nitride (AIN). How many grams of yttrium oxide are required for a 400-g batch?

3. If the required yttrium oxide content of Problem 2 is 7 mol%, how many grams of yttrium oxide are required for a 400-g batch?

4. $PbTiO_3$ can be prepared by a sol-gel technique starting with $Pb(NO_3)_2$ and $TiCl_4$. How much $TiCl_4$ must be used to yield 500 g of $PbTiO_3$?

5. Cordierite has the composition $Mg_2Al_4Si_5O_{18}$. What is the mol% MgO?

6. What is the wt% MgO in cordierite?

STUDY GUIDE

1. Fabrication of a ceramic generally requires that the starting materials be in powdered form. What are some important considerations for these powders?

2. What are some examples of powders that are used in traditional ceramics?

3. List 10 examples of applications for traditional ceramics.

4. List 10 examples of applications of modern ceramics.

5. What are some special characteristics to make powders suitable for fabrication of modern ceramics?

6. Why is control of impurities in raw materials important for ceramics destined for high-temperature, high-stress applications?

7. For electrical, magnetic, or optical applications?

8. Why is control of particle size and particle size distribution important?

9. Why is powder "reactivity" important?

10. What particle sizing method is good for sorting particles larger than 325 mesh (44 μm)?

11. What methods can produce average particle sizes less than 10 μm?

12. Picking up contamination is a major problem in reduction in particle size of a ceramic powder. What are some ways to minimize or eliminate contamination?

13. Explain the purpose or advantage of adding a milling aid during dry milling.

14. Explain some potential advantages of an attrition mill (such as a turbomill) or a vibratory mill compared to a ball mill.

15. Identify eight different processes that can yield highly controlled purity of powder to very fine particle size directly by chemical processes.

16. What is "preconsolidation," and why is it important?

17. What are four especially important categories of additives during preconsolidation, and what are their purposes?

18. What are the reasons for spray drying or granulation?

19. Describe spray drying.

20. Identify some organic binders that have been used with ceramic powders.

13 Shape-Forming Processes

The properly sized and preconsolidated powders are now ready for forming into the required shapes. Table 13.1 summarizes the major techniques for consolidation of powders and producing shapes. In this chapter we examine the major approaches in terms of the process steps and controls involved, the types of strength-limiting flaws that may result, and the range of shapes that can be produced.

13.1 PRESSING

Pressing is accomplished by placing the powder (premixed with suitable binders and lubricants and preconsolidated so that it is free-flowing) into a die and applying pressure to achieve compaction. Two categories of pressing are commonly used: (1) uniaxial and (2) isostatic. Both use powder prepared by the same procedures. Therefore, we shall discuss the procedures and the nature of the binder systems first, followed by a review of the pressing techniques. The types of problems that can be encountered in pressing and the evidence that an end-user can detect in the part will be emphasized. If the end-user can spot problems, he or she is in a better position to work with the supplier to solve the problems.

13.1.1 STEPS IN PRESSING

Like all the processes for forming ceramics, pressing involves a sequence of steps that all must be carefully controlled to achieve an acceptable product. Figure 13.1 shows a flowchart for two procedures of pressing. Let us review each procedure.

Procedure A is based on granulation to achieve a free-flowing powder. Raw materials are selected and weighed to the proper batch calculation. The powders are sized by dry milling. The sized powder is placed in a muller mixer with additions of the binder plus about 15 wt% water and mixed until homogeneous. The mixture is formed into granules by screening, running through a granulator, or

---

TABLE 13.1
Major Compaction Techniques Used for Ceramic Fabrication

Pressing	Fugitive-mold casting
Uniaxial	Gel casting
Isostatic	Electrophoretic deposition
Hot pressing [a]	**Tape Casting**
Hot isostatic pressing [a]	Doctor blade
Slip Casting	Waterfall
Drain casting	**Plastic Forming**
Solid casting	Extrusion
Vacuum casting	Roll forming
Pressure casting	Injection molding
Centrifugal casting	Compression molding

[a] Techniques that involve simultaneous compaction and densification and are discussed in Chapter 14.

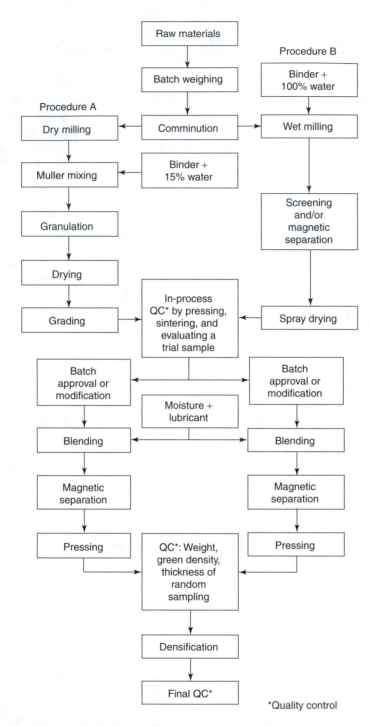

FIGURE 13.1 Typical flow sheets for fabrication by pressing. (From Ceramic Fabrication Processes, Kingery, W.D., Ed., MIT Press, Cambridge, MA, 1963. With permission.)

prepressing plus granulating. The granules are soft and damp at this stage, but become moderately hard after drying. The dried granules are graded by screening to achieve the desired size distribution. This usually involves removal of fines that are not adequately free-flowing. At this point, an in-process quality check is conducted. This typically involves pressing of a trial sample to determine

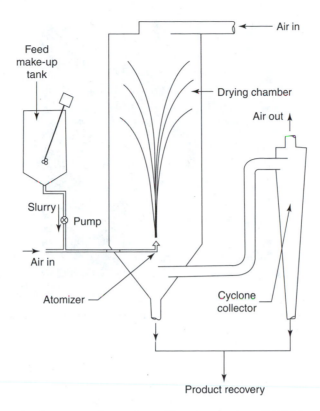

FIGURE 13.2   Schematic of one type of spray dryer. (Drawing courtesy of ASM International.)

the compaction characteristics (e.g., compaction ratio* green density, ease of release from the die) and densification characteristics (e.g., shrinkage, fired density, and key properties). Acceptable powder is then ready to be prepared for production pressing. This can involve addition of a lubricant and a little moisture and removal by magnetic separation of metal particles picked up by wear of the processing equipment. Quality is checked on random samples after pressing and again after densification.

Procedure B is based on spray drying to achieve a free-flowing powder. The weighed batch of powder plus additives is mixed with enough water to form a fluid suspension (slurry) and wet-milled to achieve homogeneous mixture and particle sizing. The slurry is passed through a screen and/or a magnetic separator to remove large particles and metallic contamination. The slurry is then spray-dried as described in Chapter 12. Figure 13.2 shows the schematic of one type of spray dryer. Figure 13.3 illustrates the morphology of typical spray-dried powder. Depending on the slurry and the spray-drying parameters, the resulting powder can consist of solid spheres, hollow spheres, or doughnut-shaped platelets. After spray drying the powder goes through the same quality control and pressing procedures as described for procedure A.

13.1.2   SELECTION OF ADDITIVES

The additives commonly required for pressing are a binder, a plasticizer, a lubricant, and/or a compaction aid. The binder provides some lubrication during pressing and gives the pressed part adequate strength for handling, inspection, and green machining. The plasticizer modifies the binder to make it more pliable. The lubricant reduces interparticle friction and die-wall friction. The compaction

---

* Compaction ratio is the ratio between the thickness of the powder in the die to the thickness after pressing. Green density is bulk density of the compact. The term "green" is commonly used to describe the porous ceramic compact prior to densification.

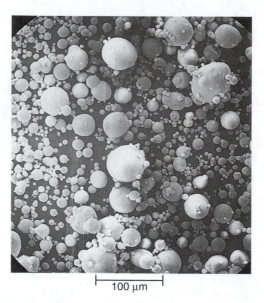

100 µm

FIGURE 13.3   Photo taken with a scanning electron microscope showing the spherical morphology of spray-dried powder. (Courtesy of Ceramatec, Inc.)

aid (which is essentially a lubricant) reduces interparticle friction. The combined effects of the additives are: (1) to allow the powder particles to slide past each other to rearrange in the closest possible packing; and (2) to minimize friction and allow all regions of the compact to receive equivalent pressure. Let us discuss these in more detail and examine some examples.

13.1.2.1   Binders and Plasticizers

Table 12.13 listed a variety of organic and inorganic materials that have been used as binders. Most binders and plasticizers are organic. They coat the ceramic particles and provide lubrication during pressing and a temporary bond after pressing. The amount of organic binder required for pressing is quite low, typically ranging from 0.5 to 5 wt%. Organic binders normally are decomposed during the high-temperature densification step and evolved as gases. Some binders leave a carbon residue, especially if fired under reducing conditions.

Inorganic binders also exist. The clay minerals such as kaolinite are a good example. Kaolinite has a layered structure and interacts with water to yield a flexible, plastic mixture. The clay minerals do not burn off during densification, but instead become part of the ceramic.

Binder selection is dependent on the type of pressing that will be conducted. Some binders such as waxes and gums are very soft and quite sensitive to temperature variations. These generally do not require moisture or lubricant additions prior to pressing, but must be handled more carefully to avoid changes in granule size that might alter flow characteristics into the pressing die or result in inhomogeneous density distribution. The soft binders also have a tendency to extrude between the die components, which can cause sticking or reduce the production rate.

Other binders can be classified as hard, that is, they produce granules that are hard or tough. These granules have the advantage that they are dimensionally stable and free-flowing and are therefore excellent for high-volume production with automated presses. However, these are generally not self-lubricating and thus require small additions of lubricant and moisture prior to pressing. They also require higher pressure to assure uniform compacts. If the starting powder agglomerates are not completely broken down into a continuous compact during pressing, artifacts of the approximate size of the agglomerates will persist through the remaining process steps and may act as large flaws, which will limit the strength.

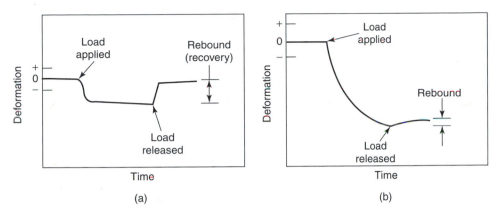

FIGURE 13.4   (a) Glassy deformation behavior below the glass transition temperature and (b) plastic behavior above the glass transition temperature. (Drawings courtesy of ASM International.)

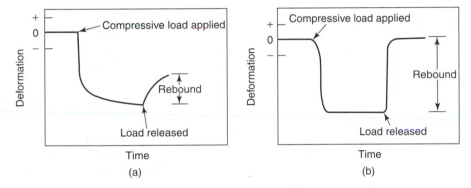

FIGURE 13.5   (a) Viscoelastic deformation behavior and (b) rubbery deformation behavior. (Drawings courtesy of ASM International.)

Dextrine, starches, lignins, and acrylates produce relatively hard granules. Polyvinyl alcohol and methyl cellulose result in slightly softer granules. Waxes, wax emulsions, and some gums produce soft granules.

The hardness and deformation characteristics of organic binders vary with temperature, humidity, and other factors. Many of these materials go through a ductile-brittle transition and behave in a brittle fashion below the transition and in a ductile fashion above the transition. The temperature at which this ductile-brittle transition occurs is referred to as the *glass transition temperature* (T_g). Typical load-deflection curves below and above the glass transition temperature are shown in Figure 13.4. Several aspects of these curves are important during pressing: (1) the total deformation; (2) the amount of rebound or recovery after the load is removed; (3) the load necessary to initiate deformation; and (4) the net deformation (permanent set). These are illustrated for a hypothetical load-deflection curve in Figure 13.5.

Now we can go back to Figure 13.4 and compare the behavior below and above the glass transition temperature T_g. Below T_g the deformation is primarily elastic and the behavior is categorized as "glassy." The total deformation is low and is completely recovered after removal of the load. This behavior provides little or no binder or lubricant capability. The material has a stronger tendency to fracture than to deform. Conversely, above T_g the deformation is large and primarily plastic. Very little rebound occurs when the load is removed. This behavior provides excellent binder and lubricant capability. The curve in Figure 13.5a illustrates behavior intermediate between glassy and plastic. This behavior is referred to as *viscoelastic*. One other type of behavior is occasionally encountered.

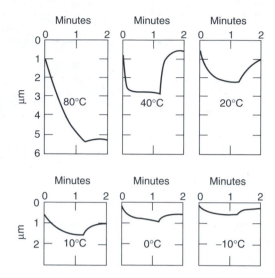

FIGURE 13.6   Deformation behavior of an alkyd organic binder at various temperatures showing transition from glassy to viscoelastic to rubbery to plastic. (Adapted from Rosen, S.L., *Fundamental Principles of Polymeric Materials*, Wiley, New York, 1982.)

This is shown in Figure 13.5b and is referred to as "rubbery." It is characterized by large elastic deformation and large rebound after the load is removed. This behavior is not favorable for pressing.

The complete range of deformation characteristics can occur for a single organic binder material over a temperature range. This is shown for an alkyd material in Figure 13.6. Between -10 and $0°C$ (15 and 32°F) the material shows glassy behavior. Between 10 and 20°C (50 and 70°F) the behavior is viscoelastic. At 40°C (105°F) the behavior is rubbery and at 80°C (175°F) the behavior is plastic.

The deformation behavior can be altered by addition of plasticizers. Let us examine polyvinyl alcohol (PVA) that is commonly used as a binder in the ceramic industry. Polyethylene glycol (PEG) and water both act as plasticizers for PVA. PVA with no plasticizer has a T_g over 60°C (140°F) and thus has glassy deformation behavior at room temperature. Relative humidity (rh) of 50% reduces T_g to below 50°C (120°F). Addition of 50% PEG at 50% rh reduces T_g to 10°C (50°F). As would be expected from these numbers, the amount of plasticizer and the pressing temperature have a significant effect on the green density achieved during pressing. This is illustrated in Figure 13.7 for Al_2O_3 powders pressed with binder-plasticizer combinations of 60% PVA–40% PEG, 80% PVA–20% PEG, and 100% PVA at different pressures and temperatures. Powder pressed at 70 MPa (10.1 ksi) at room temperature with 100% PVA binder achieved a green density of only 1.93 g/cm³ (0.069 lb/in.³) (48.5% theoretical). Powder with 60% PVA binder–40% PEG plasticizer pressed at 140 MPa (20.3 ksi) at room temperature reached a green density of 2.2 g/cm³ (0.079 lb/in.³) (55% theoretical).

Glycerol also acts as a plasticizer for PVA. The effects of glycerol plus moisture on the T_g of PVA are shown in Figure 13.8.

13.1.2.2   Lubricants and Compaction Aids

Lubricants and compaction aids are essentially the same. They reduce the friction between particles, between granules, and between the powder compact and the wall of the pressing die. This results in increased uniformity of the pressed part, improved green density, extended tool life, reduced sticking (which reduces time required for tool cleaning), and decreased pressure required to eject the part from the die. The effect on ejection pressure is illustrated in Figure 13.9. One-half percent of zinc stearate reduced the ejection pressure by a factor of four. Materials with low shear strength seem to make good lubricants. Other low-shear-strength materials besides zinc stearate that have been successful as lubricants are listed in Table 13.2.

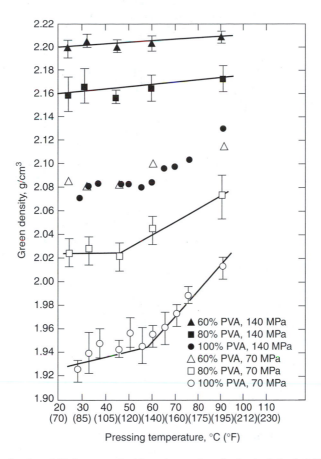

FIGURE 13.7 Green density of Al_2O_3 pressed with various ratios of polyvinyl alcohol (PVA) binder and polyethylene glycol (PEG) at different pressures and temperatures. (From Nies and Messing, *J. Am. Ceram. Soc.*, 301, 1984. With permission.)

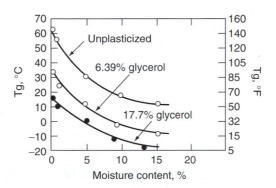

FIGURE 13.8 Effects of additions of glycerol and moisture on the glass transition temperature of a PVA binder. (From Nies and Messing, *J. Am. Ceram. Soc.*, 301, 1984. With permission.)

Compaction aids can result in substantial increase in green density and reduction in shrinkage during densification. This is illustrated for Al_2O_3 in Table 13.3. Al_2O_3 powder pressed at 34.5 MPa (5 ksi) with no compaction aid reached a green density of 2.53 g/cm³ (0.091 lb/in.³) and was fired at 1700°C (3090°F) in 1 h to a density of 3.85 g/cm³ (0.138 lb/in.³). Addition of only 2.0 wt% stearic acid increased the green density to 3.0 g/cm³ (0.107 lb/in.³) and decreased the linear shrinkage during firing to only 8.3% (which is a very low value).

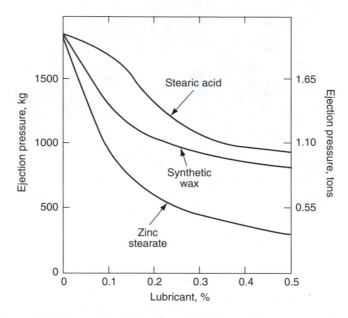

FIGURE 13.9 Effects of lubricants on decreasing die-wall friction and the pressure to eject a pressed part from a die. (Drawing courtesy of ASM International.)

TABLE 13.2
Low-Shear-Strength Materials Used as Lubricants for Pressing

Zinc stearate	Paraffin
Stearic acid	Synthetic wax
Oleic acid	Lithium stearate
Oils	Potassium stearate
Naphthenic acid	Sodium stearate
Boric acid	Ammonium stearate
Boron nitride (hexagonal)	Magnesium stearate
Graphite	Talc

TABLE 13.3
Use of Lubricants as Compaction Aids for Pressing of Al_2O_3[a]

Compaction Aid	Quantity, wt%	Green Density, g/cm³	Fired Density, g/cm³	Liner Shrinkage, %
None	0	2.53	3.85	13.1
Stearic acid	1.0	2.86	3.85	9.45
Stearic acid	2.0	3.00	3.89	8.27
Stearic acid	3.0	3.06	3.88	7.67
Oleic acid	2.0	2.97	3.90	8.63

[a] All specimens pressed at 34.5 MPa (5000 psi) and fired 1 h at 1700°C (3090°F).

Source: Hart, L.D. and Hudson, L.K., *Am. Ceram. Soc. Bull.*, 43(1), 13, 1964.

13.1.2.3   Removal of Organic Additives

Selection of binder and other additives must be compatible with the chemistry of the ceramic and the purity requirements of the application. The binder must be removed prior to densification of the ceramic. Organic binders can be removed by thermal decomposition. If reaction between the binder and the ceramic occurs below the binder decomposition temperature or if the ceramic densifies below this temperature, the final part will be contaminated or may even be cracked or bloated. If the temperature is raised too rapidly or if the atmosphere in the furnace is reducing, the binder may char rather than decompose, leaving carbon.

13.1.3   Uniaxial Pressing — Presses and Tooling

Uniaxial pressing involves the compaction of powder into a rigid die by applying pressure along a single axial direction through a rigid punch, plunger, or piston.[1,2] Most uniaxial presses are either mechanical or hydraulic. Mechanical presses typically have a higher production rate and are easy to automate. Figure 13.10 shows schematically the automated pressing sequence of a typical uniaxial mechanical press. The punches preposition in the die body to form a cavity predetermined (based on the compaction ratio of the powder) to contain the correct volume to achieve the required green dimensions after compaction. The feed shoe then moves into position and fills the cavity with free-flowing powder containing suitable binders, moisture, and lubricant. The feed shoe retracts, smoothing the powder surface as it passes, and the upper punches move down to precompress the powder. The upper and lower punches then simultaneously compress the powder as they independently move to preset positions. The upper punches retract and the lower punches eject the compact from the die body. The feed shoe then moves into position and pushes the compact away from the punches as the punches reset to accept the correct powder fill. This cycle repeats typically 6 to 100 times per minute, depending on the press and the shape being fabricated. Presses of this type generally have a capacity from 910 to 18,200 kg (1 to 20 tons), but some operate up to 91,000 kg (100 tons).

Another type of mechanical press is the rotary press. Numerous dies are placed on a rotary table. The die punches pass over cams as the table rotates, resulting in a fill, compress, and eject cycle similar to the one described for a single-stroke press. Production rates in the range of 2000 parts per minute can be achieved with a rotary press. Pressure capability is in the range 910 to 91,000 kg (1 to 100 tons).

Yet another type of mechanical press is the toggle press. It is commonly used for pressing refractory brick and is capable of exerting pressure up to about 727,000 kg (800 tons). The toggle press closes to a set volume so that the final density is controlled largely by the characteristics of the feed.

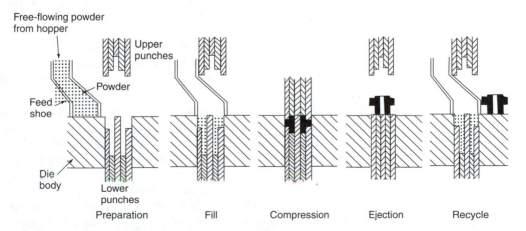

FIGURE 13.10   Schematic illustrating automated uniaxial pressing. (Adapted from Thurnauer, H., Controls required and problems encountered in production dry pressing, in *Ceramic Fabrication Processes,* Kingery, W.D., Ed., MIT Press, Cambridge, MA, 1963, pp. 62–70.)

TABLE 13.4
Uniaxial Pressing Part and Tool Classifications

Class	Definition	Type of Tooling	Typical Part Cross Sections
I	Thin, one-level parts that can be pressed from one direction	Single action	
II	Thick, one-level parts that require pressing force from both ends	Double action	
III	Two-level parts that require pressing force from both ends	Double action, multiple motion	
IV	Multiple-level parts that require pressing force from both ends	Double action, multiple motion	

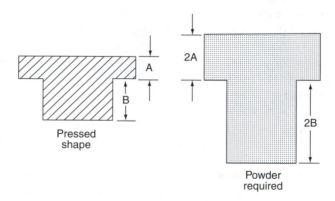

FIGURE 13.11 Schematic illustrating the different distances a punch must move to accomplish uniform compaction of the powder. Based on a powder with a compaction ratio of 2:1. (Drawing courtesy of ASM International.)

Hydraulic presses transmit pressure via a fluid against a piston. They are usually operated to a set pressure, so that the size and characteristics of the pressed component are determined by the nature of the feed, the amount of die fill, and the pressure applied. Hydraulic presses can be very large, but have a much lower cycle rate than mechanical presses.

The type of press and tooling selected is based largely on the size and shape of the part to be pressed. The parts can be divided into classes, as is done in powder metallurgy. The classes are defined in Table 13.4. Parts with a constant thickness and thin cross section can be successfully pressed with single-action, that is, with the die and bottom punch stationary and only the top punch moving. Thicker parts do not achieve uniform compaction if only pressed from one end. These require tooling where the top and bottom punches move, that is, double-action tooling. Parts with variations in cross-sectional thickness require an independent punch for each level of thickness. This again is necessary to achieve uniform compaction throughout the part. This is illustrated in Figure 13.11 for a powder with a compaction ratio of 2:1. The punch only has to travel a distance A to achieve compaction of the thin section, but most travel a distance A + B to compact the thick section. Both cannot be achieved with a single rigid punch. Two punches are required. Figure 13.12 shows the schematic of tooling needed to uniformly press a three-level part.

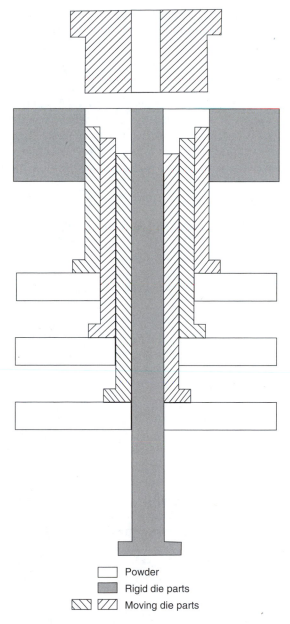

Powder

Rigid die parts

Moving die parts

FIGURE 13.12 Schematic of tooling to uniaxially press a three-level part. (Drawing courtesy of ASM International.)

13.1.3.1 Dry Pressing

Most automated pressing is conducted with granulated or spray-dried powder containing 0 to 4% moisture. This is referred to as *dry*, *semidry*, or *dust pressing*. Compaction occurs by crushing of the granules and mechanical redistribution of the particles into a close-packed array. The lubricant and binder usually aid in this redistribution and the binder provides cohesion. High pressures are normally used for dry pressing to assure breakdown of the granules and uniform compaction.

High production rates and close tolerances can be achieved with dry pressing. Millions of capacitor dielectrics approximately 0.050-cm (0.020-in.) thick are made to close tolerances and with tightly specified electrical properties. Millions of electrical substrates, packages (thin-walled insulator "boxes" for isolating miniature electronic circuits), and other parts for a wide variety of applications are made by dry pressing. Dimensional tolerances to ±1% are normally achieved in routine applications, and closer tolerances have been achieved in special cases.

13.1.3.2 Wet Pressing

Wet pressing involves a feed powder containing 10 to 15% moisture and is often used with clay-containing compositions. This feed powder deforms plastically during pressing and conforms to the contour of the die cavity. The pressed shape usually contains flash (thin sheets of material at edges where the material extruded between the die parts) and can deform after pressing if not handled carefully. For these reasons, wet pressing is not well-suited to automation. Also, dimensional tolerances are usually only held to ±2%.

13.1.3.3 Uniaxial Pressing Problems

The following are some of the problems that can be encountered with uniaxial pressing.

improper density or size
die wear
cracking
density variation.

The first two are easy to detect by simple measurements on the green compact immediately after pressing. Improper density or size are often associated with off-specification powder batches and are therefore relatively easy to resolve. Die wear shows up as progressive change in dimensions. It should also be routinely handled by the process specification and quality control.

The source of cracking may be more difficult to locate. It may be due to improper die design, air entrapment, rebound during ejection from the die, die-wall friction, die wear, or other causes. Often a crack initiates at the top edge of the part during pressure release or ejection of the part. Two mechanisms of this type cracking are illustrated in Figure 13.13. The first, shown in Figure 13.13a, occurs as pressure is released from the upper punch. The material rebounds near the top center of the compact, but is restricted momentarily at the edges due to frictional drag between the compact and the wall of the die. This results in a tensile stress concentrated at the upper edge of the compact. Cracks due to this mechanism (called endcapping) can be avoided by (1) use of a lubricant to minimize die-wall friction; (2) increasing the green strength of the part through binder selection; (3) minimizing rebound; and (4) maintaining a hold-down pressure on the upper punch during ejection.

The second mechanism is illustrated in Figure 13.13b. This also involves rebound. As the part clears the top of the die during ejection, the material rebounds to a larger cross section. This places

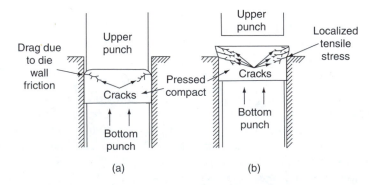

FIGURE 13.13 Mechanism of formation of laminar cracks in uniaxially pressed parts. (a) Pressure being released from upper punch and (b) material rebound at top of die. (Drawing courtesy of ASM International.)

a tensile stress in the material just above the top of the die and can result in a series of laminar cracks. This mechanism can be minimized by selecting a binder system that provides good green strength with minimum rebound.

Another important problem to be overcome in uniaxial pressing is nonuniform density.[3–5] Density variation in the green compact causes warpage, distortion, or cracking during firing. One source of density variation is the friction between the powder and the die wall and between powder particles. As shown in Figure 13.14, a uniaxial pressure applied from one end of a die full of powder

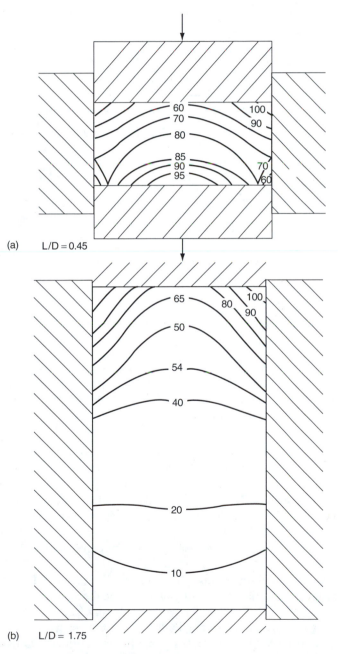

FIGURE 13.14 Pressure variations in uniaxial pressing due to die-wall friction and particle-particle friction, which lead to nonuniform density of the pressed compact. (Adapted from *Ceramic Fabrication Processes*, Kingery, W.D., Ed., MIT Press, Cambridge, MA, 1963.)

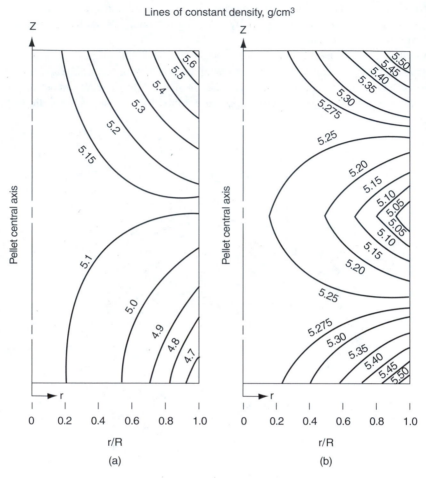

FIGURE 13.15 Decrease in pressure curves and increase in uniformity of green density by pressing compact from opposing directions with a double-acting press. (a) Single-acting press and (b) double-acting press. (From Thompson, R.A., *Am. Ceram. Soc. Bull.*, 60(2), 237–243, 1981. With permission.)

will be dissipated by friction so that a substantial portion of the powder will experience much lower than the applied pressure. These areas will compact to a lower density than the areas exposed to higher pressure. The pressure difference increases as the length-to-diameter ratio increases. During firing, the lower-density areas will either not densify completely or will shrink more than surrounding areas. Both will result in flaws that can cause rejection of the part.

Use of suitable binders and lubricants can reduce both die-wall and particle-particle friction and thus reduce density variation in the compact. Applying pressure from both ends of the die also helps. This is illustrated in Figure 13.15.

A second source of nonuniform density is nonuniform fill of the die. Powder that is heaped or otherwise nonuniformly stacked in the die will not reposition during pressing. The region with the largest amount of powder will compact to a higher green density. This region of decreased porosity will shrink less during densification, resulting in distortion of the part.

A third source of nonuniform green density is the presence of hard agglomerates (clusters of particles) in the powder or a range of hardness of the granules in a free-flowing granulated powder. The hard granules will shield surrounding softer powder or granules from exposure to the maximum pressing pressure, resulting in pore clusters that reduce strength. Sometimes the surrounding powder

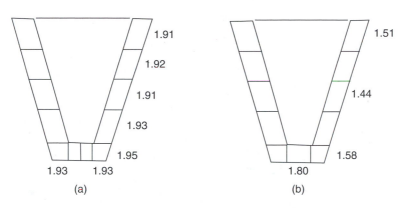

FIGURE 13.16 Improvement in green density uniformity of a thin-wall crucible achieved by isostatic pressing. (a) After isostatic pressing and (b) after die pressing. (From Gill, R.M. and Bryne, J., in *Science of Ceramics*, Vol. 4, Stewart, G.H., Ed., British Ceramic Research Association, London, 1968. With permission.)

will compact uniformly, but the hard agglomerate will trap porosity. The hard agglomerate may then shrink more than the surrounding material during densification and leave a large pore.

Note that the cause of nonuniform density is more associated with the condition of the powder loaded into the pressing die than with the pressing operation itself. Further, note that the problem may not show up until after the densification step of fabrication. This is another reminder that all of the processing steps are interlinked and that all must be coordinated and controlled to achieve the desired characteristics of the final product.

13.1.4 ISOSTATIC PRESSING

Uniaxial pressing has limitations as described in the prior section. Some of the limitations can be overcome by applying pressure from all directions instead of only one or two directions. This is referred to as *isostatic pressing* or *cold isostatic pressing* (CIP). It has also been referred to as hydrostatic pressing.[6]

Application of pressure from multiple directions achieves greater uniformity of compaction and increased shape capability. Figure 13.16 compares the green density uniformity achieved in a thin-wall crucible fabricated by isostatic vs. uniaxial pressing. The green density of the isostatically pressed part is higher and much more uniform. Green density for the isopressed part ranges from 1.91 to 1.95 g/cm³ (0.068 to 0.070 lb/in.³) compared to 1.44 to 1.80 g/cm³ (0.054 to 0.065 lb/in.³) for the uniaxially pressed part. Figure 13.17 shows the density uniformity achieved by isostatic pressing of a larger solid cylinder. Compare this with the density contours shown previously in Figure 13.15 for a uniaxially pressed solid cylinder.

Two types of isostatic pressing are commonly used: (1) wet-bag and (2) dry-bag. These are described in the following paragraphs.

13.1.4.1 Wet-Bag Isostatic Pressing

Wet-bag isostatic pressing in illustrated in Figure 13.18. The power is sealed in a water-tight die. The walls of the die are flexible. The sealed die is immersed in a liquid contained in a high-pressure chamber. The chamber is sealed using a threaded or breach lock cover. The pressure of the liquid is increased by hydraulic pumping. The walls of the die deform and transmit the pressure uniformly to

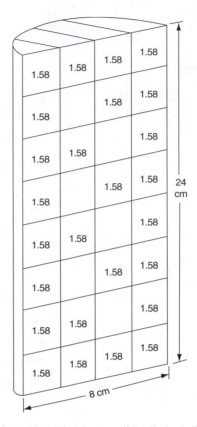

FIGURE 13.17   Density uniformity achieved in a large, solid cylinder by isostatic pressing. (From Gill, R.M. and Bryne, J., in *Science of Ceramics*, Stewart, G.H., Ed., British Ceramic Research Association, London, 1968. With permission.)

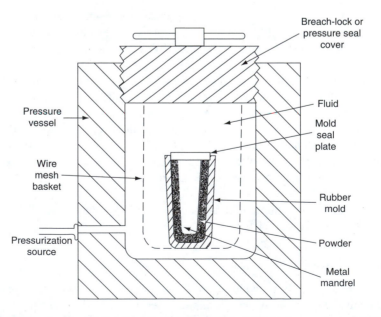

FIGURE 13.18   Schematic of a wet-bag isostatic pressing system. (© Drawing courtesy of ASM International.)

the powder, resulting in compaction. The walls of the die spring back after the pressure is removed, allowing the compact to easily be removed from the die after the cap of the die is removed.

Any noncompressible fluid can be used for isopressing. Water is commonly used, although fluids such as hydraulic oil and glycerine also work. The flexible walls of the die or mold are made of an elastomer such as rubber or polyurethane. The flexibility and wall thickness are carefully selected to allow optimum dimensional control and release characteristics. Natural rubber, neoprene, butyl rubber, nitrile, silicones, polysulphides, polyurethanes, and plasticized polyvinyl chloride have all been used.

Laboratory isostatic presses have been built with pressure capabilities among from 35 to 1380 MPa (5 to 200 ksi). However, production units usually operate at 400 MPa (58 ksi) or less.

A major concern in isostatic pressing is uniform fill of the mold. This is usually achieved by use of vibration plus free-flowing spray-dried or granulated powder. Since higher pressures are usually achieved by isostatic pressing than by uniaxial pressing and since these pressures are applied uniformly, a greater degree of compaction is achieved. This usually results in improved densification characteristics during the subsequent sintering step of processing and a more uniform, defect-free component.

As with other processes, wet-bag isopressing has advantages and disadvantages. The advantages are density uniformity, versatility, and low cost of tooling. Coupled with green machining, a wide variety and size of parts can be fabricated with minimum equipment investment. The disadvantages are long cycle time, high labor requirement, and difficulty to automate. Cycles are in minutes and tens of minutes, so production rates are low compared to uniaxial pressing.

13.1.4.2 Dry-Bag Isostatic Pressing

Dry-bag isopressing was developed to achieve increased production rate and close dimensional tolerances. Rather than immerse the tooling in a fluid, the tooling is built with internal channels into which the high-pressure fluid is pumped. This minimizes the amount of pressurized fluid required and allows the use of stationary tooling. The major challenge is constructing the tooling so that pressure is uniformly transmitted to the powder to achieve the desired shape. This is accomplished through careful positioning and shaping of the fluid channels, often by use of several different elastomer materials in a single die and by optimization of the external constraints of the die. Once a tool has been properly designed and automated, parts can be pressed at a rate of 1000 to 1500 cycles per hour.

Dry-bag isostatic pressing has been used for many years to press spark plug insulators. A schematic cross section of a pressing die is shown in Figure 13.19. Multiple dies are built into a single tool. Pressing and green machining are automated as shown in Figure 13.20 for fabrication of the zirconia electrolyte for an oxygen sensor. The rubber mold or die forms the outer circumference to the sensor. The spindle moves into the die from the bottom and forms the inner cavity of the sensor. The die is filled from the top and capped. Pressure is applied uniformly around the circumference of the die to compact the powder. After pressure release, the spindle and green compact are withdrawn from the bottom of the die. The compact is transferred automatically to a grind spindle that brings the compact in contact with a contoured grinding wheel for final shaping. The finished shape is then conveyed to a furnace for binder removal and densification.

13.1.5 Applications of Pressing

Uniaxial pressing is widely used for compaction of small shapes, especially of insulating, dielectric, and magnetic ceramics for electrical devices. These include simple shapes such as bushings, spacers, substrates, and capacitor dielectrics, and more complex shapes such as the bases or sockets for tubes, switches, and transistors. Uniaxial pressing is also used for the fabrication of tiles, bricks, grinding wheels, wear-resistant plates, crucibles, and an endless variety of parts.

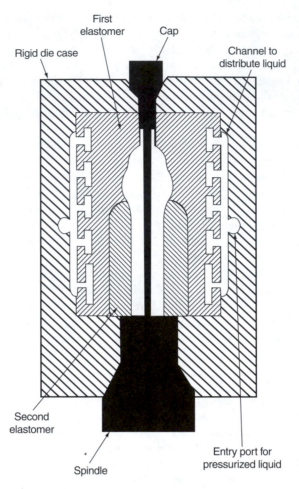

FIGURE 13.19   Schematic of a die for dry-bag isostatic pressing of a spark plug insulator. (Drawing courtesy of ASM International.)

Isostatic pressing, typically in conjunction with green machining, is used for configurations that cannot be uniformly pressed uniaxially or that require improved properties. Large components such as radomes, cone classifiers, and cathode-ray-tube envelopes have been fabricated by isostatic pressing. Bulky components for the paper industry have also been produced. Small components with a large length-to-width ratio are also fabricated by isostatic pressing and machining.

Figure 13.21 shows a variety of ceramic parts that have been fabricated by uniaxial and isostatic pressing.

13.2   CASTING

When most people hear the term "casting," they automatically think of metal casting in which a shape is formed by pouring molten metal into a mold. A limited amount of casting of molten ceramics is done in the preparation of high-density Al_2O_3 and Al_2O_3–ZrO_2 refractories and in preparation of some abrasive materials. In the latter case, casting from a melt into cooled metal plates produces rapid quenching, which results in very fine crystal size that imparts high toughness to the material. The technique of casting molten ceramic refractories is called *fusion casting*.

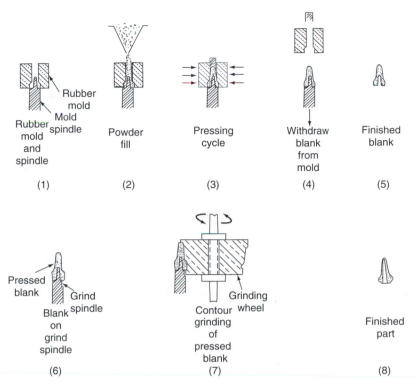

FIGURE 13.20 Automated dry-bag isostatic pressing and formed-wheel green machining of a zirconia electrolyte for an automotive oxygen sensor. (Drawings courtesy of ASM International.)

FIGURE 13.21 Ceramic parts formed by uniaxial and isostatic pressing, some with green machining. (Courtesy of Western Gold and Platinum Company, Subsidiary of GTE Sylvania, Inc.)

More frequently, the casting of ceramics is done by a room-temperature operation in which ceramic particles suspended in a liquid are cast into a porous mold that removes the liquid and leaves a particulate compact in the mold. There are a number of variations to this process, depending on the viscosity of the ceramic-liquid suspension, the mold, and the procedures used. The most common is referred to as *slip casting*. The principles and controls for slip casting are similar to those of the other particulate ceramic casting techniques. Slip casting is described in detail, followed by a brief description of other techniques.

13.2.1 SLIP CASTING

Most commercial slip casting involves ceramic particles suspended in water and cast into porous plaster molds. Figure 13.22 identifies the critical process steps in slip casting and some of the process parameters that must be carefully controlled to optimize strength or other critical properties.

13.2.1.1 Raw Materials

Selection of the starting powder is dependent on the requirements of the applications. Most applications require a fine powder, typically −325 mesh (44 μm). Applications requiring high strength require even finer powders, averaging under 5 μm (0.0002 in.), with a substantial portion under 1 μm (0.00004 in.).

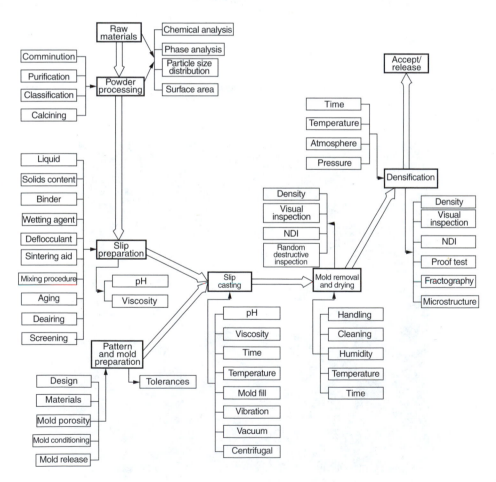

FIGURE 13.22 Critical process steps in slip casting and some of the process parameters that must be carefully controlled.

Some applications, such as kiln furniture, which must withstand cyclic thermal shock, may require a bimodal particle size with some particles considerably larger than 325 mesh.

The chemical composition is frequently an important consideration in selecting the starting powder and additives. Impurities and second phases can have pronounced effects on high-temperature properties.

13.2.1.2 Powder Processing

As discussed in Chapter 12, powder as received from the supplier does not usually meet all the specifications for the shape-forming process or application and must be processed. Processing for slip casting usually involves particle sizing to achieve a particle size distribution that will yield maximum packing and uniformity during casting. Often, particle sizing is combined in one step with addition of binder, wetting agents, deflocculants, and densification aids and with slip preparation. This is usually done by ball milling, but can also be done by vibratory milling or other processes that provide wet milling. After milling, the slip is screened and perhaps passed through a magnetic separator to remove iron contamination. Slight adjustment might be required to achieve the desired viscosity and then the slip is ready for aging, deairing, or casting.

13.2.1.3 Slip Preparation and Rheology

To understand slip preparation considerations, one needs to understand a little about rheology. Rheology is the study of the flow characteristics of matter, for example, suspensions of solid particles in a liquid, and is described quantitatively in terms of viscosity η. For low concentrations of spherical particles where no interaction occurs between the particles, the Einstein relationship applies:

$$\frac{v}{v_0} = 1 + 2.5V \tag{13.1}$$

where

v = viscosity of the suspension
v_0 = viscosity of the suspending fluid
V = volume fraction of solid particles

This idealized relationship implies that the resulting viscosity is controlled by the volume fraction of solids.

In actual systems, the volume fraction does have a major effect, but so also do particle size, particle shape, particle surface charges, and degree of agglomeration vs. dispersion. All of these are interrelated. The viscosity is essentially determined by how close particles approach each other and by the degree of attraction or repulsion between the particles. As we have discussed previously, particles of a material have incomplete bonds at the surface. These particles tend to adsorb H_2O and other molecules. It has been suggested that these adsorbed species result in a "sphere of influence" of about 20 Å around each particle. When adjacent particles approach within ~20 Å of each other, they tend to interact. When all of the particles in the suspension approach 20 Å, the force to move particles past each other will increase and the viscosity will significantly increase.

13.2.1.4 Particle Size and Shape Effects

The size and shape of particles determine the volume fraction at which particles approach the 20 Å sphere of influence. This is illustrated by calculations from Reference 7. To estimate the sensitivity of particle size, how close spherical particles of specific sizes would approach each other in a suspension containing 40 vol% of the particles was calculated:

Diameter of Spherical Particles (μm)	Mean Separation Distance for 40 vol% Solids Suspension (Å)
10	9200
1	920
0.1	92
0.05	20
0.01	9.2

The particles would all have to be about 0.05 μm to approach 20 Å at 40 vol% concentration. Increasing the volume fraction of particles causes the critical size to be larger.

The effect of particle shape was then estimated by a calculation assuming plate-shaped particles with an edge length ten times the thickness. This ratio was selected because it is representative of kaolinite clay particles that are a major constituent of slip-cast traditional ceramics. In this case, the volume fraction of ceramic plates at which the mean particle separation would be 20 Å was calculated:

Edge Length of Plate (μm)	Thickness of Plate (μm)	Volume Fraction of Solid at Which Mean Distance between Particle Surfaces Is 20 Å
10	1	30
1	0.1	30
0.1	0.01	28
0.01	0.001	17

It is apparent that shape has a large effect. The larger the deviation from spherical, the smaller the volume fraction of solids for 20 Å separation.

Particle size distribution is also important. One objective with casting is to achieve a high green density and to minimize shrinkage during densification. As was shown for pressing, this objective of close packing can be achieved best by a distribution of particle sizes. A distribution of particle sizes also helps to achieve increased solids loading in the slip. This is illustrated schematically in Figure 13.23 and by the example in Figure 13.24. The example shows the viscosity characteristics (plotted as flow rate vs. shear stress) for three particle size distributions of SiO_2 powder. The size distributions are plotted as "percent finer" vs. "equivalent spherical diameter," which is the common way of graphically displaying particle size distribution data. Compare the size range for each powder between 90% finer and 10% finer. Powder A consists in this range of particles between about 2 and 18 μm. Powder B consists in this 90 to 10% range of particles between 0.35 and 7 μm. A 50-50 mixture of A + B has a broader particle size range containing some of the finer particles from B and coarser particles from A. All three powders were successfully prepared into slips containing approximately 50 vol% solids. Note from Figure 13.24 that the finer powder B had the highest viscosity (lowest flow rate), the coarser powder had intermediate viscosity, and the mixture had the lowest viscosity. For the slips in this example, the viscosity appears to be influenced by a combination of particle size and particle packing.

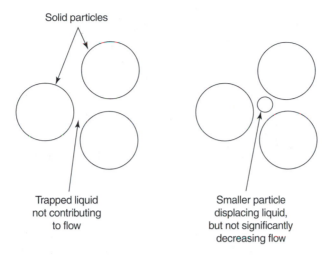

Solid particles

Trapped liquid
not contributing
to flow

Smaller particle
displacing liquid,
but not significantly
decreasing flow

FIGURE 13.23 Comparison of fluid requirements for slip containing one size of particles in suspension vs. a range of sizes. (Drawing courtesy of ASM International.)

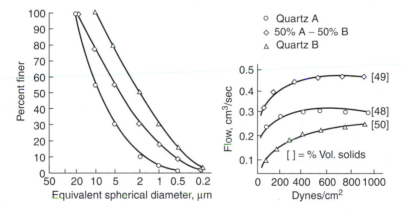

FIGURE 13.24 Effect of particle size distribution of air-elutriated SiO_2 in water using sodium silicate as a defloc-culant. (From Phelps and McLaren, in *Ceramic Processing Before Firing*, Onoda, G. and Hench, L., Eds., Wiley, New York, 1978, Chapter 17. With permission.)

An example for Al_2O_3 is illustrated in Figure 13.25. The curves on the left identify the particle size distributions. The curves on the right show the rheology behavior. The broad particle size distribution had the best rheology for casting. The 96 coarse-4 fine powder in a 59 vol% suspension showed shear-thinning behavior, that is, low viscosity at high shear rate and higher viscosity as the shear rate decreased. In other words, the slip became thinner when stirred. This behavior is also sometimes referred to as *pseudoplasticity*. Extreme cases are used for paint. The low viscosity at high shear allows the paint to flow smoothly during spraying or brushing. The high viscosity at low shear keeps the particles in suspension during storage and minimizes running once the paint has been applied. Figure 13.26 shows an example of how this behavior is controlled with the addition of different molecular weights of a polymer.

13.2.1.5 Particle Surface Effects

For high-solid-content suspensions, particle-particle attraction results in the formation of agglom-erates. In some cases, these agglomerates can act essentially like roughly spherical particles and

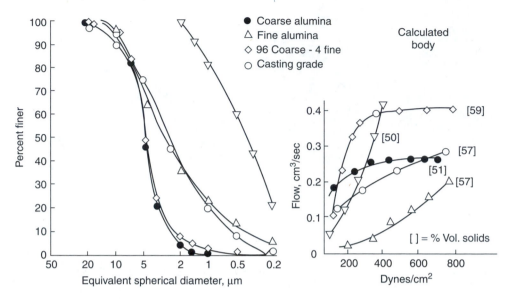

FIGURE 13.25 Rheology behavior of slips prepared with various particle size distributions of Al_2O_3. (From Phelps and McLaren, in *Ceramic Processing Before Firing*, Onoda, G. and Hench, L., Eds., Wiley, New York, 1978, Chapter 17. With permission.)

result in a decrease in viscosity. In other cases, especially for very high solids content, the agglomerates can interact with each other and increase the viscosity. The degree of agglomeration can be controlled with additives.

Dispersion and flocculation (agglomeration) of ceramic particles in a fluid are strongly affected by the electrical potential at the particle surface, adsorbed ions, and the distribution of ions in the fluid adjacent to the particle.[8–10] Thus, the chemical and electronic structure of the solid, the pH of the fluid, and the presence of impurities are all critical considerations in the preparation of a slip for casting.

Two approaches are commonly used to control and manipulate the surface characteristics of ceramic particles in a suspension: (1) electrostatic repulsion and (2) steric stabilization.

Electrostatic repulsion involves the buildup of charges of the same polarity on all the particles. Like charges repel, so the particles are held apart in the suspension by electrostatic forces. The higher the electrical charge at the surface of the particles, the better the degree of dispersion and the less agglomeration. The electrostatic forces dominate for particle separations between about 20 and 200 Å.

The charge at the surface of particles is controlled by pH of the liquid and by addition of chemicals that supply monovalent cations (Na^+, NH_4^+, Li^+) for absorption at the surface of the particles. For most oxides, dispersion can be controlled by pH using the polar properties of water and the ion concentrations of acids or bases to achieve charged zones around the particles so that they repel each other. Al_2O_3 powder can be dispersed in water by either adding an acid to achieve a low pH or a base to achieve a high pH. For instance, an Al_2O_3 slip with a specific gravity of 2.8 g/cm³ had a viscosity of 65 cP* at a pH of 4.5, but 3000 cP at a pH of 6.5. The slip at a pH of 4.5 was well dispersed and had good casting properties. It also was not extremely sensitive to changes in the solids content. Reducing the specific gravity to 2.6 g/cm³ resulted in only a factor of two decrease in viscosity. For comparison, decreasing the specific gravity of the pH 6.5 slip to 2.6 g/cm³ resulted in a tenfold decrease in viscosity.

Clay material also can be dispersed by electrostatic repulsion. Kaolinite has been studied extensively and is a good example. At pH 6 or higher, where low concentrations of sodium or lithium cations are present, kaolinite is well dispersed in water. Under these conditions, each particle has a slight negative charge and the particles repel each other. However, if aluminum or iron salts are

---

*cP = centi-Poise, the English unit for viscosity.

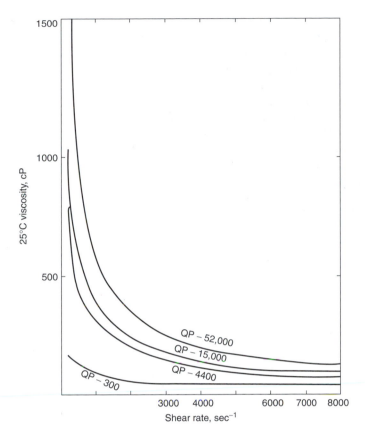

FIGURE 13.26 Addition of different molecular weights of polymers to control the degree of pseudoplasticity. (From Onoda, G., in *Ceramic Processing Before Firing*, Wiley, New York, 1978, Chapter 19. With permission.)

present in low concentration ($\sim 10^{-5}$ molar), the net charge on each particle is decreased and flocculation occurs. On the other hand, if the pH is below 6 and a $\sim 10^{-3}$ molar concentration of aluminum or ferric halides is present, the kaolinite will be dispersed. This is because the charge has been reversed under these conditions and the particles again repel each other because they have adequate levels of like charge. A similar situation exists when the pH is below 2 and monovalent anions such as chloride, nitrate, or acetate are present.

Low concentrations (0.005 to 0.3%) of certain organic and inorganic compounds have a strong dispersing effect on kaolinite suspensions. Some of these include sodium silicate, sodium hexametaphosphate (Calgon), sodium oxalate, sodium citrate, and sodium carbonate. These tend to ion exchange with ions such as calcium and aluminum, which prevent surface charge buildup and leave sodium, which allows a residual charge and causes repulsion between particles. Approximately 0.1% addition of sodium silicate reduces the viscosity by a factor of about 1000.[8]

Proper dispersion of the slip is perhaps the most important parameter in slip casting. Obtaining the optimum dispersion can be aided by the use of several pieces of equipment: a pH meter, a zeta meter, and a viscometer.

The zeta meter requires description. It consists of a thin glass tube, electrodes that attach to the ends of the tube, a DC electrical source, and a microscope mounted on a calibrated track. The tube is filled with the selected liquid (with the pH adjusted to the desired level and the desired dispersant added) containing a small number of particles of the ceramic. A particle is centered in the view of the microscope and the coordinates on the track scale recorded. A known electric field is then applied to the electrodes at the ends of the tube. A particle with a positive charge at the surface will begin moving toward the negative electrode, and vice versa for a particle with a negative charge. The motion of the particle is followed with the microscope and the coordinates on the scale recorded as a function of time. The apparent surface charge on the particle can be calculated from this information and is referred to as the *zeta potential*. This can be repeated for different values of pH and concentrations of additives and plotted on a graph to show the behavior of the material over a broad range. Optimum conditions can then be estimated for slip preparation.

An example of zeta potential vs. pH curve is shown in Figure 13.27 for silicon particles in water. The particles have a positive zeta potential below pH 4, a zero zeta potential at pH 4, and a negative zeta potential at pH above 4. The higher the absolute value of zeta potential, the greater the electrostatic repulsion between particles and the greater the degree of powder dispersion in the slip.

The curve in Figure 13.27 is similar to a typical curve for oxide ceramics. Silicon is known to have a molecular layer of SiO_2 at the surface, which apparently is dominating the rheological behavior of the particles in suspension.

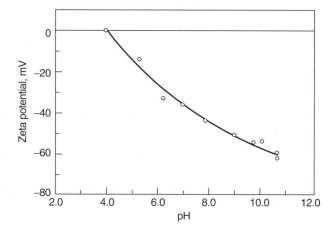

FIGURE 13.27 Zeta potential vs. pH for silicon particles in water. (Drawing courtesy of ASM International.)

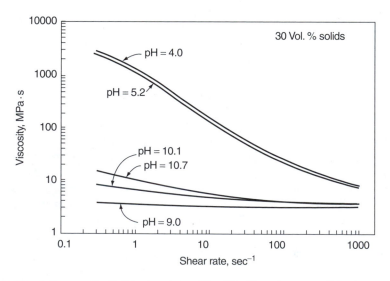

FIGURE 13.28 Viscosity vs. pH of slips containing 30 vol% silicon in water. (Drawing courtesy of ASM International.)

The silicon powder used for the zeta meter measurements was prepared in slips of varying pH using a powder concentration of 30 vol%. The viscosity of each slip was measured with a viscometer and is plotted vs. pH in Figure 13.28. High pH resulted in excellent dispersion and low viscosity. A pH of about 9 was judged to be optimum. Slips with higher solids content were then prepared and the viscosity measured at various shear rates. The results are illustrated in Figure 13.29. All of the slips were successfully cast. A viscosity of approximately 100 MPa · s (100 cP) or lower is suitable for casting.

The microstructures of the cast silicon compacts were evaluated using mercury porosimetry to estimate the total porosity and radius of interconnected pore channels. The results are summarized in Table 13.5. Sonication refers to use of an ultrasonic probe to try and break up agglomerates in the slips. Slips with low pH and near-zero zeta potential contained large agglomerates, had high viscosity, and resulted in compacts with high porosity and large pores. Slips with pH of about 7 to 9 and high negative zeta potential contained minimum agglomerates, had low viscosity, and resulted

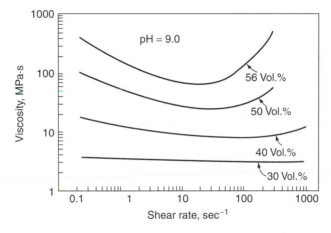

FIGURE 13.29 Viscosity of various volume fractions of silicon powder in water. (Drawing courtesy of ASM International.)

TABLE 13.5
Correlation between Zeta Potential, Viscosity, and Packing Density of Slip-Cast Silicon Powder

Solids Loading, vol%	pH	Zeta Potential, mV	Viscosity, MPa · s at 1.05⁻¹	Sonication Time, min	Total Porosity, vol%	Median Pore Radius, nm
40	4.1	0	>1000	0	46,46	310,320
40	4.1	0	>1000	15	44	290
40	6.3	−33		0	44	250
40	9.0	−50	>20	0	31	190
40	9.8	−56		0	34	200
40	10.7	−60		0	39,39	240,240
50	6.3	−33		0	42	260
50	7.8	−44		0	29	130
50	7.8	−44		10	25	110
50	9.0	−50	60	0	29	130
56	7.8	−44		0	28	105
56	7.8	−44		15	23,24	82,87
56	7.8	−44		30	22,22	73,72
56	9.0	−50	200	30	19	66

Viscosity, MPa·s vs. Shear rate, sec⁻¹ — caption values for figure:
- pH = 9.0
- 56 Vol.%
- 50 Vol.%
- 40 Vol.%
- 30 Vol.%

in compacts with low porosity and relatively small pores. Sonication effectively improved dispersion and resulted in smaller pores and lower porosity. A pH greater than 9.0 resulted in a slight increase in viscosity and porosity. This was due to too high an electrolyte strength (too many charged ions in the water) shielding the charge effect of the particles.

The above examples of Al_2O_3, kaolinite, and silicon illustrate the control of dispersion (deflocculation) and agglomeration (flocculation) that can be achieved with electrostatic repulsion. A second important approach is called *steric stabilization* or *steric hindrance*. It involves the addition of chainlike organic molecules that are adsorbed onto the ceramic particles, as shown in Figure 13.30, and provide a buffer zone around each particle. One end of the chain attaches or anchors to the ceramic and has limited solubility in the solvent. The other end extends away from the particle and is soluble in the solvent. These molecules provide a mechanical barrier to agglomeration and allow particles to approach closer than would be possible without the adsorbed molecules.

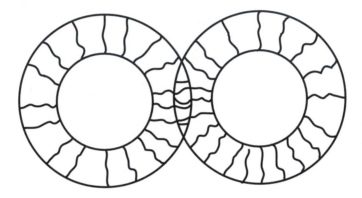

FIGURE 13.30 Adsorption of chain polymers onto the surface of ceramic particles to provide dispersion by steric hindrance. (Drawing courtesy of ASM International.)

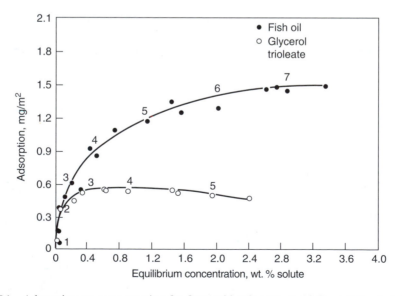

FIGURE 13.31 Adsorption vs. concentration for fatty acid polymers on Al_2O_3 particles in toluene. (From Tormey, E.S., et al., in *Advances in Ceramics, Forming of Ceramics*, Vol. 9, Mangels, J. and Messing, G., Eds., American Ceramic Society, Ohio, 1984, p. 143. With permission.)

Several factors influence steric hindrance:

1. the affinity of one end of a chain molecule to be adsorbed at the surface of the ceramic particle
2. the resistance of the tail end of the molecule to attach to the ends of adjacent molecule tails
3. the characteristics of the fluid, the length of the organic molecule

Different polymers have different affinity for different ceramics in different liquids. Figure 13.31 shows adsorption vs. concentration for two fatty acid polymers on Al_2O_3 in toluene. Note the small

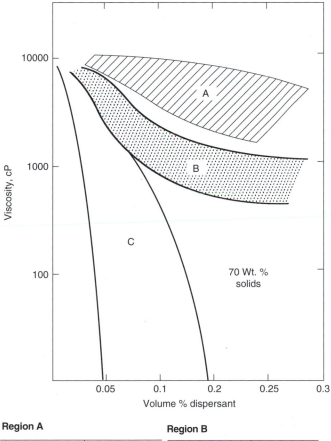

Region A

Monazoline-T	AMP-95
Monazoline-C	Alkazine-TO
Sedisperse-D	Alkazine-O
Zonyl FSN	Emerest 2423
Monazoline-O	Dispersinol-HP
Witconol H31-A	Sedisperse-F
Flourad FC-170-C	

Region B

Witcamine PA-78B	Drewfax-007
Monawet MM-80	Aerosol-OT
Aerosol AY-100	Duponol-G
Aerosol C-61	PVB
Monawet MB-45	Aerosol TR-70
Monawet MO-70	Amerlate LFA
Dispersinol-C	

Region C

Menhaden fish oil
Emphos PS–21A
Zonyl-A

FIGURE 13.32 Summary of the effect of the dispersants listed in Table 13.6 on the viscosity of slips consisting of $BaTiO_3$ in a MEK-ethanol solvent. (Adapted from Mikeska, K. and Cannon, W.R., Dispersants for tape casting pure barium titanate, in *Advances in Ceramics, Forming of Ceramics*, Vol. 9, Mangels, J. and Messing, G., Eds., American Ceramic Society, Ohio, 1984, pp. 164–183.)

amounts required to saturate the surface of the particles. For the glycerol trioleate, no further adsorption occurs above about 0.4 wt%.

Other polymers that provide steric hindrance for specific powders in specific solvents include some other fatty acids (e.g., oleic acid), amines, esters, organotitanates, branched carboxylic acids, organosilanes, polystyrene, and silanized polystyrene and are reported in Reference 11.

The adsorption characteristics and effect on viscosity for a variety of commercially available dispersants with $BaTiO_3$ powders in a methyl ethyl ketone (MEK)-ethanol solvent have been explored in Reference 12. The results are summarized in Figure 13.32. The materials in group A had little effect, those in group B had moderate effect, and those in group C had strong effect. The dispersants and their sources are listed in Table 13.6.

TABLE 13.6
Commercial Dispersants Tested for Dispersion of BaTiO₃ Tape-Casting Slip

Trade Name	Manufacturer	Identify
Monazoline-T	Mona Industries	Substituted imidazoline (1-hydroxyethyl-2-alkyl-imidazolines) from tall oil acids
Monazoline-C	Mona Industries	Substituted imidazoline (1-hydroxyethyl-2-alkyl-imidazolines) from coconut acids
Sedisperse-D	Micromeritics	Saturated aliphatic hydrocarbons
Zonyl-FSN	E. I. du Pont de Nemours & Co.	Fluorinated surfactant
Monazoline-O	Mona Industries	Substituted imidazoline (1-hydroxyethyl-2-alkyl-imidazolines) from oleic acids
AMP-95	International Minerals & Chemicals	2-amino-2-methyl-1-propanol
Sedisperse-F	Micromeritics	Saturated aliphatic hydrocarbons
Witconol H31-A	Witco Chemical Co.	Polyethylene glycol 400 monostearate
Alkazine-TO	Alkaril Chemicals	Tall oil hydroxyethylimidazoline
Fluorad FC-170-C	3M Co.	Fluorinated surfactant
Alkazine-O	Alkaril Chemicals	Oleic hydroxyethylimidazoline
Emerest 2423	Emery Industries	Glycerol trioleate
Dispersinol-HP	Arkansas Co.	Proprietary
Duponol-G	E. I. du Pont de Nemours & Co.	Proprietary
Ameriate LFA	Amerchol Corp.	Lanolin fatty acids
Poly(vinylbutyral)	Rohm and Haas	Poly(vinylbutyral)
Aerosol TR-70	American Cyanamid	Sodium bis(tridecyl)sulfosuccinate
Witcamine PA-78B	Witco Chemical Co.	Salt of fatty imidazoline
Monawet MB-45	Mona Industries	Diisobutyl sodium sulfosuccinate
Dispersinol-C	Arkansas Co.	Proprietary
Aerosol-OT-75	American Cyanamid	Sodium dioctylsulfosuccinate
Monawet MO-70	Mona Industries	Sodium dioctylsulfosuccinate
Drewfax-007	Drew Chemical	Sodium dioctylsulfosuccinate
Aerosol C-61	American Cyanamid	Ethoxylated alkylguanidine amine
Monawet MM-80	Mona Industries	Sodium dihexylsulfosuccinate
Aerosol AY-100	American Cyanamid	Sodium diisobutylsulfosuccinate
Zonyl-A	E. I. du Pont de Nemours & Co.	Ethoxylate
Fish oil	Spencer Kellogg	Fatty acid
Emphos PS-21A	Witco Chemical Co.	Phosphate ester

Source: From Mikeska, K. and Cannon, W.R., Dispersants for tape casting pure barium titanate, in *Advances in Ceramics, Forming of Ceramics*, Vol. 9, Mangels, J. and Messing, G., Eds., American Ceramic Society, Ohio, 1984, pp. 164–183, p. 169.

A simple apparatus was used for studying slip viscosity vs. volume of dispersant. The apparatus is illustrated schematically in Figure 13.33. The powder and liquid were premixed to form a dispersion containing 70 wt% solids. This was placed in a variable-speed blender and blended until homogeneous. The blender was stopped and the viscosity was measured with a rotating-disk viscometer. A known amount of dilute dispersant was added through a graduated buret, the slip was reblended, and the viscosity was measured. This procedure was repeated to provide the data for plotting viscosity vs. dispersant concentration for each dispersant. The above examples identify toluene and MEK-ethanol as fluids for slip preparation. These are referred to as *nonaqueous* (non-water-based). Another non-aqueous system uses trichloroethylene plus ethanol. Nonaqueous systems work well with steric hindrance because they are adequate solvents for the chain polymers. Some of the polymers also provide steric hindrance in an *aqueous* (water-based) system, for example, phosphate esters.

Aqueous slips using electrostatic repulsion are commonly used for slip casting. Techniques of slip preparation and slip casting are discussed in the following sections. Nonaqueous slips utilizing steric hindrance are commonly used for tape casting. Tape casting is discussed later in this chapter.

13.2.1.6 Slip Preparation

The actual physical preparation of the slip can be done by a variety of techniques. Perhaps the most common is wet ball milling or mixing. The ingredients, including the powder, binders, wetting agents, sintering aids, and dispersing agents, are added to the mill with the proper proportion of the selected casting liquid and milled to achieve thorough mixing, wetting, and (usually) particle size

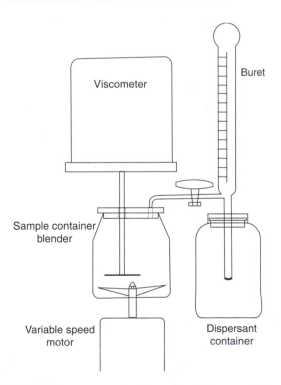

FIGURE 13.33 Schematic of simple apparatus for the study of slip viscosity vs. vol% dispersant. (From Mikeska, K. and Cannon, W.R., Dispersants for tape casting pure barium titanate, in *Advances in Ceramics, Forming of Ceramics*, Vol. 9, Mangels, J. and Messing, G., Eds., American Ceramic Society, Ohio, 1984, pp. 164–183, p. 170. With permission.)

reduction. The slip is then allowed to age until its characteristics are relatively constant. It is then ready for final viscosity checking (and adjustment, if necessary), deairing, and casting.

13.2.1.7 Mold Preparation

The mold for slip casting must have controlled porosity so that it can remove the fluid from the slip by capillary action. The mold must also be low in cost. The traditional mold material has been plaster.[13] Some newer molds, especially for pressure casting, are made of a porous plastic material.

Plaster molds are prepared by mixing water with plaster of Paris powder, pouring the mix into a pattern mold, and allowing the plaster to set. This produces a smooth-surface mold, duplicating the contours of the pattern for a complex shape. The mold is made in segments, each of which is sized so that it can be removed after slip casting without damaging the delicate casting. Plaster of Paris (hemihydrate) is partially dehydrated gypsum:

$$3CaSO_4 \cdot 2H_2O \xrightarrow[\text{(355°F)}]{180°C} 3H_2O + 2CaSO_4 \cdot 1/2H_2O$$

Gypsum   Hemihydrate

The reaction is reversible; addition of water to the hemihydrate results in precipitation of very fine needle-shaped crystals of gypsum that intertwine to form the plaster mold. The reaction is satisfied chemically by addition of 18% water, but considerably more water is necessary to provide a mixture with adequate fluidity for mold making. This extra water fills positions between the gypsum crystals during precipitation and results in very fine capillary porosity after the finished plaster mold has been dried. It is this porosity that draws the water out of the slip during slip casting. The amount of porosity can be controlled by the amount of excess water added during fabrication of the plaster mold. For normal slip casting, 70 to 80 wt% water is used.

The setting rate of plaster can be widely varied by impurities.

13.2.1.8 Casting

Once the mold has been fabricated and properly dried and an optimum slip has been prepared, casting can be conducted. Many options are available, depending on the complexity of the component and other factors:

> simple casting into a one-piece mold
> simple casting into a multipiece mold
> drain casting
> solid casting
> vacuum casting
> centrifugal casting
> pressure casting
> soluble-mold casting
> gel casting
> casting with nonabsorbing pins or mandrels inserted into the mold

Figure 13.34 illustrates schematically *drain casting*. The slip is poured into the mold and water is sucked out where the slip is in contact with the mold, leaving a close-packed deposition of particles growing into the slip from the mold walls. The slip is left in the mold until the desired thickness is built up, at which time the remaining slip is drained from the mold. Drain casting is the most common slip-casting approach. It is used for art casting (figurines), sinks and other sanitary ware, crucibles, and a variety of other products.

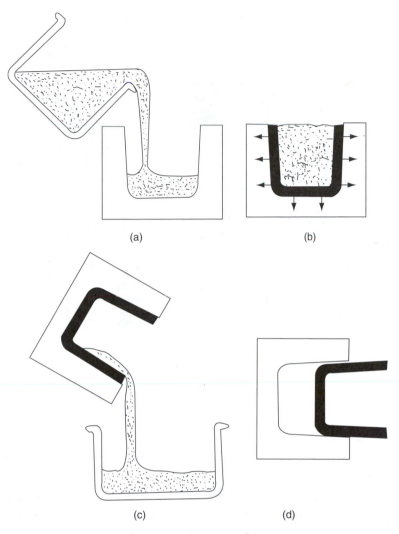

FIGURE 13.34 Schematic illustrating the drain-casting process. (a) Fill mold with slip, (b) mold extracts liquid, forms compact along mold walls, (c) excess slip drained, and (d) casting removed after partial drying.

Solid casting is identical to drain casting except that slip is continually added until a solid casting has been achieved.

Vacuum casting can be conducted either with the drain or solid approach. A vacuum is pulled around the outside of the mold. The mold can consist of a rigid permeable form or of a thin permeable membrane (like filter paper) lining a porous rigid form. Vacuum casting is commonly used in the production of porous refractory fiberboard for lining high-temperature furnaces.

Centrifugal casting involves spinning the mold to apply greater than normal gravitational loads to make sure that the slip completely fills the mold. This can be beneficial in the casting of some complex shapes.

One limitation with most slip-casting processes is the long time required to cast articles in the mold. This results in a large inventory of molds, high labor, and large floor space, all of which add to cost. Application of pressure to the slip increases the casting rate. This is referred to as *pressure casting*.[14–16] It is similar to filter pressing. In filter pressing water or other liquid is removed from a powder by pressing the powder or liquid mixture against a permeable membrane. In pressure casting the slip is pressed into a shaped permeable mold. Original pressure casting was conducted with

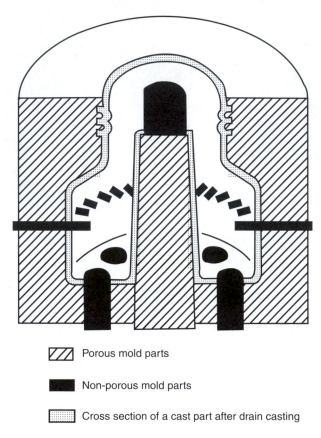

☐ Porous mold parts

■ Non-porous mold parts

▦ Cross section of a cast part after drain casting

FIGURE 13.35 Cross section of the mold arrangement to drain cast a cylinder within a cylinder with large holes through the bottom and small holes through the outer circumference. (Drawing courtesy of ASM International.)

plaster molds. However, because of the low strength of the plaster, the amount of pressure that could be applied was limited. Development of porous plastic molds allowed the pressure to be increased by ten times the values of 3 to 4 MPa (30 to 40 bar or 435 to 580 psi).

Reference 14 reports some results with pressure casting. For a porcelain composition, casting time for a 6-mm thick compact was reduced from about 45 to 15 min by increasing the pressure from 0.025 to 0.4 MPa (0.25 to 4.0 bar or 3.6 to 58 psi). Increasing the pressure to 4 MPa resulted in a casting time of about 65 sec for a 6-mm thick compact of a modified porcelain slip. A 6-mm thick compact of Al_2O_3 was cast in 100 sec and of an earthenware composition in 240 sec at a pressure of 4 MPa (40 bar or 580 psi).

Nonporous, nonabsorbing mold parts such as pins and mandrels can be used to achieve increased complexity of parts cast by the above techniques. Figure 13.35 shows schematically how a mandrel and pins were used to drain cast a complex-shaped combustor for a gas-turbine engine. The finished part is shown in Figure 13.36.

Further complexity of shape can be achieved using the *soluble-mold casting* technique.[17,18] This is a relatively new approach based on the much older technology of investment casting. It is also referred to as *fugitive wax slip casting* and is accomplished in the following steps:

1. A wax pattern of the desired configuration is produced by injection molding a water-soluble wax.
2. The water-soluble wax pattern is dipped in a nonwater-soluble wax to form a thin layer over the pattern.

FIGURE 13.36 Annular combustor for gas-turbine engine fabricated by drain casting using nonabsorbing pins and mandrels inserted into the mold. Courtesy of Garrett Turbine Engine Company (currently Honeywell Engines, Systems & Services), fabricated by Norton Company (currently Saint-Gobain/Norton Advanced Ceramics).

3. The pattern wax is dissolved in water, leaving the nonwater-soluble wax as an accurate mold of the shape.
4. The wax mold is trimmed, attached to a plaster block, and filled with the appropriate casting slip.
5. After the casting is complete, the mold is removed by dissolving in a solvent.
6. The cast shape is dried, green-machined as required, and densified at high temperature.

   The application of the fugitive-wax approach is illustrated in Figure 13.37 for the fabrication of a complex-shaped stator vane for a gas-turbine engine. The injection molding tool on the right produces the water-soluble wax stator vane pattern. The injection molding tool on the left produces the pattern for the reservoir that will hold the slip and guide it through gating channels into the stator vane mold during casting. The reservoir and vane patterns are bonded together by simple wax welding and are shown as the white wax assembly in the center of Figure 13.37. Below this is the mold produced by dipping and dissolving the pattern. Below the mold is the green casting after dissolving the mold and trimming off any material remaining in the reservoir or gating area. The stator vane discussed above required less than 1-h casting time. Some solid castings require much longer time, such as the prototype gas-turbine rotor shown in Figure 13.38. It required over 12 h. The slip must be very stable for such long casting time to avoid settling of large particles or adverse changes in viscosity.

   Other fugitive mold techniques have been developed to fabricate special shapes. One technique produces low weight, but strong ceramic foam.[19] Reticulated foam similar to a dishwashing sponge is used as the mold interior. Reticulated polymer foam of the desired pore size is cut to the desired shape and placed in a container in a vacuum chamber. A ceramic slip in poured into the container and under vacuum completely infiltrates the pores in the reticulated foam. The slip is dried and fired to burn off the polymer foam and densify the ceramic. The resulting part consists of an internal cast of the spongelike foam. Its major characteristic is continuous interconnected links of ceramic and continuous pore channels. Such a cellular structure can be very lightweight and surprisingly strong. Examples are shown in Figure 13.39. Note from the photograph that a variety of pore sizes have been achieved from several different ceramic materials. The materials are successfully used for molten metal filtration and kiln furniture and are being evaluated for removing particles from the exhaust of diesel engines.

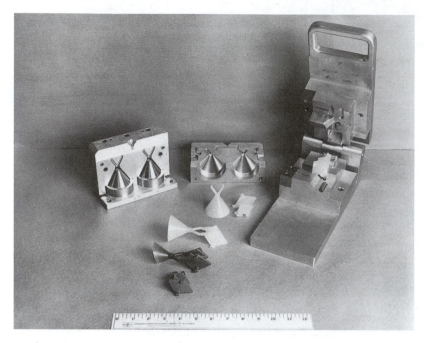

FIGURE 13.37   The fugitive-wax technique for preparing a complex-shape mold for slip casting; example of fabrication of a stator vane for a gas-turbine engine. (Courtesy of AiResearch Casting Company, Division of The Garrett Corporation, presently Honeywell Ceramic Components, Torrance, CA.)

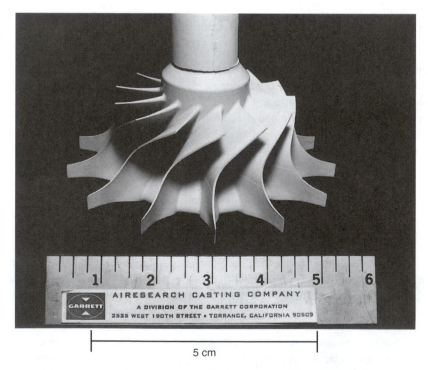

FIGURE 13.38   Prototype gas-turbine rotor fabricated by slip casting using a fugitive-wax-type process. (Courtesy of AiResearch Casting Company, Division of the Garrett Corporation, presently Honeywell Ceramic Components, Torrance, CA.)

FIGURE 13.39 Ceramic components of controlled porosity fabricated by infiltration of a ceramic casting slip into a polymer sponge structure. (Photo courtesy of Hi-Tech Ceramics, Alfred, NY.)

Some components are too complex to be fabricated in one piece by casting. An example is the turbine scroll shown in Figure 13.40a. The turbine scroll is an important component in many gas-turbine designs. It changes the direction of the hot gases coming out of the combustor to allow them to pass through the rotor. The scroll in Figure 13.40a is SiC. It was fabricated by assembling the parts shown in Figure 13.40b.[20] The shroud, sleeve, and ring were formed by isostatic pressing and green machining. The body and duct were fabricated by slip casting. The parts were successfully bonded together with a CrVTi braze developed at Oak Ridge National Laboratories (ORNL).

A final casting technique is *electrophoretic deposition* (EPD). It uses an electrostatic charge to consolidate ceramic particles from a suspension. An electrical polarity is applied to the mold that is opposite to the polarity at the surface of the ceramic particles. The ceramic particles are electrically attracted to the mold surface and deposit as a uniform compact. When the desired thickness of deposit is achieved, either the mold is removed from the container of slip or the slip is poured from the mold. Electrophoretic deposition is generally used to deposit a thin coating or to produce a thinwalled body such as a tube. It is also used to achieve very uniform deposition of spray paint onto a conductive surface.

All of the casting techniques discussed above result in a relatively weak ceramic powder compact. A technique developed at ORNL results in a much stronger compact. This technique is referred to as *gel casting*. The ceramic powder is mixed with a liquid and a polymerizable additive to form a fluid slurry similar to a casting slip. The slip is poured into a container of the desired shape. Polymerization is caused to occur before the powder in the slip has time to settle. The resulting powder compact is quite uniform and strong. However, removal of the liquid is more difficult than for conventional slip casting. Furthermore, monomers are generally toxic and require careful handling. For example, the initial material used by ORNL was acrylamid, which is a neurotoxin and has largely been discontinued because of handling concerns.

13.2.1.9 Casting Process Control

As was illustrated in Figure 13.22, careful process control is necessary in the slip-casting process. Some of the critical factors include:

constancy of properties
viscosity

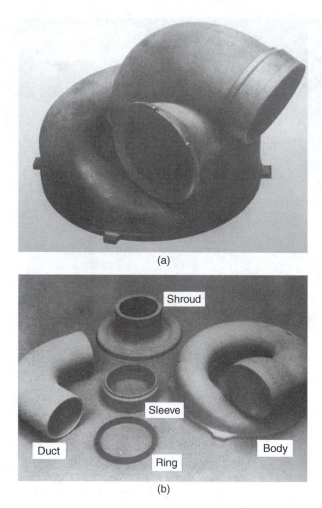

(a)

(b)

FIGURE 13.40 (a) Assembled SiC gas-turbine scroll. (b) Individual parts that were bonded together to form the scroll. (Photos courtesy of Carborundum Company for parts fabricated for Allison Gas Turbine Division of General Motors under sponsorship of the U.S. Department of Energy and administration of NASA-Lewis Research Center.)

settling rate
freedom from air bubbles
casting rate
drain properties
shrinkage
release properties
strength

Constancy of properties refers to the reproducibility of the casting slip and its stability as a function of time. The slip must be easily reproduced and preferably should not be overly sensitive to slight variations in solids content and chemical composition or to storage time. The viscosity must be low enough to allow complete fill of the mold, yet the solids content must be high enough to achieve a reasonable casting rate. Too-slow casting can result in thickness and density variations due to settling. Too-rapid casting can result in tapered walls (for a drain casting), lack of thickness control, or blockage of narrow passages in the mold.

The slip must be free of entrapped air or chemical reactions that would produce air bubbles during casting. Air bubbles present in the slip will be incorporated in the casting and may be critical defects in the final, densified part.

Once casting has been completed, the part begins to dry and shrink away from the mold. This shrinkage is necessary to achieve release of the part from the mold. If the casting sticks to the mold, it will usually be damaged during removal and rejected. Mold release can be aided by coating the walls of the mold with a release agent such as a silicone or olive oil. However, it should be recognized that the coating may alter the casting rate.

The strength of the casting must be adequate to permit removal from the mold, drying, and handling prior to the firing operation. Sometimes, a small amount (<1%) of binder is included in the slip. Organic binders such as polyvinyl alcohol work well. With the binder present, strength comparable to or greater than blackboard chalk is achieved. Such strength is adequate for handling and also for green machining if required.

13.2.1.10   Drying

Compacts fabricated by casting are saturated with the casting fluid. The fluid is trapped in all the pores, as well as forming a film on some of the particles. The fluid must be completely removed by a drying operation before the compact can be taken to high temperature for densification.[21,22] The ease of removal of the fluid is dependent on several factors: (1) the amount of porosity; (2) the size of interconnected pore channels; (3) the vapor pressure of the fluid; and (4) the thickness of the compact.

The porosity distribution is particularly important. Large pores and pore channels allow easy removal of the fluid. However, these result in low green density, shrinkage during drying, and difficulty in firing to a dense, fine-grained microstructure. Conversely, cast compacts with close-packed particles, small pores, and narrow pore channels are easy to densify to a fine-grained microstructure, but are difficult to dry. The fine channels result in large capillary pressures and stresses that can easily crack the compact if the fluid is not removed uniformly. For example, the gas-turbine rotor shown in Figure 13.38 required a dense, fine-particle-size cast compact that was very difficult to dry. The thin blades tended to dry quickly while the thick hub was still saturated. This resulted in cracking near the base of the blades during drying. Drying time greater than a week with careful control of humidity and temperature was necessary to achieve a crack-free part.

13.2.2   TAPE CASTING

Some applications such as substrates and packages for electronics and dielectrics for capacitors require thin sheets of ceramics. Tape casting has been developed to fabricate these thin sheets in large quantity and at low cost. It is similar to slip casting, except that the slip contains about 50 vol% organic binder and is spread onto a flat surface rather than being poured into a shaped mold.

13.2.2.1   Doctor Blade Process

The most common approach for tape casting is the doctor blade process.[23] A schematic of a doctor blade apparatus is illustrated in Figure 13.41. The technique consists of casting a slurry onto a moving carrier surface (usually a thin film of cellulose acetate, Teflon™, Mylar™, or cellophane) and spreading the slurry to a controlled thickness with the edge of a long, smooth blade. The slurry contains a binder system dissolved in a solvent. Enough binder is present so that a flexible tape will result when the solvent is removed. Solvent removal is achieved by evaporation. As with slip casting, the fluid must be removed slowly to avoid cracking, bubbles, or distortion. This is the purpose of the long portion of the tape-casting apparatus between the doctor blade and the take-up reel. The evaporation is achieved either by controlled heating or airflow. The dry flexible tape is rolled onto a reel to be stored for use. The uses of the tape are described later in this chapter.

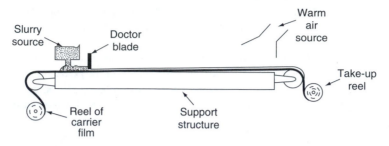

FIGURE 13.41 Schematic illustrating the doctor blade tape-casting process.

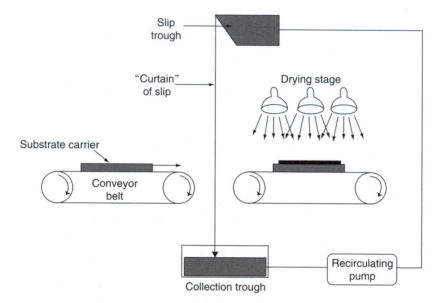

FIGURE 13.42 Schematic illustrating the "waterfall" tape-casting process. (From J. Adair, The Pennsylvania State University. With permission.)

13.2.2.2 Other Tape-Casting Processes

A second tape-casting process is the "waterfall" technique. It is illustrated in Figure 13.42. The slurry is pumped in a recirculating system to form a continuous curtain. A conveyor belt carries a flat surface through the slurry. The uniform, thin layer of slurry on the carrier is then transferred by conveyor to the drying stage. This technique has been used to form thin tape for capacitor dielectrics and thicker tape for porous electrodes for fuel cells. It is also commonly used for applying the chocolate coating to candy bars.

A third tape-casting process is the *paper-casting process*. It is illustrated in Figure 13.43. Low-ash paper is passed through a slurry. The slurry wets the paper and adheres to it. The thickness of adherence depends on the viscosity of the slurry and the nature of the paper. The coated paper passes through a drier, and the resulting tape is rolled onto a take-up reel. The paper is later removed during a firing process. This technique has been used in fabrication of honeycomb structures for heat exchangers, as shown in Figure 13.44.

13.2.2.3 Preparation of Tape-Casting Slurries

Tape-casting slurries are rheologically similar to slip-casting slurries, but contain a larger quantity of binder. In addition, the binder and plasticizer system is generally selected to be thermoplastic,

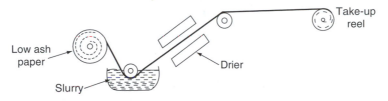

FIGURE 13.43 Schematic illustrating the paper-casting tape-forming process.

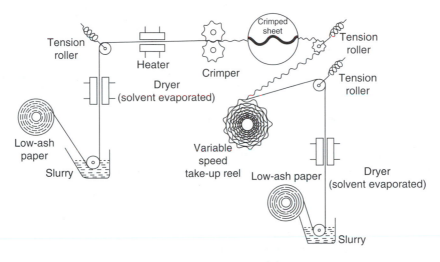

FIGURE 13.44 Use of the paper-casting process to fabricate a honeycomb structure for a heat exchanger.

that is, it can be softened by heating to moderate temperatures. This allows layers to be bonded together by lamination.

Table 13.7 describes the characteristics of the organic components commonly added to achieve an acceptable tape-casting slurry. These include a binder, a plasticizer, a dispersant, a wetting agent, and an antifoam agent. Each combination must be selected and optimized for a specific fluid (solvent). Examples of solvents are methyl ethyl ketone (MEK), alcohols, toluene, hexane, trichloroethylene, and water. Examples of binders are polyvinyl butyral, polyvinyl acetate, polyvinyl chloride, polyvinyl alcohol, polyacrylic emulsion, polystyrene, polymethacrylates, and cellulose nitrates. The criteria for the binder include: (1) forms tough, flexible film when dry; (2) volatilizes to a gas when heated and leaves no residual carbon or ash; (3) remains stable during storage, especially with no change in molecular weight; and (4) is soluble in an inexpensive, volatile, nonflammable solvent.

Tables 13.8 and 13.9 identify binder and solvent and plasticizer systems that have been reported for tape casting of Al_2O_3 for fabrication of electrically insulative substrates for integrated circuit electronic devices.[24] Figure 13.45 shows a flowchart that identifies the sequence of steps in fabrication of such a substrate by tape casting. Table 13.10 identifies a tape-casting system for $BaTiO_3$.

13.2.2.4 Applications of Tape Casting

The major applications of tape casting are for fabrication of dielectrics for multiplayer capacitors and of Al_2O_3 for substrates and multiplayer packages for integrated circuits. Figure 13.46 illustrates the use of tape casting for multilayer capacitors. Figure 13.47 shows schematically the cross section of a multilayer package for a silicon chip device. The white layers of the package are Al_2O_3. The black lines and grids are electrically conductive metal (tungsten or molybdenum). The Al_2O_3 isolates the metal circuit lines and allows miniaturization of complex circuits. Each Al_2O_3 layer is

TABLE 13.7
Functions of Organic Components of Tape-Casting Binder Systems

Binders

10,000 to 30,000 molecular weight polymers mixed with powder suspension: after drying provide flexibility and integrity to green tape

Plasticizers

Small- to medium-sized molecule that decreases cross-linking among binder molecules; this makes the tapes more pliable, e.g., polyethylene glycol

Dispersants

Typically 1000 to 10,000 molecular weight polymer molecules that aid in dispersion of ceramic particles, e.g., isooctylphenylpolyethoxyethanol

Defoaming Agents

Various organic substances used to minimize frothing usually due to addition of dispersant (one of the defoaming agent's actions may be to increase surface tension of slip), e.g., tributyl phosphate

Wetting Agents

Small- to medium-sized molecules that impart wettability to the ceramic powder by the solvent system, e.g., isooctylphenylpolyethoxyethanol

Source: Courtesy J. Adair, The Pennsylvania State University.

TABLE 13.8
Additives to Al_2O_3 for the Western Electric Company ERC-105 Doctor-Blade Process

Material	Function
MgO	Grain-growth inhibitor
Menhaden fish oil	Deflocculant
Trichloroethylene	Solvent
Ethyl alcohol	Solvent
Polyvinyl butyral	Binder
Polyethylene glycol	Plasticizer
Octylphthalate	Plasticizer

Source: Shanefield, D.J. and Mistler, R.E., Fine grained alumina substrates: I, the manufacturing process. *Am. Ceram. Soc. Bull., Part I* 53, 416–420, 1974.

TABLE 13.9
Examples of Binder-Solvent-Plasticizer Systems for Tape Casting of Al_2O_3

Thickness of Tape	Binder	Solvent	Plasticizer
>0.25 mm	3.0 Polyvinyl butyral	35.0 Toluene	5.6 Polyethylene glycol
<0.25 mm	15.0 Vinyl chloride-acetate	85.0 MEK	1.0 Butyl benzyl phthalate

Source: Mistler, R.E., Shanefield, D.J., and Runk, R.B., in *Ceramic Processing Before Firing,* Onoda, G. and Hench, L., Eds., Wiley, New York, 1978, p. 414.

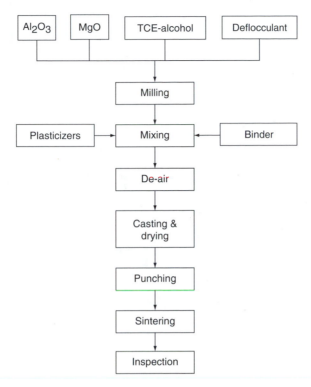

FIGURE 13.45 Flowchart for fabrication of Al_2O_3 substrates by tape casting. (From Mistler, R., et al., in *Ceramic Processing Before Firing*, Onoda, G. and Hench, L., Eds., Wiley, New York, 1978, With permission.)

TABLE 13.10

Example of a Tape-Casting System for $BaTiO_3$

Ingredient	Function	Parts by Weight
$BaTiO_3$	Ceramic	100
MEK and ethanol (50 wt% mixture)	Solvent	20
Menhaden fish oil[a]	Deflocculant	1
Santicizer 160 (Butyl benzyl phthalate)	Plasticizer	4
Carbowax 400[c] (Polyethylene glycol)	Plasticizer	4
Cyclohexanone	Homogenizer	0.7
Acryloid B-7 MEK[d] (30% solution)	Binder	13.32

[a] Defloc Z-3, Spencer Kellogg Inc., Buffalo, N.Y.

[b] Monsanto Inc., St. Louis, Mo.

[c] Rohm and Haas Co., Philadelphia, Pa.

[d] Carbide and Carbon Chemical Co., N.Y.

Source: Mackinnon, R.J. and Blum, J.B., in *Advances in Ceramics, Forming of Ceramics*, Vol. 9, Mangels, J. and Messing, G., Eds., American Ceramic Society, Ohio, 1984, pp. 150–157.

prepared from a tape-cast section. The metal patterns are screen-printed onto the tape. The layers of tape with the appropriate metal patterns are stacked and bonded together. The binders are then carefully removed at moderate temperature and the Al_2O_3 powder compact densified at high temperature. The metal circuits are trapped in the Al_2O_3 and protected from the external environment.

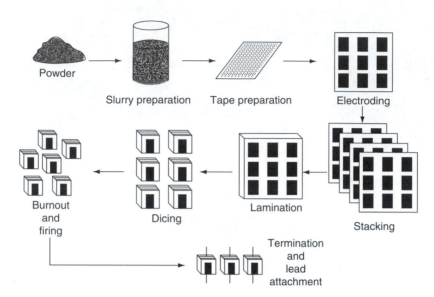

FIGURE 13.46 Process steps for fabrication of multilayer capacitors by tape casting. (Courtesy of J. Adair, The Pennsylvania State University.)

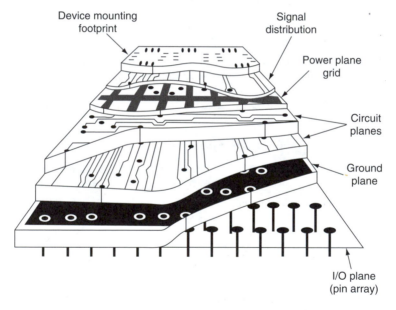

FIGURE 13.47 Schematic cross section showing the complexity of the metallized circuit patterns in a ceramic multilayer integrated circuit package. (From Ryshkewitch, E. and Richerson, D., *Oxide Ceramics*, 2nd ed., General Ceramics, Inc., N.J., 1985, p. 466. With permission.)

13.3 PLASTIC FORMING

Plastic forming involves producing shapes from a mixture of powder and additives that are deformable under pressure. Such a mixture can be obtained in systems containing clay minerals by addition of water and small amounts of a flocculant, a wetting agent, and a lubricant. In systems not containing clay, such as pure oxides and carbides and nitrides, an organic material is added in place of water or mixed with water or other fluid to provide the plasticity. About 25 to 50 vol% organic additive is required to achieve adequate plasticity for forming.

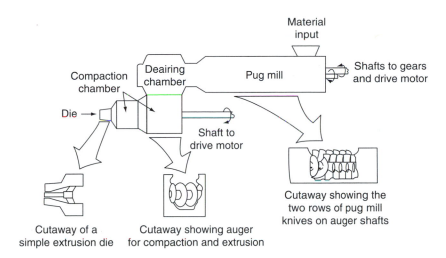

FIGURE 13.48 Schematic of an auger-type extruder.

A major difficulty in plastic-forming processes is removing the organic material prior to firing. In the case of a water-clay system, substantial shrinkage occurs during drying, increasing the risk of shrinkage cracks. In the case of organic additives, the major problems are forming a flaw-free green part and extraction of the organic. Too rapid extraction causes cracking, bloating, or distortion. Inadequate removal results in cracking, bloating, or contamination during the later high-temperature densification process.

Plastic processes are used extensively in the fabrication of traditional ceramics such as pottery and dinnerware. The compositions contain clay and have been made workable by addition of water. Modern ceramics generally do not contain clay and require organic additions to achieve plasticity.

The following sections of this chapter describe the key plastic-forming techniques that are being used or developed for shape fabrication of modern ceramics. Extrusion and injection molding are described in detail. Other techniques such as compression molding and roll forming are briefly discussed.

13.3.1 EXTRUSION

Extrusion is a plastic-forming method that has been used extensively for many years for fabrication of ceramics for furnace tubes, bricks, insulators, pipe, tile, tubular capacitors, catalyst supports, magnets, heat-exchanger tubes, and other parts with a constant cross section. The extrusion process consists of forcing a highly viscous, doughlike plastic mixture of ceramic powder plus additives through a shaped die.[22,25,26] It is analogous to squeezing frosting out of a cake decorator, although the pressures are higher and the mixture is much stiffer than frosting. The consistency is similar to that of molding or potter's clay.

13.3.1.1 Extrusion Equipment

The schematics of the two types of extruders commonly used are shown in Figures 13.48 and 13.49. One is an auger-type extruder in which the plasticized mix is forced through a shaped die by the rotation of an auger. The second type of extruder uses a piston in place of an auger. The piston-type extruder generally results in less contamination by wear.

An auger extruder consists of several sections and is capable of continuous operation. The first section is a *pug mill*. The pug mill contains two rows of blades mounted on auger shafts. The shafts rotate in opposite directions, providing a high-shear mixing action of material squeezed between the blades. The doughlike premix (ceramic powder plus liquid plus additives such as binders, plasticizers,

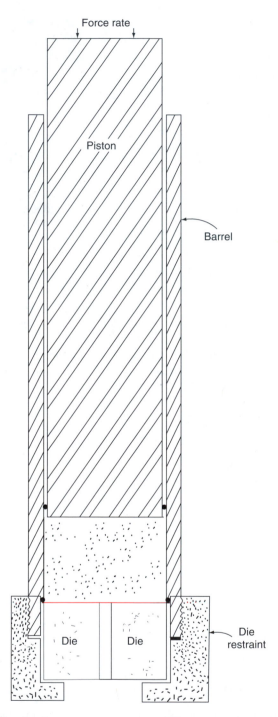

FIGURE 13.49   Schematic of a piston-type extruder. (Drawing courtesy of ASM International.)

dispersants, flocculants, lubricants and surfactants) is fed into the pug mill. The pug mill kneads the premix to provide homogeneity, to maximize plasticity, and to squeeze out excess air. The mixture then enters a *deairing chamber* that uses auger motion plus an applied vacuum (in some equipment) to remove as much air as possible. The mixture finally moves to the *compaction chamber* where auger motion or pressure from a piston precompacts the mixture to remove as much void space as possible prior to extrusion through a *shaped die*. The plasticized mixture is then forced under high pressure through the shaped die. The resulting long strands of constant-cross-section compacted ceramic are supported in trays and cut to the desired length prior to drying and firing.

13.3.1.2 Binders and Additives for Extrusion

Additives to a ceramic powder are required to achieve a mixture that has characteristics suitable for extrusion. The nature of the additives depends on whether the extrusion is conducted at room temperature with a combination of a binder and fluid or at elevated temperature with a thermosetting polymer. The following are some of the key characteristics that must be considered:

1. The mixture must be plastic enough to flow under pressure into the desired cross-section, yet rigid enough (high enough wet strength) to resist deformation due to slumping or handling.
2. The mixture must not stick to the die or other tooling and must yield smooth surfaces after extrusion.
3. The fluid and ceramic must not separate under an applied pressure.
4. The mixture must have reproducible porosity such that shrinkage during drying and firing are predictable.
5. Organics must be low-ash content to leave minimal residue during firing.

The flow characteristic (rheological behavior) is the dominant factor. As was the case for slip casting, additives and volume fraction solids are used to control the flow characteristics. Extrusion is most commonly conducted at room temperature, so we will concentrate our discussion on additives that provide flow at room temperature. The nature of the additives selected depend on the ceramic powder and the liquid. Compositions containing no clay require a substantial percentage of organic additives and either water or a solvent. Compositions containing a clay mineral such as kaolinite can be plasticized with water additions and do not require organics. Clay has a structure that absorbs water between the sheet layers, resulting in natural plasticity.

Modern ceramic compositions such as Al_2O_3, ZrO_2, and $BaTiO_3$ do not contain clay and thus must have organic additives. The major additive is the *binder*. It provides a coating over each ceramic particle to allow flow during extrusion and green strength after extrusion. Binders are available in a wide range of viscosity. Examples of water-soluble binders and their range of viscosity are identified in Table 13.11. The viscosity is based primarily on the molecular weight and the strength of molecular bonds of the organic. Examples of several systems that have been successfully extruded are listed in Tables 13.12 and 13.13.

Other additives include lubricants, surfactants (wetting agents), dispersants, flocculants, plasticizers, algicides, and antifoam agents. The *plasticizer* modifies the rheology of the binder to achieve plastic behavior at the temperature of extrusion. The *lubricant* reduces particle-particle and die-wall friction, prevents sticking and aids in achieving an acceptable surface finish. *Surfactants* enhance the wetting ability of the binder onto the ceramic particles. *Dispersants* and *flocculants* control the degree of dispersion or agglomeration of the particles. For some particle size distributions, dispersion is preferred. For others, a high degree of flocculation works better so coagulants are added. Examples are included in Table 13.14. The extrusion characteristics can be broadly varied by changing the degree of dispersion. As with slips, the dispersion characteristics can be altered by control of pH and by additives that provide electrostatic surface charges or steric hindrance.

13.3.1.3 Extrusion Steps

The steps of extrusion are similar to the steps of other processes for forming ceramic particulate compacts. They include powder sizing, batch formulation, mixing, extrusion, drying, densification, and quality control.

TABLE 13.11
Viscosity Ranges for Some Water-Soluble Binders

Binder	Very Low	Low	Medium	High	Very High
Gum arabic	•				
Lignosulfonates	•				
Lignin liquor	•				
Molasses	•				
Dextrins	•—	•			
Poly(vinylpyrrolidone)	•——	•			
Poly(vinyl alcohol)	•———	———	•		
Polyethylene oxide		•———	———	•	
Starch		•———	———	———	•	
Acrylics		•——	•		
Polyethylenimine PEI		•——|—| •		
Methylcellulose			•———|———| •	
Sodium carboxymethylcellulose			•——|——| •	
Hydroxypropylmethylcellulose			•——|——| •	
Hydroxyethylcellulose			•——	——	—	•
Sodium alginate				•—	
Ammonium alginate				•—	
Polyacrylamide				•——	— •
Scleroglucan				•	
Irish moss				•	
Xanthan gum				•	
Cationic galactomanan					•
Gum tragacanth					•
Locust bean gum					•
Gum karaya					•
Guar gum					•—•

Source: Adapted from Onoda, G., in *Ceramic Processing Before Firing*, Onoda, G. and Hench, L., Eds., Wiley, New York, 1978, p. 242.

As with other forming processes, particle size and shape and degree of agglomeration are extremely important. Fine particles (under 1 μm or 0.00004 in.) are generally easiest to extrude.

Mixing is a critical step in extrusion. All particles must be uniformly coated with the binder-liquid solution. Obtaining uniformity is difficult because the final extrudable mix is so stiff. Imagine stirring flour into honey. Without high-intensity mixing, pockets of binder and powder can remain that can cause nonuniform extrusion and also end up as strength-limiting defects in the final part. Several different mixing techniques have been used successfully. One is the "brute force" technique in which the ingredients are weighed out to their final formulations and mixed directly to the final, stiff consistency in a high-shear mixer such as a pug mill or muller. A second technique is to add excess liquid to provide improved wetting and then partially dry to the required consistency after thorough mixing. Another technique is to mix the water and powder first to form a low-viscosity mixture and then to add the binder incrementally to bring the mix uniformly up to the extrudable viscosity.

Viscosity is not the only key parameter for successful extrusion. The mix must have the proper flow characteristics when a load is applied. The options of flow behavior are illustrated in Figure 13.50 for apparent viscosity vs. shear rate. The preferred flow behaviors are *shear-thinning* or *Bingham*.

Other important factors for successful extrusion are die design and alignment. The purpose of the die is to achieve the final compaction of the ceramic mix and to form the desired cross section.

TABLE 13.12
Examples of Extrusion of Nonclay Compositions

Composition	Additives	Conditions	Shape
Graphite	50–60 parts phenol formaldehyde emulsion	–	1.6- and 2.54-cm (5/8-and 1-in.) rods
Petroleum coke	Coal-tar pitch plus heavy oil	90–110°C (195–230°F)	–
Al_2O_3–5% Cr_2O_3	Gum ghatti plus mogul starch		Round, square, and triangular tubes
30% Si–70% SiC	4% guar gum, 20% water, 3% silicone	–	–
BeO	10–18% mogul starch, 15–18% water-glyceryl mixture	–	Rods, tubes, thin-walled multicell tubes
MgO	30–40% flour paste	–	Vacuum-tube parts

Source: From Hyde, C., Vertical extrusion of nonclay composition, in *Ceramic Fabrication Processes*, Kingery, W.D., Ed., MIT Press, Cambridge, MA, 1963, pp. 107–111.

TABLE 13.13
Examples of Compositions of Extrusion Bodies

Composition (vol%)					
Refractory Alumina		**High Alumina**		**Electrical Porcelain**	
Alumina (<20 μm)	50	Alumina (<20 μm)	46	Quartz (<44 μm)	16
Hydroxyethyl cellulose	6	Ball clay	4	Feldspar (44 μm)	16
		Methylcellulose	2	Kaolin	16
Water	44	Water	48	Ball clay	16
$AlCl_3$	<1	$MgCl_2$	<1	Water	36
(pH 8.5)				$CaCl_2$	1

Source: From Reed, J.S., *Principles of Ceramic Processing*, Wiley, New York, 1988, p. 359.

TABLE 13.14
Additives Used in Nonclay Aqueous Extrusion Bodies

Flocculant and Binder	Coagulant	Lubricant
Methylcellulose	$CaCl_2$	Various stearates
Hydroxyethyl cellulose	$MgCl_2$	Silicones
Polyvinyl alcohol	$MgSO_4$	Petroleum oil
Polyacrylimides	$AlCl_3$	Colloidal talc
Polysaccharides	$CaCO_3$	Colloidal graphite

Source: From Reed, J.S., *Principles of Ceramic Processing*, Wiley, New York, 1988, p. 360.

The cross section can be very simple such as a solid rod or hollow cylinder or can be very complex such as a thin-wall honeycomb. Some examples of extruded cross sections are illustrated in Figures 13.51 and 13.52. For these sections to be complete and the extruded lengths straight, the ceramic mix must flow uniformly through the die. If it hangs up in one place or flows at a nonuniform rate,

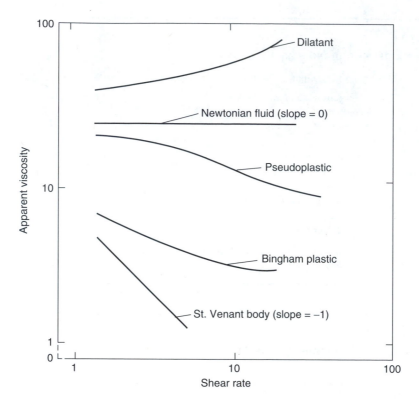

FIGURE 13.50 Options for plastic-flow behavior. (From Mutsuddy, B., *Ind. Res. & Dev.*, 78, 1983. With permission.)

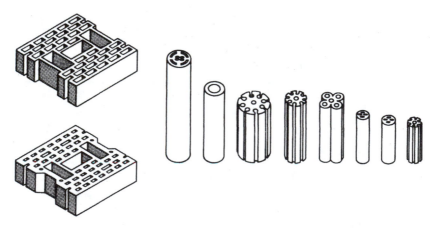

FIGURE 13.51 Examples of the types of shapes that are commonly extruded. (Drawing courtesy of ASM International.)

the extruded compact will deform as it exits the die. Major effort is required to design a die that allows uniform flow and to align the die to tight tolerances.

Extruded material contains a substantial percentage of liquid. This is removed by evaporation using the same type of careful control discussed earlier for slip-cast ceramics. In addition to liquid, the extruded compact contains a much higher binder content than a cast compact. This binder must be removed before the compact can be densified. Most binders used for extrusion can be burned off in an oxidizing atmosphere in the temperature range of 300 to 1000°C (570 to 1830°F). This burn

FIGURE 13.52 A variety of ceramic parts that have been fabricated by extrusion. (Photo courtesy of Superior Technical Ceramics Corporation, St. Albans, VT.)

off must be done slowly and carefully to allow the carbonaceous gases generated by the decomposition of the organics to escape through the pores of the compact. Too rapid a rate of burn off can crack or fracture the compact. Each binder has its unique temperature range for burn off. The allowable rate of burn off is largely determined by the particle size distribution, the degree of particle packing, and the thickness of the compact. Most organic binders will not burn off properly in a nonoxidizing atmosphere. Instead, they will decompose and leave a carbon residue. The acrylic binders are an exception. They burn out cleanly in inert and reducing atmospheres as well as oxidizing atmospheres.

13.3.1.4 Common Extrusion Defects

Extrusion is often more of an art than a science. Quality is controlled by careful inspection of extruded compacts for defects. Defects that can occur for extrusion include warpage or distortion, lamination, tearing, cracking, segregation, porosity, and inclusions.[27]

Warpage or *distortion* can occur during drying or firing due to density variations or during extrusion due to improper die alignment or die design. If the alignment or balance of the die is not correct, greater pressure on one side of the die will occur. This will cause more material to extrude from this side and result in bending of the extruded column as it exits the die.

Laminations are cracks that generally form a pattern or orientation. Examples are shown in Figure 13.53. A common cause is incomplete re-knitting as the plastic mix is cut by the auger or flows past the spider portion of the die. The spider is the portion of the die that supports any shaped channels in the die. For example, to extrude a circular tube, a solid rod of the inner diameter of the tube must be supported at the center of the die. It is generally supported by three prongs at 120° to each other that run parallel to the length of the die and are attached to the inside of the die. The material being extruded must squeeze around these prongs and reunite into a continuous hollow cylinder before leaving the die. Laminations occur if the material does not completely re-knit.

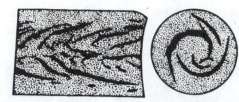

FIGURE 13.53 Drawings of the cross sections of extruded parts illustrating the appearance of severe laminations that can occur as extrusion defects. (From Robinson, G.C., Extrusion defects, in *Ceramic Processing Before Firing,* Onoda, G.Y. and Hench, L.L., Eds., Wiley-Interscience, New York, 1978. With permission.)

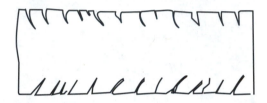

FIGURE 13.54 Drawings of the cross section of an extruded part illustrating the appearance of tearing. (From Robinson, G.C., Extrusion defects, in *Ceramic Processing Before Firing,* Onoda, G.Y. and Hench, L.L., Eds., Wiley-Interscience, New York, 1978. With permission.)

Tearing consists of surface cracks that form as the material exits the extruder. This is illustrated in Figure 13.54. The cracks extending from the surface inward result from the contact stresses and friction that are discussed earlier in this chapter. Too dry a mix with inadequate cohesiveness will tear. A mix with high rebound may also tend to tear. Die design involving a slight divergent taper at the die exit can help prevent tearing.

Lamination and tearing are two sources of cracking. Other cracks can occur due to poor mixing, shrinkage variation, and partially dried debris from a prior extrusion run.

Segregation involves a separation of the liquid and solid portions of the mix during extrusion. This can result in cracking or distortion during extrusion or during subsequent drying or firing.

The defects discussed so far are macrodefects that should be observable by visual inspection or radiography (discussed in Chapter 16). Microdefects such as pores and inclusions can also occur. Pores result from trapped air, powder agglomerates, and pockets of binder or other organic matter that burn off during firing and leave a hole in the material. Inclusions result from foreign material in the starting powder, in the binder, or picked up as contamination during mixing or extrusion. Binder solutions frequently contain contamination or undissolved binder and should be filtered if possible prior to use. Linings and surfaces of blades and augers in mixers and extruders introduce metal particles due to wear.

13.3.1.5 Applications of Extrusion

Extrusion is a low-cost method of forming large quantities of product. It has been used for many years to produce brick, furnace tubes, kiln furniture, thermocouple protection tubes, electronics substrates, tile, sewer pipe, magnets, refractories, and whiteware. More recently extrusion has been used to fabricate heat-exchanger tubes and honeycomb-shaped catalyst supports. The catalyst supports are particularly complex in shape, as illustrated in Figure 13.55. They consist of hundreds of open cells per square centimeter with wall thickness under 100 μm (0.004 in.). A unique procedure was invented for extrusion of such a complex cross section. The ceramic powder was mixed with a hydraulic-setting polyurethane resin. The mix was then extruded into a water bath at a rate that matched the rate of cure of the polyurethane. This rate was about 2 mm/sec.

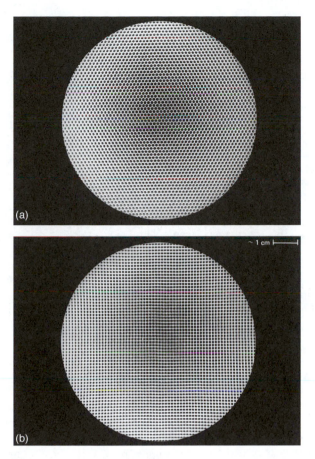

FIGURE 13.55 Cross sections of extruded honeycomb structures of cordierite for use as catalyst supports for automotive emission-control devices. (Courtesy of NGK Insulators.)

13.3.2 INJECTION MOLDING

Injection molding uses equipment[28] that has a cross section similar to an extruder, as shown in Figure 13.56. However, the process is very different from extrusion. The feed material for injection molding generally consists of a mixture of the ceramic powder with a thermoplastic polymer plus a plasticizer, wetting agent, and antifoam agent. The mixture is preheated in the "barrel" of the injection-molding machine to a temperature at which the polymer has a low-enough viscosity to allow flow if pressure is applied. A ram or plunger is pressed against the heated material in the barrel by either a hydraulic, pneumatic, or screw mechanism. The viscous material is forced through an orifice into a narrow passageway that leads to the shaped tool cavity. This helps compact the feed material and remove porosity. At the end of the passageway the strand of viscous material passes through another orifice into the tool cavity. The mixture is much more fluid at this point than an extrusion mix and could not form a self-supporting shape. The strand piles on itself until the cavity is full and the material has "knit" or fused together under the pressure and temperature to produce a homogeneous part. The shaped tool is cooler than the injection-molding mix such that the mix becomes rigid in the tool cavity. The part can be removed from the tool as soon as it is rigid enough to handle without deformation. Cycle times can be rapid, providing the potential for injection molding to be a high-volume, low-cost process for fabrication of ceramics into complex shapes.

Injection molding is used extensively in the plastics industry to make everything from garbage cans to ice cube trays to surprisingly complex constructible toys such as model boats and airplanes.[29]

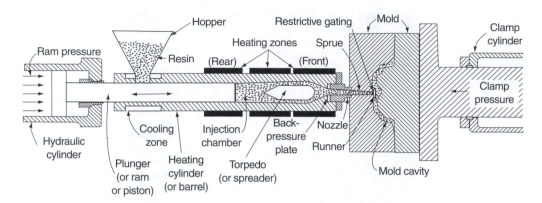

FIGURE 13.56 Cross section of a plunger- (piston-)type injection-molding machine. (From Quackenbush, C.L., et al., *Ceram. Eng. and Sic. Proc.,* 3(1–2), 31, 1982. With permission.)

Ceramic parts are made with the same injection-molding equipment, but with dies made of harder, more wear-resistant metal alloy. The ceramic powder is essentially added to the plastic as a filler. After injection molding, the plastic is then removed by careful thermal treatments.

13.3.2.1 Injection-Molding Parameters

Figure 13.57 lists the general steps in injection molding and some of the controls and inspection procedures that must be considered. The sequence of process steps is similar to that for the other forming processes we have discussed: raw material selection, powder processing, preconsolidation, consolidation, extraction of organics, and densification. The following paragraphs describe some of the process steps and parameters for injection molding.

Particle Sizing. Particle size distribution is important in injection molding, as we have determined for other forming processes, to achieve the densest packing and to minimize the amount of organic material. Both particle packing and sizing affect viscosity. It has been reported[30] that viscosity starts increasing rapidly at about 55 vol% solids for a unimodal suspension of spheres, but that the solids loading can be increased to over 70% before the viscosity starts increasing rapidly for a bimodal distribution containing about 25% fine spheres. By using a graduated particle size distribution, complex shapes of silicon powder having 76.5 vol% solids were successfully injection-molded. Those tests[31] were conducted on a plunger-type injection machine at a cylinder pressure of 13.8 MPa (2 ksi) and a temperature 10°C (50°F) above the melting temperature of the organic binder.

Preconsolidation for Injection Molding. Preconsolidation consists primarily of mixing the ceramic powder homogeneously with the organic additives. The mixing is conducted in a high-shear mixer at a temperature above the softening point of the binder and plasticizer mixture. The objective is to coast each particle with a thin layer of the polymers.[32] Once mixing is complete, the mixture is granulated or pelletized and cooled. The cooled material is hard like plastic or wax.

Consolidation. The objective of the consolidation step is to inject the ceramic powder and binder mixture such that it completely fills the die or mold without leaving porosity, cracks, or other defects. Many factors affect this and must be considered. Major factors include die design, material rheology, and injection parameters, all of which are interactive.

To understand the importance of die design, try to picture what happens during injection. The hot polymer and ceramic powder mixture is squirted into a cool die. Wherever the mixture touches the die, the polymer begins to cool and increase in viscosity. Injection is essentially a race to fill the die uniformly before the polymer becomes too rigid to flow. Variations in die design and sprue design can make a big difference in the uniformity and completeness of die fill. This is discussed later in the section on injection-molding defects.

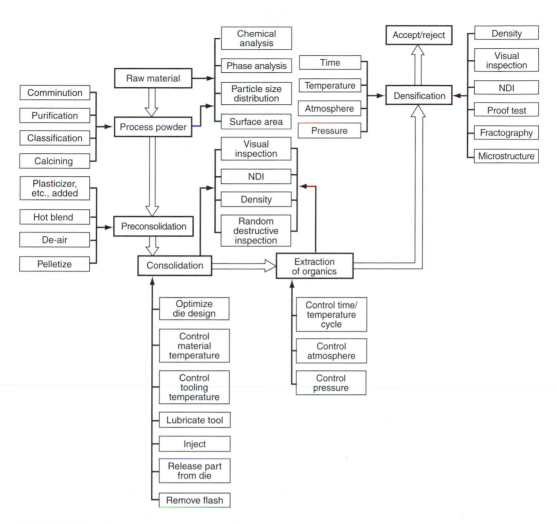

FIGURE 13.57 Flowchart for injection molding.

A second die-design factor is removal of the part after injection. For very complex shapes, the die may have to be very complex and contain removable pins and shaped inserts.

The rheology of an injection-molding mixture is determined by the vol% solids and by the nature of the binders, plasticizer, and other additives. More than one binder is often used.[33,34] The major binder is added in large quantity and provides to the mixture the general range of binder properties and injection parameters. Some of the characteristics of the major binder include: (1) to provide adequate fluidity to the powder to permit defect-free filling of the mold cavity; (2) to wet the powder; (3) to remain stable under mixing and molding conditions; (4) to provide strength to the powder compact during the initial stage of binder removal; (5) to leave a low residue during burnoff; and (6) to be commercially available at an acceptable cost.

A minor binder is often added to aid in the binder-removal cycle. The minor binder melts or decomposes at a lower temperature than the major binders. During binder removal, it volatilizes first and leaves channels or paths through which the major binder can be more easily removed at a higher temperature. Table 13.15 identifies an injection-molding system for SiC that includes more than one binder plus other additives.[35]

A plasticizer is added to increase the fluidity of the ceramic and binder mixture. Specific plasticizers work with specific binders.

TABLE 13.15
Additives for Injection Molding of SiC

Function	Options	Quantity, wt%	Characteristics
Thermoplastic resin	Acrylic	9–17	Volatilization temperature between
	Ethyl cellulose		200–400°C (390–750°F)
	Hydroxypropyl cellulose		
	Polyethylene		
	Oxidized polyethylene		
	Cellulose acetate		
	Nylon		
	Polystyrenes		
	Polybutylene		
	Polysulfone		
	Polyethylene glycol		
Wax or high-temperature volatilizing oil	Paraffin	2–3.5	Volatilization temperature between
	Mineral oils		150–190°C (300–375°F)
	Vegetable oils		
	Waxes		
Low-temperature volatilizing hydrocarbon or oil	Animal	4.5–8.5	Volatilization temperature between
	Vegetable oils		50–150°C (120–300°F)
	Mineral oils		
Lubricant or mold release	Fatty acids	1–3	
	Fatty alcohols		
	Fatty esters		
	Hydrocarbon waxes		
Thermosetting resin	Epoxy		Source of carbon; char. in range
	Polyphenylene		450–1000°C (840–1830°F)
	Phenol formaldehyde		

Source: From Ohnsorg, R., U.S. Patent 4,233,256, Nov. 11, 1980.

Other additives are used mainly as surfactants to improve the wetting characteristics between the binder and ceramic during mixing. Functions of other selected additives include deagglomeration, reduction in melt viscosity, lubrication, and die release.

The optimum binder content is generally 102 to 115% of the void volume.[34] The optimum flow characteristics are Bingham or pseudoplastic flow at an intermediate yield stress. A viscosity of less than 10^4 cP at a shear range of 100 to 1000 sec^{-1} is generally acceptable.

Table 13.16 lists a variety of binders, plasticizers, and lubricants that have been used for injection molding of ceramics. Most of the binders commonly used are thermoplastic, that is, they reversibly soften when cooled. Examples include the waxes, polypropylene, polyethylene, and polystyrene.[36] Wax-based compositions can be injected at relatively low temperature and pressure. For example, a paraffin or beeswax system can be injected at 60 to 100°C (110 to 212°F) and at pressures as low as 305 kPa (3 atm or 44.1 psi). This allows use of a very simple, inexpensive apparatus that uses compressed gases for injection and cooled aluminum for tooling. Cycle time per injection can be less than 60 sec.

Polyethylene and polypropylene systems are injection-molded at higher temperatures and pressures than wax-based compositions. In Reference 33 for example, a polypropylene-based system was molded at 225°C (435°F) and 150 MPa (~1500 atm 21.8 ksi). A screw-type injection-molding machine with steel tooling was used. Pressure was held for 120 sec with a total cycle time of 150 sec. The mold temperature was 30°C (85°F).

Limited injection molding of ceramics has been conducted with thermosetting binders such as phenolfurfural or epoxy resins. Thermosetting resins solidify by *cross-linking*, which is not a reversible

TABLE 13.16

Examples of Binders, Plasticizers, and Lubricants That Have Been Used for Injection Molding of Ceramics

Binders (Thermoplastic)
  Polypropylene
  Polyacetal polymers
  Ethylene vinyl acetate (low molecular weight)
  Atactic polypropylene (molecular weight 5000–12000)
  Styrene-butadiene copolymer
  Poly (n-butyl methacrylate)
  Polyethylene
  Polybutene
  Polystyrene
  Waxes

Binders (Thermosetting)
  Epoxy resin
  Phenol furfural
  Phenol formaldehyde
  Nylon

Binders (Water-Soluble)
  Methyl cellulose
  Hydroxypropylmethyl cellulose
  Hydroxyethyl cellulose
  Polacrylamides

Plasticizers
  Polyethylene glycol
  Other phthalates
  Beeswax
  Diethyl phthalate
  Butyl stearate
  Light oils

Lubricants
  Stearic acid
  Hydrogenated peanut oil
  Glycerol monostearate
  Paraffin wax
  Tritolyl phosphate
  Ester wax
  Oleic acid

Source: From Richerson, D.W., *Introduction to Modern Ceramics*, ASM Mat. Eng. Inst. Course 56, Lesson 11, p. 11, ASM Int., 1990.

process. The cross-linking is induced thermally or by chemical additions. For injection molding of a thermosetting resin, the mold generally is held at a higher temperature than the injection temperature. Cycle time is high to allow cross-linking to occur.

A third category of binders includes water-soluble compositions. These can be injected at or near room temperature. The challenge is achieving rigidization in the die. This is generally achieved by thermal gellation. The die is held at a temperature around. 60°C (140°F) at which the cellulose-based binder gels into a structure that is rigid enough for the part to be carefully removed from the die.

Injection-molding compositions and parameters for production forming of ceramics are generally considered proprietary by the company developing the technology. Table 13.17 identifies some injection-molding compositions that have been reported in the literature.

TABLE 13.17
Examples of Injection-Molding Batch Formulations

Ceramic powder	Additives	Reference
83 wt% Al_2O_3 or ZrO_2	16.1 wt% paraffin, 0.9 wt% oleic acid	1
91.5 wt% $BaTiO_1$	8.0 wt% paraffin, 0.5 wt% beeswax	1
87.6 wt%	4.8 wt% (15.2 vol%) Carnauba wax,	2
(68.7 vol%) Al_2O_3 plus	3.1 wt% (8.8 vol%) epoxy resin,	
	0.8 wt% (2.4 vol%) silicone oil	
3.7 wt%		
(5.0 vol%) SiO_2		
100 parts Si or Al_2O_3	20 parts atactic polypropylene, 5 parts paraffin	3
82.44 wt% Si	11.71 wt% polypropylene, 3.9 wt% wax, 1.95 wt% stearic acid	4
63.0 vol% steatite	24.3 vol% wax, 7.3 vol% epoxy resin,	5
	3.2 vol% coumarone-indene resin,	
	2.2 vol% phenol formaldehyde resin	
47 vol% SiC,	47 vol% phenol furfural-phenol formaldehyde	6
5 vol% graphite	copolymer, 1 vol% zinc stearate	

References

1. Operation and Maintenance Manual for Hot Molding of Ceramic Parts Under Low Pressure, Peltsman Corporation.
2. Newfield, S.E. and Gac, F.D., Injection Moldable Ceramics, Los Alamos Nat. Lab Rept. LA-6960, March 1978.
3. U.S. Patent 4,248,813.
4. Edirisinghe, M.J. and Evans, J.R.G., *J. Mater. Sci.*, 22, 2267–2273, 1987.
5. Strivens, *Am. Ceram. Soc. Bull.*, 42, 13–19, 1963.
6. Whalen and Johnson, *Am. Ceram. Soc. Bull.*, 60, 216–220, 1981.

Binder Removal.   Binder removal is a major step in the injection-molding process. The large volume fraction of organic materials must be removed without cracking or distortion of the ceramic powder compact. The following techniques are used: evaporation (or distillation), solvent extraction, capillary action, decomposition, and oxidation. Often these are mixed by utilizing more than one binder. One, such as a low-temperature wax, may be removed by evaporation or solvent extraction. Another, such as a thermosetting resin, will then be removed at higher temperature by oxidizing or dissociation.

Evaporation involves slow volatilization of the organic material at elevated temperature. The temperature depends on the vaporization characteristics of the specific binder composition. The temperature is increased slowly to avoid too-rapid evolution of the vapor that would cause fracture or distortion of the weak ceramic compact. The cycle time of binder removal also depends on the binder composition, but also on the particle packing (and resulting permeability) and on the cross-section thickness of the part. An 8-hr cycle may be adequate for a thin cross section (a few mm), whereas a 20-hr or even one-week cycle may be required for a thicker section (1 cm or 0.39 in.). Evaporation can be conducted under vacuum, at atmospheric pressure or at an overpressure.

Solvent extraction is accomplished by dissolving one of the binders with liquid-phase or vapor-phase solvent. This can only be done if a nonsoluble binder is also present to hold the ceramic particles together while the soluble binder is being extracted. Solvent extraction can be conducted at lower temperature than evaporation and potentially minimizes capillary and vapor stresses.

Capillary extraction involves packing the injection-molded parts in a fine powder such as activated carbon. The temperature is then increased until the binder is fluid enough to move from the part into the surrounding powder by capillary action.

Decomposition and *oxidation reactions* are used to remove the secondary binder plus any residue remaining from the major binder. These reactions occur at higher temperatures than evaporation, that is, typically above 500°C (930°F) and result in gas phases such as H_2O, CO, or CO_2. The

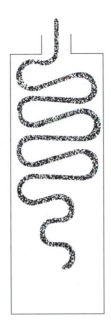

FIGURE 13.58   Schematic showing how a thin strand can inject into the mold and pile up without fully bonding to result in knit lines. (Drawing courtesy of ASM International.)

temperature must be raised slowly to allow the gases to form and diffuse out of the porous compact without building up enough pressure to cause fracture.

13.3.2.2   Injection-Molding Defects

A variety of defects can occur during the injection-molding operation. These can be divided into two categories: micro and macro. Microdefects include inclusions and microporosity due to agglomerates and improper particle size distribution. These are similar to the microdefects that can occur with any other particulate forming process.

The second category, macrodefects, can be divided into two subgroups: (1) mold-filling defects and (2) solidification defects. *Mold-filling* defects include incomplete fill, porosity, and knit lines. *Solidification* defects include void nucleation and microcracking due to residual stresses. These types of defects are defined and described in the following paragraphs.

Incomplete mold fill is easy to detect visually on an injected part. Essentially, a portion of the part is missing. This can occur for several reasons. First, the die design may not be optimum and may result in solidification before the die is completely full. This can result sometimes due to *gate freeze-off*. The gate is the opening through which material enters the mold. Since the mold is at a lower temperature, sometimes the material in the gate solidifies and blocks further injection. This can be prevented by better control of temperature and by modification of the mold design.

Porosity can occur during injection due to air entrapment. The air can either be in the ceramic and binder mix or can be trapped during mold fill. Molds are usually vented to allow a route for air to escape. Overflows are also often included in mold design to allow the initial material entering the mold to pass through the mold and accumulate in a waste cavity. Excess mold-release agent plus foreign debris in the mold will be swept into the overflow cavity where it can be discarded.

The final mold-filling defect we shall discuss is *knit lines*. Knit lines are areas where the injected material does not properly fuse together. They represent a discontinuity or a weak region in the part. They usually have a laminar or folded appearance. Some can be severe and are easily visible if they intersect the outer surface of the part. Others are very subtle and difficult to detect, even with nondestructive inspection techniques such as radiography and ultrasonics. Figure 13.58 illustrates schematically how flow lines form. The thin stream of ceramic and binder mix "jets" through the sprue into the mold cavity and begins to pile up like a strand of spaghetti. The first part that touches the mold wall begins to cool and solidify. If it solidifies too much, it will not fuse together into a single continuous compact, but instead will leave open spaces. The tendency to form knit lines can be reduced by proper mold design, especially by orientation of the gate and sprue. Figure 13.58 shows injection into a rectangular mold cavity where the gate was located on the end directed parallel to the length of the cavity. Knit lines were difficult to avoid with this geometry. Figure 13.59 shows the flow pattern with alternate sprue and gate designs. In Figure 13.59a, the gate is at the end, but is directed perpendicular to the length of the cavity. In Figure 13.59b, the gate is directed perpendicular, but placed at the center of the mold cavity. Plug flow resulted in both cases and

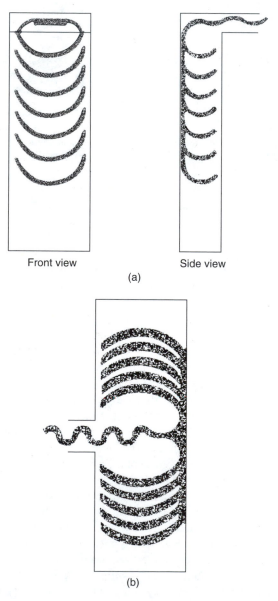

Front view   Side view

(a)

(b)

FIGURE 13.59   Alternate mold sprue and gate designs that result in plug flow rather than "jetting" and minimize formation of knit lines. (Drawing courtesy of ASM International.)

knit-line formation was minimized. This is further illustrated in Figure 13.60 for actual injection-molding trials. The "short shot" technique was used whereby injection was interrupted before the cavity was full. By conducting a sequence of short shots, a good image of the nature of mold fill for each gate configuration could be obtained.

After binder removal and densification, knit lines remain as large cracks, voids, or laminations and severely limit the strength of the part.

The short shot approach has been successfully used at Carborundum Company in developing integral radial rotors for an experimental automotive gas turbine.[37] Initial rotors were injected from the nose end. (Figure 13.61 illustrates the cross section of a radial rotor and identifies terminology that will be referred to subsequently). Short shots indicated a tendency for folds and knit lines to form in the thick region of the hub near the backface. This is illustrated in Figure 13.62. This region is exposed to the highest stresses during engine operation, so major iterative efforts were conducted to minimize the knit lines. Many parameters such as die temperature, injection pressure, hold time, and sprue bushing and nozzle diameter were systematically varied. Sixteen resulting rotors were spin tested and failed at an average speed of 80,500 rpm, significantly below the desired design speed of 86,240 rpm. Failure analysis determined that fracture initiated at surface and near-surface

FIGURE 13.60   Sequence of "short shots" showing the nature of mold fill for two different sprue and gate orientations. (Photo courtesy of Carborundum Company for work conducted under the Department of Energy Advanced Gas Turbine Program.)

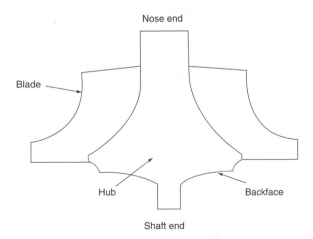

FIGURE 13.61   Schematic of the cross section of a radial turbine rotor identifying regions of the rotor. (Drawing courtesy of ASM International.)

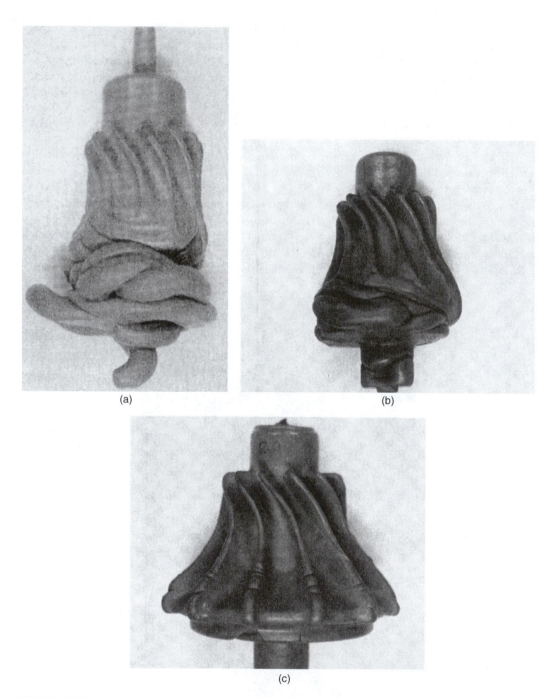

(a)

(b)

(c)

FIGURE 13.62 Sequence of short shots for injection molding of a SiC rotor from the nose end. Note the knit lines in the hub and backface regions. (Photos courtesy of Carborundum Company for parts fabricated for Allison Gas Turbine Division of General Motors under sponsorship of the U.S. Department of Energy and administration of NASA-Lewis Research Center.)

flaws in the highly stressed region of the backface. Approximately 1.25 mm was machined off the backface of 30 rotors. These failed at an average of 95,100 rpm.

Whereas the backface of the rotor is exposed to high stresses, the stresses in the nose end are relatively low. An alternate approach to machining the backface was pursued. This involved injection molding from the shaft end. A sequence of short shots for this approach is shown in Figure 13.63.

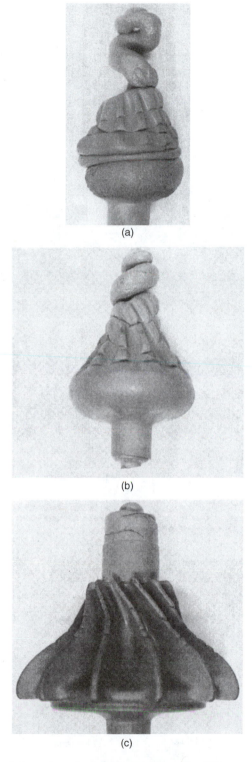

(a)

(b)

(c)

FIGURE 13.63 Sequence of short shots for injection molding of a SiC rotor from the shaft end. Note the absence of knit lines in the hub and backface region. (Photos courtesy of Carborundum Company for parts fabricated for Allison Gas Turbine Division of General Motors under sponsorship of the U.S. Department of Energy and administration of NASA-Lewis Research Center.)

Note the absence of knit lines in the hub region. Forty-two rotors fabricated by shaft-end injection were spin-tested with average failure at 96,200 rpm. An as-molded and a sintered (densified) rotor are illustrated in Figure 13.64.

Now we are ready to discuss the second subgroup of injection-molding macrodefects: solidification defects. Imagine what happens when material that is 225°C (435°F) is injected rapidly into

FIGURE 13.64 Examples of optimized SiC rotors injection-molded from the shaft end. The rotor on the left is as-molded, the one on the right is after sintering. (Photo courtesy of Carborundum Company.)

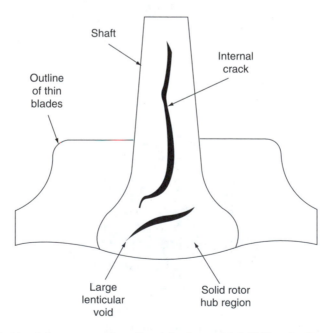

FIGURE 13.65 Sketch of the cross section of an injection-molded Si_3N_4 turbocharger rotor showing the internal void that resulted during cooling in the mold due to a combination of surface quenching and shrinkage of the interior toward the surface. (Drawing courtesy of ASM International.)

a mold that is only 30°C (85°F). The material at the surface of the part cools more quickly than the material in the interior. If the gradient is too high or the part too thick, residual stresses and even nucleation of voids can occur. This is aggravated by the high thermal-expansion coefficient of the binder polymers and the additional shrinkage that occurs when the binder goes from liquid to solid or noncrystalline to crystalline. For example, the volume change for one polypropylene system due to thermal contraction was about 2.75 vol% and due to crystallization was about 1.75 vol% for a total of about 4.5 vol%. If the outer shell is rigid and cannot shrink, while the inner material is more fluid and can reposition during further cooling, 4.5% shrinkage is adequate to form a void or crack through the center of the part. Such a void or crack is typically not visible by examining the surface of the injection-molded part and may not even be visible after densification. Figure 13.65 illustrates a large lenticular (lens-shaped) void in a Si_3N_4 turbocharger rotor that resulted primarily from this mechanism.

13.3.2.3   Applications of Injection Molding

Injection molding is usually selected for ceramics only after other processes have been rejected. It can produce a high degree of complexity, but the initial cost of tooling is very high. For example, a mold to fabricate an individual turbine blade can cost over $10,000 and a mold for a turbine rotor over $100,000. Molds for simple shapes and molds made of aluminum for low-pressure injection molding are much less expensive. As a result, the use of injection molding of ceramics is increasing.

Injection molding is presently used to manufacture a variety of parts including cores for investment (lost-wax) casting of metals, weld caps, thread guides, threaded fasteners (nut and bolt pairs), radomes, and prototype gas-turbine engine components. Drawings of complex investment casting cores for cooled metal gas-turbine blades or stator vanes are shown in Figure 13.66. During investment casting, the core is mounted in a ceramic mold. Molten superalloy is poured into the mold around the core. The ceramic mold is removed from the outside of the metal part. The injection-molded ceramic core is leached from the interior of the blade or vane to leave a complex cooling path. This substantially reduces the cost of manufacturing of internally cooled superalloy stator vanes and rotor blades for advanced gas-turbine engines.

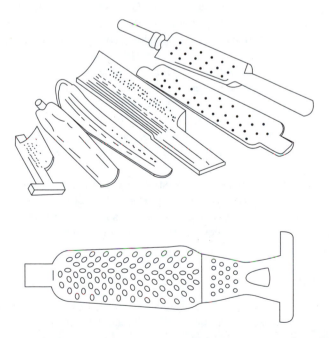

FIGURE 13.66   Drawings of injection-molded ceramic cores for investment casting of cooled rotor blades or stator vanes for gas-turbine engines. (Drawing courtesy of ASM International.)

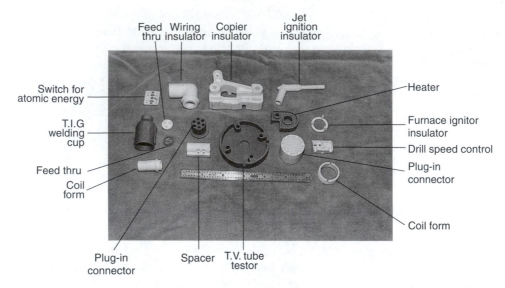

FIGURE 13.67 Examples of Al_2O_3 parts fabricated by injection molding. (Courtesy of Diamonite Division of W.R. Grace.)

Examples of other injection-molded ceramic parts are illustrated in Figures 13.67 through 13.69.

13.3.2.4 Nonthermoplastic Injection Molding

Injection molding of ceramics has traditionally been conducted with thermoplastic binders or a combination of thermoplastic and thermoset binders. Some success has also been achieved with cellulose derivatives that gel with a suitable change in temperature. Two additional approaches to injection molding have been reported in recent years. One uses polysaccharides (in particular, agar and agarose) as a gel-forming binder and water as the fluid.[38] A relatively small percentage (3 wt%) of agarose is required (compared to thermoplastic systems), so drying and binder removal are reported to be simplified.

The second new approach is identified as Quickset* injection molding.[39] It is sort of a cross between casting, injection molding, and freeze drying and appears to provide some of the benefits of each, while avoiding some of the problems. It uses a slurry (typically with a viscosity under 1000 centipoise) that is injected at typically less than 50 psi pressure into a closed cavity, nonporous mold. The pore fluid is solidified by freezing and subsequently removed by sublimation. Volume change during freezing is negligible and stresses during sublimation are substantially lower than for removal of water or thermoplastic polymers.

Quickset injection molding has been successfully accomplished with both aqueous and non-aqueous suspensions and with a variety of ceramic powders. Table 13.18 lists the properties achieved for different materials formed by the Quickset process. In addition to the excellent properties, tight dimensional tolerances are readily achieved. For example, dimensional tolerances for a SiAlON component only varied in as-fired parts by 0.09%.

13.3.3 Compression Molding

Compression molding is analogous to forging. A block of plastic mix is placed between the platens of a shaped die, and uniaxial pressure is applied until the block deforms to the shape of the die cavity.

---

* Trademark of Ceramics Process Systems, Milford, MA.

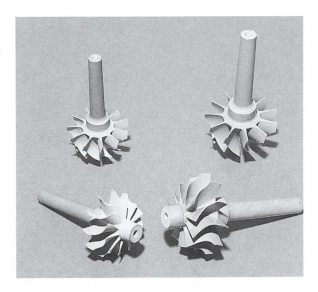

FIGURE 13.68 Prototype sintered silicon nitride turbocharger rotors fabricated by injection molding. (Courtesy of Garrett Ceramic Components Division of Allied-signal Aerospace.) Fabricated by AiResearch Casting Company (Currently Honeywell Ceramic Components, Torrance, CA.)

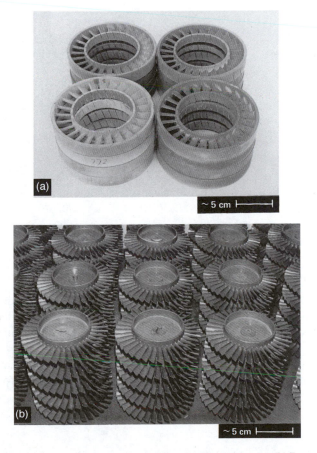

FIGURE 13.69 Complex shapes made by injection molding. (a) Integral stators. (b) Rotor blade rings. (Courtesy of Ford Motor Company, Dearborn, MI.)

TABLE 13.18

Examples of Properties Achieved by Quickset Injection Molding

Material[a]	Strength, MPa[b]	Weibull Modulus	Comments
Alumina[c]	489	10	
Zirconia[d]	850	20	$K_{1c} = 8\,MPa\text{-}m^{1/2}$
ZTA[e]	1100	22	
SiAlON[f]	968	11	UTS[h] = 650 MPa
AlN[g]	380	—	230–259 W/mK

[a] Pressureless sintered compositions.
[b] Four-point flexural strength, room temperature.
[c] Alcoa A16 Superground.
[d] Tetragonal zirconia polycrystal, 3 mol% yttria.
[e] 30 vol% 3Y-zirconia/70 vol% alumina.
[f] SiAlON 101, supplied by Vesuvius Zyalons.
[g] CPS-200 grade material.
[h] UTS = Ultimate tensile strength.

Source: From Novich, B.E., Lee, R.R., Franks, G.V., and Ouellette, D., Quickset™ injection molding of high temperature gas turbine engine components, in *Proceedings of the 27th Automotive Technology Development Contractors' Coordination Meeting*, SAE publication P-230, 1990.

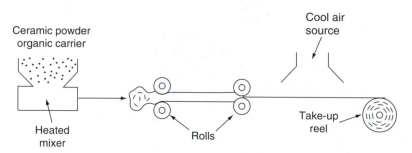

FIGURE 13.70   Schematic illustrating the roll-forming process. (Drawing courtesy of ASM International.)

Compression molding can be conducted hot or cold, depending on the nature of the binder system. It works especially well for systems containing thermosetting resins.

13.3.4   ROLL FORMING

A plastic mix is passed between two cylinders that are rotating in opposite directions as shown in Figure 13.70. The plastic mix passing between the rolls is compacted, as well as being pressed to a thickness equivalent to the spacing of the rolls. Multiple passes at diminishing roll separation can yield a constant-thickness sheet of high uniformity.

Roll forming can be conducted at room temperature using a mix equivalent to an extrusion mix or at elevated temperature using a thermoplastic polymer system. Warm roll forming has been used for many years to fabricate resin-bonded and rubber-bonded grinding wheels. It has also been used to form much thinner layers suitable for heat-exchanger fabrication.

13.3.5   JIGGERING

Jiggering is commonly used to fabricate cooking ware, electrical porcelain, and refractories. A segment of de-aired extruded mix (or other plasticized feedstock) is placed on a shaped rotating

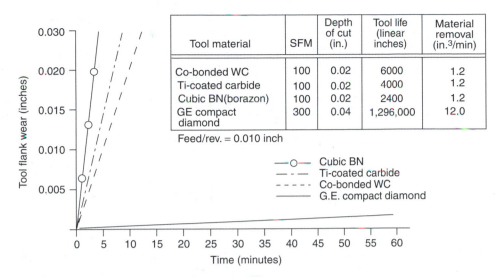

FIGURE 13.71 Tool wear for different tool insert materials for green machining of a presintered silicon compact in the fabrication of reaction-bonded Si_3N_4. (From Richerson, D.W. and Robare, M.W., Turbine component machining development, in *The Science of Ceramic Machining and Surface Finishing*, Vol. II, Hockey, B.J. and Rice, R.W., Eds., NBS Special Publication 562, U.S. Government Printing Office, Washington, DC, 1979, pp. 209–220.)

wheel. A shaped roller or template then presses the plasticized mix against the wheel to forge the shape.

13.4 GREEN MACHINING

The final shape and tolerances cannot always be achieved by pressing, casting, and plastic forming. In many cases, the final dimensions are achieved by surface grinding or other finishing operations following the densification step. This requires diamond tooling and can be very expensive. Final machining can often be avoided by machining the ceramic particulate compact prior to densification. This is referred to as *green machining*.

Unfired green material is relatively fragile, and great care is necessary in the design and fabrication of the tooling and fixturing so that the parts can be accurately and uniformly held during the various shaping operations. In addition, the machining parameters must be carefully controlled to avoid overstressing the fragile material and producing chips, cracks, breakage, or poor surface.

Holding of the compact for machining is typically accomplished either by simple mechanical gripping or by bonding or potting with a combination of beeswax and precision metal fixtures. The part must be held rigidly, but with no distortion or stress concentration.

Once a ceramic part has been secured rigidly in a fixture, machining can be conducted by a variety of methods — turning, milling, drilling, form wheel grinding, and profile grinding. Machining can be either dry or wet, depending on the binder present and whether or not the part has been bisque-fired.* In either case, the compact is abrasive and results in tool wear. A wear land on the cutting edge as little as 0.1-mm (0.0039-in.) wide can cause a buildup of cutting pressure and result in damage to the ceramic.

It is possible to machine compacts with high-speed steel or cemented carbide cutting tools, but this is not recommended for all components or all green materials. In some cases, the tool dulls so rapidly that extreme care is necessary to avoid damage to the workpiece. Figure 13.71 summarizes a green machining study comparing several cutting-tool materials. A 5° positive rake and 10° clearance

---

*Fired at a high enough temperature to form bonds at particle-particle contact points, but not at a high enough temperature to produce densification.

angle were used in the study. The compact diamond cost about 10 times as much as the tungsten carbide, but resulted in a significant cost saving in terms of increased life, less time changing inserts, and reduced risk of damage to the workpiece from a dull tool. The study was conducted with single-point turning on an engine lathe. Milling with a two-flute end mill at 61 surface meters/min (200 sfm) with compact diamond inserts showed the same life characteristics.

Green machining can also be conducted with grinding wheels containing multiple abrasive particles bonded in a resin or metal matrix. Higher surface speed, broader contact, and decreased depth of cut are characteristic of this technique, resulting usually in a better surface and less chance of damage. Excellent tool life can be achieved, especially if a diamond abrasive is used. Furthermore, coarse abrasive can be used for roughing passes and fine abrasive for finishing. Formed wheels can also be used to produce a controlled and reproducible contour, as was discussed earlier in this chapter for green machining of spark plug insulators and oxygen-sensor electrolytes.

REFERENCES

1. Thurnauer, H., Controls required and problems encountered in production dry pressing, in *Ceramic Fabrication Processes,* Kingery, W.D., Ed., MIT Press, Cambridge, Mass., 1963, pp. 62–70.
2. Whittemore, O.J., Jr., Particle compaction, in *Ceramic Processing Before Firing*, Onoda, G.Y., Jr. and Hench, L.L., Eds., Wiley, New York, 1978, pp. 343–355.
3. Kingery, W.D., Pressure forming of ceramics, in *Ceramic Fabrication Processes*, Kingery, W.D., Ed., MIT Press, Cambridge, Mass., 1963, pp. 55–61.
4. Duwez, P. and Zwell, L., AIME Tech. Publ. 2515, *Metals Trans.*, 1, 137, 1949.
5. Seelig, R.P., in *The Physics of Powder Metallurgy,* Kingston, W.E., Ed., McGraw-Hill, New York, 1950, p. 344.
6. Kingery, W.D., Hydrostatic molding, in *Ceramic Fabrication Processes,* Kingery, W.D., Ed., MIT Press, Cambridge, Mass., 1963, pp. 70–73.
7. Michaels, A.S., Rheological properties of aqueous clay systems, in *Ceramic Fabrication Processes,* Kingery, W.D., Ed., MIT Press, Cambridge, Mass., 1963, pp. 23–31.
8. Moore, F., *Rheology of Ceramic Systems*, MacLaren & Sons Ltd., London, 1965.
9. Cowan, R.E., in *Treatise on Materials Science and Technology*, Vol. 9: *Ceramic Fabrication Processes*, Wang, F.F.Y., Ed., Academic Press, New York, 1976, pp. 153–171.
10. St. Pierre, P.D.S., Slip casting nonclay ceramics, in *Ceramic Fabrication Processes,* Kingery, W.D., Ed., MIT Press, Cambridge, Mass., 1963, pp. 45–51.
11. Green, M., et al., Chemically bonded organic dispersants, in *Advances in Ceramics, Ceramic Powder Science and Technology*, Vol. 21, American Ceramic Society, Westerville, Ohio, 1987, pp. 449–465.
12. Mikeska, K. and Cannon, W.R., Dispersants for tape casting pure barium titanate, in *Advances in Ceramics, Forming of Ceramics*, Vol. 9, Mangels, J. and Messing, G., Eds., American Ceramic Society, Ohio, 1984, pp. 164–183.
13. Lambe, C.M., Preparation and use of plaster molds, in *Ceramic Fabrication Processes,* Kingery, W.D., Ed., MIT Press, Cambridge, Mass., 1963, pp. 31–40.
14. Blanchard, E.G., Pressure casting improves productivity, *Am. Ceram. Soc. Bull.*, 67(10), 1680, 1988.
15. Sheppard, L.M., Fabrication of ceramics: the challenge continues, *Am. Ceram. Soc. Bull.*, 68(10), 1815–1820, 1989.
16. Fennelly, T.J. and Reed, J.S., Mechanics of pressure casting, *J. Am. Ceram. Soc.*, 55(5), 264–268, 1972.
17. Ezis, A. and Nicholson, J.M., Method of manufacturing a slip cast article, U.S. Patent No. 4,067,943.
18. Ezis, A. and Neil, J.T., Fabrication and properties of fugitive mold slip-cast Si_3N_4, *Bull. Am. Ceram. Soc.*, 58(9), 883, 1979.
19. Sutton, W.H. and Morris, J.R., Development of ceramic foam materials for the filtration of high temperature investment casting alloys, 31st Annual Meeting of ICI, Dallas, Tex., Oct. 25, 1983.
20. Ohnsorg, R.W. and TenEyck, M.O., Fabrication of sintered alpha SiC turbine engine components, in *Ceramic Transactions*, Silicon Carbide '87, Vol. 2, Cawley, J.D. and Semler, C.E., Eds., The American Ceramic Society, Westerville, Ohio, pp. 367–386.
21. Keey, R.B., *Introduction to Industrial Drying Operations*, Pergamon Press, Elmsford, New York, 1978.

22. Reed, J.S., *Principles of Ceramic Processing*, Wiley, New York, 1988.
23. Williams, J.C., in *Treatise on Materials Science and Technology, Ceramic Fabrication Processes*, Vol. 9, Wang, F.F.Y., Ed., Academic Press, New York, 1976, pp. 173–198.
24. Shanefield, D.J. and Mistler, R.E., Fine grained alumina substrates: I, the manufacturing process. *Am. Ceram. Soc., Bull.,* Part I, 53, 416–420, 1974.
25. Janney, M., *Bibliography of Ceramic Extrusion and Plasticity*, ORNL Rept. 6363, Oak Ridge National Laboratory, Tenn.
26. Hyde, C., Vertical extrusion of nonclay composition, in *Ceramic Fabrication Processes,* Kingery, W.D., Ed., MIT Press, Cambridge, Mass., 1963, pp. 107–111.
27. Robinson, G.C., Extrusion defects, in *Ceramic Processing Before Firing,* Onoda, G.Y. and Hench, L.L., Eds., Wiley-Interscience, New York, 1978.
28. Matsuddy, B.C., Equipment selection for injection molding, *Am. Ceram. Soc. Bull.*, 68(10), 1796–1802, 1989.
29. Rubin, I., *Injection Molding of Plastics*, Wiley, New York, 1973.
30. Farris, R.J., Prediction of the viscosity of multimodal suspensions from unimodal viscosity data, *Trans. Soc. Rheol.*, 12(2), 281, 1968.
31. Mangels, J.A., in *Ceramics for High Performance Applications*, Vol. II, Burke, J.J., Lenoe, E.N., and Katz, R.N., Eds., Brook Hill Publishing Co., Mass., 1978, pp. 113–130.
32. Uhl, V.W. and Gray, J.B., *Mixing: Theory and Practice*, Vol. 1 and 2, Academic Press, New York, 1967.
33. Edirisinghe, M.J. and Evans, J.R.G., Properties of ceramic injection molding formulations, II, integrity of moldings, *J. Mater. Sci.*, 22, 2267–2273, 1987.
34. Edirisinghe, M.J. and Evans, J.R.G., Review: fabrication of engineering ceramics by injection molding, I, materials selection, *Int. J. High Tech. Ceramics*, 2, 1–31, 1986.
35. Ohnsorg, R., U.S. Patent 4,233,256, Nov. 11, 1980.
36. Renlund, G.M. and Johnson, C.A., Thermoplastic molding of sinterable silicon carbide, U.S. Patent 4,551,496, Nov. 5, 1985.
37. Ohnsorg, R., TenEyck, M., and Sweeting, T., Development of injection molded rotors for gas turbine applications, ASME paper 86-GT-45.
38. Fanelli, A.J., et al., New aqueous injection molding process for ceramic powders, *J. Am. Ceram. Soc.*, 72(10), 1833–1836, 1989.
39. Novich, B.E., Lee, R.R., Franks, G.V., and Ouellette, D., Quickset™ injection molding of high temperature gas turbine engine components, in *Proceedings of the 27th Automotive Technology Development Contractors' Coordination Meeting*, SAE publication P-230, 1990.
40. Richerson, D.W. and Robare, M.W., Turbine component machining development, in *The Science of Ceramic Machining and Surface Finishing*, Vol. II, Hockey, B.J. and Rice, R.W., Eds., NBS Special Publication 562, U.S. Government Printing Office, Washington, D.C., 1979, pp. 209–220.

ADDITIONAL RECOMMENDED READING

1. Mangels, J.A. and Messing, G.L., Eds., *Forming of Ceramics, Advances in Ceramics*, Vol. 9, American Ceramic Society, Westerville, Ohio, 1984.
2. German, R.M., Theory of Thermal Debinding, *Int. J. Powder Metall.*, Vol. 23 (No. 4), 1987, pp. 237–245.
3. Hench, L.L. and Ulrich, D.R., Eds., *Ultrastructure Processing of Ceramics, Glasses, and Composites*, John Wiley & Sons, 1984.
4. MacKenzie, J.D. and Ulrich, D.R., Eds., *Ultrastructure Processing of Advanced Ceramics*, John Wiley & Sons, 1988.
5. Messing, G.L., Mazdiyasni, K.S., McCauley, J.W., and Haber, R.A., Eds., *Ceramic Powder Processing, Advances in Ceramics,* Vol. 21, American Ceramic Society, 1987.
6. Messing, G.L., Fuller, E.R., and Hausner, H., Eds., *Ceramic Powder Science II, A and B, Ceramic Transactions*, Vol. 1, American Ceramic Society, 1988.
7. Horn, R.G., Surface Forces and Their Action in Ceramic Materials, *J. Am. Ceram. Soc.*, Vol. 73 (No. 5), 1990, pp. 1117–1135.
8. *Isostatic Pressing Technology*, Applied Science Publishers Ltd., London, 1983.
9. *Engineered Materials Handbook Vol. 4: CERAMICS AND GLASSES*, ASM International, Materials Park, Ohio, 1991.

10. Omatete, O., Janney, M., and Strehlow, R.A., Gelcasting A New Ceramic Forming Process, *Am. Ceram. Soc. Bull.*, Vol. 70 (No. 10), 1991, pp. 1641–1649.
11. Tari, G., Gelcasting Ceramics: A Review, *Am. Ceram. Soc. Bull.*, Vol. 82 (No. 4), 2003, pp. 43–46.
12. Rahaman, M.N., *Ceramic Processing and Sintering*, Marcel Dekker, New York, 1995.
13. Shanefield, K.J., *Organic Additives and Ceramic Processing,* 2nd ed., Kluwer Academic Publishers, Boston, 1996.
14. Brinker, C.J., *Sol-Gel Science*, Academic Press, Inc., San Diego, 1990.

PROBLEMS

1. One literature source lists the following shear strengths (in units of kg/cm^2) for some materials: lithium stearate, 6.0; zinc stearate, 9.3 to 20.2; palmitic acid, 12.3 to 12.7; stearic acid, 13.3 to 13.7; hard paraffin, 19.0; potassium stearate, 31.3; synthetic wax, 33.9; talc, 63.2 to 80.0; boric acid, 73.0; and graphite, 75.0. You require an effective lubricant for a pressing operation, but do not want a contaminating residue after final firing. Which of the above materials would be the best lubricant candidate?

2. Which technique is most likely to provide the highest rate of slip casting? Explain.

3. Ceramic powder batches are often evaluated by pressing small test bars for densification trials, strength, or other property measurement and examination of microstructure. Assume you are given a simple pressing die with a cavity 5-mm (0.197-in.) wide and 50-mm (1.97-in.) long and are told to press the powder at 15,000 psi. The gauge on your press reads in tons. How many tons of pressure do you apply?

4. You wish to prepare a tape-casting slip containing 50 vol% Al$_2$O$_3$ and 50 vol% polyvinyl butyral binder (PVB). Assuming that the density of Al$_2$O$_3$ is 3.98 g/cm^3 and that of PVB is 1.08 g/cm^3, how many grams of PVB are required to produce 1000 g of tape?

5. Recommend and discuss the most cost-effective forming process to achieve each of the following shapes:

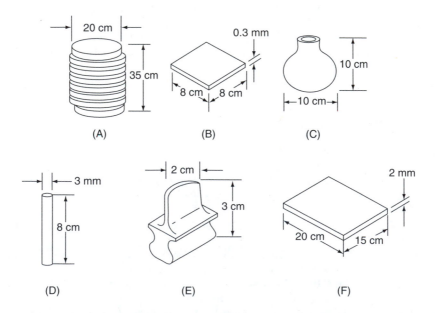

6. The deformation and rebound characteristics of a binder material are very important. The binder must have adequate flow for spray-dried granules to deform and weld together during pressing, yet must not rebound enough to cause cracking. Figure 13.6 showed the deformation behavior of an alkyd organic binder as a function of temperature. Which one of

the following temperatures would you expect to result in good pressing and die-removal characteristics for the alkyd binder? Why?

(a) $-10°C$ ($-14°F$)
(b) 0°C (30°F)
(c) 40°C (105°F)
(d) 80°C (175°F)

7. Which of the following does *not* have thermoplastic characteristics?
    (a) epoxy
    (b) polyethylene
    (c) cellulose acetate
    (d) polystyrene

8. Which of the following is likely to exhibit the least wear during green machining?
    (a) cubic boron nitride
    (b) Co-bonded tungsten carbide
    (c) diamond
    (d) tool steel

9. Which of the following is *least* likely to contribute to knit lines during injection molding?
    (a) jetting
    (b) plug flow
    (c) freeze before welding
    (d) none of the above

10. An injection-molding mix contains 19 wt% paraffin and 81 wt% Al_2O_3. What is the vol% paraffin? (Assume the following densities: paraffin = 0.91 g/cm³ and Al_2O_3 = 3.98 g/cm³.)

11. Which of the following is *not* important in achieving a straight, uniform-density extruded part?
    (a) tool design
    (b) rheology of the mix
    (c) wet strength after extrusion
    (d) all are important

STUDY GUIDE

1. What are the purposes of the additives to a powder that is to be used to form a ceramic part by pressing?
2. What are some tradeoffs in the use of a "hard" vs. a "soft" binder?
3. What factors can affect the hardness and deformation characteristics of a binder, and why is this a concern?
4. Why is "glassy" deformation behavior of a binder not favorable for compacting ceramics by pressing?
5. Why is "rubbery" deformation behavior not favorable?
6. Achieving high green density during pressing helps to achieve a fully dense fine grain size ceramic later during densification (sintering). Describe the difference in green density observed by Nies and Messing based upon pressure and the amount of plasticizer added.
7. How do lubricants and compaction aids work?
8. What benefits do they provide?
9. What are some examples of lubricants and compaction aids?
10. "Dry pressing" is used extensively for low cost production of large quantities of simple ceramic parts in automated presses. Identify typical parameters for dry pressing, and explain what happens to the powder during dry pressing.
11. What are some potential causes of cracking during uniaxial pressing?
12. Explain potential sources of nonuniform density during uniaxial pressing.

13. What is "isostatic pressing", and what are the benefits?
14. Compare wet-bag and dry-bag isostatic pressing.
15. What are some important ceramic products manufactured by dry-bag isostatic pressing?
16. Briefly describe how a ceramic part is made using slip casting.
17. What are some of the preconsolidation steps and controls that are critical to achieve a high quality part by slip casting?
18. What is a dispersant, and why is it important for slip casting?
19. Compare "electrostatic repulsion" and "steric hindrance" in achieving dispersion of ceramic particles in a slip.
20. What is the most common mold material for slip casting?
21. Casting can be done in a variety of ways. List some of these options.
22. How can slip casting achieve complex shapes?
23. "Reticulated foam" porous but strong ceramic structures have become very important for molten metal filters and for kiln furniture. How are these produced by slip casting?
24. How does "gel casting" differ from other forms of casting a slip?
25. What types of defects might occur during slip casting that would make the completed part unsuitable for use?
26. What is tape casting, and how does it differ from slip casting?
27. What are some important applications of tape casting?
28. What is extrusion, and what are the shape limitations?
29. What are some important factors for successful extrusion?
30. What are some common extrusion defects?
31. Identify some important ceramic products fabricated by extrusion.
32. How does injection molding differ from extrusion?
33. How much total binder is required for successful injection molding?
34. What are the purposes of the major binder?
35. What are the major purposes of the minor binder?
36. What other additives are often required in an injection molding mix?
37. Binder removal is a very critical step in forming a ceramic by injection molding. Discuss various mechanisms of binder removal.
38. Describe the most common defects encountered for injection molding.
39. What is "green machining"?
40. What are the benefits of green machining as an alternative to machining after the ceramic has been densified?

14 Densification

In Chapters 12 and 13 we discussed the criteria and techniques for selecting and processing ceramic powders and for forming these powders into shaped particulate compacts. In this chapter we explore the processes for densifying these particulate compacts into strong, useful ceramic components.

14.1 THEORY OF SINTERING

The densification of a particulate ceramic compact is technically referred to as sintering. Sintering is essentially a removal of the pores between the starting particles (accompanied by shrinkage of the component), combined with growth together and strong bonding between adjacent particles.[1-5] The following criteria must be met before sintering can occur:

1. A mechanism for material transport must be present.
2. A source of energy to activate and sustain this material transport must be present.

The primary mechanisms for transport are diffusion and viscous flow. Heat is the primary source of energy, in conjunction with energy gradients due to particle-particle contact and surface tension.

Although ceramic materials have been used and densified for centuries, scientific understanding and control of sintering have only developed during the past 60 to 70 years. Early controlled experiments were conducted by Muller in 1935.[6] He sintered compacts of NaCl powder for a variety of times at several temperatures and evaluated the degree of sintering by measuring the fracture strength.

Much progress in our understanding of densification has been achieved since 1935. Now sintering is studied by plotting density or shrinkage data as a function of time and by actual examination of the microstructure at various stages of sintering using scanning electron microscopy, transmission electron microscopy, and lattice imaging.

14.1.1 STAGES OF SINTERING

Sintering is often thought of in stages according to the sequence in physical changes that occur as the particles bond together and the porosity disappears. Although this is a simplified approach, it does allow the student to visualize the changes and mechanisms. The stages and primary physical changes that occur in each stage are listed in Table 14.1.

The initial stage involves rearrangement of particles and initial neck formation at the contact point between each particle. The rearrangement consists of slight movement or rotation of adjacent particles to increase the number of points of contact.

Bonding occurs at the points of contact where material transport can occur and where surface energy is highest. The changes that occur during the first stage of sintering are illustrated in Figure 14.1.[7]

The second stage of sintering is referred to as intermediate sintering. The physical changes that occur during second-stage sintering are illustrated in Figure 14.2 the size of the necks between particles grows. Porosity decreases and the centers of the original particles move closer together. This results in shrinkage equivalent to the amount of porosity decrease. The grain boundaries begin to move so that one particle (now called a grain) begins to grow while the adjacent grain is consumed. This allows geometry changes that are necessary to accommodate further neck growth and removal

---

TABLE 14.1
Stages of Sintering

1st Stage (Initial)
  Rearrangement
  Neck formation

2nd Stage (Intermediate)
  Neck growth
  Grain growth
  High shrinkage
  Pore phase continuous

3rd Stage (Final)
  Much grain growth
  Discontinuous pore phase
  Grain boundary pores eliminated

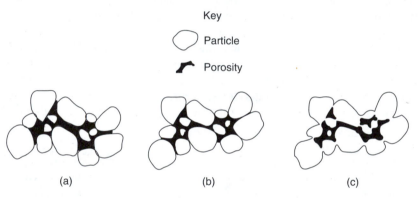

FIGURE 14.1 Changes that occur during the initial stage of sintering. (a) Starting particles, (b) rearrangement, and (c) neck formation. (Drawings courtesy of ASM International.)

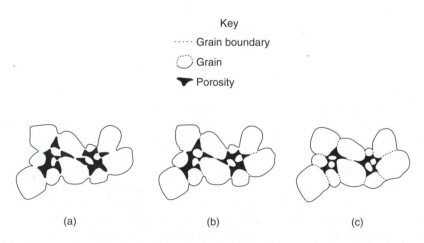

FIGURE 14.2 Changes that occur during the second stage of sintering. (a) Neck growth and volume shrinkage, (b) lengthening of grain boundaries, and (c) continued neck growth and grain boundary lengthening, volume shrinkage, and grain growth. (Drawings courtesy of ASM International.)

of porosity. Intermediate sintering continues as long as pore channels are interconnected and ends when pores become isolated. Most of the shrinkage during sintering occurs during second-stage sintering.

The third stage of sintering is referred to as final sintering.[8,9] It involves the final removal of porosity. The porosity is removed by vacancy diffusion along grain boundaries. Therefore, the pores must remain close to the grain boundaries. Pore removal and vacancy diffusion are aided by movement of grain boundaries and controlled grain growth. However, if grain growth is too rapid, the grain boundaries can move faster than the pores and leave them isolated inside a grain. As the grain continues to grow, the pore becomes further separated from the grain boundary and has decreased chance of being eliminated. Therefore, grain growth must be controlled to achieve maximum removal of porosity.

Grain growth is driven by surface energy. The forces of nature reduce surface area to a minimum to minimize surface free energy. That is why drops of water or mercury form into spheres. The same energy relationships exist in a solid material during sintering. Curved grain boundaries move in such a way that they gain a larger radius of curvature, that is, straighten out. This can only be accommodated by growth of the grains. Smaller grains have a smaller radius of curvature and more driving energy to move, change shape, and even to be consumed by larger grains. The physical changes that occur during the final stage of sintering are illustrated in Figure 14.3. The final distribution of grains and pores is referred to as microstructure.

The following sections describe in more detail the mechanisms of sintering.

14.1.2 MECHANISMS OF SINTERING

Sintering can occur by a variety of mechanisms, as summarized in Table 14.2. Each mechanism can work alone or in combination with other mechanisms to achieve densification.

14.1.2.1 Vapor-Phase Sintering

Vapor-phase sintering is important in only a few material systems and is discussed only briefly. The driving energy is the difference in vapor pressure as a function of surface curvature. As illustrated in Figure 14.4, material is transported from the surface of the particles, which have a positive radius of curvature and a relatively high vapor pressure, to the contact region between particles, which has

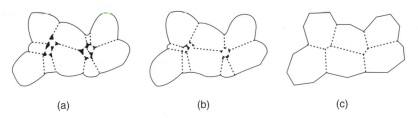

     (a)   (b)   (c)

FIGURE 14.3 Changes that occur during the final stage of sintering. (a) Grain growth with discontinuous pore phase, (b) grain growth with porosity reduction, and (c) grain growth with porosity reduction, and (c) grain growth with porosity elimination. (Drawing courtesy of ASM International.)

TABLE 14.2
Sintering Mechanisms

Type of Sintering	Material Transport Mechanism	Driving Energy
Vapor-phase	Evaporation-condensation	Differences in vapor pressure
Solid-state	Diffusion	Differences in free energy or chemical potential
Liquid-phase	Viscous flow, diffusion	Capillary pressure, surface tension
Reactive liquid	Viscous flow, solution-precipitation	Capillary pressure, surface tension

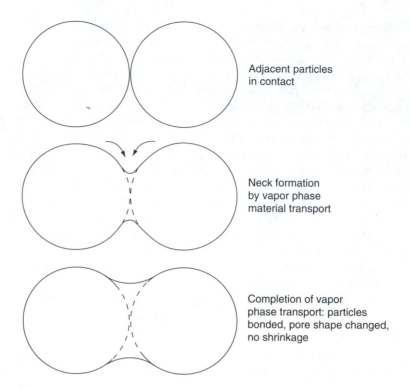

FIGURE 14.4   Schematic of vapor-phase material transport.

TABLE 14.3
Effect of Particle Size or Surface Curvature on the Pressure Difference and Relative Vapor Pressure across a Curved Surface

| Material | Surface Diameter (μm) | Pressure Difference | | Relative Vapor Pressure (P/P_0) |
		MPa	psi	
Liquid water at 25°C	0.1	2.8	418	1.02
	1.0	0.28	41.8	1.002
	10.0	0.03	4.2	1.0002
Liquid cobalt at 1450°C	0.1	67.3	9750	1.02
	1.0	6.7	975	1.002
	10.0	0.67	97.5	1.0002
Silica glass at 1700°C	0.1	12.1	1750	1.02
	1.0	1.2	175	1.002
	10.0	0.12	17.5	1.0002
Solid Al_2O_3	0.1	36.2	5250	1.02
	1.0	3.6	525	1.002
	10.0	0.36	52.5	1.0002

Source: From Kingery, W.D., Bowen, H.K., and Uhlmann, D.R., *Introduction to Ceramics*, Wiley-Interscience, New York, 1976.

---

a negative radius of curvature and a much lower vapor pressure. The smaller the particles, the greater the positive radius of curvature and the greater the driving force for vapor-phase transport. Table 14.3 shows how large an effect particle size or surface curvature can have on pressure across the curved surface and on relative vapor pressure.

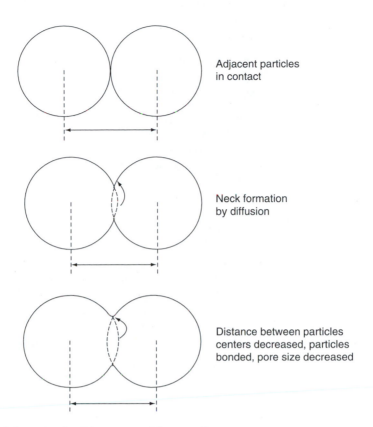

FIGURE 14.5   Schematic of solid-state material transport.

Vapor-phase transport changes the shape of the pores and achieves bonding between adjacent particles and thus increases the material strength and decreases permeability due to open porosity. However, it does not result in shrinkage and cannot produce densification. It must be accompanied by other mechanisms that provide bulk material transport or transport of pores to external surfaces.

14.1.2.2   Solid-State Sintering

Solid-state sintering involves material transport by volume diffusion as illustrated in Figure 14.5. Diffusion can consist of movement of atoms or vacancies along a surface or grain boundary or through the volume of the material. Surface diffusion, like vapor-phase transport, does not result in shrinkage. Volume diffusion, whether along grain boundaries or through lattice dislocations, does result in shrinkage.[10]

The driving force for solid-state sintering is the difference in free energy or chemical potential between the free surfaces of particles and the points of contact between adjacent particles. The general form[11] of the model of the mechanism of transport of material by lattice diffusion from the line of contact between two particles to the neck region is

$$\frac{\Delta L}{L_0} = \left(\frac{K\gamma a^3 D^* t}{kTd^n} \right)^m \tag{14.1}$$

where
$\Delta L/L_0$ = linear shrinkage (equivalent to the sintering rate)
γ = surface energy
a^3 = atomic volume of the diffusing vacancy
D^* = self-diffusion coefficient

k = Boltzmann constant

T = temperature

d = particle diameter (if we assume equal-size spherical starting particles)

t = time

K = constant dependent on geometry

The exponent n is typically close to 3 and the exponent m is generally in the range of 0.3 to 0.5.

The mathematical model agrees favorably with data for the initial stage of sintering. However, once grain growth starts, more complex models are required.

Examination of Equation (14.1) indicates that particle diameter has a major effect on the rate of sintering. The smaller the particles, the greater the rate. Although not obvious by examination of the equation, temperature also has a major effect. This is due to the exponential relationship of temperature to the diffusion coefficient.

Figure 14.6a illustrates the effects of temperature and time. Figure 14.6b shows a log-log plot of the same data. The slope of the log $\Delta L/L_0$ vs. log t line is approximately two fifths for solid-state sintering.

It is apparent from examination of Equation (14.1) and Figure 14.6 that control of temperature and particle size is extremely important, but that control of time is less important.

Finer-particle-size powder can be sintered more rapidly and at a lower temperature than coarser powder. Not apparent in the equation, but highly important to the final properties, are the uniformity of particle packing, the particle shape, and the particle size distribution.[12,13] If particle packing is not uniform in the greenware, it will be very difficult to eliminate all the porosity during sintering.

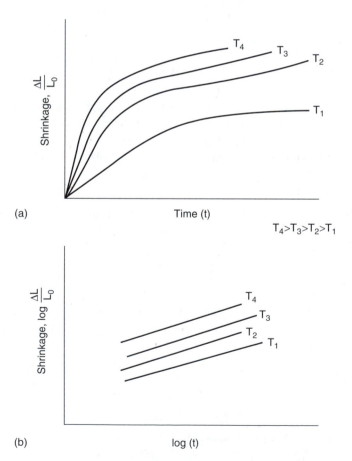

FIGURE 14.6 Typical sintering rate curves showing the effects of temperature and time.

Agglomerates are a common source of nonuniformity, as discussed in prior chapters. Nonuniformity can also result during shape forming due to gas entrapment, particle segregation (i.e., settling during slip casting), lamination, and fold lines (injection molding).

Particle shape can also be important. Too high a concentration of elongated or flattened particles can result in bridging during forming, producing a large or irregularly shaped pore that is difficult to remove during sintering.

Particle size distribution is also critical. Particles that are all of one size are difficult to pack efficiently; they form compacts with large pores and a high volume percentage of porosity. Unless very uniform close packing is achieved during compacting and grain growth occurs during densification, such compacts will undergo a high percentage of shrinkage and yet will retain significant porosity. For very fine particles (500 Å) this may be acceptable and may result in very uniform properties. However, commonly available powder has a range of particle sizes from submicron upward. Better overall packing can be achieved during compaction, but isolated pores due to bridging and agglomerates are usually quite large and result either in porosity or large grain size after sintering.

Another concern with solid-state sintering is exaggerated grain growth. Sometimes grains grow so rapidly that pores are trapped within the grains rather than moving along the grain boundaries until the pores are eliminated. This happens in pure Al_2O_3, as is illustrated in Figure 14.7 and Figure 14.8. Figure 14.7 shows how a single grain of Al_2O_3 has grown rapidly to consume surrounding grains and to trap porosity. Figure 14.8 shows a different Al_2O_3 sample in which the complete microstructure consists of large grains with porosity trapped within the grains. Porosity trapped within a grain is not easily removed during further sintering. Exaggerated grain growth has been avoided in Al_2O_3 by the addition of 0.25 wt% MgO. The resulting microstructure is depicted in Figure 14.9a. The presence of the MgO has slowed the rate of motion of the grain boundaries to allow the pores to remain on the grain boundaries until all porosity has been eliminated. The resulting Al_2O_3 has no internal pores to

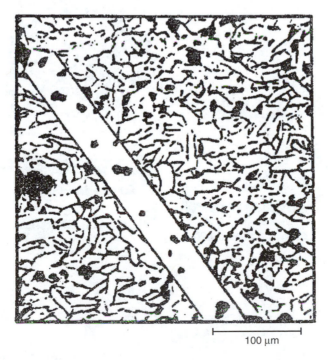

100 μm

FIGURE 14.7 Example of exaggerated grain growth in sintered Al_2O_3 involving a single grain. (Line drawing after a microstructure from Coble, R. and Burke, J., Sintering in ceramics, in *Progress in Ceramic Science*, Vol. 3, Pergamon Press, New York, 1963.)

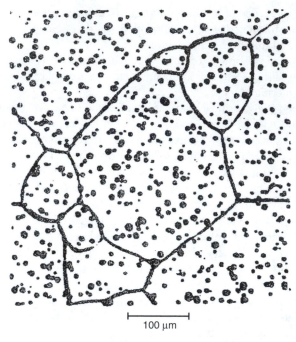

100 μm

FIGURE 14.8   Example of exaggerated grain growth in sintered Al_2O_3 where all the grains have grown large and trapped porosity within the grains. (Line drawing after a microstructure from Coble, R. and Burke, J., Sintering in ceramics, in *Progress in Ceramic Science*, Vol. 3, Pergamon Press, New York, 1963.)

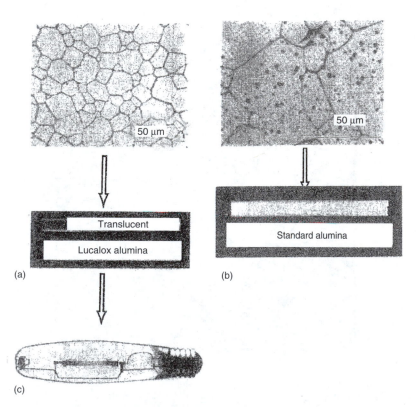

FIGURE 14.9   Comparison of the microstructure and translucency of relatively pore-free Al_2O_3 (a) with that of opaque Al_2O_3 containing pores trapped in grains. (b) Translucent Al_2O_3 tubes are used in sodium vapor lamps that provide energy efficient street lights. (Courtesy of General Electric.)

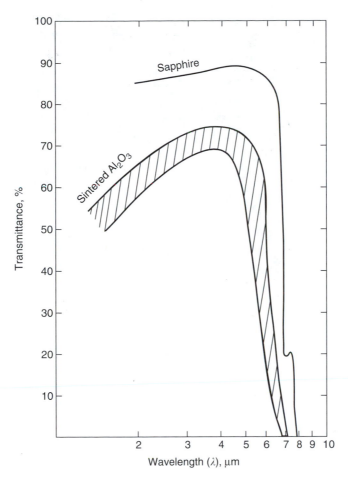

FIGURE 14.10 High degree of transparency of polycrystalline Al_2O_3 when sintered in such a way that all the porosity is eliminated. (After Amato, I., et al., *Mat. Sci. and Eng.*, 26(1), 75, 1976.)

scatter incident light and is nearly transparent (Figure 14.10). Such Al_2O_3 is used in large quantity for sodium-vapor lamps for street lighting and for translucent brackets for orthodontic braces.

Many ceramics besides Al_2O_3 have been densified by solid-state sintering, especially the relatively pure oxides. Examples include BeO, Y_2O_3, UO_2, ThO_2, ZrO_2, and doped ZrO_2. Doping can increase the number of point defects in the material and increases the rate of diffusion, thus enhancing solid-state sintering. SiC with the addition of B and C is also thought to densify by solid-state sintering. The C apparently removes SiO_2 from the surface of the SiC particles. The B has limited solid solubility in the SiC and allows a mechanism of material transfer between adjacent grains. Pure SiC particles can bond together, but not densify (no shrinkage and no removal of interparticle porosity). Figure 14.11 and Figure 14.12 illustrate SiC ceramics sintered with additions of B and C.

14.1.2.3 Liquid-Phase Sintering

Liquid-phase sintering involves the presence of a viscous liquid at the sintering temperature and is the primary densification mechanism for most silicate systems. Three factors control the rate of liquid-phase sintering:

1. particle size
2. viscosity
3. surface tension

FIGURE 14.11   Variety of SiC parts fabricated by solid-state sintering using beta-SiC powder with additions of B and C. (Photo courtesy of H. Yamauchi, Ibiden CO., Ltd., Ogaki, Japan.)

FIGURE 14.12   SiC gas-turbine components fabricated using alpha-SiC powder plus B and C sintering aids. (Photo courtesy of Carborundum Company for parts fabricated for Garrett Auxiliary Power Division of Allied-Signal Aerospace during the 1980s under sponsorship of the U.S. Department of Energy and administration of NASA-Lewis Research Center.)

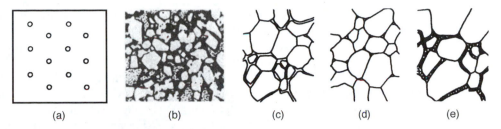

FIGURE 14.13 Schematics of types of microstructures that result from liquid-phase sintering. (a) Sintered glass, (b) crystal-liquid structure, (c) grain boundary glass, (d) glass at triple points, and (e) crystallized grain boundary. (Drawings courtesy of ASM International.)

The viscosity and surface tension are affected strongly by composition and temperatures.

Liquid-phase sintering occurs most readily when the liquid thoroughly wets the solid particles at the sintering temperature. The liquid in the narrow channels between the particles results in substantial capillary pressure,[14] which aids densification by several mechanisms:

rearranges the particles to achieve better packing
increases the contact pressure between particles, which increases the rate of material transfer by solution and precipitation, creep and plastic deformation, vapor transport, and grain growth

The magnitude of capillary pressures produced by silicate liquids can be greater than 7 MPa (1015 psi). Smaller particles result in higher capillary pressure and also have higher surface energy due to the small radius of curvature and thus have more driving energy for densification than coarser particles. Materials requiring high strength and minimum porosity are generally processed from powders having an average particle size less than 1 μm (0.00004 in.) and a surface area greater than 5 m²/g.

The rate of liquid-phase sintering is also strongly affected by temperature. For most compositions a small increase in temperature results in a substantial increase in the amount of liquid present. In some cases, this can be beneficial by increasing the rate of densification. In other cases, it can be detrimental by causing excessive grain growth (which reduces strength) or by allowing the part to slump and deform. The amount of liquid present at a selected temperature can be predicted with the use of phase equilibrium diagrams, as was discussed in detail in Chapter 6.

A range of microstructures can occur due to liquid-phase sintering.[15] Major factors that lead to microstructure variations include the particle size, the amount of liquid present at the sintering temperature, and the cooling cycle. Some examples of microstructure variations are shown schematically in Figure 14.13. Figure 14.13a might result from sintering of glass particles. Note that no grains or grain boundaries are present, but only a few remaining spherical pores. Figure 14.13b might result from densification of a composition containing a large amount of liquid and cooled rapidly enough for the liquid to solidify as a glass. Figure 14.13c and d illustrate similar situations, except with less liquid and less residual glass. In Figure 14.13c the glass is a continuous phase coating the grain boundaries and triple-point junctions between grains. In Figure 14.13d, where very little liquid was present during sintering, glass is only visible at the triple points by optical microscopy. Examination at high magnification by transmission electron microscopy would normally show a thin layer (10 to 50 nm) of glass coating the grain boundaries. As discussed in prior chapters, glass layers at grain boundaries have a major effect on the high-temperature strength, creep resistance, and stress rupture life of the material. Crystallization of the grain boundary glass improves these properties. Figure 14.13e illustrates a liquid-phase-sintered material where the liquid crystallized during cooling or by a subsequent heat treatment.

Porcelain compositions (such as compositions in the $K_2O–Al_2O_3–SiO_2$ phase diagram) are important traditional ceramics and are densified by liquid-phase sintering. Porcelain ceramic applications include china, sinks, porcelain enamels, insulators, and many ceramics used in electronics. A common porcelain body is fabricated from about 50% kaolin clay (~45% Al_2O_3, 55% SiO_2), 25% potash-feldspar, and 25% SiO_2. Enough liquid is present in this system above about 1200°C (2190°F) to achieve liquid-phase sintering. As a review, look up the $K_2O–Al_2O_3–SiO_2$ phase equilibrium diagram and draw isothermal sections at 1200, 1300, and 1400°C (2190, 2370, and 2550°F). Note the extent of the liquid plus solid regions and how they increase as the temperature increases. Increased amounts of liquid phase increase the rate of sintering.

Si_3N_4-based compositions represent an advanced family of ceramics that are densified by liquid-phase sintering. Additives are required to achieve a liquid phase. Examples of additives that have been used successfully include MgO, Al_2O_3, Y_2O_3, and rare earth oxides. These react with the SiO_2 present on the surface of the Si_3N_4 particles to form a silicate liquid. The Si_3N_4 particles dissolve in the liquid and precipitate as β-Si_3N_4 crystals. If the starting powder is α-Si_3N_4 and if the time and temperature cycle is properly selected, the β-Si_3N_4 will crystallize as elongated single crystals that impart unusually high fracture toughness to the Si_3N_4. Figure 14.14 shows the microstructures of a $Si_3N_4–Y_2O_3–Al_2O_3$ composition sintered at different temperatures. 1600°C (2910°F) was too low to achieve full densification. Much porosity is present and the original Si_3N_4 particles are still visible and of the α phase. 1750°C (3180°F) resulted in nearly complete densification and conversion of the Si_3N_4 to elongated β-Si_3N_4. 1850°C (3360°F) resulted in substantial grain growth.

14.1.2.4 Reactive Liquid Sintering

Reactive liquid sintering is also referred to as transient liquid sintering. A liquid is present during sintering to provide the same types of densification driving forces as discussed for liquid-phase sintering, but the liquid either changes composition or disappears as the sintering process progresses or after it is completed. Since the liquid phase is consumed in the reaction, the resulting material can have extremely good high-temperature properties and in some cases can even be used at temperatures above the sintering temperature.

One means of achieving reactive liquid sintering is to select starting powders or additives that go through a series of chemical combinations or reactions before the final stable compound is formed, with one or more of the intermediate compounds being liquid and the final compound being solid. Another is to use starting powders that will form a solid solution at equilibrium but will pass through a liquid stage before equilibrium is reached. A third approach (which is not really reactive liquid sintering, but gives the same results) is to liquid-phase sinter, cool to yield a glass at the grain boundaries, and then heat-treat to crystallize the glass.

An example of reactive liquid sintering has been demonstrated by using the SiC–Al_2OC system.[16] The objective was to sinter a SiC-based material at reduced temperature, yet retain good high-temperature properties. SiC (with B plus C additions) requires a temperature of about 2150°C (3900°F) and a sintering time at temperature of at least 30 min. It was hypothesized that sintering could be accomplished below 2000°C (3630°F) by using the eutectic between Al_2O_3 and Al_4O_4C in the $Al_2O_3–Al_4C_3$ system. The binary phase equilibrium diagram is shown in Figure 14.15. The eutectic is at about 1850°C (3360°F). It was further hypothesized that reactants could be selected that would initially form a liquid phase due to the eutectic and end up as the composition Al_2OC. Al_2OC has the same crystal structure as β-SiC and can form a nearly complete solid solution with SiC. If Al_2OC could be formed and taken into solid solution, the final material would contain no glassy grain boundary phase and would likely have excellent high-temperature properties. Full densification was achieved after about a 10-min hold at 2000°C (3630°F). Transmission electron microscopy (using high-resolution lattice imaging) determined that no glassy phase was present at the grain boundaries. This was verified by high-temperature creep tests.

FIGURE 14.14 Si$_3$N$_4$ containing Y$_2$O$_3$ plus Al$_2$O$_3$ sintering aids; sintered in nitrogen at (a) 1600°C (2910°F), (b) 1750°C (3180°F), and (c) 1850°C (3360°F). (From unpublished work of D.W. Richerson and I. Aksay.)

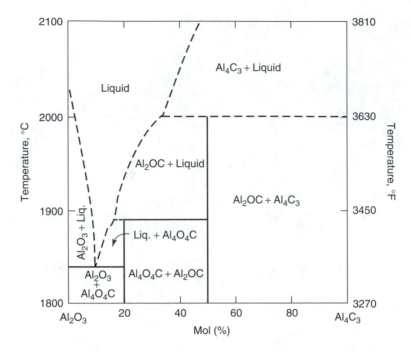

FIGURE 14.15 Phase equilibrium diagram of the system $Al_2O_3-Al_4C_3$ showing the eutectic between Al_2O_3 and Al_4O_4C at about 1830°C (3225°F). (From Foster, L.M., Lang, G., and Hurter, M.S., *J. Am. Ceram. Soc.*, 39(1), 8, 1956. With permission.)

14.1.3 CONTROL OF CONVENTIONAL SINTERING

All the sintering techniques discussed so far are referred to as conventional sintering or pressure-less sintering. Densification is controlled by composition, particle size distribution, temperature, and time at temperature. No external pressure is applied during densification. We discuss the influence of applied pressure later in this chapter.

Many other factors besides composition, particle size, temperature, and time must be controlled. These factors include atmosphere, time and temperature cycle, design of the furnace, material of construction of the furnace and heating elements, settering,* and all processing steps of the powder compact prior to sintering. Some of these factors are discussed in the following paragraphs.

14.1.3.1 Atmosphere

The atmosphere (gas composition) in the furnace can have obvious or subtle effects on sintering. Obvious effects involve oxidation-reduction conditions. Trying to sinter Si_3N_4, SiC, or other nonoxide compositions in air will result in oxidation. Conversely, trying to sinter oxide ceramics in a carbonaceous or a reducing atmosphere can lead to reduction. For example, ZrO_2 fired in a reducing atmosphere will be oxygen-deficient (ZrO_{2-x}) and will be dark gray or black in color. Residual binder can cause localized reducing conditions and cause sintering to be retarded or accelerated. Thermodynamic calculations generally can be made quickly to assess to potential for adverse interactions between the atmosphere, the powder compact, and additives present in the powder.

---

* Settering refers to the way the part being sintered is supported in the furnace.

Atmosphere can also have subtle effects. The effect on dopants is an example. Many dopants such as Fe, Ce, and Mn can exist in more than one valence (Fe^{3+}, Fe^{2+}, etc.). The atmosphere during sintering must be carefully controlled to achieve the desired valence and desired properties.

14.1.3.2 Time and Temperature Cycle

Time and temperature cycle refers to the rate of heating, peak temperature, the time at peak temperature, and the rate of cooling. Time at temperature and peak temperature influence the total densification achieved and the degree of grain growth. Cooling rate affects the amounts of residual glass in the microstructure. In the case of Mg–PSZ and Ca–PSZ (magnesia and calcia partially stabilized zirconia), the cooling rate and time at selected aging temperatures determine the nature of the strength- and toughness-controlling tetragonal or monoclinic ZrO_2 precipitates. Finally, the heating and cooling rates can affect the mechanical integrity of the part being sintered. Too-rapid heat-up or cool-down can cause cracks due to thermal shock. Too-rapid-heating can also cause cracks due to too-rapid burn-off of binders or other organic additives.

Time and temperature cycle and atmosphere must be considered simultaneously for many materials. An example is yttrium barium cuprate superconductor. The material must have the fully oxygenated composition $YBa_2Cu_3O_7$ to have optimum superconductivity.[17] $YBa_2Cu_3O_7$ requires a sintering temperature above about 900°C to achieve full densification. However, an oxygen-deficient tetragonal form is stable above 710°C. An oxygen-deficient structure is retained during normal cooling. Lengthy heat treatment below 710°C is necessary to achieve a superconductive stoichiometry near $YBa_2Cu_3O_7$. For example, one study determined that a 48-h heat treatment at 600°C was necessary to fully oxygenate a 1-mm thick slice of material.

14.1.3.3 Design of the Furnace

Proper furnace design is critical to achieve optimum sintering. Key concerns include temperature capability, temperature control, temperature uniformity, and atmosphere control. Some materials require that the temperature be held within a 10°C (18°F) range. Excess grain growth, deformation, or change in composition can occur if the furnace temperature wanders above the range. Incomplete densification can occur if the furnace temperature is below the desired range. Nonuniform temperature in the furnace can result in incomplete densification in part of the ceramic and over firing in another part, both of which can leave strength-limiting flaws. The differential shrinkage can show up as distortion such that the part is rejected by not meeting dimensional tolerances.

Figure 14.16 shows schematic cross sections of several different types of furnaces or kilns. Figure 14.16a depicts a simple resistance-heated box furnace. It consists of a chamber lined with refractory brick or fiber board. The heating elements can be horizontal rods (such as SiC "hot rods" or "globars") or can be hung vertically (such as $MoSi_2$ "Superkanthal"). Heating of a box furnace can also be achieved by combustion of natural gas. Box furnaces are generally batch-fired and the atmosphere is air or combustion-gas-modified air.

Sometimes it is desirable to sinter a material in a controlled atmosphere. A controlled atmosphere is difficult to achieve in a conventional refractory-lined box furnace. One option is to enclose the parts to be sintered in a retort (a sealed container) inserted in the chamber of the box furnace. Another option is to use a tube furnace or muffle furnace such as depicted in Figure 14.16b. This furnace consists of an impermeable ceramic tube (or other cross section) inserted through a box furnace. The ends of the tube extend beyond the walls of the furnace and are cool enough that they can be sealed, with gas manifolds to allow selected gases to be flowed into the tube. Tubes or muffles have been constructed of fused SiO_2, mullite, cordierite, Al_2O_3, SiC, stabilized ZrO_2, and other materials. Muffle furnaces are typically batch-fired, but can be adapted for continuous firing.

Continuous furnaces are generally most cost-effective for production sintering of ceramics. The key elements of a continuous furnace are illustrated in the tunnel kiln depicted in Figure 14.16c.

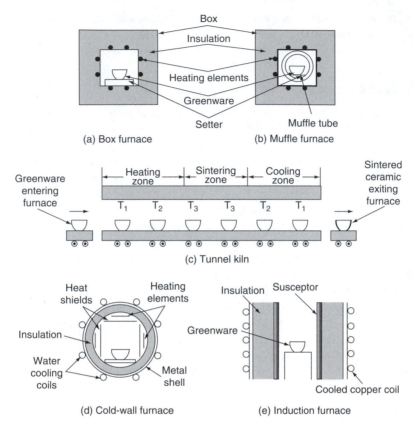

FIGURE 14.16 Examples of some of the types of furnaces used for sintering of ceramics. (Drawings courtesy of ASM International.)

These include (1) a long, refractory-lined furnace chamber; (2) a means of heating different portions of the chamber to different temperatures (to allow slow heat-up, adequate hold at sintering temperature, and slow cool-down); and (3) a conveyor system to move the ceramic parts through the furnace. Continuous furnaces can be heated by combustion or by electrical resistance and can operate with an air atmosphere or even a controlled atmosphere. For example, Al_2O_3-based multilayer ceramic packages for integrated circuits are typically fired in a continuous furnace in a hydrogen atmosphere containing a controlled dew point of H_2O. The controlled dew point allows enough oxygen at low temperatures for organic binders to be burned off. The hydrogen provides a reducing atmosphere at high temperature to keep the tungsten or molybdenum metallization from oxidizing.

Many other furnace designs exist, each with features tuned to the requirements of specific needs.[18] Figure 14.16d depicts a cold-wall furnace. It can be operated at very high temperatures ($>2000°C$ or $3630°F$), with relatively pure atmospheres, and at greater than atmospheric pressure. The furnace is enclosed in a metal container that looks like a vacuum chamber or autoclave chamber. Water-cooling tubes are attached to the outer surface of the metal. A layer of thermal insulation generally lines the interior of the metal. Plate- or ribbon-shaped heating elements (usually of a refractory metal or graphite) are mounted inside the furnace and are heated electrically.

Figure 14.16e depicts an induction furnace. The key elements of the furnace are a water-cooled copper coil and a semiconductive susceptor separated by electrical and thermal insulation. A high-frequency electrical field is applied through the coil. This produces a magnetic field that induces an electrical current in the susceptor, resulting in rapid resistance heating of the susceptor.

Table 14.4 briefly describes examples of designs of high-temperature furnaces that were available in 1990. These furnaces were typically computer-automated and were reduced energy consumption

TABLE 14.4
Examples of Some Variations in High-Temperature Furnaces[a]

Manufacturer	Type of Furnace	Application	Maximum Temperature (°C)	Special Features
Tokai Konetsu Kogyo Co., Tokyo, Japan	Multiple-control sintering furnace	Small volumes, advanced ceramics	Up to 1700	3 firing zones in 1, programmable control with up to 20 steps per program
	Pusher-type tunnel	Sintering of advanced ceramics	2100	High-purity graphite heaters and linings, automatic transfer system
	Batch-type vacuum	Same as pusher-type	2300	Fast heat-up and cool-down, plate-type graphite heaters
	Vertical and horizontal tubes	Research and development	3000	Can be equipped for measuring viscosity of oxide melts
Thermcraft, Inc., Winston-Salem, N.C.	Box furnace	Laboratory and research	Up to 1700	Pneumatically generated vertical lift doors
L&L Special Furnace Co., Aston, PA.	Car elevator	Ceramics, composites batch production	1700	Ceramic fiber insulation, easy loading, good uniformity
BTU International, Billerica, Mass.	Pusher furnace	Co-firing ceramics, Al_2O_3, AIN, electronic components	2200	Single or multi-chamber
Centorr Furnace, Suncook, N.H.	Belt furnace	Controlled atmosphere processing of materials	2200	Friction-driven belt, sight port with rotatable viewing glass
Thermal Technology, Inc., Concord, N.H.	Vacuum	Joining and heat treating of materials	2500	Fully automatic operation
Advanced Vacuum Systems, Ayer, Mass.	Vacuum (horizontal resistance)	Annealing, sintering, brazing	2500	Full access at both ends of chamber
Deltech, Inc., Denver, Colo.	Vertical tube or bottom-loading chamber	Sintering research, advanced ceramics	2000 (air atmosphere)	Zirconia heating elements
CM Furnaces, Inc., Bloomfield, N.J.	Front-loading laboratory	Same as Deltech	2000 (air atmosphere)	Rapid heat-up, fully automated, zirconia insulation, and heating elements
Vacuum Industries, Somerville, Mass.	Rotary drum laboratory vacuum batch-type horizontal	Ceramic powder processing and sintering, cofiring of Al_2O_3, AIN, BeO	2300 1900(H_2) 2200(N)	Fully programmable temperature and dewpoint controls, rectangular hotzone cross section

[a] *Note*: Based on manufacturer's data. This is not a complete list.

Source: From Sheppard, L.M., Firing technology heats up for the 90s, *Am. Ceram. Soc. Bull.*, 69(10), 1674–1689, 1990.

TABLE 14.5

Comparison of a Countertravel Kiln and a Tunnel Kiln

	Countertravel Kiln	Tunnel Kiln
Effective capacity (kg bricks/h)	4167	4167
Gross fuel consumption (kcal/h)[a]	833,333	1,250,000
Flue gas volume (m_n^3/h)	1042	8334
Hydrocarbons in the exhaust (kg/h)	0	8334
Concentration (mg/m_n^3/h)	0	1000
CO_2 emission (m_n^3/h)	≈100	≈150
Costs of flue gas cleaning (DM)[b]	≈35,000	≈230,000

[a] 1 kcal/h = 4.184×10^3 W.

[b] DM = Deutsch mark.

Source: From Sheppard, L.M., Firing technology heats up for the 90s, *Am. Ceram. Soc. Bull.*, 69(10), 1674–1689, 1990.

designs. A major factor in reduced energy consumption is the use of fibrous insulation. Fibrous insulation can reduce heat loss through the lining to less than 5% and result in an energy saving of 30 to 50%.[18] Furnaces with fibrous insulation can also be heated and cooled more rapidly than brick-lined furnaces. Another energy-saving innovation is the countertravel kiln of Fritz Werner GmbH. It consists of a tunnel kiln with two parallel tunnels rather than one. Ceramic ware passes through the adjacent tunnels in opposite directions. The sintered ware gives off enough heat in its cooling zone to preheat the green ware in the adjacent tunnel. The countertravel design reduces fuel, flue gas volume, and pollutants as shown in Table 14.5.

The sintering furnaces discussed so far heat the ceramic particulate compact externally by radiant and convective heat transfer. The heat then must reach the interior of the compact primarily by conduction. Since the compact is highly porous, the thermal conductivity is very low. To avoid a large temperature gradient from the surface to the interior of the compact (and thus to avoid the potential for thermal shock fracture or uneven densification), slow increase of the furnace temperature is required. Studies during recent years have shown that many ceramics can be heated to sintering temperature by microwave energy.[19,20] Microwaves are electromagnetic energy with a wavelength in the range of 100 µm to 1 mm (frequencies of 300 MHz to 300 GHz). Ceramics with a low dielectric loss are transparent to microwaves. Ceramics with a moderate to high dielectric loss couple with the microwaves and convert the microwave energy to heat energy. Since the microwave energy penetrates deeply into the ceramic (up to about 20 cm, depending on the wavelength), thick sections of a susceptible ceramic can be heated uniformly and rapidly to the sintering temperature. For small parts, complete heating and densification can occur within a few minutes if the part is insulated at the surface to avoid heat loss from the surface by infrared radiation.

Another technique that has demonstrated rapid sintering of small ceramic samples is plasma sintering. A plasma is a high-temperature, ionized electrically conducting gas. The temperature in a plasma is typically in the range 4000 to 10,000°C. Al_2O_3, ZrO_2, TiO_2, SiC, beta-alumina, and MgO have been plasma-sintered.[21,22]

14.1.4 SINTERING PROBLEMS

A variety of conditions can result in improper sintering and have a deleterious effect on the material properties. Normally, the manufacturer will detect these problems either during processing or during quality-control inspection. However, sometimes defective or inferior material is not detected by the manufacturer and is shipped to the user, where the defect does not show up until the component fails prematurely in service. Under these circumstances the source of the problem and a feasible solution must be found quickly. The responsible engineer will have a distinct advantage if he or she knows generally how the ceramic was processed and knows what possible problems to look for. The following paragraphs describe some of the problems that can occur during sintering and some of the artifacts in the ceramic component that will help the engineer to identify the cause.

14.1.4.1 Warpage

Warpage is a common problem and usually is detected before the part is put into service. It increases reject rate and hence the cost per part. It also can cause delays if it arises intermittently. Warpage usually results from inadequate support during sintering or from density variations in the green-ware. The former can be corrected by shifting the orientation of the part in the furnace or by supporting the part with saggers (refractory, nonreactive ceramic pieces that restrict the component from deforming during sintering). The latter can be corrected only by solving the problem in an earlier processing step that caused the inhomogeneity. The two sources of warpage can usually be distinguished from each other by dimensional inspection or by examination of a polished section of the microstructure. Warpage due to sagging will not show variations in thickness or microstructure across the cross section, but warpage due to density variation will.

14.1.4.2 Overfiring

Overfiring is another of the more common sintering problems with ceramics. It can cause warpage, reaction with surrounding furnace structures, bloating, or excessive grain growth. The first three are usually easy to detect visually. Excessive grain growth is more difficult to detect during routine inspection and may require preparation of a polished surface, etching to accentuate the grain structure, and examination by reflected light microscopy.

However, the presence of large grains is readily visible on a fracture surface at low magnification and can provide the engineer with valuable insight into the cause of a component failure. As discussed in Chapter 8, an increase in grain size usually results in a decrease in strength. This is true even if only a portion of the grains have increased size. Sometimes over firing results in exaggerated grain growth, whereby a few grains preferentially grow very large compared to other grains in the microstructure and compared to the optimum grain size required for the intended application.

14.1.4.3 Burn-Off of Binders

As discussed in Chapter 13, binders are often added to the ceramic powder prior to compaction. These are usually organic and can leave a carbon residue in the ceramic during sintering if the time and temperature and atmosphere parameters are not properly controlled. If large percentages of binders are present, such as in injection-molded ceramics, the binder may have to be removed very slowly as a gas or liquid by thermal decomposition or capillary extraction. Too-rapid removal results in formation of cracks in the component.

The author once conducted experiments on sintering compacts of glass powders. A variety of binders were evaluated in glass compositions having a wide range in melting temperature. If the glass started to soften before the binder was completely burned off, discoloration would result. In one case the binder subsequently decomposed to produce a gas after the glass had partially sintered and expanded the glass into a porous foam having many times the volume of the original compact.

Proper binder removal is normally accomplished by slowly raising the temperature to a level at which the binder can volatilize, and holding at this temperature until the binder is gone. The temperature can then be safely increased to the sintering temperature. However, if the temperature is increased before the binder has completely volatilized, the portion remaining will char and leave a residue of carbon.

In some materials, the carbon will be relatively inert, but in others, it can cause severe chemical reactions during sintering. In one case the carbon resulted in localized reducing conditions in the core of a part, causing bloating and severe dark discoloration. The surface of the same part was white and sintered properly.

14.1.4.4 Decomposition Reactions

Ceramics are frequently prepared using a different starting composition than the final composition. For instance, carbonates, sulfates, nitrates, or other salts are often used rather than the oxides, even though the final product is an oxide. There are a variety of reasons for doing this. The salts are often purer or more reactive or can be mixed more uniformly. However, during sintering the salt must decompose to the oxide and react with other constituents to form the desired final composition. If the salt does not decompose early enough, the component can be damaged by gas evolution. If the salt does not decompose completely, an off-composition or inhomogeneous condition can result. The degree of sensitivity of a component to this is dependent on the sintering temperature, the time-temperature schedule, and the decomposition temperature and kinetics of the salt. Problems are usually not encountered with hydrates and nitrates because they have low decomposition temperatures. Carbonates tend to have higher decomposition temperatures, but usually do not pose a problem if the sintering temperatures is above 1000°C (1832°F). Some sulfates do not completely decompose until 1200 to 1300°C (2200 to 2372°F).

14.1.4.5 Polymorphic Transformations

Polymorphic transformations do not usually cause problems during sintering, but can cause problems during cool-down after sintering if a substantial volume change is involved. A good example is ZrO_2. ZrO_2 transforms with a substantial volume change back and forth between monoclinic and tetragonal polymorphs in the temperature range 800–1100°C. No problem occurs during heat-up because the individual ZrO_2 particles are not constrained. However, after sintering the original particles are now grains that are solidly bonded to adjacent grains and are thus restrained. They are also randomly oriented. Now, when the component goes through the transformation, the grains are not free to move. Very high internal stresses result at the grain boundaries and many cracks are initiated, significantly weakening the material.

The problem with ZrO_2 has been resolved by controlled additions of CaO, MgO, or Y_2O_3 that produce a cubic form of ZrO_2 that does not undergo a transformation. Modifications of ZrO_2 are discussed in Chapter 20.

Many ceramic materials undergo polymorphic transformations that can decrease the strength of sintered material either during cool-down or by further thermal cycling. Quartz and cristobalite forms of SiO_2 both have displacive transformations accompanied by substantial volume change.

An engineer has several ways of determining if a material is susceptible to damage by polymorphic transformation. First, he or she can determine what crystallized compositions are present in the material by x-ray diffraction analysis. Then the engineer can look up phase equilibrium diagrams for the material and its constituents. These diagrams will show if polymorphic phases are present. Unfortunately, the equilibrium diagram provides no information about the volume change during transformation. This can be obtained by thermal expansion measurement.

14.2 MODIFIED DENSIFICATION PROCESSES

Sintering is a process by which a particulate compact is transformed into a ceramic article that has adequate strength or other characteristics to satisfy the needs of an application. A variety of processes other than conventional sintering exist that can achieve a functional ceramic part. Many of these are listed in Table 14.6. The remainder of this chapter defines and briefly discusses each of these processes.

14.2.1 Modified Particulate Processes

The three modified particulate processes are overpressure sintering, hot pressing, and hot isostatic pressing. These processes are extensions of conventional sintering in which pressure is applied simultaneously to temperature. The application of pressure increases the energy at particle-particle contacts and thus enhances the normal mechanisms of sintering.

14.2.1.1 Overpressure Sintering

Some materials have a high enough vapor pressure at the sintering temperature to result in vaporization or decomposition rather than densification. Si_3N_4 is a good example. Si_3N_4 at high temperature ($\sim 1850°C$ or $3360°F$ and above) has the following equilibrium reaction:

$$Si_3N_4(s) \rightleftarrows 3Si(l) + 2N_2(g) \qquad (14.2)$$

TABLE 14.6
Alternate Densification or Bonding Processes

Modified Particulate Processes
    Overpressure sintering
    Hot isostatic pressing
    Hot pressing
Chemical Processes
    Reaction
    Pyrolysis
    Cementitious bonding
Melt Processing
    Casting
    Drawing
    Spraying
    Blowing
    Quenching
    Devitrification
    Crystallization
Vapor Processing
    Chemical vapor deposition (CVD)
    Physical vapor deposition (PVD)
Infiltration
    Melt
    Vapor
    Liquid
    Sol
    Polymer
Metal-gas reaction
Ceramic superconductor fibers and films
Lanxide™ process

If Si_3N_4 powder is heated to 1900°C (3450°F) in an open container, N_2 gas will form and go into the atmosphere; the above relationship will proceed from left to right. The volatilization will inhibit or prevent sintering. Conducting the sintering in a closed system with an overpressure of N_2 drives the equilibrium from right to left and allows the Si_3N_4 to densify rather than volatilize.[23–25]

14.2.2 HOT PRESSING

Hot pressing is analogous to sintering except that pressure and temperature are applied simultaneously.[26–28] Hot pressing is often referred to as pressure sintering. The application of pressure at the sintering temperature accelerates the kinetics of densification by increasing the contact stress between particles and by rearranging particle positions to improve packing. It has been established that the energy available for densification is increased by greater than a factor of 20 by the application of pressure during sintering, providing several processing and property advantages:

1. reduces densification time
2. can reduce densification temperature, often resulting in less grain growth than would occur with pressureless sintering
3. minimizes residual porosity
4. results in higher strength than can be achieved through pressureless sintering, due to the minimization of porosity and grain growth
5. can reduce the amount of sintering aid and result in improved high-temperature properties
6. can be conducted starting with a loose powder so that no binders or other organic additives are required

Keys to hot pressing are equipment design and die design. Figure 14.17 shows a simple schematic of a typical uniaxial hot-pressing setup. It consists of a furnace surrounding a high-temperature die with a press in-line to apply a controlled load through the die pistons. The type of furnace is dependent on the maximum temperature and uniformity of the hot zone required. Induction heating, with water-cooled copper coils and a graphite susceptor, is most commonly used and has a temperature capability greater than 2000°C (3630°F). The furnace must either be evacuated or backfilled with N_2, He or Ar during operation to minimize oxidation of the graphite. Furnaces with graphite or other resistance heating elements can also be used for hot pressing.

The source of pressure is usually a hydraulic press with a water-cooled platen attached to the ram. However, this does not provide adequate cooling to extend the ram into the furnace, so blocks of graphite or other refractory material are used. Obviously, the size of the press is dependent on the size of the part being hot pressed and the pressure required. Most hot pressing is done in the range of 6.9 to 34.5 MPa (1 to 5 ksi).

The die material is perhaps the most important element of the hot press. It must withstand the temperature, transient thermal stresses, high hot-pressing loads, and be chemically inert to the material being hot pressed. Graphite is the most widely used die and piston material. It has high-temperature capability, its strength increases with temperature, and it has low coefficient of friction. It does not react with most materials and can be coated with a boundary layer such as boron nitride (BN) to prevent direct contact with material with which it might interact. As with the graphite susceptor, though, graphite does oxidize and must be used under a protective environment.

Refractory metal dies such as molybdenum, tantalum, and the molybdenum alloy TZM have been used in limited cases. However, they are expensive, have high reactivity, and deform easily at high temperatures. TZM coated with $MoSi_2$ or a composite die consisting of a molybdenum jacket surrounding an Al_2O_3 liner has been recommended.[27] This latter approach takes advantage of the strength of the molybdenum and the abrasion resistance, creep resistance, and moderate thermal expansion coefficient of the Al_2O_3.

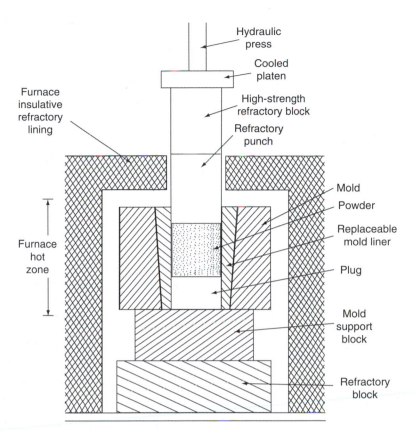

FIGURE 14.17 Schematic showing the essential elements of a hot press. (Drawing courtesy of ASM International.)

Superalloys have also been used for hot-pressing dies for ceramics, but only at temperatures below 900°C (1650°F) and loads below 104 MPa (15 ksi). A major problem with these materials is high thermal expansion. If the expansion of the die is higher than that of the material being hot pressed, the die will essentially shrink fit around the material during cooling and make ejection extremely difficult.

Ceramic dies, especially Al_2O_3 and SiC, have been used successfully for hot pressing. They have reasonably low thermal expansion, are nonreactive, and have excellent resistance to galling and abrasion. Al_2O_3 can be used to approximately 1200°C (2190°F), dense SiC to about 1400°C (2550°F).

Reactivity is a special concern of die assemblies. Many of the carbides, ferrites, and other materials are very susceptible to property alteration through variations in stoichiometry and must be hot pressed under very controlled conditions. Graphite dies are often lined with a "wash" or spray coating of BN or Al_2O_3. Dies used for ferrites and some other electronic ceramics are often lined with ZrO_2 or Al_2O_3 powder.

The nature of the powder to be hot pressed is equally important to correct selection of the die material. the same type of fine-grained powders suitable for pressureless sintering are usually acceptable for hot pressing. In most cases a densification aid or a grain-growth inhibitor is added to achieve maximum density and minimum grain size. Table 14.7 summarizes sintering aids and grain-growth modifiers for a variety of oxides, carbides, nitrides, and borides. Specific references on the hot pressing of each of these ceramic materials are listed in References 11 and 29.

Hot pressing is typically conducted at approximately half the absolute melting temperature of the material,[27] which is usually a lower temperature than that at which the material can be densified by

TABLE 14.7

Densification Aids and Grain-Growth Modifiers

Material	Densification Aids	Grain-Growth Inhibitors	Modifiers, Enhancers
Al_2O_3	LiF	Mg, Zn, Ni, W, BN, ZrB_2	H_2, Ti, Mn
MgO	LiF, NaF	MgFe, Fe, Cr, Mo, Ni, BN	Mn, B
BeO	Li_2O	Graphite	–
Si_3N_4	MgO, Y_2O_3, $BeSiN_2$	–	–
SiC	B, Al_2O_3, Al	–	–
TaC, TiC, WC	Fe, Ni, Co, Mn	–	–
ZrB_2, TiB_2	Ni, Cr	–	–
ThO_2	F	Ca	–
ZrO_2	–	H_2, Cr, Ti, Ni, Mn	–
$BaTiO_3$	–	Ti, Ta, Al/Si/Ti	–
Y_2O_3	–	Th	–
$Pb(ZrTi)O_3$	–	Al, Fe, Ta, La	–

Source: From Reed, J.S., *Principles of Ceramic Processing*, Wiley, New York, 1988 and Hodge, E.S., Hot isostatic pressing improves powder metallurgy parts, *Mat. Des. Eng.*, 61, 92–97, 1965.

pressureless sintering. Time at temperature is also reduced. The reduced temperature and time at temperature combine to minimize grain growth, thus providing better potential for improved strength.

Powder to be hot pressed can be loaded directly into the die or can be precompacted separately into a powder preform or compact that is then loaded into the die. Loading powder directly into the die is the most common procedure. However, the problems with this method are the difficulty in achieving uniformity and the pickup of contamination. Another disadvantage involves the low packing density of the loose powder and the resulting increase in the stack height to achieve a given part thickness. This reduces the number of parts that can be produced in a hot-pressing run and also increases the die-wall friction. Increased die-wall friction increases the variation of pressure within the compact and increases the chances for nonuniformity in the final part. The author encountered another problem with loose powder die loading while scaling up the hot pressing of Si_3N_4 from 7.6-cm (3-in.)-diameter development samples to 15.2-cm (6-in.)-diameter pilot production billets. The 7.6-cm (3-in.) samples had a strength of 897 MPa (130,000 psi) and were uniform across the diameter. The early 15.2-cm (6-in.) billets were near theoretical density around the edges, but of decreased density in the interior. The overall density was within specification. The strength also appeared within specification since it was being measured on material sliced from the edge of the part. From all appearances the billets were of equivalent quality to the smaller development samples and were acceptable for delivery to a customer. However, when further testing was conducted, which included an evaluation of the billet interior, it was found that this region had density below specification and a strength of less than 690 MPa (100,000 psi). The source of the problem turned out to be a combination of loose powder loading and nonuniform temperature distribution. The loose powder had a very low thermal conductivity such that the edges in close proximity to the graphite die heated up faster than the interior and began to sinter. This physically shifted material from the center toward the edge and ultimately resulted in the density and strength gradient. The lesson is that flaws may result in a part that are not readily detectable, but if the engineer is aware of the mechanisms of processing and of some of the things that can go wrong, he or she will have a better chance of solving a problem that occurs or producing a quality-control specification that will minimize such occurrences. The problem was resolved by precompacting the powder better and by modifying the time and temperature profile during hot pressing. Recurrence was prevented by initiating a more rigid density specification and strength certification procedure.

TABLE 14.8
Comparison of Densities and Strengths Achieved by Hot Pressing vs. Sintering

Material	Sintering Aid	Density (% theoretical)	RT MOR[a] MPa	RT MOR[a] kpsi	1350°C MOR MPa	1350°C MOR kpsi
Hot-pressed Si_3N_4[b]	5% MgO	98	587	85	173	25
Sintered Si_3N_4[b]	5% MgO	~90	483	70	138	20
Hot-pressed Si_3N_4[c]	1% MgO	>99	952	138	414	60
Sintered Si_3N_4[d]	$BeSiN_2 + SiO_2$	>99	560	81	–	–
						–
Sintered Si_3N_4[e]	6% Y_2O_3	~98	587	85	414	60
Hot-pressed Si_3N_4[e]	13% Y_2O_3	>99	897	130	669	97

[a] Room-temperature modulus of rupture.
[b] Terwilliger, G.R., *J. Am. Ceram. Soc.*, 57(1), 48–49, 1974.
[c] Richerson, D.W., *Am. Ceram. Soc. Bull.*, 52, 560–562, 569, 1973.
[d] Greskovich, C.D. and Palm, J.A., U.S. DOE Conference 791082, 1979, pp. 254–262.
[e] Data from Quackenbush, C.L., GTE Laboratories, Waltham, MA.

14.2.2.1 Unique Hot-Pressed Properties

Hot pressing permits achieving near-theoretical density and very fine grain structure, which result in optimization of strength. It also permits reduction of the amount of sintering aid required to obtain full density. This can result in orders-of-magnitude improvement in high-temperature properties such as creep and stress rupture life.

Table 14.8 compares the properties of several sintered and hot-pressed Si_3N_4 compositions. Similar differences exist between sintered and hot-pressed varieties of other materials such as Al_2O_3, SiC, spinel, and mullite.

Hot pressing can cause preferred orientation of the grain structure of some materials and result in different properties in different directions. This occurs predominantly when powders with a large aspect ratio such as rods or needles are used. It can also occur due to flattening of agglomerates or laminar distribution of porosity perpendicular to the direction of hot pressing. Figure 14.18 illustrates the strength variations measured for specimens cut from various orientations from a hot-pressed Si_3N_4 billet. The strength was greatest in the plane perpendicular to the direction of hot pressing. This was thought to be due to a combination of preferred orientation of Si_3N_4 grains and laminar density contours.

Preferred orientation has also been encountered in hot-pressed Al_2O_3 reinforced with SiC whiskers. The whiskers have a high length-to-diameter ratio (usually over 20:1) and orient perpendicular to the hot-pressing direction. Test bars cut from a plane perpendicular to hot pressing break across the whiskers and have high strength (>600 MPa or 87 ksi) and toughness (7 MPa · m$^{1/2}$ or 6.37 ksi · in.$^{1/2}$). Bars cut from a plane parallel to the hot-pressing direction break parallel to the whiskers and have much lower strength (typically <400 MPa or 58 ksi) and toughness (~3.5 to 4.0 MPa · m$^{1/2}$ or 3.18 to 3.64 ksi · in.$^{1/2}$).

Strength test specimens are normally cut from the plane perpendicular to the hot-pressing direction. This is usually the strongest direction (if anisotropy is present) and may give the engineer false confidence in the material. The engineer should be aware that the strength and other properties in the other directions may be inferior and adjust the material qualification testing accordingly.

14.2.2.2 Hot-Pressing Limitations

The major limitation of hot pressing is shape capability. Flat plates, blocks, or cylinders are relatively easy to hot press. Long cylinders, nonuniform cross sections, and intricate or contoured

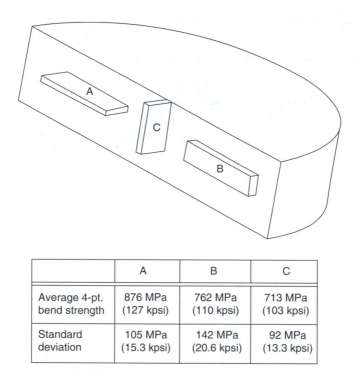

	A	B	C
Average 4-pt. bend strength	876 MPa (127 kpsi)	762 MPa (110 kpsi)	713 MPa (103 kpsi)
Standard deviation	105 MPa (15.3 kpsi)	142 MPa (20.6 kpsi)	92 MPa (13.3 kpsi)

FIGURE 14.18   Variations in the strength of hot-pressed Si_3N_4 as a function of direction.

shapes are difficult and often impossible by conventional uniaxial techniques. Figure 14.19 and the following paragraphs describe the nature of the problem.

The starting powder goes into the die as a relatively uniform stack of powder or as a uniform preform. During densification the powder or preform will compact in the axial direction of pressure application until the porosity has been eliminated and near-theoretical density achieved. The amount of compaction required to go from the loose powder or preform to the pore-free part is referred to as the compaction ratio. The compaction ratio for a well-compacted preform usually ranges from 2:1 to 3:1 and can be even higher for loose powder having a very fine particle size. For instance, one batch of Si_3N_4 powder had a compaction ratio of 8:1.

Figure 14.19a illustrates the shape of a preform having a compaction ratio of 2:1 that would be required to make a fully dense part of an arbitrary nonuniform cross section. The shape of the preform is different than the final shape and the required movement of the graphite die punches is greater for thick sections than for thin sections. For instance, in the example in Figure 14.19, the total shrinkage in the thick section of the part is four times greater than the shrinkage in the thin section, even though the percentage is the same in each case. And this is only for a minimal compaction ratio of 2:1. The shrinkage difference is greater for higher compaction ratios. How can one design rigid graphite tooling to accommodate the differences in distance and still achieve the required shape? Usually, it cannot be done. One either has to make the preform a different shape than the graphite tooling [as shown in Figure 14.19b] and hope the preform does not break up prematurely and alter the powder distribution, or one has to load loose powder to fill the die cavity [as shown in Figure 14.19c] and hope that the powder will redistribute to the required distribution during hot pressing.

The latter approach has been used with success for some materials and shapes and is worth trying because the tooling is usually not prohibitively expensive. The former approach requires two sets of tooling and has not yet been developed, but may also be worth considering.

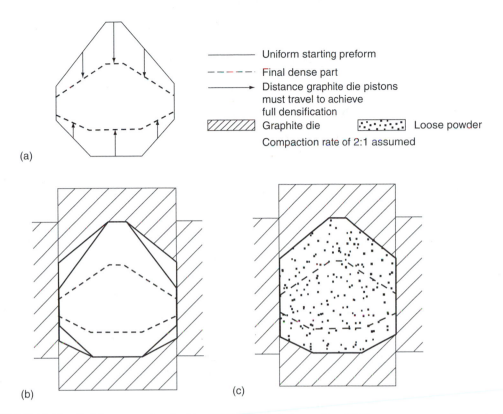

FIGURE 14.19 Problems associated with complex-shape hot pressing. (a) If we assume a uniform starting preform or powder stack, the preform will have to shrink different amounts to achieve uniform final density and the required shape. (b) A preform with the correct powder distribution has a different shape than the die. (c) Loose powder fill requires a powder redistribution during hot pressing.

Another approach to uniaxial hot pressing of shapes having nonuniform cross section is the use of nonrigid tooling. Two concepts are illustrated in Figure 14.20. The first is referred to as pseudoisostatic hot pressing. A preform is prepared by cold pressing, slip casting, or another approach. The dimensions are selected (with knowledge of the compaction ratio for the specific material) such that the required shape will result after densification. The preform is embedded in loose powder in the hot-press die cavity. The loose powder is selected so that it will not densify and will not chemically react with the preform being hot-pressed. Hexagonal boron nitride and graphite powders have both been used successfully and work especially well because of their self-lubricating character and excellent chemical stability. During hot pressing the loose powder transmits pressure from the die punches to the preform. A true isostatic pressure distribution is not achieved, but enough pressure is apparently transmitted to the preform to allow densification. Most shapes hot pressed by this approach have achieved near-theoretical density but have undergone some distortion during pressing. However, once the distortions are accounted for in the preform, near net shape can probably be reproducibly achieved.

Figure 14.20b illustrates the second nonrigid tooling approach. In this case three preforms having the same compaction ratio are required. The center preform will densify to become the required shaped part. The other two preforms are simply conforming layers between the flat die punch surfaces and the contoured part surface. A nonreactive boundary layer such as boron nitride is placed between the preforms so that they can be separated after hot pressing. This approach simulates hot

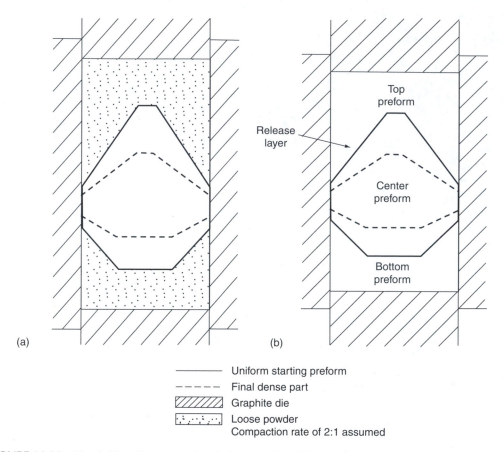

FIGURE 14.20 Nonrigid tooling approaches for hot pressing. (a) Pseudoisostatic approach. (b) Multiple preforms simulating a flat-plate approach.

pressing of a flat plate of uniform cross section and results in uniform density and properties in the finished part.

Some ceramic materials have been successfully hot-forged using a uniaxial hot-pressing apparatus. Examples include a Si_3N_4 material and a zirconia material. The Si_3N_4 material contained enough liquid phase to readily creep during forging. The zirconia material was a specific, fine-grained composition that actually deformed superplastically.

14.2.3 HOT ISOSTATIC PRESSING

Most of the limitations of hot pressing result from the uniaxial pressure application. Techniques have been developed to hot press from multiple directions. This is referred to as hot isostatic pressing (HIP) and is analogous to cold isostatic pressing.[30] Apparatus for HIP consists of a high-temperature furnace enclosed in a water-cooled autoclave capable of withstanding internal gas pressures up to about 32,000 kg/mm² (45,419 ksi) and providing a uniform hot-zone temperature up to about 2000°C (3630°F). Pressurization gas is either argon or helium. Heating is usually by molybdenum or graphite resistance-heated elements. A schematic of a HIP apparatus is shown in Figure 14.21.[31]

To achieve densification of a ceramic preform, the preform must first be evacuated and then sealed in a gas-impermeable envelope. If any high-pressure gas leaks into the preform, pressure is equalized and the preform cannot hot press. The earliest HIP studies encapsulated the ceramic in tantalum or other metal, depending on the temperature required for densification. However, this

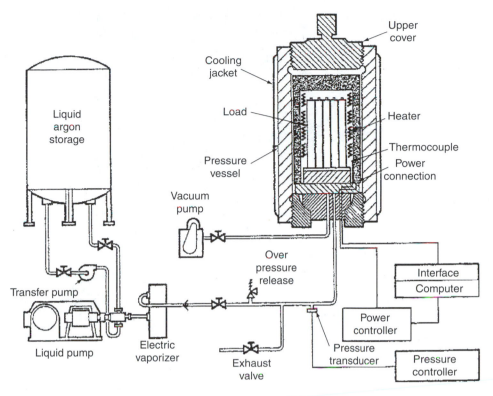

FIGURE 14.21 Schematic of hot isostatic pressing apparatus. (From Clauer, A.H., Meiners, K.E., and Boyer, C.B., Hot Isostatic Pressing MCIC Rept. MCIC-82-46, Metals and Ceramics Information Ctr., Battelle, Columbus, OH, Sept. 1982. With permission.)

severely limited the shape capability. Later studies were directed toward the use of glass encapsulation. The glass was applied as a preformed envelope, which was sealed around the part under vacuum and then collapsed at high temperature to conform to the ceramic preform shape. This worked for relatively simple shapes. ASEA in Sweden and Battelle in the United States subsequently developed techniques to apply the glass as a particulate coating.[32] The preform with this coating was placed in the HIP autoclave and evacuated. The temperature was raised until the glass softened and formed a continuous layer on the surface. The pressure and temperature were then increased to the levels required to accomplish densification of the ceramic preform. Battelle has demonstrated that individual gas-turbine rotor blades of Si_3N_4 could be fabricated by this approach. ASEA fabricated small (12.7 cm or 5 in.) integral axial rotors of Si_3N_4 by this approach, as well radial rotors of Si_3N_4 for turbochargers. Later ASEA established production capability and also licensed the HIP and glass encapsulation technology to other companies. Figure 14.22 shows Si_3N_4 gas-turbine rotors fabricated by the ASEA process.

HIP has the potential of resolving some of the major limitations of uniaxial hot pressing. It makes possible net-shape forming because the pressure is equally applied from all directions. This also results in greater material uniformity by eliminating die-wall friction effects and preferred orientation, resulting in higher strength and Weibull modulus. Also, much higher pressures and temperatures can be used, making possible more complete densification and greater flexibility in selection of composition. For instance, the higher pressure and temperature may permit densification of compositions containing less sintering aid and having dramatically improved stress rupture life and oxidation resistance.

HIP has been used to improve the strength and wear resistance of some MnZn and NiZn ferrites, yttrium iron garnet, and $BaTiO_3$, especially for applications such as magnetic recording heads.

FIGURE 14.22 Si_3N_4 gas-turbine rotors fabricated by pressure casting before and after glass encapsulation and hot isostatic pressing. Dense rotors are over 13 cm in diameter. (Photo courtesy of Garrett Ceramic Components Division of Allied-Signal Aerospace Company, currently Honeywell Ceramic Components.)

The ceramic is first sintered to closed porosity and is then further densified by HIP without a requirement for encapsulation.[33] This is sometimes referred to as sinter-HIP. This technique also has been used to achieve improved strength in SiC,[34] in transformation-toughened ZrO_2, in Al_2O_2–TiC cutting tools, and in Al_2O_3–SiC whisker-reinforced composites.

14.2.4 CHEMICAL PROCESSES

Chemical processes can produce strongly bonded ceramic materials by direct chemical interactions, by chemical reactions induced by temperature, and by a combination of chemical reaction and conventional sintering. Three processes are briefly discussed in this chapter: (1) reaction, (2) cementitious bonding, and (3) pyrolysis.

14.2.4.1 Chemical Reaction

Many options exist for chemical reactions that can yield a ceramic material. These generally involve mixing the reactants together as powders, compacting the powders into the desired shape, and heating to a temperature at which the powders react. In some cases, such as combustion synthesis,[35] the reaction is exothermic (gives off heat) and produces enough heat to bond the particles together. Examples of combustion synthesis reactions are listed in Table 14.9.

Other options involve reaction between compacted solid particles and either a gas or a liquid. Two important examples are reaction-sintered Si_3N_4 and reaction-sintered SiC.[36–39]

Reaction-sintered Si_3N_4 is also referred to as reaction-bonded Si_3N_4 or simply RBSN. RBSN is fabricated from silicon powder. The silicon powder is processed to the desired particle size distribution and formed into the required shape by pressing, slip casting, injection molding, or another suitable process. The compacted Si shape is then placed in a furnace under a nitrogen or mixed nitrogen and hydrogen or nitrogen and helium atmosphere and heated initially to about 1200 to 1250°C (2190 to 2280°F). The nitrogen permeates the porous Si Compact and begins to react with the Si to form Si_3N_4. Initially, α-Si_3N_4 fibers grow from the Si particles into the pores. As the reaction progresses, the temperature is slowly raised to approximately 1400°C (2550°F), near the melting temperature of Si. As the temperature increases, the reaction rate increases and primarily β-Si_3N_4 is formed. Great care is necessary in controlling the rate of temperature increase and nitrogen flow. The reaction of N_2 and Si is exothermic and, if allowed to proceed too fast, will cause the silicon to melt and ball up into Si particles too large to nitride or which exude out of the surface of the part. A typical nitriding cycle in which the exotherm (heat rise and fall vs. time) is controlled and no

---

TABLE 14.9

Examples of Materials Prepared by the Combustion Synthesis Method

Borides	CrB, HfB$_2$, NbB$_2$, TaB$_2$, TiB$_2$, LaB$_6$, MoB$_2$
Carbides	TiC, ZrC, HfC, NbC, SiC, Cr$_3$C$_2$, B$_4$C, WC
Carbonitrides	TiC-TiN, NbC-NbN, TaC-TaN
Cemented carbides	TiC-Ni, TiC-(Ni, Mo), WC-Co, C$_3$C$_2$-(Ni, Mo)
Chalcogenides	MoS$_2$, TaSe$_2$, NbS$_2$, WSe$_2$
Composites	TiC-TiB$_2$, TiB$_2$-Al$_2$O$_3$, B$_4$C-Al$_2$O$_3$, TiN-Al$_2$O$_3$
Hydrides	TiH$_2$, ZrH$_2$, NbH$_2$
Intermetallics	NiAl, FeAl, NbGe, TiNi, CoTi, CuAl
Nitrides	TiN, ZrN, BN, AlN, Si$_3$N$_4$, TaN (cubic and hexagonal)
Silicides	MoSi$_2$, TaSi$_2$, Ti$_3$Si$_2$, ZrSi$_2$

Source: From Munir, Z.A., Synthesis of high temperature materials by self-propagating combustion methods, *Am. Ceram. Soc. Bull.*, 67(2), 342–349, 1988.

---

exuding occurs is on the order of 7 to 12 days, depending on the volume of material in the furnace and the green density of the starting Si compacts.

Approximately 60% weight gain occurs during nitriding, but less than 0.1% dimensional change. This makes possible excellent dimensional control. Bulk densities up to 2.8 g/cm^3 (0.1 lb/in.^{3}) have been achieved (compared to a theoretical density for Si$_3$N$_4$ of about 3.2 g/cm^3 or 0.11 lb/in.^{3}).

The earliest reaction-bonded Si$_3$N$_4$ had a density of about 2.2 g/cm^3 (0.07 lb/in.^{3}) and a strength under 138 MPa (20 ksi). By the late 1970s, RBSN had been developed with density of 2.8-g/cm^3 and four-point flexure strength in the range of 345 MPa (50 ksi).

Another advantage of the reaction-sintered Si$_3$N$_4$ is its creep resistance. No sintering aids are added to achieve densification, so no glassy grain boundary phases are present. Strength is retained to temperatures greater than 1400°C (2550°F) and the creep rate is very low. In addition, reaction-sintered Si$_3$N$_4$ has a relatively low elastic modulus and coefficient of thermal expansion and a relatively high thermal conductivity (considering its porosity). These properties, combined with a moderately high strength, give RBSN good thermal shock resistance and make it a feasible candidate for such applications as welding nozzle tips and some prototypes gas-turbine static-structure components.

The primary disadvantage of RBSN is its porosity. The porosity is interconnected and can result in internal oxidation and accelerated surface oxidation at high temperature. The internal oxidation appears to affect the thermal stability of a component. For instance, Ford Motor Company observed that experimental turbine-engine parts fractured when the weight gain due to oxidation reached about 2%. The mechanism was not reported, but could have been associated with internal stresses induced by the thermal expansion mismatch between the Si$_3$N$_4$ and the cristobalite formed during oxidation. To avoid oxidation, a post-sintered RBSN was developed.[40,41] A sintering aid such as Y$_2$O$_3$ was added to the Si prior to nitriding. After nitriding, the 80%-dense RBSN was heated to around 1900°C (3450°F) in nitrogen (preferably with an overpressure) and sintered by a liquid-phase mechanism to near-theoretical density (typically less than 1% porosity). This material had a strength greater than 700 MPa (101.5 ksi) and very good oxidation resistance. The oxidation characteristics of RBSN vs. sintered RBSN are shown in Figure 14.23.

Reaction-sintered SiC is also referred to as reaction-bonded SiC and siliconized SiC. It is processed from an intimate mixture of SiC powder and carbon. This mixture is formed into the desired shape and exposed at high temperature to molten Si. The silicon wicks through the porous powder compact and reacts with the carbon to form in-situ SiC, which bonds together the original SiC particles. The remaining pores are filled with Si. The resulting material is basically a nonporous Si-SiC composite

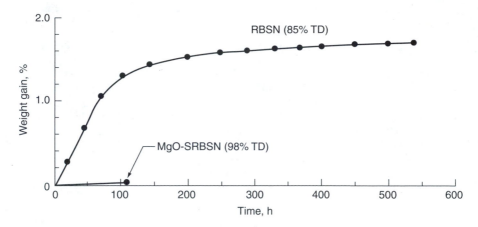

FIGURE 14.23 Comparison of the rate of oxidation of porous, reaction-bonded silicon nitride (RBSN) vs. nonporous sintered RBSN. (From Mangels, J., Sintered reaction-bonded silicon nitride, *Cer. Eng. and Sci. Proc.*, 2(7–8), 596, 1981. With permission.)

TABLE 14.10

Properties of Reaction-Sintered SiC Materials Illustrating Variations Achievable by Varying Microstructure and Silicon Content

Material	Vol% Silicon	Young's Modulus (E)		Flexure Strength (σ)	
		GPa	psi	MPa	ksi
NC-435[a]	~20	349	50.7×10^6	394	57.2
Refel[b]	–	396	57.5×10^6	309	44.9
Type TH[c]	15–20	393	57×10^6	482	70.0
Type THL[c]	55–60	303	44×10^6	331	48.0
Type F[c]	75–80	200	29×10^6	207	30.0

[a] Norton Company, Worcester, MA.

[b] British Nuclear Fuels, Ltd., U.K.

[c] General Electric Company, Schenectady, NY.

Source: Data compiled from Larsen, D.C., Property Screening and Evaluation of Ceramic Turbine Engine Materials, Final Rep. Period July 1, 1975–August 1, 1979, AFMLTR-79-4188, Oct. 1979, and Hillig, W.B., Mehan, R.L., Morelock, C.R., DeCarlo, V.J., and Laskow, W., Silicon and silicon carbide composites, *Am. Ceram. Soc. Bull.*, 54(12), 1054–1056, 1975.

and can have a broad range of strength and elastic modulus, depending on the particle size distribution and the percent of Si. Table 14.10 summarizes data for several reaction-sintered SiC materials.

Reaction-sintered SiC materials have a relatively flat strength vs. temperature curve nearly up to the melting temperature of silicon, at which point the strength drops off rapidly. A typical strength vs. temperature curve is shown in Figure 14.24 compared to reaction-bonded Si_3N_4, hot-pressed Si_3N_4*, and sintered SiC.

Reaction-sintered SiC has similar advantages to reaction-bonded Si_3N_4 for complex-shape fabrication; it undergoes dimensional change of less than 1% during densification. The initial

---

* Note that the reference reports data for materials developed in the mid-1970s. Substantial improvements have occurred particularly for the Si_3N_4 materials. Si_3N_4 materials are presently available with room temperature flexure strength over 800 MPa and 1375°C flexure strength over 500 MPa.

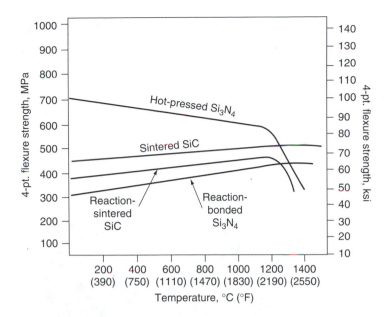

FIGURE 14.24 Typical strength vs. temperature behavior of reaction-sintered SiC compared to other mid-1970s vintage SiC and Si₃N₄ materials. (Compiled from Larsen, D.C., Property Screening and Evaluation of Ceramic Turbine Engine Materials, Final Rep. Period July 1, 1975–Aug. 1, 1979, AFML-TR-79-4188, Oct. 1979.)

shape can be formed by casting, plastic molding, pressing, extrusion, and any of the other processes applicable to ceramics. It is especially suitable for the plastic processes such as extrusion and compression molding. The binder can be a thermosetting resin such as a phenolic. Instead of having to remove the binder after molding, as is required with other ceramics, the binder is simply charred to provide the carbon source for reaction with the silicon.

The siliconizing process is accomplished by several different techniques. The most common are illustrated in Figure 14.25. Both of these are conducted at temperatures above about 1500°C (2730°F) in a vacuum chamber. The Si is molten and is able to travel by capillary pressure into the pores where it can react with the C to form SiC. The pack technique surrounds the SiC–C preform with granular Si plus inert carrier or dilution particles such as Al_2O_3. After siliconization, the pack is removed by mechanical means such as sand blasting. The wick technique avoids contact of the surfaces of the preform with the Si or other materials. The molten Si is wicked into the preform through a single surface or through a protrusion that is cut off after siliconization.

Another interesting method of fabricating reaction-sintered SiC is to start with woven carbon fibers or felt. Laminae of carbon fiber weave are laid up much the same as fiberglass to form the desired shape. This can then be reacted with molten silicon to form a composite of SiC–Si or C–SiC–Si. By controlling the type of fibers and the weave, a complete range of composites with varying SiC-to-Si ratios can be engineered. Those with high Si have low elastic modulus (~206 GPa or 30×10^6 psi) and relatively low strength (<206 MPa or 30 ksi); those with high SiC have high elastic modulus (>344 GPa or 50×10^6 psi) and high strength (~482 MPa or 70 ksi). The range of composites can be further expanded by pitch impregnation or deposition of pyrolytic carbon or glassy carbon prior to siliconizing.

14.2.5 CEMENTITIOUS BONDING

All of the densification processes discussed so far involve high-temperature operations to achieve a strong, useful part. Another important approach is cementitious bonding (also called chemical

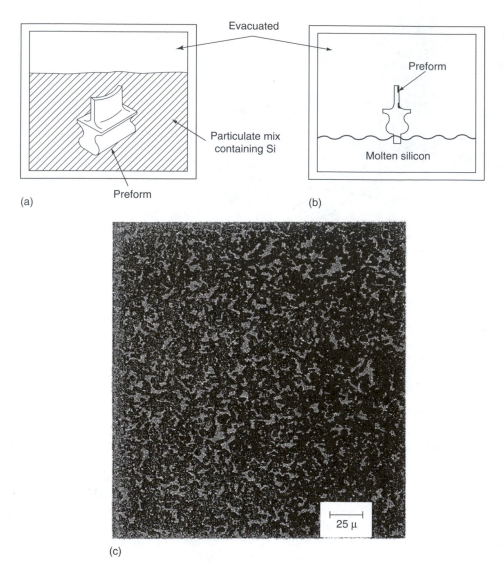

FIGURE 14.25 (a) and (b) Molten Si infiltration of SiC plus C preforms to produce dense, reaction-sintered SiC. (c) Typical microstructure of Si–SiC material. Residual Si is light-colored phase. (Photo courtesy of Carborundum Company, Niagara Falls, NY, currently a part of Saint-Gobain Advanced Ceramics.)

bonding), in which an inorganic ceramic adhesive bonds together an aggregate of ceramic particles. The adhesion results primarily from hydrogen bonding. The resulting materials are not of high strength, but are adequate for many wear-resistance, building, and refractory applications. Major advantages are that the cement can be poured, troweled, or gunned into place, has little dimensional change during setting, and can be repaired on-site.

Many different cements have been developed, ranging from common concrete to very-high-temperature furnace linings. The cements can be classified according to the mechanism of bond formation: hydraulic bonds, reaction bonds, and precipitation bonds.[42]

Hydraulic cements set by interaction with water. The most common hydraulic cement is Portland cement, which is primarily an anhydrous (without water) calcium silicate. It is slightly soluble in water and sets by a combination of solution-precipitation and reaction with water to form a hydrated (i.e., water-incorporating) composition. The reaction is exothermic and care must be taken

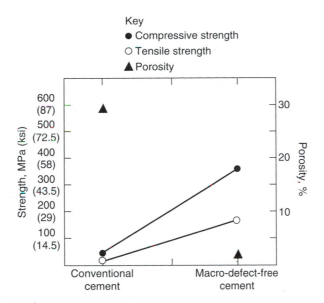

FIGURE 14.26 Increases in strength and decreases in porosity achieved with advanced macro-defect-free (MDF) cements. (Compiled from data of Roy, D., The Pennsylvania State University.)

to ensure that adequate water is initially present and that the heat of reaction does not dry out the cement prematurely. This explains why a competent cement contractor keeps the surface of freshly laid concrete damp.

The ratio of water to cement in the initial mix has a primary effect on the final strength of the cement. As long as adequate water is added for hydration and for workability, the lower the water-to-cement ratio, the higher the resulting strength.

Calcium aluminate cements are also hydraulic setting, but have much higher temperature capability than Portland cement and are thus used for refractory applications such as furnace linings. Other hydraulic cements include natural lime-silica cements, barium silicate and barium aluminate cements, slag cements, and some ferrites.

Similar to hydraulic cements are gypsum cements such as plaster of Paris and Keene's cement. They set by a hydration reaction but are much more soluble than the hydraulic cements and recrystallize to a highly crystalline structure that has little adhesion. Rather than being used for bonding aggregates, plaster of Paris is used alone to make wallboard, plaster molds for slip casting and metal casting, and decorative knick-knacks and statuettes.

Major progress has occurred in recent years in achieving hydraulic cements of decreased porosity and increased strength. One approach involves adding ultrafine SiO_2 ("microsilica") particles to Portland cement. This is referred to as DSP (densified with small particles) cement. The SiO_2 particles are much smaller than the cement particles and fill void spaces between the cement particles. Less water is required to prepare a flowable mix. The decreased water and better particle packing result in decreased total porosity and a very large decrease in pore size. Strength can be increased by a factor of 4 and permeability to liquid absorption nearly eliminated. The life of road, bridge, and building structures can be greatly increased, especially those that contain metal reinforcing bars which are attacked in conventional concrete by corrosion.

A second approach to improved hydraulic cements is called macro-defect-free (MDF) cement. This approach utilizes organic additives to the cement to achieve better dispersion and particle packing. It also results in a dramatic increase in strength and decrease in permeability with little increase in cost.

Figure 14.26 illustrates some of the improvements that have occurred in advanced cements.[43]

Reaction cements are formed by a chemical reaction between two constituents other than water. One of the most common is monoaluminum phosphate, formed by the reaction of aluminum oxide powder with phosphoric acid. This cement sets in air at room temperature, but is usable over a very broad temperature range. Above 310°C (590°F) the cement is dehydrated to form the metaphosphate, which is then stable up to a very high temperature. In fact, the strength increases as the temperature is increased.

Most metal oxides form phosphate cements when reacted with phosphoric acid.[44] In addition to Al_2O_3, the following metal oxides have been shown to form phosphate cements: BeO, CdO, Fe_2O_3, Y_2O_3, ZnO, ZrO_2, CuO, CoO, ThO_2, V_2O_5, and SnO. Compositions based on ZnO have been widely used for dental cements.

Most of the ceramic cements are porous and brittle and require an aggregate (filler) to provide durability. An interesting reaction cement that has some resiliency is magnesium oxychloride cement. It has been used for floors, building facings, signs, and a variety of other applications.

Precipitation cements are primarily gels formed by precipitating colloidal (uniformly dispersed) suspensions by adjusting the pH or ion concentration. Sodium silicate is perhaps the best known and most widely used precipitation cement. It is inexpensive and its composition can be controlled to achieve setting by drying, heating, or chemical means. Chemical setting is achieved by addition of acid salts, especially sodium silicofluoride. The setting rate can be controlled by the amount and grain size of the silicofluoride and the amount of water in the cement. Organic materials such as esters (ethyl acetate) and alcohols also precipitate alkali silicate cements.

Another important precipitation cement is prepared from ethyl orthosilicate. It is precipitated by condensation polymerization. Reaction can be accelerated by addition of magnesia or by heating and also can be catalyzed by acids or bases.

Precipitation cements are used extensively in applications where acid resistance is critical. They are also used for some abrasion-resistance applications, for bonding low-temperature refractories, and for forming of foundry molds for metal casting.

14.2.6 PYROLYSIS

Pyrolysis refers to the thermal decomposition of a polymer. Polymers that contain only hydrogen and carbon thermally decompose to H_2O and CO_2 in air and to carbon in an inert or reducing atmosphere. Some types of carbon fibers and carbon-carbon composites are fabricated in this way.[45] Some other polymers contain atoms besides H and C. These other atoms remain as reactants or residue during the pyrolysis to yield ceramic compounds.[46,47] Examples are shown in Figure 14.27. The siloxanes yield SiO_2, the silazanes yield SiC or Si_3N_4, and the metalorganic polymers can yield a variety of ceramics. Other polymers can yield boron nitride, boron carbide, or mixed compositions.

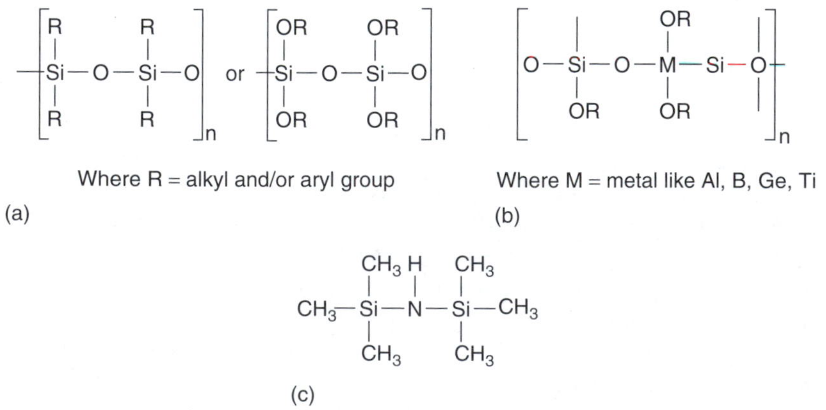

Where R = alkyl and/or aryl group

(a)

Where M = metal like Al, B, Ge, Ti

(b)

(c)

FIGURE 14.27 Types of polymers that can be pyrolyzed to yield ceramic or glass products. (a) Siloxanes, (b) metal-organic polymers, and (c) silazanes (hexamethyldisilazane shown.)

High-strength SiC fibers have been achieved by pyrolysis of polycarbosilane polymers.[48] The fibers* have tensile strength of 2.7 GPa (391.5 ksi), elastic modulus of 200 GPa (29 × 10[6] psi), and elongation of 1.5 to 2%. These fibers have been successfully used to reinforce aluminum, glass, SiC, and lithium aluminum silicate glass ceramics. Figure 14.28 shows the strength improvements that have been achieved.[49]

Fabrication of ceramics from polymer precursors has a number of problems that have limited commercial application: (1) poor yield of ceramic; (2) large volume of gas evolved during pyrolysis; (3) large shrinkage during pyrolysis; and (4) high cost and limited availability of polymers. The poor yield, high gas evolution, and high shrinkage can be visualized by examining the polymer structures shown previously in Figure 14.27. Note that the bulk of the polymers shown are carbon and hydrogen, most of which are lost as gaseous compositions during pyrolysis. For example, hexaphenylcyclotrisilazane ($C_{72}H_{66}Si_6N_6$) theoretically loses 78% by weight during pyrolysis. However, polymers can be synthesized that might result in a greater yield of ceramic. For instance, methylcyclotetrasilazane ($C_1H_{14}Si_4N_4$) theoretically would have only a 7.3% weight loss.

14.2.7 MELT PROCESSING

Glass compositions and many polycrystalline and single-crystal ceramics are processed from a melt. Melt processing offers many options for shape forming and generally results in very low porosity. Some of the important processes that involve a melt include casting, drawing, spraying, blowing, quenching, devitrification, and crystallization.

14.2.7.1 Casting, Drawing, and Blowing

Casting, drawing, and blowing are commonly utilized to form glass shapes. Casting is also used to produce dense bricks or blocks of Al_2O_3 or ZrO_2–Al_2O_3–SiO_2. These are highly resistant to high-temperature corrosion and are used to line glass-melting furnaces. The Al_2O_3 or ZrO_2–Al_2O_3–SiO_2 is melted in an arc furnace at temperatures well above 2000°C (3630°F). High-toughness abrasives

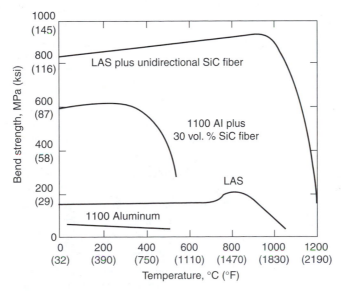

FIGURE 14.28 Strength improvement of lithium aluminum silicate (LAS) glass ceramic and 1100 aluminum alloy by reinforcement with Nicalon™ SiC Fiber. (Drawing courtesy of ASM International.)

* Nicalon™ from the Nippon Carbon Company.

are also made by casting from a melt. In this case, the melt is quenched very rapidly to achieve very fine grain size, high strength, and high toughness.

Drawing is used to produce long, thin fibers. Glass fibers for fiberglass and insulation are made by drawing (also called spinning). Some high-quality glass fiber for fiber optics is also made by drawing.

14.2.7.2 Spraying

A variety of techniques and apparatus have been developed to melt ceramic particles and spray the molten droplets onto a surface. This molten-particle deposition is commonly referred to as flame spray or plasma spray. Almost any oxide, carbide, boride, nitride, or silicide that does not sublime or decompose can be applied by molten-particle techniques. Coatings are most often applied, but free-standing parts can also be made by using a removable mandrel or form.

The first widely used molten-particle approach was the oxyacetylene powder gun, more frequently referred to as the flame-spray gun. Ceramic powder is aspirated (sucked in) into the oxyacetylene flame and melts. The molten particles exit the gun through a nozzle and strike the substrate to be coated at a velocity of about 45 m/sec (150 ft/sec). By moving either the substrate or the gun, a uniform coating can be built up having approximately 10 to 15 vol% porosity and a surface finish of 150 to 300 µin. rms.

A similar approach is the oxyacetylene rod gun. Instead of using ceramic powder, a sintered rod of the coating material is fed into the oxyacetylene flame. Molten ceramic at the tip of the rod is carried to the substrate by bursts of air traveling at about 180 m/sec (590 ft/sec). Surface finishes are similar to those produced by the powder gun, but porosity is usually lower (6 to 10%) because the particles are completely molten and the impact velocity is higher.

The oxyacetylene guns are widely used. Another widely used approach is the arc-plasma gun. A high-intensity, direct-current arc is maintained in a chamber. Helium or argon is passed through the chamber, heated by the arc, and expelled through a water-cooled copper nozzle as a high-temperature, high-velocity plasma. Ceramic particles are injected into the plasma, where they are melted and directed against the substrate. Velocities as high as 450 m/sec (1500 ft/sec) have been obtained, yielding coatings with porosity as low as 3% and surfaces with a finish in the range 75 to 125 µin. rms. The major difficulty of the arc-plasma gun is the temperature of the plasma. Ceramic substrates may have to be preheated to avoid thermal shock damage and metal substrates may have to be cooled to avoid melting.

Deposition rates for molten-particle spray are much higher than for vapor deposition approaches (chemical or physical vapor deposition), in the kilograms per hour range rather than in grams or milligrams per hour. However, for some applications this is not enough. One technique has been developed that has demonstrated 2 to 4 kg/min (4.4 to 8.8 lb/min) and is projected to scale up to 200 kg/min (440 lb/min). The ceramic powder is mixed with fuel oil to form a slurry and is burned in oxygen in a water-cooled gun. Al_2O_3, mullite, and SiO_2 have been successfully sprayed by this method to form refractory linings for high-temperature furnaces such as the oxygen converters in steel mills.

Molten-particle spray techniques have been used extensively to deposit wear-resistant and chemically resistant coatings on a wide variety of metal and ceramic products. One interesting example is the spraying of chromium oxide (Cr_2O_3) on the propeller shafts of large seagoing ships. The chromium oxide greatly reduces erosive wear, provides a good surface to seal against (after surface grinding to achieve a suitable surface finish), and inhibits seawater corrosion.

Coatings are also applied to provide thermal protection. Stabilized ZrO_2 has a very low thermal conductivity and emissivity and is applied to stainless steel and superalloy parts as a thermal barrier coating. Although some other oxide ceramics have similar thermal properties, ZrO_2 was selected because it has a coefficient of expansion similar to that of the metals. This is one case for a ceramic in which a high coefficient of thermal expansion is beneficial.

An important advantage of molten-particle spray techniques is that a wide range of sizes and shapes of substrates can be coated. On-site repairs are often feasible. Flame spray has great versatility and the student should be aware of the various techniques, sources, and capabilities.

14.2.7.3 Devitrification

Devitrification refers to conversion of a glass or another noncrystalline solid into a polycrystalline solid. It sometimes occurs spontaneously during cooling or during application. For example, long-term use of fused SiO_2 above 1000°C (1830°F) results in the formation of regions of cristobalite by devitrification.

Devitrification is used commercially to produce high-quality glass ceramics.[50,51] An example is Corningware™. Corningware is initially fabricated from a lithium aluminum silicate (LAS) glass by conventional glass-forming techniques. The glass is then heat-treated at a temperature at which many tiny crystalline nuclei form throughout the glass. The glass containing crystal nuclei is then increased to a temperature at which the nuclei grow to crystallize all the remaining glass. The result is a very uniform, fine-grained, nonporous ceramic. In the case of LAS, the glass ceramic has a very low coefficient of thermal expansion and excellent thermal shock resistance.

14.2.8 CRYSTALLIZATION

Many ceramic compositions are utilized as single crystals that are carefully crystallized from a melt. A variety of techniques have been developed for growing single crystals from a melt.[52-55] Important examples include the:

1. Verneuil process
2. Czochralski technique
3. EFG technique
4. gradient furnace technique
5. skull melting technique
6. flux technique

The Verneuil process is a flame fusion technique developed between 1886 and 1902 by Auguste Victor Louis Verneuil. A controlled flow of fine Al_2O_3 powder is gravity-fed through a hydrogen-oxygen torch or an oxygen-gas torch at temperatures above 2050°C (3720°F). The Al_2O_3 particles fuse and deposit on an Al_2O_3 pedestal. The position of the pedestal and temperature of the torch are carefully controlled to nucleate and sustain growth of a single crystal. As the Al_2O_3 accumulates and the crystal grows, the pedestal is slowly lowered until a rounded, elongated single-crystal "boule" has formed. Boules as large as 9 cm (3.54 in.) in diameter reportedly have been produced and the total world production in one year was approximately 200,000 kg. Most were machined into "jewel" bearings for watches and instruments. However, this market has decreased since the commercial introduction of electronic watches.

A variety of different colored synthetic gemstones have been produced by the Verneuil process by doping the Al_2O_3 with different impurities. A ruby red color results from Cr additions and a sapphire blue color from Ti plus Fe. Orange is achieved with Ni + Cr + Fe, yellow with Ni, yellow-green with Ni + Fe + Ti, green with Co + V + Ni, and purple with Cr + Ti + Fe.

The Verneuil process does not produce high-optical-quality crystals. The crystals have inclusion, voids, growth rings, distortions, and residual stress and are not suitable for precision optical applications.

The Czochralski process produces much-higher-optical-quality single-crystal Al_2O_3, which is relatively free of inclusions, growth rings, and dislocations. This material is suitable for laser and other optical applications. Al_2O_3 is melted in an iridium crucible. A seed crystal is touched to the

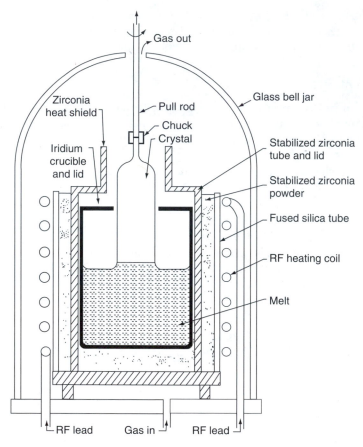

FIGURE 14.29   Apparatus for the Czochralski method for growth of single crystals. (From Nassau, K., *Gems Made by Man*, Chilton Book Co., Radnor, PA, 1980. With permission.)

surface of the melt and slowly withdrawn as illustrated in Figure 14.29. Large sizes are feasible. Crystals reportedly have been grown from a 15-cm (5.9-in.)-diameter crucible.

EFG stands for "edge-defined film-fed growth" technique. The EFG technique is a modification of the Czochralski process and is illustrated in Figure 14.30. A molybdenum die is immersed in the molten alumina. A capillary hole allows molten alumina to be wicked by surface tension to the top of the die. A seed crystal is touched to the surface of the melt at the top of the die and slowly pulled away to start single-crystal growth. A film of the melt spreads over the surface of the die and the growing crystal assumes the cross-sectional shape defined by the sharp edges of the die. The EFG technique can produce shapes similar to those that are produced by extrusion.

The EFG process was developed as an outgrowth of studies to grow single-crystal filaments. Filaments as fine as 0.005 cm (0.002 in.) in diameter have been grown. Continuous filaments 0.025 cm (0.009 in.) in diameter were grown at a rate of 60 m/h (197 ft/h). Twenty-five filaments were grown simultaneously. These had a tensile strength of 2758 MPa (400 ksi). Single-hole and multihole single-crystal tubing over 1.5-m (5-ft) long also have been produced, as well as plates up to 1.25-cm (0.5-in.) thick, 30-cm (11.8-in.) wide, and 30-cm (11.8-in.) long. Some shapes produced by the EFG process are shown in Figure 14.31.

The need for large, high-quality single crystals for lasers, laser windows, substrates, transparent armor, and other applications led to the development of the gradient furnace or heat-exchanger technique. A seed crystal is placed in the bottom of a molybdenum crucible and covered with alumina chips. Helium gas is impinged through a tungsten tube onto the bottom of the crucible directly under the seed crystal. The crucible is heated to melt all of the alumina except the seed crystal,

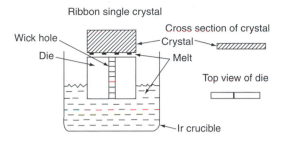

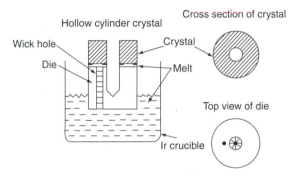

FIGURE 14.30 Schematic illustrating single crystal growth by the edge-defined film-fed (EFG) technique. (Drawing courtesy of ASM International.)

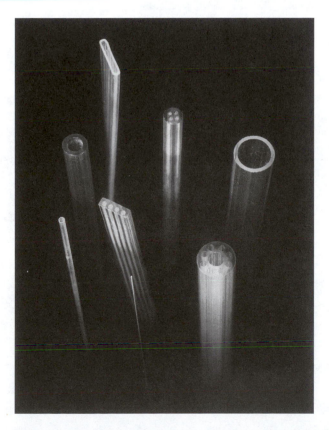

FIGURE 14.31 Examples of single-crystal configurations achieved with the edge-defined film-fed growth technique. (Courtesy of Saphikon Division of Tyco Laboratories, Inc.)

which is kept cooler by the helium flow. The temperature of the melt and seed crystal are then controlled by adjusting the furnace temperature and helium flow to initiate crystallization of the melt on the seed crystal. Continued control of the temperature gradient of the melt and growing crystal results in crystallization of the complete contents of the crucible into a single crystal.

Sapphire crystal free of light scatter have been grown in the 30.5-cm (12-in.) diameter, 12.7-cm (5-in.) thick size range. An example is illustrated in Figure 14.32. Average dislocation densities less than $1 \times 10^3/cm^2$ can be achieved.

FIGURE 14.32   Single-crystal sapphire grown by the heat-exchanger method. (Courtesy of Crystal Systems, Inc., Salem, MA.)

Skull melting involves melting of a powder in a cooled crucible. The powder in the interior of the crucible melts, but the powder in contact with the cooled walls of the crucible does not. This unmelted shell or "skull" essentially forms a crucible of the same composition as the melt.

The skull melting process has been adapted to growth of single crystals of stabilized zirconia as illustrated in Figure 14.33. Zirconia powder, plus the CaO or Y_2O_3 stabilizer, plus metallic zirconium are placed in a water-cooled copper crucible. The zirconium metal is added to make the mix electrically conductive at room temperature. The crucible is surrounded by an induction-heating coil and energy is applied with a radio-frequency generator. The system is designed to obtain induction heating of the zirconium metal first, followed by the zirconia once the temperature is high enough for it to become electrically conductive. The zirconium metal oxidizes and becomes part of the ZrO_2 charge. Adequate cooling is applied to the crucible to prevent the adjacent zirconia from melting. Thus, a thin layer of sintered zirconia (skull) remains lining the crucible walls, preventing

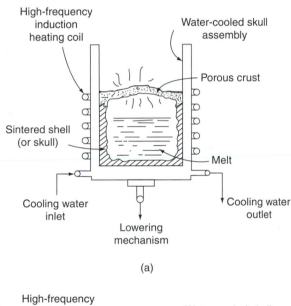

(a)

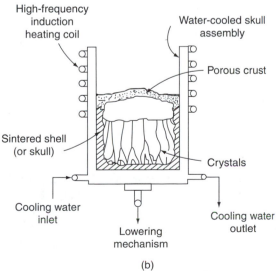

(b)

FIGURE 14.33 Schematics illustrating growth of single crystals of zirconia by the skull-melting technique. (a) During melting and (b) after crystal growth. (From Elwell, D., *Man-Made Gemstones*, Ellis Horwood Ltd., Chichester, England, 1979. With permission.)

contact of the crucible with the molten zirconia. When a uniform melt is achieved inside the crucible, crystal growth is initiated either by gradually reducing the power input or by slowly withdrawing the crucible from the induction-heating coil. Columnar crystals grow from the base of the skull.

Cubic zirconia single crystals have optical properties relatively close to those of diamond. This includes a high refractive index of 2.17 to 2.18 (compared to 2.42 for diamond) and a dispersion of 0.060 (compared to 0.044 for diamond). As a result, faceted cubic zirconia has brilliance similar to that of faceted diamond and has become an important gemstone. Approximately 50 million carats (5000 carats per kg) reportedly were produced in 1980.

14.2.9 VAPOR PROCESSING

Vapor processing involves deposition of a solid from a vapor phase. Vapor processing is generally classified into two categories: (1) chemical vapor deposition (CVD) and (2) physical vapor deposition (PVD).

Chemical vapor deposition is typically conducted by passing a mixture of gases across a heated surface. The temperature is selected such that the gases will react or decompose to form a solid when they come into contact with the surface. For example, vapor phase CH_3SiCl_4 plus a hydrogen gas react in the temperature range of 1000 to 1400°C (1830 to 2550°F) to form a SiC coating on the heated substrate. Other examples are identified in Table 14.11.[56]

Deposition rate is low for CVD, usually less than 250 µm (0.01 in.)/h. The resulting coating is very fine-grained and impervious, is of high purity, and generally has higher hardness than is achieved by conventional particulate ceramic fabrication processes.

CVD in capable of producing high-quality ceramic materials. High-optical-quality glass fiber is produced for fiber optic telecommunications. SiC fibers are produced that have a tensile strength of 3.5 GPa (507.5 ksi) and an elastic modulus of 407 GPa (59×10^6 psi). High-strength, high-toughness composites have been fabricated from these fibers, as shown in Figure 14.34 for a HfB_2-SiC composite. The composite had ultimate tensile strength (in four-point bending) of >1000 MPa (145 ksi) and fractured in a nonbrittle mode. Coatings are also produced by CVD. Coatings of CVD Al_2O_3 and AlN are applied to WC cutting tools. These ceramic coatings, even though only a few microns thick, greatly increase the life of the cutting tools.

The second category of vapor processing is PVD. In this case, material is vaporized from one surface and deposited on another. This can be done by evaporation and condensation or by use of a high-energy source to remove atoms one at a time from a target and deposit them on a substrate. The energy source can be a beam of electrons or ions.

14.2.10 INFILTRATION

Another technique to densify a ceramic is to fill or partially fill the porosity of a particulate compact or a porous ceramic. This is referred to as infiltration. Infiltration can be achieved with a liquid (such as a solution), a melt, a vapor, a polymer, or a sol. The following describe examples of the use of infiltration.

Many water-soluble or solvent-soluble chemicals can be decomposed to form a ceramic material. An example is chromic acid (H_2CrO_4). Porous Al_2O_3 has been infiltrated with liquid chromic acid, dried, and heat-treated to convert the chromic acid to Cr_2O_3. The resulting Cr_2O_3 is very fine-grained and has excellent wear resistance. By a sequence of infiltrations and heat treatments, the porous Al_2O_3 (or alternate substrate material) can be densified to nearly full density. Because of the ultrafine grain size of the Cr_2O_3, the resulting infiltrated material has excellent strength, wear resistance, and tribology characteristics.

Melt infiltration can be conducted with metallic and nonmetallic melts. One example has already been discussed, that is, infiltration of SiC–C with Si. Other options that have been tried include silicate melts and CaF_2.

TABLE 14.11
Ceramic Materials Produced by Chemical Vapor Deposition

Coating	Chemical Mixture	Deposition Temperature °C	Method[a]	Application[b]
		Carbides		
TiC	$TiCl_4$-CH_4-H_2	900–1000	CCVD	wear
	$TiCl_4$-$CH_4(C_2H_2)$-H_2	400–600	PACVD	elec.
HfC	$HfCl_4$-CH_4-H_2	900–1000	CCVD	wear, cor./ox.
ZrC	$ZrCl_4$-CH_4-H_2	900–1000	CCVD	wear, cor./ox.
	$ZrBr_4$-CH_4-H_2	>900	CCVD	wear, cor./ox.
SiC	CH_3SiCl_3-H_2	1000–1400	CCVD	wear, cor./ox.
	SiH_4-C_2H_2	200–500	PACVD	elec., cor.
B_4C	BCl_3-CH_4-H_2	1200–1400	CCVD	wear
B_4C	B_2H_6-CH_4	400	PACVD	wear, elec., cor
W_2C	WF_4-CH_4-H_2	400–700	CCVD	wear
Cr_7C_3	$CrCl_2$-CH_4-H_2	1000–1200	CCVD	wear
Cr_3C_2	$Cr(CO)_6$-CH_4-H_2	1000–1200	CCVD	wear
TaC	$TaCl_5$-CH_4-H_2	1000–1200	CCVD	wear, elec.
VC	VCl_2-CH_4-H_2	1000–1200	CCVD	wear
NbC	$NbCl_5$-CCl_4-H_2	1500–1900	CCVD	wear
		Nitrides		
TiN	$TiCl_4$-N_2-H_2	900–1000	CCVD	wear
	$TiCl_4$-N_2-H_2	250–1000	PACVD	elec.
HfN	$HfCl_4$-N_2-H_2	900–1000	CCVD	wear, cor./ox.
	Hfl_4-NH_3-H_2	>800	CCVD	wear, cor./ox.
Si_3N_4	$SiCl_4$-NH_3-H_2	1000–1400	CCVD	wear, cor./ox.
	SiH_4-NH_3-H_2	250–500	PACVD	wear, cor./ox.
	SiH_4-N_2-H_2	300–400	PACVD	elec.
BN	BCl_3-NH_3-H_2	1000–1400	CCVD	wear
	BCl_3-NH_3-H_2	25–1000	PACVD	elec.
	$BH_3N(C_2H_2)_3$-Ar	25–1000	PACVD	elec.
	$B_3N_3H_6$-Ar	400–700	CCVD	elec., wear
	BF_3-NH_3-H_2	1000–1300	CCVD	wear
	B_2H_6-NH_3-H_2	400–700	PACVD	elec.
ZrN	$ZrCl_4$-N_2-H_2	1100–1200	CCVD	wear, cor./ox.
	$ZrBr_4$-NH_3-H_2	>800	CCVD	wear. cor./ox.
TaN	$TaCl_5$-N_2-H_2	800–1500	CCVD	wear
AlN	$AlCl_3$-NH_3-H_2	800–1200	CCVD	wear
	$AlBr_3$-NH_3-H_2	800–1200	CCVD	wear
	$AlBr_3$-NH_3-H_2	200–800	PACVD	elec., wear
	$Al(CH_3)_3$-NH_3-H_2	900–1100	CCVD	elec., wear
VN	VCl_4-N_2-H_2	900–1200	CCVD	wear
NbN	$NbCl_5$-N_2-H_2	900–1300	CCVD	wear, elec.
		Oxides		
Al_2O_3	$AlCl_3$-CO_2-H_2	900–1100	CCVD	wear, cor./ox.
	$Al(CH_3)_3$-O_2	300–500	CCVD	elec., cor.
	$Al[OCH(CH_3)_2]_3$-O_2	300–500	CCVD	elec., cor
	$Al(OC_2H_5)_3$-O_2	300–500	CCVD	elec., cor
SiO_2	SiH_4-CO_2-H_2	200–600	PACVD	elec., cor.
	SiH_4-N_2O	200–600	PACVD	elec.
TiO_2	$TiCl_4$-H_2O	800–1000	CCVD	wear. cor.
	$TiCl_4$-O_2	25–700	PACVD	elec.
	$Ti[OCH(CH_3)_2]_4$-O_2	25–700	PACVD	elec.

(continued)

TABLE 14.11

(continued)

Coating	Chemical Mixture	Deposition Temperature °C	Method[a]	Application[b]
ZrO_2	$ZrCl_4$-Co_2-H_2	900–1200	CCVD	wear, cor./ox.
Ta_2O_5	$TaCl_5$-O_2-H_2	600–1000	CCVD	wear, cor./ox.
Cr_2O_3	$Cr(CO)_4$-O_2	400–600	CCVD	wear
		Borides		
TiB_2	$TiCl_4$-BCl_3-H_2	800–1000	CCVD	wear, cor/ox.
MoB	$MoCl_5$-BBr_3	1400–1600	CCVD	wear, cor.
WB	WCl_4-BBr_3-H_2	1400–1600	CCVD	wear, cor.
NbB_2	$NbCl_5$-BCl_3-H_2	900–1200	CCVD	wear, cor.
TaB_2	$TaBr_5$-BBr_3	1200–1600	CCVD	wear, cor.
ZrB_2	$ZrCl_4$-BCl_3-H_2	1000–1500	CCVD	wear, cor./ox.
HfB_2	$HfCl_4$-BCl_3-H_2	1000–1600	CCVD	wear, cor.

[a] CCVD = conventional CVD; PACVD plasma-assisted CVD.

[b] Wear = wear-resistant coatings; elec. = electronics; cor. = corrosion-resistant coatings; ox. = oxidation-resistant coatings.

Source: From Stinton, D.P., Besmann, T.M. and Lowden, R.A., Advanced ceramics by chemical vapor deposition techniques, *Am. Ceram. Soc. Bull.*, 67(2), 350–355, 1988, p. 351.

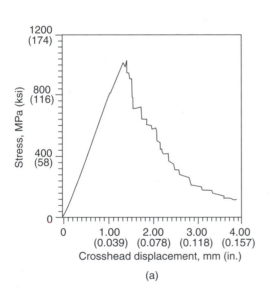

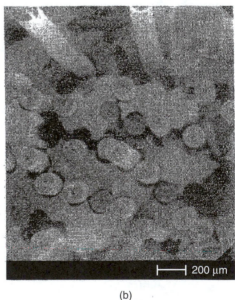

FIGURE 14.34 (a) Load-deflection curve and (b) fracture surface for HfB_2-CVD SiC fiber composite. (Courtesy of D.W. Richerson, Ceramatec, Inc.)

Vapor infiltration is currently being used to densify carbon-carbon composites and is being developed to densify ceramic-ceramic composites. Both of these involve chemical vapor infiltration (CVI). CVI is analogous to CVD, except that the deposition occurs inside a porous preform rather than only on a surface. Carbon-carbon starts with stacked layers of coarse cloth woven loosely from high-strength carbon fibers. Carbon is vapor-deposited on the carbon fibers to bond them together

FIGURE 14.35 Components fabricated by the Lanxide process. (Photo courtesy of the Lanxide Company, Newark, DE.)

and to fill a portion of the porosity. Full densification cannot be achieved. However, the material is still strong and tough. Similar composites have been fabricated with a CVI SiC matrix and a variety of fibers (carbon, SiC, mullite, borosilicate glass).

Polymer infiltration is similar to both liquid and melt infiltration. Some polymers are liquid at room temperature. Others must be heated to achieve low enough viscosity for infiltration. Others must be dissolved in a solvent. The infiltration is normally conducted in a vacuum chamber. Solvents must then be evaporated and the polymer pyrolyzed to yield the desired composition. As with other polymer pyrolysis, a large volume of gases is evolved. Several sequences of infiltration and pyrolysis are required, each filling a little more of the porosity.

Sol gel infiltration is similar to other liquid infiltration. The preform is vacuum-infiltrated with the sol, gelled to immobilize the sol, dried to remove water, and heat-treated. Repeated infiltration cycles are required to achieve density approaching 90% of theoretical.

14.2.11 METAL-GAS REACTION

Metal-gas reactions are a common method of synthesizing powders. Several processes have also been developed for forming a hard ceramic coating or an insulative ceramic coating on metals. Examples include Al_2O_3 on Al (for capacitors) and nitrides and carbides on Ti and other metals. However, relatively few examples exist for the use of metal-gas reactions to produce dense bulk ceramics. The reason is that the initial layer of ceramic that forms on the metal provides an effective diffusion barrier that limits further penetration of the gas. One example of recent interest is the Lanxide process.*

The Lanxide process involves the reaction of a molten metal with a gas to form a ceramic-metal composite on the surface of the metal.[57] As mentioned before, oxidation of a metal generally forms a solid ceramic layer that limits further reaction. This has been altered in the Lanxide process by additives that promote wetting of the ceramic by the alloy and reduce grain boundary stability of the ceramic. As the ceramic crystals grow away from the melt, some edges remain wetted by the metal and act as a wick. This metal is continuously made available to react with the gas such that the growth rate is independent of the thickness.

The initial material fabricated by the Lanxide process was $Al–Al_2O_3$. Additions of Mg and Si to the Al metal provided the wetting action to allow sustained reaction of the metal with oxygen.

---

*Process developed by the Lanxide Company, Newark, DE.

Growth was achieved in the temperature range of 900 to 1400°C (1650 to 2550°F). The Al_2O_3 content increased with increasing reaction temperature, resulting in a range of controllable microstructures and properties. Composites produced at 900°C (1650°F) had a bend strength of 300 MPa (43.5 ksi) and a fracture toughness of 7.8 MPa · $m^{1/2}$ (7.09 ksi · $in.^{1/2}$). Composites produced at 1000°C (1830°F) had a bend strength of 390 MPa (56.5 ksi) and a toughness of 5.4 MPa · $m^{1/2}$ (4.91 ksi · $in.^{1/2}$). Processing at 1150°C (2100°F) resulted in a strength of 525 MPa (76.1 ksi) and toughness of 4.7 MPa · $m^{1/2}$ (4.27 ksi · $in.^{1/2}$). Some parts fabricated by the Lanxide process are shown in Figure 14.35.

Other composites reported to have been fabricated by the Lanxide process include TiN–Ti, ZrB_2–Zr, TiB_2–Ti, and B_4C–B.

REFERENCES

1. Kuczynski, G.C., Hooten, N.A., and Gibson, C.F., Eds., *Sintering and Related Phenomena*, Gordon and Breach, New York, 1967.
2. Burke, J.E. and Rosolowski, J.H., *Sintering, in Treatise on Solid State Chemistry: Reactivity of Solids*, Vol. 4 (Hannay, N.B., Ed.), Plenum, New York, 1976.
3. Kingery, W.D., Bowen, H.K., and Uhlmann, D.R., *Introduction to Ceramics*, Wiley-Interscience, New York, 1976.
4. Yan, M.F., Sintering of ceramics and metals, in *Advances in Powder Technology*, Chin, G.Y., Ed., ASM International, Materials Park, Ohio, 1982.
5. Kuczynski, G.C., Ed., *Materials Science Research, Sintering Processes*, Vol. 13 (Proceedings of the 5th International Conference on Sintering and Related Phenomena), Plenum, New York, 1980.
6. Muller, H.G., Zur Natur der Rekristallisationsvorgange, *Z. Phys.*, 96, 279, 1935.
7. Richerson, D.W., Ceramic fabrication — densification, Lesson 12 from Introduction to Modern Ceramics, ASM Materials Engineering Institute Course 56, ASM International, Materials Park, Ohio, 1990.
8. Hsueh, C.H., Evans, A.G., and Coble, R.L., Microstructural development during final/intermediate stage sintering, I. pore/grain boundary separation, *Acta Metall.*, 30, 1269–1280, 1982.
9. Spears M.A. and Evans, A.G., Microstructural development during final/intermediate stage sintering, II. grain and pore coarsening, *Acta Metall.*, 30, 1281–1290, 1982.
10. Coble, R.L., Sintering crystalline solids. I. intermediate and final state diffusion models and II. experimental test of diffusion models in powder compacts, *J. Appl. Phys.*, 32(5), 787–799, 1961.
11. Reed, J.S., *Principles of Ceramic Processing*, Wiley, New York, 1988.
12. Yeh, T.-S. and Sacks, M.D., Effect of green microstructure on sintering of alumina, in *Ceramic Transactions*, Vol. 7, Handwerker, C.A., Blendell, J.E., and Keyser, W.A., Eds., American Ceramics Society, Westerville, Ohio, 1990, pp. 309–311.
13. Occhionero, M.A. and Halloran, J.W., The influence of green density upon sintering, in *Sintering and Heterogeneous Catalysis*, Kuczynski, G.C., Miller, A.E., and Sargent, G.A., Eds., Plenum, New York, Materials Science Research, Vol. 16, 1984.
14. Cahn, J.W. and Heady, R.B., Analysis of capillary forces in liquid-phase sintering of jagged particles, *J. Am. Ceram. Soc.*, 53(7), 406–409, 1970.
15. Courtney, T.H., Microstructural evolution during liquid phase sintering: I and II, *Met. Trans.*, A (8A), 679–689, 1977.
16. Cutler, R.A. and Jackson, T.B., Liquid phase sintered silicon carbide, in *Ceramic Materials and Components for Engines*, Tennery, V.J., Ed., American Ceramics Society, Westerville, Ohio, 1989, pp. 309–318.
17. Clarke, D.R., Shaw, T.M., and Dimos, D., Issues in the processing of cuprate ceramic superconductors, *J. Am. Ceram. Soc.*, 72(7), 1103–1113, 1989.
18. Sheppard, L.M., Firing technology heats up for the 90s, *Am. Ceram. Soc. Bull.*, 69 (10), 1674–1689, 1990.
19. Sutton, W.H., Brooks, M.H., and Chabinsky, I.J., Eds., *Microwave Processing of Materials*, Vol. 124, Materials Research Society, 1988.
20. Sheppard, L.M., Manufacturing ceramics with microwaves: the potential for economic production, *Am. Ceram. Soc. Bull.*, 67(10), 1656–1661, 1988.

21. Kemer, E.L. and Johnson, D.L., Microwave plasma sintering of alumina, *Am. Ceram. Soc. Bull.*, 64(8), 1132–1136, 1985.

22. Upadhya, K., An innovative technique for plasma processing of ceramics and composite materials, *Am. Ceram. Soc. Bull.*, 67(10), 1691–1694, 1988.

23. Priest, H.F., Priest, G.L., and Gazza, G.E., Sintering of Si_3N_4 under high nitrogen pressure, *J. Am. Ceram. Soc.*, 60, 81, 1977.

24. Greskovich, C., Preparation of high density Si_3N_4 by gas pressure sintering process, *J. Am. Ceram. Soc.*, 64(12), 725–730, 1981.

25. Tani, E., Kishi, K., Umebayashi, S., and Nishijima, M., Gas pressure sintering of composite of Si_3N_4-SiC system, *J. Ceram. Soc. Jpn.*, 95, 916–920, 1987.

26. Vasilos, T. and Spriggs, R.M., The hot pressing of ceramics, *Proc. Brit. Ceram. Soc.*, 3, 195–221.

27. Leipold, M.H., in *Treatise on Materials Science and Technology, Ceramic Fabrication Processes*, Vol. 9, Wang, F.F.Y., Ed., Academic Press, New York, 1976, pp. 95–134.

28. Vieira, J.M. and Brook, R.J., Kinetics of hot pressing: the semilogarithmic law, *J. Am. Ceram. Soc.*, 67(4), 245–249, 1984.

29. Hodge, E.S., Hot isostatic pressing improves powder metallurgy parts, *Mat. Des. Eng.*, 61, 92–97, 1965.

30. Wills, R.R. and Brockway, M.C., Hot isostatic pressing of ceramics, *Proc. Brit. Ceram. Soc.*, 31(6), 1981.

31. Clauer, A.H., Meiners, K.E., and Boyer, C.B., Hot Isostatic Pressing MCIC Rept. MCIC-82-46, Metals and Ceramics Information Ctr., Battelle, Columbus, Ohio, Sept. 1982.

32. Larker, H., Adlerborn, J., and Bohman, H., Fabricating of dense silicon nitride parts by hot isostatic pressing, SAE Paper 770335, 1977.

33. Hardtl, K.H., Gas isostatic hot pressing without molds, *Am. Ceram. Soc. Bull.*, 54, 201, 1975.

34. TenEyck, M.O., Ohnsorg, R.W., and Groseclose, L.E., Hot isostatic pressing of sintered alpha silicon carbide turbine components, ASME paper 87-GT-161.

35. Munir, Z.A., Synthesis of high temperature materials by self-propagating combustion methods, *Am. Ceram. Soc. Bull.*, 67(2), 342–349, 1988.

36. Washburn, M.E. and Coblenz, W.S., Reaction-formed ceramics, *Am. Ceram. Soc. Bull.*, 67(2), 356–363, 1988.

37. Parr, N.L., Silicon nitride, a new ceramic for high temperature engineering and other applications, *Res. (Lon.)*, 13, 261–269, 1960.

38. Popper, P., The preparation of dense self-bonded silicon carbide, in *Special Ceramics*, Popper, P., Ed., Academic Press, New York, 1960, pp. 209–219.

39. Bonnell, D.A. and Tien, T.Y., Eds., *Preparation and Properties of Silicon Nitride Based Materials*, Vol. 47, Materials Science Forum, Translation of Technical Publication, Zurich, Switzerland, 1989.

40. Giachello, A. and Popper, P., Post-sintering of reaction bonded silicon nitride, *Ceram. Int.*, 5(3), 110–114, 1979.

41. Mangels, J., Sintered reaction-bonded silicon nitride, *Cer. Eng. and Sci. Proc.*, 2(7–8), 589–603, 1981.

42. Wygant, J.F., Cementitious bonding in ceramic fabrication, in *Ceramic Fabrication Processes*, Kingery, W.D., Ed., MIT Pres, Cambridge, Mass., 1963, pp. 171–188.

43. Shah, S.P. and Young, J.F., Current research at the NSF Science and Technology Center for Advanced Cement-Based Materials, *Am. Ceram. Soc. Bull.*, 69(8), 1319–1331, 1990.

44. Kingery, W.D., Fundamental study of phosphate bonding in refractories, *J. Am. Ceram. Soc.*, 33(8), 239–250, 1950.

45. Buckley, J.D., Carbon-carbon, an overview, *Am. Ceram. Soc. Bull.*, 67(2), 364–368, 1988.

46. Wynne, K.J. and Rice, R.W., Ceramics via polymer pyrolysis, *Ann. Rev. Mat. Sci.*, 14, 297–334, 1984.

47. Wills, R.R., et al., Siloxanes, silanes, and silazanes in the preparation of ceramics and glasses, *Am. Ceram. Soc. Bull.*, 62(8), 1983.

48. Yajima, S., Special heat-resisting materials from organometallic polymers, *Am. Ceram. Soc. Bull.*, 62(8), 1983.

49. Prewo, K.M. and Brennan, J.J., Properties of silicon carbide fiber–glass composites, *J. Mat. Sci.*, 17, 1201–1206, 1982.

50. Stookey, S.D., Catalyzed crystallization of glass in theory and practice, *Ind. Eng. Chem.*, 51(7), 805–808, 1959.

51. Beall, G.H., Properties and process development in glass-ceramic materials, in *Glass: Current Issues*, Wright, A.F. and Dupuy, J., Eds., North-Holland, New York, 1985, pp. 31–48.

52. Paladino, A.E. and Roiter, B.D., Czochralski growth of sapphire, *J. Am. Ceram. Soc.*, 47(9), 465, 1980.
53. LaBelle, H.E., Jr., EFG, the invention and application to sapphire growth, *J. Cryst. Growth* 50, 8–17, 1980.
54. Schmid, F. and Viechnicki, D., Growth of sapphire disks from the melt by a gradient furnace technique, *J. Am. Ceram. Soc.*, 53(9), 528–529, 1970.
55. Nassau, K., *Gems Made by Man*, Chilton Book Co., Radnor, Pa., 1980.
56. Stinton, D.P., Besmann, T.M. and Lowden, R.A., Advanced ceramics by chemical vapor deposition techniques, *Am. Ceram. Soc. Bull.*, 67(2), 350–355, 1988.
57. Newkirk, M.S., et al., Preparation of Lanxide™ ceramic matrix composites: matrix formation by the directed oxidation of molten metals, *Ceram. Eng. and Sci. Proc.*, 8(7–8), 879–885, 1987.

ADDITIONAL RECOMMENDED READING

1. Handwerker, C.A., Blendell, J.E., and Kaysser, W.A., Eds., Sintering of Advanced Ceramics, Ceramic Transactions, Vol. 7, American Ceramic Society, Westerville, Ohio, 1990.
2. Fulrath, R.M. and Pask, J.A., Eds., *Ceramic Microstructures*, John Wiley & Sons, New York, 1968.
3. Rahaman, M.N., *Ceramic Processing and Sintering*, Marcel Dekker, New York, 1995.
4. Firing/Sintering: Densification, Section 4 pp. 242–312 in ASM Engineered Materials Handbook Vol. 4: CERAMICS AND GLASSES, ASM Int., Materials Park, Ohio, 1991.
5. Pejovnik, S. and Kristic, M.M., Eds., *Sintering — Theory and Practice*, Elsevier Scientific, 1984.
6. German, R.M., *Liquid Phase Sintering*, Plenum Press, New York, 1985.

PROBLEMS

1. A compact of silicon particles has a bulk density of $1.57 \, g/cm^3$. The compact is nitrided to achieve complete conversion of the Si to Si_3N_4. What is the bulk density of the resulting Si_3N_4 part?

2. Tetramethylcyclotetrasilazane, $C_4H_{20}Si_4N_4$, can be pyrolyzed to yield a combination of silicon nitride and silicon carbide. Assuming that all the Si and N and a portion of the C go to form Si_3N_4 plus SiC, and that the rest of the C and all the H form volatile species, what is the theoretical weight loss during pyrolysis?

3. What type of cement is magnesium oxychloride?
   (a) reaction
   (b) hydraulic
   (c) precipitation
   (d) sol–gel

4. Ten wt% (Al_2O_3 plus Al_4C_3) are added to SiC to achieve liquid-phase sintering. To achieve the maximum amount of liquid at about 1835°C (3335°F), how much Al_2O_3 is required for a 1000-g batch? (Hint: Use Figure 14.15.)

STUDY GUIDE

1. What is the meaning of the term sintering?
2. What criteria must be met before sintering can occur?
3. What are the primary mechanisms of sintering and the source of energy?
4. Describe the physical changes that occur during each stage of sintering.
5. What are the key mechanisms of densification during the final stage of sintering?
6. Explain the factors that lead to grain growth.
7. Explain the driving force for solid state sintering.
8. What are some of the key factors that influence the rate of solid state sintering?
9. Explain how the presence of a liquid phase influences sintering.

10. How can the microstructure vary for liquid phase sintered ceramics, and how can phase equilibrium diagrams be used to predict the microstructure?
11. Identify some ceramic materials that have been densified by liquid phase sintering.
12. What is reactive liquid sintering, and how might it provide a benefit?
13. What parameters or factors must be controlled during sintering to achieve a high quality, reproducible product?
14. What are some problems or defects that can result during sintering?
15. What are some advantages and disadvantages of hot pressing?
16. How does hot isostatic pressing resolve some of the limitations of uniaxial hot pressing?
17. Describe how silicon nitride and silicon carbide materials can be fabricated by reaction-sintering processes.
18. Compare hydraulic, reaction, and precipitation cementitious bonding.
19. What are several ways to prepare a ceramic from a melt?
20. Briefly describe or compare six different methods of growing single crystals.
21. What other methods can be used to produce or densify ceramics?

15 Final Machining

Some ceramic parts can be fabricated to net shape by the methods described in Chapter 14. However, more frequently, machining of some of the surfaces is required to meet dimensional tolerances, achieve improved surface finish, or remove surface flaws. This machining can represent a significant portion of the cost of fabrication and thus should be minimized and conducted as efficiently as possible.

15.1 MECHANISMS OF MATERIAL REMOVAL

Ceramic materials are difficult and expensive to machine due to their high hardness and brittle nature. Machining must be done carefully to avoid brittle fracture of the component. Most ceramics cannot be successfully machined with the type of cutting tools used for metal because these tools are either not hard enough to cut the ceramic or because they apply too great a local tensile load and cause fracture. The tool must have a higher hardness than the ceramic being machined and must be of a configuration that removes surface stock without overstressing the component.

Ceramic material can be removed by mechanical, thermal, or chemical action. Mechanical approaches are used most commonly and are discussed first. They can be divided into three categories: mounted abrasive, free abrasive, and impact.

15.1.1 MOUNTED-ABRASIVE MACHINING

Mounted-abrasive tools consist of small, hard, abrasive particles bonded to or immersed in a softer matrix. The abrasive particles can be diamond, SiC, Al_2O_3, Al_2O_3–ZrO_2, or other hard ceramic material, and the matrix can be rubber, organic resin, glass, or a crystalline ceramic composition softer than the abrasive particles. Good examples are the wide variety of grinding wheels used extensively in home workshops and industry. For machining very hard ceramics such as Al_2O_3, Si_3N_4, and SiC, diamond is usually the most efficient abrasive, mounted in a matrix of soft metal or organic resin.

Mounted-abrasive tools can be made in a wide variety of configurations and compositions. Coarse abrasives are used for rough machining, where rapid stock removal is desired. Finer abrasives are used for final machining, where close tolerances and smooth surface finishes are required.

Stock removal is achieved by moving the tool in relation to the ceramic workpiece while simultaneously applying pressure. The abrasive particles are small and irregular in shape such that a sharp corner of the particle is usually in contact with the ceramic. This small contact area produces high localized stress concentration and the particle plows a microscopic groove across the surface of the ceramic. The larger the abrasive, the larger the groove and the greater the depth of damage in the ceramic. This will be discussed in more detail later in a section on effects of machining on material strength, including suggested procedures to minimize strength reduction.

Schematics of commonly used mounted-abrasive tooling are illustrated in Figure 15.1.[1]

15.1.2 FREE-ABRASIVE MACHINING

Free-abrasive machining consists of the use of loose abrasive and is usually used for achieving the final surface finish with very fine particle size abrasive. Lapping is the most commonly used free abrasive approach. The fine abrasive is placed on a soft material such as cloth or wood, which is then moved in relation to the ceramic being lapped. Because of the fine abrasive size, material-removal rate is very slow. However, very smooth surfaces with flatness measured in wavelengths of

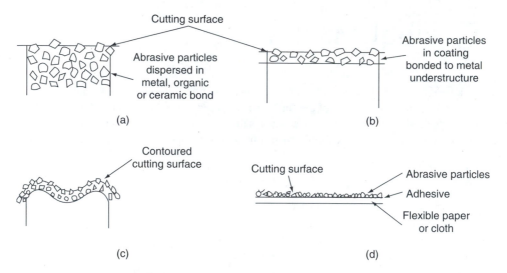

FIGURE 15.1 Schematics of types of mounted abrasive tooling. (a) Conventional grinding wheel, (b) coated abrasive grinding wheel, (c) formed grinding wheel, and (d) abrasive paper or cloth. (Drawing courtesy of ASM International.)

light can be achieved. These surfaces result in very low friction and have been used in bearing and seal applications.

Free abrasive is also used in trepanning. This is a technique used for drilling circular holes in a ceramic. The tool consists of a thin-walled hollow cylinder of a soft metal such as brass. The loose abrasive is placed between the tool and the workpiece together with a coolant such as water or oil, and pressure is applied simultaneously with rotation to achieve grinding action.

15.1.3 Impact Abrasive Machining

The two previous machining methods remove stock essentially by sliding motion. Material can also be removed by impact. This is generally achieved by accelerating loose abrasive particles to high velocity such that they cause local impact damage when they strike the workpiece. Important approaches include sandblasting,[2] water-jet machining,[3] and ultrasonic machining.[4]

Two types of sandblasting (abrasive blasting) are commonly used: conventional sandblasting and wheelabrasion. In conventional sandblasting, the abrasive particles are accelerated by compressed air and directed through a nozzle at high velocity against the workpiece. In wheel-abrasion, the abrasive particles are accelerated by a paddle wheel rotating at high velocity. The abrasive particles are fed into the wheel at the center, accelerated along the length of the paddles, and essentially thrown at high velocity against the workpiece.

Sandblasting is used for cleaning the surface of ceramics and as a test method for evaluating the wear resistance of materials. It is also used in sculpture and for fabrication of tombstones, but is not used often for fabrication of ceramic components for engineering applications. The primary drawbacks are difficulty in achieving close tolerances or a uniform surface.

Wheel-abrasion is used primarily for cleaning and peening of metal surfaces.

Rate of material removal for a ceramic workpiece increases with particle size, hardness, and velocity of the abrasive, and for angles of impingement approaching 90°. Al_2O_3, SiC, and SiO_2 are the most commonly used abrasives.

Water-jet machining is similar to sandblasting, except that the abrasive is carried in water rather than air and is focused through a nozzle to a small-diameter cutting point. The abrasive is accelerated by water pumped at around 413 MPa (60 ksi). Water-jet machining is gaining in popularity as a low-cost method of accurately cutting metals and ceramics.

Ultrasonic machining uses impact in a much different way than sandblasting and water-jet machining. Very small abrasive particles are suspended in a water slurry that flows over the surface of a shaped tool. The tool is vibrated at high frequency, which accelerates each abrasive particle over a very short distance to strike the workpiece. Impact occurs only where the tool is in close proximity to the workpiece, so that close tolerances can be achieved by control of the tool and abrasive dimensions. No pressure is applied between the tool and abrasive so that very little strength-limiting damage to the ceramic results. Boron carbide is frequently used as the abrasive because it is harder than most other ceramics (except diamond), is less expensive than diamond, and is available in narrow-size ranges.

The volume removal of material by ultrasonic machining is low compared to mounted abrasive techniques. However, ultrasonic machining reproduces the shape of the tool and can produce complex shapes to close tolerances. It can emboss a shape or contour into a surface. It can cut shaped holes or cross sections analogous to those made with a cookie cutter. Figure 15.2 shows the schematic of an ultrasonic machining apparatus. Figure 15.3 illustrates examples of shapes machined ultrasonically.

15.1.4 CHEMICAL MACHINING

Chemical machining is used primarily to achieve improved surface finish and thus increased strength or decreased friction. It is achieved by immersing the surface to be machined into a liquid in which the ceramic is soluble. Most silicate glass compositions can be etched or chemically machined with hydrofluoric acid (HF). Al_2O_3 can be etched with molten $Na_2B_3O_7$. These treatments are often referred to as chemical polishing because they produce such a smooth surface.

15.1.4.1 Photoetching

Some glass compositions can be chemically machined into very complex geometries using photoetching.[5] One such glass contains Ce_2O_3 and Cu_2O. A mask or photo negative is placed on the glass and irradiated with ultraviolet light. In the unmasked areas the Cu_2O is reduced by the Ce_2O_3 by the reaction $Ce^{3+} + Cu^+ \rightarrow Ce^{4+} + Cu$. The glass is then exposed to a controlled heat treatment in which the Cu particles act as nucleation sites for localized crystallization. The crystallized material can be etched in hydrofluoric acid at a rate 15 times the rate of the original glass.

Some very intricate configurations have been produced by this photoetching technique. An example is a 600-mesh sieve.

15.1.4.2 Electrical Discharge Machining

Electrical discharge machining (EDM) can be performed only with electrically conductive materials. A shaped tool is held in close proximity to the part being machined, retaining a constant predetermined gap with the use of a servomechanism that responds to change in the gap voltage. A dielectric liquid is flowed continuously between the tool and workpiece. Sparks produced by electrical discharge across this dielectric erode the ceramic by a combination of vaporization, cavitation, and thermal shock produced by the intense local heating.[6,7]

EDM has been used successfully with conductive carbides, silicides, borides, and nitrides. The advantages are that no mechanical load is applied during EDM and that holes, recesses, and outer dimensions can be produced in the same types of shapes that could be formed in metal by stamping or in dough by a cookie cutter. The disadvantages are the slow rate of cut, the limitation to conductive materials, and the relatively poor surface finish achieved. The surface is typically pitted and microcracked and results in substantial strength reduction.

15.1.4.3 Laser Machining

Only a few studies have been reported on laser machining of ceramics. Lumley[8] and Brody and Molines[9] report the use of laser machining to score Al_2O_3 electronic substrates to allow them to be fractured to the desired size. The mechanism of material removal appeared to be localized thermal shock spalling.

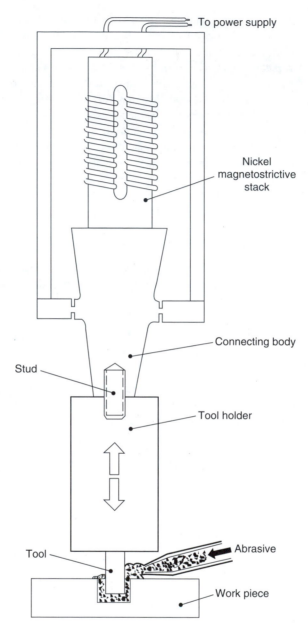

FIGURE 15.2 Schematic of an ultrasonic machining setup. (From Moreland, M.A., in *Ceramics Applications in Manufacturing*, Richerson, D.W., Ed., Society of Manufacturing Engineers, Dearborn, MI, 1988. With permission.)

Copley et al.[10] have reported machining of SiC, Si_3N_4, and SiAlON using a CO_2 laser. In this case material removal was apparently by evaporation, since all three of these materials decompose rather than melt. A hot-pressed Si_3N_4 cylinder containing 1/4 in. × 20 threads was laser-machined from rectangular stock. A SiAlON cylinder was also machined. Surface finish measurement determined that the maximum nonsmoothness from peak to valley was 7.5 μm. This suggests that laser machining of ceramics may be a feasible approach and should be evaluated further. In particular, the effects on material strength should be assessed.

FIGURE 15.3   Examples of shapes fabricated by ultrasonic machining. (Courtesy of Bullen Ultrasonics, Inc., Eaton, OH.)

15.2   EFFECTS ON STRENGTH

To understand the effects of machining on the strength of a ceramic material, we must examine the interactions that occur at the tool-workpiece interface and define the flaws that are initiated in the ceramic. First, let us consider a single mounted-abrasive particle plowing a furrow in a ceramic workpiece. Material directly in the path of the abrasive particle sees very high stress and temperatures and is broken and deformed. Material adjacent to the abrasive particle is placed in compression and may also deform plastically. After the abrasive particle passes, this material rebounds and either cracks or spalls off, due to the resulting tensile stresses.[11,12] Thus, the size of the machining groove for most ceramics is larger than the size of the abrasive particle.

Figure 15.4 shows schematically the types of cracks that can form adjacent to the grinding groove.[13–17] The median crack is parallel to the direction of grinding and perpendicular to the surface, and results from the high stresses at the bottom of the grinding groove. Because it is parallel to the direction of grinding, it has also been called a *longitudinal crack*. It is usually the deepest crack and produces the greatest strength reduction.

Lateral cracks are parallel to the surface and extend away from the plastic zone. They result from the high tensile stress that exists at the edge of the plastic zone and extend as the material relaxes immediately after the abrasive particle passes. Lateral cracks tend to curve toward the surface and often result in a chip spalling off. Because lateral cracks are parallel to the surface, they do not result in stress concentration during subsequent mechanical loading and thus do not significantly reduce the strength of the material. However, they do account for a substantial portion of stock removal during grinding.

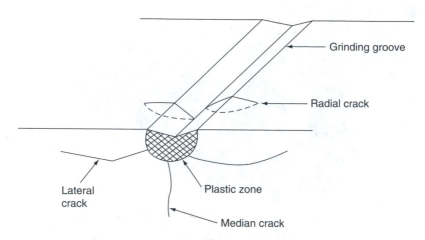

FIGURE 15.4 Schematic showing the cracks and material deformation that occurs during grinding with a single abrasive particle.

Radial cracks normally result from single-particle impact or indentation and extend radially from the point of impact. They are perpendicular to the surface, but are usually shallow and do not degrade the strength as much as a median crack. The cracks shown in Figure 15.4, which are perpendicular to the grinding groove, are analogous to radial cracks. There can be many of these along the length of the grinding groove. They have been referred to in the literature as transverse, chatter, or crescent cracks, but their mechanism of formation has generally not been discussed. It is most likely that they are initiated by the high tensile stress that arises at the trailing edge of the contact of the abrasive particle and the workpiece. This biaxial stress mechanism for high-friction situations was discussed in Chapter 9.

15.2.1 EFFECT OF GRINDING DIRECTION

Most ceramics are machined with tools containing many abrasive particles rather than just one single point. However, it is likely that the resulting surface flaws are similar for both types of tools and that the median and radial cracks control the strength. Which flaw controls the strength depends on the orientation of the grinding grooves to the direction of stress application. This is shown schematically in Figure 15.5 for specimens loaded in bending.

As the load is applied and the specimens begin to bend, stress concentration will occur at the tips of cracks perpendicular to the stress axis but not at cracks parallel to the stress axis. Thus, for specimens ground in the longitudinal direction, stress concentration will occur at the radial (transverse) cracks, and for specimens ground in the transverse direction, stress concentration will occur at the median (longitudinal) cracks. Since the median cracks are usually the most severe, one would expect the strength to be lowest for transverse grinding, where the grooves and median cracks are perpendicular to the tensile stress axis. This can be seen in Table 15.1.

From Table 15.1 it is obvious that substantial differences in load-bearing capability for a ceramic component can result, depending on the orientation of grinding with respect to the stress distribution in the component. This anisotropy of strength is an important consideration for an engineer designing a component that must withstand high stress.

15.2.2 EFFECTS OF MICROSTRUCTURE

The microstructure of the ceramic material has a pronounced effect on the rate of machining and on the residual strength after machining. Rice[11] reports that fine-grained ceramics require higher grinding

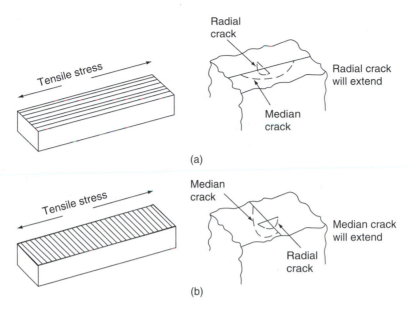

FIGURE 15.5 Grinding direction and crack distribution vs. tensile stress axis.

TABLE 15.1

Strength vs. Grinding Direction Orientation with Respect to Tensile Stress Axis

Material	Longitudinal Grinding		Transverse Grinding	
	MPa	psi	MPa	psi
Hot-pressed Si_3N_4	669	97,000	428	62,000
Soda-lime glass	97	14,100	68	9,900
Mullite	319	46,300	259	37,600
MgF_2	87	12,600	53	7,700
B_4C	374	54,200	154	22,300

Source: From Richerson, D.W., Schuldies, J.J., Yonushonis, T.M., and Johansen, K.M., ARPA/Navy ceramic engine materials and process development summary, in *Ceramics for High Performance Applications*, *II*, Burke, J.J., Lenoe, E.N., and Katz, R.N., Eds., Brook Hill, Chestnut Hill, MA, 1978, pp. 625–650. (Available from MCIC, Battelle Columbus Labs., Columbus, OH) for Si_3N_4 data and Rice, R.W. and Mecholsky, J.J., The nature of strength-controlling machining flaws in ceramics, in *The Science of Ceramic Machining and Surface Finishing*, *II*, Hockey, B.J. and Rice, R.W., Eds., NBS Special Publication 562, U.S. Government Printing Office, Washington, DC, 1979, pp. 351–378. for data for other materials.

force and longer time to slice or machine. This is shown for several ceramic materials in Figure 15.6. Uniformly distributed porosity increases the rate of machining, but also decreases the smoothness of surface finish that can be achieved.

The degree of strength reduction resulting from machining is dependent on a comparison of the size of flaws initially present in the ceramic to those produced by machining. Machining has very little effect on the strength of ceramics containing high porosity or large grain size because the

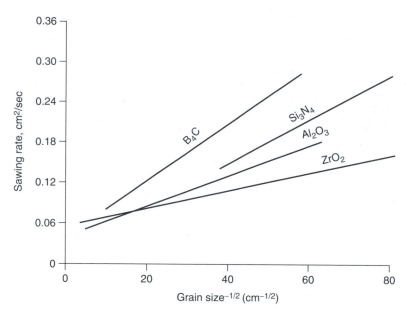

FIGURE 15.6 Effect of machining rate on grain size. (From Rice, R.W., Machining of ceramics, in *Ceramics for High Performance Applications*, Burke, J.J., Gorum, A.E., and Katz, R.N., Eds., Brook Hill, Chestnut Hill, MA, 1974, pp. 287–343.)

flaws introduced during machining are no larger than the microstructure flaws initially present. On the other hand, the strength of fine-grained ceramics such as most Si_3N_4 and Al_2O_3 can be reduced significantly.

15.2.3 EFFECTS OF GRINDING PARAMETERS

The parameters selected for machining a ceramic have a large effect on the rate of machining and tool wear and on the resulting properties of the ceramic. Table 15.2 summarizes general trends associated with variations in individual machining parameters. It should be emphasized that these are just trends and that they may not hold true for all ceramic materials and all levels of variation. Also, most of these parameters are interactive and may have a different effect when combined than when considered individually.

Large abrasive particles result in a greater depth of grinding damage and are used for roughing operations. Finer abrasives are used to remove the subsurface damage produced by the large abrasive during roughing and to achieve the final tolerances and surface finish. Table 15.3 shows the relative surface finish one might expect for surface grinding with different diamond grit sizes.[20]

To achieve maximum strength, grinding is usually done in several steps, decreasing the abrasive size in each step and removing enough surface stock to remove subsurface damage resulting from the prior step. Sometimes, intermediate machining steps are skipped to save time and decrease cost. The desired tolerances and surface finish can be achieved, but the strength requirements may not, and the component may fail in service.

Harder ceramic materials are more difficult to cut and grind than softer materials and require higher force.[21] This results in increased wheel wear, higher interface temperatures, and greater danger of damage to the ceramic.

However, fracture toughness of the ceramic workpiece also affects the difficulty of machining and the degree of machining damage. Figure 15.7 compares the normal force required to surface

TABLE 15.2
Effects of Machining Parameters

Parameter Variation	Effect on Machining Rate[a]	Effect on Tool Wear[a]	Effect on Material Strength[a]
Increasing abrasive size	Increases rate	Depends more on other factors	Decreases strength
Increasing downfeed or table speed	Increases rate	Increases wear	Decreases strength
Increasing wheel speed	Increases rate	Usually decreases wear	Increases strength
Use of lubricants and coolants	May increase rate	Usually decreases wear	Usually increases strength
Increasing abrasive hardness	Increases rate	Decreases wear	Depends more on other factors
Changing abrasive concentration and wheel bond	Different optimum for different materials and configurations	Can be optimized	Can be optimized

[a] Relative to an arbitrary baseline.

TABLE 15.3
Diamond Grit Size vs. Expected Surface Finish

| Grit size | Particle Size | | Expected Surface |
	μ	in.	Finish (RA)
60	250	0.010	100
80	177	0.007	64
150	105	0.004	32
180	88	0.0035	32
220	74	0.0030	20
320	44	0.0017	12
400	37	0.0015	8
600	30	0.0012	4
1200	15	0.0006	2

Source: From Chand, R.H., Costello, K.P., and Payne, T.M., Diamond Wheels put finish on ceramics, *Ceram. Indus.*, 42–44, 1990.

grind hot-pressed Si_3N_4 (HPSN), transformation toughened ZrO_2, ferrite, Al_2O_3–TiC, and Co-bonded WC. The ferrite has the lowest hardness and toughness, the ZrO_2 relatively low hardness and high toughness, the Al_2O_3–TiC moderate hardness and low toughness, the Si_3N_4 moderate hardness and moderate toughness, and the Co-bonded WC moderate hardness and high toughness.

Increased wheel speed decreases the required force and usually results in lower tool wear, lower temperatures, and less surface damage. The use of water or other suitable lubricant or coolant provides similar benefits. Table 15.4 identifies surface speed vs. wheel rpm for various-diameter grinding wheels.

15.2.4 OPTIMIZATION OF GRINDING

Optimization of grinding requires a systems approach in which the grinding machine, the grinding wheel, the grinding parameters, and the characteristics and specifications of the ceramic workpiece

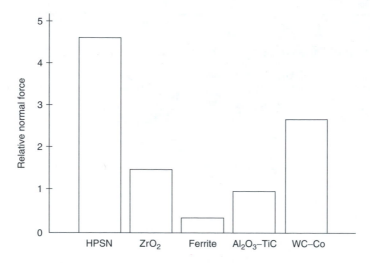

FIGURE 15.7 Relative normal force to surface grind ceramics of various hardness and toughness. (Adapted from Subramanian, K., Ramanath, S., and Matsuda, Y., Diamond abrasives cut path to volume production, *Ceram. Indus.*, 30–32, 1990.)

TABLE 15.4
Surface Speed vs. rpm for Various-Diameter Grinding Wheels

Wheel Diameter (in.)	Surface Speed (ft/min)	Wheel (rpm)
5	4500	3450
6	5300	3450
7	6300	3450
8	7250	3450
10	4600	1750
12	5500	1750
14	6400	1750
16	7300	1750

Source: From Chand, R.H., Costello, K.P., and Payne, T.M., Diamond Wheels put finish on ceramics, *Ceram. Indus.*, 42–44, 1990.

---

are all considered.[22] Table 15.5 lists some of the key factors for each of these. Tables 15.6 through 15.8 provide information about diamond concentration terminology, a description of types of grinding wheels, and recommendations for specific grinding wheels for specific ceramic workpieces.

Optimization of grinding minimizes cost, as well as reducing the strength degradation of the ceramic. However, acceptable strength and surface finish cannot always be achieved by grinding. Postmachining procedures have been developed to obtain further improvement in the properties. These include the following:

lapping
annealing
oxidation
chemical etching
surface compression
flame polishing

TABLE 15.5
Key Factors for Optimization of Machining Ceramics

Machine	High rigidity and stiffness
	Low vibration levels
	Effective coolant system
	Precision positioning
	Precision movements
	On-machine dynamic balancing
	Truing and dressing systems
	Preferably multiaxis CNC capability
Grinding wheel	Shape or design of wheel (straight, cup, etc.)
	Composition of abrasive (diamond, cubic BN, etc.)
	Type of abrasive (single-crystal, polycrystalline, porous, etc.)
	Grit size of abrasive
	Concentration of abrasive
	Characteristics of bond or matrix (hardness, thermal conductivity, stiffness, porosity, etc.)
Grinding parameters	Power
	Normal load
	Wheel speed
	Rate of downfeed
	Table speed
	Method of holding ceramic workpiece
	Coolant parameters
	Truing, dressing procedures
Ceramic workpiece	Hardness, toughness, strength
	Grain size, porosity
	Thermal properties
	Elastic properties
	Melting, creep, decomposition properties
	Final shape and tolerances required
	Final strength required

TABLE 15.6
Definition of Diamond Concentration Terminology

Standard Diamond Concentration	Vol. % of Diamond	Carats/in.^{3}
100	25.00	72
75	18.75	54
50	12.50	36
25	6.25	18

Source: From Chand, R.H., Costello, K.P., and Payne, T.M., Diamond Wheels put finish on ceramics, *Ceram. Indus.*, 42–44, 1990.

Each of these approaches either removes surface and subsurface flaws resulting from machining or reduces the stress concentration due to the flaw.

15.2.4.1 Lapping

Lapping involves the use of very fine free abrasive particles suspended in a slurry and applied to the workpiece surface with a tool surface such as metal, cloth, or wood. It is equivalent to

TABLE 15.7

Features of the Bond or Matrix Options for Diamond Wheels

Bond Type	Features
Resin	Readily available
	Easy to true and dress
	Moderate "freeness" of cut
	Applicable for a range of operations
	First selection for learning the use of diamond wheels
Vitrified	Free cutting
	Easy to true
	Needs no dressing (if selected and trued properly)
	Controlled porosity to enable coolant flow to the grinding zone and chip removal
	Intricate forms can be crush-formed on the wheels
	Suitable for creep feed or deep grinding, inside-diameter grinding, or high-conformity grinding
	Potentially longer wheel life than resin bond
	Excellent under oil as coolant
Metal	Very durable
	Excellent for thin slot, groove, cutoff, simple form, or slot grinding
	High stiffness
	Good form holding
	Good thermal conductivity
	Potential for high-speed operation
	Generally requires high grinding forces and power
	Difficult to true and dress
Electroplated	Single abrasive layer plated on a premachined steel preform
	Extremely free cutting
	High material-removal rate
	Form wheels are easily produced
	From accuracy depends on perform and plating accuracy
	High abrasive content
	Generally not truable
	Poor surface finish compared to bonded abrasive wheels

Source: From Subramanian, K., Precision finishing of ceramic components with diamond abrasives, *Am. Ceram. Soc. Bull.*, 67(6), 1026–1029, 1988.

mechanical polishing used in preparation of metallographic samples for microstructure examination. The size of the abrasive determines the final surface finish that can be achieved. However, as mentioned before, the strength does not necessarily increase as the surface smoothness increases. To achieve strength increase, machining and lapping must be done in a diminishing abrasive size sequence such that each step successfully removes the worst surface damage produced by the prior step. A suitable sequence might be to rough machine with 200-grit diamond, finish machine with 320- and 600-grit diamond, rough lap with 30- and 9-μm Al_2O_3 or CeO_2, and finish lap with 3-, 0.3-, and 0.06-μm Al_2O_3 or CeO_2. Final lapping can also be done with diamond.

The degree of lapping or surface finishing is obviously dependent on the cost restraints and the criticality of the application. Because of the fine grain structure and hardness of advanced ceramic materials, excellent surface finish and tolerances can be achieved. For example, the following

TABLE 15.8

Examples of Grinding Wheel and Coolant Options Successfully Used with Various Ceramic Materials

Material to be Ground	Suggested Wheel Specification		Coolant pH Level
	Rough	Finish	
Hot-pressed	180-grit	320-grit	High alkaline
B_4C	Resinoid	Resinoid	High flow
	100 conc.	125 conc.	
Hot-pressed	150	320	High alkaline
Si_3N_4	R	R	High flow
	100	100	
Hot-pressed	120	320	Alkaline
SiC	R	R	
	75	75	
Sintered	220	320	High alkaline
SiC	R	R	
	100	100	
High-density	220	320	Neutral
Al_2O_3,	R	R	
Zr_2O_3	75	50	
SiC Whisker	150	320	Neutral
Reinforced	R	R	
Al_2O_3	75	75	
Al_2O_3	80	180	Neutral
	Metal	Metal	
	50	50	

Source: From Chand, R.H., Costello, K.P., and Payne, T.M., Diamond Wheels put finish on ceramics, *Ceram. Indus.*, 42–44, 1990.

capabilities have been reported for dense Al_2O_3 by Western Gold and Platinum Co. (WESGO) in their Technical Circular No. L-779:

flat lapping to half a light band (0.000006 in. or 0.000152 mm flatness)
parallelism to 0.000010 in. (0.00025 mm)
dimensional tolerances of 0.000010 in. (0.00025 mm)
cylindrical outside diameters between 0.060 in. (1.524 mm) and 3 in.
    (76.2 mm) lapped to 0.000005 in. (0.000127 mm) dimensional tolerance
    and roundness to 0.000005 in. (0.000127 mm)
stepped diameters lapped to 0.000050 in. (0.00127 mm) concentricity
outside-inside mating diameters lapped to 0.000050 in. (0.00127 mm)
    tolerance of mating clearance
cylindrical inside diameters between 0.060 in. (1.524 mm) and 2 in.
    (50.8 mm) lapped to 0.000005 in. (0.000127 mm) dimensional tolerance
    and roundness to 0.000005 in. (0.000127 mm)
blind holes and intersecting holes lapped to 0.000005 in. (0.000127 mm) dimensional tolerance
    and roundness to 0.000005 in. (0.000127 mm)
radial and spherical concave and convex surfaces lapped to within 0.000010 in. (0.00025 mm)
    of true radius and 0.000010 in. (0.00025 mm) roundness

Most applications do not require such precise lapping. Some of the more critical applications in terms of surface finish or tolerances include optical glass, laser ceramics, bearings, seals, some papermaking components, and thread guides.

15.2.4.2 Annealing

Since ceramic materials are normally processed at very high temperature, internal stresses often result during cooldown. Sometimes these residual stresses can improve strength, but often they reduce strength. Annealing at high temperature followed by slow cooldown can often relieve these stresses. Annealing can also relieve surface stresses resulting from machining and actually heal subsurface flaws such as median cracks. Annealing can also crystallize glass phases and achieve improved strength or stability.

15.2.4.3 Oxidation

Oxidation has been demonstrated to increase the strength of machined hot-pressed Si_3N_4[18] and other nitrides and carbides. In the case of Si_3N_4, the reaction $Si_3N_4 + 3O_2 \rightarrow 3SiO_2 + 2N_2$ can increase the strength by completely removing the depth of surface containing the residual machining cracks or by rounding the crack tip and reducing stress concentration. This approach was used effectively for improving the load-bearing capability of hot-pressed Si_3N_4 rotor blades.[18] The blade machining process resulted in transverse grinding perpendicular to the tensile stress axis such that the median crack was strength-controlling. The as-machined strength was about 428 MPa (62,000 psi). Oxidation at 960°C (1800°F) for 50 h increased the strength to 635 MPa (92,000 psi).

15.2.4.4 Chemical Etching

Chemical etching has been used for many years for removing machining or other surface damage, especially the use of hydrofluoric acid with glass. Strength increases greater than tenfold have been achieved both for glass compositions and for oxide ceramics.

15.2.4.5 Surface Compression

Placing the surface in compression obviates the effects of surface flaws by preventing concentration of tensile stresses at the crack tip. The surface compressive stress must first be exceeded by an applied tensile stress before the stress concentration will begin to build up and lead to crack propagation.

Perhaps the most common example of surface compression is in safety glass. Surface compression is achieved in glass most commonly by either ion exchange or quenching. In the former case, glass is exposed at elevated temperature to positive ions that are larger than those initially in the glass. Since the glass structure is expanded at the elevated temperature, these larger ions are able to trade places with smaller ions near the surface. When the glass is cooled, these ions no longer fit, but are trapped in the structure and result in surface compression. An example is that of exchanging potassium ions for sodium ions.

Ion exchange can also be applied to crystalline materials. The exchange ion can cause compression by size difference or by producing a surface composition with a lower coefficient of thermal expansion.[22]

Quenching has also been known for a long time as a method of improving strength. Kirchner[24] discusses in detail the use of quenching to strengthen Al_2O_3, TiO_2, spinel, steatite ($MgSiO_3$), forsterite (Mg_2SiO_4), SiC, and Si_3N_4. Quenching Al_2O_3 rods from 1600°C into a silicone oil resulted in an increase in the average strength from 331 MPa (47,900 psi) to 694 MPa (100,600 psi).

For quenching to be effective, the material must be heated to a temperature such that some plasticity is present. A temperature of 1500 to 1600°C is appropriate for Al_2O_3, but 1900 to 2000°C is required for SiC. During quenching the surface cools very rapidly and is placed in compression as the interior cools more slowly.

TABLE 15.9
Surface Preparation vs. Tensile Strength for Single-Crystal Al$_2$O$_3$

Surface Preparation	Tensile Strength	
	MPa	psi
As-machined	440	60,000
Polished by centerless grinding	590	90,000
Annealed in oxygen after polishing by centerless grinding	1,040	150,000
Chemical polished with molten borax	6,860	1,000,000
Flame polished	7,350	1,100,000
Pristine whiskers	15,900	2,300,000

Source: Stokes, R.J., Effect of surface finishing on mechanical and other physical properties of ceramics, in *The Science of Ceramic Machining and Surface Finishing*, Schneider, S.J. and Rice, R.W., Eds., NBS Special Publication 348, U.S. Government Printing Office, Washington, DC, 1972, pp. 348–352.

Surface compression can also be achieved with glazes (glass surface coatings). The glaze composition is selected to have a lower coefficient of expansion than the matrix material. For instance, Kirchner used a glaze with a thermal expansion of 5.3×10^{-6} per °C for coating Al$_2$O$_3$ with a thermal expansion of 6.5×10^{-6} per °C. Glazes can be used in conjunction with quenching or ion exchange to achieve additional benefits. Kirchner achieved a strength of 767 MPa (111,200 psi) for Al$_2$O$_3$ that was glazed and quenched, compared to 331 MPa (47,900 psi) for as-received material.

15.2.4.6 Flame Polishing

Flame polishing is used primarily for reducing the size and quantity of surface flaws in small-diameter rods or filaments, especially of sapphire or ruby (single-crystal Al$_2$O$_3$). Flame polishing is conducted by rotating the rod or filament and passing it through a H$_2$-O$_2$ flame such that the thin surface layer melts. Noone and Heuer[25] report bend strengths for flame-polished ruby and sapphire in the range 4000 to 5000 MPa (580,000 to 724,000 psi) compared to approximately 300 MPa (43,500 psi) for as-ground specimens. Stokes[26] compares the tensile strength of flame-polished single-crystal Al$_2$O$_3$ with other surface preparations. His results are summarized in Table 15.9.

Stokes[26] also discussed the effects of machining on properties other than strength. The shape of the hysteresis loop in magnetic ferrites is significantly changed by near-surface stresses resulting from machining. Polishing, annealing, and etching procedures are routinely used to attain reproducibly the desired loop shape. Electrical and optical properties are also affected strongly by machining and surface condition.

15.3 ADDITIONAL SOURCES OF INFORMATION

Stinton[27] provided additional discussion on machining of ceramics and includes a bibliography of 90 references.

REFERENCES

1. Richerson, D.W., Ceramic fabrication: finishing and quality control, Lesson 13 in *Introduction to Modern Ceramics*, ASM Materials Engineering Institute Course 56, ASM International, Materials Park, Ohio, 1990.
2. Kulischenko, W., *Abrasive Jet Machining*, SME Tech. Paper MR 76-694, 1976.

3. Haylock, R., Waterjet cutting, in *Ceramics Applications in Manufacturing*, Richerson, D.W., Ed., Society of Manufacturing Engineers, Dearborn, Michigan, 1988, pp. 163–175.

4. Moreland, M.A. and Moore, D.O., Versatile performance of ultrasonic machining, *Am. Ceram. Soc. Bull.*, 67(6), 1045–1052, 1988.

5. Stookey, S.D., Chemical machining of photosensitive glass, *Ind. Eng. Chem.*, 45, 115–118, 1953.

6. Lee, D.W. and Feick, G., The techniques and mechanisms of chemical, electrochemical and electrical discharge machining of ceramic materials, in *The Science of Ceramic Machining and Surface Finishing*, Schneider, S.J. and Rice, R.W., Eds., NBS Special Publication 348, U.S. Government Printing Office, Washington, D.C., 1972, pp. 197–211.

7. Petrofes, N.F. and Gadalla, A.M., Electrical discharge machining of advanced ceramics, *Am. Ceram. Soc. Bull.*, 67(6), 1048–1052, 1988.

8. Lumley, R.M., Controlled separation of brittle materials using a laser, *Am. Ceram. Soc. Bull.*, 48(9), 850–854, 1969.

9. Brody, C.J. and Molines, J.L., The evolution, process and effects of laser machined alumina, in *Ceramics Applications in Manufacturing*, Richerson, D.W., Ed., Society of Manufacturing Engineers, Dearborn, Michigan, 1988, pp. 142–150.

10. Copley, S.M., Bass, M., and Wallace, R.G., Shaping silicon compound ceramics with a continuous wave carbon dioxide laser, in *The Science of Ceramic Machining and Surface Finishing*, *II*, Hockey, B.J. and Rice, R.W., Eds., NBS Special Publication 562, U.S. Government Printing Office, Washington, D.C., 1979, pp. 283–292.

11. Rice, R.W., Machining of ceramics, in *Ceramics for High Performance Applications*, Burke, J.J., Gorum, A.E., and Katz, R.N., Eds., Brook Hill, Chestnut Hill, Mass., 1974, pp. 287–343. (Available from MCIC, Battelle Columbus Labs., Columbus, Ohio.)

12. Gielisse, P.J. and Stanislao, J., *Dynamic and Thermal Aspects of Ceramic Processing*, University of Rhode Island, Kingston, Naval Air Systems Command Contract Rept., Nov. 15, 1969–Nov. 15, 1970 (AD 728 011), Dec. 1970.

13. Evans, A.G., Abrasive wear in ceramics: an assessment, in *The Science of Ceramic Machining and Surface Finishing*, *II*, Hockey B.J. and Rice, R.W., Eds., NBS Special Publication 562, U.S. Government Printing Office, Washington, D.C., 1979, pp. 1–14.

14. Hagan, J.T., Swain, M.V., and Field, J.E., Nucleation of median and lateral cracks around Vickers indentations in soda-lime glass, in *The Science of Ceramic Machining and Surface Finishing*, *II*, Hockey, B.J. and Rice, R.W., Eds., NBS Special Publication 562, U.S. Government Printing Office, Washington, D.C., 1979, pp. 15–21.

15. Kirchner, H.P., Gruver, R.M., and Richard, D.M., Fragmentation and damage penetration during abrasive machining of ceramics, in *The Science of Ceramic Machining and Surface Finishing*, *II*, Hockey, B.J. and Rice, R.W., Eds., NBS Special Publication 562, U.S. Government Printing Office, Washington, D.C., 1979, pp. 23–42.

16. Van Groenou, A.B., Maan, N., and Veldkamp, J.B.D., Single-point scratches as a basis for understanding grinding and lapping, in *The Science of Ceramic Machining and Surface Finishing*, *II*, Hockey, B.J. and Rice, R.W., Eds., NBS Special Publication 562, U.S. Government Printing Office, Washington, D.C., 1979, pp. 43–60.

17. Evans, A.G., Ed., *Fracture in Ceramic Materials: Toughening Mechanisms, Machining Damage, Shock*, Noyes Publications, Park Ridge, N.J., 1984.

18. Richerson, D.W., Schuldies, J.J., Yonushonis, T.M., and Johansen, K.M., ARPA/Navy ceramic engine materials and process development summary, in *Ceramics for High Performance Applications*, *II*, Burke, J.J., Lenoe, E.N., and Katz, R.N., Eds., Brook Hill, Chestnut Hill, Mass., 1978, pp. 625–650. (Available from MCIC, Battelle Columbus Labs., Columbus, Ohio.)

19. Rice, R.W. and Mecholsky, J.J., The nature of strength-controlling machining flaws in ceramics, in *The Science of Ceramic Machining and Surface Finishing*, *II*, Hockey, B.J. and Rice, R.W., Eds., NBS Special Publication 562, U.S. Government Printing Office, Washington, D.C., 1979, pp. 351–378.

20. Chand, R.H., Costello, K.P., and Payne, T.M., Diamond Wheels put finish on ceramics, *Ceram. Indus.*, 42–44, 1990.

21. Subramanian, K., Ramanath, S., and Matsuda, Y., Diamond abrasives cut path to volume production, *Ceram. Indus.*, 30–32, 1990.

22. Barks, R., Subramanian, K., and Ball, K.E., Eds., *Machining of Advanced Ceramic Materials and Components*, American Ceramic Society, Westerville, Ohio, 1987.

23. Subramanian, K., Precision finishing of ceramic components with diamond abrasives, *Am. Ceram. Soc. Bull.*, 67(6), 1026–1029, 1988.

24. Kirchner, H.P., *Strengthening of Ceramics*, Marcel Dekker, New York, 1979.

25. Noone, M.J. and Heuer, A.H., Improvements in the surface finish of ceramics by flame polishing and annealing techniques, in *The Science of Ceramic Machining and Surface Finishing*, Schneider, S.J. and Rice, R.W., Eds., NBS Special Publication 348, U.S. Government Printing Office, Washington, D.C., 1972, pp. 213–232.

26. Stokes, R.J., Effect of surface finishing on mechanical and other physical properties of ceramics, in *The Science of Ceramic Machining and Surface Finishing*, Schneider, S.J. and Rice, R.W., Eds., NBS Special Publication 348, U.S. Government Printing Office, Washington, D.C., 1972, pp. 348–352.

27. Stinton, D.P., *Assessment of the State of the Art in Machining and Surface Preparation of Ceramics*, ORNL/TM-10791, Oak Ridge National Laboratory, Tenn., Nov. 1988.

ADDITIONAL RECOMMENDED READING

1. Subramanian, K. and Ramanath, S., Principles of Abrasive Machining, pp. 315–328 in *ASM Engineered Materials Handbook Vol. 4: Ceramics and Glasses*, ASM Int., Materials Park, Ohio, 1991.

2. Ratterman, E. and Cassidy, R., Abrasives, *ibid*, pp. 329–335.

3. Subramanian, K. and Kopp, R.N., Production Grinding Methods and Techniques, *ibid*, pp. 336–350.

4. Indge, J.H. and Wolters, P., Lapping, Honing, and Polishing, *ibid*, pp. 351–358.

5. Moreland, M.A., Ultrasonic Machining, *ibid*, pp. 359–362.

6. Singh, P.J., Abrasive Fluid Jet Machining, *ibid*, pp. 363–366.

7. Capp, A.O., Jr., Laser Beam Machining, *ibid*, pp. 367–370.

8. Faulk, N., Electrical Discharge Machining, *ibid*, pp. 371–376.

9. Hahn, R.S. and King, R.I., *Handbook of Modern Grinding Technology*, Chapman and Hall, 1986.

PROBLEMS

1. Which ceramic would you expect to be the least difficult to cut or grind? Assume that each has less than 1% porosity. Why?
   (a) silicon nitride
   (b) aluminum oxide
   (c) zirconium oxide
   (d) boron carbide

2. Based on the data in Table 15.1, compare the size of the radial and median cracks produced in Si_3N_4 by the specific surface-grinding procedure used to prepare the test bars.

3. Which abrasive would you expect to be most efficient for grinding SiC?
   (a) Al_2O_3
   (b) SiC
   (c) diamond
   (d) B_4C

4. Which machining approach is *least* likely to cause strength-reducing damage?
   (a) cutting
   (b) grinding
   (c) water-jet
   (d) ultrasonic

5. Relatively weak, porous ceramic molds are used for investment casting of complex metal parts such as gas-turbine rotors. How would you remove remnants of the ceramic mold without damaging the intricate metal part?
   (a) grinding
   (b) sandblasting
   (c) hammering
   (d) lapping

6. Which of the following is *not* a potential objective of annealing?
    (a) relieve residual stresses
    (b) allow equilibrium phases to form
    (c) crystallize grain boundary glass
    (d) achieve final densification

STUDY GUIDE

1. Explain why machining of ceramic materials requires a different approach than machining of metals.
2. Biaxial contact stress was discussed in Chapter 9. How might biaxial contact stress play a role in machining of ceramics?
3. Identify some examples of impact abrasive machining.
4. What is EDM, and what are its limitations for machining of ceramics?
5. Describe the near-surface damage that occurs in a ceramic material when an abrasive particle is gouged across the surface.
6. Describe how surface grinding can affect the strength of a ceramic material.
7. What factors might influence the effectiveness, cutting rate, and surface finish for mounted abrasive machining?
8. Identify examples of postmachining procedures.
9. What are the primary reasons for postmaching procedures?
10. Identify some potential benefits of annealing.
11. How does surface compression provide a benefit?
12. Describe several ways of achieving surface compression.

16 Quality Assurance

Quality assurance (QA) is required throughout processing of any material or product, and ceramics are no exception. The degree of QA is determined by the criticality of the application. Most applications require a specification or a written manufacturing procedure and one or more certification tests to assure that the manufacturing procedure has been followed and the specification met. More critical or demanding applications may require destructive sampling, proof testing, or nondestructive inspection (NDI). In this chapter these various aspects of QA are explored from the perspective of both the manufacturing engineer and the applications engineer.

16.1 IN-PROCESS QA

The fundamental step in in-process QA is the preparation of a formal written manufacturing procedure. This document describes each operation required in the process from procurement of raw materials through final shipping. It also normally defines the paperwork that must accompany the part through the process, lists the signatures that are required on this paperwork to certify that each operation has been followed as specified, and describes the procedure that must be followed if a process change is considered. The signed paperwork that follows the part through the process is kept on file as traceability for that part and as a record that the part was processed by the designated fixed process.

In-process QA starts with procurement of raw materials. It may consist simply of checking the chemical analysis and particle size distribution analyses submitted for the material by the supplier to ensure that they are within the specification of the manufacturing operation document. Or it may involve additional analyses, depending on how critical the raw material characteristics are to achieving the desired characteristics of the final product. For electrical, magnetic, optical, and many structural applications, the purity and particle size aspects of the raw materials are extremely important and careful QA is justified.

In-process QA continues into the next step of the fabrication process — powder processing. Again, purity and particle sizing are usually most important, so that QA consists of chemical analysis (such as emission spectroscopy, x-ray fluorescence, infrared spectroscopy, x-ray diffraction, or atomic absorption) and particle size distribution analysis (such as x-ray sedimentation, Coulter counter, BET surface area, and screening).

The nature of in-process QA changes with the shape-forming and densification steps of the manufacturing process. There is less interest in the bulk chemistry and more interest in isolated forming defects, dimensions, and properties. Forming defects such as cracks, pores, inclusions, laminations, and knit lines can be detected by visual inspection and NDI. Dimensions can be determined by gauges, shadow graphs, and other standard inspection instruments.

Electrical, magnetic, optical, and physical properties can be determined by direct measurement, either on each part or by random sampling. Mechanical properties are usually not as easy to qualify. The final product shape is usually not of a configuration readily tested for mechanical properties. The options are to process appropriate mechanical property test shapes with the product, to cut test bars out of random samples of the product, or to conduct a proof test on the product.

In-process QA has two major objectives: (1) to maintain the process under control with the aim of achieving maximum yield and quality and (2) to reject unacceptable components and to accept good components. The accept and reject criteria result from experience, primarily joint experience between the manufacturer and the user. This usually leads to a specification prepared by the user and accepted by the manufacturer, plus a QA or certification procedure prepared by the manufacturer and accepted by the user.

16.2 SPECIFICATION AND CERTIFICATION

At some point in their career, most engineers will have to prepare or use a specification. Specifications are required not only for ceramic components, but also for metal and organic components and for systems. In fact, most products manufactured have a specification and certification procedure to assure that the specification is met.

Table 16.1 and Table 16.2 illustrate how a specification and certification procedure are integrated into a manufacturing operation procedure to produce the hypothetical ceramic component shown in

---

TABLE 16.1
Specification for the Part Shown in Figure 16.1

1.0 Material composition
2.0 Properties
    2.1 Density >98.5% of theoretical
    2.2 Hardness >2200 kg/mm^2 Knoop$_{500}$
    2.3 Average room temperature MOR[a] for eight specimens must be >280 MPa (40,000 psi)
    2.4 Average 1000°C (1832°F) MOR[a] for four specimens must be >207 MPa (30,000 psi)
    2.5 Stress rupture life for same specimen configuration under 140 MPa (20,000 psi) load at 1000°C is >100 hour
3.0 Dimensions
    3.1 ID 1.000 + 0.002–0.000 in. with a 2-μin. surface finish
    3.2 OD$_1$ 1.500 ± 0.005 in.
    3.3 OD$_2$ 3.000 ± 0.005 in.
    3.4 h_1 0.750 ± 0.010 in.
    3.5 h_2 0.500 ± 0.002 in. with surface flat and parallel to 0.001 in.

[a] Specimen cross section 0.32 × 0.64 cm (0.125 × 0.25 in.) to be tested in three-point bending over a span of 3.8 cm (1.5 in.).

TABLE 16.2
Integration of QA and Certification into the Manufacturing Process

Process Steps	In-Process and Final QA	Certification Tests
Starting powder	Chemical analysis X-ray diffraction Particle size distribution	Chemical analysis
Powder processing	Chemical analysis Particle size distribution	Particle size distribution
Shape-forming	Visual and dimensional Radiography Green density	Visual and dimensional
Densification	Visual NDI Final bulk density Critical property measurement	Visual Final bulk density Critical property measurement
Final machining	Dimensional	Dimensional

Figure 16.1. The objective is to conduct certification tests at each step in the process that will allow rejection of faulty parts as early as possible. Of course, there will be a trade-off based on the economics and service conditions of the part being manufactured. More in-process QA and certification testing can be justified for an expensive or especially critical component than for a very low-cost component or one that has a loose specification.

The specification in Table 16.1 is quite tight both in terms of properties and dimensions and thus will require substantial certification. The component is apparently for a high-temperature application where strength and creep resistance are critical. Since creep and high-temperature strength are often controlled by composition and impurity levels, the earliest opportunity in the process for QA and certification is chemical analysis of the starting powder, or at least review of the supplier's analysis. The component manufacturer typically has background experience in the levels of variation in the chemical composition that can be tolerated and has an in-house specification (separate from the component specification of the user). Therefore, chemical analysis becomes an effective certification test for the manufacturer, even though the chemical composition is not listed in the user's specification.

A similar situation exists for the processed powder. The manufacturer has an in-house specification relating parameters such as particle size distribution to properties such as strength and hardness and can then use a certification test to accept or reject a batch of processed powder.

The next step in the process is shape-forming. Emphasis now shifts to dimensions. Generally, one or two dimensions are most critical for a component and can be measured quickly in the green state. This will be an automatic procedure if a portion of the shaping is achieved by green machining.

After densification the certification procedure usually shifts back to properties as shown in Table 16.2. Based on the specification in Table 16.1, bulk density would be measured first (for parts not showing any visual defects). If the density was acceptable, the other critical properties listed in the specification would be measured as part of the certification testing.

The final process step is often finish machining. The tolerances in the hypothetical component are tight and would require accurate dimensional measurements as part of the certification procedure.

Components passing all the certification tests would then be delivered to the customer, usually along with written documentation of the certification test results.

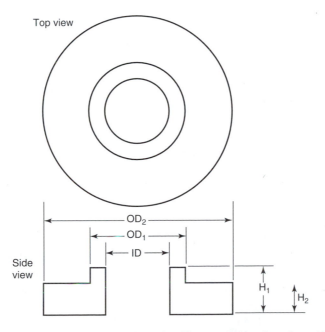

FIGURE 16.1 Hypothetical ceramic component used to illustrate integration of specification and certification into the manufacturing operation. Dimensions and tolerances are listed in Table 16.1.

16.3 PROOF TESTING

Many components are *proof-tested*, i.e., exposed to conditions comparable to or exceeding those of the service environment. Proof tests can be conducted for critical electrical, mechanical, or other properties using parameters known from prior application experience to control the acceptability or life of the component in the application. Proof tests are applied to metals, organics, and ceramics. For example, superalloy metal rotors for small gas-turbine engines are typically spin-tested to a predetermined overspeed prior to being installed into engines. Similarly, ceramic grinding wheels are proof-tested by spin testing. Ceramic electrical insulators for high-voltage application are usually electrically proof-tested to make sure that they will not undergo dielectric breakdown.

To be a true proof test, the test condition must exactly simulate the service conditions, including temperature and stress distributions. This is extremely difficult to achieve without actually putting the component in real service. Hence, many tests referred to as proof tests really are not. Instead, most of these tests only expose the material or component to the most critical aspect of the service environment and are more properly referred to as *qualification* or *screening tests*.

A major concern with proof or qualification testing is that no damage is done to the material that might reduce its service life. This is especially important with ceramics in mechanical loading. The ceramic can withstand the proof load, but be damaged subcritically either by the load or during unloading. This is especially likely to occur when a room-temperature overload condition is applied to qualify a part for a high-temperature application. Damage that will not cause failure at room temperature may cause failure at elevated temperature or grow subcritically and cause failure after only a short time.

Proof or qualification testing can dramatically improve the reliability of a ceramic component, especially for applications requiring high mechanical strength.[1,2] As shown in Chapter 8, the strength of a ceramic material is dependent on the type of loading and the volume and area under stress. This, plus the wide flaw size distribution in a single ceramic component, results in a significant probability of failure over a wide range of applied load. A properly applied proof test can truncate this distribution and substantially reduce the probability of failure under service conditions. This is illustrated in Figure 16.2 using a Weibull plot of probability of failure versus applied stress. Without proof testing, 8 components out of 1000 would fail at the service load of 50,000 psi (345 MPa) and 3 of 100 at an accidental overload condition of 60,000 psi (414 MPa). By proof testing at 55,000 psi (379 MPa), the strength distribution is truncated and no failures would occur at the service load of 50,000 psi and less than 1 in 100 at an overload condition of 60,000 psi.

16.4 NONDESTRUCTIVE INSPECTION

NDI involves examination of a component or test specimen by instrumentation that detects surface or subsurface flaws in the material but does not result in material damage. Such techniques are also referred to as nondestructive evaluation (NDE) and nondestructive testing (NDT) and are applicable to ceramics, metals, and organics. In this section the administration of various NDI techniques and the nature of their strengths and limitations are reviewed. Emphasis is on techniques that show the most promise for ceramics.

16.4.1 PENETRANTS

Penetrants are used extensively for the detection of surface flaws. Usually, a three-step procedure is used: (1) the part is soaked in a fluorescent dye, (2) the part is dried or cleaned in a controlled fashion to remove the dye from smooth surfaces but not from surface defects, (3) the part is examined under ultraviolet light. Surface defects such as cracks and porosity that retain dye show up brilliantly under the ultraviolet light.

The use of penetrants for inspection is widespread and is frequently included as part of a specification. Penetrants are categorized into classes according to their sensitivity and are usually identified in a specification only according to their sensitivity category.

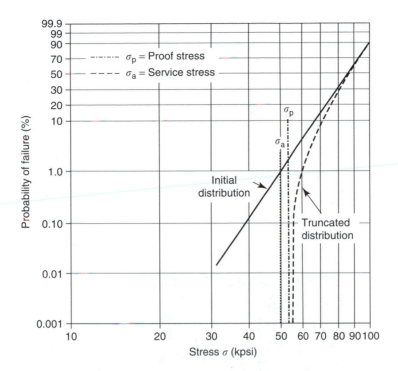

FIGURE 16.2 Truncation of the material strength distribution to achieve improved reliability and increased operating margin. (Adapted from Wynn, L.P., Tree, D.J., Yonushonis, T.M., and Solomon, R.A., Proof testing of ceramic components, in *Proceedings of the 1977 DARPA/NAVSEA Ceramic Gas Turbine Demonstration Engine Program Review*, Fairbanks, J.W. and Rice, R.W., Eds., MCIC Rept. MCIC-78-36, 1978, pp. 493–516.)

Penetrants are effective for most metals and for nonporous ceramics. If a ceramic has open porosity, the penetrant will usually enter all the pores and result in fluorescence of the whole part, preventing detection of other surface flaws.

Penetrants are valuable for detecting surface-connected cracks and regions of abnormally low density, but do not detect internal defects. As ceramics enter applications with higher stress, techniques such as x-ray radiography and ultrasonics may be beneficial.

16.4.2   X-RAY RADIOGRAPHY

16.4.2.1   Conventional X-Ray Radiography

Conventional x-ray radiography consists of passing a beam of x-rays through the part being examined to expose a sheet of film, as illustrated schematically in Figure 16.3a. The material absorbs a portion of the x-rays. If the material is of constant thickness and contains no flaws, the film adjacent to the part will be uniformly exposed and will be uniformly gray in color after developing, surrounded by fully exposed black where the film was not shielded by the part. If the part is of a nonuniform thickness, the thicker portions will absorb more of the x-rays and the film in this region will be less exposed (lighter gray in color). Density variations in the material will show up in a similar manner, higher density showing up lighter and lower density showing up darker. Similarly, as shown in Figure 16.3b, a void or hole will show up darker and a high-density inclusion that absorbs x-rays more than the matrix will show up lighter. Obviously, if a positive photographic print is made from the film negative, the opposite will be true; voids will appear lighter than the matrix and a high-density inclusion will appear darker.

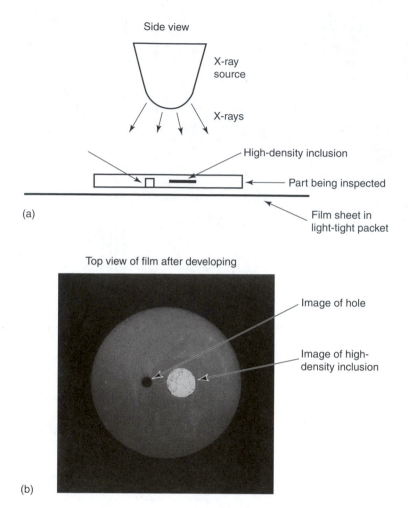

FIGURE 16.3 (a) Schematic of conventional x-ray radiography setup. (b) Resulting image on the developed negative.

The size of defect that can be detected by x-ray radiography depends on a combination of factors:

the thickness of the part and its x-ray absorption
the size of the flaw compared to the thickness of the part
the difference in x-ray absorption between the flaw and the part
the orientation of the flaw

Elements of low atomic weight and structures of low packing density have relatively low absorption of x-rays and can thus be inspected by x-ray radiography in greater thicknesses than materials having high absorption, owing to their content of high-atomic-weight elements. Al_2O_3, graphite, Si_3N_4, SiC, and other similarly low-atomic-weight ceramics have low absorption for x-rays and can be inspected over a broad thickness range. WC has high atomic weight, and most metals have close packing and therefore have higher absorption for x-rays, resulting in restriction of the thickness that can be inspected.

Other factors can also affect x-ray absorption. Different elements preferentially absorb specific wavelengths of x-rays.

The difference in x-ray absorption between the part and the inclusions or voids it contains determines the sensitivity or resolution capability of the material. Small voids are difficult to detect in materials like Si_3N_4 and SiC because the difference in absorption between the void and the material is relatively small. On the other hand, inclusions of WC or iron in Si_3N_4 or SiC can be resolved even in very small sizes because the x-ray absorptions are so different.

The resolution capability or specified requirement is often stated in terms of the thickness of the defect compared to the thickness of the part. Inspection to a 2T level means that the technique or apparatus used must be able to detect flaws larger than 2% of the thickness of the part. Kossowsky[3] reported the following limits of resolution for defects in hot-pressed Si_3N_4 using conventional x-ray radiography: voids (holes or cracks), 3% of thickness; BN-filled cavity, 4% of thickness; steel particle, 0.7% of thickness; and WC particle, 0.5% of thickness. Similar sensitivities were reported by Richerson et al.[4] using a microfocus x-ray.

The final factor affecting detection sensitivity is the orientation of the flaw. A tight crack perpendicular to the x-ray source will not be detected, whereas one that is parallel or at a low angle to the source will have a better chance of detection. It is for this reason that radiographs are normally taken with the part in several different orientations. Also, two radiographs are usually taken for each orientation, so that artifacts due to film blemishes, developing, or surface contamination can be distinguished from defects in the material.

Conventional x-ray radiography is widely used for all types of materials and is a quick, convenient, and cost-effective way of detecting internal flaws in components. A major advantage of radiography is that it can be used effectively with complex shapes, which is a limiting factor for most other NDI techniques. However, as with other NDI techniques, radiography requires standards in order to quantify the size and type of defect being evaluated.

16.4.2.2   Microfocus X-Ray Radiography

Microfocus x-ray radiography is a relatively new technique made possible by development at Magnaflux of their MXK-100M x-ray tube, which has a small focal spot of 0.05 mm (0.002 in.). This tube provides improved resolution and geometric sharpness. Because of the fine focal spot, closer working distances are possible and direct radiographic enlargements up to 36× can be achieved without reduction in sensitivity due to parallax and secondary radiation effects.[5,6]

Microfocus x-ray is especially useful for complex shapes where a small region is especially critical, such as the leading and trailing edges of rotor blades and stator vanes for gas-turbine engines. The MXK-100M tube is small and portable (weighs under 4 kg) and can be easily maneuvered to inspect the desired region at the appropriate orientation.

The microfocus x-ray tube has also been modified to permit panoramic radiography[5] of a hollow cylindrical object.

16.4.2.3   Image Enhancement

Once a photographic negative has been produced by radiography, it must be examined with back lighting and judgments made as to which indications on the film represent defects in the material. Such film interpretation is very subjective and requires an experienced individual. For instance, Figure 16.4 shows radiographs of graphite, iron, and tungsten carbide in hot-pressed Si_3N_4. Which coloration differences actually represent flaws? Are there four graphite inclusions in Figure 16.4a?

Image-enhancement technology, developed initially for evaluation of satellite photos, has been adapted to radiography. In general, the radiograph film is back-lighted and the image picked up by a detector or device that can divide the image into discrete elements (pixels) and enter the digitized data into a computer. In the study conducted by Schuldies and Spaulding,[6] the image was divided into an array of 480 × 512 pixels and each pixel assigned a gray-level value ranging from 0 (black) to 255 (white). This is far more gray levels than can be distinguished by the human eye.

Once the image data are digitized and stored in the computer, a variety of computer programs can be run to achieve greater contrast and thus enhance the image. The enhanced image is displayed on a black-and-white TV screen that can be photographed to provide a permanent record.

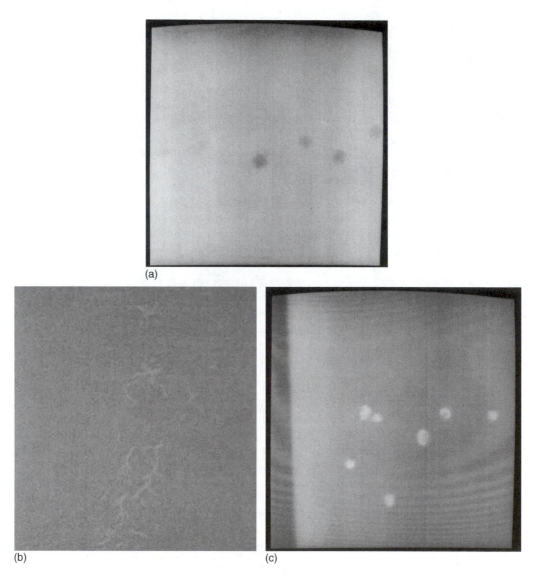

(a)

(b)   (c)

FIGURE 16.4 Microfocus x-ray radiographs of inclusions seeded in hot-pressed Si_3N_4. (a) 500-μm (0.02-in.) graphite inclusions. (b) 250-μm (0.01-in.) iron inclusions. (c) 500-μm (0.02-in.) WC inclusions. (Courtesy of Garrett Turbine Engine Company, Phoenix, AZ, Division of Allied-Signal Aerospace, currently Honeywell Engines, Systems & Services.)

Figure 16.5 shows a block schematic of the image-enhancement system. Besides the computer enhancement, visual enhancement can be achieved directly with the scanner to obtain improved contrast. Color enhancement can also be used in which gray levels are replaced by colors. In addition, a video cursor is tied into the computer and display system to permit physical distance measurements and digital readout.

Image enhancement is a significant aid to interpretation of radiographs. It helps reduce the subjectivity of the operator and provides more objective evaluation of the radiograph. However, it should be recognized that film anomalies such as graininess, scratches, water marks, and processor defects will also be enhanced to the same degree as images of defects in the part, so it still ends up that the individual has to make judgments.

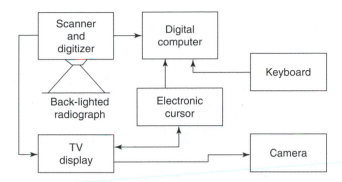

FIGURE 16.5 Schematic of the image-enhancement system.

Figure 16.6 through 16.8 are examples of the benefits of image enhancement. Figure 16.6 illustrates the enhancement of the radiograph shown previously in Figure 16.4 for graphite inclusions in Si_3N_4. The edge-enhancement computer algorithms have dramatically isolated the inclusions. Now we can answer the question we raised before. There are more than four graphite inclusions. If only the fifth inclusion had been present, it would have been missed by the film examiner without the aid of enhancement.

Figure 16.7 illustrates the enhancement of the radiograph shown previously in Figure 16.4b for iron inclusions in Si_3N_4. The iron appears to be present as dendrites. Since it was initially seeded in the material as equiaxed particles,[7] the iron evidently recrystallized or reacted with the Si_3N_4 matrix during hot pressing. This provides us with information about the ceramic processing that is not normally available from NDI data. The circular inclusion in the bottom photo of Figure 16.7 turned out to be WC. It had been seeded as a near-spherical particle and did not change its shape during processing, suggesting that it was more stable or inert during processing than the iron was.

Figure 16.8 illustrates the enhancement of the radiograph shown in Figure 16.4c for WC inclusions in Si_3N_4. As in the other examples, enhancement has substantially increased the detectability of the inclusions.

16.4.3 COMPUTED TOMOGRAPHY

Microfocus radiography and image enhancement were developed for ceramics during the 1970s primarily in an effort to locate flaws in silicon nitride and silicon carbide components being evaluated for prototype gas turbine engines. However, these film radiography techniques were not adequately effective for complex shapes such as gas turbine engine rotors. Advances in computed tomography (CAT-scan or simply CT) in the medical field began to be applied to ceramics in the 1980s and 1990s.[8,9] As shown earlier in Figure 3.51, a thin microfocus beam of x-rays is passed through the ceramic object. The intensity getting through for each tiny segment of the object is detected by a scintillator and photo diode array or a sensitive amorphous silicon or amorphous selenium detector. The x-ray source and the detector are rotated around the object so that thousands of data points are gathered from all directions. All of these data go into a powerful computer such that images of cross sections of the interior of the object can be reconstructed and printed or displayed on a TV monitor (as shown earlier in Figure 3.52).

16.4.4 ULTRASONIC NDI

Ultrasonic NDI is another important technique for detecting subsurface flaws in materials. A simple schematic illustrating the basics of the technique is shown in Figure 16.9. The part to be inspected is immersed in water. A piezoelectric transducer in close proximity to the surface of the part is stimulated by an electric current to emit acoustic waves of a known amplitude and wavelength. These

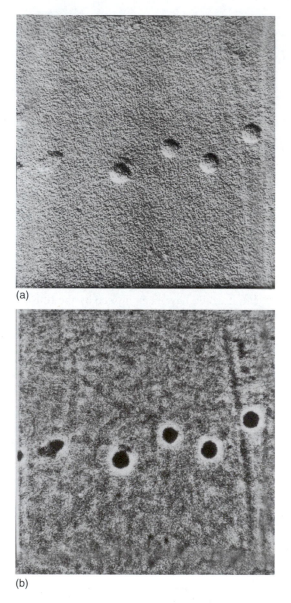

(a)

(b)

FIGURE 16.6   Image enhancement of 500-μm (0.02-in.) graphite inclusions in hot-pressed Si_3N_4. (Courtesy of Garrett Turbine Engine Company, Phoenix, AZ, Division of Allied-Signal Aerospace, currently Honeywell Engines, Systems & Services.)

waves pass into and through the part. Each discontinuity (the surfaces plus any internal defects) perturbs the acoustic waves, resulting either in scattering or in reflected secondary waves. A receiver picks up the secondary waves and the electronics of the equipment converts the signal into a graphical representation. The receiver can either be opposite the emitting transducer and pick up the transmitted waves, or be on the same side and pick up reflected waves. The latter approach is commonly used and is referred to as the *pulse-echo technique*. As was the case with radiography, flaw orientation affects detectability and inspection should be conducted in multiple directions if possible.

Ultrasonic inspection is most easily conducted on material having a flat surface and a constant cross section. The transducers are scanned across the part and results are plotted with a pen-type XY-recorder. The electronics are adjusted so that an electronic window eliminates the wave reflections for the two surfaces. This approach is known as *C-scan*. Since the surface reflections must be filtered out,

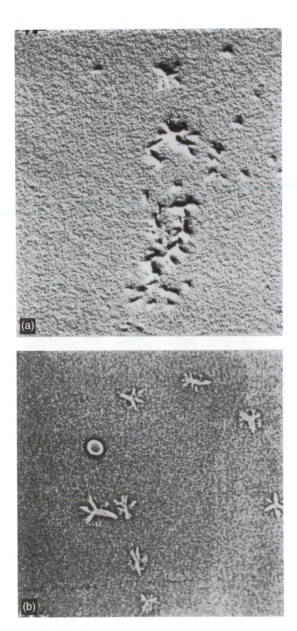

FIGURE 16.7   Image enhancement of 250-μm (0.01-in.) iron inclusions in hot-pressed Si_3N_4. (Courtesy of Garrett Turbine Engine Company, Phoenix, AZ, Division of Allied-Signal Aerospace, currently Honeywell Engines, Systems & Services.)

C-scan does not detect near-surface flaws. The closeness to the surface that can be evaluated depends partially on how accurately the electronic window can be set. It is also affected by transducer noise and electronic signal damping limitations. Surface irregularities and variations in cross-sectional thickness further restrict the ability of the operator to set the window close to the surface and make C-scan of complex shapes very difficult.

Figure 16.10 shows the *C-scan* printout for a 0.64-cm (0.25-in.)-thick flat plate of hot-pressed Si_3N_4 containing various sizes of inclusions and voids.[10] The resolution of both inclusions and voids is quite good. However, such success was not achieved on the first attempt. A variety of transducers and electronic gating procedures were tried before optimum conditions were defined. This re-emphasizes the importance of standards containing defects of known size and composition. The

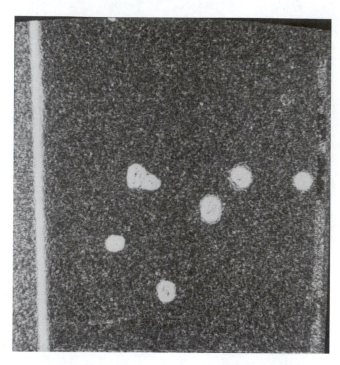

FIGURE 16.8   Image enhancement of 500-μm (0.02-in.) WC inclusions in hot-pressed Si_3N_4. (Courtesy of Garrett Turbine Engine Company, Phoenix, AZ, Division of Allied-Signal Aerospace, currently Honeywell Engines, Systems & Services.)

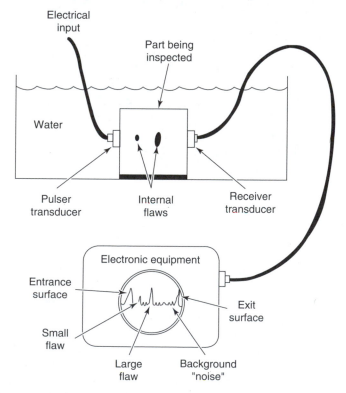

FIGURE 16.9   Schematic illustrating the basic principles of ultrasonic NDI.

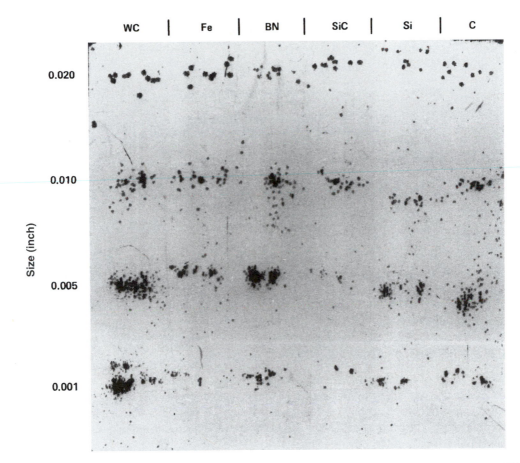

FIGURE 16.10   Ultrasonic C-scan with a 25-MHz transducer of a 0.64-cm (0.25-in.)-thick hot-pressed Si_3N_4 plate. (Courtesy of Garrett Turbine Engine Company, Phoenix, AZ, Division of Allied-Signal Aerospace, currently Honeywell Engines, Systems & Services.)

Si_3N_4 plate had originally been prepared as a standard with seeded defects specifically to evaluate and optimize the resolution capabilities of different NDI techniques.[7] Without standards, the operator has difficulty optimizing equipment parameters and interpreting printout results.

The ultrasonic C-scan inspection detected more of the defects in the hot-pressed Si_3N_4 plate than could be detected by other techniques available in the late 1970s. X-ray radiography had shown up only the high-density WC and Fe inclusions and the largest graphite inclusion. Neutron radiography showed only the BN.

A limitation of ultrasonics is the loss in intensity by the scattering of the waves as they pass through the material. This is called *attenuation* and limits the thickness of the part that can be inspected. Attenuation is accentuated by porosity or other microstructural features that cause scattering (second-phase distributions, microcracking, etc.). Attenuation is also affected by the frequency of the transducer. Increasing the frequency increases the sensitivity of detecting smaller flaws, but also increases scattering and decreases the thickness that can be effectively inspected.

One approach for improvement in the resolution and shape inspection capability for internal defect detection of ceramics was computer-aided ultrasonics. Resolution sensitivity is reduced by system noise (from the transducer and electronics) and material noise (wave scattering by microstructure and surfaces). Seydel[11] has showed that both sources of noise could be reduced significantly by digitizing the ultrasonic pulses and using a minicomputer for signal averaging. A simple schematic of a computerized system is shown in Figure 16.11.

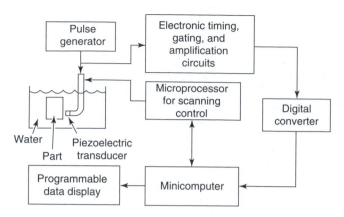

FIGURE 16.11   Schematic of computer-aided ultrasonic NDI system.

Another ultrasonic method has potential for detecting surface or near-surface flaws. The transducer is placed at a low angle to the surface. The acoustic waves travel along the surface rather than penetrating the interior of the part and interact with surface or near-surface discontinuities. Since the strength of ceramic materials is so sensitive to surface flaws, this may be a useful method to consider.

16.4.5   OTHER NDI TECHNIQUES

By this time, hopefully you recognize that NDI of ceramics is generally expensive and not as reliable or definitive as desired. Efforts are continually in progress to decrease cost and improve resolution. Many techniques have been explored, but will not be discussed here: laser holographic interferometry, acoustic holography,[12] high-frequency ultrasonics,[13] microwave[14] acoustic emissions, laser scattering,[15] air-coupled ultrasonics, and thermal imaging.[16]

16.5   QUALITY PROBLEM SOLVING AND IMPROVEMENT

Quality is obviously a key issue to both the manufacturer and to the end user. Whether you are a manufacturer or an end user, sometime in your career you will likely be in a position of trying to solve a quality problem. The objective of this section is to introduce you to statistical process control (SPC) and specific techniques or tools that you can use to understand and solve a quality problem or to make continuous improvements in quality.[17–23] First let us look at the nature of variation in a process, material, or product. Then let us look at techniques for effectively identifying sources of variation and maintaining control over a process or product. Since the discussion of SPC tools in this chapter represents a brief introduction, you may choose to check References 17 to 23 for further information.

16.5.1   NATURE OF VARIATION IN A FABRICATION PROCESS

Variation can occur in two forms. One is "normal or natural variation," which is the normal statistical variation that one would expect from any action. For example, if you flip a coin you have two options: heads or tails, each of which has a 0.5 probability of occurring on any one flip. In practice, however, if you flip a coin 10 times you will rarely get 5 heads and 5 tails. And every time you do a sequence of 10 flips, you will get a different combination of heads and tails. Figure 16.12 illustrates the results of 10 sets of 10 flips. There appears to be enormous variation. Note, though, that the total number of heads after 100 flips was 52 and tails was 48, close to the statistical average of 0.5. This is a graphic example of natural variation.

The second category of variation is "abnormal or caused variation," which is variation outside the normal statistical variation and has a distinct cause. The cause may be an equipment problem,

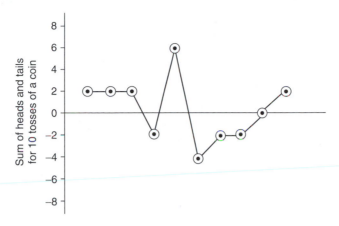

FIGURE 16.12   Results of ten sets of ten coin flips where each head is valued at $+1$ and each tail is valued at -1. Illustrates the wide range in natural variation. (From Richerson, D.W., Testing for Design, Material, and Fabrication Optimization, Chapter 9 (pp. 387–425) in *Mechanical Testing Methodology for Ceramic Design and Reliability*, Cranmer, D.C. and Richerson, D.W., Eds., Marcel Dekker, Inc., New York, 1998. With permission.)

a material problem, an operator error, or even a change in the environment such as temperature or humidity. Remember the behavior of organic binders such as polyvinyl alcohol discussed in an earlier chapter? Slight variations in humidity, temperature, and amount of plasticizer had a large effect on the deformation characteristics of the binder and on the resulting properties of the material.

Both natural variation and abnormal variation are important in achieving control in the fabrication process of ceramics. The challenge is to recognize which variations are normal and which are abnormal, to eliminate the abnormal variations, and to minimize the natural variations. We accomplish this through a combination of observation, gathering of reliable characterization data (in-process and post-process), and use of SPC tools.

16.5.2   SPC Tools and Techniques

As shown in Figure 16.13, SPC tools are used for problem identification and for problem analysis and control. The items on the left are used to identify the problem, those on the right to analyze and control the problem, and those in the middle for both. Subsequent paragraphs define and describe briefly each tool.

16.5.2.1   Flow Chart

The flow chart is a simple pictorial representation of all the steps and decision points in a process. Figure 16.14 is a very simplified illustration of the key elements of a flow chart. The flow chart has multiple benefits:

It can provide a detailed procedure for the operators to follow.
It can include a listing of the required key measurements.
It can clearly indicate decision points and even the criteria to be satisfied for a go-forward decision.

16.5.2.2   Check Sheet

A check sheet is a simple tabular form for keeping track of the frequency of process events such as causes of rejection. As shown in Figure 16.15, one column can list the events or key measurements and other columns list the frequency of occurrence of each. For this hypothetical example, one can quickly see that the most frequent defect causing rejection was "wrong weight" and the next most frequent was "wrong length." This very simple summary of information quickly gives the investigator the priority of problems to solve and some guidance on where to start.

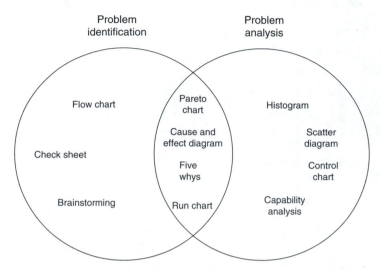

FIGURE 16.13   Tools that are useful for problem identification and problem analysis/solution for statistical process control (SPC). (From Richerson, D.W., Testing for Design, Material, and Fabrication Optimization, Chapter 9 (pp. 387–425) in *Mechanical Testing Methodology for Ceramic Design and Reliability*, Cranmer, D.C. and Richerson, D.W., Eds., Marcel Dekker, Inc., New York, 1998. With permission.)

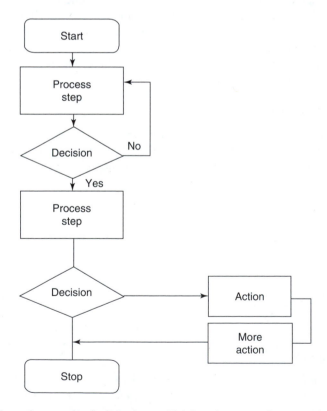

FIGURE 16.14   Schematic example of a flow chart, which is an important document that shows the sequence of steps of a process and key decision points. (From Richerson, D.W., Testing for Design, Material, and Fabrication Optimization, Chapter 9 (pp. 387–425) in *Mechanical Testing Methodology for Ceramic Design and Reliability*, Cranmer, D.C. and Richerson, D.W., Eds., Marcel Dekker, Inc., New York, 1998. With permission.)

Defect	Day					Total
	10/1	10/2	10/3	10/4	10/5	
Wrong length	⊬⊬⊬	||||	⊬⊬⊬ |	||||	⊬⊬⊬	24
Wrong OD	|		|	|		3
Wrong ID	|||	||	|||	|||	||	13
Wrong weight	⊬⊬⊬ ||	⊬⊬⊬ |||	⊬⊬⊬ |	⊬⊬⊬ ||	⊬⊬⊬ ||||	37
Wrong surface			|	|		2
Total	16	14	17	16	16	79

FIGURE 16.15 Hypothetical example of a check sheet, a tool that tracks how often different events happen. (From Richerson, D.W., Testing for Design, Material, and Fabrication Optimization, Chapter 9 (pp. 387–425) in *Mechanical Testing Methodology for Ceramic Design and Reliability*, Cranmer, D.C. and Richerson, D.W., Eds., Marcel Dekker, Inc., New York, 1998. With permission.)

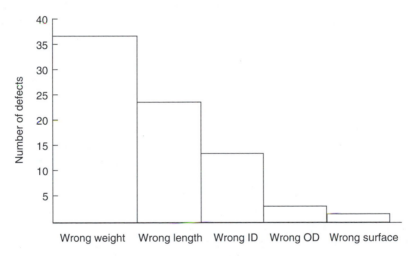

FIGURE 16.16 Example of a Pareto chart, a tool that graphically displays the frequency of occurrence of events. (From Richerson, D.W., Testing for Design, Material, and Fabrication Optimization, Chapter 9 (pp. 387–425) in *Mechanical Testing Methodology for Ceramic Design and Reliability*, Cranmer, D.C. and Richerson, D.W., Eds., Marcel Dekker, Inc., New York, 1998. With permission.)

A variation of the check sheet is to list the key measurements versus each fabrication run or lot. This allows a quick comparison of numerical values over a series of runs or lots.

16.5.2.3 Pareto Chart

A Pareto chart is a simple graphical format for displaying information from a check sheet or other compilation of frequency of events. Figure 16.16 illustrates the data from the check sheet in Figure 16.15 organized into a Pareto chart.

16.5.2.4 Brainstorming

Brainstorming is a technique for bringing together a group of persons to quickly gather ideas to help identify a problem, the cause of a problem, and potential solutions. It is best done with a facilitator

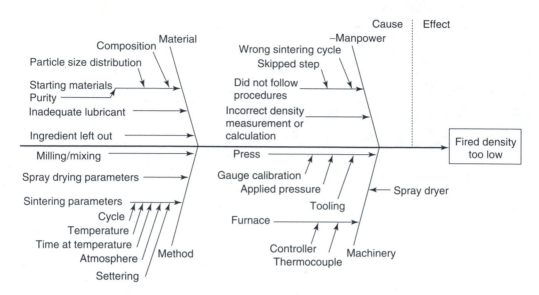

FIGURE 16.17 Hypothetical example of a fishbone diagram, a tool that is helpful in exploring the interrelationships in a process, especially in seeking cause and effect ideas. (From Richerson, D.W., Testing for Design, Material, and Fabrication Optimization, Chapter 9 (pp. 387–425) in *Mechanical Testing Methodology for Ceramic Design and Reliability*, Cranmer, D.C. and Richerson, D.W., Eds., Marcel Dekker, Inc., New York, 1998. With permission.)

(leader or moderator) and the following procedure. First arrange a meeting of a diverse group of people that may have ideas. For example, if there is a processing problem, it might be important to include engineers, technicians, equipment operators, and quality control personnel. The facilitator then poses a very specific question such as "What factors might cause the viscosity of the slip to vary?" Each member of the group spends several minutes privately writing down all the ideas they can think of that respond to the question. Then the facilitator or moderator requests one idea verbally from each person in turn and writes the idea as stated on a board or flip chart. This gives each person an equal minimum-stress opportunity to present his or her ideas. New ideas stimulated by those listed are also recorded. This procedure continues, each person in turn, until all the ideas have been listed. No discussions, criticism or interruptions are allowed until all the items are listed. Then discussions begin, primarily to categorize and prioritize the ideas and to agree to a path of action.

16.5.2.5   Cause and Effect Diagram (Fishbone Diagram)

The cause and effect diagram is another way to get a group of people or even individuals to gather ideas and to start to analyze the source or sources of a problem. A hypothetical example is shown in Figure 16.17 where the "effect" is "fired density too low." The cause might be in manpower, the material, the method, or the machinery (the four m's). Each category is represented by a major line intersecting the primary horizontal line. Specific ideas are then written on horizontal lines extending from each category line. Organizing the ideas from individuals and a group into categories often can guide a team in the right direction to isolate the cause of a problem and come up with a solution.

16.5.2.6   Five-Whys Diagram

The five-whys diagram is another tool to get people to think about the causes of a problem and especially to delve into secondary sources that may not be obvious. The concept is illustrated in Figure 16.18. The problem or event is listed in the center. The team then must come up with five reasons why the problem or event might have occurred. Then they do the same with each of these five reasons and again with each of these.

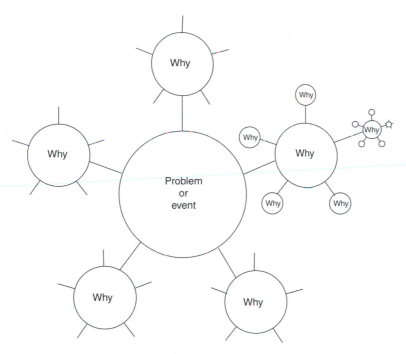

FIGURE 16.18 Schematic of a "five whys" diagram, another tool for stimulating thinking about the possible causes of an event or problem. (From Richerson, D.W., Testing for Design, Material, and Fabrication Optimization, Chapter 9 (pp. 387–425) in *Mechanical Testing Methodology for Ceramic Design and Reliability*, Cranmer, D.C. and Richerson, D.W., Eds., Marcel Dekker, Inc., New York, 1998. With permission.)

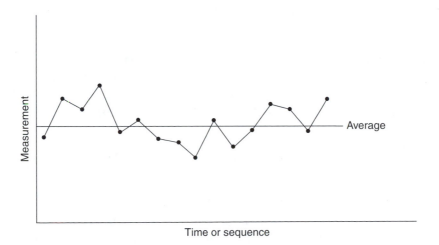

FIGURE 16.19 Example of a run chart, a tool that illustrates variations from batch-to-batch or run-to-run and is useful in monitoring a step in a process or the final output of the process. (From Richerson, D.W., Testing for Design, Material, and Fabrication Optimization, Chapter 9 (pp. 387–425) in *Mechanical Testing Methodology for Ceramic Design and Reliability*, Cranmer, D.C. and Richerson, D.W., Eds., Marcel Dekker, Inc., New York, 1998. With permission.)

16.5.2.7 Run Charts and Control Charts

The tools discussed so far are excellent for trying to identify the source of a problem, but SPC requires numerical, data-based tools to help monitor and control a process. Two basic tools are the run chart and the control chart. As shown in Figure 16.19, the run chart is simply a plot of the variation of a

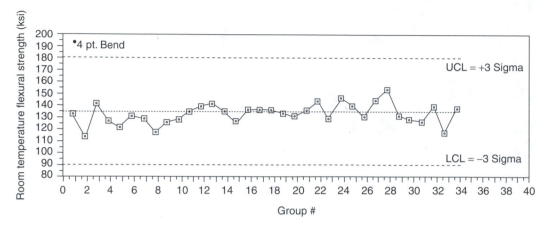

FIGURE 16.20 Actual example of a control chart (for an advanced silicon nitride material), a tool that helps show if a process step or product is within normal variation or has a problem that must be immediately addressed. (From Richerson, D.W., Testing for Design, Material, and Fabrication Optimization, Chapter 9 (pp. 387–425) in *Mechanical Testing Methodology for Ceramic Design and Reliability*, Cranmer, D.C. and Richerson, D.W., Eds., Marcel Dekker, Inc., New York, 1998. With permission.)

key control measurement for each run or process batch compared to the average for that measurement. Normally it shows the natural variation of the process step. The control chart, though, goes a step further and places an upper specification limit (USL) and a lower specification limit (LSL) above or below which the process step deviates from natural variation and becomes abnormal variation. As long as the data point for the key measurement lies between the USL and LSL, the process is in control. If the data point for a run falls outside these limits, a problem or "special cause" exists and must immediately be investigated. Figure 16.20 shows an example of a control chart. Note that the process was maintained in control throughout the 34 process batches.

The control chart allows the manufacturer to statistically monitor his process, but it does not assure that the product will meet the customer's specification. A "capability analysis" is necessary in addition to the control analysis. The capability analysis simply compares the natural variation of the process with the specification limits of the customer, as illustrated schematically in Figure 16.21. The Gaussian curve is usually based on ± three times the standard deviation, which is the 6σ standard often referred to by quality control personnel. Stated in a different way, the specification limit is USL-LSL and the capability index (C_p) is USL-LSL divided by 6σ. If 6σ = (USL-LSL), C_p = 1 and the variation would fit exactly within the specification limits. The desired production condition is for $C_p > 1$.

16.5.3   USE OF SPC TOOLS FOR CONTINUOUS IMPROVEMENT

Manufacturing processes are developed through a long sequence of laboratory research and development and subsequent scale-up. At the time of first production, the process is rarely optimum. There are invariably many ways that the process and product can be improved such as better raw materials, parameters that increase yield, optimization of process parameters, reduction in the number of key measurements required to control the process, better consistency of operators, fine tuning of equipment, and improved equipment. Many ceramic processes have a yield of less than 50% when first released to production. Implementing a philosophy and SPC procedures for continuous improvement provides a tremendous opportunity for cost reduction and customer satisfaction.

Effective use of SPC tools depends upon high quality measurements. The first step in use of SPC is therefore selecting the appropriate key measurements and to verify that the measurement systems (equipment, procedures, personnel training) are under control and reproducibly yield data that are accurate. The second step is to bring the fabrication process under control by eliminating

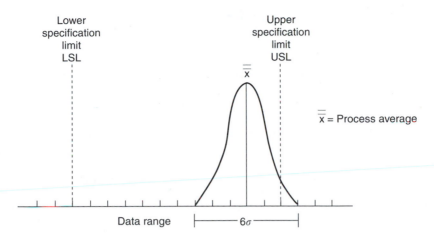

FIGURE 16.21 Example of a comparison of the process variation based on $\pm$ three times the standard deviation (6σ) to the customer specification limits. Variation is small, but still overlaps with one of the specification limits and produces rejects. (From Richerson, D.W., Testing for Design, Material, and Fabrication Optimization, Chapter 9 (pp. 387–425) in *Mechanical Testing Methodology for Ceramic Design and Reliability*, Cranmer, D.C. and Richerson, D.W., Eds., Marcel Dekker, Inc., New York, 1998. With permission.)

special causes of abnormal variation. The third step is to establish run charts and control charts for each key measurement and compare these using capability analysis with the specification. Then you can prepare a plan for continuous process improvement to reduce normal variation and to shift the normal distribution to the optimum position within the product specification (to give the greatest margin of safety or performance).

16.5.4 QUALITY ASSURANCE PERSPECTIVE OF THE END USER

If you are an end user, especially if your expertise is other than ceramics, your perspective for quality assurance is different. Your concern will be that you receive a product from the ceramic manufacturer that will meet the reliability that you require. To do this you will typically have to establish a specification and work with the manufacturer to define a certification procedure to assure quality. You also will probably want to conduct an audit of the manufacturer's quality assurance system. All of these tasks can best be achieved if you have basic knowledge of SPC and quality control techniques. In addition, if a part fails in your application you will need to use some of the same tools to investigate the nature of the failure and to locate a solution.

REFERENCES

1. Wynn, L.P., Tree, D.J., Yonushonis, T.M., and Solomon, R.A., Proof testing of ceramic components, in *Proceedings of the 1977 DARPA/NAVSEA Ceramic Gas Turbine Demonstration Engine Program Review*, Fairbanks, J.W. and Rice, R.W., Eds., MCIC Rept. MCIC-78-36, 1978, pp. 493–516.
2. Wiederhorn, S., Reliability, life prediction, and proof testing of ceramics, in *Ceramics for High Performance Applications*, Burke, J.J., Gorum, A.E., and Katz, R.N., Eds., Brook Hill, Chestnut Hill, Mass., 1974, pp. 635–663. (Available from MCIC, Battelle Columbus Labs., Columbus, Ohio.)
3. Kossowsky, R., Defect detection in hot-pressed Si_3N_4, in *Ceramics for High Performance Applications*, Burke, J.J., Gorum, A.E., and Katz, R.N., Eds., Brook Hill, Chestnut Hill, Mass., 1974, pp. 665–685. (Available from MCIC, Battelle Columbus Labs., Columbus, Ohio.)
4. Richerson, D.W., Schuldies, J.J., Yonushonis, T.M., and Johansen, K.M., ARPA/Navy ceramic engine materials and process development summary, in *Ceramics for High Performance Applications*, Burke, J.J., Gorum, A.E., and Katz, R.N., Eds., Brook Hill, Chestnut Hill, Mass., 1974, pp. 625–650. (Available from MCIC, Battelle Columbus Labs., Columbus, Ohio.)

5. Cassidy, D.J., NDE techniques used for ceramic turbine rotors, in *Ceramics for High Performance Applications, II*, Burke, J.J., Lenoe, E.N., and Katz, R.N., Eds., Brook Hill, Chestnut Hill, Mass., 1978, pp. 231–242. (Available from MCIC, Battelle Columbus Labs., Columbus, Ohio.)

6. Schuldies, J.J. and Spaulding, W.H., Radiography and image enhancement of ceramics, in *Proceedings of the 1977 DARPA/NAVSEA Ceramic Gas Turbine Demonstration Engine Program Review*, Fairbanks, J.W. and Rice, R.W., Eds., MCIC Rep. MCIC-78-36, 1978, pp. 403–428.

7. Schuldies, J.J. and Richerson, D.W., NDE approach, philosophy and standards for the DARPA/NAVSEA ceramic turbine program, in *Proceedings of the 1977 DARPA/NAVSEA Ceramic Gas Turbine Demonstration Engine Program Review*, Fairbanks, J.W. and Rice, R.W., Eds., MCIC Rept. MCIC-78-36, 1978, pp. 381–402.

8. Ellingson, W.A., Ikeda, Y., and Goebbels, J., Nondestructive evaluation/characterization, Chapter 23 (pp. 493–519) in *Ceramic Gas Turbine Component Development and Characterization*, van Roode, M., Ferber, M.K., and Richerson, D.W., Eds., ASME Press, New York, 2003.

9. Goebbels, J., Heidt, H., Kettschau, A., and Reimers, P., Computed Tomography, pp. 97–100 in *NBS/BAM 1986 Symposium on Advanced Ceramics*, Hsu, S.M. and Czichos, H., Eds., NIST Special Publication 766, 1989.

10. Schuldies, J.J. and Derkacs, T., Ultrasonic NDE of ceramic components, in *Proceedings of the 1977 DARPA/NAVSEA Ceramic Gas Turbine Demonstration Engine Program Review*, Fairbanks, J.W. and Rice, R.W., Eds., MCIC Rept. MCIC-78-36, 1978, pp. 429–448.

11. Seydel, J.A., Improved discontinuity detection in ceramic materials using computer-aided ultrasonic nondestructive techniques, in *Ceramics for High Performance Applications*, Burke, J.J., Gorum, A.E., and Katz, R.N., Eds., Brook Hill, Chestnut Hill, Mass., 1974, pp. 697–709. (Available from MCIC, Battelle Columbus Labs., Columbus, Ohio.)

12. Brenden, B.B., Recent developments in acoustical imaging, *Mat. Res. and Stand.*, *MTRSA II* (9), 16, 1971.

13. Kino, G.S., Khuri-Yakub, B.T., Murakami, Y., and Yu, K.H., Defect characterization in ceramics using high frequency ultrasonics, in *Proceedings of the DARPA/AFML Review of Progress in Quantitative NDE*, AFML-TR-78-25, 1979, pp. 242–245.

14. Bahr, A.J., Microwave techniques for the nondestructive evaluation of ceramics, in *Proceedings of the DARPA/AFML Review of Progress in Quantitative NDE*, AFML-TR-78-25, 1979, pp. 236–241.

15. Sun, J.G., Ellingson, W.A., Steckenrider, J.S., and Ahuja, S., Application of optical scattering methods to damage detection in ceramics, in *Machining of Ceramics and Composites*, Jahanmir, S., Ramulu, M., and Koshy, P., Eds., Marcel Dekker, Inc., New York, 1999, pp. 669–699.

16. Sun, J.G., Petrak, K.R., Pillai, T.A.K., Deemer, C., and Ellingson, W.A., Nondustructive evaluation and characterization of damage and repair for continuous fiber ceramic composites panels, *Ceram. Eng. and Sci. Proc.*, 19(3), 615–622, 1998.

17. Deming, W.E., *Out of the Crisis*, MIT, Ctr. Advanced Engineering Study, Cambridge, Mass., 1986.

18. Diamond, W.J., *Practical Experiment Designs for Engineers and Scientists*, Van Nostrand Reinhold, New York, 1981.

19. Bennett, C.A. and Franklin, N.L., *Statistical Analysis in Chemistry and the Chemical Industry*, Marbern House, Mass., 1988.

20. Grant, E.L. and Leavenworth, R.S., *Statistical Quality Control*, McGraw-Hill, New York, 1980.

21. Scherkenbach, W.W., *The Deming Route to Quality and Productivity, Roadmaps and Roadblocks*, Mercury Press, MD, 1986.

22. Ishikawa, K., *What is Total Quality Control? The Japanese Way*, Prentice-Hall, Inc., Englewood Cliffs, NJ, 1985.

23. Richerson, D.W., Testing for Design, Material, and Fabrication Optimization, Chapter 9 (pp. 387–425) in *Mechanical Testing Methodology for Ceramic Design and Reliability*, Cranmer, D.C. and Richerson, D.W., Eds., Marcel Dekker, Inc., New York, 1998.

ADDITIONAL RECOMMENDED READING

1. Proceedings of the Joint Conference on Nondestructive Testing of High Performance Ceramics, Aug. 25–27, 1987, Boston, Mass., American Ceramic Society, Westerville, Ohio, 1987.

2. Cartz, L., *Nondestructive Testing*, ASM Int., Materials Park, Ohio, 1995.

PROBLEMS

1. X-ray radiography has been successful for detection of some types of internal defects in ceramics. Assume that a ceramic material 2.5-mm thick can be inspected by x-ray radiography to a 2T level. What is the minimum-size defect that should be routinely detected in this material by this method?

2. Identify important certification tests for a ceramic starting powder. For a powder compact. For a sintered part.

3. Which of the following would you select as the most economical method to examine a nonporous ceramic for surface cracks?
   (a) dye penetrant
   (b) x-ray radiography
   (c) ultrasonics
   (d) holography

STUDY GUIDE

1. How would you define quality assurance?

2. What are some of the important elements of quality assurance for manufacturing of ceramic components?

3. What is "in-process QA" and what are its key objectives?

4. What is a "formal written manufacturing procedure" document?

5. What are the key contents of the manufacturing procedure document?

6. Process flow charts were shown in Chapter 13 for pressing (Figure 13.1), slip casting (Figure 13.22), and injection molding (Figure 13.57). How might you modify these process flow charts to produce a process control "formal written manufacturing procedure" document?

7. Explain the purpose and content of a specification.

8. Explain the purpose and content of a certification.

9. List a few of the critical parameters or characteristics of ceramic processing that are likely to need in-process measurement and control by a specification or certification procedure.

10. Once a component has been manufactured, what are some of the options for post-processing QA?

11. Explain how a proof test provides increased assurance of quality.

12. Explain the meaning of "nondestructive inspection" (NDI).

13. What is a penetrant, and what might be its limitations for inspection of ceramic materials?

14. Explain the limitations or difficulties of inspecting a shape of nonuniform thickness with conventional X-ray radiography.

15. Explain why computed tomography is more effective at locating defects in a complex ceramic shape.

16. Briefly explain ultrasonic NDI.

17. What are some of the limitations or difficulties of ultrasonic NDI for ceramics?

18. Identify and briefly compare two forms of variation that can occur in ceramic processing.

19. How might someone minimize natural variations in a process?

20. What are some sources of abnormal variation in a process?

21. Describe a couple simple ways of identifying the major sources of rejection of parts from a manufacturing process.

22. Once a problem has occurred that is causing an abnormally high level of rejects, what are several tools that can be used to try to locate the source of the problem?

23. Many engineers call a meeting to brainstorm and simply open the meeting to discussion. Why might the technique described in this chapter be more efficient and effective?
24. How does one go beyond a run chart to actually achieve statistical process control?
25. List some things that can be done to improve a material or process by decreasing the normal variation.
26. Explain why high quality measurements are critical to successful SPC.
27. Why should an end user be familiar with ceramic processing and SPC?

Part IV

Design with Ceramics

Part II discussed the relationships among atomic bonding, crystal structure, and properties of ceramics as compared to other materials. It was shown that the intrinsic properties are controlled largely by the nature of the bonding and structure, but that the extrinsic or actual properties are controlled by such factors as structural defects, impurities, and fabrication flaws.

Part III reviewed the fabrication processes for ceramic materials and components, defined potential sources in these fabrication processes of property-limiting flaws, and described techniques for detecting and limiting the occurrence of these flaws.

The objective of Part IV is to apply the property, fabrication, and inspection principles learned in Parts II and III to the selection and design of ceramic components for advanced engineering applications. Chapter 17 discusses design considerations, such as requirements of the application, property limitations, fabrication limitations, cost limitations, and reliability requirements. Chapter 18 considers design approaches. The approach is normally based on the design considerations and can range from empirical to deterministic to probabilistic. Chapter 19 explores the importance and techniques of failure analysis. If a ceramic component fails, often the only means of determining whether the failure was design-oriented or material-oriented is by examination of the fractured pieces. Chapter 20 identifies methods of increasing the toughness of ceramics to improve their resistance to fracture.

The final chapter, Chapter 18, reviews a range of ceramic applications. Emphasis is on the criteria for selecting the best material for each application based on the design considerations and design approaches described in Chapters 14 and 15.

17 Design Considerations

The selection of a material and a fabrication process for a component for an engineering application is governed by a variety of factors, not just the material properties. The shape and cost limitations of the fabrication process must be considered. The requirements of the application, including such factors as load distribution, environment, and tolerances, must be considered, as also must be the reliability requirements, such as life expectancy, the risk of premature failure, and the effects of premature failure on the rest of the system and personnel.

17.1 REQUIREMENTS OF THE APPLICATION

The first step in the design of a ceramic component or any other component is to define clearly and prioritize the requirements of the application. Usually, one or two characteristics will be most critical and allow an initial selection of candidate materials. For instance, a primary characteristic of a wear-resistant material is hardness. However, if wear resistance is required in a severe chemical environment or at high temperature, other characteristics become critical and must be considered on an equal or nearly equal basis to hardness. Table 17.1 lists some of the design characteristics that an engineer must consider for an application.

To get a better feeling for the thought process that an engineer goes through in defining and prioritizing the critical design requirements for an application, consider two examples: a grinding wheel and a gas-turbine rotor. These both rotate at high speeds and must have similar design requirements. Right? Not necessarily.

The grinding wheel and rotor do have some important requirements in common:

1. They must have suitable strength to remain intact at their respective design speeds.
2. They must have an acceptable margin of safety as defined by industry and government standards.
3. They must be fabricated such that they are in balance when rotating.

TABLE 17.1
Examples of Design Characteristics That Must Be Considered

Load	Tolerances
Stress distribution	Surface finish
Attachment	Stability to radiation
Interfaces	Life requirement
Friction	Safety requirements
Chemical environment	Toxicity
Temperature	Pollution
Thermal shock	Electrical property requirements
Creep	Magnetic property requirements
Strain tolerance	Optical property requirements
Impact	Cost
Erosion	Quantity

Other critical design requirements of the grinding wheel include controlled surface breakdown to expose fresh abrasive grains, impact resistance, low cost, and adaptability to mass production. Other critical design requirements for the rotor include high strength and oxidation and corrosion resistance at high temperature, resistance to extreme thermal shock, and complex shape fabrication to close tolerances. The differences in design requirements result in very different material design selections with corresponding design, manufacturing, and quality-control choices. The grinding wheel is best made from a composite material with hard abrasive particles bonded by a softer matrix. The turbine rotor requirements have been met for a limited number of applications by monolithic (noncomposite) silicon nitride fabricated by overpressure sintering. Further material and design development are required before ceramic rotors will be reliable enough for most turbine engine applications.

Design requirements can be determined in many ways. For existing applications, in which an alternative material is being sought to achieve benefits such as lower cost, longer life, or improved performance, a specification usually exists defining quantitatively the critical design requirements. This can be a good starting point. However, one must remember that ceramics have different properties than other materials and that modification or redesign may be necessary. The engineer should especially consider thermal expansion mismatch (if the component is to be used over a temperature range) and the implication of point loading or flexural loading.

For new applications, design requirements will either have to be assumed based on the best estimates of service conditions, estimated by analogy with similar applications, determined experimentally, or predicted analytically. This can result in a multiphase program in which the first phase would be design analysis and material property screening; the second phase would then be fabrication of prototypes; the third phase, component testing, will overlap with the second phase and allow iteration back and forth between prototype fabrication, component testing, and redesign.

17.2 PROPERTY LIMITATIONS

The second step in the design of a ceramic component is to compare the properties of candidate ceramic materials with the requirements of the application. This is usually hampered by lack of property data at the design conditions, especially if an adverse service environment is involved. However, an initial set of candidates that have the closest fit with the design requirements can usually be defined. Examples of design requirements for some applications and candidate ceramics with the needed combination of properties are listed in Table 17.2. These candidates can then be included in screening tests to isolate the best candidate.

The method and extent of property evaluation vary according to the nature of the application. Some materials may clearly satisfy the property requirements so that no measurements are necessary. Such is the case in many room-temperature wear-resistance applications, where technical ceramics such as polycrystalline sintered Al_2O_3 or hot-pressed B_4C have strengths a factor of 10 higher than design loads and more than adequate hardness. In this case, factors such as cost and large quantity availability are usually more important and determine the final selection.

In other applications, extensive property and QA measurements are required. This is especially true of electrical and magnetic ceramics, where properties must be precisely controlled. It is also true of optical applications, where index of refraction, absorption, and color more often must be controlled to a tight tolerance.

Various approaches can be pursued in evaluating the suitability of a material's properties. If the shape is simple and the part can be fabricated quickly and inexpensively, it may be best to make the part to print and test it directly in the system being developed. This has the potential of leading to commercialization with a minimum of time and development cost. However, the engineer must carefully assess the consequences of a failure during this testing. Will a failure damage much more costly components in the system? Will it endanger personnel or facilities? Will initial test parts be

TABLE 17.2
Examples of Design Requirements of Various Applications and Ceramics with Properties That Match the Requirements

Application	Requirements of the Applications	Candidate Ceramics	Key Properties
Seal	Wear resistance, high surface finish, low friction	Al_2O_3, SiC, Si_3N_4	Hardness, low porosity, machinability to high surface finish
Turbine stator	Thermal shock resistance, oxidation resistance, high-temperature stability, complex aerodynamic shape	Si_3N_4, SiC	High strength, moderate to low thermal expansion, moderate to high thermal conductivity, can be fabricated to complex shape
Heating element	Produces heat when electric current passes through, long-term stability at temperature and to thermal cycling	$MoSi_2$, SiC, C, doped ZrO_2	Semiconductor level of electrical resistivity, high-temperature stability in the furnace atmosphere
Rotary heat exchanger	Continuously exchanges heat between hot exhaust and cold inlet gasses, high surface area honeycomb shape, high-thermal-shock resistance	LAS, NZP, cordierite	Low thermal expansion, long-term stability in high-temperature exhaust environment
Heat sink for IC and transistor devices	Rapid heat dissipation, low electrical losses, compatibility with Si and GaAs chips and metal interconnects	BeO, AlN, diamond	High electrical resistance, low dielectric constant, high thermal conductivity
Furnace insulation	Withstand high temperatures, retain heat within furnace	Porous fire bricks, fiber-board, "wool"	Low thermal conductivity, high-temperature stability
Miniature capacitor	High charge-storage capability	Various titanates and mixed zirconate-titanates	High polarizability, high dielectric constant, low dielectric loss
High-speed, high-load bearing	Resistance to high contact loads, wear resistance, low friction, heat dissipation	Si_3N_4	High toughness, hardness, and strength; very low porosity, machinable to high surface finish, moderate to high thermal conductivity
Segments of watch band	Impact resistance, abrasion resistance, attractive appearance	PSZ, TZP (transformation-toughened zirconia materials)	High toughness and strength, moderate hardness, machinable to high surface finish, ability to be produced in a range of colors

Source: From Richerson, D.W., Design and fracture analysis, Lesson 14 in Introduction to Modern Ceramics, ASM Materials Engineering Institute Course 56, ASM International, Materials Park, OH, 1990.

of high-enough quality to provide a meaningful component test, or should material development and property verification be conducted first?

For many advanced applications, no existing material is clearly suitable. In fact, at the current time, we are design-limited in most advanced materials applications. This means that engineers have already identified approaches to improving overall systems, but do not have materials with acceptable properties. Therefore, these applications are dependent on material development; often the project engineer has the responsibility to complete this development. An example where material development is required is for heat engine components.

Ceramics are currently being evaluated for gas-turbine components to allow increased operating temperatures. By increasing operating temperatures from current metal-limited levels of 1800 to 2100°F to 2500°F or greater, fuel savings from 10 to 25% could be achieved. The feasibility has been demonstrated,[2-4] but present ceramic materials do not yet have the predictably reproducible strength and environmental stability to provide long-term reliability.[5]

As discussed in previous chapters, property limitations frequently result from fabrication limitations. The property-controlling material defects occur during the various steps of processing. Often, design needs can be met simply by increased care during processing. Sometimes this can be achieved by a minor modification in the processing specification. Other times, iterative development will be necessary.

Another factor that affects properties is the quantity of parts being manufactured. Industry experience has shown that part-to-part variation is usually high in prototype or small production quantities, but decreases substantially when high-volume production is reached.

17.3   FABRICATION LIMITATIONS

Comparison of the design requirements with the property limitations dictates the fabrication requirements. At this point, two primary questions will be asked: (1) Will existing fabrication experience and technology achieve the required properties? (2) Can existing fabrication experience and technology achieve the required configuration in the necessary quantity at an acceptable cost?

If the answer to the first question is "yes," the engineer can concentrate on the second question. If the answer is "no," then the following options need to be considered.

1. Achieve the required properties by improvement of an existing commercial material or fabrication process.
2. Continue development of an emerging or developmental material or fabrication process.
3. Develop a new material fabrication process or material system (such as a composite).

Obviously, the difficulty, time, and cost will increase substantially if item 2 or 3 is the only feasible option rather than item 1. It is the engineer's responsibility to assess which level of development is required and whether the program resources are adequate to implement the development. Many programs have failed or experienced substantial cost overruns because an engineer did not make an adequate assessment of the material property and fabrication limitations.

Shape capability is the next critical fabrication concern. Once shape and tolerances have been defined for the application by the design analysis, the engineer must evaluate the fabrication approaches and manufacturing sources. This is usually best done by direct discussion with the material suppliers; however, finding the appropriate supplier to talk to is the first step. The following are potential sources of information:

1. *Thomas Register of Products and Services*
2. publications such as the *American Ceramic Society Bulletin, Journal of the American Ceramic Society*, and *Ceramic Industry*; both the *Bulletin* and *Ceramic Industry* publish a yearly directory of suppliers of materials, services, and finished components

3. the library, especially reference periodicals such as *Ceramic Abstracts, Chemical Abstracts*, and *Engineered Materials Abstracts*

4. special information services such as Chemical Abstracts at Columbus, Ohio; Materials and Ceramics Information Center at Battelle Columbus Laboratories; National Technical Information Service, Springfield, Virginia; and the Materials Information Department of ASM International at Materials Park, Ohio

5. ceramics consultants, usually listed in the classified ads of monthly ceramics publications and in the *Directory of Consultants and Translators for Engineered Materials* available from ASM International

6. faculty at universities that offer degrees in ceramic engineering or materials science, such as Massachusetts Institute of Technology, Pennsylvania State University, University of Washington, University of Utah, Ohio State University, Iowa State University, University of Illinois, University of California at Los Angeles, University of California at Berkeley, Alfred University, Virginia Polytechnic Institute, University of Michigan, Rutgers, and University of Arizona

7. research institutes such as Battelle Columbus Laboratories and IIT Research Institute

8. annual meeting and exposition of the American Ceramic Society and exhibitions and conferences sponsored by ASM International

9. the internet

The first contact with a supplier involves a description of the required component together with critical considerations such as service environment, quantity required, and key properties. If this first discussion is encouraging, a set of prints or drawings is sent to the potential supplier for further evaluation. This is usually followed by meetings during which the final procurement decision is made and program details are negotiated.

An engineer with a knowledge of the various ceramic fabrication processes has a pronounced advantage in evaluating the fabrication limitations associated with a new design. Processes such as uniaxial pressing and extrusion are very good for reproducibly fabricating large quantities of simple parts. Injection molding can produce more complex parts in large quantity, but greater care is necessary in tool design and quality control because of the increased likelihood of fabrication flaws. Slip casting can also produce complex parts, but in lesser quantity than pressing or injection molding. For high-strength, high-reliability requirements, hot pressing might be considered, but one must remember the difficulties and cost of achieving complex shape by this process.

A development program is usually required to fabricate a new ceramic component. A typical flowchart is shown in Figure 17.1. The steps usually consist of tool design, tool fabrication, fabrication of initial parts, evaluation of the dimensions and integrity of these parts, tool redesign and

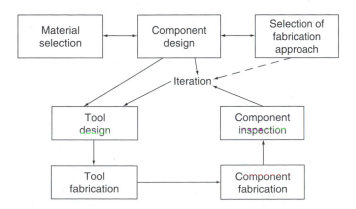

FIGURE 17.1   Schematic of program approach to develop fabrication capability for a new ceramic part.

rework as required, fabrication of parts, inspection, testing of the parts in the application or a simulation rig, and iteration as required. Frequently, prototype parts will be made by a different fabrication process than the one intended for production. This is done to minimize program cost, especially tool cost. Once feasibility has been demonstrated and a workable design configuration verified, it is much easier to justify large capital outlay for expensive production tooling. For instance, a complex injection-molding tool can cost more than $100,000. The program would be quite expensive if several retoolings were required to determine the optimum component design. However, making the initial parts by cold isostatic pressing and green machining or by slip casting could limit initial tooling cost to perhaps $10,000.

17.4 COST CONSIDERATIONS

Cost is an important design consideration and must be evaluated concurrently with other factors such as property, fabrication, and reliability requirements. Although it must be realized that initial prototype parts will be expensive and that adequate information may not be available to project production costs, an initial cost analysis should be conducted and a strategy defined for obtaining the necessary information and achieving the ultimate production-cost objectives. There have been many programs in which an engineer has ignored cost considerations and ultimately made a component work, only to find that there is no way of reducing the system cost to a marketable level. Conversely, there have been other programs not started or terminated prematurely because an engineer took high prototype costs too seriously and did not adequately evaluate production-cost projections.

Cost projection has many pitfalls. The individual engineer should not assume the whole responsibility, but should seek other individuals with as much experience as possible. The optimum consultant would be a person who has solved and commercialized a different ceramic component of the same material for a similar application. In cost projection there is no equivalent for experience and technical understanding of the specific material and process.

17.5 RELIABILITY REQUIREMENTS

The reliability requirements are also part of the initial requirements of the application and may be ultimately written into a specification or a warranty. The term "reliability" is really rather ambiguous and varies dramatically depending on the application.[3] For instance, the heat shield tiles on the space shuttle must be 100% reliable for the time required. If only one tile fails, burnthrough could result and lead to destruction of the vehicle. On the other hand, breakage of a household floor or wall tile causes some inconvenience, but does not jeopardize life or equipment. A similar comparison could be made between the glass windows in a deep-sea submergence vehicle and those in an automobile or in a house. Each has its own definition of reliability.

The following are some of the factors that must be considered when evaluating reliability requirements:

1. the acceptable failure rate for the application
2. the type of warranty for the system and its subcomponents
3. expectations of the potential customer
4. safety requirements defined by industry or government regulations

Reliability for mechanical applications generally is determined by a comparison of the design requirements (particularly the peak stress) with the material properties (particularly the strength). This is illustrated in Figure 17.2. Figure 17.2a represents a condition in which the strength distribution of the material does not overlap the estimated stress distribution in the application. If we assume that the stress estimates are relatively accurate, the material should perform reliably in the application. Figure 17.2b represents a condition in which there is some overlap between the strength distribution of the material and the stress distribution of the application. One would question the reliability of the material for this application. However, the material could still be reliable if the region

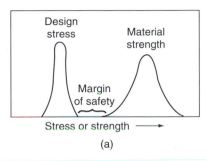
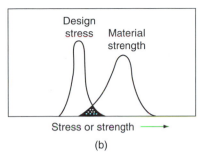

FIGURE 17.2   Comparison of the design stress distribution vs. the material strength distribution to estimate mechanical reliability. (a) Likely to have acceptable reliability and (b) likely to have marginal or unacceptable reliability. (From Richerson, D.W., Design and fracture analysis, Lesson 14 in *Introduction to Modern Ceramics*, ASM Materials Engineering Institute Course 56, ASM International, Materials Park, OH, 1990. With permission.)

of high stress corresponded with an equivalent high-strength region of the material. This could be accomplished through proof testing.

17.6   SUMMARY

The probability of success of a new ceramic component can be effectively increased by using a systematic design approach that first quantitatively defines the requirements of the application and then evaluates candidate materials in terms of property and fabrication limitations, as well as cost and reliability requirements. The probability of success can be further improved by an iterative, overlapping program in which close liaison is maintained between designers and manufacturers throughout the development and demonstration program.

REFERENCES

1. Richerson, D.W., Design and fracture analysis, Lesson 14 in *Introduction to Modern Ceramics*, ASM Materials Engineering Institute Course 56, ASM International, Materials Park, Ohio, 1990.
2. Harper, J.E., ARPA/NAVAIR ceramic gas turbine engine demonstration program, in *Ceramics for High Performance Applications, III*, Burke, J.J., Lenoe, E.N., and Katz, R.N., Eds., Plenum, New York, 1982.
3. Metcalfe, A.G., in *Ceramics for High Performance Applications, III*, Burke, J.J., Lenoe, E.N., and Katz, R.N., Eds., Plenum, New York, 1982.
4. Katz, R.N., High-temperature structural ceramics, *Science*, 208, 841–847 (May 23, 1980).
5. *Reliability of Ceramics for Heat Engine Applications*, prepared by the Committee on the Reliability of Ceramics for Heat Engine Applications, Natl. Acad. Sci. Publ. NMAB-357, Washington, D.C., 1980.

STUDY GUIDE

1. What factors must be considered when selecting a material and fabrication process for a component for an engineering application?
2. Explain how you would proceed to define the requirements of an application.
3. What are some reasons why an alternate material might be sought for an existing application?
4. How do you determine if a material is a suitable candidate for an application?
5. What are some key fabrication concerns when trying to match a material to an application?
6. How do you locate potential candidate suppliers for a new ceramic component?
7. What factors must be considered when assessing reliability requirements?

18 Design Approaches

In Chapter 17 we discussed briefly some of the important considerations of component design in general and ceramic design in particular. In this chapter we discuss in more detail the design approaches for ceramics. For the purposes of this discussion, design approaches can be divided roughly into five categories:

1. empirical
2. deterministic
3. probabilistic
4. linear elastic fracture mechanics
5. combined

18.1 EMPIRICAL DESIGN

Empirical design is a trial-and-error approach that emphasizes iterative fabrication and testing and deemphasizes mathematical modeling and analysis. It can be the most effective approach in cases where a ceramic is already in use and is only being modified and in cases where mechanical loads are minimal. It can also be the optimum approach when the available property data for the candidate ceramic material are too limited for the more analytical approaches. Finally, empirical design may be the only approach, or may be required in addition to analytical approaches, where the survival of a component is strongly affected by environmental factors such as chemical attack or erosion.

Historically, most ceramic design has been empirical, especially with traditional ceramics. Only recently, with the advent of ceramics in demanding structural applications, has it become necessary to use analytical approaches.

18.2 DETERMINISTIC DESIGN

Deterministic design is a standard "safety-factor" approach. The maximum stress in a component is calculated by finite-element analysis or closed-form mathematical equations.[1] A material is then selected that has a strength with a reasonable margin of safety over the calculated peak component stress. The margin of safety is usually determined from prior experience, so that this approach is really a combination of analytical and empirical.

The deterministic approach is routinely used with the design of metals. It works well, largely because metals have relatively low property scatter. Figure 18.1 depicts a typical strength distribution for a metal. Such a curve is obtained by categorizing all the measured strength data into short ranges such as 900 to 910 MPa (130.5 to 132 ksi), 910 to 920 MPa (132 to 133.4 ksi), and 920 to 930 MPa (133.4 to 135 ksi) and plotting the number (frequency) of data points in each category vs. the strength range. For example, Figure 18.1 indicates that approximately 100 data points were in the vicinity of 1000 MPa (145 ksi), but less than 10 around 900 MPa (130.5 ksi). Several other observations can be made regarding Figure 18.1:

1. The curve is tall and thin and represents a material having relatively low strength scatter. Nearly all of the test bars failed between 900 and 1100 MPa (130.5 and 159.5 ksi). No test bars failed below 850 MPa (123 ksi) or above 1150 MPa (167 ksi).

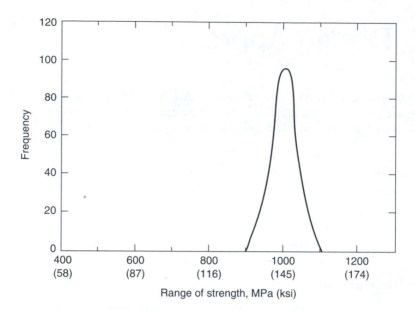

FIGURE 18.1   Typical strength distribution for a metal. (Drawing courtesy of ASM International.)

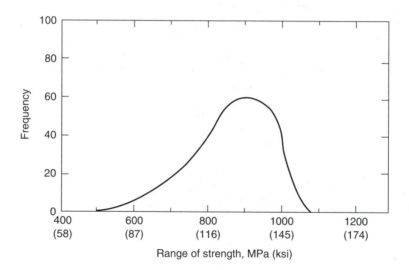

FIGURE 18.2   Typical strength distribution for a high-strength ceramic material. (Drawing courtesy of ASM International.)

2.   The curve is symmetrical (normal Gaussian bell-shaped distribution). Although the curve does not have to be symmetrical for a metal, it is more commonly close to symmetrical than for a ceramic. If the curve is symmetrical, the average strength corresponds to the position of the peak of the curve, i.e., 1000 MPa (145 ksi) for the curve shown.

Often, metals can be designed within a small margin of their ultimate strength and used with the confidence that they will not fail prematurely. This is not true with many structural ceramic materials. Ceramics generally have wide strength scatter and the measured strength is affected by the volume and area of material under stress (as discussed in detail in prior chapters). Figure 18.2

depicts a typical strength distribution curve for a high-strength ceramic material. The following can readily be observed:

1. The curve is broad, indicating large strength scatter.
2. The curve is not symmetrical, but instead is skewed toward the low-strength side and has a long tail extending all the way down to 500 MPa (72.5 ksi).
3. Because the curve is nonsymmetrical, the peak of the curve does not correspond to the average strength.

Now let us compare Figures 18.1 and 18.2 with Figure 17.2 and imagine that an average design stress of 700 MPa (101.5 ksi) is applied. Which material has the greater design margin? The answer is the metal. Because of the narrow strength distribution of the metal, a significant safety margin exists between the design stress and the minimum strength. No safety margin exists for the ceramic; the design stress and strength curves have substantial overlap. As a result, the metal can be designed by the deterministic approach and the ceramic cannot (at least for the example cited).

Normal statistics[2] are commonly used for deterministic design. The average or mean strength $\overline{\sigma}$ identifies the typical strength of the material. It is obtained by adding all the measured strength values and dividing by the number of tests N. As expressed in Equation (18.1):

$$\overline{\sigma} = \frac{\displaystyle\sum_{i=1}^{N} \sigma_i}{N} \tag{18.1}$$

In Equation (18.1), the expression $\sum_{i=1}^{N} \sigma_i$ is summation notation. It is the statistical shorthand way of saying, "add all the measured strength values of the individual test bars together." This sum of all the strength values, when divided by the number of strength values (i.e., number of tests N), then is the average strength $\overline{\sigma}$. The degree of strength scatter (i.e., deviation of the individual strength value from the average strength) is identified by the standard deviation S, which is defined in Equation (18.2):

$$S = \left[\frac{\displaystyle\sum_{i=1}^{N} (\sigma_i - \overline{\sigma})^2}{N} \right]^{1/2} \tag{18.2}$$

where
  σ_i = strength of individual test bars
  N = the number of test bars

Again, $\sum_{i=1}^{N} (\sigma_i - \overline{\sigma})^2$ in Equation (18.2) is summation notation. It is statistical shorthand that says, "add all the squares of the deviations between the individual strength values (σ_i) and the average strength ($\overline{\sigma}$)." This sum, when divided by the number of strength values (i.e., number of tests N) and raised to the 1/2 power (i.e., the square root is taken), then is the standard deviation of the selected group of strength values.

18.3 PROBABILISTIC DESIGN

Empirical and deterministic design approaches may be adequate for most ceramic applications, but are limited in cases where high stresses or complex stress distributions are present. In such cases, a probabilistic approach that takes into account the flaw distribution and the stress distribution in the material may be required.[3,4]

18.3.1 WEIBULL STATISTICS

Currently, the most popular means of characterizing the flaw distribution is by the Weibull[5] approach. It is based on the weakest link theory, which assumes that a given volume of ceramic under a uniform stress will fail at the most severe flaw. It thus presents the data in a format of probability of failure F vs. applied stress σ, where F is a function of the stress and the volume V or area S under stress

$$F = f(\sigma, V, S) \tag{18.3}$$

Weibull proposed the following relationship for ceramics:

$$f(\sigma) = \left(\frac{\sigma - \sigma_\mu}{\sigma_0} \right)^m \tag{18.4}$$

where σ is the applied stress, σ_μ the threshold stress (i.e., the stress below which the probability of failure is zero), σ_0 a normalizing parameter (often selected as the characteristic stress, at which the probability of failure is 0.632), and m the Weibull modulus, which describes the flaw size distribution (and thus the data scatter). The probability of failure as a function of volume is

$$F = 1 - \exp\left[-\int_v \left(\frac{\sigma - \sigma_\mu}{\sigma_0} \right)^m dV \right] \tag{18.5}$$

This results in the shape of curve shown in Figure 18.3. Such a curve can easily be plotted from experimental data by estimating F by $n/(N + 1)$, where n is the ranking of the sample and N the total number of samples. This is plotted vs. the measured strength value for each value of n as shown in Table 18.1 and Figure 18.4 for a hypothetical set of data.

The curve in Figure 18.4 provides only an approximation of the probability of failure and does not yield the m value. Plotting $\ln [1/(1 - F)]$, calculated using Equation (18.5), vs. $\ln \sigma$ results in a straight line of slope m, as shown in Figure 18.5. This form of the Weibull curve is used extensively in depicting the reliability or predicted reliability of materials or components. Figure 18.6 shows the data distribution resulting from bend strength testing of 30 reaction-bonded Si_3N_4 specimens plotted in this fashion. The m value determined from the slope is 5.5. This is a relatively low value and indicates substantial scatter.

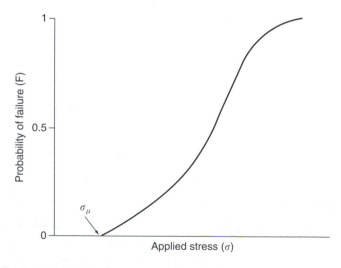

FIGURE 18.3 Typical Weibull distribution.

TABLE 18.1
Organization of Experimental Data to Plot Weibull Curve

Number of Ordered Data	Measured Strength σ (MPa)	Estimated Probability of Failure, $F \sim \dfrac{n}{N+1}$
1	178	0.1
2	210	0.2
3	235	0.3
4	248	0.4
5	262	0.5
6	276	0.6
7	296	0.7
8	318	0.8
9	345	0.9

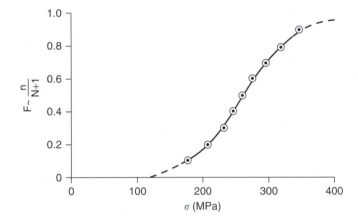

FIGURE 18.4 Weibull curve plotted from experimental data in Table 18.1.

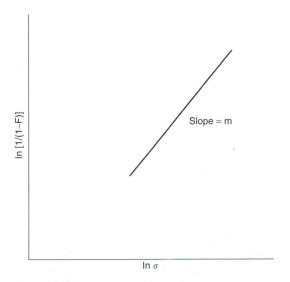

FIGURE 18.5 Format of Weibull plot commonly used to present probability of fracture data for ceramics.

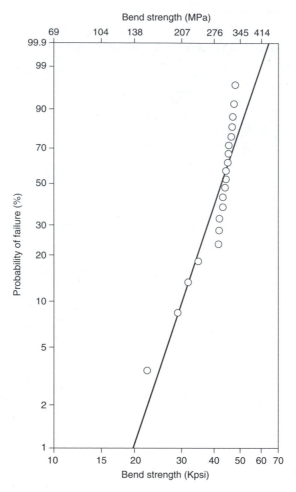

FIGURE 18.6   Example of Weibull curve generated from strength test data for reaction-bonded Si$_3$N$_4$. (Data from Johansen, K.M., Richerson, D.W., and Schuldies, J.J., Ceramic Components for Turbine Engines, AiResearch Rep. 21-2794(08), prepared under contract F33615-77-C-5171, 1980, Appendix A, p. 67.)

Equation (18.4) and Equation (18.5) represent three-parameter Weibull functions where σ_μ, σ_0, and m are the three parameters. Usually, a two-parameter form is used for ceramics, where the threshold stress σ_μ is set equal to zero. Thus, the equation becomes

$$F = 1 - \exp\left[-\int_v \left(\frac{\sigma}{\sigma_0} \right)^m dV \right] \tag{18.6}$$

Cracks initiate and propagate in ceramics under tensile loading rather than compressive loading, so that only the volume or area of material under tension is of concern in the Weibull equation. Therefore, if the full volume is under uniform uniaxial tension, the two-parameter equation becomes

$$F = 1 - \exp\left[-V \left(\frac{\sigma}{\sigma_0} \right)^m \right] \tag{18.7}$$

If the loading is three-point or four-point bending, the effective volume under tensile stress is substantially lower. For three-point bending, the effective volume is equal to $V/2(m + 1)^2$ and for four-point bending is $V(m + 2)/4(m + 1)^2$. For an m of 10, the effective volumes for three-point and four-point loading are, respectively, only 0.004 and 0.025 of the beam volume under load.

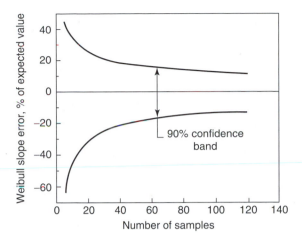

FIGURE 18.7 Potential error in the Weibull slope calculation compared to the number of samples tested. (From McLean, A.F. and Fisher, F.A., Brittle material design, high temperature gas turbine, Army Materials and Mechanics Research Center-CTR-77-20, Interim Report, August 1977, pp. 111–114. With permission.)

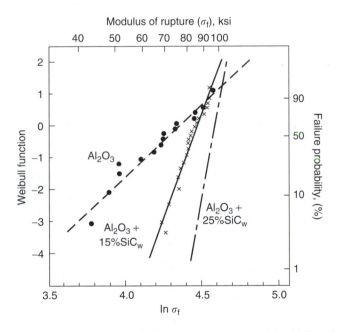

FIGURE 18.8 Example of the use of Weibull plots to compare materials. (Adapted from Rhodes, J.F., Rootare, H.M., Springs, C.A., and Peters, J.E., data presented at the 88th Annual Meeting of the American Ceramic Society, Chicago, Ill. April, 28, 1986.)

A substantial number of test samples are required to determine an accurate value for the Weibull slope m.[4] This is illustrated in Figure 18.7. Testing of only 10 samples can result in an error of + or −40% in the m value. Over 60 samples are required to obtain 90% confidence in the m value.

An example of the use of Weibull curves to compare ceramic materials and predict reliability is shown in Figure 18.8. The line on the left is for 100% Al_2O_3 and represents a characteristic strength of 530 MPa (76.9 ksi) and a Weibull modulus (slope) of $m = 4.6$. The middle line is for a composite of 85% Al_2O_3–15% SiC whiskers and represents a characteristic strength of 595 MPa (86.3 ksi) and a Weibull modulus of $m = 13.4$. The line on the right is for a composite of 75% Al_2O_3–25%

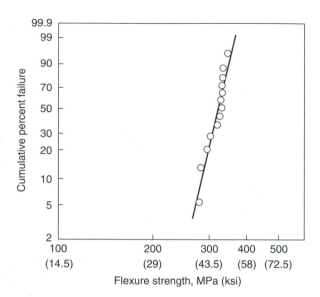

FIGURE 18.9 Improved RBSN material resulting from process modification to eliminate the lower flaw population causing the bimodal distribution in Figure 15.6. (Drawing courtesy of ASM International.)

SiC whiskers and represents a characteristic strength of 680 MPa (98.7 ksi) and a Weibull modulus of $m = 22.5$. Which of the three materials would you predict to be the most reliable (least likely to fail)? The answer is the 75% Al_2O_3–25% SiC whisker composite. Note the distribution of the strength data points. This material has a maximum slightly over 100 ksi and a minimum around 90 ksi. The 85% Al_2O_3–15% SiC whisker composite has a maximum of about 95 ksi and a minimum of about 70 ksi. The 100% Al_2O_3 has about the same maximum, but a very low minimum slightly under 310 MPa (45 ksi).

Data does not always fit a Weibull plot. An example was shown previously in Figure 18.6 for a reaction-bonded Si_3N_4 material. The data would be fit better by two separate Weibull curves, one of low slope through the bottom four data points and one of higher slope through the upper 16 data points. This is referred to as a bimodal distribution. Examination of the fracture surface of each test bar revealed that a separate type of flaw population controlled the two regions. The fractures for the four lowest points all occurred at relatively large white linear flaws. All the rest occurred at smaller, dark-colored regions. The white linear flaws were identified as present in the silicon powder compact before nitriding. The others all developed during nitriding. The cause of the white linear flaws was hypothesized and the process modified. As shown in Figure 18.9, these flaws were eliminated in subsequent material to achieve a monomodal Weibull distribution and a substantial increase in Weibull modulus and characteristic strength.

The above is a good example of the use of Weibull curves and fracture analysis to help improve a ceramic material. Another example is shown in Figure 18.10. In this case, completely different processes are compared rather than just process modifications.

18.3.2 Use of the Weibull Distribution in Design

Plotting the Weibull curve from experimental data provides useful information about the probability of failure vs. applied stress for the material, but does not provide an assessment of failure probability of the component. To do this, the Weibull distribution for the material must be integrated with the stress distribution for the component. This can be done conveniently with the use of finite-element analysis.[7–9] An example of a finite-element model is shown in Figure 18.11. The material strength

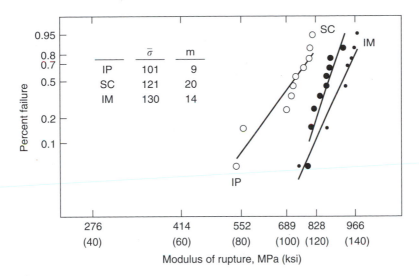

FIGURE 18.10 Weibull plots comparing the strength of a sintered Si_3N_4 fabricated by various techniques. IP is isostatically pressed, SC is slip-cast, and IM is injection-molded. (From Pasto, A., Neil, J., and Quackenbush, C.L., paper presented at International Conference on Ultrastructure Processing of Ceramics, Glasses and Composites, Gainesville, FL, Feb. 13–17, 1983. With permission.)

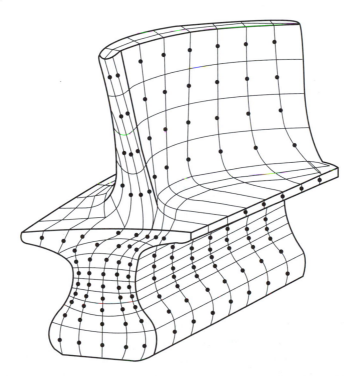

FIGURE 18.11 Finite-element analysis model for a ceramic rotor blade for a gasturbine engine. (Drawing courtesy of ASM International.)

data in the form of a Weibull probability curve is compared with each finite element to determine the probability of failure of that specific element. The probabilities, both volume and surface, for all the elements are then summed to determine the probability of failure for the component. This can also be plotted on a Weibull curve as a function of some component operation parameter. For

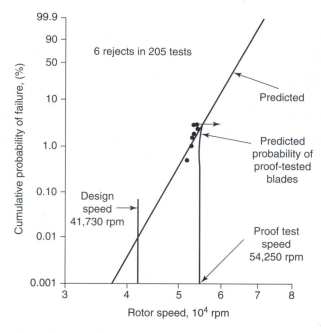

FIGURE 18.12 Rotor blade spin proof-test results that show good correlation between predicted and actual probability of failure for hot-pressed Si_3N_4 rotor blades. (From Richerson, D.W., Design with ceramics for heat engines, paper presented at U.S./Japan Seminar on Structural Ceramics, Seattle, WA, Aug. 13–15, 1984. With permission.)

instance, Figure 18.12 shows a Weibull curve that predicts the reliability of hot-pressed Si_3N_4 rotor blades (of the design shown in the finite-element model in Figure 18.11) run in a specific gas-turbine engine application as a function of rotation speed. These blades were proof-tested at 30% overspeed in a spin-test rig. The probabilistic design predicted a 3% failure rate. Six of the 205 rotor blades tested failed, providing excellent verification of the probabilistic design methodology.

18.3.3 ADVANTAGES OF PROBABILISTIC DESIGN

The primary advantage of a probabilistic approach is that it allows designing closer to the properties of the material. For instance, in a component with high but localized stress, a material can be successfully used with a low probability of failure, even though its measured property scatter band overlaps the peak design stress. The probability is low that one of the more severe flaws in the material will be in the region of peak stress. Deterministic design in this case would probably conclude that the ceramic material was not acceptable.

Use of probabilistic design allows a trade-off in material selection between high strength and low scatter. For instance, an application requiring 0.99 reliability could be satisfied by a material with a characteristic strength of 120 MPa and an m of 8 or a material with a strength of 94 MPa and an m of 10 or a material with a strength of only 61 MPa and an m of 16. The advantage of probabilistic design is that these trade-offs are considered and can be integrated into the design analysis. This visibility and flexibility is not available in empirical and deterministic approaches.

18.3.4 LIMITATIONS OF PROBABILISTIC DESIGN

Probabilistic design is limited primarily by inadequacy in defining peak stresses and stress distributions in the component and in defining the true strength-flaw size distribution in the candidate material. Stresses in the component arise from thermal and mechanical loading. The accuracy of predicting the magnitude and distribution of these stresses is restricted by the accuracy of defining the boundary conditions. For instance, the local heat-transfer conditions of both the component and the environment plus all the effects of geometry (e.g., heat sinks) and thermal conductivity plus material anisotropy plus boundary-layer effects must all be accurately defined or assumed before the thermal stress can be accurately calculated. Similarly, precise loads and load application angles plus contact areas and friction coefficients plus the effects of geometry and tolerances must all be accurately defined or assumed before the mechanical design stresses can be accurately calculated. Substantial inaccuracies in boundary conditions for metals can be tolerated because both thermal and mechanical stresses can redistribute slightly due to the metals' ductility. Such is not the case with ceramics. Stresses are concentrated and cannot be adequately predicted unless the boundary conditions are accurately defined.

There are also pitfalls in defining the true strength-flaw size distribution in ceramics. Because of the large size range of flaws in a ceramic part and the resulting wide scatter in strength values, a large number of specimens must be strength-tested to determine adequately the Weibull parameters.

A further problem in accurately defining the m value is the effect of bimodal or multimodal flaw size distributions. The Weibull relationship in Equation (18.5) assumes a uniform, random, unimodal flaw size distribution. As long as such a uniform unimodal distribution is present in the material, the calculated m value makes sense and can be used by routine procedures for probabilistic design. However, if the distribution is multimodal, but the data are used as a single distribution in calculating the Weibull modulus, the resulting m value will not be as suitable for design. This is especially true when using the three-parameter Weibull approach.

18.4   LINEAR ELASTIC FRACTURE MECHANICS APPROACH

Linear elastic fracture mechanics is a useful approach to the design of ceramics and other brittle materials. It takes into account that the material (and thus the component) contains flaws and it treats fracture in terms of fracture toughness, stress intensity factors, and flaw size rather than ultimate strength or yield strength.

Fracture is a dynamic process that involves crack formation and crack propagation. If the criteria and characteristics for crack initiation and propagation for a material are known quantitatively, these factors can become the basis for component design by fracture mechanics. This involves the following:

1. analysis of the stress and temperature distribution in the component
2. analysis of the time each region within the component will be exposed to a specific stress and temperature over the life of the component
3. identification of the number of operation cycles required for the application
4. use of fast-fracture data and either a deterministic or probabilistic design approach to determine whether the material will withstand the steady-state and peak stresses imposed by the application; and
5. use of stress rupture, creep, oxidation-corrosion, cyclic fatigue and other data relevant to slow crack growth and formation of new defects to aid in prediction of the life of the component in the specific application.

The intent of this section is not to describe the theory and techniques of fracture mechanics, but rather to alert the engineer to its existence and provide some references. The technology is still evolving and the uses expanding. Early work on the fundamentals of fracture mechanics was conducted by Griffith,[10] Irwin,[11] and Williams.[12] More recent work relating to ceramics has been reported in References 13 through 15. General discussion and reviews are available in References 16 through 19.

18.5 COMBINED APPROACHES

It is apparent from the prior discussions in this chapter that a variety of design approaches are available. The approach selected will depend on the severity of the application, the timing, the available budget, and the existing database. Usually, the final approach will be a combination of empirical, deterministic, or probabilistic and fracture mechanics. As an example, empirical studies may be conducted to screen candidate materials in parallel with deterministic or probabilistic analysis. Once a configuration has been selected and is in the test phase, fracture mechanics and probabilistic analyses may be used to predict life and reliability.

REFERENCES

1. Boresi, A.P., Sidebottom, O.M., Seely, F.B., and Smith, J.O., *Advanced Mechanics of Materials*, 3rd ed., Wiley, New York, 1978.
2. Lipson, C. and Sheth, N.J., *Statistical Design and Analysis of Engineering Experiments*, McGraw-Hill, New York, 1973.
3. Paluszny, A. and Wu, W., Probabilistic aspects of designing with ceramics, *J. Eng. for Power*, 99(4), 617–730, 1977.
4. McLean, A.F. and Hartsock, L.L., Design with structural ceramics, in *Structural Ceramics, Treatise on Materials Science and Technology*, Vol. 29, Wachtman, J.B., Jr., Ed., Academic Press, San Diego, California, 1989, pp. 27–98.
5. Weibull, W., A statistical distribution function of wide applicability, *J. Appl. Mech.*, 18(3), 293–297, Sept. 1951.
6. Johansen, K.M., Richerson, D.W., and Schuldies, J.J., Ceramic Components for Turbine Engines, AiResearch Rep. 21-2794(08), prepared under contract F33615-77-C-5171, 1980, Appendix A, p. 67.
7. DeSalvo, G.J., Theory and Structural Design Applications of Weibull Statistics, WANL-TME-2688, 1970.
8. Tree, D.J. and Kington, H.L., Ceramic component design objectives, goals and methods, in *Ceramic Gas Turbine Demonstration Engine Program Review,* Fairbanks, J.W. and Rice, R.W., Eds., MCIC Rep. MCIC-78-36, 1978, pp. 41–75.
9. Gyekenyesi, J.P., Scare—a Post-Processor Program to MSC/Nastran for the Reliability Analysis of Structural Ceramics Components, NASA TM87188, Cleveland, Ohio, 1986.
10. Griffith, A.A., The phenomena of rupture and flow in solids, *Philos. Trans. R. Soc. Lond. Ser. A.*, 221(4), 163–198, 1920–1921.
11. Irwin, G.R., Analysis of stresses and strains near the end of a crack traversing a plate, *J. Appl. Mech.*, 24(3), 361–364, 1957.
12. Williams, M.L., On the stress distribution at the base of a stationary crack, *J. Appl. Mech.*, 24(1), 109, 1957.
13. Davidge, R.W., McLaren, J.R., and Tappin, G., Strength-probability-time (SPT) relationships in ceramics, *J. Mater. Sci.*, 8(12), 1699–1705, 1973.
14. Weiderhorn, S.M., Evans, A.G., Fuller, E.R., and Johnson, H., Application of fracture mechanics to space-shuttle windows, *J. Am. Ceram. Soc.*, 57(7), 319–323, 1974.
15. Bradt, R.C., Hasselman, D.P.H., and Lange, F.F., Eds., *Fracture Mechanics of Ceramics,* Vols. 1 and 2 (1973), Vols. 3 and 4 (1978), Plenum, New York.
16. Wachtman, J.B., Jr., Highlights of progress in the science of fracture of ceramics and glass, *J. Am. Ceram. Soc.*, 57(12), 509–519, 1974.
17. Knott, J.F., *Fundamentals of Fracture Mechanics*, Butterworth, Kent, England, 1973.
18. Tetelman, A.S. and McEvily, A.J., *Fracture of Structural Materials*, Wiley, New York, 1967.
19. Dukes, W.H., *Handbook of Brittle Material Design Technology*, AGARD-ograph 152, Dec. 1970.

ADDITIONAL RECOMMENDED READING

1. Cranmer, D.C. and Richerson, D.W., Eds., *Mechanical Testing Methodology for Ceramic Design and Reliability*, Marcel Dekker, Inc., New York, 1998.
2. van Roode, M., Ferber, M.K., and Richerson, D.W., Eds., *Ceramics Gas Turbine Design and Test Experience*, ASME Press, New York, 2002.

PROBLEMS

Problems 18.1 through 18.6 are based on the data listed in the following:

Strength Data (in MPa)

Si_3N_4	Y-TZP	Al_2O_3	$Al_2O_3-SiC_w$
710	925	272	584
835	840	410	572
740	1048	402	615
680	1012	320	567
815	970	371	586
764	937	315	602
615	865	418	593
632	1071	351	595
716	992	335	582
682	914	380	562
790	980	248	578
702	894	307	604
734	917	480	585
602	968	218	574
728	957	385	598

1. What is the average strength of the Y-TZP material?
2. What is the standard deviation of the Al_2O_3 material?
3. What is the standard deviation of the $Al_2O_3-SiC_w$ material?
4. What is the ranking of the Si_3N_4 data point 716 MPa?
5. Which of these four hypothetical materials has the highest Weibull modulus?
6. Estimate the probability that the Si_3N_4 material will fracture at 716 MPa.
7. From the data in Figure 18.8, estimate the design stress that would result in a 10% probability of failure for the Al_2O_3 material.
8. From the data in Figure 18.8, estimate the design stress that would result in a 10% probability of failure for the $Al_2O_3-15\%$ SiC material.
9. Which of the following is least likely to increase the structural reliability of a ceramic?
   (a) increased Weibull modulus
   (b) increased toughness
   (c) increased strength
   (d) increased grain size

Problems 18.10 through 18.14 are based on the following strength data (MPa) for a reaction-sintered SiC material:

310	264
273	286
278	292
337	302
295	318

10. What is the average strength of the reaction-sintered SiC?
11. What is the standard deviation of the reaction-sintered SiC?
12. What is the rank of the test bar that fractured at 302 MPa?
13. Estimate the probability of a test bar of the material fracturing at an applied stress of 302 MPa.
14. Estimate the flaw size that caused the weakest reaction-sintered SiC test bar to fracture at 264 MPa. Assume that the fracture toughness is 2.5 MPa $\cdot$ m$^{1/2}$ and that the dimensionless term associated with the crack configuration and load geometry is 0.75.

STUDY GUIDE

1. What are the key distinctions between empirical, deterministic, and probabilistic design?
2. Why does deterministic design work well for metals, but probabilistic design sometimes is required for ceramics?
3. What are "Weibull statistics"?
4. What is the significance of the Weibull modulus?
5. The samples used to plot Figure 18.7 were 4-point flexure samples. Why were so many samples required to minimize the error in Weibull slope calculation? You may need to review Chapter 8.
6. Examine Figure 18.8. Which of the three materials would be easiest to design with? Why?
7. What are the advantages of probabilistic design?
8. What are the limitations of probabilistic design?
9. What is "linear elastic fracture mechanics"?

19 Failure Analysis

Failure analysis is extremely important in engineering, especially with ceramics, because it is the only means of isolating a failure-causing problem. In particular, failure analysis helps determine whether failure or damage occurred due to a design deficiency or a material deficiency. Until this has been determined, effort cannot be efficiently directed toward finding a solution. The result is usually a "shotgun" approach that includes a little design analysis, a little empirical testing, and a little material evaluation and often ends up only in a repeat of the test or operating conditions that initially caused failure.

Much of the shotgun approach can often be avoided by fracture analysis. Fracture analysis or fractography is the examination of the fractured or damaged hardware in an effort to reconstruct the sequence and cause of fracture. The path a crack follows as it propagates through a component provides substantial information about the stress distribution at the time of failure. Features on the fracture surfaces provide further information, especially the position at which the fracture initiated (fracture origin), the cause of fracture initiation (impact, tensile overload, thermal shock, material flaw, etc.), and even the approximate local stress that caused the fracture. The primary objective of this chapter is to acquaint the reader with these fracture surface features and the techniques to interpret the cause of fracture in ceramic components.

19.1 FRACTOGRAPHY

The first step is to determine where the fracture initiated. Often, simply reconstructing the pieces will pinpoint the fracture origin and may even give useful information about the cause of fracture. After assembling the pieces, look for places where a group of cracks come together or where a single crack branches. Preston[1] has shown that the angle of forking is an indicator of the stress distribution causing fracture. Examples are shown schematically in Figure 19.1.

The frequency of crack branching provides qualitative information about the amount of energy available during fracture. To branch, a crack must reach a critical speed. For glass, the critical speed is typically a little greater than half the speed of sound in the specific glass. At the instant of crack initiation, the crack velocity is zero, but quickly accelerates. The rate of acceleration is a function of the energy available either due to the stress applied or to energy stored in the part (such as residual stresses or prestresses, as in tempered glass). The more energy, the more rapidly the crack will reach its critical branching velocity and the more branching that will occur. A baseball striking a window will cause much more branching than a BB, due to the larger applied energy. Tempered glass will break into many fragments due to release of the high stored energy. On the other hand, a thermal-shock fracture may not branch at all, especially if it initiates from a localized heat source and propagates into a relatively unstressed or compressively stressed region of the component. In this case, the fracture will tend to follow a temperature or stress contour and will have a characteristic wavy or curved appearance, as shown in Figure 19.2 for a thermally fractured ceramic setter plate for a furnace.

19.1.1 LOCATION OF THE FRACTURE ORIGIN

The pattern of branching will often lead the engineer to the vicinity of the fracture origin. The engineer will then have to examine the fracture surfaces in this region, often under a low-power optical

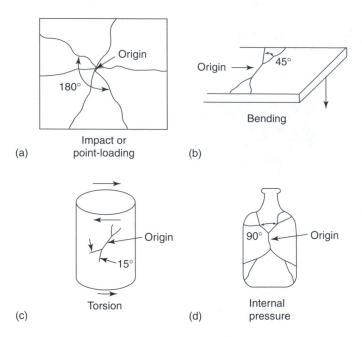

FIGURE 19.1   Information available by examining crack direction and crack branching. (Adapted from Frechette, V.D., *Fracture of Heliostat Facets*, presented at the ERDA Solar Thermal Projects Semiannual Review, Seattle, WA, Aug. 23–24, 1977.)

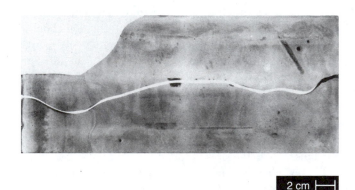

FIGURE 19.2   Thermal shock fracture showing lack of branching.

binocular microscope, to locate the precise point at which fracture initiated. This point of origin can be a flaw such as a pore or inclusion in the material, a cone-shaped Hertzian surface crack resulting from impact, a crack in a surface glaze, an oxidation pit, intergranular corrosion, a position of localized high stress, or a combination of these. Location and examination of the fracture origin will help determine which of these factors is dominant and provide specific guidance in solving the fracture problem.

As mentioned before, a fracture begins at zero velocity at the fracture origin and then accelerates as it travels through the part. As it does, it interacts with the microstructure, the stress field, and even acoustic vibrations and leaves distinct features on the fracture surface that can be used for locating the fracture origin.[2] The most important features include hackle, the fracture mirror, and Wallner lines.

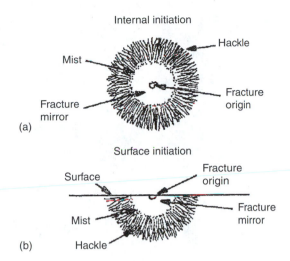

FIGURE 19.3 Schematic showing the typical fracture features that surround the fracture origin.

19.1.1.1 The Fracture Mirror and Hackle

When a crack initiates at an internal flaw, it travels radially in a single plane as it accelerates. The surface formed is flat and smooth and is called the *fracture mirror*. When the crack reaches a critical speed, intersects an inclusion, or encounters a shift in the direction of principal tensile stress, it begins to deviate slightly from the original plane, forming small radial ridges on the fracture surface. The first of these are very faint and are referred to as *mist*. Mist is usually visible on the fracture surfaces of glass, but may not be on crystalline ceramics. The mist transitions into larger ridges called *hackle*. Hackle is also referred to as *river patterns* because the appearance is similar to the branching of a river into tributaries and the formation of deltas. The hackle region transitions into macroscopic crack branching such that the remaining portion of the fracture surface is often on a perceptibly different plane than the mirror and hackle. Sometimes, this gives the appearance that the fracture origin is either on a step or a pedestal.

Figure 19.3a shows schematically the fracture mirror, mist, and hackle for a fracture that initiated in the interior of a part. The mirror is roughly circular and the fracture origin is at its center. Note that lines drawn parallel to the hackle will intersect at or very near the fracture origin. Similarly, Figure 19.3b shows the fracture features for a crack that started at the surface of a part.

The hackle lines surrounding the mirror result from velocity effects and are sometimes called *velocity hackle*. Another form of hackle, called *twist hackle*, usually forms away from the mirror and results from an abrupt change in the tensile stress field, such as going from tension to compression. Twist hackle points in the new direction of crack movement and appears more as parallel cracks than as ridges and does not have to point to the fracture origin. Twist hackle is an important feature for deducing the stress distribution in the ceramic at the time of fracture.

The size of the fracture mirror is dependent on the material characteristics and the localized stress at the fracture origin at the time of fracture. Studies by Terao,[3] Levengood,[4] and Shand[5] on glass suggest that the fracture stress σ_f times the square root of the mirror radius r_m equals a constant A for a given material:

$$\sigma_f \, r_m^{1/2} = A \qquad\qquad (19.1)$$

Kirchner and Gruver[6] determined that this relationship also provides a good approximation for polycrystalline ceramics, as long as the mirror is clearly visible and can be measured accurately.

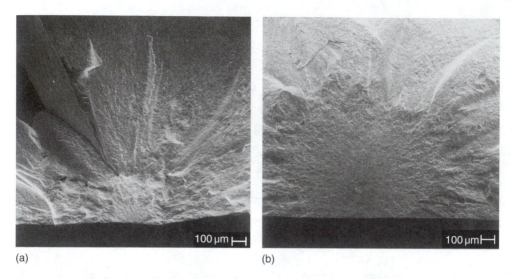

FIGURE 19.4   Examples of typical fracture mirrors for high-strength polycrystalline ceramics. (a) Initiation at a surface flaw in hot-pressed silicon nitride. (b) Initiation at an internal flaw in reaction-sintered silicon nitride.

They obtained values of A ranging from 2.3 for a glass to 9.1 for a sintered Al_2O_3 to 14.3 for a hot-pressed Si_3N_4.

It is therefore possible to determine the stress causing failure of a ceramic component by comparing the mirror size with a graph of $r_m^{1/2}$ vs. σ_f for the material. The graph can be compiled from bend strength or tensile strength data using scanning electron microscopy (SEM) of the fracture surfaces of the test bars to determine the mirror radius.

The stress causing the failure can also be estimated using the Griffith equation (described in Chapter 8) by measuring the flaw size on the fracture surface. Because inclusions and pores and other flaws are not symmetrical and their boundaries are often not well defined, stress estimates based on flaw size are only approximate. If a knowledge of the local stress at the fracture origin is needed, perhaps it should be calculated both from the mirror size and the flaw size and then the most appropriate value selected by good engineering judgment.

Figure 19.4 shows examples of fracture mirrors and flaws on the fracture surfaces of strength test specimens. The photomicrographs were taken by a scanning electron microscope. Since the ceramic specimens were not electrical conductors, a thin layer of gold was applied to the surface by sputtering to avoid charge buildup, which would result in poor resolution.

The mirror size cannot always be measured. If the fracture-causing stress is low and the specimen size is small, the mirror may cover the whole fracture surface. If the material has very coarse grain structure or a bimodal grain structure, the mirror and other fracture features may not be visible or distinct enough for measurement. Figure 19.5 shows examples of fracture surfaces with indistinct fracture features.

19.1.1.2   Wallner Lines

Sonic waves are produced in a material during fracture. As each succeeding wave front overtakes the primary fracture crack, the principal stress is momentarily perturbed. This results in a series of faint arc-shaped surface lines that are termed *Wallner lines*. The curvature of each line shows the approximate shape of the crack front at the time it was intersected by the sonic wave and provides information about the direction of crack propagation and the stress distribution. The direction is from the concave to the convex side of the Wallner lines. The stress distribution is inferred from the

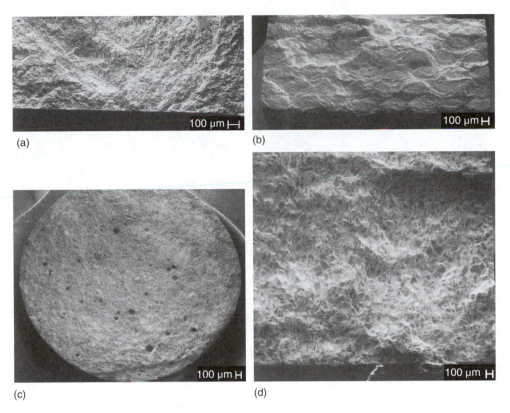

FIGURE 19.5   Examples of fracture surfaces with indistinct fracture features. (a) Sintered silicon carbide. (b) Silicon carbide-carbon-silicon composite. (c) Porous lithium aluminum silicate. (d) Bimodal grain distribution reaction-sintered silicon carbide.

distance of each portion of a single line from the origin. If the stress distribution were of uniform tension, each portion of a line would be about equidistant from the origin. If a stress gradient were present, the distance of various portions of the Wallner line from the origin would vary, being farthest where the tensile stress was highest. These effects are shown schematically in Figure 19.6.

Wallner lines are not always present. For high-energy fractures, where the fracture velocity is high and the surface is rough, Wallner lines often cannot be distinguished. In very slow crack velocities, such as those that occur in subcritical crack growth, Wallner lines are not present because the sonic waves are damped and gone before the crack has propagated appreciably.

19.1.1.3   Other Features

Other fracture features besides the mirror, hackle, and Wallner lines are useful in the interpretation of a fracture. These include arrest lines, gull wings, and cantilever curl.

An *arrest* line occurs when the crack front temporarily stops. The reason for crack arrest is usually a momentary decrease in stress or a change in stress distribution. When the crack starts moving again, its direction invariably has changed slightly, leaving a discontinuity. This line of discontinuity looks a little a Wallner line, but is usually more out of plane and more distinct. It is also called a *rib mark*. Arrest lines or rib marks provide essentially the same information as Wallner lines, i.e., the direction of crack movement and the stress distribution. Twist hackle frequently is present after an arrest line.

The gull wing is a feature that occurs due to the crack intersecting a pore or inclusion. As the crack travels around the inclusion, two crack fronts result. These do not always meet on the same

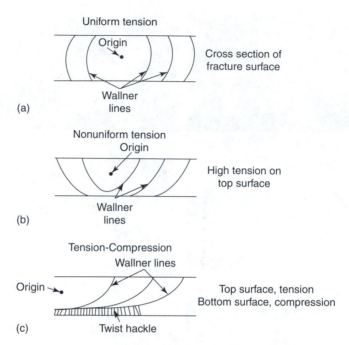

FIGURE 19.6 Relationship of Wallner lines on a fracture surface to the stress distribution at the time of fracture. (From Frechette, V.D., *Fracture of Heliostat Facets*, presented at the ERDA Solar Thermal Projects Semiannual Review, Seattle, WA, Aug. 23–24, 1977. With permission.)

plane on the opposite side of the inclusion or pore, resulting in a ridge where the two link up and again become a single crack front. In some cases, the ridge is immediately in the wake of the inclusion or pore and looks like a tadpole. In other cases, two ridges resembling a gull wing form in the wake.

Another very useful feature is referred to as the *cantilever curl* or *compression lip*. It occurs when the material is loaded in bending. The fracture initiates on the tensile side perpendicular to the surface and exits on the compressive side not perpendicular to the surface. This is illustrated in Figure 19.7. If the part were fractured under pure tension, the crack would be straight through the thickness and would thus exit at 90°. This information can be valuable in diagnosing the cause of fracture. For instance, thermal fractures of plate-shaped parts typically approach a stress state of pure tension near the point of origin and will not result in a compression lip. These same parts fractured mechanically will normally have some bend loading and will thus have a compression lip. Another example is a part containing prestressing or residual internal stresses. The crack will not pass straight through the thickness, but instead will follow contours consistent with the stress fields it encounters. A third example is strength testing. A problem with testing a ceramic in uniaxial tension is avoiding parasitic bend stresses. Examination of the fracture surface for signs of cantilever curl after tensile testing will help determine if pure tension was achieved or not.

Figure 19.8 shows the cross sections of typical specimens tested in four-point bending. Note the variations in the shape of the compression lip.

19.1.2 Techniques of Fractography

The techniques of fractography are relatively simple and the amount of sophisticated equipment minimal. Often, the information required to explain the cause of fracture of a component can be obtained with only a microscope and a light source. In fact, sometimes an experienced individual can explain the fracture just by examining the fracture surfaces visually. At other times, a variety of

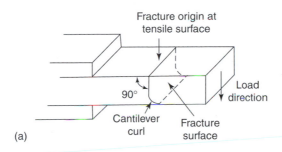

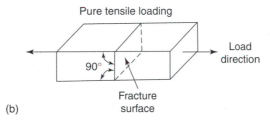

FIGURE 19.7 Difference in the fracture contour through the specimen thickness for bend loading vs. pure tensile loading.

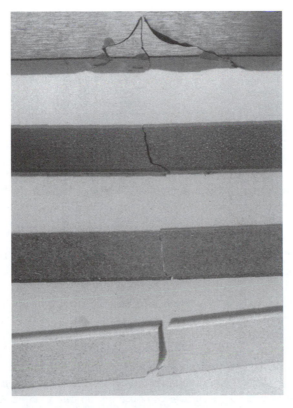

FIGURE 19.8 Examples showing cantilever curl in four-point bend specimens. Specimens 0.32-cm (0.125-in.) thick.

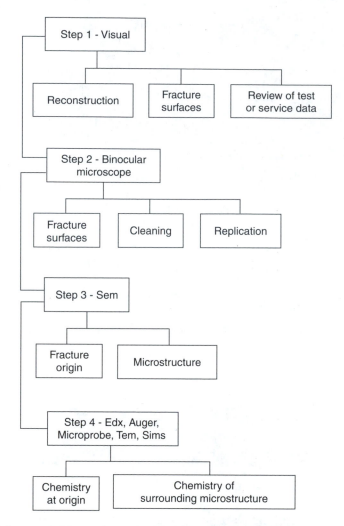

FIGURE 19.9   Major sequential steps in conducting fracture analysis.

steps and techniques including sophisticated approaches such as SEM, electron microprobe, and Auger analysis are required.[8–10] When extensive fractography is necessary, the steps and procedures are as shown schematically in Figure 19.9.

Step 1 involves visual examination of the fractured pieces and review of data regarding the test or service conditions under which the hardware failed. These data usually provide some hypotheses to guide the evaluation. Reconstructing the broken pieces and sketching probable fracture origins and paths are also helpful.

A primary objective of visual examination is to locate the point of fracture origin. This can be done with the use of Wallner lines, hackle, and the fracture mirror, as described previously.

Visual examination determines the extent of additional evaluation that will likely be required. It also determines if cleaning procedures are necessary prior to microscopy. Generally, one must be careful not to handle or damage the fracture surface. The origin and features can be fragile and key information explaining the fracture lost by improper handling. For instance, fingerprints can be mistaken for Wallner lines. Debris on the surface can obscure the true fracture surface.

Cleaning can sometimes be accomplished with compressed air, but the source must be considered. Some compressors mix small amounts of oil with the air, which could produce a thin surface film that would later result in interpretation difficulties during SEM analysis. Compressed air should not be unless it is known to be clean.

Ultrasonic cleaning in a clean solvent such as acetone or methyl alcohol is frequently used. Caution and judgment must be exercised, however, because ultrasonic cleaning is quite vigorous due to the cavitation action at the fluid-specimen interface and can damage the fracture surface. For instance, if fracture initiated at a low-density region or a soft inclusion, this material might be removed during ultrasonic cleaning and either prevent interpretation or lead to an erroneous interpretation.

It is apparent that cleaning and handling should be avoided unless absolutely necessary. Before trying other cleaning approaches, try a soft camel's-hair brush.

Step 2 involves examination of the fracture surfaces under a low-power microscope. Usually, a binocular microscope with magnification up to $40\times$ is adequate. Sometimes higher power and special lighting or contrast features are required. Stereo photography may also be useful, since it accentuates the fracture surface features.

Preparation of replicas can also be useful; a replica often provides better resolution of the fracture features than examination of the original part. Several methods are available for preparing replicas. A room-temperature method makes use of cellulose acetate, acetone, and polyvinyl chloride (PVC). A thin sheet of cellulose acetate is placed on a piece of PVC and submerged in acetone for about 15 sec. It is then pressed against the fracture surface while being flooded with acetone and held for about 5 min. The acetone is then allowed to dry and the cellulose acetate replica peeled off.

Replicas can also be prepared with PVC alone. The specimen is heated and PVC pressed on with a Teflon rod. After cooling, the PVC replica is peeled off. This technique is quick, but should not be used if there is a chance that the heating of the specimen will alter the fracture surface.

The fracture origin can usually be located by low-power optical microscopy using either the original specimen or a replica and an assessment made as to whether the fracture resulted from a material flaw or some other factor. However, details of the fracture origin, such as the nature of the flaw and interaction between the flaw and the microstructure, require higher magnification. Optical microscopes do not have adequate depth of focus at high magnification, so the scanning electron microscope must be used.

Examination of the fracture surface, in particular, the fracture origin, by SEM is the third step in fractography. The scanning electron microscope provides an extremely large depth of focus (compared to the optical microscope) and a range in magnification from around $10\times$ to well over $10,000\times$. Most fractography is conducted between $25\times$ and $5000\times$.

Scanning electron microscopy shows the differences between the fracture origin and the surrounding material and helps the engineer to develop a hypothesis of the cause of failure. One can easily detect if the fracture initiated at a machining groove or at a material pore or inclusion. SEM shows if the surface region is different from the interior and provides visual evidence of the nature or cause of the differences.

As with other techniques, SEM requires interpretation and must be used with caution. Because of the large depth of focus, it is difficult sometimes to differentiate between a ridge and a depression or to determine the angle of intersection between two surfaces. This can be better appreciated by comparing a single SEM photomicrograph with a stereo pair of the same surface. Only after doing this does one understand how easy it is to misinterpret a feature on an SEM photomicrograph.

Difficulty in interpretation is inevitable but can be minimized if the engineer is present during the SEM analysis. This is especially true with respect to artifacts. An *artifact* is defined as extraneous material on the surface of the specimen. It can be a particle of dust or lint, a chunk of debris resulting from the fracture or handling, or a smeared coating resulting from oil contamination. If the engineer suspects that a feature is an artifact, he or she can instruct the SEM operator to look at it from different view angles and to examine surrounding areas in an effort to be sure.

As noted, SEM can usually locate the fracture origin and provide a photo with a calibrated scale that allows accurate measurement of the size and shape of the flaw and the size of the fracture mirror. The engineer can then use Equation (19.1) and Equation (8.11) to estimate the magnitude of the tensile stress that caused failure.

Once the fracture origin has been located, evaluation of the localized chemistry is often desirable. This leads to step 4 of fractography, the use of sophisticated instruments to conduct microchemical analysis. Most SEM units have an energy-dispersive x-ray (EDX) attachment that permits chemical analysis of the x-rays that are emitted when the SEM electron beam excites the electrons within the material being examined. Each chemical element gives off x-rays under this stimulation that are characteristic of that element alone. These are detected by the EDX equipment and displayed as peaks by peripheral equipment. Comparison of the peak height of each element provides a semiquantitative chemical analysis of the microstructural feature being viewed.

Electron microprobe works on the same principle as EDX, except an alternate x-ray detection mechanism is used that provides better resolution and detects a wider range of elements. EDX cannot detect the lower-atomic-number elements.

Auger analysis provides additional features for chemical analysis of microstructural features. It can remove material by sputtering while it is conducting chemical analysis and can thus determine if the chemistry changes as we progress inward from the fracture surface. Auger analysis is especially useful in cases where oxidation, corrosion, slow crack growth, or other intergranular effects are suspected. In some cases, fracture of specimens can be conducted in the Auger apparatus under an inert atmosphere prior to conducting the chemical analysis. This allows examination of a fresh fracture surface that has not had a chance to pick up contamination from the atmosphere and handling.

In addition to analysis of the fracture surface, other material tests can help determine the cause of failure. These include surface and bulk x-ray diffraction analysis, reflected-light microscopy of polished specimens, and in occasional cases, transmission electron microscopy (TEM).

There is no guarantee that the four steps of fractography will explain the cause of a failure and suggest a solution. However, it is still the most effective technology available and should be used routinely.

19.1.3 DETERMINING FAILURE CAUSE

As mentioned before, determining the cause of failure is critical. It is obviously important in liability suits, where responsibility for failure must be established, but it is also important for other reasons:

    to determine if failure is resulting from design or material limitations
    to aid in material selection or modification
    to guide design modifications
    to identify unanticipated service problems such as oxidation or corrosion
    to identify material or material-processing limitations and suggest direction for improvement
    to define specification requirements for materials and operating conditions

Figure 19.10 shows schematically how fractography interacts with other sources of information to determine the cause of failure and lead to action in the right direction to achieve a solution. It is imperative that all sources of data be considered and that the source of fracture be isolated to determine as quickly as possible whether it is design- or materials-oriented.

The following paragraphs review some of the common causes of fracture and describe the typical fracture surfaces that result.

19.1.3.1 Material Flaws

As discussed in Chapter 8, flaws in a ceramic material concentrate the stress. When this concentrated stress at an individual flaw reaches a critical value that is high enough to initiate and extend

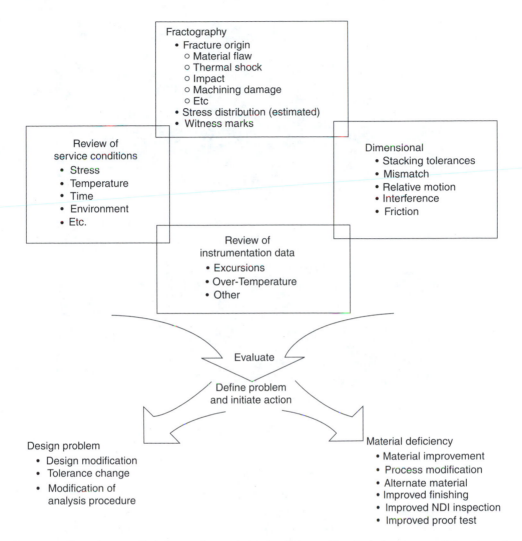

FIGURE 19.10 Failure analysis interaction to determine if the problem is design- or materials-oriented and to define a plan of action to solve the problem.

a crack, fracture occurs. Therefore, the first thing to look for on the fracture surface is whether there is a material flaw at the fracture origin. If a flaw such as a pore or inclusion is present, the engineer can do the following:

Compare the nature of the flaw with prior certification and service experience to determine if the flaw is intrinsic to the normal baseline material or if it is abnormal. If the latter is the case, a problem in the material fabrication process probably exists, and the help of the component manufacturer should be solicited.

Measure the flaw size and/or mirror size and estimate the fracture stress σ_f using Equation (8.12) and Equation (19.1). Compare this with the stresses estimated by design analysis and with fracture stress distribution projected for the material from prior strength certification testing. If the calculated fracture stress is within the normal limits specified for the material, a design problem should be suspected. If the fracture stress is below the normal limits specified for the material, a material problem is indicated.

If the flaw is at or near the surface, assess whether the flaw is intrinsic to the material or resulted from an extrinsic source such as machining, impact, or environmental exposure. Comparison of the

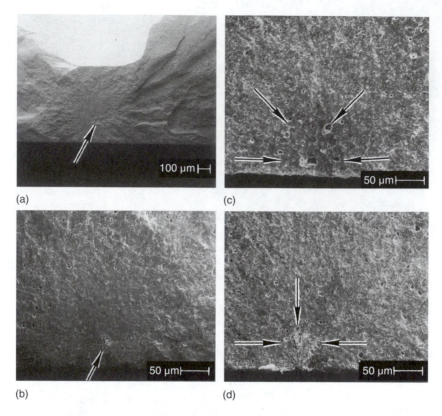

FIGURE 19.11 SEM photomicrographs of fracture-initiating material flaws in reaction-bonded Si_3N_4 (RBSN) that are typical and consistent with the normal microstructure and strength. Arrows point to the fracture origins.

chemical and physical nature of the flaw with the surrounding baseline microstructure will help in making this assessment.

Figures 19.11 through 19.14 illustrate the types of intrinsic material flaws that can cause fracture and compares whether they are normal or abnormal for the material. Although the types of flaws vary depending on the material, the fabrication process, and the specific step in the process in which they formed, it is relatively easy to distinguish between normal and abnormal flaws and thus to determine if errors in processing contributed to the failure.[11]

19.1.3.2 Machining Damage

Surface flaws resulting from machining are a common source of failure in ceramic components, especially in applications where high bend loads or thermal loads are applied in service. The flaws resulting from machining were discussed in Chapter 15, the most important being the median crack and the radial cracks.

The median crack is elongated in the direction of grinding and is like a notch, in that fracture usually initiates over a broad front. A broad mirror usually results, but no flaw is readily visible because of the shallow initial flaw depth and because the median crack is perpendicular to the surface. The principal tensile stress is usually distributed such that the crack will extend in a plane perpendicular to the surface. If the flaw is perpendicular to the surface to start with, it will be in the same plane as the fracture and will be difficult to differentiate from the rest of the fracture mirror.

Figure 19.15 shows examples of specimens that fractured at transverse grind marks such that the median crack was likely the strength-determining flaw. Note the length of the surface involved

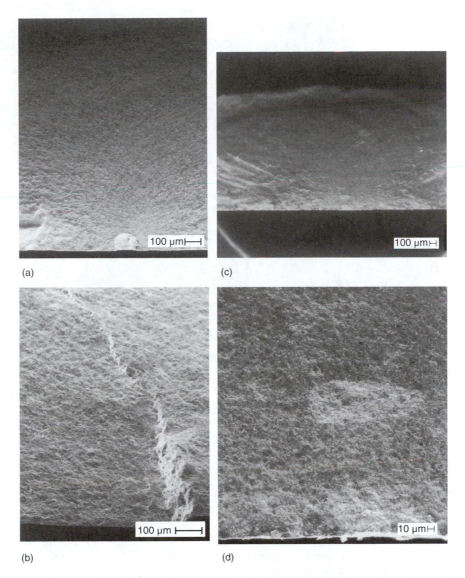

(a)

(c)

(b)

(d)

FIGURE 19.12 SEM photomicrographs of abnormal fracture-initiating material flaws in RBSN traceable to improper processing prior to nitriding. (a) Large pore in slip-cast RBSN resulting from inadequate de-airing. (b) Crack in greenware prior to nitriding. (c) and (d) Low-density regions in slip-cast RBSN resulting from agglomerates in the slip.

in the fracture origin and the lack of a distinguishable flaw. Note also that the fracture origin is at a grinding groove and is elongated parallel to the direction of grinding.

Radial cracks produced by machining are roughly perpendicular to the direction of machining, are usually shallower than the median cracks, and are usually semicircular rather than elongated. The resulting fracture mirror is similar to one produced by a small surface pore or inclusion, but a well-defined flaw is not visible. However, by examining the intersection of the fracture surface with the original specimen surface at the center of the semicircular mirror under high magnification with the scanning electron microscope, the source can usually be seen and interpreted as machining damage. There are two features to look for. First, check to see if a grinding groove, especially an unusually deep one, intersects the fracture surface at the origin. Second, look for a small slightly

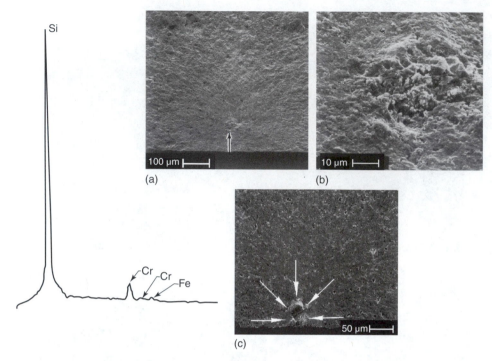

FIGURE 19.13 SEM photomicrographs of abnormal fracture-initiating material flaws in RBSN traceable to the nitriding process. (a) and (b) Porous aggregate rich in Cr and Fe, resulting from reaction of the silicon during the nitriding cycle with stainless steel contamination picked up during powder processing; energy-dispersive x-ray analysis shown. (c) Large aggregate of unreacted silicon resulting from localized melting due to local exothermic overheating. Arrows identify the fracture origins.

out-of-plane region at the origin. This could indicate machining damage, but could also have other interpretations, such as a contact damage or simply a tensile overload.

Machining damage often limits the strength of fine-grained ceramics and determines the measured strength distribution of certification specimens. Rice et al.[12] reported that hot-pressed Si_3N_4 specimens machined in the transverse direction fractured at machining flaws 98% of the time and that bars machined in the longitudinal direction failed at machining flaws greater than 50% of the time. This is similar to results of the author and his co-workers for both Si_3N_4 and SiC.[13–15]

19.1.3.3 Residual Stresses

Many ceramics have residual stresses, which usually result from the surface cooling down faster than the interior after sintering or are due to chemical differences between the surface and interior. Often, the interior is under residual tension and the exterior is under compression. This can provide a strengthening effect if the material is used in service in the as-fired, prestressed condition. However, most components require some finish machining. Once the compressive surface zone has been penetrated, the component is substantially weakened, often to the point of spontaneous crack initiation during machining.

Frechette[16] recounts a case in which simple blanks of boron carbide cracked during machining. At the time, the cause of cracking was unknown, but improper machining was the primary suspect. Fracture analysis showed that the crack initiated at the point of machining and entered the material roughly perpendicular to the surface. However, the crack then quickly changed direction and propagated through the interior of the material parallel to the surface, and finally changed direction again and exited through the bottom surface. The fracture path and fracture surface features (Wallner lines

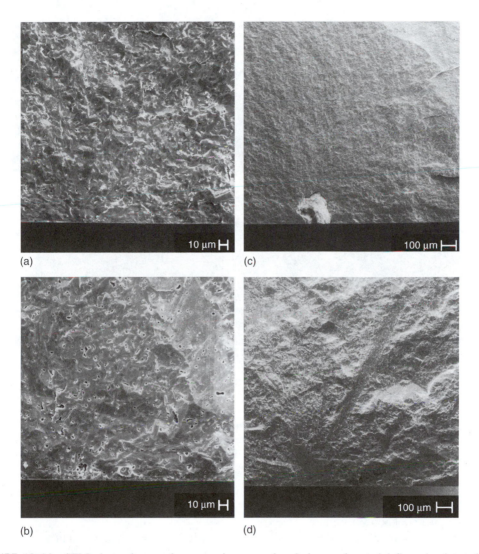

FIGURE 19.14 SEM photomicrographs comparing normal and abnormal material flaws in sintered SiC. (a) and (b) Typical microstructure of high-strength material. (c) Large pore resulting from powder agglomeration during powder preparation and shape forming. (d) Large grains resulting from improper control of temperature during sintering.

and hackle) indicated that the surfaces were in compression and the interior was in tension and suggested that the problem was not linked to machining practice or material defects. Review of the processing history showed that the boron carbide was allowed to freely cool after hot pressing from 1950 to 1800°C, followed by slow, controlled cooling thereafter. The free cooling was within the creep temperature range of the material, resulting in an effect comparable to tempering of glass, i.e., formation of surface compression and internal tension. Based on this hypothesis, derived from fracture surface analysis alone, the hot-pressing operation was modified to permit slow cooling from 1950 to 1750°C. This eliminated the residual stress condition and permitted machining without cracking.

19.1.3.4 Thermal Shock

Little work has been reported on identification of thermal-shock fractures, and much additional work is needed. Therefore, the comments here will be fairly general, and the reader should be aware

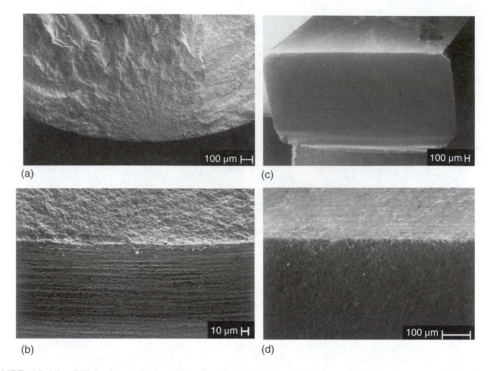

FIGURE 19.15 SEM photomicrographs showing fractures initiating at transverse machining damage. (a) The fracture surface of a tensile specimen of hot-pressed silicon nitride that had been machined circumferentially. (b) The intersection of this fracture surface with the machined surface, illustrating that the fracture origin is parallel to the grinding grooves. (c) and (d) The same situation for reaction-sintered silicon nitride.

that the distinguishing features described may not always occur and may not be the best ones in a specific case. As in all other interpretive studies, the best approach is to apply known principles, thoroughly analyze all available data and options, and the make a judgment decision, rather than depending on a cookbook procedure.

Thermal-shock fractures tend to follow a wavy path with minimal branching and produce rather featureless fracture surfaces. This appears to be especially true for weak- or moderate-strength materials, including both glass and polycrystalline ceramics. It is common for thermal-shock cracks not to propagate all the way through the part. These characteristics suggest that a nonuniform stress field is present during thermal fracture; i.e., the fracture initiates where the tensile stress is highest, follows stress or temperature contours (thus the wavy path), and stops when the stress drops below the level required for further extension or when a compressive zone is intersected. These conditions would also suggest a low crack velocity, which would account for lack of branching and lack of fracture surface features.

In line with the foregoing general considerations, several actual cases of fracture or material damage due to severe thermal transients are now analyzed.

The first case is the water quench thermal-shock test described in Chapter 9 for comparing the relative temperature gradient required to cause damage in simple rectangular test bars of different ceramic materials. The test bars are heated in a furnace to a predetermined temperature and then dropped into a controlled-temperature water bath. The bars generally do not break due to the quench. Instead, a large number of microcracks are produced. These become critical flaws that lead to fracture during subsequent bend testing; the size of the flaws is dependent on the material properties and the ΔT, and the residual strength is dependent on the size of these cracks. Further discussion of this test procedure and the results on specific materials is available in References 17 and 18.

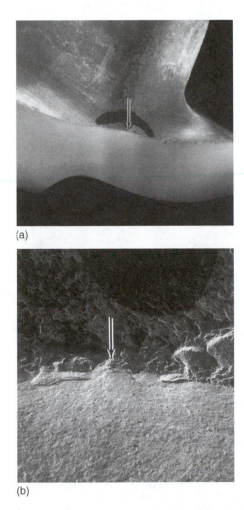

(a)

(b)

FIGURE 19.16 SEM photomicrograph showing a typical featureless thermal-shock fracture surface. (a) Overall surface at low magnification. (b) Fracture origin at higher magnification. (Courtesy of Garrett Turbine Engine Company, Phoenix, AZ, Division of Allied-Signal Aerospace, currently Honeywell Engines, Systems & Services.)

Ammann et al.[19] conducted cyclic fluidized bed thermal-shock tests on wedge-shaped specimens of hot-pressed Si_3N_4. They reported that multiple surface cracking occurred and that the cracks grew in depth as a function of the total number of cycles.

Oxyacetylene torch thermal-shock testing and gas-turbine rig testing of reaction-bonded Si_3N_4 stator vanes by the author and his co-workers[20] have shown that thermal-shock fractures of actual components under service conditions can be quite varied. Figure 19.16a shows a typical featureless fracture surface. The origin is shown by the arrow and the crescent-shaped ink mark and occurred in a region predicted by three-dimensional finite-element analysis to have the peak thermal stress during rapid heat-up. No material flaw or machining damage is visible, even at high magnification, as shown in Figure 19.16b. The crack apparently initiated at the material surface where the stress was maximum and propagated at moderate velocity without branching or making any abrupt changes in direction.

Figure 19.17a shows a thermal-shock failure that initiated at the trailing edge of a stator vane airfoil. Finite-element analysis also determined this position to be under high thermal stress during

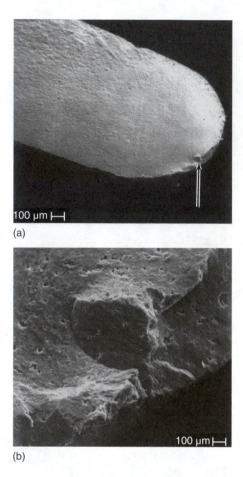

FIGURE 19.17   SEM photomicrograph of a thermal-shock fracture initiating at a material flaw. (a) Overall surface at low magnification. (b) Preexisting crack at fracture origin. (Courtesy of Garrett Turbine Engine Company, Phoenix, AZ, Division of Allied-Signal Aerospace, currently Honeywell Engines, Systems & Services.)

heat-up and steady-state service conditions. However, in this case the thermal-shock fracture initiated at a preexisting material flaw, as shown more clearly in Figure 19.17b at higher magnification. The flaw was a penny-shaped crack nearly normal to the surface of the airfoil, but out of plane to the principal tensile stress. The fracture initiated at this crack due to the thermal stress and then quickly changed direction to follow the plane of maximum tensile stress. Hackle marks can be seen in the upper part of Figure 19.17a that point to the vicinity of the origin.

The fracture surface shown in Figure 19.17 is not what one would expect for thermal-shock conditions. It has relatively well-defined fracture features and could just as easily have been interpreted as a mechanical overload. Similarly, the fracture surface in Figure 19.16 looks very much like fractures resulting from contact loading (discussed later). How does the engineer make the distinction? At our current state of knowledge, he or she does not, at least not based solely on fracture surface examination. The engineer needs other inputs such as stress analysis, controlled testing (such as the calibrated oxyacetylene torch thermal-shock tests), and a thorough knowledge of the service conditions. The engineer then needs to evaluate all the data concurrently and use his or her best judgment.

19.1.3.5   Impact

Impact can cause damage or fracture in two ways: (1) localized damage at the point of impact and (2) fracture away from the point of contact due to cantilevered loading. The former will have

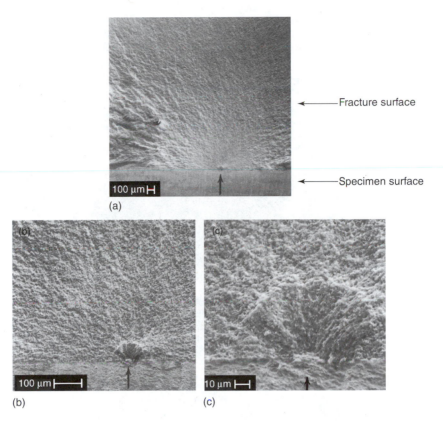

Fracture surface

Specimen surface

(a)

(b)

(c)

FIGURE 19.18 Typical Hertzian cone crack resulting from impact and acting as the flaw that resulted in fracture under subsequent bend load. Shown at increasing magnification from (a) to (c). (Courtesy of Garrett Turbine Engine Company, Phoenix, AZ, Division of Allied-Signal Aerospace, currently Honeywell Engines, Systems & Services.)

distinctive features and can usually be identified as caused by impact. The latter will appear like a typical bend overload (with a compressive lip on the exit end of the fracture) and can only be linked to impact by supporting data, such as location of the local damage at the point of impact.

The damage at the point of impact may be so little as to resemble a scuff mark on the ceramic or a smear of the impacting material (both cases referred to as a *witness mark*) or it may be as severe as complete shattering.[21] The degree depends on the relative velocity, mass, strength, and hardness of the impacting bodies. A baseball striking a plate glass window causes shattering; a BB striking a window only causes a series of concentric cone-shaped (Hertzian) cracks intersected by radial cracks. However, if the window shattered by the baseball could be reconstructed, it would also have conical and radial cracks.

The conical cracks are typical of impact and are thus a strong diagnostic fracture feature.[22] Figure 19.18a shows a fracture surface of a polycrystalline ceramic where impact damage occurred first and was followed by fracture due to a bend load. The origin is indicated by an arrow and is easily located by observing the hackle lines and fracture mirror. Figures 19.18b and c show the origin at higher magnification. A distinct cone shape is present; the apex is at the surface where the impact occurred, and the fracture flared out as the crack penetrated the material.

Figure 19.19 shows another example of fracture due to impact. In this case, a ceramic rotor blade rotating at 41,000 rpm struck a foreign object and was fractured by the impact. Note the Hertzian crack extending from the fracture origin. Also note that additional damage is present at the origin, possibly a series of concentric cracks, providing evidence that the degree of impact was severe.

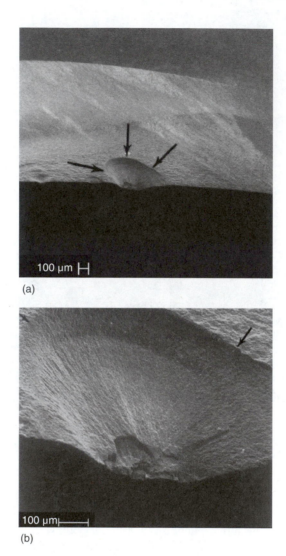

(a)

(b)

FIGURE 19.19 Impact fracture of a ceramic rotor blade showing Hertzian cone crack. (Courtesy of Garrett Turbine Engine Company, Phoenix, AZ, Division of Allied-Signal Aerospace, currently Honeywell Engines, Systems & Services.)

19.1.3.6 Biaxial Contact

Biaxial contact refers to a situation in which normal and tangential forces are being applied simultaneously at a ceramic surface. Examples where this might occur are numerous and include the following:

surface grinding
sliding contact (such as in bearings, seals, and very many other applications)
applications involving a shrink fit
interfaces where the materials have different thermal expansion coefficients and operate under
    varying temperatures
any high-friction interface

Tensile stress concentration at a biaxially loaded interface was discussed briefly in Chapter 9 and is explored in more detail by Finger[23] and Richerson et al.[24] The fracture surface can be quite varied. If

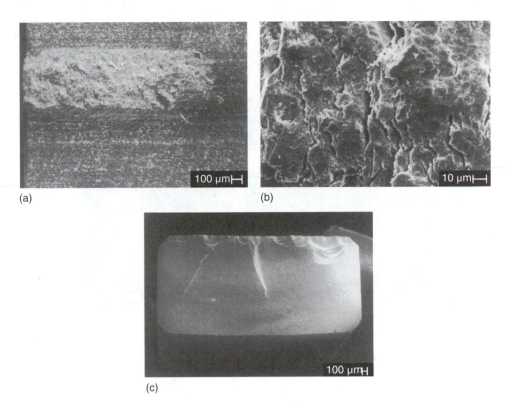

(a)

(b)

(c)

FIGURE 19.20 (a) and (b) Surface cracks resulting from relative movement between two contact surfaces under a high normal load and with a high coefficient of friction. (c) Typical multiple chipping resulting from contact loading and visible on a fracture surface.

the contact is concentrated at a point, a Hertzian cone crack can form and the resulting fracture surface will be similar to the one in Figure 19.18. However, if the contact is more spread out, as in high-friction sliding contact, the damage will be spread over a larger surface area and will result in a relatively featureless fracture surface similar to the one shown in Figure 19.16 for a thermal-shock fracture.

Distinguishing features that differentiate a contact failure from a thermal-shock failure are just beginning to be defined[23] and appear to include the following:

Contact cracks tend to enter the material surface at an angle other than 90°.

An extreme case is shown in Figure 19.20a for a ceramic specimen with a high-interference shrink fit axially loaded in tension.

Multiple parallel cracks often occur during a contact failure and can either be seen on the surface adjacent to the fracture origin or show up as chips obscuring the fracture origin (Figure 19.20b).

Surface witness marks are often present at the fracture origin in cases of contact-initiated failure (Figure 19.21a).

One or more clamshell shapes, faint Hertzian cones, or pinch marks occur at the origin of some contact-initiated fracture surfaces (Figure 19.21b).

19.1.3.7 Oxidation-Corrosion

Some types of oxidation or corrosion are easy to detect because they leave substantial surface damage that is clearly visible to the naked eye. In this case, the objective is to identify the mechanism of attack and find a solution. In other cases, especially where the oxidation or corrosion is isolated along grain boundaries, the presence and source of degradation may be more difficult to detect. In this case, the degree of attack may only be determined by strength testing, and the cause may be

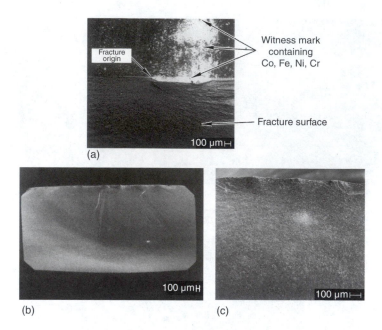

(a)

(b)   (c)

FIGURE 19.21 (a) Witness mark on the surface of the ceramic adjacent to the fracture origin, suggesting fracture due to contact loading. (b) Multiple cone features resulting from a contact fracture. (c) Multiple cone features resulting from a contact fracture.

ascertained by controlled environment exposures and/or sophisticated instruments such as Auger spectroscopy, which can detect slight chemical variations on a microstructural level.

Let us first examine some examples of oxidation and corrosion in which visible surface changes have occurred. Figure 19.22 shows the surface and fracture surface of NC-132 hot-pressed $Si_3N_4^*$ after exposure in a SiC resistance-heated, oxide-refractory-lined furnace for 24 h at 1100°C (2012°F)[25] Figure 19.22a shows the complete cross section of the test bar. The fracture origin is at the surface on the left side of the photo and is easily located by the hackle marks and the fracture mirror (the dark spots on the fracture surface are artifacts that accidentally contaminated the surface in preparing the sample for SEM). The specimen surface appears at low magnification to have many small spots that were not present prior to the oxidation exposure. At higher magnification (Figure 19.22b), these spots appear to be blisters or popped bubbles and one is precisely at the fracture origin. Still higher magnification (Figure 19.22c) reveals that a glass-filled pit is at the base on the center of the blister. It also reveals that a surface layer less than 5 μm thick covers the specimen and that this layer appears to be partially crystallized.

By simply examining the specimen surface, especially the intersection of the oxidized surface and the fracture surface, we have obtained much insight into both the nature and sequence of oxidation. What else can we do to obtain further information? We can compare the strength of the oxidized specimen with that of unoxidized material. In this specific case, the oxidation exposure resulted in a reduction in strength from 669 MPa (97,000 psi) to 497 MPa (72,000 psi). We can also compare x-ray diffraction and chemical analyses for the original surface, the oxidized surface, and the bulk material. In this case, the oxidized surface contained much more Mg and Ca than the original surface or the bulk material. Energy-dispersive x-ray (EDX) analysis verified that the glassy material in the pit also had high concentrations of Mg and Ca. X-ray diffraction revealed crystallized cristobalite (SiO_2) plus magnesium silicate and calcium magnesium silicate phases in the oxide layer. No sign of Mg or Ca contamination was detected in the furnace.

---

* Manufactured by the Norton Company, Worcester, MA (currently Saint-Gobain Advanced Ceramics).

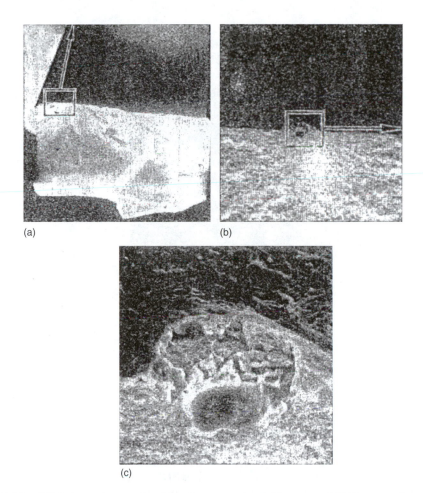

(a)   (b)

(c)

FIGURE 19.22   SEM photomicrographs of the fracture surface of hot-pressed Si_3N_4 exposed to static oxidation for 24 h at 1100°C (2012°F). (a) Overall fracture surface showing hackle marks and fracture mirror (the irregular dark spots on the fracture surface are artifacts). (b) Higher magnification showing the fracture mirror with an oxidation corrosion pit at the origin. (c) Higher magnification showing the nature of the pit and the surface oxidation layer. Specimen size 0.64 × 0.32 cm. (From Yonushonis, T.M. and Richerson, D.W., Strength of reaction-bonded silicon nitride, in *Ceramic Gas Turbine Demonstration Engine Program Review*, Fairbanks, J.W. and Rice, R.W., Eds., MCIC Rept. MCIC-78-36, 1978, pp. 219–234. With permission.)

Simultaneous evaluation of all the data led to a plausible hypothesis of the mechanism of oxidation degradation. Mg and Ca, present as oxide or silicate impurities in the Si_3N_4, were diffusing to the surface, where they reacted with SiO_2 that was forming simultaneously at the surface from reaction of the Si_3N_4 with oxygen from the air. The resulting silicate compositions apparently locally increased the solubility or oxidation rate of the Si_3N_4. The reason for the formation of isolated pits was not determined, but could have resulted from impurity segregation or other factors and would have required additional studies to determine.

A similar example of static oxidation for reaction-bonded $Si_3N_4^*$ is illustrated in Figure 19.23. In this case, the exposure was for 2 h at 1350°C (2462°F) plus 50 h at 900°C (1652°F)[11] Only isolated pits were present on the surface and these appeared to occur where small particles of the furnace

---

* RBN-104 reaction-bonded Si_3N_4 from the AiResearch Casting Co., Torrance, CA (currently Honeywell Ceramic Components).

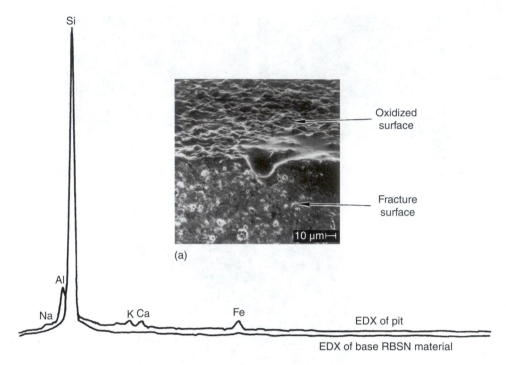

(a)

FIGURE 19.23   SEM photomicrograph of the fracture-initiating oxidation-corrosion pit on the surface of reaction-bonded Si_3N_4. The EDX graph shows the relative concentration of chemical elements in the glassy region at the base of the pit. (Courtesy of Garrett Turbine Engine Company, Phoenix, AZ, Division of Allied-Signal Aerospace, currently Honeywell Engines, Systems & Services.)

lining had contacted the specimen during exposure. The EDX analysis included in Figure 19.23 was taken in the glassy region at the base of the pit, showing that Al, Si, K, Ca, and Fe were the primary elements present and again indicating a propensity for Si_3N_4 to be corroded by alkali silicate compositions. However, it should be noted that the size of the pit is much smaller than in the prior example and resulted in only a small strength decrease.   .

Figure 19.24 and Figure 19.25 show examples of more dramatic corrosion of hot-pressed and reaction-bonded Si_3N_4,[25] resulting from exposure to the exhaust gases of a combustor burning jet fuel and containing a 5-ppm addition of sea salt. Exposure consisted of 25 cycles of 899°C (1650°F) for 1.5 h, 1121°C (2050°F) for 0.5 h, and a 5-min air quench. At 899°C (1650°F), Na_2SO_4 is present in liquid form and deposits along with other impurities on the ceramic surface. The EDX analyses taken in the glassy surface layer near its intersection with the Si_3N_4 documents the presence of impurities such as Na, Mg, and K from the sea salt, S from the fuel, and Fe, Co, and Ni from the nozzle and combustor liner of the test rig. An EDX analysis for the Si_3N_4 on the fracture surface about 20 μm beneath the surface layer is also shown in Figure 19.25. Only Si is detected (nitrogen and oxygen were outside the range of detection for the EDX equipment used), indicating that the corrosion in this case resulted from the impurities in the gas stream plus the surface oxidation.

The strength of the hot-pressed Si_3N_4 exposed to the dynamic oxidation with sea salt additions decreased to an average of 490 MPa (71,000 psi) from a baseline of 669 MPa (97,000 psi). The reaction-bonded material decreased to 117 MPa (17,000 psi) from a baseline of 248 MPa (36,000 psi). Repeating the cycle with fresh specimens and no sea salt resulted in an increase to 690 MPa (100,000 psi) for the hot-pressed Si_3N_4 and only a decrease to 207 MPa (30,000 psi) for the reaction-bonded Si_3N_4.

The examples presented so far for oxidation and corrosion have had distinct features that help distinguish the cause of fracture from other mechanisms, such as impact or machining damage.

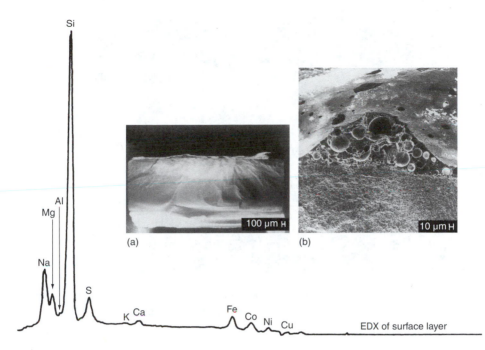

FIGURE 19.24 SEM photomicrograph of hot-pressed Si_3N_4 that was exposed to combustion gases with sea salt additions, showing that fracture initiated at the base of the glassy surface buildup. EDX analysis shows the chemical elements detected in the glassy material adjacent to the Si_3N_4. (Courtesy of Garrett Turbine Engine Company, Phoenix, AZ, Division of Allied-Signal Aerospace, currently Honeywell Engines, Systems & Services.)

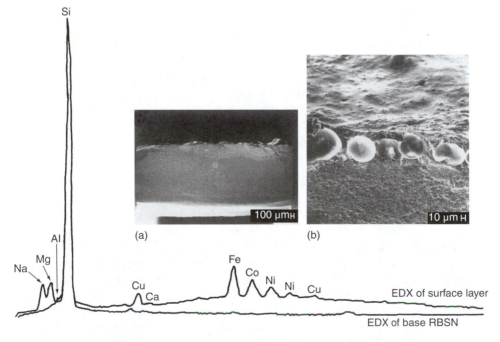

FIGURE 19.25 SEM photomicrograph of reaction-bonded Si_3N_4 that was exposed to combustion gases with sea salt additions, showing that fracture initiated at the base of the glassy surface buildup. EDX analysis shows the chemical elements detected in the glassy material adjacent to the Si_3N_4. (Courtesy of Garrett Turbine Engine Company, Phoenix, AZ, Division of Allied-Signal Aerospace, currently Honeywell Engines, Systems & Services.)

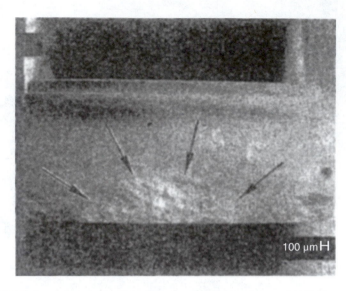

100 μm ⊢

FIGURE 19.26 SEM photomicrograph of the fracture surface of a low-purity Si_3N_4 material sintered with MgO and showing slow crack growth. Region of slow crack growth identified by arrows.

Some corrosion-initiated fractures are more subtle. The corrosion or oxidation may only follow the grain boundaries and be so thin that it is not visible on the fracture surface. Its effects may not even show up in room-temperature strength testing since its degradation mechanism may only be active at high temperature. How do we recognize this type of corrosion? The following suggestions may be helpful:

Prepare a polished section of the cross section and try various etchants; this may enhance the regions near the surface where intergranular corrosion is present.

Conduct EDX, microprobe, or Auger analysis scans from the surface inward to determine if a composition gradient is present.

Use high-magnification SEM of the fracture surface to look for differences between the microstructure near the surface and in the interior; if the fracture surface near the specimen surface is intergranular and near the interior is intragranular, grain boundary corrosion is a possibility.

Conduct controlled exposures under exaggerated conditions in an effort to verify if the material is sensitive to attack.

19.1.3.8 Slow Crack Growth

Slow or subcritical crack growth was discussed in Chapter 9 and a typical fracture surface was shown in Figure 9.8. The region of subcritical growth appears rough and intergranular and is very distinctive.

Slow crack growth can occur under sustained loading, as in the example in Chapter 9, and also under relatively fast loading, depending on the nature of the material, the temperature, the atmosphere, and the load. Figure 19.26 shows the fracture surface of a sintered Si_3N_4 material* (which was developed for low- to moderate-temperature applications) after four-point bend testing at 982°C (1800°F) at a load rate of 0.05 cm/min (0.02 in./min). In spite of the rapid loading, substantial slow crack growth occurred. Examination of the fracture surface quickly tells the engineer that this material is not suitable for high-temperature application under a tensile load.

---

* Kyocera International, San Diego, CA.

Sometimes, examination of the fracture surface by EDX or other surface chemical analysis technique can help identify the cause of slow crack growth. Specifically, the roughened region is analyzed separately from the rest of the fracture surface and bulk material. Chemical elements present in greater concentration in the slow-crack-growth region are probably associated with the cause. The tensile stress at fracture can also be approximated by assuming that the flaw size is equivalent to the slow-crack-growth region and using the Griffith Equation (8.11). However, it should be realized that this is only an approximation and that the reported elastic modulus and fracture energy values for the material, when used in the Griffith equation, may not be good approximations for the material under slow-crack-growth conditions.

There are other limitations to the information available from the fracture surface. The size of the slow-crack-growth region provides no information about the time to failure, the rate of loading, or the mode of loading (cyclic vs. static).

19.2 SUMMARY

Fractography is a powerful tool to the engineer in helping to determine the cause of a component or system failure. Well-defined features usually present on the fracture surface of a ceramic provide the engineer with useful information regarding the place where fracture initiated, the cause of fracture, the tensile stress at the point of failure, and the nature of the surrounding stress distribution. This information helps the engineer to determine if the failure was design- or material-initiated and provides direction in finding a solution. It can also help in achieving process or product improvement. Finally, it can help determine legal liability for personal or property damage.

REFERENCES

1. Preston, F., Angle of forking of glass cracks as an indicator of the stress systems, *J. Am. Ceram. Soc.*, 18, 175, 1935.
2. Mecholsky, J.J., Freiman, S.W., and Rice, R.W., in *Fractography and Failure Analysis,* Strauss, D.M. and Cullen, W.H., Jr., Eds., ASTM STP 645, American Society of Testing, Philadelphia, Pa., 1978, pp. 363–379.
3. Terao, N., *J. Phys. Soc. Jap.*, 8, 545, 1953.
4. Levengood, W.C., Effect of origin flaw characteristics on glass strength, *J. Appl. Phys.*, 29, 820, 1958.
5. Shand, E.B., Breaking stress of glass determined from dimensions of fracture mirrors, *J. Am. Ceram. Soc.*, 42, 474, 1959.
6. Kirchner, H.P. and Gruver, R.M., Fracture mirrors in polycrystalline ceramics and glass, in *Fracture Mechanics of Ceramics*, Vol. I, Bradt, R.C., Hasselman, D.P.H., and Lange, F.F., Eds., Plenum, New York, 1974, pp. 309–321.
7. Frechette, V.D., *Fracture of Heliostat Facets*, presented at the ERDA Solar Thermal Projects Semiannual Review, Seattle, Washington, Aug. 23–24, 1977.
8. Rice, R.W., Fractographic identification of strength-controlling flaws and microstructure, in *Fracture Mechanics of Ceramics*, Vol. I, Bradt, R.C., Hasselman, D.P.H., and Lange. F.F., Eds., Plenum, New York, 1974, pp. 323–345.
9. Marcus, H.L., Harris, J.M., and Szalkowsky, F.J., Auger spectroscopy of fracture surfaces of ceramics, in *Fracture Mechanics of Ceramics*, Vol. I, Bradt, R.C., Hasselman, D.P.H., and Lange, F.F., Eds., Plenum, New York, 1974, pp. 387–398.
10. Johari, O. and Parikh, N.M., in *Fracture Mechanics of Ceramics*, Vol. I, Bradt, R.C., Hasselman, D.P.H., and Lange, F.F., Eds., Plenum, New York, 1974, pp. 399–420.
11. Johansen, K.M., Richerson, D.W., and Schuldies, J.J., *Ceramic Components for Turbine Engines*, Phase II Final Rept., AiResearch Rept. 21-2794(08), Feb. 29, 1980, prepared under Air Force contract F33615-77-C-5171.
12. Rice, R.W., Freiman, S.W., Mecholsky, J.J., and Ruh, R., Fracture sources in Si_3N_4 and SiC, in *Ceramic Gas Turbine Demonstration Engine Program Review*, Fairbanks, J.W. and Rice, R.W., Eds., MCIC Rept. MCIC-78-36, 1978, pp. 665–688.

13. Richerson, D.W., Yonushonis, T.M., and Weaver, G.Q., Properties of silicon nitride rotor materials, in *Ceramic Gas Turbine Demonstration Engine Program Review*, Fairbanks, J.W. and Rice, R.W., Eds., MCIC Rept. MCIC-78-36, 1978, pp. 193–217.

14. Yonushonis, T.M. and Richerson, D.W., Strength of reaction-bonded silicon nitride, in *Ceramic Gas Turbine Demonstration Engine Program Review*, Fairbanks, J.W. and Rice, R.W., Eds., MCIC Rept. MCIC-78-36, 1978, pp. 219–234.

15. Carruthers, W.D., Richerson, D.W., and Benn, K., *3500 Hour Durability Testing of Commercial Ceramic Materials*, Interim Rept., NASA CR-159785, July 1980.

16. Frechette, V.D., Fractography and quality assurance of glass and ceramics, in *Quality Assurance in Ceramic Industries*, Frechette, V.D., Pye, L.D., and Rase, D.E., Eds., Plenum, New York, 1979, pp. 227–236.

17. Davidge, R.W. and Tappin, G., Thermal shock and fracture in ceramics, *Trans. Br. Ceram. Soc.*, 66, 8, 1967.

18. Weaver, G.Q., Baumgartner, H.R., and Torti, M.L., Thermal shock behavior of sintered silicon carbide and reaction-bonded silicon nitride, in *Special Ceramics*, Vol. 6, Popper, P., Ed., British Ceramic Research Association, Stoke-on-Trent, England, 1975, pp. 261–281.

19. Ammann, C.L., Doherty, J.E., and Nessler, C.G., The thermal fatigue behavior of hot pressed silicon nitride, *Mater. Sci. Eng.*, 22, 15–22, 1976.

20. Johansen, K.M., Lindberg, L.J., and Ardans, P.M., *Ceramic Components for Turbine Engines*, 8th Interim Rept., AiResearch Rept. 21-2794, 10, June 5, 1980, prepared under Air Force contract F33615-77-C-5171.

21. *Ceramic Gas Turbine Engine Demonstration Program*, Interim Rept. 11, AiResearch Rept. 76-212188(11), Nov. 1978, prepared under contract N00024-76-C-5352, pp. 3–47 to 3–64.

22. Wimmer, J.M. and Bransky, I., Impact resistance of structural ceramics, *Ceram. Bull.*, 56(6), 552–555, 1977.

23. Finger, D.G., *Contact Stress Analysis of Ceramic-to-Metal Interface*, Final Rept., contract N00014-78-C-0547, Sept. 1979.

24. Richerson, D.W., Carruthers, W.D., and Lindberg, L.J., Contact stress and coefficient of friction effects on ceramic interfaces, in *Surfaces and Interfaces in Ceramic and Ceramic-Metal Systems*, Pask, J.A. and Evans, A.G., Eds., Plenum, New York, 1981.

25. Richerson, D.W. and Yonushonis, T.M., Environmental effects on the strength of silicon nitride materials, in *Ceramic Gas Turbine Demonstration Engine Program Review*, Fairbanks, J.W. and Rice, R.W., Eds., MCIC Rept. MCIC-78-36, 1978, pp. 247–271.

ADDITIONAL RECOMMENDED READING

1. Section 9: Failure Analysis, pp. 629–673 in *Engineered Materials Handbook, Vol. 4: CERAMICS AND GLASSES*, ASM Int., Materials Park, Ohio, 1991.

2. Mecholsky, J.J., Jr. and Powell, S.R., Jr., Editors, *Fractography of Ceramic and Metal Failures*, ASTM STP 827, American Society for Testing and Materials, 1984.

3. Varner, J.R. and Fréchette, V.D., Eds., *Advances in Ceramics, Vol. 22, Fractography of Glasses and Ceramics*, The American Ceramic Society, 1986.

4. Fréchette, V.D., *Failure Analysis of Brittle Materials, Vol. 28, Advances in Ceramics*, The American Ceramic Society, 1990.

5. Varner, J.R. and Fréchette, V.D., Eds., *Fractography of Glasses and Ceramics II, Vol. 17, Ceramic Transactions*, The American Ceramic Society, Westerville, Ohio, 1991.

6. Varner, J.R. and Quinn, G.D., *Fractography of Glasses and Ceramics IV, Ceramic Transactions, Vol. 122*, American Ceramic Society, Westerville, Ohio, 2001.

7. *Metals Handbook, Vol. 9: Failure Analysis and Prevention*, 9th ed., ASM Int., Materials Park, Ohio, 1986.

8. Brooks, C. and Choudhury, A., *Failure Analysis of Engineering Materials*, McGraw-Hill, New York, 2002.

STUDY GUIDE

1. What information is often available on the fracture surface of a ceramic?
2. What is the "fracture mirror"?

3. What is a "river pattern"?
4. How are the river pattern and mirror used to help determine the cause of failure?
5. What is the Griffith equation, and how can it be used with information from the fracture surface to estimate the stress that caused the fracture?
6. What is a "Wallner line"?
7. What key information is provided by "cantilever curl"?
8. Identify four key steps in fractography of a ceramic part.
9. Identify five reasons why determination of the cause of failure might be important.
10. Why is it so important to determine whether a failure is due to a design problem or to a material deficiency?
11. How does one determine if a failure resulted from a material flaw?
12. What is a "median crack" and why is it often difficult to see on a fracture surface?
13. How are stress distribution and the contour of the fracture surface related for a thermal shock fracture?
14. What are some features that might distinguish a contact-initiated fracture from other causes of fracture?
15. If a fracture is influenced by oxidation or corrosion, it is important that an engineer identify the mechanism of attack and find a solution. How might this be done using the fracture surface?
16. How can slow crack growth be diagnosed and perhaps understood by study of the fracture surface?

20 Toughening of Ceramics

Prior chapters characterized ceramics as brittle and susceptible to catastrophic fracture. This behavior was attributed to the ease of both crack initiation and crack propagation.[1,2] The ease of crack initiation was shown to result because of a high degree of stress concentration of an applied load at very small microstructural or surface flaws in the ceramic material. The ease of crack propagation was identified to result because of the low fracture toughness of most ceramic materials.

The flaw sensitivity and the low fracture toughness of typical ceramics provide a challenge to achieving reliability of these ceramics in structural applications. Major efforts have occurred since the late 1960s to improve reliability through the fabrication process by reducing the size of microstructural and surface flaws. Many examples were presented in prior chapters including use of ultrafine starting powders, improvements in particle packing during the consolidation process, and minimization of porosity through advanced sintering and HIP techniques. Additional major effort since the mid–1970s has been directed toward improving reliability by increasing the fracture toughness of ceramics to inhibit crack propagation. Substantial progress has been achieved with approaches such as self-reinforced microstructures, particle dispersions, whisker dispersions, transformation toughening, long-fiber reinforcement, ductile phase reinforcement, and prestressing.

This chapter reviews progress in development of the above approaches to reduce the flaw-sensitivity of ceramics, to avoid catastrophic failure by improving resistance to fracture propagation, and thus to improve reliability for structural and thermal applications. The first part of the chapter discusses the mechanisms for achieving toughening in a ceramic material. The remainder of the chapter reviews progress with the application of these mechanisms to specific ceramic materials.

20.1 TOUGHENING MECHANISMS

Before we discuss potential toughening mechanisms for ceramics, let us first review what happens during fracture of a typical polycrystalline ceramic. Visualize a ceramic that consists of an assemblage of grains, grain boundary phases, pores, and likely isolated inclusions and surface scratches or cracks. These microstructural features represent a range in the size of flaws. If the sample is being loaded in uniform tension, fracture will initiate at the largest flaw in the gage section. Assume that we start at zero load and increase the load slowly until we reach the critical stress for the largest flaw. Below the critical stress, we are essentially storing energy elastically within the material. At the critical stress, a crack initiates at the critical flaw. The stored energy is now available to concentrate at the tip of this new crack and drive it through the ceramic. A typical ceramic has no mechanism to prevent this, so that crack rapidly propagates through the ceramic and results in catastrophic (break-up into two or more pieces), brittle fracture.

If we are to avoid this brittle fracture mode, the challenge is to build into the ceramic microstructure mechanisms that either allow the material to withstand the concentration of stored energy at the crack tip or to delocalize (spread out) the energy. Table 20.1 lists mechanisms that have the potential to display one or both of these functions. The following sections describe each of these mechanisms.

20.1.1 MODULUS TRANSFER

Modulus transfer generally involves high elastic modulus fibers in a lower elastic modulus matrix. Examples of composites that use modulus transfer as a primary toughening mechanism include polymers reinforced with glass or carbon fibers, metals reinforced with boron or SiC fibers, and concrete

---

TABLE 20.1
Potential Mechanisms for Achieving Toughening in Ceramics

Modulus transfer
Prestressing
Crack deflection or impediment
Bridging
Pullout
Crack shielding
Energy dissipation

---

reinforced with carbon fibers. In each of these examples the fiber has much higher elastic modulus (and thus cohesive strength) than the matrix. A stress applied to the composite is "transferred" from the matrix to the fibers such that the high-modulus, high-strength fibers carry the load.

Several factors control the degree of toughening that can be achieved by modulus transfer: (1) difference in modulus between the fiber and matrix, (2) strength of the fibers, (3) volume fraction of the fibers and architecture of the fiber distribution, (4) length of the fibers, and (5) interfacial bond between the fibers and matrix.

20.1.1.1 Effect of Elastic Modulus

The larger the difference in elastic modulus between the fibers and matrix and the higher the strength of the fibers, the greater the stress that can be carried by the fibers. Generally, it is preferable to use fibers that have at least double the elastic modulus of the matrix.

The strength of the fibers is determined by the cohesive strength of the atomic bonding and by the size of microstructural and surface defects. Ceramic fibers have been synthesized with tensile strength exceeding 2.5 GPa (>360,000 psi). This is substantially higher than the strength of typical polycrystalline ceramics. However, care is required during composite fabrication to avoid surface damage that would reduce the strength of the fibers.

20.1.1.2 Effect of Volume Fraction and Architecture

The volume fraction of fibers that can be successfully built into a matrix has a substantial affect on the toughness and strength that can be achieved in the composite. This is illustrated in Figure 20.1 for a unidirectional SiC fiber in an aluminum matrix. The strength of the matrix alone is less than 75 MPa. Thirty vol% fibers increases the strength to about 600 MPa and 35 vol% increases the strength to about 700 MPa. Note that the strength is increased over the whole temperature range.

Large volume fraction is desirable as long as the fibers do not interact in such a way that they are damaged and lose strength. Figure 20.2 illustrates an example of a nondamaging fabrication technique for unidirectional fiber reinforcement of aluminum. Single-strand filaments of SiC are wound onto a mandrel. The fibers are spaced at the desired spacing and wound until a single layer is present on the mandrel. A thin layer of aluminum metal is then plasma-sprayed over and between the fibers to form a uniform thickness of fiber-reinforced aluminum sheet. The sheet is removed from the mandrel, stacked with additional layers, and diffusion-bonded at elevated temperature and pressure to form a unidirectional fiber-reinforced composite. A typical cross section is shown in Figure 20.3.

This "drum winding" process has also been used for preparation of a variety of ceramic matrix composites. Rather than plasma spraying, ceramic particles and an organic binder are dispersed in a liquid to form a slurry and either sprayed or painted onto the fibers on the mandrel. An alternative is to pass the fiber through the slurry to pre-coat the fibers before winding onto the mandrel. When the slurry has dried, the resulting "tape" is cut off the mandrel, layered in the desired thickness (number of plies) and orientation, and densified by infiltration, hot pressing or HIP.

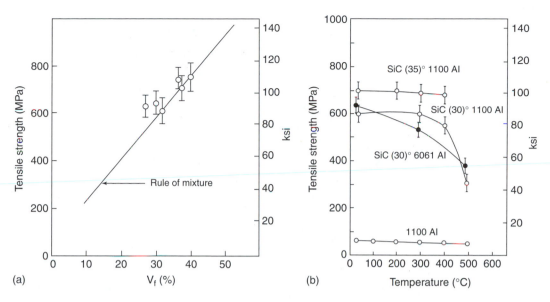

FIGURE 20.1 Strength improvement of Al alloys by reinforcement with unidirectional SiC fibers. (a) Tensile strength of unidirectional vs. fiber volume fraction. (b) Tensile strength as a function of test temperature. (From Yajima, S., *Revue de Chemie Minerale*, 118, 1981 and Yajima, S., *J. Mater Sci.*, 116, 1981. With permission.)

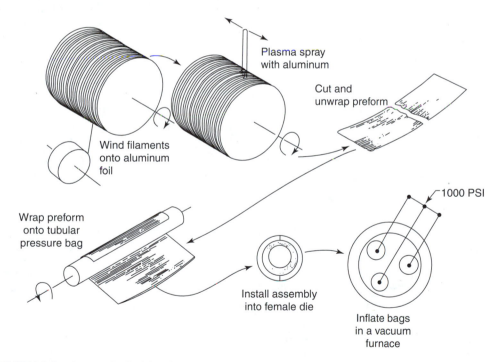

FIGURE 20.2 One method of forming a ceramic-fiber, metal-matrix composite in which the fibers are oriented unidirectionally. (Courtesy of Textron Specialty Materials, Lowell, MA.)

Unidirectional fibers provide toughening and strengthening only in the direction parallel to the fibers. Such composites are referred to as having one-dimensional architecture. Strength and toughening in the other directions are no better than for the matrix. Such composites often fail in interlaminar shear rather than in tension. To increase the resistance to shear failure, fibers are built into the

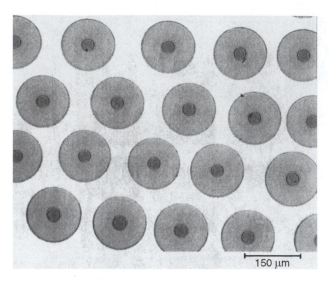

FIGURE 20.3 Optical photomicrograph of polished cross section showing unidirectional SiC filaments in a metal matrix. (Courtesy of Textron Specialty Materials, Lowell, MA.)

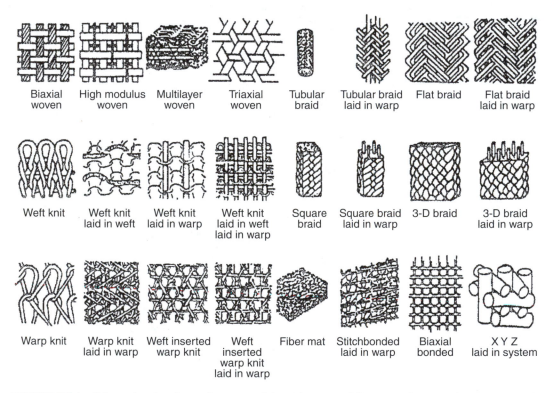

FIGURE 20.4 Schematics showing various potential architectures of fibers and filaments for use in ceramic matrix composites. (From Ko, F.K., *Am. Ceram. Soc. Bull.*, 68(2), 401–414, 1989. With permission.)

composite in additional directions.[3] Examples of some possible architectures are shown schematically in Figure 20.4. The two-dimensional architecture can be approximated with fibers woven into a fabric. Multidirectional architectures result in a decreased number of fibers in any single direction, so that the strength in the strongest direction is not as high as for unidirectional fibers. However, the strength in the minimum direction is increased for three-dimensional and higher composites to yield improved resistance to failure by the interlaminar shear mode.

20.1.1.3 Effect of Fiber Length

Fiber length also affects the toughening and strengthening capability of fibers in a composite. The examples discussed so far have involved "continuous" fibers in which the fiber essentially extends from one end of the composite to the opposite end. Modulus transfer can occur with shorter "chopped" fibers, but achieving a controlled architecture or distribution of chopped fibers is difficult. The minimum length of fiber that will yield modulus transfer depends on the relative moduli of the fiber and matrix and on the degree of bonding between the fiber and matrix. Some studies have indicted that modulus transfer can occur down to an aspect ratio of about 8:1 (length of fiber:diameter of fiber).

20.1.1.4 Effect of Interfacial Bond

As the term "modulus transfer" implies, the stress applied to the composite must be transferred from the matrix to the fibers. This transfer requires a reasonable level of bond or friction between the fibers and matrix. Too-weak a bond can result in shear at the fiber-matrix interface and reduce the amount of modulus transfer. However, this can lead to alternate toughening mechanisms that are particularly important for ceramic matrix composites where there is not a large modulus difference between the matrix and the fibers.

20.1.2 Prestressing

Ceramics generally fracture in tension, i.e., in a crack-opening stress mode. Prestressing involves placing a portion of the ceramic under a residual compressive stress. A crack cannot start or extend as long as the ceramic is prestressed in compression. Tensile fracture will only occur after a large enough load is applied to exceed the compressive prestress and to build up a tensile stress large enough to initiate a crack at a critical flaw.

A compressive prestress can be achieved by many approaches.[4] One approach is to place the surface in compression by quenching, ion exchange, or layering. A second approach is to achieve compression throughout a matrix by using fibers. We will address the latter approach first.

Compression can be achieved with fibers in two ways. One involves a multistep process consisting of (1) elastically stretching the fibers in tension, (2) surrounding the fibers with matrix, and (3) releasing the applied tension on the fibers. The fibers elastically contract and pull the surrounding matrix into compression. The fibers retain tensile prestress equivalent to the compressive prestress that results in the surrounding matrix. It is apparent that this technique will only work if the fibers are very strong and if there is reasonable bond or mechanical interlocking between the fibers and matrix. A practical example of this technique is the use of metal rods to prestress concrete.

The second fiber-prestressing technique uses a thermal expansion mismatch between the fibers and matrix. The fiber is selected to have a higher coefficient of expansion than the matrix. The matrix is formed around the fibers at high temperature. The fibers contract more than the matrix during cooling and pull the matrix into a compressive prestress. The fibers retain an equivalent level of tensile prestress.

Surface prestressing was saved until last because it is more commonly used than fiber prestressing. Surface prestressing can be achieved by quenching, ion exchange, or layering. Quenching achieves a compressive surface stress in the following way. The ceramic or glass is heated to a temperature where creep or viscous flow can occur. The surface is then quenched more quickly than the interior, forming a rigid case around the still flexible interior. Initially, the hot interior can deform to conform to the stress applied by the quenched rigid case. However, as the interior cools below the range of creep or viscous flow, it then shrinks and pulls the lower temperature surface into compression. This technique is commonly used to produce a surface compressive layer on glass to achieve safety glass.

Safety glass can also be achieved by ion exchange. The glass is exposed at an elevated temperature to larger ions than were originally present in the glass. Some of these larger ions exchange places with smaller ions in the glass. By proper selection of the ions and careful control of temperature, the depth of penetration of the exchanged ions can be controlled. When the glass is cooled, the larger ions near the surface result in a compressive prestress at and near the surface. The degree of prestress can be controlled by the depth of ion exchange compared to the original cross section of the glass. Compressive surface stresses have also been achieved for polycrystalline ceramics by this technique.

A final technique for achieving surface compression is layering. Corelle dishes by Corning represent a good commercial example of this technique. Corelle consists of a laminate in which the surface layers have a different coefficient of thermal expansion than the interior. During cooling from the fabrication temperature, a large compressive prestress results at the surface. Dropping a Corelle dish on the floor generally does not result in high enough tensile stress to exceed the compressive surface stress. As a result, a Corelle dish is much more resistant to in-use fracture than a conventional stoneware or porcelain dish.

TABLE 20.2

Typical Fracture Toughness Values for Ceramic-Based Materials

Material	K_{Ic} MPa $\cdot$ m$^{1/2}$ (ksi $\cdot$ in.$^{1/2}$)
Glass	0.7 (0.6)
Single Crystals	
NaCl	0.3 (0.27)
Si	0.6 (0.54)
MgO	1 (0.91)
ZnS	1 (0.91)
SiC	1.5 (1.36)
Al_2O_3	2 (1.82)
WC	2 (1.82)
Polycrystalline Ceramics	
Al_2O_3	3.5–4.0 (3.18–3.64)
SiC	3.0–3.5 (2.73–3.18)
Stabilized ZrO_2	2 (1.82)
RBSN	2.5 (2.3)
Sintered Si_3N_4	4–6 (3.64–5.46)
Transformation-Toughened Ceramics	
Mg–PSZ	9–12 (8.19–10.92)
Y–TZP	6–9 (5.46–8.19)
Ce–TZP	10–16 (9.1–14.56)
Al_2O_3–ZrO_2	6.5–13.0 (5.91–11.83)
Dispersed-Particle Ceramics	
Al_2O_3–TiC	4.2–4.5 (3.82–4.09)
Si_3N_4–TiC	4.5 (4.09)
Dispersed-Whisker Ceramics	
Al_2O_3–SiC whiskers	6–9 (5.46–8.19)
Fiber-Reinforced Ceramics	
SiC in borosilicate glass	15–25 (13.65–22.75)
SiC in LAS	15–25 (13.65–22.75)
SiC in CVD SiC	8–15 (7.28–13.65)

20.1.3 CRACK DEFLECTION OR IMPEDIMENT

Fracture toughness is strongly affected by the microstructure of a ceramic[5–8] and by the path that a crack follows as it propagates through the material. If the crack follows a planar, smooth path, the new surface area produced is a minimum and the fracture energy term in the equation is low. This is the situation for single crystals and glass. In a single crystal, the crack can follow a cleavage plane and meet no obstructions. The K_{Ic} is very low. Although a glass has no crystal structure or cleavage planes, the crack follows a similar planar path due to the homogeneity of the glass and the absence of microstructure. As a result, glass also has a very low fracture toughness.

Most ceramics in commercial use are polycrystalline. Each grain generally has a different crystal orientation than adjacent grains. A crack passing through a polycrystalline ceramic does not follow a smooth planar path. It follows grain boundaries around some grains and fractures other grains. The total new surface generated as the crack propagates is greater than was the case for a single crystal or glass. This results in greater fracture surface energy and greater fracture toughness. For comparison, glass usually has a fracture toughness under $1\,MPa \cdot m^{1/2}$, single crystals generally range from 0.3 to 2.0, and equiaxed polycrystalline ceramics range from about 2 to $4\,MPa \cdot m^{1/2}$. Examples are listed in Table 20.2, along with toughness values for ceramics toughened by other mechanisms that are discussed later in this chapter.

Ceramic microstructures can be modified to achieve increased crack deflection and impediment and increased fracture toughness. Approaches include controlled grain boundary phases, multimodal grain structures, elongated or fibrous grain structures, and dispersions of foreign particles, plates, whiskers, or chopped fibers.

The microstructure can be modified either by addition of a second phase of the desired morphology or by inducing growth of the desired morphology during sintering. The pros and cons of each approach are identified in Table 20.3. Specific examples of each are described later.

The amount of toughness improvement by dispersed particles depends on the volume fraction of the particles and on the shape of the particles. Theoretical and experimental studies indicate that the toughness can be improved (compared to an equiaxed microstructure of the matrix ceramic) by a factor of 2 for spherical particles, a factor of 3 for disk-shaped particles, and a factor of 4 for rod-shaped

---

TABLE 20.3
Pros and Cons of *In Situ* vs. Added Reinforcement Phase

Mode of Microstructure Modification	Pros	Cons
Reinforcement formation during sintering	Ease of achieving near-theoretical density	Potential difficulty in control of the morphology and distribution of the reinforcement
	Potential for achieving near-net shape by conventional sintering	Limited material options that result in reinforcement
	Potential for low cost	
	Relatively stable microstructure	
Addition of reinforcement	Precise control over percent addition and over morphology	Inhibition of sintering of matrix, leading to low-volume fraction reinforcement or substantial residual porosity
	Wide range of options for matrix and reinforcement	Generally requires hot pressing or liquid-enhanced sintering
		Difficult to achieve low cost
		Potential health hazard of whisker materials

particles.[8] This is illustrated in Figure 20.5 as a function of the volume fraction of dispersed particles. Experimental results provided good correlation. For example, 30 vol% disk-shaped particles in a glass ceramic resulted in about a 2.5 times increase in toughness.

Another benefit that has been achieved by particle dispersion is increased abrasion resistance through increased hardness. Addition of TiC to Al_2O_3 increases the hardness from about 90 to 95 Rockwell A. Addition of TiC to Si_3N_4 increases the hardness from about 91 to 94 Rockwell A. The addition of TiC to Si_3N_4 also results in a material that can be machined by electrical discharge techniques, which provides new shape-forming options.

20.1.4 CRACK BRIDGING

Crack bridging involves a feature of the microstructure that extends across a crack behind the crack tip and imposes a closing force on the crack.[7] Inhibiting crack opening reduces the stress intensity at the tip of the crack and stops or slows crack propagation. Crack bridging is a major toughening mechanism for ceramics reinforced with long fibers. This is illustrated in Figure 20.6. The crack has extended through the matrix beyond the fibers, but the fibers near the crack tip have adequate strength and elastic modulus to not break. Additional energy (stress) must be applied to elongate, pull out, or break fibers before the stress intensity at the crack tip is high enough for the crack to further propagate.

Crack bridging has been identified for fibrous microstructure ceramics, for some coarse-grain ceramics, for ceramics reinforced with platelets or whiskers, and for ceramics reinforced with long or continuous fibers.

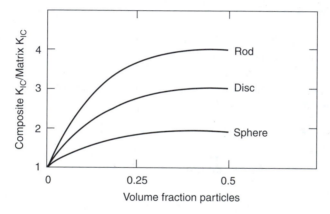

FIGURE 20.5   Predicted relative change in fracture toughness vs. shape of dispersed particles. (Courtesy of A.G. Evans, University of California at Santa Barbara.)

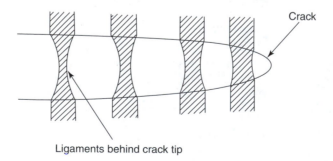

FIGURE 20.6   Increase in fracture toughness by the bridging of cracks by long fibers. The degree of necking of the fibers is highly exaggerated. (Drawing courtesy of ASM International.)

Crack bridging can also occur if a ductile phase is present. The ductile phase elongates rather than immediately fracturing and inhibits crack opening.

20.1.5 Pullout

Pullout consists of a fiber, particle, or grain that debonds from the adjacent microstructure and pulls out as the crack opens. Energy that would normally cause crack propagation is partially expended by debonding and by friction as the fiber, particle, or grain slides against adjacent microstructure features. This can effectively increase fracture toughness. Pullout often accompanies bridging.

Pullout appears to be an important mechanism in achieving optimum toughening in fiber-reinforced ceramic matrix composites. The nature of the fiber-matrix interface is critical. If the bonding is too strong to allow pullout, the fracture travels directly through the fiber and no pullout or bridging occurs. The material exhibits low toughness. If the fiber-matrix interface bond is weak enough to allow crack deflection or debonding, the material can exhibit toughening.[9–11]

20.1.6 Crack Shielding

Crack shielding is a stress-induced microstructural change that results in a reduction in stress at the crack tip. The effect occurs in a zone around the crack tip and extending back along the crack (referred to as the wake). The thickness of the zone and the extent of the wake both affect the degree of stress shielding at the crack tip.

Several types of crack shielding have been identified and studied: microcracking, ductile zone, and transformation zone.[12–15] Microcracking results in crack shielding by locally reducing the elastic modulus and by spreading the applied stress over many cracks rather than one primary crack. A ductile zone results in crack shielding by allowing plastic deformation around the crack tip. The nature of crack shielding by a transformation zone is described in the following paragraphs.

Transformation toughening is a relatively new approach to achieve high toughness and strength in ceramics.[16–21] The key ceramic material is zirconium dioxide (ZrO_2). ZrO_2 goes through a martensitic phase transformation from the tetragonal-to-monoclinic crystal form while cooling through a temperature range from approximately 1150°C (2100°F) to 1000°C (1832°F). By control of composition, particle size, and the heat treatment cycle, zirconia can be densified at high temperature and cooled such that the tetragonal phase is maintained as individual grains or as precipitates to room temperature. An example involving tetragonal precipitates in a cubic ZrO_2 matrix is shown in Figure 20.7.

FIGURE 20.7 Transmission electron micrograph of optimally aged, transformation-toughened ZrO_2–MgO showing the oblate spheroid precipitates of tetragonal ZrO_2 in a MgO-stabilized ZrO_2 cubic matrix. (Courtesy of A.H. Heuer, Case Western Reserve University.)

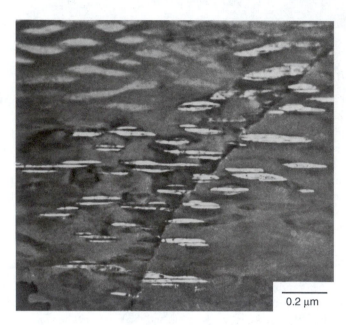

0.2 µm

FIGURE 20.8 Dark field transmission electron micrograph of optimally aged, transformation-toughened ZrO_2–MgO showing twinned monoclinic precipitates adjacent to a crack and tetragonal precipitates away from the crack. (Courtesy of A.H. Heuer, Case Western Reserve University.)

The tetragonal phase is not in equilibrium at room temperature and would normally spontaneously transform to the stable monoclinic phase. However, such a transformation involves an increase in volume, as shown earlier in Figure 7.17. If the grain or precipitate size is small enough (less than about 0.5 µm), the strength of adjacent grains prevents the transformation from occurring by preventing the necessary volume expansion. When a stress is applied to the zirconia and a crack tries to propagate, the metastable tetragonal zirconia grains adjacent to the crack tip can now expand and transform to the stable monoclinic crystal form. This is illustrated in Figure 20.8. Precipitates that have transformed to monoclinic can be distinguished from untransformed precipitates in the TEM photomicrograph by the presence of twinning. Note that only the precipitates near the crack have transformed. This martensitic transformation is accompanied by a 3% volume increase of these grains or precipitates adjacent to the crack, which places the crack in compression and stops it from propagating. To extend the crack further requires additional tensile stress. The result is a ceramic that is very tough and strong and that has been appropriately referred to as "ceramic steel."

Pure ZrO_2 does not have transformation-toughening behavior. Additives are required to stabilize such behavior. These additives are CaO, MgO, Y_2O_3, CeO_2, and rare earth oxides. Too much addition fully stabilizes the ZrO_2 in a cubic crystal structure, which also does not have transformation-toughening behavior because it does not go through the tetragonal-to-monoclinic transformation. Toughening requires the presence of the metastable tetragonal state. The range of addition to achieve the metastable tetragonal state and toughening is shown for various ZrO_2-based compositions in Figure 20.9. The peak of the curve for each material corresponds to the maximum tetragonal content. Monoclinic content increases to the left of the peak and cubic to the right of the peak, each resulting in a decrease in toughness and strength. PSZ in Figure 20.9 stands for partially stabilized zirconia. DCB, ICL, and NB identify the method that was used to measure fracture toughness. DCB stands for the double-cantilever beam, ICL the indentation crack length, and NB the notched beam. Note that the composition zones for achieving peak toughness are relatively narrow.

Transformation toughening is not limited to ZrO_2. Very small grains of ZrO_2 can be added to another ceramic such as Al_2O_3 and be retained as tetragonal during cooling. These grains will then

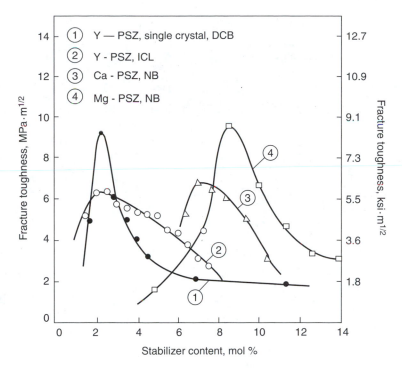

FIGURE 20.9 Fracture toughness vs. composition for transformation-toughened zirconia materials. (From Claussen, N., Transformation toughened ceramics, in *Ceramics in Advanced Energy Technologies*, Krockel, H., et al., Eds., Reidel, D., Dordrecht, 1984, pp. 51–86. With permission.)

transform near a crack tip and inhibit crack propagation. Several criteria are necessary before transformation toughening can be achieved by addition of partially stabilized ZrO_2 particles to a host ceramic: (1) ZrO_2 particles not dissolved by host; (2) particle size of the ZrO_2 typically under 0.5 μm; and (3) host microstructure strong enough to retain the ZrO_2 particles in the tetragonal form during cooling. Al_2O_3, Mullite, spinel, Si_3N_4, cordierite, β''-alumina, MgO, and glass have been successfully used as hosts for ZrO_2 to achieve transformation toughening.

Figure 20.10 schematically illustrates options for transformation toughening and some of the microstructures that have been achieved. Specific materials are discussed later in this chapter.

20.1.7 ENERGY DISSIPATION

Energy dissipation is a consequence of some of the toughening mechanisms discussed so far. An example is fiber pullout. Before the fiber can pull out, debonding must occur along a length of the fiber. If the fiber is not positioned parallel to the direction of the initial crack, the crack temporarily changes direction, producing debonding for some distance along the fiber. This results in a delocalization in the stress and essentially a mode of energy dissipation. As the fiber pulls out, the friction of the fiber against the matrix results in additional energy dissipation.

Other mechanisms have been identified that increase toughness directly by energy dissipation. One is referred to as ferroelastic domain switching.[22,23] The term ferroelastic is analogous to ferromagnetic or ferroelectric. A ferromagnetic material contains domains that can change direction under the influence of a magnetic field. A ferroelastic material contains domains that can change direction under the influence of a stress field. Energy that would normally contribute to crack formation and extension is instead partially dissipated as heat by domain reorientation. Ferroelastic domain switching has been demonstrated in tetragonal zirconia. Figure 20.11 illustrates ferroelastic domains in yttria-stabilized tetragonal zirconia samples.

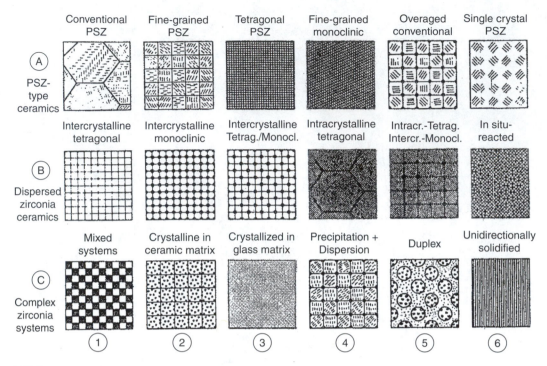

FIGURE 20.10 Classification of transformation-toughened ceramics based on microstructural feates (From Claussen, N., Transformation toughened ceramics, in *Ceramics in Advanced Energy Technologies*, Krockel, H., et al., Eds., Reidel, D., Dordrecht, 1984, pp. 51–86. With permission.)

20.2 EXAMPLES OF TOUGHENED CERAMICS

Tremendous progress has occurred during recent years in increasing the toughness of ceramic materials. The following sections describe specific examples.

20.2.1 SELF-REINFORCED CERAMICS

Self-reinforcement refers to achieving a microstructure imparting enhanced toughness during sintering or heat treating without adding a second phase. This has been achieved with silicon nitride and sialon compositions through a relatively conventional sintering cycle, with a zirconium diboride-zirconium-zirconium carbide material through a reaction mechanism, and with zirconia materials through chemical additions. These are discussed in this section. Transformation-toughened ceramics also qualify as self-reinforced, but are discussed in a subsequent section.

20.2.1.1 Self-Reinforced Si_3N_4

Hot-pressed Si_3N_4 was determined in the early 1970s to have unusually high fracture toughness for a monolithic ceramic. Values from 4 to 5.5 MPa $\cdot$ m$^{1/2}$ were observed. Lange[24] suggested that the high toughness resulted from a microstructure consisting of elongated grains. The elongated grains formed during liquid phase sintering by a solution-precipitation mechanism. The starting alpha-Si_3N_4 powder was dissolved by the liquid phase (consisting of sintering aids plus SiO_2 coating the alpha-Si_3N_4 powder) and reprecipitated as acicular beta-Si_3N_4 grains.

A number of studies[25–30] have been conducted in efforts to optimize the microstructure and toughness. These studies have generally shown that chemisty, temperature, and time at temperature are

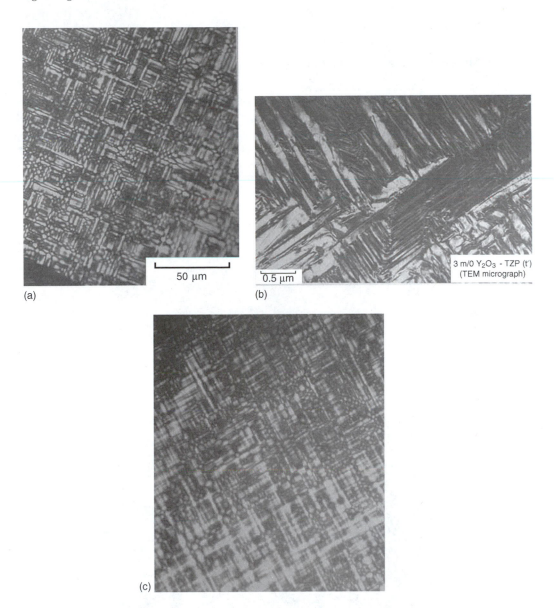

FIGURE 20.11 Ferroelastic domains in ZrO$_2$ stabilized to the tetragonal phase with 3 mol% Y$_2$O$_3$. (a) Transmission optical photomicrograph. (b) Transmission electron microscope photomicrograph of domains in one grain of a polycrystalline sample. (c) Transmission optical photomicrograph of domains in a single-crystal sample (×400). (Courtesy of A.V. Virkar and Jan Fong Jue, University of Utah.)

key parameters. Table 20.4 summarizes some of the results. Figure 20.12 shows a photomicrograph of the microstructure of a self-reinforced Si$_3$N$_4$ material with high-aspect-ratio beta-Si$_3$N$_4$ grains. This material had average fracture toughness of 10.5 MPa · m$^{1/2}$, as measured by the chevron notch technique.

The toughening mechanisms for self-reinforced Si$_3$N$_4$ have been identified primarily as crack deflection and bridging. Highest toughnesses have been achieved for large grain size and for compositions containing substantial grain boundary phase. Mechanical properties above 1200°C were initially limited by the properties of the grain boundary phase. However, major improvements were achieved during the 1990s as shown earlier in Figure 9.8.

TABLE 20.4

Results of Studies to Increase the Fracture Toughness of Si_3N_4 by Self-Reinforcement

Composition (wt%)	Sintering Temperature (°C)	Total Time at Temperature (hr)	Bulk Density (g/cm³)	Approximate Average Grain Diameter (μm)	Flexural Strength (MPa)	Fracture Toughness (MPa·m$^{1/2}$)	Ref.
$1Al_2O_3$–$5CeO_2$	1800/2000	2	3.25	>1 μm	650	9.0	25
$3Al_2O_3$–$5CeO_2$	1800/2000	2	3.27	>1 μm	735	7.9	25
$2Al_2O_3$–$5Y_2O_3$	1800/2000	3	3.25	>1 μm	785	8.2	25
$5CeO_2$–$4MgO$–$1SrO$–$3ZrO_2$	1600	–	3.25	Finer	990	5.8	26
$6Y_2O_3$–$4MgO$–$0.5ZrO_2$	1600	–	3.26	Coarser	1020	7.4	26
A Si_2N_4 Glass	–	–	3.27	0.52	960	6.5	27
B S_3N_4 Glass	–	–	3.38	0.54	810	8.3	27
C Si_3N_4 Glass	–	–	3.18	0.72	900	9.7	27
6.4 vol% Y_2O_3	–	As-sintered	3.22	Fine	860	5.8	28
6.4 vol% Y_2O_3	–	Heat-treated	3.22	Coarse	690	8.3	28
AS-700	–	–	3.31	>1 μm	550	10.6	29
Y_2O_3–MgO–CaO	–	–	–	>1 μm	–	10.5	30
Y_2O_3–MgO–ZrO_2							
A Hot-pressed	–	–	–	<1 μm	900	10–14	30
B Sintered	–	–	–	<1 μm	800	8.5	30

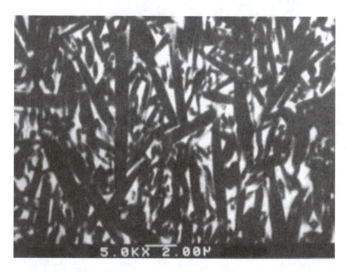

FIGURE 20.12 Photomicrograph of pressureless sintered Si_3N_4 material with composition and sintering parameters selected to achieve a fibrous self-reinforced microstructure. This material had a composition in the Si_3N_4–Y_2O_3–MgO–CaO system, contained about 15% glass, and had an average fracture toughness of 10.5 MPa · m$^{1/2}$. (Courtesy of A. Pyzik, Dow Chemical Company.)

20.2.1.2 Self-Reinforced ZrC

High-toughness platelet-reinforced ZrC/Zr matrix composites have been achieved by a reactive densification process.[31,32] Boron carbide powder is compacted with a binder using conventional techniques such as pressing, molding, or casting. The preform is placed in a graphite mold with a controlled amount of Zr metal. The material is heated to 1850 to 2000°C in an inert atmosphere. The Zr becomes molten, infiltrates the preform, and reacts with the B_4C to form ZrB_2 plus ZrC. The quantity of Zr

FIGURE 20.13 Crack deflection by dispersed particles of ZrB_2 in a matrix of ZrC and Zr. (From Claar, T.D., Johnson, W.B., Andersson, C.A., and Schiroky, G.H., *Ceram. Eng. Sci. Proc.*, 10(7–8), 599–609, 1989. With permission.)

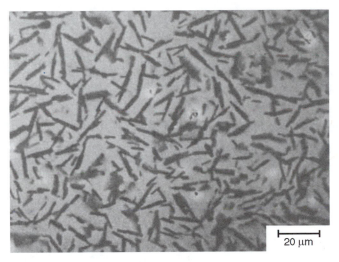

FIGURE 20.14 Aluminate platelets in situ reinforcement in a transformation-toughened ZrO_2. (Courtesy of R.A. Cutler, Ceramatec, Inc.)

can be varied either to be completely consumed by the reaction or to be retained as a residual metal phase to over 30% by volume. The ZrC crystallizes from the reaction as equiaxial grains, whereas the ZrB_2 forms platelets.

The strength, toughness, and other properties can be varied by the amount of residual Zr metal. A sample with 2.5 vol% Zr had a mean four-point flexural strength of 870 MPa and a Weibull modulus of about 28. A sample with 24.2 vol% Zr had a mean strength of 888 MPa and a Weibull modulus of about 68. Fracture toughness values ranged from about 11 MPa $\cdot$ m$^{1/2}$ for about 1 vol% Zr to over 20 MPa $\cdot$ m$^{1/2}$ for about 30 vol% Zr.

Three toughening mechanisms were identified as the primary sources of high toughness: crack deflection, bridging, and crack clamping plus energy dissipation (through plastic deformation of the Zr metal phase). Figure 20.13 shows the path of propagation of a crack through the microstructure, illustrating the high degree of crack deflection.

20.2.1.3 Aluminate Platelet-Reinforced Transformation-Toughened ZrO_2

Platelets of $SrO \cdot 6Al_2O_3$ can form in CeO_2-doped transformation-toughened ZrO_2 during sintering[33] for a range of mixtures of powders of $SrZrO_3$, Al_2O_3, and coprecipitated ZrO_2–12 mol% CeO_2. The platelets are typically 0.5 μm wide by 5 to 10 μm long. The microstructure is shown in Figure 20.14. Material containing 2.0 wt% $SrZrO_3$ and 30 vol% Al_2O_3 and sintered at 1500°C had a four-point flexure strength of 726 MPa and a fracture toughness of 11.2 MPa $\cdot$ m$^{1/2}$. Material containing 2 wt% $SrZrO_3$ and 15 wt% Al_2O_3 and sintered at 1550°C had a strength of 519 $\pm$ 20 MPa and toughness of 19.2 ± 2.8 MPa $\cdot$ m$^{1/2}$. This material exhibited permanent deformation during room-temperature testing.

20.2.1.4 La-β Alumina-Reinforced Transformation-Toughened ZrO$_2$

Fujii et al.[34] have demonstrated that elongated grains of La$_2$O$_3 \cdot$ 11Al$_2$O$_3$ (La beta alumina) can form in CeO$_2$-doped transformation-toughened ZrO$_2$ during densification and can provide additional toughening. A composition consisting of 80% CeO$_2$-doped ZrO$_2$, 5% La-β alumina, and 15% alumina was sintered to near-theoretical density at 1600°C for 4 hr. The ZrO$_2$ powder contained 12 mol% CeO$_2$. The La was added as lanthanum oxalate. The resulting self-reinforced material had average flexural strength of 910 MPa and toughness (measured by single-edge precracked beam) of 11.0 MPa $\cdot$ m$^{1/2}$.

20.2.2 TRANSFORMATION-TOUGHENED CERAMICS

The mechanism of transformation toughening was described in a prior section of this chapter. The key is to retain grains or precipitates of the tetragonal ZrO$_2$ phase metastably to room temperature. This is done differently, depending on the additives in the ZrO$_2$.

20.2.2.1 Transformation-Toughened ZrO$_2$

Early studies were conducted in the CaO–ZrO$_2$ system. The phase diagram for the CaO–ZrO$_2$ system is shown in Figure 20.15. Toughening in this system has been achieved in the composition range of roughly 6 to 11 mol% CaO. Note from Figure 20.15 that this corresponds to the position of the T$_{ss}$ + C$_{ss}$ (tetragonal solid solution plus cubic solid solution) phase field. To achieve high

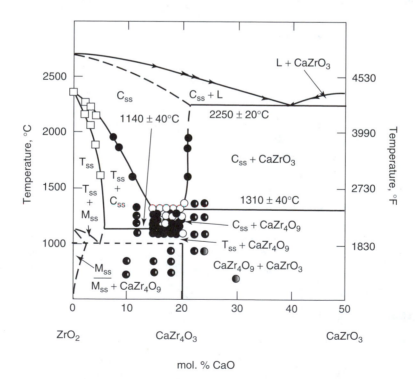

FIGURE 20.15 ZrO$_2$–CaO binary phase equilibrium diagram (C$_{ss}$ = cubic solid solution, T$_{ss}$ = tetragonal solid solution, M$_{ss}$ = monoclinic solid solution). (From Stubican, V.S. and Ray, S.P., *J. Am. Ceram. Soc.*, (1–2), 534–537, 1977. With permission.)

toughness in this system, the following procedure is used: (1) a composition such as 10 mol% CaO, 90 mol% ZrO_2 is selected; (2) the powder is compacted into the desired shape; (3) the compact is densified at a temperature just above the $T_{ss} + C_{ss}$ field in the C_{ss} field to achieve a polycrystalline microstructure of uniform cubic solid solution, i.e., about 1800 to 1850°C (3270 to 3330°F) for the 10% CaO composition; (4) the solutioned material is quenched to about 1300°C (2370°F) (at this temperature, the material is supersaturated; tetragonal precipitates form in the cubic ZrO_2 grains); (5) the material is aged at about 1300°C (2370°F) until the precipitates have reached optimum size; and (6) the material, now consisting of tetragonal precipitates about 0.3 μm long in large cubic grains, is cooled to room temperature.

Similar procedures are used in the MgO–ZrO_2 system. The phase diagram in Figure 20.16 identifies that the crucial $T_{ss} + C_{ss}$ phase field is roughly between 3 and 12 mol% MgO. Solution and densification have been achieved at about 2000°C (3630°F), followed by aging at 1500°C (2730°F) for 1 h. Figure 20.7 shows the shape and size of optimum tetragonal ZrO_2 precipitates in a MgO–ZrO_2 cubic grain.

Formation of tetragonal precipitates has also been achieved in single-crystal ZrO_2 using CaO, MgO, and Y_2O_3 additions. The single crystal is grown at high temperature in the cubic solid solution phase field and then heat-treated in the cubic plus tetragonal solid solution phase field to form the tetragonal precipitates in the cubic single crystal.

Although tetragonal precipitates can be achieved using Y_2O_3 additions, an alternate approach is generally used that yields a different microstructure. The Y_2O_3–ZrO_2 phase equilibrium diagram is illustrated in Figure 20.17. In this case, very fine powder (<0.3 μm) containing about 2 to 3 mol% Y_2O_3 and 97 to 98 mol% ZrO_2 can be densified completely in the tetragonal phase field to yield a fine-grained microstructure consisting almost totally of tetragonal grains. Each grain in this material

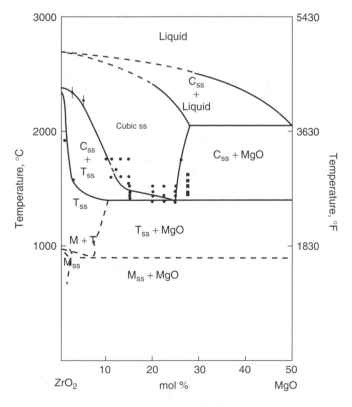

FIGURE 20.16 ZrO_2–MgO binary phase equilibrium diagram. (From Viechnicki, D. and Stubican, V.S., *J. Am. Ceram. Soc.*, 48(6), 292–297, 1965. With permission.)

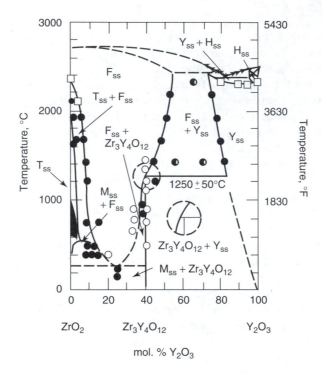

FIGURE 20.17 ZrO₂–Y₂O₃ binary phase equilibrium diagram. (From Stubican, V.S., Hink, R.C., and Ray, S.P., *J. Am Ceram. Soc.*, 61(1–2), 17–21, 1978. With permission.)

can transform near a crack tip to inhibit propagation of the crack. A similar microstructure can be achieved with CeO_2 additions.

Specific terminology has evolved to describe the different transformation-toughened ZrO_2 materials based on additive and microstructure. Material with tetragonal precipitates is referred to as partially stabilized zirconia, or simply PSZ. The additive is normally identified by a prefix, i.e., Ca–PSZ, Mg–PSZ, and Y–PSZ. If enough stabilizer is added, the material only contains cubic phase and is referred to as *fully stabilized zirconia*. The Y_2O_3–ZrO_2 and CeO_2–ZrO_2 materials with essentially all tetragonal phase are called *tetragonal zirconia polycrystal*, i.e., Y–TZP and Ce–TZP.

Figure 20.9 showed fracture toughness vs. composition for ZrO_2 with additions of Y_2O_3, CaO, and MgO. Table 20.5 lists some published property data. Each composition has its own characteristics such that one material may be optimum for one application and another for a different application. Fracture toughness typically ranges from 5 to 10 MPa · m$^{1/2}$ (4.55 to 9.1 ksi · in.$^{1/2}$), although some higher values have been reported. Strength (measured in four-point bending) is typically in the range 600 to 1200 MPa (87 to 174 ksi), but isolated values over 2000 MPa (290 ksi) have been reported.

Transformation-toughened zirconia materials have great potential for wear-resistance applications. They can withstand considerably more handling and in-service abuse than conventional ceramics. The zirconia compositions have a coefficient of thermal expansion very similar to cast iron, ranging from about 9.5–11.0 × 10^{-6}/°C. The reasonable thermal expansion match simplifies attachment of the ceramic to metal systems.

Transformation-toughened ceramics also have limitations. The major limitation is temperature. For toughening to occur, the tetragonal phase must be able to transform to the monoclinic phase. Above about 1150°C (2100°F), all of the tetragonal phase is in equilibrium, so no toughening occurs. A portion of the tetragonal phase is stable all the way down to 650°C (1200°F), as shown in Figure 20.18. As a result, transformation-toughened materials have a progressive reduction in strength and toughness as the temperature is increased. This is illustrated in Figure 20.19.

TABLE 20.5

Examples of Property Data for Transformation-Toughened Zirconia

Material	Composition	Tetragonal Phase Content[b] %	K_{Ic}[a] MPA·m$^{1/2}$ (ksi·in.)$^{1/2}$	Test Technique[c]	Strength, MPa (ksi)	Grain Size (µm)
Mg–PSZ	3 wt% MgO	40	9.5 (8.64)[d]	NB	650 (94.3)	60
Mg–PSZ	9 mol% MgO	40	8.95 (8.14)	NB	600 (87)	80
Ca, Mg–PSZ	10 mol% (CaO + MgO)	10	5.6 (5.09)	NB	360 (52.2)	60
Ca–PSZ	3.7 wt% CaO	40	6.8 (6.18)[e]	NB	645 (93.5)	80
Y–TZP	2.5 mol% Y$_2$O$_3$	97	6.4 (5.82)	DCB	700 (101.5)	1

[a] The toughness of fully stabilized (cubic) ZrO$_2$ ranges between 2 and 3 MPa·m$^{1/2}$ (1.82 and 2.73 ksi·in.$^{1/2}$).

[b] Rest cubic, traces of monoclinic.

[c] NB: notched beam, DCB: double-cantilever beam, ICL: indentation crack length.

[d] As-notched.

[e] Notched and etched; as-notched: 9.6 MPa·m$^{1/2}$ (8.73 ksi·in.$^{1/2}$).

Source: Adapted from Claussen, N., Transformation toughened ceramics, in *Ceramics in Advanced Energy Technologies*, Krockel, H., et al., Eds., Reidel, D., Dordrecht, 1984, pp. 51–86.

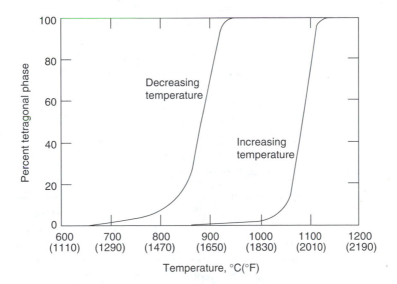

FIGURE 20.18 Thermal hysteresis of the tetragonal-to-monoclinic transformation of ZrO$_2$. (From Wolten, G.M., *J. Am. Ceram. Soc.*, 46(9), 418–422, 1963. With permission.)

Many of the transformation-toughened zirconia materials are sensitive to thermal aging. Mg–PSZ and Ca–PSZ exhibit changes in precipitate size and grain size at temperatures roughly above 1000°C. This results in an increase in the monoclinic content, a decrease in the tetragonal content, and a resulting reduction in strength and toughness. For example, in one study the strength loss of four Mg–PSZ materials ranged from 30 to 70% due to 1000-h aging in air at 1000°C. A Y–TZP material in the same study decreased in strength only 7%. Another study showed that 50% of the tetragonal precipitates transformed to monoclinic during 250-h aging at 1000°C and 100% transformed during 200-h aging at 1100°C. Less than 10% monoclinic conversion occurred during the same aging trials for Y–TZP. Even after aging for 250-h at 1300°C, less than 40% conversion had occurred.

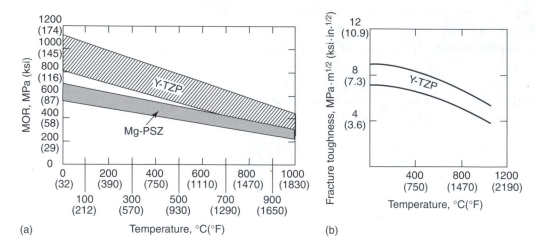

FIGURE 20.19   Strength and toughness of transformation-toughened zirconia vs. temperature. (a) Strength curve, (Based on data from Hecht, N.L., Jang, S.D., and McCullum, D.E., University of Dayton Research Institute, Ohio.) (b) Toughness curve. (Based on data from Oda, I. and Yamamoto, N., NGK Insulators, Nagoya, Japan.)

The above examples illustrate that Y–TZP has better high-temperature stability than Mg–PSZ. However, other studies have shown that some Y–TZP materials are unstable at lower temperatures. Claussen[19] describes an example in which the strength of a Y–TZP material dropped from 900 to 250 MPa after exposure to 300°C for 500 hr. Sato and Shimada[35] showed that the strength loss resulted from conversion of tetragonal grains to monoclinic during low-temperature aging and that the conversion was enhanced by the presence of water vapor. Grain size was also a factor. Material with a grain size of 1.0 μm degraded, whereas material with a grain size of 0.4 or 0.2 μm did not degrade. Addition of Al_2O_3 also restrained strength loss during low-temperature aging. No strength loss was measured after 1000-h low-temperature aging for Y–TZP containing 5 to 30 wt% Al_2O_3.[36]

20.2.2.2   Other Transformation-Toughened Ceramics

As discussed earlier in this chapter, very fine particles of ZrO_2 can be added to other materials and result in transformation toughening of these materials. Some examples are listed in Table 20.6. Especially good results have been achieved with Al_2O_3 as the matrix. Strength greater than 1000 MPa and toughness greater than 10 MPa · $m^{1/2}$ have been reported for Al_2O_3 containing only 16 vol% ZrO_2. In addition, the resistance to thermal-shock damage and to slow crack growth are increased for zirconia-toughened alumina (ZTA). Rice[37] reported that Al_2O_3 with 14 vol% ZrO_2 had a critical ΔT for thermal-shock damage of about 700 to 900°C, compared to 300°C for Al_2O_3. Furthermore, the strength dropped from 700 to about 500 MPa for the Al_2O_3–14% ZrO_2 compared to a drop from 320 MPa to less than 150 MPa for Al_2O_3 without ZrO_2 addition.

20.2.3   Particulate-Reinforced Ceramics

Addition of a dispersion of particles of a material that does not react with the matrix material generally increases toughness by crack deflection. If the particles are irregular in shape or much larger in grain size than the matrix, some bridging can occur. If the particles are significantly different in thermal expansion coefficient than the matrix, some toughening by microcrack formation can occur.

The presence of nonreactive second-phase particles generally inhibits densification unless a liquid phase is present. Full densification can generally be achieved by hot pressing or HIP and in some cases by conventional sintering. Fine grain size generally results. Table 20.7 lists some examples of

TABLE 20.6

Examples of Property Data for Other Ceramics Toughened by Additions of Zirconia

Matrix	Vol% ZrO_2 (additive)	Tetragonal ZrO_2 at RT[a], %	K_{Ic} (matrix), $MPa \cdot m^{1/2}$	Test Technique[a]	Strength (matrix), MPa
Al_2O_3	15	20	9.6 (5.4)	NB-a	480 (550)
Al_2O_3	16	100	15.0 (5.2)	NB-a	1200 (400)
Al_2O_3	7.5 (0.12 wt% MgO)	95	6.8 (5.2)	NB-b	600 (460)
Al_2O_3	11.5	70	6.5 (4.2)	DCB	750 (480)
Al_2O_3	29.5 (2 mol% Y_2O_3)	100	7.4 (4.9)	ICL	950 (600)
ZnO	36	0	3.3 (1.6)	Nb-b	n.d.
ThO_2	13	n.d.	3.5 (1.5)	ICL	n.d.
Spinel	17.5	50	4.6 (2)[b]	NB-b	500 (200)[b]
Mullite	23	30	4.5 (3)[b]	NB-a	400 (270)[b]
Si_3N_4	15 (10 eq% Al)	30 cub.	8.5 (5.6)	NB-a	950 (670)
Si_3N_4	10 (1 eq% Al)	30 cub.	7.2 (5.5)	NB-b	700 (610)
SiC	15 (ZrO_2)	0	5.9 (4.7)	ICL	n.d.

[a] Rest monoclinic.

[b] Assumed.

[c] NB: notched beam, a: as-notched, b: annealed at 1250°C (2280°F), DCB: double-cantilever beam, ICL: identation crack length.

Source: From Claussen, N., Transformation toughened ceramics, in *Ceramics in Advanced Energy Technologies*, Krockel H., et al., Eds., Reidel, D., Dordrecht, 1984, pp. 51–86.

TABLE 20.7

Examples of Particulate-Reinforced Ceramics

Material	Flexural Strength (MPa)	Fracture Toughness $MPa \cdot m^{1/2}$	Ref.
Baseline hot-pressed AY6 Si_3N_4	773	4.6	38
AY6 + 10 vol% 8 μm SiC	950	5.0	38
AY6 + 20 vol% 8 μm SiC	763	4.8	38
AY6 + 30 vol% 8 μm SiC	885	4.9	38
Baseline hot-pressed Si_3N_4	760	4.2	39
Si_3N_4 + 30 vol% 12 μm SiC	520	6.7	39
Baseline hot-pressed Si_3N_4	–	5.0	40
Si_3N_4 + 20 vol% 40 μm SiC	–	8.0	40
Baseline Al_2O_3	420	4.0	41
Al_2O_3–5 vol% TiB_2	650	6.5	41
Sintered Al_2O_3–30 wt% TiC	480	4.4	42
Hot-pressed Al_2O_3–30 wt% TiC	583	4.5	42
Sinter-HIP Al_2O_3–30 wt% TiC	638	4.4	42
Al_2O_3–10 vol% 30 μm Ba-mica	–	8.6	43
Baseline SiC	360	≈3	44
Sintered SiC–16 vol% TiB_2	478	6.8–8.9	44

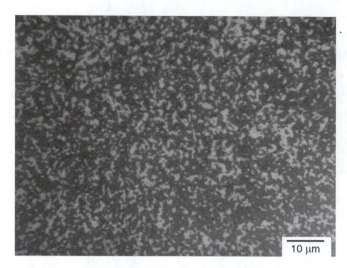

10 μm

FIGURE 20.20   Reflected light optical photomicrograph of a pressureless sintered post-HIPed Al$_2$O$_3$–30 wt% TiC particulate composite. (Courtesy of R. Cutler, Ceramatec, Inc., Salt Lake City, UT.)

particulate-reinforced ceramics. Figure 20.20 shows the microstructure of a Al$_2$O$_3$–30 wt% TiC material successfully densified by a combination of pressureless sintering and unencapsulated HIP.[42]

20.2.4 WHISKER-REINFORCED CERAMICS

Whiskers are usually single crystals that have grown preferentially along a specific crystal axis. Whiskers typically range in size from about 0.5 to 10 μm in diameter and a few microns to a few centimeters in length. A note of caution is in order. The smaller whiskers can become lodged in the lungs and represent a health hazard. Thus, they should be handled with proper precautions such as a hood, respirator, and careful cleaning of work areas.

Whiskers generally grow under vapor-solid or vapor-liquid-solid conditions that result in a small defect size in the whiskers. For example, defect size in some SiC whiskers was estimated to range from 0.1 to 0.4 μm.[45] Defects identified were voids, inclusions, and surface roughness. Such small defect size results in very high strength. Tiegs et al.[45] calculated strengths in the range of 9 to 17 GPa for various SiC whiskers. Lowden[46] listed strength and other properties for a variety of ceramic whiskers. These are included in Table 20.8, along with comparison data for some matrix materials to which whiskers have been added. Figure 20.21 illustrates various SiC whiskers as viewed by SEM.

Whiskers have been added to a wide variety of ceramics in an effort to achieve increased toughness. In most cases, hot pressing has been required to achieve near-theoretical density of the matrix. The whiskers are fully dense and rigid and thus do not shrink during sintering. As a result, they inhibit the material transport of the matrix powder necessary for sintering to reach completion. Hot pressing provides the force to reorient whiskers and to force the matrix material to move large enough distances to permit full densification of the matrix around the whiskers. The reorientation of the whiskers during hot pressing produces a microstructure with many of the whiskers aligned perpendicular to the direction of hot pressing. This microstructure results in anisotropy in the strength and toughness. The strength and toughness are strongly enhanced perpendicular to the length of the fibers, but decrease to values not much better than the matrix parallel to the whiskers. Increased randomness of whisker orientation and decreased anisotropy of strength and toughness have been demonstrated with conventional sintering with <15% whisker addition, with hot isostatic pressing, and with liquid phase sintering.

TABLE 20.8

Comparison of the Properties of Some Whisker Materials and Potential Matrix Materials

	Density (g/cm³)	Tensile Strength (MPa)	Young's Modulus (GPa)	Thermal Expansion (10⁻⁶/°C)	Whisker Diameter (μm)
Whiskers					
SiC whiskers	3.2	21,000	840	4.8	0.1–0.6
VLS SiC whiskers	3.2	8,400	580	5.5	6
Alumina	3.9	21,000	430	8.6	
Beryllia	2.9	13,000	345	9.5	
Boron carbide	2.5	14,000	480	4.5	
Graphite	1.7	21,000	700	2.0	
Silicon nitride	3.2	14,000	380	2.8	
Matrix Glasses and Ceramics					
Borosilicate glass	2.3	100	60	3.5	
Soda-lime glass	2.5	100	60	8.9	
Lithium aluminosilicate glass-ceramic	2.0	150	100	1.5	
Magnesium aluminosilicate glass-ceramic	2.7	170	120	2.5–5.5	
Silica glass	2.3	120	70	0.5–8.6	
Mullite	2.8	185	145	5.3	
Si_3N_4	3.2	410	310	2.8	
SiC	3.2	310	440	4.8	
CVD SiC	3.2	1,200	440	5.5	
Al_2O_3	3.9	300	400	8.5	
ZrO_2	5.7	140	250	7.6	
HfO_2	9.7	69	565	5.9	
ZrC	6.8	210	480	6.7	
B_4C	2.5	490	450	4.5	
TiB_2	4.5	1,000	570	8.1	

Source: Adapted from Lowden, R.A., *Characterization and Control of the Fiber-Matrix Interface in Ceramic Matrix Composites*, ORNL Rept. TM-11039. March 1989.

Major efforts have been conducted with SiC whiskers in Al_2O_3, Si_3N_4 and $MoSi_2$. Table 20.9 summarizes some of the results.[47–54] Specific examples, key properties, toughening mechanisms, and important applications are discussed in the following sections.

20.2.5 Al_2O_3 REINFORCED WITH SiC WHISKERS

Much of the exploratory work of whisker toughening of ceramics was conducted with SiC whiskers in Al_2O_3.[55,56] This was first made possible when a source of SiC whiskers became available. The story of the development and eventual commercialization of these SiC whiskers is quite interesting. Whiskers were first produced on a research basis at the University of Utah by J.G. Lee and Dr. Ivan B. Cutler.* They were studying ways of utilizing solid waste materials, specifically rice hulls. Chemical analysis showed that rice hulls contained about 14% SiO_2, in addition to the normal carbon content one would expect for an organic material. The University of Utah team hypothesized that they could

---

* Lee, J.G. and Cutler, I.B., Formation of SiC from rice hulls, *Am. Ceram. Soc. Bull.*, 54(2), 195–198, 1975.

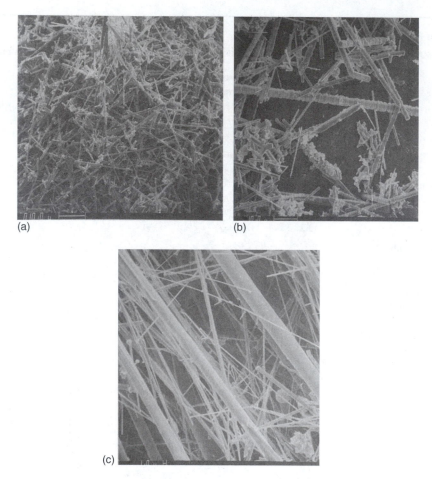

FIGURE 20.21 Examples of SiC whiskers showing variations in size and smoothness. The Arco (presently Advanced Composite Materials Corp.) and Tateho whiskers in (a) and (b) were synthesized by vapor-solid techniques. The Los Alamos National Labs (LANL) whiskers in (c) were synthesized by a vapor-liquid-solid process that used molten metal catalyst balls to transfer reactants from the vapor phase to the growing whisker. Some of these metal spheres can be seen still attached to the end of some of the whiskers.

heat the rice hulls to around 1500°C in a nonoxidizing atmosphere and produce SiC powder. When they conducted their experiments to evaluate their hypothesis, they got SiC, but much of it was in the form of whiskers rather than particles. This technology was licensed and refined and ultimately ended up with Arco Chemical (presently Advanced Composite Materials). Individuals at Arco Chemical collaborated with researchers at Oak Ridge National Labs (ORNL) to evaluate the whiskers in Al_2O_3 and other matrix materials.

The Arco Chemical and ORNL efforts demonstrated that substantial improvements could be achieved in Al_2O_3 with SiC whisker reinforcement. As shown in Table 20.9, the room-temperature toughness of Al_2O_3 was doubled and the strength increased by over 50%. These improvements persisted to elevated temperatures, as illustrated in Figure 20.22. The presence of the whiskers also resulted in improved thermal-shock resistance and Weibull modulus. Baseline Al_2O_3 showed a severe drop in strength after being heated to 200°C and quenched in water. Whisker-reinforced Al_2O_3 survived quenches from above 900°C.[57] Baseline Al_2O_3 had a Weibull modulus of 4.6. Al_2O_3 with 15% SiC whiskers had a Weibull modulus of 13.4 and Al_2O_3 with 25% SiC whiskers had a Weibull modulus of 22.5.[47] Which material would you expect to have the highest reliability? Whisker-reinforced Al_2O_3 also exhibited a lower creep rate than baseline Al_2O_3.

TABLE 20.9
Examples of Whisker-Reinforced Ceramics

Material	Flexural Strength (MPa)	Fracture Toughness MPa · m$^{1/2}$	Ref.
Hot-pressed Al$_2$O$_3$	480	3.8	47
Hot-pressed Al$_2$O$_3$–15 vol% SiC whiskers[a]	570	5.8	47
Hot-pressed Al$_2$O$_3$–30 vol% SiC whiskers[a]	660	6.9	47
Hot-pressed Al$_2$O$_3$	–	4.5	48
Hot-pressed Al$_2$O$_3$–10 vol% SiC whiskers[a]	420	7	48
Hot-pressed Al$_2$O$_3$–30 vol% SiC whiskers[a]	660	8.6	48
Sinter-HIP Al$_2$O$_3$–20 vol% SiC whiskers[b]	539	7.8	49
Hot-pressed Si$_3$N$_4$	780	4.7	50
Hot-pressed Si$_3$N$_4$–30 vol% SiC whiskers[a]	970	6.4	50
Hot-pressed Si$_3$N$_4$	660	7.1	51
Hot-pressed Si$_3$N$_4$–30 vol% SiC whiskers[c]	450	10.5	51
Reaction-bonded/hot-pressed Si$_3$N$_4$–5 wt% SiC whiskers[d]	871	10.4	52
Hot-pressed Si$_3$N$_4$	649	3.9	53
Hot-pressed Si$_3$N$_4$–30 vol% Si$_2$N$_4$ whiskers[e]	659	8.6	53
Hot-pressed Si$_3$N$_4$–30 vol% BN-coated Si$_3$N$_4$ whiskers[e]	428	9.2	53
Hot-pressed MoSi$_2$	150	5.3	54
Hot-pressed MoSi$_2$–20 vol% SiC whiskers[e]	310	8.2	54
Hot-pressed MoSi$_2$–20 vol% SiC whiskers[f]	283	6.8	54

[a] Arco Chemical Co., Greer, S.C. (now Advanced Composite Materials), 0.1 to 1 μm in diameter.

[b] Tokai Carbon, Tokyo, Japan, 25 μm long, 0.5 μm in diameter.

[c] Los Alamos National Lab., Los Alamos, N.M., 100 to 200 μm long, 5 μm in diameter.

[d] American Matrix, Knoxville, Tenn., 1.5 μm in diameter.

[e] Ube Industries Ltd., Ube City, Japan, 1.0 μm in diameter, 30 μm long.

[f] XPW2 SiC whiskers from J.M. Huber Corp., Borger, TX, 1 to 5 μm long, 0.1 μm in diameter.

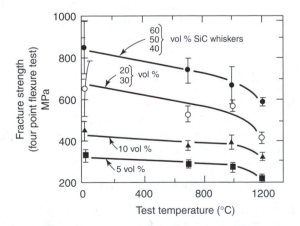

FIGURE 20.22 Strength vs. temperature for hot-pressed composites of Al$_2$O$_3$ containing various-volume fractions of SiC whiskers. (From Tiegs, T.N. and Becher, P.F., Alumina-SiC whisker composites, in *Proceedings of the 23rd Auto. Tech. Devt. Contractors Coord. Meeting*, SAE P-165, 1985.)

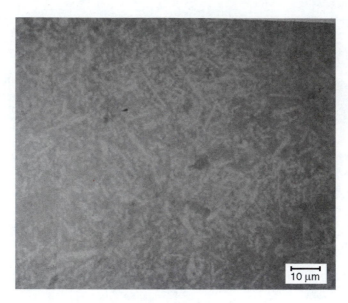

FIGURE 20.23   Photo taken through a reflected-light optical microscope of a polished section of a hot-pressed Al_2O_3–SiC whisker composite cutting tool insert. The cutting tool sample was from the Greenleaf Corp. The sample was polished and photographed at Ceramatec.

Further examination of Table 20.9 and the references from which the data were extracted indicates that different composite properties result for different whisker sources and composite processing procedures. Whiskers from different sources have significant variations. Some are relatively smooth. Others have growth steps and look like bamboo or in severe cases like they have been threaded. Some consist of isolated whiskers. Others are reticulated and have whiskers growing out of the side of other whiskers. Even the composition of "SiC" whiskers varies. Some have a carbon-rich surface, some a surface composition similar to "black glass" (Si–O–C). Different whiskers contain different levels of inclusions, voids, or other defects and thus have different strength.

The differences in whisker characteristics have a pronounced effect on processing and properties of the composites. Surface chemistry affects the degree of dispersion of the whiskers with the Al_2O_3. Reticulation also affects dispersion by forming mechanically linked clusters (agglomerates) of whiskers. Surface chemistry affects the interface bonding between the whiskers and matrix. This plus the presence of growth steps influences the degree of whisker pullout, bridging, and stress delocalization. Defects in whiskers limit the strength of the whiskers. As a result, present SiC whisker-reinforced materials are not optimum. Toughness values greater than $15\,MPa \cdot m^{1/2}$ are possible with further development.[45]

Figure 20.23 shows the microstructure of a typical Al_2O_3 matrix and SiC whisker composite. Figure 20.24 illustrates some examples of components fabricated from whisker-Al_2O_3 matrix composites. A major success story has been the use of hot-pressed whisker-Al_2O_3 material for cutting tool inserts for high-speed turning and milling of some metals.

20.2.6 Si_3N_4 Reinforced with SiC and Si_3N_4 Whiskers

Table 20.9 includes some strength and toughness values for hot-pressed Si_3N_4 containing SiC or Si_3N_4 whiskers. In general, the room-temperature values are no better than what has been achieved with self-reinforced Si_3N_4 prepared by overpressure sintering. High-temperature strength and creep resistance may benefit by whisker reinforcement, but further study is necessary. Buljan et al.[50] reported $970\,MPa/6.4\,MPa \cdot m^{1/2}$ strength/toughness for 30 vol% SiC whiskers at room temperature, 820/7.5 at 1000°C, and 590/7.7 at 1200°C.

FIGURE 20.24 Examples of components fabricated of Al_2O_3–SiC whisker composite compositions. (Courtesy of Advanced Composite Materials Corporation, Greer, SC.)

20.2.7 MoSi₂ AND MoSi₂-WSi₂ REINFORCED WITH SiC WHISKERS

$MoSi_2$ is an intermetallic material that has a melting temperature of about 2030°C and excellent oxidation resistance. It is used as a heating element for high-temperature furnaces and can operate above 1700°C in air. $MoSi_2$ has potential for structural use, but has two primary problems: (1) low toughness at room temperature and (2) high rate of creep at elevated temperature. The source of both types of behavior is a brittle-ductile transition at about 1000°C. Below this transition, $MoSi_2$ is brittle and has low toughness (about 5.3 MPa · $m^{1/2}$). Above this transition, $MoSi_2$ becomes increasingly ductile. For example, at 1200°C $MoSi_2$ has a yield stress of 139 MPa that drops to 19 MPa at 1400°C and 8 MPa at 1500°C.[58] Addition fo 20 vol% large-diameter ($\approx$0.5 µm) SiC whiskers increases the room-temperature toughness to 8.2 MPa · $m^{1/2}$, the 1200°C yield strength to 247 MPa, and the 1400°C yield strength to 43 MPa. Addition of 20 vol% small-diameter ($\approx$0.5 µm) SiC whiskers increases the room temperature toughness to 6.8 MPa, the 1200°C yield strength to 386 MPa, the 1400°C yield strength to 89.5 MPa, and the 1500°C yield strength to over 50 MPa.[58] Addition of 50 vol% WSi_2 into solid solution in the $MoSi_2$ plus 20 vol% small-diameter whiskers increases the 1200°C yield strength to 590 MPa, the 1400°C yield strength to 125 MPa, and the 1500°C yield strength to about 70 MPa.[58]

20.2.8 FIBER-REINFORCED CERAMICS

The prior sections have shown that dispersions of particles or whiskers can increase toughness and strength of ceramics. This generally results in a material of increased reliability and reduced flaw sensitivity. However, the strain-to-failure is still about the same as for a monolithic ceramic, so the material continues to fracture in a brittle, catastrophic mode. Reinforcing a ceramic with long fibers can increase the distance over which a toughening mechanism acts and lead to enough strain-to-failure that fracture is no longer catastrophic.

A variety of high strength fibers have become commercially available during the past 20 to 30 years. To achieve high strength, the defect size in the fiber must be very small. The small size has been achieved by several processing approaches: (1) drawing a homogeneous glass, (2) drawing a homogeneous polymer and decomposing (pyrolyzing) the polymer to yield a ceramic composition fiber, (3) drawing a fiber using a sol–gel process and heat treating the fiber to convert it chemically to an oxide, and (4) depositing a material such as boron or silicon carbide onto a small diameter carbon or tungsten fiber by chemical vapor deposition (CVD). Strength over 4000 MPa and elastic

TABLE 20.10

Typical Properties of Glass Fibers

Designation	Composition	Strength (MPa)	Elastic Modulus (GPa)	Density (g/cm³)	Thermal Expansion (ppm/K)	Elongation (%)	Diameter (μm)
E-glass	Calcium aluminoborosilicate	3445	81	2.62	5.4	4.9	3–20
S-glass	Magnesium aluminosilicate	4585	89	2.50	1.6	5.7	3–20
C-glass	Soda lime borosilicate	3310	–	2.56	6.3	4.8	3–20
Astroquartz II	99.95% silica	3450	69	2.2	0.54	–	9

Source: Adapted from Miller, D.W., Glass Fibers, in *Engineering Materials Handbook, Vol. 1: Composites*, ASM Int., Materials Park, OH, 1987, pp. 45–48 and Johnson, D.D. and Sowman, H.G., Ceramic Fibers, *ibid.*, pp. 60–65.

modulus of 500 GPa have been demonstrated with various materials. The following paragraphs briefly review the properties of some fiber materials.[59]

20.2.8.1 Glass Fibers

Glass fibers are produced by extruding glass at high temperature through a die and then rapidly drawing the glass into a fiber 3 to 20 μm in diameter. Multiple fibers are drawn simultaneously, typically through a platinum alloy bushing containing 204 orifices. An organic coating is applied to each fiber to provide abrasion protection (surface scratches are defects that reduce the strength of the fiber) and in some cases to act as a coupling agent later in the fabrication of a polymer matrix composite. This coating is referred to as a "size." Following coating with the size, the fibers are gathered into strands typically ranging from 500 to 6000 filaments. These are generally referred to as "tow" or "roving."

Table 20.10 summarizes the compositions and properties of several important glass fibers.[60,61] Note that the strength is very high, but the elastic modulus is relatively low. The elastic modulus is still high compared to polymer matrix materials but low compared to ceramic matrix materials. Because of this, plus the fact that the glass typically reacts with a ceramic at the densification temperature required for the ceramic, glass fibers are not an important source of reinforcement for ceramic matrix composites. However, glass fibers could be compatible with chemically bonded ceramics (cements, adhesives) and provide some benefits.

E-glass is a general-purpose fiber and is especially useful when strength and high electrical resistivity are required. S-glass was developed for high tensile strength, and C-glass was developed for high corrosion resistance in acidic environments.

20.2.8.2 Carbon Fibers

Carbon fibers are produced from polymer precursors. A variety of precursors have been used, including polyacrylonitrile (PAN), rayon, and pitch. The viscosity of the polymer is controlled by solvent or temperature to allow the polymer to be pulled through an orifice. The polymer is cured as it exits the orifice. The fiber is then stabilized by an oxidation exposure at an intermediate temperature to allow it to subsequently be taken to higher temperatures without deforming. As the temperature is progressively increased, the polymer decomposes. Oxygen, nitrogen, and hydrogen are deleted from the structure and carbon–carbon bonds progressively form a graphene-layer network.[62] The strength and modulus can be controlled by the extent of graphene formation and network connection, which is primarily controlled by the peak temperature and time at temperature.[63] Fibers treated at about 1500°C have relatively small crystallites and exhibit moderate elastic modulus. Fibers treated above 2000°C have larger crystallites and exhibit higher elastic modulus. The crystallites in both categories of carbon fibers are oriented along the length of the fibers. This results in high strength and modulus along the length of the fiber but low strength and modulus transverse to the fiber (radial direction).

Several important characteristics can be noted for carbon fibers by examining the data in Table 20.11. One is the anisotropy in the properties in the axial (parallel to the length of the fibers) and transverse (perpendicular to the length of the fibers) directions. The modulus is much higher in the axial direction due to the strong C–C bonds in this direction. The modulus in the transverse direction is quite low due to the presence of van der Waals bonds rather than C–C bonds. This also affects the thermal expansion and thermal conduction properties. The thermal expansion is actually negative in the axial direction and is moderate to high and positive in the transverse direction. The thermal conduction is high axially and low in the transverse direction.

Another characteristic to note is the difference between the PAN-based fibers and the pitch-based. PAN-based fibers can have very high strength but intermediate modulus. Pitch-based fibers have lower strength but can have very high elastic modulus (725 GPa). Pitch-based fibers with a strength of 3100 MPa have recently been developed.

Two other characteristics to note from Table 20.11 are the density and the elongation. The specific gravity and bulk density of the carbon fibers are very low. Linked with low specific gravity polymers, very low bulk density composites are possible with very high specific strength and specific modulus. This has led to broad use in aerospace and sports applications. The elongation for carbon fibers is high compared to that of monolithic ceramic materials but low compared to that of glass fibers.

From a composite fabrication perspective, carbon fiber technology is relatively mature. Many fiber options are available so that the desired composite can be engineered. Carbon fibers have been successfully used for reinforcement of glass and ceramic matrices. The carbon is relatively nonreactive with the ceramic and debonds adequately during stressing to allow maximum toughening by bridging and pull-out. The big deficiency with carbon fibers in ceramic matrices is temperature capability in an oxidizing atmosphere. Carbon fibers oxidize readily in the 600 to 800°C range. Ceramic composites for use at and above these temperatures must either have the fibers completely embedded in the matrix with no direct exposure to the atmosphere or have a protective coating. Carbon–carbon composites for the space shuttle are a good example. They have the original surface converted to SiC, followed by a tetraorthosilicate overlayer as well as an internal oxidation inhibitor.

20.2.8.3 Oxide Fibers

Table 20.12 summarizes the properties of some oxide continuous filament fibers.[61,64–67] The Nextel and Sumitomo fibers are fabricated using a sol–gel process. The precursor is an organoalumino compound such as an alkyl or alkoxide of aluminum [$M(OR)_n$, where M is aluminum, R is an organic compound, and $n = 3$, the valence of aluminum]. For example, 3M has reported that some of the Nextel fibers use [$Al(OH)_2(OOCCH_3) \cdot (1/3)H_3BO_3$] as a precursor.[65] The viscosity of the precursor is adjusted by heat-induced polymerization plus solvents or evaporation to the range of about 100 to 150 Pa · s. This is extruded through a spinneret containing holes of the desired size (around 100 μm). The fiber is allowed to gel as it exits the spinneret. This precursor fiber is then heat treated at temperatures up to about 1400°C to remove water, pyrolyze the polymer, and convert it to a fine-grained oxide ceramic. The fiber shrinks in diameter to about 10 μm during this pyrolysis and calcination and sintering step. The resulting fibers are very smooth and have a uniform microstructure with very small defect size.

A wide variety of oxide compositions can be achieved by the sol–gel approach. Most of those available commercially are alumina- or Mullite-based or are aluminoborosilicates.

Another technique for producing oxide fibers is to prepare a slurry of alumina powder with organic additives to control the viscosity within a range in which dry spinning can be accomplished. This approach was developed by DuPont for fabrication of their fiber FP and PRD-166 fiber.[64] Fiber

TABLE 20.11
Typical Properties of Carbon Fibers

Fiber Designation	Axial Tensile Strength (MPa)	Axial Tensile Modulus (GPa)	Transverse Tensile Modulus (GPa)	Density (g/cm³)	Axial Thermal Expansion (ppm)	Transverse Thermal Expansion (ppm)	Elongation (%)	Diameter (μm)
PAN Precursor Fibers								
Low modulus	3300	230	40	1.76	−0.7	10	1.4	7–8
Intermediate modulus	5300–6800	270	21	1.8	−0.5	7	2.0–2.5	6
High modulus	5500	320	–	1.9	–	–	0.9	7
Mesophase Pitch Precursor Fibers								
Low modulus	1400	160	–	1.9	–	–	0.9	11
High modulus	1700	380	21	2.0	−0.9	7.8	0.4	10
Very high modulus	2200	725	–	2.15	−1.6	–	0.3	10
Rayon precursor fibers	1000	41	–	1.6	–	–	2.5	8.5
Isotropic pitch precursor	700	55	–	1.6	–	–	1.4	10

Source: Adapted from Diefendorf, R.J., Carbon/Graphite Fibers, in *Engineering Materials Handbook, Vol. 1: Composites*, ASM Int., Materials Park, OH, 1987, pp. 48–53.

TABLE 20.12
Compilation of Data for Oxide Continuous Filament Ceramic Fibers

Manufacturer	Designation	Composition (wt%)	Use Temperature (°C)	Tensile Strength (MPa)	Tensile Modulus (GPa)	Density (g/cm³)	Diameter (μm)	Thermal Expansion (ppm/°C)
3 M	Nextel 312	62 Alumina, 24 silica, 14 boria	1200	1720	152	2.7–2.9	11	3.5
	Nextel 440	70 Alumina, 28 silica, 2 boria	1430	2060	186	3.05	11	–
	Nextel 480	70 Alumina, 28 silica, 2 boria	1430	2070	220	3.05	11	–
	Nextel 550	73 Alumina, 27 silica	–	2240	193	3.03	11	–
	Nextel 610	>99 Alumina, 0.2–0.3 silica, 0.4–0.7 Fe_2O_3	–	1900	370	3.75	11	–
	Nextel 720	85 Alumina, 15 silica	–	2070	260	3.4	11	6
Sumitomo	Sumica	85 Alumina, 15 silica	1250	2600	250	3.2	8	8.8
Mitsui Mining	ALMAX	>99.5 Alumina	–	1760	324	3.6	10	7
Saphikon	Single-crystal	100 Alumina	–	3100	380	3.8	70–250	–
DuPont	FP	Alumina	–	1380	380	–	20	–
	PRD-166	Alumina-zirconia	–	2108	380	–	20	–

Source: Johnson, D.D. and Sowman, H.G., Ceramic Fibers, in *Engineering Materials Handbook, Vol. 1: Composites*, ASM Int, Materials Park, OH, 1987, pp. 60–65; Diefendorf, R.J., Carbon/Graphite Fibers, *ibid.*, pp. 48–53; Donnet, J.B. and Bansal, R.C., *Carbon Fibers*, Marcel Dekker, New York, 1984; Chawla, K.K., *Ceramic Matrix Composites*, Chapman and Hall, London, 1993; Sowman, H.G., in *Sol–Gel Technology*, Noyes, Park Ridge, NJ, 1988, p. 162; Saitow, Y., et al., SAMPE Annual Meeting, 1992; and Anon., *High Performance Synthetic Fibers for Composites*, NMAB-458, National Academy Press, Washington, DC, 1992.

FP was 98% theoretical dense α-alumina. PRD-166 fiber was alumina containing a dispersion of 15 to 20% very fine (0.1 μm) yttria-stabilized zirconium oxide particles. The purpose of the zirconia was to inhibit grain growth and reduce high temperature creep. These fibers are presently not commercially available. However, an alumina fiber designated Almax fabricated by a similar process is available from Mitsui Mining in Japan.[64,66]

The fibers spun using slurry and sol–gel precursors are all available in diameters in the 9 to 12 μm range as continuous multifilament yarn with a fine-grained microstructure. Long single-crystal alumina fibers are also commercially available. These range in diameter from about 70 to 250 μm and are single crystals grown from a melt. They are fabricated by Saphikon by the edge-defined film-fed growth (EFG) process. A molybdenum die is immersed in molten alumina. A capillary hole allows molten alumina to be wicked by surface tension to the top of the die. A seed crystal is touched to the surface of the melt at the top of the die and slowly pulled away to start single-crystal growth. Filaments can be grown at a rate of about 60 m/h. The cross section of the fiber can be controlled by the die design.

Oxide fibers have been used extensively woven into cloth as insulation and firewalls. They also have been fabricated into composites by chemical vapor infiltration and by sol infiltration. Examples are presented later. The major limitations of oxide fibers have been grain growth and creep at elevated temperatures and avoidance of interaction with the matrix during composite fabrication.

20.2.8.4 Nonoxide Fibers

A wide variety of nonoxide compositions have been synthesized into fibers.[67] Three primary approaches have been used: (1) chemical conversion of a precursor fiber, (2) chemical vapor deposition on a precursor fiber, and (3) thermal decomposition (pyrolysis) of a fiber spun from a polymer.

Chemical conversion of precursors has typically used either carbon fibers or boric oxide fibers. SiC, B_4C, Mo_2C, and NbC have been formed by reacting carbon fibers with a mixture of hydrogen and the chloride of the cation (such as boron trichloride). Boron carbide fibers have been fabricated with elastic moduli of 207 to 482 GPa and tensile strength of 2070 to 2760 MPa (300–400 ksi).[67] BN fibers have been synthesized by chemical conversion of melt spun boric oxide fibers using NH_3. The resulting fibers are 4 to 6 μm in diameter and have 2 to 3% elongation prior to fracture. However, the strength and modulus are relatively low, 345 to 862 MPa and 27.6 to 68.9 GPa, respectively.[67] The BN fibers are stable up to about 2500°C in an inert atmosphere but begin to oxidize at about 850°C in air.

The most widely used fibers for ceramic matrix composites have been SiC-based compositions synthesized either by chemical vapor deposition or by polymer pyrolysis. The properties of some of these fibers are summarized in Table 20.13. The fabrication and key characteristics of several of these fibers are discussed in the following paragraphs.

Textron SCS-6 Fibers.   The SCS-6 fibers consist of high purity β-SiC deposited on a 33 μm pitch-derived carbon fiber precoated with a thin layer of pyrolytic carbon.[68] The outer surface of the fiber is graded over about 4 μm from pure stoichiometric SiC internally to slightly carbon rich at the surface.[64] The overall diameter of each fiber is about 140 μm, resulting in fibers that are not very flexible. These fibers cannot be woven into a two-dimensional (2-D) cloth or wound around a small radius. Thus they are used as monofilaments, usually formed into unidirectional tapes by coating with particles of the matrix material and carefully laying up on a surface or winding on a large diameter mandrel. The tapes are then cut to the desired shape and stacked to form a 2-D architecture that is subsequently densified. An example of this type composite was shown earlier in Figure 20.3, clearly illustrating the cross section of the fibers.

The SCS-6 fibers are fabricated by passing the carbon fiber through a CVD reactor. The carbon fiber is heated resistively and passed through a mixture of hydrogen and chlorinated alkyl silanes. The silanes are decomposed when they contact the hot precursor fiber, resulting in deposition of SiC on the surface of the fiber.

As shown in Table 20.13, the SCS-6 fibers have high strength >3400 MPa and high modulus (400 GPa) at room temperature. They have been successfully integrated into a variety of matrix

TABLE 20.13
Compilation of Data for Nonoxide Continuous Filament Ceramic Fibers

Manufacturer	Designation	Composition	Tensile Strength (MPa)	Tensile Modulus (GPa)	Density (g/cm³)	Average Diameter (μm)	Thermal Expansion (ppm/°C)	Elongation (%)
Textron	SCS-6	CVD SiC on carbon core	>3400	428	3.0	140	4.1 at 500°C 5.1 at 1300°C	1.0
	SCS-9	CVD SiC on carbon core	>2800	315	2.8	75	4.1 at 500°C 5.1 at 1300°C	1.0
British Petroleum	Sigma	CVD SiC on W core	3600	400		100		
Nippon Carbon	Ceramic grade Nicalon	58 Si–31 C–11 O	2900	190	2.55	15	3.9	1.5
	Hi-Nicalon	63.7 Si–35.8 C–0.5 O	2800	270	2.74	14		1.0
Ube Industries	Tyranno Lox M	Si–Ti–C–O	320	180	2.4	8–12	3.1	1.75
Dow Corning	HPZ	Si–N–C–O	2800	180	2.4	10–12	4.1	1.5
	SiC	>95% β-SiC	260	450	3.1–3.2	8–10		
Carborundum	α-SiC	>95% α-SiC	1500–1750	400	3.15	18–20	4	0.38–0.44

FIGURE 20.25   SEM photomicrograph of as-received Nicalon fibers. (From Richerson, D.W., unpublished IR&D studies conducted at Garrett Turbine Engine Company, Phoenix, AZ, 1979–1982.)

materials and have demonstrated good stability in air at higher temperatures than those at which carbon or polymer derived SiC fibers are stable.

Textron SCS-9 Fibers.   The SCS-9 fibers are fabricated in the same fashion as the SCS-6 fibers, but to a smaller diameter of about 75 μm. These fibers have slightly lower strength >2800 MPa) and modulus (315 GPa) but are more flexible than the thicker SCS-6 fibers. Thus they provide greater versatility of fabrication. However, they still appear to be restricted to fabrication by monofilament approaches.

Nippon Carbon SiC-Based Fibers.   Nippon Carbon introduced a polymer-derived fiber called Nicalon in the mid-1970s.[69,70] The fiber was melt spun from polycarbosilane, cured in air or ozone, and pyrolyzed in vacuum to yield a Si–C–O–H composition. Nicalon fibers are illustrated in Figure 20.25[71] Nicalon had tensile strength of 2000 to 2500 MPa, tensile modulus of 180 to 200 GPa, and elongation of 1.5 to 2.0%. The fibers were 13 to 15 μm in diameter and available in multifilament yarn consisting of about 500 filaments per tow. This yarn was very flexible and could be processed like carbon yarn (weaving, filament winding, etc.).

Nicalon was used for many of the early studies of ceramic matrix composites. Its primary limitation was temperature capability. The microstructure was very fine grained (SiC grain radius of about 1.7 nm), and the composition was a combination of SiC, free carbon, and SiO_2. The composition and microstructure began to change above about 600°C, resulting in degradation in properties.[64]

The ceramic grade (HVR grade) Nicalon and Hi-Nicalon shown in Table 20.13 are improved grades of Nicalon. They retain their strength to higher temperature. Hi-Nicalon is a low oxygen fiber and has especially improved stability. Exposure for 10 h at 1400°C in argon resulted in no visual degradation, compared to substantial degradation in earlier Nicalon compositions and in Tyranno Lox M.[72] The low oxygen content of Hi-Nicalon was achieved by curing with an electron beam in an inert atmosphere rather than in air or ozone.

Ube Tyranno Lox M. Tyranno is also synthesized by pyrolysis of a polymer. The starting polymer is poly (titano carbosilane). The resulting fiber contains oxygen and about 1.5 to 4.0% titanium.[73] As shown in Table 20.13, the strength is high (3200 MPa at room temperature), but the modulus is relatively low, similar to the moduli of other polymer-derived SiC-based fibers.

Dow Corning HPZ. HPZ is a multifilament fiber derived from pyrolysis of hydridopolysilazane precursor fibers. The HPZ fibers are oblong in cross section, have an amorphous structure, and have a typical composition of 57 wt% Si, 28% N, 10% C, and 4% O.[74]

Dow Corning SiC. These monofilament fibers are also polymer-derived and are nearly stoichiometric SiC.[75] They have moderately high strength (2600 MPa) and high modulus (450 GPa). The material has potential for application at higher temperature than prior polymer-derived SiC-based fibers. The fibers retained 96% of their strength after aging in argon for 10 h at 1550°C. Fast fracture strength at 1400°C in air was 1550 MPa.

Carborundum Sintered α-SiC Fibers. The final fiber material listed in Table 20.13 is an α-SiC fiber developmented at Carborundum. A fiber preform was fabricated by extrusion of SiC powder mixed with suitable plasticizers. The plasticizers were carefully removed and the fiber was fired at high temperature. The fiber densified to near-theoretical density to result in a polycrystalline fiber approximately 18 to 20 µm in diameter. The strength of the fiber was determined by the grain size and the microstructural flaws. Fibers with average grain size of about 7 µm had a strength of about 1000 MPa; fibers with average grain size of 2 to 3 µm had strength approaching 1500 MPa.[76] The best fibers produced had strength of about 1500 to 1750 MPa. The polycrystalline fibers attracted interest because of their potential for low cost. However, they did not perform adequately in ceramic composites. The fiber surface was too rough because of the relatively large crystal size and inhibited debonding and pullout.

20.2.9 EXAMPLES OF CERAMIC MATRIX COMPOSITES REINFORCED WITH CERAMIC FIBERS

The following paragraphs describe some specific examples of the use of ceramic fibers to reinforce a variety of ceramic matrices.

20.2.9.1 Cement Matrix Composites

Perhaps the first widely used ceramic matrix composite in modern times was asbestos fibers added to hydraulic cement,[77] which was developed around 1900. Later, cements reinforced with fibers of steel, glass, nylon, polypropylene, carbon or Kevlar were developed.[78,79]

One of the earliest successful ceramic matrix composites consisted of glass-fiber-reinforced cement (GFRC). The glass fibers have higher elastic modulus and strength than the cement and are able to sustain increased tensile loading, even after cracks have initiated in the cement matrix. The fibers are not bonded to the matrix and thus provide toughening by crack deflection, microcracking, pullout, bridging, and stress delocalization.

The Kajima Corporation and Kurhea Chemical Industries of Tokyo, Japan have developed cement reinforced with 1 to 3% discontinuous carbon fibers.[80] The fibers alter the fracture behavior and instill high toughness and improved reliability. The load-deflection behavior for cement with 2% chopped carbon fibers is illustrated in Figure 20.26. Note the large increase in strength and strain to failure for the reinforced cement compared to the unreinforced cement. The reinforced cement has been successfully used for exteriors of buildings in Japan. It is 33% lighter than concrete, requires less steel framing, and takes less time to install. Considering all these factors, it is less expensive than conventional concrete.

20.2.9.2 Glass Matrix Composites

Work on glass matrix composites with discontinuous carbon fibers was reported by Sambell et al.[81] in the early 1970s. The composites were prepared by doctor blade tape casting followed by hot

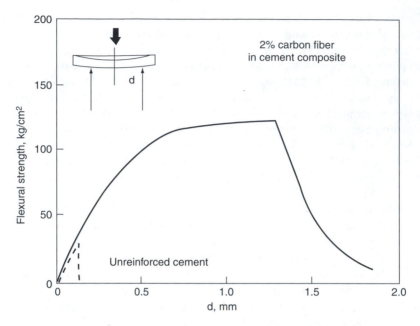

FIGURE 20.26 Comparison of the load-deflection behavior of unreinforced cement and a cement matrix composite containing 2% chopped carbon fibers. (From Prewo, K.M., Glass and ceramic matrix composites present and future, in *High Temperature/High Performance Composites*, MRS Symposium Proceedings, Vol. 120, Lemkey, F.D., et al., Eds., Materials Research Society, Pittsburgh, PA, 1988, pp. 145–156. With permission.)

pressing of stacked layers. The viscous behavior of the glass at the hot-pressing temperature (1150°C) allowed enough mobility of the matrix under moderate pressure to achieve full density without damaging the fibers. Furthermore, densification was at a low-enough temperature that the fibers were not chemically degraded or strongly bonded to the matrix. Substantial increase in toughness (as indicated by work of fracture measurements) was achieved for a Pyrex matrix. Pure Pyrex had a work of fracture value of less than $20 J/m^2$. This was increased to $125 J/m^2$ for 10 vol% aligned carbon fibers, to $280 J/m^2$ for 20 vol% fibers, and to $460 J/m^2$ for 30 vol% fibers.[81] Strength was decreased by 10% for 10 vol% fibers, but increased for 20 vol% and 30 vol% fibers by about 20 and 45%, respectively.[81]

Extensive work was conducted on glass matrix composites at United Technologies Research Center during the 1970s and 1980s.[82,83] Initial work was conducted with carbon fibers, followed by studies with SiC fibers as they became available. Some of the carbon fiber results are summarized in Table 20.14. The high volume fraction (50 to 60%) of fibers was achieved by a drum-winding technique. The tow of fiber filaments was passed through a slurry containing the matrix powder, a binder, and other additives selected to provide optimum wetting of the slurry onto each filament. The coated, wet fiber was then wound onto a drum to form a single-layer tape. This tape was dried, removed from the drum, cut into the desired shape, stacked in a graphite die, and hot-pressed.

Figure 20.27 shows the strength vs. temperature behavior of a composite of the HMS carbon fibers in 7740 borosilicate glass. Note that a strength of over 600 MPa is maintained to 600°C. The strength of unreinforced 7740 glass is typically below 100 MPa. The strength of the composite was affected by oxygen. Testing in air at 530°C resulted in a decrease in strength of 10 to 35%. Thus, the carbon fiber and glass composites are limited in use temperature.

Efforts in the late 1970s and early 1980s evaluated various SiC fibers in borosilicate and high silica glass matrices. Investigators hoped that the SiC fibers would result in composites with improved high-temperature capability. Both monofilament (Textron) and multifilament (Nicalon) fiber were evaluated. Both resulted in strength vs. temperature curves identical in shape to the one shown in Figure 20.27, but displaced vertically to slightly higher strength. A Textron monofilament

TABLE 20.14

Examples of Graphite Fiber-Reinforced 7740 Borosilicate Glass Composites Containing 50 to 60 Vol% Unidirectional Fibers

Fiber Type	Fiber Density (g/cm³)	Fiber Modulus (GPa)	Fiber Tensile Strength (MPa)	Composite Tensile Strength (MPa)	Composite Modulus (GPa)
Hercules HTS	1.66	257	2830	370	–
Hercules HMS	1.80	350	2700	689	200
Thornel 300s	1.75	228	2650	498	–
Celanese DG-102	1.95	530	1725	342	–

Source: Adapted from data in Prewo, K.M., Bacon, J.F., and Dicus, D.L., Graphite fiber reinforced glass matrix composites for aerospace applications, *SAMPE Proc.*, 24, 61–71, 1979.

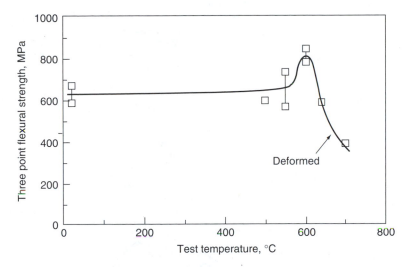

FIGURE 20.27 Flexural strength vs. temperature for HMS carbon fiber and 7740 glass composite. (From Prewo, K.M., Bacon, J.F., and Dicus, D.L., Graphite fiber reinforced glass matrix composites for aerospace applications, *SAMPE Proc.*, 24, 61–71, 1979. With permission.)

and 7740 glass composite with 65 vol% fibers had flexure strength of 830 MPa at 22°C, 930 MPa at 350°C, and 1240 MPa at 600°C.[83] A Nicalon® multifilament and 7740 glass composite with 50 vol% fibers had flexural strength of 800 MPa at 22°C and about 900 MPa at 600°C.[84] Toughness measurement by notched beam indicated K_{Ic} for the monofilament and glass composite of 18.9 MPa · m$^{1/2}$ and for the multifilament and glass composite of 11.5.

The use of SiC fibers in a glass matrix substantially improved the stability in an oxidizing atmosphere in the 500 to 600°C range. SiC fiber and 7740 glass exhibited no strength reduction after 500-h exposure at 540°C.

20.2.9.3 Glass-Ceramic Matrix Composites

The two primary limiting factors for glass matrix composites were softening of the glass above 1000°C and oxidation at lower temperatures. Glass ceramics in the lithium aluminosilicate (LAS) system were evaluated to potentially increase the temperature capability of the matrix. As discussed in an earlier chapter, glass ceramics are formed as a glass and heat-treated to yield a fine-grained crystalline microstructure. Whereas a glass exhibits softening behavior, the crystallized glass does not.

Early efforts with LAS produced composites with low strength and toughness that fractured in a brittle fashion. The matrix and fiber interacted during hot pressing to produce strong interfacial bonding. Cracks were not deflected by the fibers and no bridging and pullout occurred, leading to the brittle fracture behavior. TiO_2 was added to LAS as a nucleating agent and was judged to contribute to strong interfacial bonding. Other additives were tried. A combination of ZrO_2 and Nb_2O_5 resulted in weak interfacial bonding and excellent composite behavior. A thin layer of NbC at the fiber-matrix interface was determined to be the source of the weak interfacial bonding.[10] The LAS with ZrO_2 and Nb_2O_5 added is referred to as LAS III.

Figure 20.28 illustrates the flexural strength and toughness of Nicalon SiC filaments in LAS III, as measured in argon.[85] Strength, toughness, and temperature capability have all been dramatically

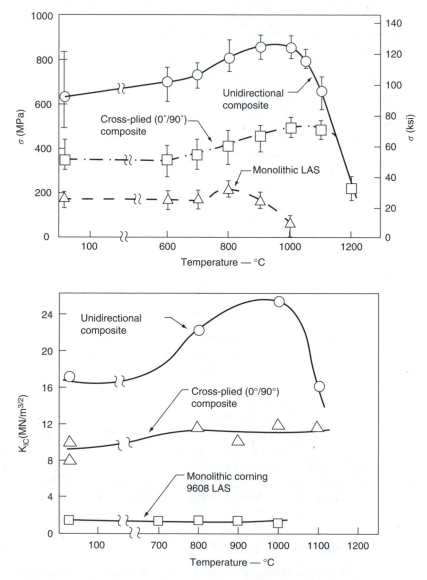

FIGURE 20.28 Flexural strength and toughness of Nicalon SiC multifilament yarn in LAS matrix. (From Brennan, J.J. and Prewo, K.M., *J. Mater. Sci.*, 17, 7371–7383, 1982. With permission.)

increased compared to unreinforced (monolithic) LAS. Figure 20.29 shows the fracture surface of a typical test sample. Note the high degree of fiber pullout. This indicates the presence of debonding, pullout, and bridging of fibers consistent with high composite toughness. Samples tested in air exhibited approximately 25% lower strength at 700°C and 50% lower strength at 1000°C compared to argon testing. This correlated with a 20-fold increase in interfacial bond strength at 1000°C compared to room temperature.[86]

20.2.9.4 SiC Matrix Composites Fabricated by Chemical Vapor Infiltration

Hot pressing is suitable for glass matrix and glass-ceramic matrix composites because glass is viscous at a reasonable temperature and can squeeze into the spaces between fibers. SiC does not melt and requires very high temperature to pressureless sinter or hot press (generally above 2000°C). This is too high a temperature for existing SiC-based fibers to withstand without severe degradation. Furthermore, hot pressing is not amenable to making parts of complex shape or large size. To circumvent these problems, SiC chemical vapor infiltration (CVI) into a SiC fiber preform was developed.

Early work in CVI was conducted by Societé Europeenne de Propulsion (SEP) in France using an isothermal process analogous to that used for fabrication of carbon–carbon composites.[87] In the isothermal process, fiber preforms are supported in a reaction chamber and heated to a desired temperature (around 1000°C). Reactant gases are passed through the chamber. These infiltrate into the fiber preform and react to form a pure SiC layer on each fiber. The process is continued until the pores narrow enough such that further infiltration either ceases or is very slow. Weeks of deposition time are required for a preform containing about 40 vol% fiber to be densified to about 70 to 80% theoretical density.

Initial attempts to CVI SiC into a SiC fiber preform did not result in composite fracture behavior, but instead resulted in brittle fracture. The bond interface between the SiC matrix and SiC fiber was too strong. SEP applied a thin CVD coating of pyrolytic graphite onto the SiC fiber prior to CVI SiC infiltration.[88] This resulted in a much weaker interface bond, allowed toughening by pullout and bridging, and yielded noncatastrophic composite fracture behavior. Early SEP material had

(a)

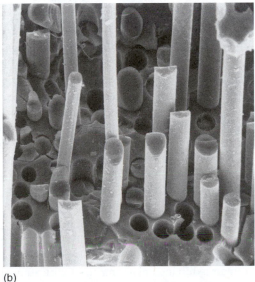

(b)

FIGURE 20.29 Fracture surface of continuous-fiber-reinforced ceramic matrix composite showing fiber pullout. (a) 15× overview and (b) 500× closeup. (Courtesy of J. Brennan, United Technologies Research Center.)

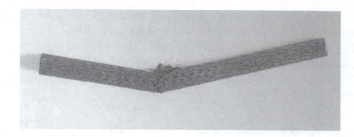

FIGURE 20.30 Test bar of SEP SiC/SiC composite fractured in bending. (From Richerson, D.W., unpublished IR&D studies conducted at Garrett Turbine Engine Company, Phoenix, AZ, 1979–1982.)

(a)

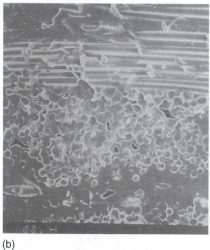

(b)

FIGURE 20.31 Fracture surface (as viewed by SEM) of early SEP SiC/SiC composite stress rupture tested at 1200°C and 138 MPa. Composite failed in a brittle fashion in less than 24 h. Note that no fiber pullout occurred. (From Richerson, D.W., unpublished IR&D studies conducted at Garrett Turbine Engine Company, Phoenix, AZ, 1979–1982.)

a flexure strength of 300–400 MPa[71] and toughness of about 25 MPa · m$^{1/2}$. A fractured sample of this early 1980s vintage material is shown in Figure 20.30.

Other tests during the early 1980s identified that the properties of CVI SiC/SiC with carbon interface degraded when the composite was exposed to an oxidizing environment roughly above 700°C, especially while under tensile stress. For example, a test bar in a flexural stress rupture test at 1200°C and 138 MPa failed in a brittle mode with no fiber pullout (as shown in Figure 20.31a) in less than 24 h.[71] Microstructure examination (Figure 20.31b) showed that the carbon layer had been removed and replaced with a silica layer that appeared to bond the fibers to the matrix and not allow debonding and pullout during fracture. Similar results were observed for a test bar exposed to static oxidation with no stress for 24 h, as shown in Figure 20.32.[71]

Many other studies during the 1980s and 1990s provided further understanding of the mechanisms and kinetics[89–94] of high temperature degradation of CVI SiC/SiC composites and other SiC-based composites. The design limits relevant to both stress and oxidation are now better understood, and material modifications have been developed to extend composite life and temperature capability in an oxidizing environment. Modifications include (1) use of a CVD SiC seal coat on the surface of the composite as an oxidation barrier, (2) additions to the matrix that inhibit oxidation, (3) alternate interface layers such as boron nitride, layered BN/SiC, and Si-doped BN, and (4) oxide "environmental barrier coatings."[95–101]

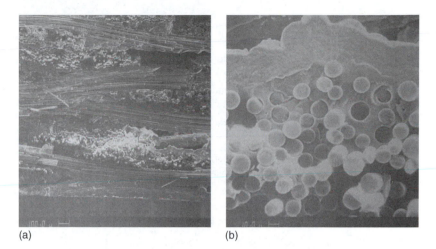

(a)   (b)

FIGURE 20.32 Fracture surface (as viewed by SEM) of early SEP SiC/SiC composite tested at room temperature following 24-h static oxidation exposure at 1000°C. Note the small degree of fiber pullout. (From Richerson, D.W., unpublished IR&D studies conducted at Garrett Turbine Engine Company, Phoenix, AZ, 1979–1982.)

The development of CVI SiC/SiC ceramic matrix composites is a good case study of identifying multiple engineering challenges, conducting studies to understand the challenges, and performing iterative research and development to overcome the difficult challenges. For example,[95] baseline CVI SiC/SiC composite fabricated by DuPont Lanxide Composites (currently GE Power Systems Composites, LLC) under a SEP license failed in less than one hour in a stress rupture test at 100 MPa and 1000°C. By 1993 an improved material with Hi-Nicalon fibers failed in about 3.3 h at 100 MPa and 1250°C. Using the Hi-Nicalon fibers in a DuPont Lanxide Composites proprietary "enhanced" SiC matrix increased the life in the stress rupture test at 100 MPa up to 7 h at 1400°C. By 1998 enhanced CVI SiC/SiC with Hi-Nicalon fibers survived 540 h tensile stress rupture testing at 1204°C and 110 MPa. However, by this time testing in an actual turbine engine environment identified the silica volatilization problem that was discussed in Chapter 9, so a whole new effort of research and development was required to solve this problem. In this case the solution was a surface layer of an oxide environmental barrier coating.[95,99,100] This example illustrates the frustration to the end user of taking years to come up with a solution, but also illustrates the fun for materials engineers to develop the solution.

As mentioned before, the conventional isothermal CVI process takes many days and still results in about 20% porosity. A "forced CVI" process has been developed at Oak Ridge National Laboratories (ORNL) that reduces the deposition time typically to less than 24 h.[102] The fiber preform is compressed into a graphite holder as illustrated in Figure 20.33. The holder and preform are heated such that a gradient exists from top to bottom. The fibers near the top are heated hot enough to allow deposition of SiC from reactant gases introduced through the bottom of the holder. The hot zone moves from the top of the holder toward the bottom of the holder as SiC deposition continues. This results in a high rate of deposition without blocking of near-surface pore channels such as occurs with isothermal CVI. Because of the reduced deposition time, reaction temperature can be increased to around 1200°C to further decrease deposition time.

Composites 85 to over 90% theoretical density have been achieved by the forced CVI process starting with 40% dense cloth preforms. Room-temperature load-displacement curves are shown in Figure 20.34 for Nicalon and Tyranno fiber-reinforced composites. The strength measured in four-point flexure was 470 MPa for the Nicalon-reinforced composite and 390 MPa for the Tyranno-reinforced composite.[102]

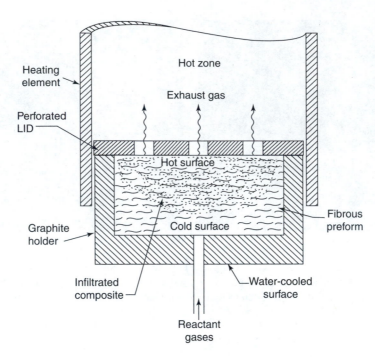

FIGURE 20.33 Schematic illustrating the forced-flow thermal-gradient CVI process. (From Stinton, D.P., Lowden, R.A., and Krabill, R.H., *Mechanical Property Characterization of Fiber-Reinforced SiC Matrix Composites*, ORNL/TM-11524, April 1990. With permission.)

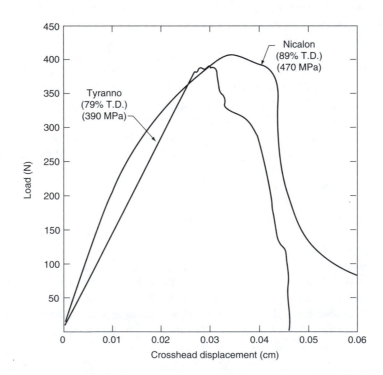

FIGURE 20.34 Room-temperature load-displacement curves for CVI SiC–SiC composites tested in four-point bending. Density and flexure strength are indicated for the two materials. (T.D. = theoretical density.) (From Stinton, D.P., Lowden, R.A., and Krabill, R.H., *Mechanical Property Characterization of Fiber-Reinforced SiC Matrix Composites*, ORNL/TM-11524, April 1990. With permission.)

20.2.9.5 SiC Matrix Composites Fabricated by Si Melt Infiltration

An alternate process, referred to as melt infiltrated ceramic matrix composites (MI-CMC), has been developed to fabricate SiC/SiC composites.[103–105] Fabrication involves the following steps:

1. Coat the fibers with a thin layer of a material or materials that will prevent degradation of the fiber during infiltration and will result in a weak interface bond between the matrix and fibers (similar to the coatings used for the CVI process).
2. Coat the fibers with a slurry containing particles of carbon and SiC plus a polymer binder and use drum winding or filament winding; or prepare a fiber preform by layup of fabric and add the slurry by infiltration.
3. Dry the preform and bring it into contact with molten silicon at about 1500°C. The molten silicon will infiltrate throughout the preform by capillary action and will react with the carbon to form SiC. All of the remaining pores will be filled with silicon.

Advantages compared to the CVI process include zero porosity, greatly reduced fabrication time, and increased thermal conductivity. As of 2004, MI-CMC combustor liners with environmental barrier coatings (EBCs) have survived greater than 15,000 h in an industrial gas turbine engine (approximately 5000 kW), and EBC-coated turbine shroud segments have successfully operated for about 5000 h in a much larger utility scale turbine.

20.2.9.6 SiC Matrix Composites Fabricated by Preceramic Polymer Infiltration

Some polymers contain silicon and carbon and decompose when heated in an inert atmosphere to SiC. Other contain nitrogen in addition to the silicon and carbon and yield Si–C–N compositions. These polymers can be used via multiple infiltration and pyrolysis (heating to decomposition) steps to build up a matrix within a fiber preform. Most of the development on this process was conducted during the 1990s by Dow Corning Corporation and was given the name Polymer Impregnation and Pyrolysis (PIP).[106] The advantage is that techniques and equipment developed over many years for making polymer matrix composites are directly applicable. The disadvantages are that multiple infiltration and pyrolysis cycles are required, that matrix cracks are usually present after composite fabrication, and that 5 to 10% porosity remains.

20.2.9.7 Oxide Matrix Composites Fabricated by Infiltration

One major problem with SiC/SiC composites is oxidation. An obvious way to avoid this problem is to develop a composite that is comprised of oxide fibers and matrix. Table 20.12 listed a variety of oxide fibers that have been developed, many of which are commercially available. These fibers can be formed into a preform and the matrix added by sol–gel or CVI techniques. Most of the work has been done with sol–gel.[106–109] The sol, which generally consists of alkoxide compounds with catalysts, dissolved salts, or other additives, is infiltrated into the preform. A gel forms by a hydrolysis reaction, followed by dehydration, polymerization, and calcining or pyrolysis to yield tiny crystalline particles. Most sols have very low yield of ceramic particles (as low as about 5%), so multiple cycles of infiltration, drying, and firing are required. However, if the sol is filled with ceramic particles to form a slurry, as few as one cycle can achieve nearly 80% of theoretical density.

The matrix produced by sol infiltration is typically quite weak. Fitzer and Gadow[107] reported that a matrix of 90% SiO_2–10% GeO_2 had strength of only 20 MPa. However, when reinforced with 40 volume % of alumina FP fibers, the strength increased to 230 MPa. The weak, porous matrix actually performs an important function. It allows fracture in the matrix at the fiber-matrix interface and results in toughening by debonding, pullout, and bridging. Stronger, less porous oxide matrices have been achieved, but an interface coating was required to allow debonding and composite mechanical behavior.

Oxide–oxide composites were evaluated briefly for gas turbine engine combustor liners in the mid-1990s, but did not perform well. The materials available at that time did not withstand the temperature >1100°C) and thermal stress conditions. Significant improvements have been achieved in recent years as reported by Szweda et al.[106] and has survived in a turbine engine test for >15,000 h. Strength, creep rupture, and cyclic fatigue results are summarized in Table 20.14 for their material comprised of Nextel 720 aluminosilicate fibers in a porous aluminosilicate matrix.

20.2.9.8 Si_3N_4 Matrix Composites

The SiC/SiC and oxide/oxide ceramic matrix composites appear to be progressing towards commercial applications. Work began on silicon nitride matrix composites in the early 1980s, but none of these materials appear at this time to be candidates for commercial products.

Studies with Si_3N_4 concentrated on SiC fibers. Early studies explored hot-pressing and reaction-bonding methods of Si_3N_4 fabrication.[71] Hot pressing achieved full densification, but was conducted

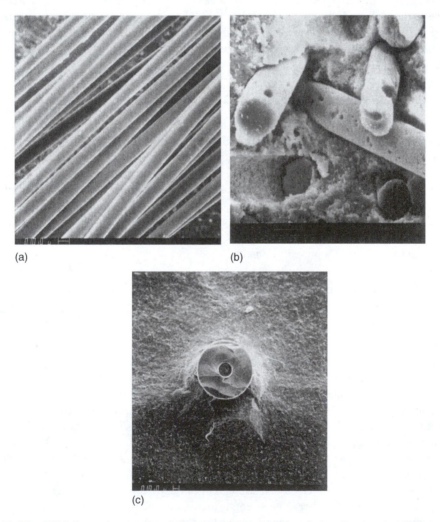

(a)   (b)

(c)

FIGURE 20.35   SEM photomicrographs of (a) degraded Nicalon fibers in reaction-bonded Si_3N_4 matrix, and (b) visually nondegraded Textron (AVCO) SCS-6 CVD monofilament fiber in reaction-bonded Si_3N_4. (From Richerson, D.W., unpublished IR&D studies conducted at Garrett Turbine Engine Company, Phoenix, AZ, 1979–1982.)

at a high enough temperature (>1700°C) that the properties of the fibers were degraded. Polymer-derived multifilament fibers such as Nicalon were completely destroyed. CVD monofilament fibers survived, but did not result in substantial improvement. For example, Shetty et al.[110] reported a slight toughness increase, but about 50% three-point bend strength decrease.

Reaction sintering was conducted at substantially lower temperature (1400°C), but for a long time at that temperature. As shown in Figure 20.35, polymer-derived (Nicalon) fibers severely degraded, but the CVD monofilament (Textron SCS-6) fibers showed no visual signs of degradation.[71] Subsequent studies at NASA and Norton demonstrated that careful layup of unidirectional monofilaments resulted in substantial improvements in strength and toughness compared to unreinforced reaction-bonded Si$_3$N$_4$.[111,112] This is illustrated in Figure 20.36 and Figure 20.37. The Norton NC-350 unreinforced RBSN fractured in a brittle fashion at about 350 MPa. The Norton composite fractured in a nonbrittle mode and had an ultimate tensile strength (as measured in bending) of about 650 MPa. The NASA SiC fiber-reinforced RBSN exhibited similar behavior. Note that the first cracks formed in the matrix at around 250 MPa. The stress at which first matrix cracking occurs is important. Ceramic matrix composite stressed above this level appear to be susceptible to both cyclic fatigue and oxidation and corrosion degradation.

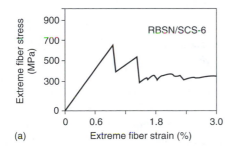
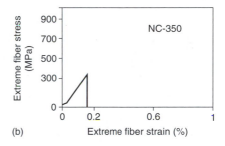

FIGURE 20.36 Comparison of the fracture behavior of Norton reaction-bonded Si$_3$N$_4$ material (a) with fiber reinforcement (RBSN/SCS-6 SiC unidirectional monofilaments) and (b) without fiber reinforcement (NC-350). (From Corbin, N.D., Rossetti, G.A., Jr. and Hartline, S.D., *Ceram. Eng. Sci. Proc.*, 7(7–8), 958–968, 1986. With permission.)

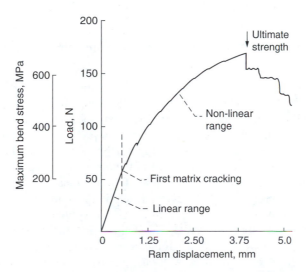

FIGURE 20.37 Fracture behavior of SiC monofilament fiber-reinforced reaction-bonded Si$_3$N$_4$ from NASA. (From Bhatt, R.T., Mechanical properties of SiC fiber-reinforced reaction-bonded Si$_3$N$_4$ composites, in *Proceedings 23rd Auto. Tech. Devt. Cont. Coord.* Meeting, SAE, P-165, 1985. With permission.)

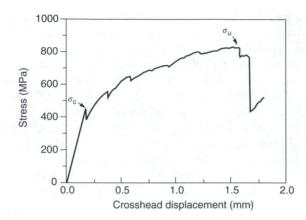

FIGURE 20.38 Stress-displacement curve for Si₃N₄-mullite matrix composite containing 35 vol% unidirectional monofilament SiC fibers. (From Bright, J.D., Flinders, R.M., Shetty, D.K., Griffin, C.W., and Richerson, D.W., *High Strength Silicon Nitride Based SiC Continuous Fiber-Reinforced Ceramic Matrix Composites*, Paper 34-SVI-91, American Ceramics Society 93rd Annual Meeting, April 28–May 2, 1991.)

The stress for first matrix cracking and relative ease or difficulty of fiber pullout is strongly affected by the difference in thermal expansion between the matrix and fiber. This concept was applied to development of a hot-pressed Si₃N₄-based ceramic matrix composite at Ceramatec that had high first matrix cracking stress.[113] A stress-displacement curve is shown in Figure 20.38. First matrix cracking occurred at about 450 MPa, followed by substantial fiber debonding, pullout, and bridging. The ultimate bend strength was about 908 MPa. The composite consisted of a matrix of 95 vol% Si₃N₄ and 5 vol% Mullite with 35 vol% unidirectional Textron SCS-6 SiC monofilament fiber. The Mullite provided an improved thermal expansion match between the matrix and fiber that helped increase first matrix cracking stress. The Mullite also helped decrease the hot-pressing temperature and time (1500°C, 30 min) to minimize interfacial bond between the fibers and matrix. The composite also exhibited excellent high-temperature properties. It retained nonbrittle composite fracture behavior to 1400°C and survived stress rupture testing for over 1000 h at 1000°C in air at an applied flexural stress of 400 MPa.[114]

20.2.9.9 Other Fiber-Reinforced Ceramic Matrix Composites

Many additional studies have been conducted in recent years with oxide and nonoxide fibers in oxide and nonoxide matrices. This section identifies some of the combinations and processes that have been tried and provides a few references to give the reader a starting point in locating further information.

Matrix Formation by Gas-Metal Reaction. The Lanxide directed metal oxidation (DIMOX) process (discussed in an earlier chapter) has been adapted to produce fiber-reinforced composites.[115] The fiber preform is placed between the molten alloy and the reaction gas source as shown schematically in Figure 20.39. The dense reaction product of the metal and gas grows into and through the fiber preform. The growth temperature is low so that fiber degradation does not occur and so that interface bonding can be minimized. No pressure is applied and no shrinkage occurs during processing, so that the architecture of the fiber preform and the volume fraction can be controlled. A composite of Nicalon woven fabric in an Al₂O₃-based matrix had flexure strength of about 450 MPa at room temperature and 350 MPa at 1200°C. Toughness values ranged from 18 to 23 MPa · m^{1/2}.[116] A composite of Nicalon woven fabric in an AIN-based matrix had similar strength and toughness.

Oxide Fibers in CVD SiC Matrix. 3M Company Nextel oxide fibers, as well as Nicalon SiC fibers, have been formed by braiding into tubes and other configurations and bonded into a composite with CVD SiC. Thin-walled tubes have been fabricated that have exceptional thermal-shock

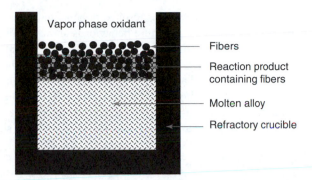

FIGURE 20.39 Fabrication of a fiber-reinforced ceramic matrix composite using the Lanxide directed metal oxidation process. (From Newkirk, M.S., Lesher, H.D., White, D.R., Kennedy, C.R., Urquhart, A.W., and Claar, T.D., *Ceram. Eng. Sci. Proc.*, 8(7–8), 879–885, 1987. With permission.)

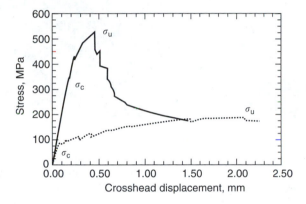

FIGURE 20.40 Stress-displacement curves showing extremes in the fracture behavior of SiC monofilament fiber-NZP matrix composites. (From Griffin, C.W., Limaye, S.Y., Richerson, D.W., and Shetty, D.K., *Ceram. Eng. Sci. Proc.*, 11(9–10), 1577–1591, 1990. With permission.)

resistance. Scoville et al.[117] described tests in which a composite tube was heated internally with a high-temperature torch and survived spraying of a stream of water locally onto the hot surface.

Diboride Matrix Composites. Composites have successfully been fabricated using unidirectional CVD SiC monofilament fibers in matrices consisting of diboride-SiC-C compositions.[118,119] A HfB_2-based composite had flexural strength of 1100 MPa, compared to 780 MPa for a ZrB_2-based composite and 329 MPa for a TiB_2-based composite. All fractured in a noncatastrophic composite mode.

NZP Matrix Composites. NZP was discussed in an earlier chapter as a family of low thermal expansion materials with the [NZP] (sodium zirconium phosphate) structure. Some of the most interesting compositions involve crystal chemical substitutions of divalent ions (Ca, Mg, Ba, and Sr) in the sodium position and Si substitutions for a portion of the phosphorus. Through the proper composition manipulation, one can select a bulk polycrystalline thermal expansion coefficient from slightly negative all the way to $+5 \times 10^{-6}/°C$. The ability to vary thermal expansion characteristics without major chemical variation has allowed an interesting study of the effects of different thermal expansion mismatch between SiC fiber and the surrounding matrix.[120] The study was conducted with CVD SiC monofilament fibers that have a radial expansion of about $2.3 \times 10^{-6}/°C$.[121] Figure 20.40 shows the range of composite fracture characteristics that were observed. The bottom dotted line represents samples in which the expansion of the matrix and fiber were about equal. The matrix fractured at low stress and easily slipped along the fibers. This resulted in an extreme case of bridging with the fibers carrying essentially all of the load. This fracture behavior is shown in Figure 20.41. The solid line in Figure 20.40 illustrates the behavior of samples with maximum thermal expansion mismatch where the expansion of the matrix was higher than that of the fiber. Note that the first matrix cracking stress is high and that the ultimate strength has a distinct peak before decreasing to around 200 MPa.

The two types of behavior depicted in Figure 20.40 correlate well with the fiber-matrix interfacial sliding friction stress and the fiber-matrix thermal expansion mismatch. This is illustrated in Figure 20.42. A high mismatch (high matrix expansion) results in high sliding-friction stress.

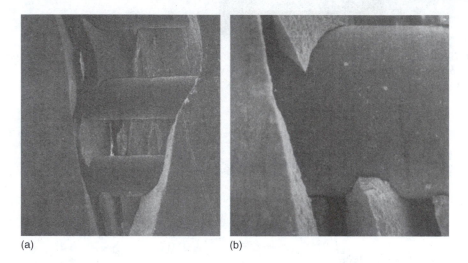

(a)   (b)

FIGURE 20.41   Side view of fractured SiC-NZP composite showing high degree of debonding, pullout, and bridging. (From Griffin, C.W., Limaye, S.Y., Richerson, D.W., and Shetty, D.K., *Ceram. Eng. Sci. Proc.*, 11(9–10), 1577–1591, 1990. With permission.)

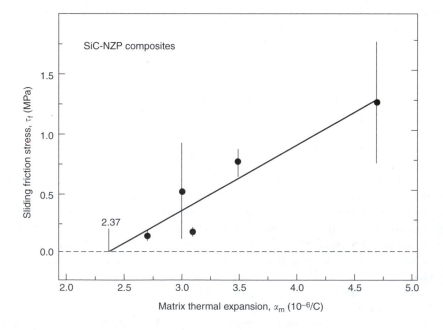

FIGURE 20.42   Linear correlation between sliding friction stress (τ_f) and coefficient of thermal expansion of the matrix (α_m) in SiC-NZP composites. (From Griffin, C.W., Limaye, S.Y., Richerson, D.W., and Shetty, D.K., *Ceram. Eng. Sci. Proc.*, 11(9–10), 1577–1591, 1990. With permission.)

A close match results in low sliding-friction stress. The curve extrapolates to zero frictional stress at $2.37 \times 10^{-6}/°C$, which is very close to the radial thermal expansion coefficient reported for the SiC fiber.

20.2.10 COMPOSITES WITH SURFACE COMPRESSION

Surface compression can be an effective mode of increasing the resistance of a ceramic to fracture. The examples of safety glass and Corelle were discussed earlier. Surface compression has also been achieved in Si_3N_4–SiC and Al_2O_3–ZrO_2 materials. The use of a layering approach to achieve maximum level of compressive stress is discussed in this section.

The Si_3N_4–SiC composite consisted of an interior that was 60% Si_3N_4–40% SiC and a surface that was 10% Si_3N_4.[122] The Si_3N_4 had a coefficient of thermal expansion of $3.17 \times 10^{-6}/°C$ compared to $3.87 \times 10^{-6}/°C$ for the 60% Si_3N_4–40% SiC. During cooling from the hot-pressing temperature, the lower expansion surface was pulled into compression by the higher expansion interior. For a layered composite with the inner layer comprising 60% of the thickness and the surface layers each 20%, the flexural strength was 1200 MPa. This compared to 990 MPa for the Si_3N_4 material alone and 827 MPa for the 60 Si_3N_4–40 SiC material alone. The compressive stress in the Si_3N_4 surface layer could be increased by decreasing the thickness of the layer.

Surface compression was achieved in the ZrO_2–Al_2O_3 system by a volume change due to a polymorphic transformation.[123] The interior consisted of Al_2O_3 containing 15 vol% ZrO_2 stabilized with Y_2O_3 to retain the tetragonal polymorphic phase during cooling. The outer layer consisted of Al_2O_3 containing 15 vol% ZrO_2 without the Y_2O_3 to stabilize the tetragonal phase. During cooling, the ZrO_2 in the outer layer transformed from tetragonal to monoclinic, accompanied by a substantial volume increase. This volume increase of the ZrO_2 particles placed the surface in compression. Figure 20.43 illustrates the improvements in strength and reliability that were achieved. Nonlayered material had average strength of 451 MPa and Weibull modulus of 9.9. Material with a surface layer 375-μm thick had average strength of 825 MPa and Weibull modulus of 16.1. The layered composite was remarkably resistant to strength degradation due to surface damage.[124] This is illustrated in Figure 20.44. Vicker's hardness indentations up to 1000 N load resulted in strength reduction of less than 10% to about 600 MPa. A load of only 100 N decreased the strength of unlayered material to about 200 MPa compared to an as-fabricated strength of about 400 MPa.

20.2.11 FIBROUS MONOLITH

High strength fibers such as the Nicalon SiC fibers have been very expensive. In the late 1980s a concept referred to as the "fibrous monolith" was devised to make composites with behavior similar to continuous fiber reinforced composites, but without requiring expensive fibers.[125–128] A fibrous monolith is fabricated in several steps. First the ceramic powder (such as silicon nitride) is mixed with a polymer and extruded into a thin strand. This strand is coated with a nonreactive barrier layer (such as BN). The coated strands are then laid up in a die in the desired architecture and hot pressed. The ceramic powder comprising each strand densifies, but is prevented from bonding to adjacent strands by the barrier layer. The resulting microstructure consists of long parallel cells of a flattened hexagonal shape separated by the barrier layer and behaves very much like wood when a stress is applied. The fracture is noncatastrophic with extensive crack deflection, bridging, and pullout.

Fibrous monolith microstructures have been achieved with a variety of compositions.[126] Most of the early development was conducted at the University of Michigan and at Advanced Ceramics Research (Tucson, AZ) with SiC/graphite, SiC/BN, and Si_3N_4/BN. The SiC/BN fibrous monolith was reported to have flexural strength in the range of 300–375 MPa and very high work of fracture of about 2400 J/m².[127] A sample exposed to 1400°C in air for 250 h retained its noncatastrophic fracture behavior and had work of fracture of about 2200 J/m². Silicon nitride containing about 15% BN had flexural strength of about 400 MPa for uniaxial cell architecture and about 200 MPa for woven (0/90) and chopped "fiber" architectures.[128]

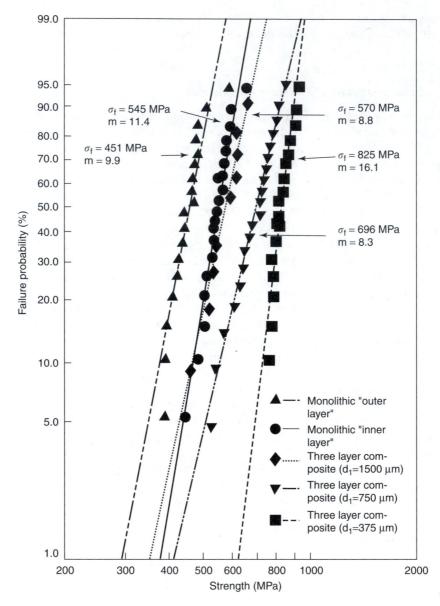

FIGURE 20.43 Increase in the reliability of an Al_2O_3–ZrO_2 material by inducing a surface compressive stress through layering. (From Cutler, R.A., Bright, J.D., Virkar, A.V., and Shetty, D.K., *J. Am. Ceram. Soc.*, 70(10), 714–718, 1987. With permission.)

20.3 SUMMARY

Enormous progress has been made since the mid-1970s in the understanding of microstructure vs. properties and in the subsequent development of ceramic-based materials with increased toughness. Aluminum oxide materials reinforced with TiC particles and with SiC whiskers have been in production for cutting tool inserts since the mid-1980s. Silicon nitride materials with self-reinforced microstructures have achieved considerable success for bearings, cutting tool inserts, a wide variety of wear resistant parts, and for prototype engine components. Transformation toughened zirconia and transformation toughened alumina have also become important as wear parts, tooling for metal forming, and grinding media.

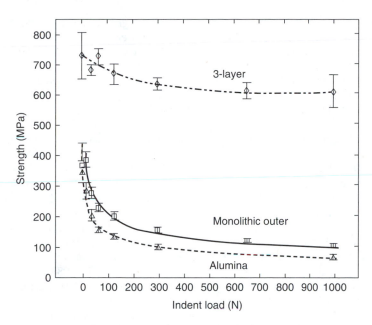

FIGURE 20.44 Strength vs. surface damage (indentation load of Vickers indentor) for Al_2O_3, monolithic Al_2O_3–ZrO_2 and compressive-surface-strengthened Al_2O_3–ZrO_2 layered composite. (From Hansen, J.J., Cutler, R.A., Shetty, D.K., and Virkar, A.V., *J. Am. Ceram. Soc.*, 71(12), C501–505, 1988. With permission.)

While the toughness was substantially increased with particulate and whisker reinforced ceramics, *in situ* reinforced silicon nitride, and transformation toughened ceramics, these materials still fractured in a brittle mode. A wide range of high strength ceramic fibers have been synthesized and incorporated in various ceramic matrices to achieve composites that do not fracture in a brittle mode, but instead have considerable strain-to-failure. Although many challenges (cost and technology) remain before these continuous fiber reinforced ceramic matrix composites achieve widespread application, these materials appear very promising and represent a good opportunity for materials engineers to participate in rewarding development programs.

REFERENCES

1. Bradt, R.C., Hasselman, D.P.H., and Lange, F.F., Eds., *Fracture Mechanics of Ceramics*, Vol. 1 and 2 (1973), Vol. 3 and 4 (1978), Plenum, New York.
2. Evans, A.G., Ed., *Fracture in Ceramic Materials*, Noyes Publications, Park Ridge, N.J., 1984.
3. Ko, F.K., Architecture for ceramic-matrix composites, *Am. Ceram. Soc. Bull.*, 68(2), 401–414, 1989.
4. Kirchner, H.P., *Strengthening of Ceramics*, Marcel Dekker, New York, 1979.
5. Rice, R.W., Microstructure dependence of mechanical behavior of ceramics, in *Treatise on Materials Science and Technology*, Vol. 11, MacCrone, K.K., Ed., Academic Press, New York, 1977, pp. 199–381.
6. Rice, R.W., Mechanism of toughening in ceramic matrix composites, *Ceram. Eng. Sci. Proc.*, 2(7–8), 661–701, 1981.
7. Becher, P.F., Microstructural design of toughened ceramics, *J. Am. Ceram. Soc.*, 74(2), 255–269, 1991.
8. Faber, K.T. and Evans, A.G., Crack deflection processes-I, theory, and II, experiment, *Acta. Metall.*, 31(4), 565–584, 1983.
9. Rice, R.W., Spann, T.R., Lewis, D., and Coblenz, W., The effect of ceramic fiber coatings on the room-temperature mechanical behavior of ceramic-fiber composites, *Ceram. Eng. Sci. Proc.*, 5(7–8), 614–624, 1984.

10. Brennan, J.J., Interfacial characterization of glass and glass-ceramic matrix/Nicalon® SiC fiber composites, in *Tailoring Multiphase and Composite Ceramics*, Tressler, R.E., Messing, G.L., Pantano, C.G., and Newnham, R.E., Eds., Plenum, New York, 1986, pp. 549–560.

11. Cranmer, D.C., Fiber coating and characterization, *Am. Ceram. Soc. Bull.*, 68(2), 415–419, 1989.

12. Claussen, N., Steeb, J., and Pabst, R.F., Effect of induced microcracking on the fracture toughness of ceramics, *Am. Ceram. Soc. Bull.*, 56(6), 559–562, 1977.

13. Magley, D.J., Winholtz, R.A., and Faber, K.T., Residual stresses in a two-phase microcracking ceramic, *J. Am. Ceram. Soc.*, 73(6), 1641–1644, 1990.

14. Sigl, L.S., Mataga, P.A., Dalgleish, B.J., McMeeking, R., and Evans, A.G., On the toughening of brittle materials reinforced with a ductile phase, *Act. Metall.*, 36(4), 945–953, 1988.

15. Marshall, D.B., Shaw, M.C., Dauskardt, R.H., Ritchie, R.O., Readey, M., and Heuer, A.H., Crack-tip transformation zones in toughened zirconia, *J. Am. Ceram. Soc.*, 73(9), 2659–2666, 1990.

16. Heuer, A.H. and Hobbs, L.W., Eds., *Science and Technology of Zirconia, Advances in Ceramics*, Vol. 3, American Ceramics Society, Westerville, Ohio, 1981.

17. Claussen, N., Ruhle, M., and Heuer, A.H., Eds., *Science and Technology of Zirconia II, Advances in Ceramics*, Vol. 11, American Ceramics Society, Westerville, Ohio, 1984.

18. Somiya, S., Yamamoto, N., and Hanagida, H., Eds., *Science and Technology of Zirconia III, Advances in Ceramics*, Vol. 24, American Ceramics Society, Westerville, Ohio, 1988.

19. Claussen, N., Transformation toughened ceramics, in *Ceramics in Advanced Energy Technologies*, Krockel, H., et al., Eds., Reidel, D., Dordrecht, 1984, pp. 51–86.

20. Cannon, W.R., Transformation toughened ceramics, in *Structural Ceramics, Treatise on Materials Science and Technology*, Vol. 29, Wachtman, J.B., Jr., Ed., Academic Press, San Diego, Calif., 1989, pp. 195–228.

21. Evans, A.G. and Cannon, R.M., Toughening of brittle solids by martensitic transformations, *Acta. Metall.*, 34(5), 761–800, 1986.

22. Virkar, A.V. and Matsumoto, R.L.K., Ferroelastic domain switching as a toughening mechanism in tetragonal zirconia, *J. Am. Ceram. Soc.*, 69, 224–226, 1986.

23. Virkar, A.V., Jue, J.F., Smith, P., Mehta, K., and Prettyman, K., The role of ferroelasticity in toughening of brittle materials, *Phase Trans.*, 35(1), 27–46, 1991.

24. Lange, F.F., Relation between strength, fracture energy and microstructure of hot pressed Si_3N_4, *J. Am. Ceram. Soc.*, 56(10), 518, 1973.

25. Tani, E., Umebayashi, S., Kishi, K., Kobayashi, K., and Nishijima, M., Gas-pressure sintering of Si_3N_4 with concurrent addition of Al_2O_3 and 5 wt.% rare earth oxide: high fracture toughness Si_3N_4 with fiber-like structure, *Am. Ceram. Soc. Bull.*, 65(9), 1311–1315, 1986.

26. Matsuhiro, K. and Takahashi, T., Physical properties of sintered silicon nitride controlled by grain boundary chemistry and microstructure morphology, in *MRS International Meeting on Advanced Materials*, Vol. 5, Materials Research Society, 1989.

27. Matsuhiro, K. and Takahashi, T., The effect of grain size on the toughness of sintered Si_3N_4, *Ceram. Eng. Sci. Proc.*, 10(7–8), 807–816, 1989.

28. Whalen, P.J., Gadsaska, C.J., and Silvers, R.D., The Effect of microstructure on the high-temperature deformation behavior of sintered silicon nitride, *Ceram. Eng. Sci. Proc.*, 11(7–8), 633–649, 1990.

29. Li, C.W. and Yamanis, J., Super-tough silicon nitride with R-curve behavior, *Ceram. Eng. Sci. Proc.*, 11(7–8), 632–645, 1989.

30. Pyzik, A. and Meenan, B.J., *Sintering of Self-Reinforced Silicon Nitride*, Paper 46-C-91, 93rd Annual Meeting American Ceramics Society, Cincinnati, Ohio, April 30, 1991.

31. Johnson, W.B., Claar, T.D., and Schiroky, G.H., Preparation and processing of platelet-reinforced ceramics by the directed reaction of zirconium with boron carbide, *Ceram. Eng. Sci. Proc.*, 10(7–8), 588–598, 1989.

32. Claar, T.D., Johnson, W.B., Andersson, C.A., and Schiroky, G.H., Microstructure and properties of platelet-reinforced ceramics formed by the directed reaction of zirconium with boron carbide, *Ceram. Eng. Sci. Proc.*, 10(7–8), 599–609, 1989.

33. Cutler, R.A., Mayhew, R.J., Prettyman, K.M., and Virkar, A.V., High-toughness Ce-TZP/Al_2O_3 ceramics with improved hardness and strength, *J. Am. Ceram. Soc.*, 74(1), 179–186, 1991.

34. Fujii, T., Muragaki, H., Hatano, H., and Hirano, S., *Microstructure Development and Mechanical Properties of Ce-TZP/La-Alumina Composites*, Paper 29-SVI-91, 93rd Annual Meeting of American Ceramics Society, Cincinnati, Ohio, April 29, 1991.

35. Sato, T. and Shimada, M., Transformation of yttria-doped tetragonal ZrO_2 polycrystals by annealing in water, *J. Am. Ceram. Soc.*, 68(6), 356–359, 1985.

36. Tsukama, K. and Shimada, M., Thermal stability of yttria-partially stabilized zirconia (Y-PSZ) and Y-PSZ/alumina composites, *J. Mater. Sci. Lett.*, 4, 857–861, 1985.

37. Rice, R.W., Capabilities and design issues for emerging tough ceramics, *Am. Ceram. Soc. Bull.*, 63(2), 256–262, 1984.

38. Buljan, S.T., Baldoni, J.G., Neil, J., and Zilberstein, G., *Dispersoid-Toughened Silicon Nitride Composites*, Final Rept. ORNL/Sub/85-22011/1, Sept. 1988.

39. Rolfson, B., Richerson, D., and Hurford, A.C., *High Thermal Conductivity Dispersion-Strengthened Silicon Nitride*, Final Rept. DOE SBIR ER 80367, March 10, 1987.

40. Kodama, H. and Miyoshi, T., Fabrication and properties of Si_3N_4 composites reinforced by SiC whiskers and particles, *Ceram. Eng. Sci. Proc.*, 10(9–10), 1072–1082, 1989.

41. Liu, J. and Ownby, P.D., Enhanced mechanical properties of alumina by dispersed titanium diboride particulate inclusions, *J. Am. Ceram. Soc.*, 74(1), 241–243, 1991.

42. Cutler, R.A., Hurford, A.C., and Virkar, A.V., Pressureless-sintered Al_2O_3-TiC composites, in *Science of Hard Materials*, Vol. 3, Sarin, B.K., Ed., Elsevier, New York, 1988, pp. 183–192.

43. McCauley, J.W., A systematic framework for fabricating new ceramic matrix composites, *Ceram. Eng. Sci. Proc.*, 2(7–8), 648, 1981.

44. McMurtry, C.H., Boecker, W.D.G., Seshardri, S.G., Zanghi, J.S., and Garnier, J.E., Microstructure and material properties of SiC-TiB_2 particulate composites, *Am. Ceram. Soc. Bull.*, 66(2), 325–329, 1987.

45. Tiegs, T.N., Allard, L.F., Becher, D.F., and Ferber, M.F., Identification and development of optimum silicon carbide whiskers for silicon nitride matrix composites, in *Proceedings 27th Auto. Tech. Devt. Contractors Coord. Meeting*, SAE P-230, 1990, pp. 167–172.

46. Lowden, R.A., *Characterization and Control of the Fiber-Matrix Interface in Ceramic Matrix Composites*, ORNL Rept. TM-11039. March 1989.

47. Rhodes, J.F., Rootare, H.M., Springs, C.A., and Peters, J.E., *Hot-pressed Al_2O_3-SiC Whisker Composites*, paper presented at the 88th Annual Meeting of the American Ceramics Society, Chicago, Ill., April 28, 1986.

48. Tiegs, T.N. and Becher, P.F., Alumina-SiC whisker composites, in *Proceedings of the 23rd Auto. Tech. Devt. Contractors Coord. Meeting*, SAE P-165, 1985.

49. Griffin, C.W., Hurford, A.C., Virkar, A.V., and Richerson, D.W., Properties of pressureless sintered alumina matrix composites containing up to 30 vol.% SiC whiskers, *Ceram. Eng. Sci. Proc.*, 10(7–8), 695–706, 1989.

50. Buljan, S.T., Baldoni, J.G., and Huckabee, M.L., Si_3N_4-SiC composites, *Am. Ceram. Soc. Bull.*, 66(2), 347–352, 1987.

51. Shalek, P.D., Petrovic, J.J., Hurley, G.F., and Gac, F.D., Hot pressed SiC whisker/Si_3N_4 matrix composites, *Am. Ceram. Soc. Bull.*, 65(2), 351–352, 1986.

52. Shih, C.J., Yang, J.-M., and Ezis, A., Processing of SiC whisker-reinforced Si_3N_4 composites, *Ceram. Eng. Sci. Proc.*, 10(9–10), 1064–1071, 1989.

53. Neergaard, L.J. and Homeny, J., Mechanical properties of beta-silicon nitride whisker/silicon nitride matrix composites, *Ceram. Eng. Sci. Proc.*, 10(9–10), 1049–1062, 1989.

54. Carter, D.H., Petrovic, J.J., Honnell, R.E., and Gibbs, W.S., SiC-$MoSi_2$ composites, *Ceram. Eng. Sci. Proc.*, 10(9–10), 1121–1129, 1989.

55. Wei, G.C. and Becher, P.F., Development of SiC-whisker reinforced ceramics, *Am. Ceram. Soc. Bull.*, 64(2), 298–304, 1985.

56. Tiegs, T.N. and Becher, P.F., Sintered Al_2O_3-SiC-whisker composites, *Am. Ceram. Soc. Bull.*, 66(2), 339–342, 1987.

57. Data sheets from Advanced Composite Materials Corp., Greer, S.C.

58. Petrovic, J.J. and Honnell, R.E., SiC reinforced $MoSi_2$/WSi_2 alloy matrix composites, *Ceram. Eng. Sci. Proc.*, 11(7–8), 734–744, 1990.

59. Richerson, D.W., Ceramic Matrix Composites, Chapter 19, in *Composites Engineering Handbook*, Mallick, P.K., Ed., Marcel Dekker, Inc., New York, 1997, pp. 983–1038.

60. Miller, D.W., Glass Fibers, in *Engineering Materials Handbook, Vol. 1: Composites*, ASM Int., Materials Park, Ohio, 1987, pp. 45–48.

61. Johnson, D.D. and Sowman, H.G., Ceramic Fibers, *ibid.*, pp. 60–65.

62. Diefendorf, R.J., Carbon/Graphite Fibers, *ibid.*, pp. 48–53.

63. Donnet, J.B. and Bansal, R.C., *Carbon Fibers*, Marcel Dekker, New York, 1984.

64. Chawla, K.K., *Ceramic Matrix Composites*, Chapman and Hall, London, 1993.

65. Sowman, H.G., in *Sol–Gel Technology*, Noyes, Park Ridge, NJ, 1988, p. 162.

66. Saitow, Y., et al., SAMPE Annual Meeting, 1992.

67. Anon, *High Performance Synthetic Fibers for Composites*, NMAB-458, National Academy Press, Washington, D.C., 1992.

68. Schoenberg, T., Boron and Silicon Carbide Fibers, in *Engineered Materials Handbook, Vol. 1: Composites*, ASM Int., Materials Park, Ohio, 1987, pp. 58–59.

69. Yajima, S., Okamura, K., Hayashi, J., and Omori, M., *J. Am. Ceram. Soc.*, 59, 324, 1976.

70. Yajima, S., Special heat-resisting materials from organometallic polymers, *Am. Ceram. Soc. Bull.*, 62(8), 1983.

71. Richerson, D.W., unpublished IR&D studies conducted at Garrett Turbine Engine Company, Phoenix, Ariz., 1979–1982.

72. Takeda, M., Imai, Y., Ichikawa, H., Ichikawa, T., Kasai, N., Seguchi, T., and Okamura, K., *Ceram. Eng. Sci. Prod.*, 14(7–8) 540–547, 1993.

73. Yamamura, T., Ishikawa, T., Shibuya, M., Hisayuki, T., and Ukamura, K., Development of a new continuous Si-Ti-C-O fiber-bonded ceramic material, *J. Mater. Sci.*, 23, 2589, 1988.

74. LeGrow, G.E., Lim, T.F., Lipowitz, J., and Reaoch, R.S., Ceramics from hydridopolysilazane, *Am. Ceram. Soc. Bull.*, 66, 363, 1987.

75. Lipowitz, J., Barnard, T., Bujalski, D., Rabe, J., Zank, G., Zangril, A., and Xu, Y., Fine-diameter polycrystalline SiC fibers, *Compos. Sci. Tech.*, 51, 167–171, 1994.

76. Srinivasan, G. and Venkateswaran, V., Tensile strength evaluation of polycrystalline SiC fibers, *Ceram. Eng. Sci. Proc.*, 14(7–8) 563–572, 1993.

77. Marikunte, S. and Shah, S.P., Cement Matrix Composites, in *Composite Engineering Handbook*, Chapter 20, Mallick, P.K., Ed., Marcel Dekker, Inc., New York, 1997, pp. 1039–1066.

78. Balaguru, P.N. and Shah, S.P., *Fiber-Reinforced Cement Composites*, McGraw-Hill, New York, 1992.

79. Bentur, A. and Mindess, S., *Fiber Reinforced Cementitious Composites*, Elsevier, New York, 1990.

80. Prewo, K.M., Glass and ceramic matrix composites present and future, in *High Temperature/High Performance Composites*, MRS Symposium Proceedings, Vol. 120, Lemkey, F.D., et al., Eds., Materials Research Society, Pittsburgh, Pa., 1988, pp. 145–156.

81. Sambell, R.A.J., et al., Carbon fiber composites with ceramic and glass matrices, *J. Mat. Sci.*, 7, 663–681, 1972.

82. Prewo, K.M., Bacon, J.F., and Dicus, D.L., Graphite fiber reinforced glass matrix composites for aerospace applications, *SAMPE Proc.*, 24, 61–71, 1979.

83. Prewo, K.M. and Brennan, J.J., High-strength silicon carbide fiber-reinforced glass matrix composite, *J. Mater. Sci.*, 15(2), 463–468, 1980.

84. Prewo, K.M. and Brennan, J.J., Silicon carbide yarn reinforced glass matrix composite, *J. Mater. Sci.*, 17, 1201–1206, 1982.

85. Brennan, J.J. and Prewo, K.M., Silicon carbide fiber reinforced glass-ceramic matrix composites exhibiting high strength and toughness, *J. Mater. Sci.*, 17, 7371–7383, 1982.

86. Luh, E.Y. and Evans, A.G., High temperature mechanical properties of a ceramic matrix composite, *J. Am. Ceram. Soc.*, 70(7), 466, 1987.

87. Lamicq, P.J., Bernhart, G.A., Dauchier, M.M., and Mace, J.C., SiC/SiC composite ceramics, *Am. Ceram. Soc. Bull.*, 65(2), 336–337, 1986.

88. Lamicq, P.J. and Jamet, J.F., Thermostructural CMCs: an Overview of the French Experience, in *Ceramic Transactions Vol. 57: High Temperature Ceramic-Matrix Composites I*, Evans, A.G. and Naslain, R., Eds., The American Ceramic Society, Westerville, Ohio, 1995, pp. 1–11.

89. Mah, T-I., Mendiratta, M.G., Katz, A.P., and Mazdiyasni, K.S., Recent developments in fiber-reinforced high temperature ceramic composites, *Am. Ceram. Soc. Bull.*, 66, 304–308, 1987.

90. Luthra, K.L., Oxidation of ceramic composites, *Ceram. Trans.*, 10, 183–195, 1990.

91. Tortorelli, P.F., Jijhawan, S., Riester, L., and Lowden, R.A., Influence of fiber coatings on the oxidation of fiber-reinforced SiC composites, *Ceram. Eng. Sci. Proc.*, 14(7–8), 358–366, 1993.

92. Jacobson, N.S., Smialek, J.L., Fox, D.S., and Opila, E.J., Durability of Silica-Protected Ceramics in Combustion Atmospheres, in *Ceramic Transactions Vol. 57: High Temperature Ceramic-Matrix*

Composites I, Evans, A.G., and Naslain, R., Eds., The American Ceramic Society, Westerville, Ohio, 1995, pp. 157–170.

93. Eckel, A.J., Cawley, J.D., and Parthasarathy, T.A., Oxidation kinetics of a continuous carbon face in a nonreactive matrix, *J. Am. Ceram. Soc.*, 78, 972–980, 1995.

94. Jacobson, N.S., et al., Corrosion issues for ceramics in gas turbines, in *Ceramic Gas Turbine Component Development and Characterization*, Chapter 26, van Roode, M., Ferber, M.K., and Richerson, D.W., Eds., ASME Press, New York, 2003, pp. 607–640.

95. Landini, D.J., Fareed, A.S., Wang, H., Craig, P.A., and Hemstad, S., Ceramic Matrix Composites Development at GE Power Systems Composites, LLC, Chapter 14, pp. 259–276, *ibid.*

96. Bender, B.A., Lewis D., III, Coblenz, W.S., and Rice, R.W., Electron microscopy of ceramic fiber-ceramic matrix composites—comparison with processing and behavior, *Ceram. Eng. Sci. Proc.*, 5, 513–529, 1984.

97. Lowden, R.A. and Stinton, D.P., *The Influence of the Fiber-Matrix Bond on the Mechanical Behavior of Nicalon/SiC Composites*, ORNL Rep. ORNL/TM-10667, Dec. 1987.

98. Hatton, K., Landini, D., Hemstad, S., and Robinson, R.C., *SiC Fiber Reinforced Ceramic Matrix Composites for Turbine Engine Applications*, ASME Paper 2000-GT-72, presented at the International ASME TURBO EXPO 2000, Land, Sea & Air, Munich, Germany, May 8–11, 2000.

99. Eaton, H.E., Linsey, G.D., More, K.L., Kimmel, J.B., Price, J.R., and Miriyala, N., *EBC Protection of SiC/SiC Composites in Gas Turbine Combustion Environment*, ASME Paper 2000-GT-0.631, *ibid.*

100. Lee, K.N., Fritze, H., and Ogura, Y., Coatings for Engineering Ceramics, in *Ceramic Gas Turbine Component Development and Characterization*, Chapter 27, van Roode, M., Ferber, M.K., and Richerson, D.W., Eds., ASME Press, New York, 2003, pp. 641–664.

101. Lee, K.N., Miller, R.A., and Jacobson, N.S., New generation of plasma-sprayed mullite coatings on silicon carbide, *J. Am Ceram. Soc.*, 78(3), 705–710, 1995.

102. Stinton, D.P., Lowden, R.A., and Krabill, R.H., *Mechanical Property Characterization of Fiber-Reinforced SiC Matrix Composites*, ORNL/TM-11524, April 1990.

103. Luthra, K.L., Singh, R.N., and Brun, M.K., Toughened silcomp composites—process and preliminary properties, *Am. Ceram. Soc. Bull.*, 72(7) 79–85, 1993.

104. Corman, G.S., Luthra, K.L., and Brun, M.K., Silicon Melt Infiltrated Ceramic Composites—Processes and Properties, in *Ceramic Gas Turbine Component Development and Characterization*, Chapter 16, van Roode, M., Ferber, M.K., and Richerson, D.W., Eds., ASME Press, New York, 2003, pp. 291–312.

105. Kameda, T., Suyama, S., Itoh, Y., Nishida, K., Ikeda, I., Hijikata, T., and Okamura, F., Development of continuous fiber-reinforced reaction sintered silicon carbide matrix composite for gas turbine hot parts application, Chapter 18, pp. 331–351 in *ibid.*

106. Szweda, A., Easlser, T.E., Jurf, R.A., and Butner, S.C., Ceramic matrix composites for gas turbine applications, Chapter 15, pp. 277–289 in *ibid.*

107. Fitzer E. and Gadow, R., Fiber-reinforced ceramics, *Proc. ASM Whisker and Fiber Toughened Ceramics Conf.*, 1988, pp. 165–192.

108. Goettler, R.W., Pak, S.S., and Long, W.G., Continuous fiber ceramic composites for industrial applications, 39th Int. SAMPE Symp., Anaheim, CA, Apr. 11–14, 1994.

109. Fitzer, E. and Gadow, R., Fiber-reinforced composites via the sol–gel route, in *Tailoring Multiphase and Composite Ceramics, Mater. Sci. Res.* Vol. 20, Tressler, R.E., Messing, G.L., Pantano, C.G., and Newnham, R.E., Eds., Plenum, New York, 1985, pp. 571–608.

110. Shetty, D.K., Pascucci, M.R., Matsuddy, B.C., and Wills, R.R., SiC monofilament-reinforced Si_3N_4 matrix composites, *Ceram. Eng. Sci. Proc.*, 6(7–8), 632–645, 1985.

111. Corbin, N.D., Rossetti, G.A., Jr., and Hartline, S.D., Microstructure/property relationships for SiC filament-reinforced RBSN, *Ceram. Eng. Sci. Proc.*, 7(7–8), 958–968, 1986.

112. Bhatt, R.T., Mechanical properties of SiC fiber-reinforced reaction-bonded Si_3N_4 composites, in *Proceedings 23rd Auto. Tech. Devt. Cont. Coord. Meeting*, SAE, P-165, 1985.

113. Bright, J.D., Flinders, R.M., Shetty, D.K., Griffin, C.W., and Richerson, D.W., *High Strength Silicon Nitride Based SiC Continuous Fiber-Reinforced Ceramic Matrix Composites*, Paper 34-SVI-91, American Ceramics Society 93rd Annual Meeting, April 28–May 2, 1991.

114. Bright, J.D., Flinders, R.M., Shetty, D.K., and Swab, J.J., *High Temperature Fracture and Stress Rupture Properties of a SiC (Filament) Reinforced Si_3N_4 Matrix Composite*, Paper 89-SVI-91, American Ceramics Society 93rd Annual Meeting, April 28–May 2, 1991.

115. Newkirk, M.S., Lesher, H.D., White, D.R., Kennedy, C.R., Urquhart, A.W., and Claar, T.D., Preparation of Lanxide™ ceramic matrix composites: matrix formation by the directed oxidation of molten metals, *Ceram. Eng. Sci. Proc.*, 8(7–8), 879–885, 1987.

116. Fareed, A.S., Sonuparlak, B., Lee, C.T., Fortini, A.J., and Schiroky, G.H., Mechanical properties of 2-D Nicalon fiber reinforced Lanxide™ aluminum oxide and aluminum nitride matrix composites, *Ceram. Eng. Sci. Proc.*, 11(7–8), 782–794, 1990.

117. Scoville, A.N., Reagan, P., and Huffman, F.N., Evaluation of SiC matrix composites for high temperature applications, *Adv. Mater. Mfg. Proc.*, 3(4), 643–668, 1988.

118. Richerson, D.W., Griffin, C.W., and Stuffle, K., *Hypersonic Ramjet Leading Edge Materials Development*, Final Rep., Phase I contract N60921-88-C-0109, NSWC SBIR, Oct. 14, 1988.

119. Richerson, D.W., Stuffle, K.L., Griffin, C.W., and Martin, C., *Development of Continuous Fiber Reinforced Group IVB Diboride Composites*, presented at Proceedings of NASA/DOD Composites Meeting, Cocoa Beach, Fla., Jan. 18–21, 1989.

120. Griffin, C.W., Limaye, S.Y., Richerson, D.W., and Shetty, D.K., Correlation of interfacial and bulk properties of SiC-monofilament-reinforced sodium-zirconium-phosphate composites, *Ceram. Eng. Sci. Proc.*, 11(9–10), 1577–1591, 1990.

121. Goettler, R.W. and Faber, K.T., Interfacial shear stresses in fiber reinforced glasses, *Comp. Sci. Technol.*, 37, 129–147, 1989.

122. Torti, M.L. and Richerson, D.W., High strength composite ceramic structure, U.S. Patent, 3,911,188, Oct. 7, 1975.

123. Cutler, R.A., Bright, J.D., Virkar, A.V., and Shetty, D.K., Strength improvement in transformation-toughened alumina by selective phase transformation, *J. Am. Ceram. Soc.*, 70(10), 714–718, 1987.

124. Hansen, J.J., Cutler, R.A., Shetty, D.K., and Virkar, A.V., Indentation fracture response and damage resistance of Al_2O_3–ZrO_2 composites strengthened by transformation-induced residual stresses, *J. Am. Ceram. Soc.*, 71(12), C501–505, 1988.

125. Coblenz, W.S., Fibrous Monolithic Ceramic and Method for Production, U.S. Patent 4, 772, 524, Sept. 20, 1988.

126. Baskaran, S., Nunn, S.D., Popovic, D., and Halloran, J.W., Fibrous Monolithic Ceramics: I. Fabrication, Microstructure and Indentation Behavior, *J. Am. Ceram. Soc.*, 76(9), 2209–2216, 1993.

127. Baskaran, S. and Halloran, J.W., *J. Am. Ceram. Soc.*, 77(5), 1249–1255, 1994.

128. Popovic, D., Danko, G., Stuffle, K., King, B.H., and Halloran, J.W., Relationship Between Architecture, Flexural Strength and Work of Fracture for Fibrous Monolith Ceramics, Symp. Adv. Ceram. Compos., Am. Ceram. Soc., Cocoa Beach, FL, January 1995.

PROBLEMS

1. A Matrix contains 140-μm diameter fibers that are oriented unidirectionally and uniformly separated by an average of 100 μm. Estimate the vol% fibers.

2. Estimate the wt% of fibers for Problem 1 if the matrix is cordierite with a bulk density of 2.6 g/cm^3 and the fiber is pure SiC with a density of 3.17 g/cm^3.

3. What if the SiC fibers in Problem 2 consist of CVD SiC on a 15-μm core of carbon? What will the wt% fiber be (assuming that the carbon has a density of 2.2 g/cm^3)?

4. A whisker is reported to have a tensile strength of 320,000 psi. What is this in GPa? In kg/mm^2?

5. If the composite in Problem 20.1 has a Si_3N_4 matrix with a thermal expansion coefficient of 3.2×10^{-6}/°C and SiC fibers with a radial expansion of 2.3×10^{-6}/°C.

   (a) Estimate the degree of thermal expansion mismatch if the composite were fabricated at 1700°C.

   (b) Estimate the residual stress (assuming an elastic modulus of 44×10^6 psi for the Si_3N_4).

   (c) Would the fiber be in tension or compression?

   (d) How about the matrix?

   (e) What effect would this have on matrix cracking?

   (f) On ease of fiber pullout?

(g) What would the stress characteristics be if the matrix were a cordierite-based composition with a bulk thermal expansion coefficient of $1.2 \times 10^{-6}/°C$ (assume a fabrication temperature of 1250°C and an elastic modulus of 20×10^6 psi)?

STUDY GUIDE

1. Explain why most ceramics are brittle and susceptible to catastrophic fracture.
2. Identify seven approaches for increasing the resistance of a ceramic to brittle fracture.
3. List seven mechanisms that can increase toughness in a ceramic-based material.
4. Briefly describe the factors that control the degree of toughening that can be achieved by "modulus transfer".
5. Explain why modulus transfer is especially effective for polymer matrix composites, but not for most ceramic matrix composites.
6. What is meant by "fiber architecture" and what is the difference between one-dimensional and two-dimensional architectures?
7. Explain how the directional properties of a composite can be designed by control of fiber architecture and fiber loading.
8. Identify some approaches for achieving variations in fiber architecture.
9. What is achieved with "drum winding"?
10. Describe a couple ways that "prestressing" can be achieved to increase the resistance of a ceramic to fracture.
11. Explain the relationship between the fracture energy term, the microstructure, and the roughness of the fracture surface in achieving toughening by crack deflection.
12. List some approaches to achieve crack deflection or impediment in a ceramic material.
13. Explain why rod-shaped and disc-shaped dispersed particles result in higher fracture toughness than spherical dispersed particles.
14. Reinforcement can be achieved either by adding particles, whiskers, or fibers before sintering or by carefully manipulating composition and other factors to achieve a self-reinforced structure during sintering. What are some advantages and disadvantages for each of these approaches?
15. Explain how crack bridging and pullout can combine to increase the fracture toughness of a ceramic material.
16. Explain how microcracking can result in "crack shielding".
17. Explain how transformation toughening increases the toughness of a ceramic by "crack shielding".
18. Can transformation toughening be achieved in materials other than zirconia? Explain.
19. Explain how self-reinforcement is achieved in silicon nitride.
20. Describe an example of how a self-reinforced composite structure can be achieved by a reactive densification process.
21. With reference to the phase equilibrium diagram, explain how a transformation toughened ceramic is achieved in the zirconia-magnesia system.
22. With reference to the phase equilibrium diagram, explain how a transformation toughened ceramic is achieved in the zirconia-yttria system.
23. Explain the differences in the microstructure for the zirconia-magnesia vs. zirconia-yttria transformation toughened materials.
24. Explain a couple critical limitations of transformation toughened zirconia materials.
25. Compare the strength of SiC whiskers with polycrystalline SiC and alumina.
26. Describe the benefits of adding SiC whiskers to alumina.
27. What is the benefit of reinforcing a ceramic with long fibers rather than particles or whiskers?
28. What is the meaning of the term "sizing"?
29. What is the meaning of the terms "tow" and "roving"?

30. What are the major property differences between glass fibers and carbon fibers?
31. Explain why most carbon fibers have a large difference in properties in the axial and radial directions.
32. Identify a major limitation or challenge of using carbon fibers in ceramic matrix composites.
33. How are most oxide fibers fabricated?
34. Identify several methods (processes) for production of long strands of SiC or Si-C-N fibers.
35. Compare the fracture behavior and strength of unreinforced cement with cement containing only 2% carbon fibers.
36. Describe the improvement at room temperature and high temperature by adding SiC fibers to LAS.
37. Describe how a ceramic matrix composite is achieved by chemical vapor infiltration.
38. Why did initial attempts to make a SiC/SiC composite by CVI result in a brittle material, and what was the initial solution?
39. Why did the solution described in question 38 have a limitation at high temperature in an air environment?
40. Identify some R&D directions that were pursued to resolve the problem identified in question 39.
41. What are some advantages and disadvantages of fabricating ceramic composites by the CVI process?
42. Describe the Si melt infiltration process.
43. What are some potential advantages of the MI process and resulting material compared to CVI SiC/SiC process and material?
44. What is a potential advantage of an oxide/oxide ceramic matrix composite?
45. Describe the typical fabrication process for an oxide/oxide CMC.
46. Explain some possible reasons why silicon nitride matrix composites have not progressed as far towards commercial applications as have SiC/SiC and oxide/oxide composites.
47. Why might a ceramic matrix composite have much better fatigue life at a stress below the first matrix cracking stress compared to a stress above the first matrix cracking stress?
48. Describe how the thermal expansion mismatch between the fibers and the matrix can influence the fracture behavior of a ceramic matrix composite.

Appendix A: Glossary

Abrasion   Typically refers to surface wear by solid particles.

Abrasive   Hard, mechanically resistant material used for grinding or cutting.

Agglomerates   Groups of ceramic particles that adhere to each other in porous clusters that are typically detrimental during the ceramic fabrication process.

Amorphous   Noncrystalline, without long-range order.

Anion   Negative ion.

Anisotropic   Properties vary in different crystallographic directions of a material.

Anneal   A heat treatment usually applied to improve homogenization or to minimize residual stresses in a glass, ceramic, or metal.

Ball milling   A technique used to reduce the particle size of ceramic powders and to achieve homogeneous mixing of powders, suspensions, and organic additives such as binders.

Bend strength   Common mode of strength measurement for ceramics in which a rectangular cross-section bar of the ceramic is supported at opposite ends and an increasing load is applied at the midspan until the bar fractures.

Binders   Additives to a ceramic powder that provide a temporary bond between the particles when the powder is compacted and usually burn off during a later step in the fabrication process; allow handling and "green machining" of the powder compact.

Brittle fracture   Typical fracture mode of a glass or ceramic; occurs when an externally applied load induces a critical stress; accompanied by the formation of a crack that rapidly travels through the ceramic with no ductile deformation of the adjacent ceramic.

Calcining   A heat treatment at a temperature well below the densification temperature; used for various reasons such as particle coarsening, removal of chemically bonded water, and decomposition of salts (carbonates, nitrates, sulfates, etc.) to oxides.

Casting   Generic term referring to a process by which a complex shape is formed by pouring a fluid material into a shaped mold; see *cement casting, investment casting,* and *slip casting.*

Cement casting   Casting of a ceramic composition that hardens to a rigid structure by chemical bonding.

Cementitious bonding   Formation of a rigid ceramic structure at moderate to low temperature by chemical reactions such as hydration, dehydration, solution/precipitation, and gelation; examples: plaster and Portland cement.

Cermet   A composite structure consisting of ceramic particles bonded together by a ductile metal, e.g., cobalt-bonded tungsten carbide.

Comminution   Generic term referring to reduction in particle size.

Composition   The total chemical content of a material including all crystalline and noncrystalline constituents.

Compound   Combination of two or more chemical elements to form a single crystalline constituent having a defined chemical composition.

Congruent melting   The melting behavior of a compound that remains solid until the melting temperature is reached and then goes directly to a liquid with no change in composition.

Coordination number	The number of nearest neighbor atoms or ions surrounding an atom or ion; 12 for close-packed atoms, 8 for an ion at the center of a cube, 6 for an ion at the center of an octahedron, 4 for an ion at the center of a tetrahedron, 3 for an ion surrounded by three ions in a plane, and 2 for ions sharing the corners of a tetrahedron or octahedron.
Coordination polyhedron	A grouping of ions consisting of anions at the corners of a polyhedron with a cation in the interior of the polyhedron; building block within a ceramic crystal structure; includes simple cubes, octahedrons, and tetrahedrons, both in undistorted and distorted forms.
Cordierite	An important magnesium aluminum silicate ceramic that has low thermal expansion and low dielectric constant.
Corrosion	Deterioration and material removal by chemical attack.
Creep	Deformation of a ceramic material over a period of time due to the combined influence of temperature plus an applied load.
Cristobalite	An important high-temperature form of silicon dioxide that undergoes a high-volume-change reversible phase transformation in the temperature range 200 to 270°C (390 to 520°F).
Crystal	A physically uniform solid with long-range repetitive order in the arrangement of the atoms.
Deflocculant	An additive to a fluid suspension of ceramic particles that causes agglomerates or clusters of particles to break up into a uniform dispersion of individual particles.
Densification	Change during the firing process (sintering) of a ceramic from a loosely bonded, porous compact of individual particles to a strongly bonded, nonporous structure.
Densification aids	Chemical additions that allow densification to occur more easily, at a reduced temperature, or with a greater degree of control.
Density	A measurement that identifies the amount of porosity in a ceramic material.
Derivative structure	Modification of a crystal structure by substitution of ions, stuffing, ordering, non-stoichiometry, or distortion to produce a closely related, but crystallographically distinct new crystal structure.
Dielectric constant	A relative measurement of the degree of polarization (shift of positive charge toward the negative electrode and negative charge toward the positive electrode) that occurs when a material is placed in an electric field.
Dielectric strength	The capability of a material to withstand an electric field without breaking down and allowing an electric current to pass.
Diffusion	Motion of atoms, ions, or vacancies through a material; strongly influences densification, creep, and electrical properties.
Dipole	A system or object whose one end has a negative charge and the other a positive charge.
Dislocation	A linear defect in the stacking of atoms in a crystal.
Domains	Small crystalline areas of aligned ferromagnetic or ferroelectric atoms.
Ductility	Permanent (plastic) deformation before fracture.
Elasticity	Nonpermanent (elastic) deformation.
Elongation	The amount of permanent strain (deformation) prior to fracture.
Elutriation	Procedure for sizing of ceramic particles based on the rate of settling from a fluid suspension.
Encapsulation	Sealing a porous ceramic compact in an impervious but flexible container to allow an increase in densification by the application of an external isostatic pressure; used for cold isostatic pressing and hot isostatic pressing.
Energy bands	Permissible energy levels for valence electrons.

Energy gap	Nonpermissible energy levels for valence electrons.
Equiaxed	Powder or grain shapes with approximately equal dimensions.
Equilibrium	Condition where all phases are in their most stable form for the given temperature and pressure; dynamic balance, lowest free energy.
Erosion	Mechanical abrasion by solids suspended in a fluid.
Etching	Chemical surface corrosion, usually conducted in a controlled fashion on a polished surface of a material sample to reveal details of the microstructure.
Eutectic	The lowest melting composition in a material system, which at a specific temperature crystallizes directly from the melt to the solid and produces a distinctive microstructure.
Extrusion	A fabrication procedure that forms a constant cross-sectional shape by forcing a material under high pressure through a shaped orifice; used for producing particulate compacts of ceramics.
Failure analysis	Observation of fracture surfaces and correlation with measured data, stress analysis, and other information to determine the cause and mechanism of failure of a ceramic or other material.
Fatigue (cyclic)	The tendency for a material to fail under cyclic stresses.
Fatigue (static)	The tendency for a material to fail under sustained stress; see *stress corrosion* and *stress rupture*.
Ferrites	Ceramic compositions having magnetic characteristics.
Ferroelasticity	Spontaneous alignment of crystallographic domains under the influence of an applied load to produce permanent deformation, resulting in a hysteresis in the stress–strain curve.
Ferroelectricity	Spontaneous alignment of electric dipoles within a material under the influence of an electric field, resulting in a hysteresis loop when the direction of electric field is switched.
Ferromagnetism	Spontaneous alignment of magnetic dipoles within a material under the influence of a magnetic field, resulting in a hysteresis loop when the direction of the magnetic field is switched.
Fracture mirror	A feature on the fracture surface of a ceramic that enables the observer to locate the position where the fracture initiated and to determine the cause of fracture.
Fracture toughness	A material characteristic and measurement that relates to the resistance of the material to propagation of a crack.
Framework structure	A crystalline structure with primary atomic bonding in all three directions.
Free energy	Energy available for chemical reaction.
Glass	An amorphous material with three-dimensional primary atomic bonding.
Grain	A distinct unit in the microstructure of a ceramic material that usually consists of a single crystal.
Grain boundary	A discontinuity separating adjacent grains in the microstructure of a ceramic.
Grain growth	The average increase in the size of the grains in a ceramic microstructure under the influence of temperature.
Granulation	Compaction of groups of ceramic particles into large agglomerates or granules to achieve a free-flowing condition or to provide precompaction.
Green machining	Grinding, turning, milling, and other machining operations conducted on unfired compacts of ceramic particles to minimize the machining required after the densification process.
Greenware	The generic term used to refer to the unfired compact of ceramic particles.
Hardness	Resistance of a material to penetration of its surface.
Heat capacity	The energy required to raise the temperature of a material.

Hole	Vacancy in the atomic structure of a crystal or in an electronic structure.
Hot isostatic pressing	A method used to densify a material, whereby heat and pressure are imposed simultaneously and the pressure is applied from all directions via a pressurized gas such as argon or helium.
Hot pressing	A method used to densify a material, whereby heat and pressure are applied simultaneously and the pressure is typically applied unidirectionally via rigid tooling.
Hydration	Chemical reaction that consumes water, such as in some cements.
Hydrogen bridge	A weak form of bonding between molecules (such as water) that contain hydrogen and have a charge polarization because they are nonsymmetrical.
Inclusions	Foreign particles present as an undesirable impurity in a ceramic and typically resulting in a reduction in strength.
Incongruent melting	The behavior of a compound that does not melt directly, but instead decomposes into a liquid plus a different solid compound.
Index of refraction	Ratio of the velocity of light in a vacuum to the velocity in the material.
Injection molding	Technique of forming ceramic greenware into a complex shape by injecting a mixture of ceramic particles and a plastic organic carrier under pressure and temperature into a shaped mold.
Interstitial site	Octahedral and tetrahedral open spaces within a close-packed arrangement of atoms or ions in which a cation can fit.
Ion	An atom with a positive charge because it has had electrons removed or a negative charge because it has had electrons added.
Ionic bonding	One of the primary types of atomic bonding in ceramics in which electrons are transferred from one atom to another to leave two oppositely charged ions that are attracted by coulombic forces.
Ionic radius	Assuming a spherical model for ions, the ionic radius is the approximate distance between the center of an ion in a specific crystal structure and the outer diameter of the nearest neighbor ions.
Isostatic pressing	Method of compacting a powder in which the powder is sealed in a flexible container, is placed in a liquid-filled autoclave, and the pressure of the liquid is increased to compact the powder uniformly from all directions.
Isotropic	Having the same properties in all directions.
Kiln	Term referring to a furnace in which ceramics are fired.
Knit line	A critical type of material defect that commonly occurs during injection molding.
Lapping	A surface finishing operation used to achieve a fine polish and close tolerances.
Lattice	The space arrangement of atoms in a crystal.
Lever rule	Method for calculating the percentage of the different phases for any point on a phase equilibrium diagram.
Liquidus	The line on the phase equilibrium diagram above which only liquids are stable and below which some solid is present.
Lubricant	An additive to ceramic powders to achieve an improvement in flow during pressing and other green forming operations.
Matrix	The continuous phase in a composite in which a second phase is dispersed.
Melting temperature	The temperature at which a ceramic goes from a solid crystal structure with long-range order to a liquid with only short-range order.
Mesh	A term used to describe the screen size for particle measurement.
Metallic (electronic) bonding	The type of primary bonding in metals that involves a nondirectional sharing of electrons between many atoms.

Micron   Abbreviation of the unit of measurement micrometer, equal to 1/1000 of a millimeter; commonly used to describe the size of ceramic powders and the features in a ceramic microstructure.

Microstructure   The microscopic assemblage of grains, grain boundaries, amorphous phases, pores, and inclusions that make up a polycrystalline ceramic.

Miller indices   A notation that has been defined to identify the various atomic planes in a crystal structure.

Mixture   A combination of phases where there is no chemical interaction between the phases.

Modulus of elasticity   Proportionality constant between elastic stress and elastic strain; can be thought of simply as the amount of stress required to produce unit elastic strain.

Modulus of rupture   Breaking strength in a nonductile solid as measured by bending.

Mullite   An important aluminum silicate ceramic.

Normal   Refers to a direction perpendicular to a surface or cross section.

Notch sensitivity   A reduction in properties by the presence of a stress concentration; typical of brittle materials.

Nucleation   The start of growth of a new phase.

Ordering   Positioning of host and substitution ions in an ordered, repetitious pattern rather than in a random arrangement.

Oxidation   Interaction of oxygen gas with a surface to produce a different compound.

Phase   A physically homogeneous part of a material system.

Piezoelectricity   Mechanical distortion when an electrical current is applied and conversely, an electric current resulting when pressure is applied.

Plastic deformation The ability of a material to be permanently deformed without fracture.

Plasticizer   An additive to a binder, ceramic extrusion mix, or injection-molding mix that increases the workability.

Poisson's ratio   The ratio of the transverse contracting strain to the longitudinal elongational strain when a tensile stress is applied to a material.

Polarization   Displacement of the centers of positive and negative charge.

Polycrystalline   A bulk ceramic made up of many individual grains.

Polymorphism   Different crystal structures at different temperatures or pressures for a single compound.

Porosity   Open spaces between grains or trapped in grains in a ceramic microstructure.

Preconsolidation   Treatments of sized ceramic powders to prepare them for compacting into a shape.

Pressing   Compaction of powders through the application of pressure.

Quench   Rapidly cool a material typically to retain at room temperature a structure that otherwise is only stable at high temperature.

Radius ratio   Ratio of the ionic radius of a cation divided by the ionic radius of an anion; used to estimate which coordination polyhedron the cation will fit into.

Reaction sintering   Use of a change in composition to enhance sintering.

Refractories   General term referring to ceramics used for high-temperature furnace linings and related kiln furniture.

Refractory   General term for a heat-resistant material.

Residual stresses   Internal stresses in a material often resulting from thermal or mechanical straining.

Rheology   Study of flow characteristics, especially for suspensions of ceramic particles in a liquid or resin carrier.

Screening   Sizing of particles by using a mesh containing openings of a controlled size.

Shear	Relative displacement by sliding.
Shrinkage	Decrease in physical dimensions of a particulate compact during the drying and sintering processes.
Sintering	Densification of a particulate ceramic compact involving a removal of the pores between the starting particles (accompanied by equivalent shrinkage) combined with coalescence and strong bonding between adjacent particles.
Sintering aid	A chemical additive that enhances sintering.
Slip	For solid structures, a relative displacement along a structural direction usually caused by shear stress; for ceramic fabrication, a suspension of ceramic particles in a fluid as used for slip casting.
Slip casting	A method of forming a particulate ceramic compact to a complex shape, whereby a liquid suspension of ceramic particles is poured into a porous mold, the liquid is removed through the pores, and the ceramic particles deposit in the shape of the mold.
Slow crack growth	Slow extension of a crack in a ceramic material at a lower load than is required for typical brittle fracture.
Softening point	Temperature well below the melting point at which a glass begins to soften and is susceptible to viscous flow under an applied stress.
Solidification	Freezing of a melt.
Solid solution	Substitution of one ion for another ion in a crystal structure without a change in the structure other than an increase or decrease in the size of the unit cell.
Solidus	The temperature in a phase equilibrium diagram below which no liquids are present.
Solute	The minor component of a solution.
Solvent	The major component of a solution.
Solvus	The curve on a phase equilibrium diagram that defines the limits of solid solubility.
Specific heat	The ratio of heat capacity of a material to the heat capacity of water.
Spray drying	A preconsolidation step in ceramic fabrication that produces a uniform, free-flowing powder suitable for compaction by pressing.
Stoichiometry	Refers generally to the composition of a material and specifically to the relative atomic proportions of cations and anions. A stoichiometric ceramic contains the exact ratio of cations to anions as defined by the ideal chemical formula (e.g., ZrO_2, WC, and Al_2O_3). A nonstoichiometric ceramic has a deficiency of either cations or anions (e.g., $Zr_{0.85}$ $Ca_{0.15}$ $O_{1.85}$, or $WC_{0.95}$).
Strength	Resistance of a ceramic material to crack initiation and subsequent fracture.
Stress	Force per unit area.
Stress corrosion	Combined chemical and stress effects on crack growth.
Stress rupture	Time-dependent fracture resulting from a constant stress, usually at elevated temperature.
Stuffing	Substituting into a crystal structure a cation of lesser charge for one of greater charge and stuffing (adding) an additional cation into a vacant hole or channel in the structure to achieve charge balance.
Surface energy	Energy necessary to produce two new surfaces as a crack travels through material.
Thermal conductivity	The rate of travel of heat through a material.
Thermal expansion	Change in dimensions of a material resulting from a change in temperature.
Thermal shock	Stresses induced in a material because of a rapid temperature change or a thermal gradient.

Toughness   Resistance of a material to extension of a crack present in the material.

Transformation   The change induced by temperature or pressure of one crystallographic form to another for a specific material composition; see *polymorphism*.

Unit cell   The smallest repetitive volume that comprises the complete pattern of a crystal.

Vacancy   An unfilled lattice site in a crystal structure.

Valence   The charge on an ion based on the number of electrons transferred or shared within a specific structure.

Van der Waals forces Weak, secondary atomic bonds arising from structural polarization.

Viscosity   Coefficient of resistance to flow.

Vitreous   Glasslike.

Weibull statistics   A statistical approach used to describe the probabilistic fracture behavior of ceramics.

Appendix B: Effective Ionic Radii for Cations and Anions

Ionic Radius Depends on Charge & Size of Ion (Units (Å))

Ion	Radius for Coordination Number [CN]				Other (shown in superscript)
	[4]	[6]	[8]	[12]	
Ac^{3+}		(1.30)			
Ag^+	1.16[#]	1.29	1.44		0.81[2]
Ag^{2+}		(1.01)			
Ag^{3+}	0.79[#]				
Al^{3+}	0.53*	0.67*			0.62[5]
Am^{3+}		1.14			
Am^{4+}		(1.04)	1.09		
As^{3+}		(0.70)			
As^{5+}	0.475*	0.64			
At^{7+}		(0.74)			
Au^+		(1.49)			
Au^{3+}	0.84[#]	(0.97)			
B^{3+}	0.26*				0.16[3]
Ba^{2+}		1.50	1.56	1.74	
Be^{2+}	0.41*				0.31[3]
Bi^{3+}		1.16	1.25		1.13[5]
Bi^{5+}		(0.86)			
Bk^{3+}		1.10			
Bk^{4+}			1.07		
Br^-		(1.82)			
Br^{7+}	0.40				
C^{4+}					0.06[3]
Ca^{2+}		1.14	1.26*	1.49	
Cd^{2+}	0.94	1.09	1.21	1.45	
Ce^{3+}		1.15	1.28	1.43	1.29[9]
Ce^{4+}		0.94[†]	1.11		
Cf^{3+}		1.09			
Cl^-		(1.67)			
Cl^{5+}					0.26[3]
Cl^{7+}	0.34				
Cm^{3+}		1.12			
Cm^{4+}			1.09		
$Co^{2+}LS$		0.79			
$Co^{2+}HS$	0.71	0.885*			
$Co^{3+}LS$		0.665			
$Co^{3+}HS$		0.75			
$Cr^{2+}LS$		0.87			
$Cr^{2+}HS$		0.96			
Cr^{3+}		0.755*			

(continued)

Ion	Radius for Coordination Number [CN]				Other (shown in superscript)
	[4]	[6]	[8]	[12]	
Cr^{4+}	0.58	0.69			
Cr^{5+}	0.49		0.71		
Cr^{6+}	0.44				
Cs^+		1.84	1.90†	2.02	1.95[10]
Cu^+		(1.08)			0.60[2]
Cu^{2+}	0.76#	0.87			0.79[5]
D^+					0.04[2]†
Dy^{3+}		1.052	1.17		
Er^{3+}		1.030	1.14		
Eu^{2+}		1.31	1.39		
Eu^{3+}		1.087	1.21		
F^-	1.17	1.19			1.145[2]
					1.16[3]
$Fe^{2+}LS$		0.75			
$Fe^{2+}HS$	0.77	0.92*			
$Fe^{3+}LS$		0.69			
$Fe^{3+}HS$	0.63*	0.785*			
Fr^+		(1.92)			
Ga^{3+}	0.61*	0.76*			
Gd^{3+}		1.078	1.20		1.18[7]
Ge^{2+}		(0.85)			
Ge^{4+}	0.54*	0.68*			
H^+					−0.24[1]
					−0.04[2]
Hf^{4+}		0.85	0.97		
Hg^+					1.11[3]
Hg^{2+}	1.10	1.16	1.28		0.83[2]
Ho^{3+}		1.041	1.16		
I^-		(2.06)			
I^{5+}		1.09†			
I^{7+}		(0.62)			
In^{3+}		0.94*	1.063		
Ir^{3+}		0.87†			
Ir^{4+}		0.77			
K^+		1.52	1.65†	1.74†	
La^{3+}		1.185	1.32	1.46†	
Li^+	0.73	0.88			
Lu^{3+}		1.001	1.11		
Mg^{2+}	0.72	0.86*	1.03		
$Mn^{2+}LS$		0.81			
$Mn^{2+}HS$		0.97*	1.07		
$Mn^{3+}LS$		0.72			
$Mn^{3+}HS$		0.785*			
Mn^{4+}		0.68			
Mn^{6+}	0.41				
Mn^{7+}	0.40				
Mo^{3+}		0.81			
Mo^{4+}		0.79			
Mo^{5+}		0.77			
Mo^{6+}	0.56*	0.74*			0.85[7]
N^{3+}		(0.28)			
N^{5+}					0.02[3]

(continued)

Ion	Radius for Coordination Number [CN]				Other (shown in superscript)
	[4]	[6]	[8]	[12]	
Na^+	1.13[†]	1.16	1.30[†]		1.46[9][†]
Nb^{2+}	0.85[#]				
Nb^{3+}		0.84			
Nb^{4+}		0.83			
Nb^{5+}	0.46[†]	0.78			0.80[7]
Nd^{3+}		1.123	1.26		
NH_4^+		(1.63)			
Ni^{2+}		0.83*			
$Ni^{3+}LS$		0.70			
$Ni^{3+}HS$		0.74			
Np^{2+}		1.24			
Np^{3+}		1.16			
Np^{4+}		(1.07)	1.12		
Np^{7+}		(0.83)			
O^{2-}	1.24	1.26	1.28		1.21[2]
					1.22[3]
Os^{4+}		0.77			
P^{3+}		(0.56)			
P^{5+}	0.31*				
Pa^{3+}		(1.25)			
Pa^{4+}		(1.10)	1.15		
Pa^{5+}		(1.01)	1.05		1.09[9]
Pb^{2+}	1.08[#]	1.32	1.45	1.63	1.53[11]
Pb^{4+}		0.915	1.08		
Pd^+					0.73[2]
Pd^{2+}	0.78[#]	1.00			
Pd^{3+}		0.90[†]			
Pd^{4+}		0.76			
Pm^{3+}		1.11			
Po^{4+}			1.22		
Po^{6+}		(0.79)			
Pr^{3+}		1.137	1.28		
Pr^{4+}		0.92	1.10		
Pt^{2+}	0.74[#]	(0.92)			
Pt^{4+}		0.77			
Pu^{3+}		1.15			
Pu^{4+}		0.94[†]	1.10		
Ra^{2+}		(1.55)	1.62	1.78	
Rb^+		1.63	1.74	1.87	
Re^{4+}		0.77			
Re^{5+}		0.66[†]			
Re^{6+}		0.66			
Re^{7+}	0.54	0.71			
Rh^{3+}		0.805			
Rh^{4+}		0.755			
Ru^{3+}		0.82			
Ru^{4+}		0.76			
S^{2-}		(1.70)			
S^{4+}		(0.49)			
S^{6+}	0.26*				
Sb^{3+}	0.91[#]				0.94[5]
Sb^{5+}		0.75			

(continued)

Ion	Radius for Coordination Number [CN]				Other (shown in superscript)
	[4]	[6]	[8]	[12]	
Sc^{3+}		0.885*			
Se^{2-}		(1.84)			
Se^{4+}		(0.62)			
Se^{6+}	0.43				
Si^{4+}	0.40*	0.54*			
Sm^{3+}		1.098	1.23		
Sn^{2+}		(1.05)	1.36		
Sn^{4+}		0.83*			
Sr^{2+}		1.27	1.39	1.54	
Ta^{3+}		0.81			
Ta^{4+}		0.80			
Ta^{5+}		0.78	0.83		
Tb^{3+}		1.063	1.18		1.16[7]
Tb^{4+}		0.90	1.02		
Tc^{4+}		0.78			
Tc^{7+}		(0.68)			
Te^{2-}		(2.07)			
Te^{4+}		(0.82)			0.66[3]
Te^{6+}		(0.68)			
Th^{4+}		1.14	1.18		1.23[9]
Ti^{2+}		1.00			
Ti^{3+}		0.81			
Ti^{4+}		0.745*			0.67[5]
Tl^{+}		1.64	1.74	1.90	
Tl^{3+}		1.025	1.14		
Tm^{3+}		1.020	1.13		
U^{3+}		1.18			
U^{4+}		(1.09)	1.14*		1.19[9]
U^{5+}		0.90			1.10[7]
U^{6+}	0.62	0.87			0.59[2]
					1.02[7]
V^{2+}		0.93			
V^{3+}		0.78			
V^{4+}		0.73			
V^{5+}	0.495	0.68			0.60[5]*
W^{4+}		0.79			
W^{6+}	0.56*	0.74*			
Y^{3+}		1.040*	1.155*		1.24[9]
Yb^{3+}		1.008	1.12		
Zn^{2+}	0.74*	0.89*	1.04		
Zr^{4+}		0.86	0.98		

* Radii considered particularly reliable.

† Radii considered doubtful.

Radii so marked refer to a square-planar arrangement and for Pb^{3+} and Sb^{3+} to a pyramidal environment. All other radii in this column are tetrahedral radii.

Most of the radii given here are the "CR" radii derived by Shannon and Prewitt. Cation radii in parentheses are mostly Ahren's radii to which 0.12 Å has been added to make them more compatible with the Shannon–Prewitt radii. Anion radii in parentheses represent Ahren's radii (for Cl^-, Br^-, and I^-) and Pauling radii (for S^{2-}, Se^{2-}, and Te^{2-}) from which 0.14 Å has been substracted. LS = low spin. HS = high spin.

Source: After O. Muller and R. Roy, *The Major Ternary Structural Families*, pp. 5–7, Springer-Verlag, Berlin, 1974.

Appendix C: The Periodic Table of the Elements

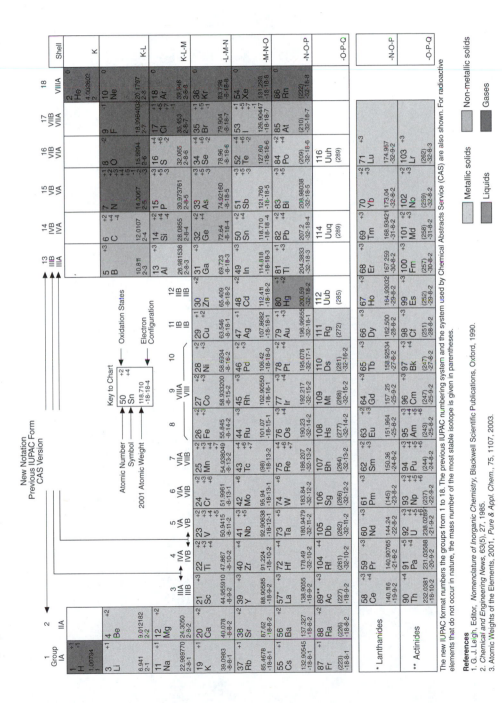

The new IUPAC format numbers the groups from 1 to 18. The previous IUPAC numbering system and the system used by Chemical Abstracts Service (CAS) are also shown. For radioactive elements that do not occur in nature, the mass number of the most stable isotope is given in parentheses.

References
1. G. J. Leigh, Editor, *Nomenclature of Inorganic Chemistry*, Blackwell Scientific Publications, Oxford, 1990.
2. *Chemical and Engineering News*, 63(5), 27, 1985.
3. Atomic Weights of the Elements, 2001, *Pure & Appl. Chem.*, 75, 1107, 2003.

Index